Digital Imaging Primer

Alan Parkin

Digital Imaging Primer

Springer

Alan Parkin
London
UK

Additional material to this book can be downloaded from http://extras.springer.com.

ISBN 978-3-662-50048-4 ISBN 978-3-540-85619-1 (eBook)
DOI 10.1007/978-3-540-85619-1

Springer Heidelberg New York Dordrecht London

Softcover re-print of the Hardcover 1st edition 2016

Printed on acid-free paper

Springer-Verlag GmbH Berlin Heidelberg is part of Springer Science+Business Media
(www.springer.com)

For Lily 1926–2008

Preface

This book is about making pictures via a computer. Making pictures, that is to say making marks on a surface to represent things seen or imagined, is an ancient pursuit. Since the earliest times, techniques for making pictures have been in continuous development, branching out in new directions as new tools became available. Over the past 50 years we have seen just such a branching-out in the use of computers for imaging, overlapping, and often over-topping the techniques inherited from the past. Personal computers and their peripherals are modern tools for an ancient task.

Nearly all the material in this book can be found in one form or another on the Internet. However, there seems to be room for an integrated treatment which is accessible at entry level, practicable with modest equipment, and fairly deep, leading on to further explorations in many directions. It is offered to scientists, engineers, and students with a computer background but no particular knowledge of digital imaging, and to artists, authors, and amateurs wanting a well-founded introduction to digital imaging and a well-integrated reference source. Readers of any level can dip in wherever they feel inclined, and skip whatever they already know.

To this end, the prerequisites are merely possession of current standard equipment and a working knowledge of its use. Standard equipment is taken as: a personal computer with a Microsoft Windows operating system, Microsoft PAINT image editor, and Microsoft QuickBASIC programming language; plus a simple scanner, digital camera, and desk-top printer.

The first half of the book (Parts I–IV) considers the roots of digital imaging, which lie in common knowledge drawn from other fields: the arithmetic of numbers, the geometry of location in space and a plane, the optics of lenses and the eye, the measurement of light, the psycho-physics of perception, and the phenomena of complexity.

The second half of the book (Parts V–VIII) covers practical imaging. Here the primary classification is by bit-depth, and within that by the natural flow from input, through processing and storage, to output and viewing.

Programs are introduced as brief descriptions throughout. In the e-book, source code in Microsoft QuickBASIC for every program can be downloaded. These programs are really a collection of sub-programs which are offered as bare working demonstrations, to be improved, developed, or rewritten to taste. QuickBASIC is free, easily learned, and can be transposed up to more powerful languages by any interested reader.

The References in each chapter are to some fundamental works, and to some current literature, for further pursuit.

Let us honour the cascade of universities, commercial enterprises, and curious persons who have made digital imaging available to all. Special thanks are due to Dr. Claus Ascheron and his team at Springer for bringing this book into being, and for their personal kindness throughout. And may you, gentle reader, enjoy a happy lifetime among the pixels.

London and Hydra Alan Parkin

Contents

Part II Location in Digital Imaging

Part VIII 24-bit Imaging

Symbols

Symbols used in this book are here separated into four lists: units, identifiers, variables, and acronyms. Each list is in alphabetic order. Each entry shows the symbol, its usage, and a pointer to further explanation (usually first occurrence). These lists of symbols may be a helpful supplement or alternative to the Contents and the subject Index in accessing a particular topic.

Units

B	Byte (Sect. 2.7)
b	Bit (Sect. 2.7)
bpp	Bits per pixel (Sect. 26.7)
cd	Candela (Sect. 15.5)
cd/m^2	Candela per square metre (Sect. 15.8)
dpi	Printer dots per inch (Sect. 28.3)
dpp	Printer dots per pixel (Sect. 28.3)
GB	Gigabyte (Sect. 4.2)
Gb	Gigabit (Sect. 4.2)
JND	Just noticeable difference (Sect. 17.3)
kB	Kilobyte (Sect. 4.4)
kb	Kilobit (Sect. 4.4)
lm	Lumen (Sect. 15.5)
lm/w	Lumens per Watt (Sect. 15.5)
lppd	Line-pairs per degree (Sect. 10.5)
lppi	Line-pairs per inch (Sect. 10.5)
lppmm	Line-pairs per mimillimetre (Sect. 10.5)
lppul	Line-pairs per unit length (Sect. 10.5)
lx	Lux (Sect. 15.7)
MB	Megabyte (Sect. 4.4)
Mb	Megabit (Sect. 4.4)

Mpx	Megapixel (Sect. 29.4)
nibble	Half-byte (Sect. 4.2)
nm	Nanometre (Sect. 15.3)
ppi	Pixels per inch (Sect. 22.18)
ppm	Pixels per metre (Sect. 10.3)
ppmm	Pixels per millimetre (Sect. 10.5)
px	Pixel (Sect. 5.2)
rad	Radian (Sect. 15.4)
spd	Samples per degree (Sect. 10.6)
spi	Samples per inch (Sect. 22.18)
sr	Steradian (Sect. 15.9)
TB	Terabyte (Sect. 4.4)

Identifiers

$\mathbf{A}^{\mathrm{I}}$	Inverse of matrix (Sect. 3.11)
$\mathbf{A}^{\mathrm{T}}$	Transpose of matrix (Sect. 3.11)
A, B, …	Member of a set (Sect. 1.1)
A, B, …, F	Hexadecimal digits for 10–15 dec (Sect. 2.4)
A, B, C, …	Point (Sect. 6.2)
AB, AC, …	Line (Sect. 6.2)
ABC, ABD, …	Plane (Sect. 6.2)
C	Centre of lens (Sect. 9.1)
CIE V(λ)	Standard eye sensitivity (Sect. 15.3)
$\det\lvert\mathbf{A}\rvert$	Determinant of matrix (Sect. 3.16)
F, F'	Focal points of lens (Sect. 9.2)
M	Matrix (Sect. 3.11)
MOD	Integer remainder (Sect. 3.7)
N	Nodal point of eye (Sect. 10.2)
N	Set of natural numbers (Sect. 1.2)
O	Coordinate origin (Sect. 6.2)
O, U, X	Coordinate base elements (Sect. 6.2)
OX, OY, OZ	Coordinate axes (Sect. 6.2)
OXY, OXZ, OYZ	Coordinate planes (Sect. 6.2)
O1, O2,…, O8	Octants of space (Sect. 6.2)
P	Principal point of eye (Sect. 10.3)
Q	Set of rational numbers (Sect. 1.4)
Q1, Q2,…, Q4	Quadrants of plane (Sect. 6.3)
R	Set of real numbers (Sect. 1.6)
sRGB	Standard RGB colour space (Sect. 16.5)
U	Unit matrix (Sect. 3.11)
V	Vector (Sect. 3.10)

$\mathbf{Z}$	Set of integers (Sect. 1.3)
1BW	1-bit colour-table (Sect. 19.7)
4CO	4-bit colour-table (Sect. 30.2)
4GM	4-bit grey-match colour-table (Sect. 30.4)
8CO	8-bit colour-table (Sect. 37.2)
8GM	8-bit grey-match colour-table (Sect. 37.3)
8GS	8-bit grey-scale colour-table (Sect. 37.4)
.BMP	Bitmap file extension (Sect. 4.4)
.FON	Bitmapped font file extension (Sect. 5.3)
.GIF	Graphics Interchange File extension (Sect. 5.7)
.JPG	JPEG file extension (Sect. 4.4)
.OTF	OpenType Font file extension (Sect. 5.3)
.PDF	Portable Document file extension (Sect. 5.7)
.PNG	Portable Network Graphic file extension (Sect. 4.4)
.TIF	Tagged Image File extension (Sect. 4.4)
.TTF	TrueType Font file extension (Sect. 5.3)
.TXT	ASCII Text file extension (Sect. 4.5)
$+$	Addition (Sect. 1.2)
$-$	Subtraction (Sect. 1.3)
$\times$	Multiplication (Sect. 1.2)
$\backslash$	Integer division (Sect. 3.6)
$/$	Rational division (Sect. 1.4)
$\pm$	Plus-or-minus (Sect. 8.6)
$=$	Equal (Sect. 1.1)
$< >$	Not equal (Sect. 1.1)
$>$	Greater than (Sect. 1.2)
$<$	Less than (Sect. 1.2)
$\neg$	Boolean negation (NOT) (Sect. 3.18)
$\vee$	Boolean disjunction (OR) (Sect. 3.18)
$\wedge$	Boolean conjunction (AND) (Sect. 3.18)
$\oplus$	Group operation (Sect. 23.12)
$\surd$	Square root (Sect. 1.5)
∞	Infinity (Sect. 1.4)

Variables Taking Numeric Values

a	Sample/pixel linear width (Sect. 10.4)
a	Amplitude of sinusoid (Sect. 40.3)
B	Brightness sensation (Sect. 15.3)
B	Bandwidth of channel (Sect. 40.3)
b	Binocular overlap (Sect. 11.2)
(B, G, R)	Blue, Green, Red triple (Sect. 16.2)

Symbol	Meaning
c	Coefficient of grating (Sect. 40.6)
C, M, Y, K	Cyan, Magenta, Yellow, Black (Sect. 16.6)
C_{R}	Contrast ratio (Sect. 29.6)
C, λ, A, A'	Plane projection parameters (Sect. 8.1)
C, π, A, A'	Space projection parameters (Sect. 8.5)
d	Optical density (Sect. 22.14)
d	Print dot width (Sect. 28.3)
DP	Dioptric power of lens (Sect. 9.1)
E	Illuminance (Sect. 15.7)
E_{A}	Ambient illuminance (Sect. 29.7)
F	Luminous flux (Sect. 15.5)
F	Frequency (Sect. 18.4)
f	Focal length of lens (Sect. 9.2)
F/n	F-number of camera lens (Sect. 46.6)
F_{S}	Linear sampling frequency (Sect. 10.4)
F_{LP}	Linear line-pair frequency (Sect. 10.4)
gmv	Grey-match value (Sect. 32.6)
H	Entropy (Sect. 18.2)
I	Luminous intensity (Sect. 15.5)
$I, J, \ldots$	Integer variable (Sect. 1.7)
I_2	1-bit colour index (Sect. 19.7)
I_{16}	4-bit colour index (Sect. 30.1)
I_{256}	8-bit colour index (Sect. 22.8)
L	Luminance (Sect. 15.8)
M	Luminous emittance (Sect. 15.8)
m	Magnification of lens (Sect. 9.2)
n, n'	Refractive index (Sect. 10.3)
P	Count of pixels in width (Sect. 5.2)
p	Probability (Sect. 18.2)
Q	Count of pixels in height (Sect. 5.2)
(R, G, B)	Red, Green, Blue triple (Sect. 16.2)
$r, s, \ldots$	Real variable (Sect. 1.7)
rf	Reflection factor (Sect. 15.8)
S	Sensation (Sect. 10.1)
S	Count of samples in width (Sect. 22.9)
s	Surface area (Sect. 15.5)
s	Binocular separation (Sect. 11.2)
T	Half-tone threshold (Sect. 39.17)
v	Voltage into ADC (Sect. 22.7)
v, v'	Gaussian object, image dist (Sect. 9.2)
w	Object linear width (Sect. 10.4)
(X, Y, Z)	Integer Cartesian coordinates (Sect. 6.3)
(x, y, z)	Real Cartesian coordinates (Sect. 6.2)
$(X : Y : Z : W)$	Integer homogeneous coordinates (Sect. 6.8)
$(x : y : z : w)$	Real homogeneous coordinates (Sect. 6.8)

(Y, M, C, K)	Yellow, Magenta, Cyan, Black (Sect. 16.6)
z, z'	Newtonian object, image dist (Sect. 9.2)
$\alpha, \beta, \ldots$	Angle (Sect. 8.10)
γ	Encoding/decoding exponent (Sect. 22.12)
δ	Angle of deviation (Sect. 12.11)
δ	Printer dot gain (Sect. 35.8)
θ	Plane angle (Sect. 15.4)
λ	Wavelength (Sect. 15.3)
μ	Mean (Sect. 18.9)
σ	Standard deviation (Sect. 18.9)
Φ	Phase difference (Sect. 23.17)
Φ_S	Angular sampling frequency (Sect. 10.4)
Φ_{LP}	Angular line-pair frequency (Sect. 10.4)
χ(PQRS)	Cross-ratio of P, Q, R, S (Sect. 8.2)
ω	Object angular width (Sect. 10.4)
ω	Solid angle (Sect. 15.4)

Acronyms

ADC	Analog-to-Digital Converter (Sect. 22.7)
ASCII	American Standard Code for Information Interchange (Sect. 2.5)
CFL	Compact fluorescent lamp (Sect. 15.5)
CIE	La Commission Internationale de l'Éclairage (Sect. 5.3)
CRT	Cathode Ray Tube (Sect. 27.5)
DAC	Digital-to-Analog Converter (Sect. 27.4)
DCT	Discrete Cosine Transform (Sect. 40.13)
EMR	Electro-Magnetic Radiation (Sect. 15.1)
HDR	High Dynamic Range (Sect. 46.15)
IDCT	Inverse Discrete Cosine Transform (Sect. 40.13)
LCD	Liquid Crystal Display (Sect. 27.5)
L, M, S	Long, Middle, Short cone types (Sect. 17.10)
OLED	Organic Light-Emitting Diode (Sect. 27.5)
SD	Secure Digital (Sect. 46.14)
SI	Système Internationale (Sect. 4.4)
SPD	Spectral Power Distribution (Sect. 15.2)
TRC	Tone Reproduction Curve (Sect. 17.5)
V, E, F	Vertices, Edges, Faces (Sect. 7.9)

Part I
Numbers and Alphanumerics in Digital Imaging

Chapter 1
Number Set

Several sets of numbers are used in digital imaging. Natural numbers are what we use for everyday counting, as one, two, three, ... Integer numbers are all the whole numbers, including the natural numbers and zero and negative numbers. Rational numbers also include fractions, where one integer is divided by another and the result terminates exactly, or repeats without end. Irrational numbers also include fractions which do not terminate or repeat, such as the square-root of 2. Real numbers include all of the above.

1.1 Set

A set is a collection of objects A, B, C, ..., (the members of the set), with an equivalence relation = between them. If A and B are alike in some way, we write A = B, other wise A <> B. Equivalence is:

- Reflexive, that is, A = A, B = B, C = C, ...for every member of the set,
- Symmetric, that is, if A = B then B = A,
- Transitive, that is, if A = B and B = C, then A = C.

1.2 Natural Number

The set of natural numbers **N** is constructed from five axioms[1] [1–3]:

[1]Landau's formulation, following Dedekind and Peano. Edmund Georg Landau (1877–1938) was Professor at Göttingen and colleague of Klein and Hilbert. Richard Dedekind (1831–1916) was Gauss's last pupil, and Lecturer at Braunschweig. Guiseppe Peano (1856–1932) was Professor at Turin: he devised the logical symbols used by Whitehead and Russell in *Principia Mathematica*.

A. Parkin, *Digital Imaging Primer*, DOI 10.1007/978-3-540-85619-1_1

I: 1 is a natural number. That is, our set is not empty; it contains an object called 1 (read "one").

II: For each A there exists exactly one natural number, called the *successor* of A, which will be denoted by A'.

III: We always have $A' <> 1$. That is, there exists no number whose successor is 1.

IV: If $A' = B'$ then $A = B$. That is, for any given number there exists either no number or exactly one number whose successor is the given number.

V: (Axiom of Induction): Let there be given a set **N** of natural numbers such that,

- 1 belongs to **N**,
- If A belongs to **N** then so does A'.

It follows that **N** contains all the natural numbers.

The set of natural numbers is *ordered*, since for any two natural numbers A and B, one of the following holds:

A is greater than B: $A > B$,
or A is equal to B: $A = B$,
or A is less than B: $A < B$.

We define an *addition* operation + on natural numbers, where

- $A + 1 = A'$. That is, adding 1 to any number gives the successor of that number.
- $A + B' = (A + B)'$. That is, adding the successor of any number to any number gives the successor of the sum.

By repeated induction on B we get $A + B =$ the Bth successor of A. We also define a *multiplication* operation $\times$ on natural numbers, where

- $A \times 1 = A$. That is, multiplying any number by 1 gives the same number.
- $A \times B' = (A \times B) + A$. That is, multiplying any number by the successor of any number gives the same as *A* added to the product of the two numbers.

By repeated induction on B we get $A \times B =$ A added to itself B times.

The set of natural numbers **N** is *closed* with respect to addition and multiplication; that is, every sum and product is also in **N**.

1.3 Integer Number

The set of *integer numbers* **Z** (for German Zahlen) is constructed from ordered pairs of natural numbers (A,B), and an equivalence relation $\equiv$ between such pairs, where

$$(A,B) \equiv (C,D) \text{ if } A + D = B + C\,.$$

This is another way of saying that B subtracted from A is the same as D subtracted from C.

We define an *equivalence class* [(A,B)] as that class which has (A,B) as a member; that is, all the pairs which are equinumerous with (A,B). An *integer number* is an equivalence class [(A,B)], and we write:

- Z if A $>$ B (a positive integer),
- 0 if A $=$ B (zero, neither positive nor negative),
- $-Z$ if A $<$ B (a negative integer).

For example:

$$\begin{aligned}
&\dots \\
2 &= [(3,1)] = [(4,2)] = \dots = [(N+2,N)] \\
1 &= [(2,1)] = [(3,2)] = \dots = [(N+1,N)] \\
0 &= [(1,1)] = [(2,2)] = \dots = [(N,N)] \\
-1 &= [(1,2)] = [(2,3)] = \dots = [(N,N+1)] \\
-2 &= [(1,3)] = [(2,4)] = \dots = [(N,N+2)] \\
&\dots
\end{aligned}$$

Addition, multiplication and subtraction between equivalence classes is defined as

$$\begin{aligned}
[(A,B)] + [(C,D)] &= [(A+C, B+D)] \\
[(A,B)] \times [(C,D)] &= [(AC+BD, AD+BC)] \\
[(A,B)] - [(C,D)] &= [(A+D, B+C)] \,.
\end{aligned}$$

Subtraction is the inverse of addition. The set of integer numbers **Z** is closed with respect to addition, multiplication and subtraction; that is, every sum, product and difference of integers is also in **Z**.

The natural numbers are a subset of the integers.

1.4 Rational Number

The set of *rational numbers* **Q** (for Quotient) is constructed from (ordinary) division of one integer by another integer. Given integer A and integer B, we get the rational number A/B. This can also be represented as a *fraction* with an *integer part* (possibly 0) and a *fractional part* which terminates or repeats (possibly 0). For example:

$$\begin{aligned}
1/1 &= 1.0\\
1/2 &= 0.5\\
1/3 &= 0.333\ldots\\
&\ldots\\
2/1 &= 2.0\\
2/2 &= 1.0\\
2/3 &= 0.666\ldots\\
&\ldots
\end{aligned}$$

Between any two rational numbers we can insert further rational numbers without limit. For example, between $1/1 = 1.0$ and $1/2 = 0.5$ we have $1/3 = 0.333\ldots$, $1/4 = 0.25$, $1/5 = 0.2$, ...

Division is the inverse of multiplication. The set of rational numbers **Q** is closed with respect to addition, multiplication, subtraction, and division; that is, every sum, product, difference, and quotient of rationals is also in **Q**.

The integers are a subset of the rationals; for example

$$1 = 1/1 = 2/2 = 3/3\ldots;\, 2 = 2/1 = 4/2 = 6/3\ldots.$$

An *arithmetic progression* is a succession of numbers with a constant difference between neighbours, such as the integers:

$$-\infty, \ldots, -3, -2, -1, 0, 1, 2, 3, \ldots, \infty\,.$$

A *harmonic progression* is a succession of reciprocals of numbers in an arithmetic progression:

$$-0, \ldots, -1/3, -1/2, -1, \infty, 1, 1/2, 1/3, \ldots, 0\,.$$

1.5 Irrational Number

An *irrational number* has a fractional part which does not terminate or repeat; for example:

$$\begin{aligned}
\sqrt{2} &= 1.4142\ldots\\
\pi &= 3.1416\ldots
\end{aligned}$$

Between any two rational numbers we can insert further irrational numbers without limit.

1.6 Real Number

The integers, rationals, and irrationals together constitute the *continuum of real numbers* **R**.

1.7 Number Variables

We shall generally use italic capitals $I, J, K, \ldots$ for variables taking integer values, and italic lower-case $r, s, t, \ldots$ for variables taking real values.

References

1. Dedekind R (1888) Was sind und was sollen die Zahlen? Vieweg, Braunschweig. English edition: Dedekind R (1963) Essays on the theory of numbers. Dover, New York
2. Landau E (1930) Grundlagen der Analysis. Akademischhe Verlagagesellschaft MBH, Leipzig. English edition: Landau E (1951) Foundations of analysis. Chelsea, New York
3. Peano G (1889) Arithmetices Principia, Nova Methodo Exposita. Bocca, Turin (written in Peano's own Latino Sine Flexione)

Chapter 2
Number Input

Numbers are variously represented by symbols in decimal, binary, and hexadecimal notation, and by ASCII codes. Decimal representation is what humans habitually use; binary and hexadecimal representations are what most computers use; and ASCII code links human to computer representations. Any such representation can be converted to any of the others. A digital computer represents binary notation internally as binary hardware. A flowchart shows the flow of information input to, processed by, stored by, and output from a computer.

2.1 Representation of Numbers

Numbers are abstract objects which platonists believe exist outside time and space, and nominalists believe exist only as human constructs. Either way, we have to represent numbers by physical *symbols*. Different representations are used for various purposes:

- Ten decimal digit symbols, for ordinary human use.
- Two binary digit symbols, for computer use.
- Sixteen hexadecimal digit symbols, for computer use.
- 128 ASCII codes, representing text characters as well as numbers, for bridging human and computer use.

2.2 Decimal Representation of Integers

Decimal representation is what we use for everyday counting and arithmetic. In decimal or *base-ten* representation there are ten digits:

$$0, 1, 2, 3, 4, 5, 6, 7, 8, 9$$

A. Parkin, *Digital Imaging Primer*, DOI 10.1007/978-3-540-85619-1_2

used for zero and its first nine successors. Then

- the successor of 9 is 10 (that is, the base itself)
- the successor of 10 is 11 (that is, base + 1)
- the successor of 11 is 12 (that is, base + 2)

  ...
- the successor of 19 is 20 (that is, base × 2)
- the successor of 20 is 21 (that is, (base × 2) + 1)
- the successor of 21 is 22 (that is, (base × 2) + 2)

  ...
- the successor of 99 is 100 (that is, base × base)
- the successor of 100 is 101 (that is, (base × base) + 1)
- the successor of 101 is 102 (that is, (base × base) + 2)

  ...

In such a *position* representation, each position $P = 0, 1, 2, \ldots$(counting from right to left) gives a *weight* of $(\text{base})^{\text{P}}$ to its digit.

There is also a matching sequence of decimal numbers, each prefixed with a "–", representing the negative integers.

2.3 Binary Representation of Integers

In binary or *base-two* representation there are two digits:

0, 1

used for zero and its first successor. Then

- the successor of 1 is 10 (that is, the base itself)
- the successor of 10 is 11 (that is, base + 1)
- the successor of 11 is 100 (that is, base × base)
- the successor of 100 is 101 (that is, (base × base) + 1)
- the successor of 101 is 110 (that is, (base × base) + 10)
- the successor of 110 is 111 (that is, (base × base) + 11)
- the successor of 111 is 1000 (that is, base × base × base)
- the successor of 1000 is 1001 (that is, (base × base × base) + 1)

  ...

And so on, each power of the base represented by one further digit to the left. There is also a matching sequence of binary numbers, each prefixed with a "–", representing the negative integers.

2.4 Hexadecimal Representation of Integers

In hexadecimal or *base-sixteen* representation there are sixteen digits:

0, 1, 2, 3, 4, 5, 6, 7, 8, 9, A, B, C, D, E, F

used for zero and its first fifteen successors. Then

- the successor of F is 10 (that is, the base itself)
- the successor of 10 is 11 (that is, base + 1)
- the successor of 11 is 12 (that is, base + 2)

  ...
- the successor of 1 F is 20 (that is, base × 2)
- the successor of 20 is 21 (that is, (base × 2) + 1)
- the successor of 21 is 22 (that is, (base × 2) + 2)

  ...
- the successor of FF is 100 (that is, base × base)
- the successor of 100 is 101 (that is, (base × base) + 1)
- the successor of 101 is 102 (that is, (base × base) + 2)

  ...

And so on, each power of the base represented by one further digit to the left. There is also a matching sequence of hexadecimal numbers, each prefixed with a "–", representing the negative integers.

Hexadecimal is a convenient abbreviation for binary, since each hexadecimal digit 0 to F corresponds to one of the 16 possible combinations of four binary digits 0 and 1.

2.5 ASCII Representation of Characters

The American Standard Code for Information Interchange or *ASCII*[1] [2] defines 128 codes representing 95 *printable characters*, including the ten decimal digits, plus 33 *control codes*. The control codes 0–31 and 127 decimal are now virtually obsolete.

The following tables show ASCII character-to-code and code-to-character conversions. "spc" means space.

```
ASCII PRINTABLE CHARACTER TO HEXADECIMAL
Enter by character (e.g. X) to find hexadecimal code 58
(which is binary 1011000). The first 32 codes, hex 00
to 1F, and the last code, hex 7F, are non-printable
control characters
```

[1]First developed by a committee of the American Standards Association (ASA) in 1963 as a voluntary standard agreed by the main manufacturers. Revisions were issued in 1967 (when ASA had become USASI), 1986 (when USASI had become ANSI), and 1992. It was made mandatory for computers bought by the US Government in 1969.

```
Char|spc  !   "   #   $   *   &   '   (   )   *   +   ,   -   .   /
Hex | 20 21  22  23  24  25  26  27  28  29  2A  2B  2C  2D  2E  2F

Char|  0  1   2   3   4   5   6   7   8   9   :   ;   <   =   >   ?
Hex | 30 31  32  33  34  35  36  37  38  39  3A  3B  3C  3D  3E  3F

Char|  @  A   B   C   D   E   F   G   H   I   J   K   L   M   N   O
Hex | 40 41  42  43  44  45  46  47  48  49  4A  4B  4C  4D  4E  4F

Char|  P  Q   R   S   T   U   V   W   X   Y   Z   [   \   ]   ^   _
Hex | 50 51  52  53  54  55  56  57  58  59  5A  5B  5C  5D  5E  5F

Char|  `  a   b   c   d   e   f   g   h   i   j   k   l   m   n   o
Hex | 60 61  62  63  64  65  66  67  68  69  6A  6B  6C  6D  6E  6F

Char|  p  q   r   s   t   u   v   w   x   y   z   {   |   }   ~
Hex | 70 71  72  73  74  75  76  77  78  79  7A  7B  7C  7D  7E

HEXADECIMAL TO ASCII PRINTABLE CHARACTER
Enter by hexadecimal code (e.g. 58) to find character X

Hex | 20 21  22  23  24  25  26  27  28  29  2A  2B  2C  2D  2E  2F
Char|spc  !   "   #   \   *   &   '   (   )   *   +   ,   -   .   /

Hex | 30 31  32  33  34  35  36  37  38  39  3A  3B  3C  3D  3E  3F
Char|  0  1   2   3   4   5   6   7   8   9   :   ;   <   =   >   ?

Hex | 40 41  42  43  44  45  46  47  48  49  4A  4B  4C  4D  4E  4F
Char|  @  A   B   C   D   E   F   G   H   I   J   K   L   M   N   O

Hex | 50 51  52  53  54  55  56  57  58  59  5A  5B  5C  5D  5E  5F
Char|  P  Q   R   S   T   U   V   W   X   Y   Z   [   \   ]   ^   _

Hex | 60 61  62  63  64  65  66  67  68  69  6A  6B  6C  6D  6E  6F
Char|  `  a   b   c   d   e   f   g   h   i   j   k   l   m   n   o

Hex | 70 71  72  73  74  75  76  77  78  79  7A  7B  7C  7D  7E
Char|  p  q   r   s   t   u   v   w   x   y   z   {   |   }   ~

ASCII PRINTABLE CHARACTER TO DECIMAL
Enter by character (e.g. X) to find decimal code 88

Char|spc  !   "   #  \$   *   &   '   (   )   *   +   ,   -   .   /
Dec | 32 33  34  35  36  37  38  39  40  41  42  43  44  45  46  47
```

```
Char|  0  1  2  3  4  5  6  7  8  9  :  ;  <  =  >  ?
Dec | 48 49 50 51 52 53 54 55 56 57 58 59 60 61 62 63

Char|  @  A  B  C  D  E  F  G  H  I  J  K  L  M  N  O
Dec | 64 65 66 67 68 69 70 71 72 73 74 75 76 77 78 79

Char|  P  Q  R  S  T  U  V  W  X  Y  Z  [  \  ]  ^  _
Dec | 80 81 82 83 84 85 86 87 88 89 90 91 92 93 94 95

Char|  `  a  b  c  d  e  f  g  h  i  j  k  l  m  n  o
Dec | 96 97 98 99 00 01 02 03 04 05 06 07 08 09 10 11

Char|  p  q  r  s  t  u  v  w  x  y  z  {  |  }  ~
Dec | 12 13 14 15 16 17 18 19 20 21 22 23 24 25 26
```

ASCII HEXADECIMAL TO CHARACTER
Enter by hexadecimal column head (e.g. 50) and row stub (e.g. 8) to find hexadecimal 58 codes character X

```
  | 00  10  20  30  40  50  60  70
----------------------------------
0 | nul dle spc 0   @   P   `   p
1 | soh dc1 !   1   A   Q   a   q
2 | stx dc2 "   2   B   R   b   r
3 | etx dc3 #   3   C   S   c   s
4 | eot dc4 $   4   D   T   d   t
5 | enq nak %   5   E   U   e   u
6 | ack syn &   6   F   V   f   v
7 | bel etb '   7   G   W   g   w
8 | bs  can (   8   H   X   h   x
9 | ht  em  )   9   I   Y   i   y
A | nl  sub *   :   J   Z   j   z
B | vt  esc +   ;   K   [   k   {
C | np  fs  ,   <   L   \   l   |
D | cr  gs  -   =   M   ]   m   }
E | so  rs  .   >   N   ^   n   ~
F | si  us  /   ?   O   _   o  del
```

ASCII DECIMAL TO CHARACTER
Enter by decimal column head (e.g. 080) and row stub (e.g. 8) to find decimal 88 codes character X

```
  | 000 010 020 030 040 050 060 070 080 090 100 110 120
-------------------------------------------------------
0 | nul nl  dc4 rs   (   2   <   F   P   Z   d   n   x
1 | soh vt  nak us   )   3   =   G   Q   [   e   o   y
2 | stx np  syn spc  *   4   >   H   R   \   f   p   z
3 | etx cr  etb  !   +   5   ?   I   S   ]   g   q   {
4 | eot so  can  "   ,   6   @   J   T   ^   h   r   |
5 | enq si  em   #   -   7   A   K   U   _   i   s   }
6 | ack dle sub  $   .   8   B   L   V   `   j   t   ~
7 | bel dc1 esc  %   /   9   C   M   W   a   k   u  del
8 | bs  dc2 fs   &   0   :   D   N   X   b   l   v
9 | ht  dc3 gs   '   1   ;   E   O   Y   c   m   w
```

2.6 Correspondence Between Representations

Clearly, the integers in any of the representations correspond 1–1 with the integers in any of the other notations, as shown in Table 2.1.

2.7 Computer Representation of Numbers

Within a digital computer, binary representation of numbers is represented by hardware elements capable of two states (off/on, negative/positive, ...), corresponding to binary digit symbols (0 and 1). One such hardware element is called a *bit*.[2]

- One bit holds one binary digit 0 or 1
- One *nibble*[3] = four bits holds four binary digits 0000 to 1111, = one hexadecimal digit 0–F, = decimal 0–15
- One *byte*[4] = 8 bits holds eight binary digits 00000000 to 11111111, = two hexadecimal digits 00–FF, = decimal 0–255. Seven bits of a byte hold one ASCII code
- One *word* = 32 bits holds the block of data which the computer reads for input, processes, stores to file, or writes for output. Current personal computers usually have a 32-bit word, larger machines 64-bit

[2]Contraction of binary digit, originated about 1943 by John Wilder Tukey (1915–2000), Princeton University and Bell Laboratories.

[3]Half a byte.

[4]Term originated in 1956 by Werner Buchholz, IBM. Other lengths of byte have been used, but are now obsolete.

Table 2.1 Correspondence between integers in binary, decimal, hexadecimal, and ASCII notation

Bin.	0	1	10	11	100	101	110	111	1000	1001	1010	1011	1100	1101	1110	1111
Dec.	0	1	2	3	4	5	6	7	8	9	10	11	12	13	14	15
Hex.	0	1	2	3	4	5	6	7	8	9	A	B	C	D	E	F
ASCII	48	49	50	51	52	53	54	55	56	57						

Notice that each hexadecimal digit abbreviates a four-binary-digit combination (assumed to be zero-padded to the left)

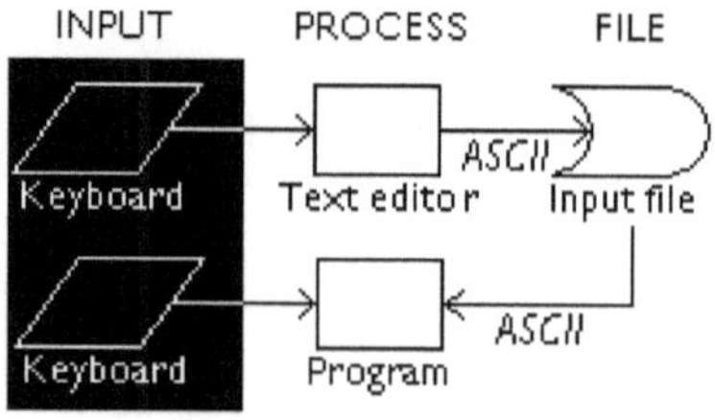

Fig. 2.1 Flowchart for number input to a computer system. A user may input alphanumerics via the keyboard of a text editor (such as Microsoft Notepad), which stores them as ASCII code in a named data file, which can then be used for input to a program. Or a user may input alphanumeric data directly to a program

2.8 Flowchart

A flowchart[5] as used in this book traces the broad flow of information which enters a computer system (input), is changed within the system (process), is stored within the system (file), and is delivered from the system (output) [3]. A sloped box shows an input device (such as a keyboard, mouse, scanner or camera) or an output device (such as a display screen or printer). A rectangle shows a computer process (such as a program). A box with dished ends shows a storage device (such as a hard disk or a removable memory stick) where information is held in an agreed structure as a file. An arrow shows the flow from box to box.

The flowchart in Fig. 2.1 shows alphanumeric input to a program, either from an ASCII keyboard to a stored file, or directly from a keyboard to a program.

The program in Sect. 3.22 shows an example of simple input from a keyboard.

2.9 Development of Computer Input

Very early computers accepted input by punched cards[6] or by punched paper tape.[7]

General input of programs, data, and controls to current PCs is usually from a standard keyboard, with 101 keys, 48 of which show printable characters, which a *shift* function extends to the 95 ASCII printable characters.

Arithmetic input can be made from a small arithmetic keypad arranged rather like a pocket calculator.

---

[5]Flowcharts for industrial motion study were used by Frank Gilbreth (1868–1924) and others from 1921. Flowcharts for computer processes were originated by Herman Goldstine (1913–2004) and John von Neumann (1903–1957).

[6]From IBM's long experience with the Hollerith 80-column punched card, where a hole (= 1) or no-hole (= 0) is detected electro-mechanically.

[7]From teleprinter experience with 5-track tape, where a hole or no-hole is detected opto-mechanically.

ASCII is a seven-bit code, covering 128 values 0 to 127. An eight-bit byte covers 256 values 0 to 255, so computer manufacturers used the upper 128 values for various purposes. In 1981 IBM introduced *code-page 437* for European accented characters and line graphic elements; and other code-pages followed. The usual keyboard access to ASCII 128-255 is by holding down the Alt key and entering the code on the numeric keypad [1]; for example the square-root sign is Alt+251.

The Universal Character Set (UCS) [4] assigns names and codes to the characters used by practically all world languages, and by HTML, and special symbols. The coding is by UCS Transformation Format (UTF-8) [5], which defines a unique code U+hhhh (h = hexadecimal digit) for every character. The ASCII characters are a sub-set of UCS: so, for example, the ASCII decimal digits 0 to 9 are UCS codes U+0030 to U+0039.

A full (or reduced) keyboard can be replaced by an image on a touch-screen. A cable connection from keyboard to computer can be replaced by a wireless connection.

There is steady development of Optical Character Recognition (OCR) software for printed or typed sources, as an alternative to a keyboard.

Speech recognition software is sometimes used for alphanumeric input.

References

1. Alt-Codes.net (2013) List of Alt-key Codes. http://www.alt-codes.net/. Accessed 1 Feb 2014
2. ANSI INCITS 4-1986 (R2007) 7-Bit American National Standard Code for Information Interchange (7-Bit ASCII). http://webstore.ansi.org/. Accessed 1 Feb 2014
3. ISO 5807:1985, Information processing—documentation symbols etc. http://www.iso.org/iso/home/standards.htm. Accessed 1 Feb 2014
4. ISO/IEC10646:2012, Information technology—Universal Coded Character Set (UCS). http://standards.iso.org/ittf/PubliclyAvailableStandards/index.html. Accessed 1 Feb 2014
5. Kuhn M (2011) UTF-8 and unicode FAQ. http://www.cl.cam.ac.uk/~mgk25/unicode.html. Accessed 1 Feb 2014

Chapter 3
Number Process

An algorithm is a recipe for using a set of simple operations to achieve a required result. Six elementary operations are sufficient to do all of integer arithmetic and integer matrix arithmetic. Real arithmetic is significantly different from integer arithmetic. A Boolean algebra is a binary structure which is fundamental to digital computer hardware, and is defined here with its operations and rules. To get a computer to do arithmetic, some programming language must be chosen. In this book, Microsoft QuickBASIC is used for examples throughout. A simple example program is given here, which inputs two integers, does some arithmetic, and displays the result. Several other current programming languages are listed.

3.1 Algorithm

An *algorithm*,[1] is a sequence of elementary instructions which produce a required result. A person, or a machine, capable of performing a small set of simple steps can thereby arrive at quite complex goals. For a person, an algorithm can be presented in a loose pseudo-code, as for example the following conversion from decimal to binary notation:

```
begin
   comment variables r a b c are decimal integer
   comment variables d e are binary integer
   r = 0
   comment r is a row counter
   a = the given decimal representation
   b = a \ 2
   c = a MOD 2
```

[1]Earlier *algorism* derived from the Latin form of the Persian savant's name Al-Khwarizmi (ca. 780–850).

A. Parkin, *Digital Imaging Primer*, DOI 10.1007/978-3-540-85619-1_3

Table 3.1 Example of converting given decimal 123 to binary 1111011

R	A	B	C	D	E
0	**123**	61	1	1	1
1	61	30	1	10	11
2	30	15	0	000	011
3	15	7	1	1000	1011
4	7	3	1	10000	11011
5	3	1	1	100000	111011
6	1	0	1	1000000	**1111011**

```
        d = c * 10^r
        e = d
        do until b = 0
            r = r + 1
            a = b
            b = a \ 2
            c = a MOD 2
            d = c * 10^r
            e = e + d
        loop
        comment e is the required binary representation
    end
```

Table 3.1 shows an example of the progressive steps of this algorithm, given decimal number 123. In practice, decimal-to-binary conversion is normally done within a programming language, or by a pocket calculator.

An inverse algorithm to convert binary to decimal is

```
    begin
        comment variables r a b c are binary integer
        comment variables d e are decimal integer
        r = 0
        comment r is a row counter
        a = the given binary representation
        b = a \ 1010
        c = a MOD 1010
        d = c * 10^r
        e = d
        do until b = 0
            r = r + 1
            a = b
            b = a \ 1010
            c = a MOD 1010
```

```
        d = c * 10^r
        e = e + d
    loop
    comment e is the required decimal representation
end
```

Table 3.2 shows an example of the progressive steps of this algorithm, given binary number 1111011.

An algorithm to convert decimal to hexadecimal is

```
begin
    comment variables r a b c are decimal integer
    comment variables d e are hexadecimal integer
    r = 0
    comment r is a row counter
    a = the given decimal representation
    b = a \ 16
    c = a MOD 16
    d = c * 10^r
    e = d
    do until b = 0
        r = r + 1
        a = b
        b = a \ 16
        c = a MOD 16
        d = c * 10^r
        e = e + d
    loop
    comment e is the required hexadecimal representation
end
```

Table 3.3 shows an example of the progressive steps of this algorithm, given decimal number 123. In practice, decimal-to-hexadecimal conversion is normally done within a programming language, or by a pocket calculator.

Table 3.2 Example of converting given binary 1111011 to decimal 123

R	A	B	C	D	E
0	**1111011**	1100	11	3	3
1	1100	1	10	20	23
10	1	0	1	100	**123**

Table 3.3 Example of converting given decimal 123 to hexadecimal 7B

R	A	B	C	D	E
0	**123**	7	11	B	B
1	7	0	7	70	**7B**

Table 3.4 Example of converting given hexadecimal 7B to decimal 123

R	A	B	C	D	E
0	**7B**	7	B	11	11
1	7	0	7	112	**123**

An inverse algorithm to convert hexadecimal to decimal is

```
begin
   comment variables r a b c are hexadecimal integer
   comment variables d e are decimal integer
   r = 0
   comment r is a row counter
   a = the given hexadecimal representation
   b = a \ 10
   c = a MOD 10
   d = c * 16^r
   e = d
   do until b = 0
      r = r + 1
      a = b
      b = a \ 10
      c = a MOD 10
      d = c * 16^r
      e = e + d
   loop
   comment e is the required hexadecimal representation
end
```

Table 3.4 shows an example of the progressive steps of this algorithm, given hexadecimal number 7B.

3.2 Integer Arithmetic

Integer arithmetic is any calculation done by applying six elementary operations to given integers. The elementary operations are

- Integer addition.
- Integer subtraction.
- Integer multiplication.
- Integer division.

- Integer remainder.
- Integer power.

Since each operation gives an integer result from integer values, the sequence of integers is *closed* under integer arithmetic. There are no fractions, and no further numbers can be inserted between any two successive integers.

3.3 Integer Addition

Given any two integers A and B, we can *add* them to get the *integer sum* C:

$$A + B = C\,.$$

In decimal representation, we add digit-wise from right to left. For each digit position, if the digit sum exceeds the base, 10, we set the excess and *carry* 1 to the next position at left. In binary representation, the entire addition table is shown in Table 3.5.

3.4 Integer Subtraction

Given any two integers A and B, we can *subtract* B from A to get the *integer difference* D:

$$A - B = D\,.$$

Subtraction is the inverse of addition. In decimal representation, we subtract digitwise from right to left; for each position, when the second term exceeds the first we *borrow* 1 from the next position at left. In binary representation, the entire subtraction table is shown in Table 3.6.

Table 3.5 Binary addition table

A	B	$A + B$	Carry
0	0	0	0
0	1	1	0
1	0	1	0
1	1	0	1

Table 3.6 Binary subtraction table

A	B	$A - B$	Borrow
0	0	0	0
0	1	1	1
1	0	1	0
1	1	0	0

3.5 Integer Multiplication

Given any two integers A and B, we can *multiply* them to get the *integer product* E:

$$A \times B = E\,.$$

Multiplication is repeated addition of A, B times over. Multiplication of variables is sometimes written without an interposed symbol as AB; or a number and a variable as $2A$. In decimal representation, we multiply digit-wise from right to left; for each position, when the product exceeds the base we set the excess and carry 1 or more to the next position at left.

In binary representation, there is no carrying, and the entire multiplication table is shown in Table 3.7.

3.6 Integer Division

Given any two integers A and B, $B \neq 0$, we can *integer-divide* A by B to get the *integer quotient* F:

$$F = A \backslash B\,,$$

(sometimes written as $F = A$ DIV B). Integer division is repeated subtraction: of B from A, then B from the remainder, ..., each subtraction adding 1 to F, until the remainder is less than B, when the remainder is discarded. $A \setminus 0$ is undefined. In binary representation, the entire integer-division table is shown in Table 3.8.

Table 3.7 Binary multiplication table

A	B	$A \times B$
0	0	0
0	1	0
1	0	0
1	1	1

Table 3.8 Binary integer-division table

A	B	$A \setminus B$
0	0	Undefined
0	1	0
1	0	Undefined
1	1	1

3.7 Integer Remainder

Given any two integers A and B, $B \neq 0$, we can get the *integer remainder* G when A is integer-divided by B:

$$A \text{ MOD } B = G .$$

In binary representation, the entire remainder table is shown in Table 3.9.

3.8 Integer Power

Given any two integers A and B, we can raise A to the Bth *non-negative integer power* H:

$$A^{\mathrm{B}} = H .$$

Raising to a power, or *evolution*, is repeated multiplication of A by itself, B times over. The zeroth power of any number is 1, and the first power of any number is the number itself. Negative powers are excluded from integer arithmetic, since $A^{-\mathrm{B}} = 1/A^{\mathrm{B}}$, a rational number.

In binary representation, the entire power table is shown in Table 3.10. Some further numerical structures and operations used in digital imaging are defined below.

Table 3.9 Binary remainder table

A	B	A MOD B
0	0	Undefined
0	1	1
1	0	Undefined
1	1	0

Table 3.10 Binary power table

A	B	A^{B}
0	0	1
0	1	0
1	0	1
1	1	1

Table 3.11 Binary rational-multiplier table

I	A	B	J	K
0	0	0	Undefined	
0	0	1	0	0
0	1	0	Undefined	
0	1	1	0	0
1	0	0	Undefined	
1	0	1	0	0
1	1	0	Undefined	
1	1	1	1	0

3.9 Rational Multiplier

A *rational multiplier* is an ordered pair of integers $A \setminus B$, $B \neq 0$, where A is understood as a multiplier and B as an integer-divisor. Given any integer I and any rational multiplier $A \setminus B$, we can first multiply $I \times A$ then integer-divide the product by B to get the integer quotient J and the integer remainder K:

$$I \times (A \setminus B) = J\,,$$
$$I \text{ MOD}(A \setminus B) = K\,.$$

In binary representation, the entire rational-multiplier table is shown in Table 3.11. A rational multiplier is a notational convenience in integer arithmetic.

3.10 Integer Vector

An *integer vector* $\mathbf{V}$ is a one-dimensional array of N integers V_{S}, each with an integer *subscript* $1 \leq S \leq N$:

$$\mathbf{V} = \begin{pmatrix} V_1 & V_2 & \ldots & V_{\mathrm{N}} \end{pmatrix}$$

Each subscript is the address of an element in the array. For example, if a vector

$$\mathbf{V} = \begin{pmatrix} V_1 & V_2 & V_3 \end{pmatrix}$$

contains the elements:

2 4 6

then V_1 addresses 2, V_2 addresses 4, and V_3 addresses to 6.

The value held in an element is, of course, independent of the value of its subscript address.

3.11 Integer Matrix

An *integer matrix* **A** is a two-dimensional array of integers A, each with two integer subscripts, the first showing the element's *row* and the second showing the element's *column* in the array:

$$\mathbf{A} = \begin{pmatrix} A_{11} & A_{12} & \dots & A_{1N} \\ A_{21} & A_{22} & \dots & A_{2N} \\ \vdots & \vdots & \vdots & \vdots \\ A_{M1} & A_{M2} & \dots & A_{MN} \end{pmatrix}$$

Each pair of subscripts is the address of an element in the array. For example, if a matrix

$$\mathbf{A} = \begin{pmatrix} A_{11} & A_{12} & A_{13} \\ A_{21} & A_{22} & A_{23} \\ A_{31} & A_{32} & A_{33} \end{pmatrix}$$

contains the elements

$$\begin{matrix} 2 & 4 & 6 \\ 3 & 6 & 9 \\ 4 & 8 & 12 \end{matrix}$$

then A_{11} addresses 2, A_{12} addresses 4, A_{13} addresses 6; A_{21} addresses 3, A_{22} addresses 6, ...; and A_{33} addresses 12.

The value held in an element is, of course, independent of the value of its subscripts address.

A matrix with R rows and C columns is said to be of *order* $R \times C$. It contains R *row vectors* and C *column vectors*.

Two matrices **A** and **B** are equal if they are of the same order and each element of **A**, A_{RC}, is equal to the corresponding element of **B**, B_{RC}.

A *null* or *zero* matrix has every element 0.

A *unit* or *identity* matrix **U** is square, $R = C$, and has elements $= 1$ on the *main diagonal* and elements $= 0$ elsewhere:

$$\mathbf{U} = \begin{pmatrix} 1 & 0 & \dots & 0 \\ 0 & 1 & \dots & 0 \\ \vdots & \vdots & 1 & \vdots \\ 0 & 0 & \dots & 1 \end{pmatrix}$$

The *transpose* of a matrix switches rows to columns and columns to rows. For example, if

$$\mathbf{A} = \begin{pmatrix} 1 & 2 & 3 \\ 4 & 5 & 6 \end{pmatrix}$$

then the transpose is

$$\mathbf{A}^{\mathrm{T}} = \begin{pmatrix} 1 & 4 \\ 2 & 5 \\ 3 & 6 \end{pmatrix}$$

3.12 Matrix Addition

Given any two matrices of the same order **A** and **B**, we can *add* them to get the *sum matrix* **C**:

$$\mathbf{A} + \mathbf{B} = \mathbf{C}$$

where corresponding elements are added: $C_{\mathrm{NM}} = A_{\mathrm{NM}} + B_{\mathrm{NM}}$. For example:

$$\begin{pmatrix} 1 & 2 & 3 \\ 4 & 5 & 6 \end{pmatrix} + \begin{pmatrix} 7 & 8 & 9 \\ 10 & 11 & 12 \end{pmatrix} = \begin{pmatrix} 8 & 10 & 12 \\ 14 & 16 & 18 \end{pmatrix}$$

3.13 Matrix Subtraction

Given any two matrices of the same order **A** and **B**, we can *subtract* **B** from **A** them to get the *difference matrix* **C**:

$$\mathbf{A} - \mathbf{B} = \mathbf{C}$$

where corresponding elements are subtracted: $C_{\mathrm{NM}} = A_{\mathrm{NM}} - B_{\mathrm{NM}}$. For example:

$$\begin{pmatrix} 1 & 2 & 3 \\ 4 & 5 & 6 \end{pmatrix} - \begin{pmatrix} 7 & 8 & 9 \\ 10 & 11 & 12 \end{pmatrix} = \begin{pmatrix} -6 & -6 & -6 \\ -6 & -6 & -6 \end{pmatrix}$$

3.14 Scalar Multiplication

A *scalar* is a 1×1 matrix, that is, a single integer. Given a scalar K and a matrix **A**, we can *multiply* them to get the *scalar product* matrix:

$$K\mathbf{A}$$

where each element A_{NM} is multiplied by K. For example:

$$7\begin{pmatrix}1&2&3\\4&5&6\end{pmatrix}=\begin{pmatrix}7&14&21\\28&35&42\end{pmatrix}$$

3.15 Matrix Multiplication

Given two matrices **A** and **B** where the number of columns of **A** is the same as the number of rows of **B**, we can *multiply* them to get the *product* matrix **C**:

$$\mathbf{AB}=\mathbf{C}$$

If **A** is of order $R \times C$ and **B** is of order $C \times S$, then **C** is of order $R \times S$. We find C_{NM} as the sum of the products of the Cth row of **A** with the Rth column of **B**; that is

$$C_{\text{CR}} = A_{\text{C1}}B_{\text{1R}} + A_{\text{C2}}B_{\text{2R}} + \cdots + A_{\text{CS}}B_{\text{SR}}$$

For example:

$$\begin{pmatrix}1&2&3\\4&5&6\end{pmatrix}\begin{pmatrix}7&8\\9&10\\11&12\end{pmatrix}=\begin{pmatrix}58&64\\139&154\end{pmatrix}$$

where

$$\begin{aligned}(1\times 7)+(2\times 9)+(3\times 11)&=7+18+33=58\\(1\times 8)+(2\times 10)+(3\times 12)&=8+20+36=64\\(4\times 7)+(5\times 9)+(6\times 11)&=28+45+66=139\\(4\times 8)+(5\times 10)+(6\times 12)&=32+50+72=154\,.\end{aligned}$$

Integer multiplication is *commutative*: $AB = BA$. However, matrix multiplication is *not commutative*: $\mathbf{AB} \neq \mathbf{BA}$. For example:

$$\begin{pmatrix}7&8\\9&10\\11&12\end{pmatrix}\begin{pmatrix}1&2&3\\4&5&6\end{pmatrix}=\begin{pmatrix}35&54&69\\45&68&87\\59&82&105\end{pmatrix}$$

In **AB** we say that **A** is *post-multiplied* by **B**; and in **BA** we say that **A** is *pre-multiplied* by **B**.

3.16 Integer Matrix Inversion

The *determinant* of a square matrix **A** is a single number det | A |. The determinant of a 1×1 matrix is just that number:

$$\det \left| A \right| = A$$

The determinant of a 2×2 matrix is

$$\det \begin{vmatrix} A & B \\ C & D \end{vmatrix} = AD - BC$$

where A, B, C, D are particular integer values. The determinant of a 3×3 matrix of the special kind used for coordinate transformations (see Chap. 18) is also

$$\det \begin{vmatrix} A & B & 0 \\ C & D & 0 \\ E & F & 1 \end{vmatrix} = AD - BC$$

The *inverse* of a square matrix **A** is a square matrix $\mathbf{A^1}$ such that

$$\mathbf{AA^1} = \mathbf{A^1A} = \mathbf{U}$$

that is, the product is the unit matrix (so inversion is the matrix analogue of division).

A square matrix has an *integer inverse* if and only if its determinant is 1 or -1. Then

$$\mathbf{A} = \begin{pmatrix} A & B & 0 \\ C & D & 0 \\ E & F & 1 \end{pmatrix} \qquad \mathbf{A^1} = \det|A| \begin{pmatrix} D & -B & 0 \\ -C & A & 0 \\ CF - DE & BE - AF & AD - BC \end{pmatrix}$$

3.17 Real Arithmetic

The fundamental difference between integer arithmetic and real arithmetic is that integer division truncates the quotient to the integer part only, whereas real division continues the quotient to any number of fractional places. We may sometimes wish to estimate the *truncation error* incurred by integer arithmetic, compared with real arithmetic.

Let n be a real number, given to an arbitrary number of fractional places, and let N be its truncated integer part. The *absolute truncation error* e is the difference

$$e = n - N$$

The *relative truncation error* r is the ratio of the absolute error e to the real value n:

$$r = e/n$$

and the *percentage truncation error* p is

$$p = 100r$$

A real number n may be exact, or may be approximate to so many fractional places. But the greatest fractional part of n that can be truncated is $0.999\ldots$, that is, less than 1. So the truncation errors are limited by

- $e < (N + 1) - N = 1$,
- $r < 1/(N + 1)$,
- $p < 100/(N + 1)$.

As integer N increases in value, its absolute truncation error remains constant, and its relative and percentage truncation errors decrease, as shown in Table 3.12.

In real arithmetic, practical computations perforce use only values with a limited fractional length, that is, approximations to the (in general unlimited) real values. If we start with values with known possible errors, the absolute error of a sum/difference of two or more reals is the sum of the absolute errors of the terms. The relative or percentage error of a product/quotient of reals is the sum of the relative or percentage errors of the factors. In lengthy computations with reals there can be significant loss of accuracy.

In integer arithmetic, all values have fractional length zero. The values of integers, and sums, differences, and products of integers, are exact. Approximation occurs only in the truncation of a quotient of integers, where the absolute error is less than 1 and the relative error is less than 1/quotient (a real number). In lengthy computations with integers the number of divisions measures the possible loss of accuracy.

Table 3.12 Truncation error limits

N	e	r	p
1	1	1/2	50
10	1	1/11	9.09
100	1	1/101	0.99
1000	1	1/1001	0.10
10000	1	1/10001	0.01

3.18 Boolean Algebra

A *Boolean algebra*[2] [1] is a set of elements $A, B, C, \ldots$ where

- each element has value 0 or 1,
- a unary operation *negation* $\neg$ is defined,
- two binary operations *disjunction* $\vee$ and *conjunction* $\wedge$ are defined.

Negation (NOT) is defined by Table 3.13. Element A has two possible values, 0 or 1. When A has value 0, $\neg A$ has value 1; and when A has value 1, $\neg A$ has value 0.

Disjunction (OR) is defined by Table 3.14. $A \vee B$ is 1 if either A or B or both are 1. It is like binary addition (Table 3.5) without the carry.

Conjunction (AND) is defined by Table 3.15. $A \wedge B$ is 1 if both A and B are 1. It is like binary multiplication (Table 3.7).

A Boolean algebra has the following axioms:

$$\begin{aligned} A \vee B &= B \vee A \text{ (commutative)} \\ A \vee (B \wedge C) &= (A \vee B) \wedge (A \vee C) \text{ (distributive)} \\ A \vee 0 &= A \text{ (identity)} \\ A \vee \neg A &= 1 \text{ (negation)}\,. \end{aligned}$$

and with *duality* under interchange of $\vee$ and $\wedge$, and 0 and 1. Thus we have immediately the dual axioms:

$$\begin{aligned} A \wedge B &= B \wedge A \text{ (commutative)} \\ A \wedge (B \vee C) &= (A \wedge B) \vee (A \wedge C) \text{ (distributive)} \\ A \wedge 1 &= A \text{ (identity)} \\ A \wedge \neg A &= 0 \text{ (negation)}\,. \end{aligned}$$

Table 3.13 Boolean negation

A	0	1
$\neg A$	1	0

Table 3.14 Boolean disjunction

A	0	0	1	1
B	0	1	0	1
$A \vee B$	0	1	1	1

Table 3.15 Boolean conjunction

A	0	0	1	1
B	0	1	0	1
$A \wedge B$	0	0	0	1

[2]George Boole (1815–1864), Professor of Mathematics at Queen's College, Cork 1849–1864.

Table 3.16 The 16 Boolean functions of two values 0 and 1

Symbol	Values				Expansion	Name
A	0	0	1	1		
B	0	1	0	1		
0	0	0	0	0	$A \wedge \neg A$	Contradiction
$A \wedge B$	0	0	0	1	$A \wedge B$	Conjunction
$A \neg \rightarrow B$	0	0	1	0	$A \wedge \neg B$	Non-implication
A	0	0	1	1	A	Left projection
$A \neg \leftarrow B$	0	1	0	0	$\neg A \wedge B$	Converse non-implication
B	0	1	0	1	B	Right projection
A XOR B	0	1	1	0	$(A \vee B) \wedge \neg(A \wedge B)$	Exclusive disjunction
$A \vee B$	0	1	1	1	$A \vee B$	(Inclusive) Disjunction
A NOR B	1	0	0	0	$\neg A \wedge \neg B$	Joint denial
A XNOR B	1	0	0	1	$(A \rightarrow B) \wedge (B \rightarrow A)$	Equivalence
$\neg B$	1	0	1	0	$\neg B$	Right complementation
$A \leftarrow B$	1	0	1	1	$A \vee \neg B$	Converse implication
$\neg A$	1	1	0	0	$\neg A$	Left complementation
$A \rightarrow B$	1	1	0	1	$\neg A \vee B$	(Material) implication
A NAND B	1	1	1	0	$\neg A \vee \neg B$	Non-conjunction
1	1	1	1	1	$A \vee \neg A$	Tautology

When A and B have the values shown in the heading, each function takes the values shown in the body

A *theorem* is any equation derivable from the axioms. The dual of a theorem is also a theorem.

De Morgan's[3] *laws* are a dual pair of theorems:

$$\neg(A \vee B) = (\neg A) \wedge (\neg B)$$
$$\neg(A \wedge B) = (\neg A) \vee (\neg B)\,.$$

Thus a negation of a disjunction can be replaced by a conjunction of negations, and a negation of a conjunction can be replaced by a disjunction of negations.

[3] Augustus De Morgan (1806–1871), first Professor of Mathematics at London, 1828–1831 and 1836–1866 (resigned twice on matters of principle).

A *Boolean function* is any combination of negations, disjunctions and conjunctions. The two values 0 and 1 produce 16 Boolean functions, as shown in Table 3.16 [5].

3.19 Computer Representation of Arithmetic

The idea of casting certain deep problems in mathematics in the form of an algorithm of instructions, executed by an imaginary machine without ingenuity or insight, was put forward by Turing[4] in 1936 [11]. At about the same time, Church[5] developed a recursive system of symbolic logic (λ-calculus) which would do much the same thing [2]. In 1945–1946 Turing and von Neumann[6] independently proposed a practical machine with a memory to store instructions and data, a unit to process instructions, and a unit to do arithmetical and logical operations. All subsequent digital computers have been based on this stored-program architecture.

Shannon[7] had the fundamental insight that Boolean algebra could be used to design electrical circuits which would perform logical, hence binary numerical, operations [9]. Shannon's scheme is based on representing binary 0 by a low voltage on a line, and binary 1 by a high voltage. A *logic gate* is a circuit which performs an elementary Boolean operation:

- A NOT gate, given low-voltage input 0, outputs high-voltage 1; given input 1, outputs 0.
- An OR gate, given two inputs, outputs 1 if either or both inputs are 1, else 0.
- An AND gate, given two inputs, outputs 1 if both inputs are 1, else 0.

Such gates can be variously combined, according to Boolean algebra, to provide circuits which represent all 16 Boolean functions. It turns out that a dual pair of gates, performing NAND and NOR, are sufficient to perform binary integer arithmetic.

Number operations can be implemented by electro-mechanical relays.[8] The earliest research computers used relays, then thermionic valves, then transistors. Digital computers now use semiconductor gates in integrated circuits, which are very fast,

---

[4]Alan Mathison Turing (1912–1954), doctorate under Church at Princeton University 1936–19388, designed computing machines for code-breaking at Bletchley Park 1939–1945, designed the ACE stored-program machine at National Physical Laboratory, Teddington 1946, software for Manchester Mark 1 machine 1949.

[5]Alonzo Church (1903–1995), Princeton University 1929–1967, University of California at Los Angeles 1967–1990.

[6]John von Neumann (1903–1957), Princeton University Institute for Advanced Study.

[7]Claude Elwood Shannon (1916–2001) worked with Vannevar Bush on the analogue electromechanical differential analyser at Massachusetts Institute of Technology 1938–1940; researched at Bell Laboratories 1941–1972 where he developed a general theory of information, with the bit as the unit [10]; Professor at MIT 1958–1978.

[8]George Robert Stibitz (1904–1995), Bell Laboratories 1930–1964, Dartmouth Medical School 1964–1970, built adders and other devices from relays in 1937 and onwards. In later life he turned to very interesting computer art.

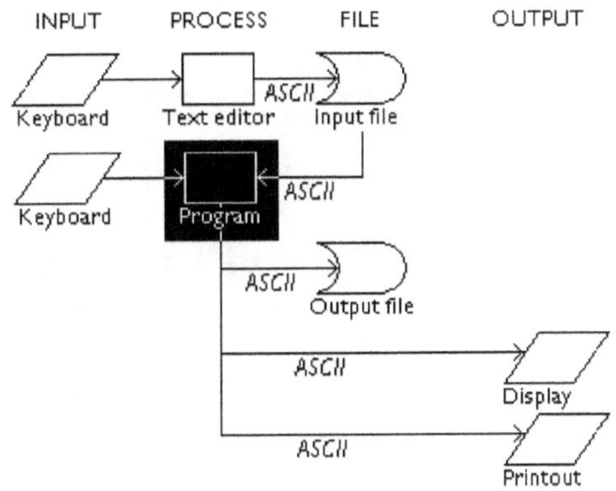

Fig. 3.1 Flowchart for number process in a computer system. A program may read number input directly from a keyboard or from a data file, perform arithmetic processes, and write the results to an ASCII file and/or a display and/or a printout

very small, use very low power, and are immune from the low-level noise which plagues analogue computing systems.

3.20 Computer Process

The flowchart (Fig. 3.1) shows a number process.

3.21 Programming Language

A digital computer works internally with very low-level instructions. It is possible, but inconvenient, to set up programs at this machine level. Normal practice is to use a high-level programming language which hidden software translates down to machine code. To do arithmetic by computer:

- The user writes (in a chosen language) a program of arithmetic operations to be performed, and inputs it as ASCII characters from a keyboard.
- The computer compiles and stores the program as a program file.
- The user runs the program.
- The computer reads any numerical data input from file or keyboard.
- The computer processes the data according to the program and outputs the results.

In this book, all the programming examples are written in Microsoft QuickBASIC 4.5 [4], the compiled version of QBasic [3], which runs in a DOS shell under Windows on a PC. It is available at [7], with tutorials at [8]. QBasic is a development from the original Beginner's All-purpose Symbolic Instruction Code.[9] It is chosen as the most

[9]Written by John Kemeny (1926–1992) and Thomas Kurtz (*1928) at Dartmouth College, New Hampshire, USA, in 1964.

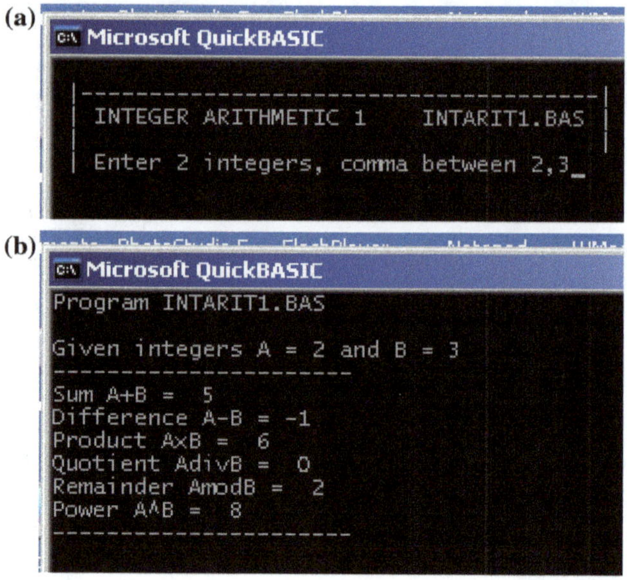

Fig. 3.2 Program INTARIT1. **a** Program screen asks for input. **b** Program screen shows output

accessible and widespread entry-level programming language, learned by many and scorned by some.

3.22 Program with Display Output

For example, the following simple program is written in QuickBASIC and stored as a program file INTARIT1.BAS. When run, the screen asks for input as shown in Fig. 3.2a then displays the results on-screen as shown in Fig. 3.2b.

```
REM  **  Program INTARIT1 to do arithmetic on two
REM  **     integers, input from keyboard, output
REM  **     to screen.
REM  **  Input numbers from keyboard
         CLS
         PRINT " "
         PRINT "|--------------------------------------|"
         PRINT "| INTEGER ARITHMETIC 1     INTARIT1.BAS |"
         PRINT "|                                      |"
         INPUT "| Enter 2 integers, comma between ", A, B
         PRINT "|                                      |"
         PRINT "|--------------------------------------|"
```

```
REM  **  Calculate
         C = A + B
         D = A - B
         E = A * B
         F = A \ B
         G = A MOD B
         H = A ^ B
REM  **  Output results to screen
         CLS
         PRINT "Program INTARIT1.BAS"
         PRINT " "
         PRINT "Given integers "; A; "and "; B
         PRINT "------------------------------"
         PRINT "Sum A+B = "; C
         PRINT "Difference A-B = "; D
         PRINT "Product AxB = "; E
         PRINT "Quotient AdivB = "; F
         PRINT "Remainder AmodB = "; G
         PRINT "Power A^B = "; H
         PRINT "------------------------------"
         END
```

3.23 Development of Programming Languages

The earliest prototype computers were programmed by plug-board.

The earliest stored programs were machine-coded addresses for operation, source register, and destination register.

Very many programming languages have been written, for different purposes. The best become very popular, elicit strong loyalties, and spawn descendants; others do not. Some notable programming-language debuts were

1957 FORTRAN (engineering and scientific),
1958 ALGOL (scientific),
1959 COBOL (business),
1964 BASIC (educational and applications),
1972 C (systems and applications),
1991 PYTHON (educational and applications),
1995 JAVA (applications and web).

The ranking by usage in 2013 on one reckoning [6] was

- 1 C
- 2 JAVA
- 5 BASIC

- 7 PYTHON
- 17 COBOL
- 18 FORTRAN
- not in the top 100 ALGOL.

Most languages have similar repertoires of

- Number types (integer, long integer, floating-point, ...)
- Arithmetic operations (add, subtract, multiply, ...)
- Array structures (vector, matrix, ...)
- Statements (DO, FOR, EXIT, ...).

Most computing is done by very high-level application packages, often written in C or Java.

References

1. Boole G (1854) An investigation of the laws of thought: on which are founded the mathematical theories of logic and probabilities. http://www.gutenberg.org/files/15114/15114-pdf.pdf. Accessed 1 Feb 2014
2. Church A (1936) An unsolvable problem in elementary number theory. Am J Math 58:345–363
3. Jamsa KA (1991) MS-DOS QBasic. Microsoft Press, Redmond
4. Kantaris N (1994) Programming in QuickBASIC. Bernard Babani, London
5. Knuth DE (2008) The art of computer programming, Fascicle 0, Introduction to combinatorial algorithms and Boolean functions, vol 4. Addison-Wesley, Reading
6. Language Popularity Index (2013) http://www.sourceforge.net/projects/lang-index. Accessed 1 Feb 2014
7. QuickBasic 4.5 Download (1990) http://www.phatcode.net/downloads.php?id=172. Accessed 1 Feb 2014
8. QBasic Tutorial (2013) http://chortle.ccsu.edu/QBasic/index.html. Accessed 1 Feb 2014
9. Shannon CE (1937) A symbolic analysis of relay and switching circuits. Trans Am IEE 57(12):713–723
10. Shannon CE, Weaver W (1948) A mathematical theory of communication. University Illinois Press, Urbana
11. Turing AM (1936) On computable numbers, with an application to the Entscheidungsproblem. Proc Lond Math Soc 42:230–265

Chapter 4
Alphanumeric File

A digital computer stores information in a file. There are various types of file, for different purposes. An ASCII file is the normal way to input a program, and to input, store, and output numerical data. The structure of an ASCII number file is described. An example of an arithmetic program is given, where the input is from a file and the output is to a file, with the appearance of the files in a text editor and in a hexadecimal dump.

4.1 Flowchart

The flowchart (Fig. 4.1) is for general guidance in this chapter.

4.2 Memory

In computing, a *memory* is a physical device which stores information [2]. The device contains a large number of simple binary elements, each of which may be set to one of two possible states. Volatile memory holds information only while the device is receiving power (such as the registers in a central processing unit); permanent memory retains information without power (such as the files in a fixed hard disk or in a removable flash drive).

Figure 4.2 shows the standard structures used in memory. A *bit* occupies one binary element of memory, and so can represent either of two possible values, 0 or 1. Four bits constitute a *nibble*, which represents one of $2^4 = 16$ possible values. Two nibbles constitute a *byte*, representing one of $2^8 = 256$ possible values. Four bytes constitute a *32-bit word*.

A. Parkin, *Digital Imaging Primer*, DOI 10.1007/978-3-540-85619-1_4

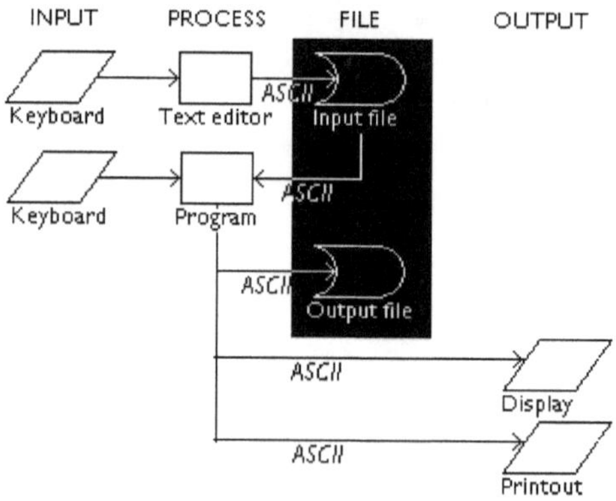

Fig. 4.1 Flowchart for ASCII file. A user may write an alphanumeric file, via a text editor. An alphanumeric file with a suitable structure may be read as data input, and/or written as output, by a program

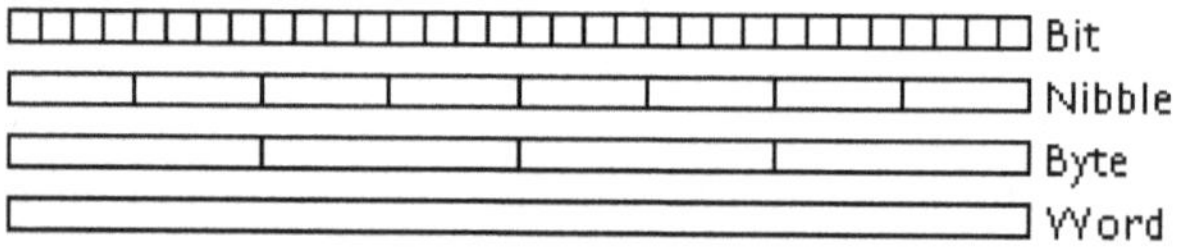

Fig. 4.2 Memory structures. The element is a bit, set to either of two possible states. Four bits constitute a nibble, set to any of 16 possible states. Eight bits or two nibbles constitute a byte, set to any of 256 possible states. A machine word is usually 32 b (4 B) or 64 b (8 B)

4.3 Development of Memory

There is continual development of cheaper, faster, and larger memory, in the existing technologies and emergent (such as Millipede memory [5]).

A *serial access memory* requires the medium to be read in sequence from beginning to end; while a *random access memory* (RAM) can access any part of the medium. A *fixed* memory is built-in to a computer; while a *removable* memory can be removed from a computer for security or for transfer elsewhere.

Various media have been used:

- Very early computers, serial removable punched cards or paper tape.
- From 1951, serial removable magnetic tape.
- From 1955, serial fixed magnetic hard disk (and more recently removable hard disk).
- From 1970s, RAM fixed semiconductor integrated circuit.
- From 1976, serial removable 5–1/4-in. magnetic floppy disk.
- From 1980s, serial removable 3–1/2-in. magnetic floppy disk.
- From 1980s, serial removable read-only optical compact disk (CD-R).
- From 1987, serial removable read/write optical compact disk (CD-RW).
- From 1998, universal serial bus (USB) removable flash drive.
- From 2000, secure digital (SD) removable flash card.

A typical current personal computer has a magnetic hard disk with capacity up to perhaps 100 GB, an optical CD drive with capacity 650 MB, and a USB port for a flash drive with capacity up to 500 GB or more.

4.4 File

In computing, a *file* is a well-defined structure of information which is stored in memory.

The size of a file is measured in *kilobytes* (kB), *megabytes* (MB), *gigabytes* (GB), or *terabytes* (TB). The standard SI prefixes k, M, G, T are customarily applied to the non-SI b for bit or B for byte. Since the early days of computing a kilobyte could mean either $10^3 = 1{,}000$ B or $2^{10} = 1{,}024$ B; and similarly a megabyte could mean $10^6 = 1{,}000{,}000$ B or $2^{20} = 1{,}048{,}576$ B; a gigabyte could mean $10^9 = 1{,}000{,}000{,}000$ B or $2^{30} = 1{,}073{,}741{,}824$ B; and a terabyte could mean $10^12 = 1{,}000{,}000{,}000{,}000$ B or $2^{40} = 2{,}147{,}483{,}648$ B. Telecommunication rates are usually measured in *kilobits per second* (kb/s), always meaning $10^3 = 1{,}000$ b/s, or *megabits per second* (Mb/s), meaning $10^6 = 1{,}000{,}000$ b/s, or *gigabits per second* (Gb/s), meaning $10^9 = 1{,}000{,}000{,}000$ b/s.

The main kinds of information stored in files are

- *programs*, which are sequences of alphanumeric commands for the computer to perform arithmetical or logical operations on data,
- *data*, which are collections of numeric or logical values ready to be transformed by programs.

A file is identified in a computer operating system by an alphanumeric *file-name*. In Microsoft Windows, the file-name [3] may be preceded by a *path* of nested *folder names*, and is normally followed by a three-character *file extension* [1, 4]. The extension identifies a broad file type, such as .TXT for text or .EXE for an executable program or .BAS for a BASIC program.

For image files in bitmap formats, identified by extension .BMP, see Chaps. 26, 33, 41, and 48. Bitmap files may be converted to, and from, compressed formats (such as .JPG, .TIFF, .PNG, ...) or to vector graphics formats (such as AutoDesk .DWG).

To write a file, an output device sets a sequence of storage elements to represent information of one kind or another. To read the file, an input device detects the settings of the sequence of elements, and hence recovers the information. A file is written and read computer-word by computer-word, and within that byte by byte.

In Microsoft QuickBASIC the available field types are

- *binary digit*: $2^1 = 2$ values 0 or 1,
- *hexadecimal digit*: $2^4 = 16$ values 0, 1, 2, ..., 9, A, B, ..., F,
- *ASCII character*: $2^7 = 128$ values coded 0 to 127 decimal (0 to FF hex),
- *integer*: $2^{16} = 65,536$ values $-32{,}768$ to 32,767,

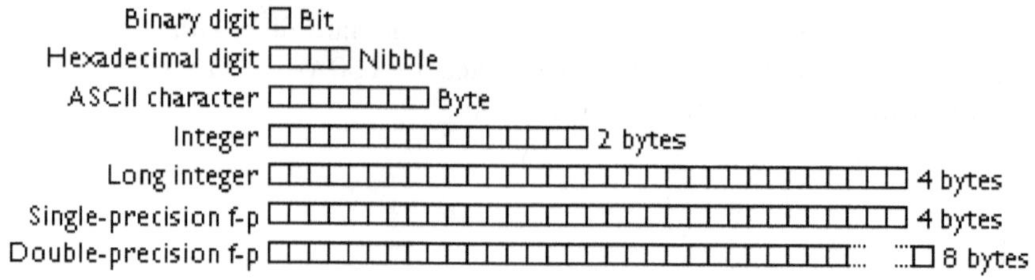

Fig. 4.3 Data types and their physical representation in a file

- *long integer*: $2^{32} = 4,294,967,296$ values −2,147,483,648 to 2,147,483,647,
- *single-precision floating-point*: exponential notation ±3.4E−37 to ±3.4E+38, that is seven significant figures,
- *double-precision floating-point*: exponential notation ±1.8E−307 to ±1.8E+308, that is 15 significant figures.

Figure 4.3 shows the physical representation of the data types in a file.

4.5 ASCII File

An *ASCII file* (or plain-text file) is a common way to input, store, and output alphanumeric data. An ASCII file has an assigned name string, usually with the extension .TXT, preceded by a path to its storage folder or directory. The file contains a sequence of bytes, each containing one ASCII code representing a numeric, alphabetic, or control character (see Sect. 2.5). An ASCII file can be written and read by a simple text editor, such as Microsoft NOTEPAD, or by a program which does other things as well.

4.6 Program with ASCII File Input and Output

For example, program INTARIT2 takes input from a previously-prepared ASCII file and sends output to an ASCII file. The calculations are the same as program INTARIT1 in Chap. 3.

```
REM  **  Program INTARIT2 to do arithmetic on two
REM  **     integers, input from file, output
REM  **     to file.
REM  **  Input numbers, and filename for text output
         CLS
         PRINT ""
         PRINT "|-------------------------------------|"
```

(a)

(b)

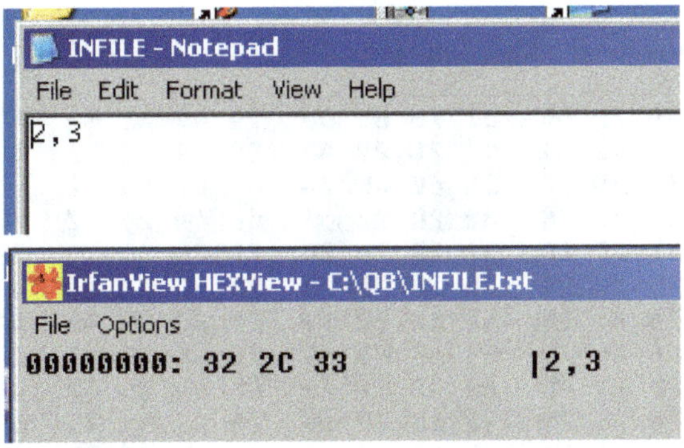

Fig. 4.4 Program INTARIT2 input file. **a** Input file INFILE.TXT as seen in Notepad. **b** INFILE.TXT as seen in hexadecimal dump. On the left is a (hex) line count, then 3 B. Each byte contains two hexadecimal digits, which convert to eight binary digits. The byte-wise character correspondence is shown on the right. Notice the comma separator (hexadecimal 2C)

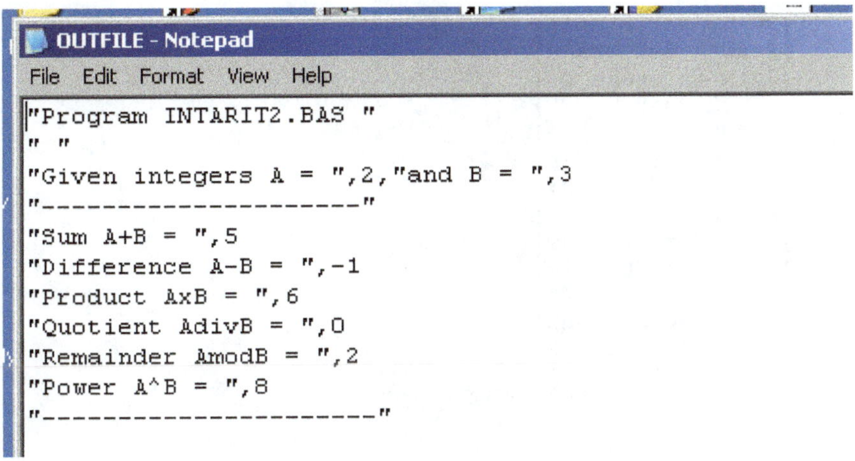

Fig. 4.5 Program INTARIT2 output file OUTFILE.TXT as seen in Notepad

```
        PRINT "| INTEGER ARITHMETIC 2     INTARIT2.BAS |"
        PRINT "|                                      |"
        INPUT "| Enter input file inc.TXT", INFILE$
        INPUT "| Enter output file inc. TXT", OUTFILE$
        PRINT "|                                      |"
        PRINT "|--------------------------------------|"
REM  ** Read text file
        OPEN INFILE$ FOR INPUT AS #1
        INPUT #1, A, B
        CLOSE #1
REM  ** Calculate
```

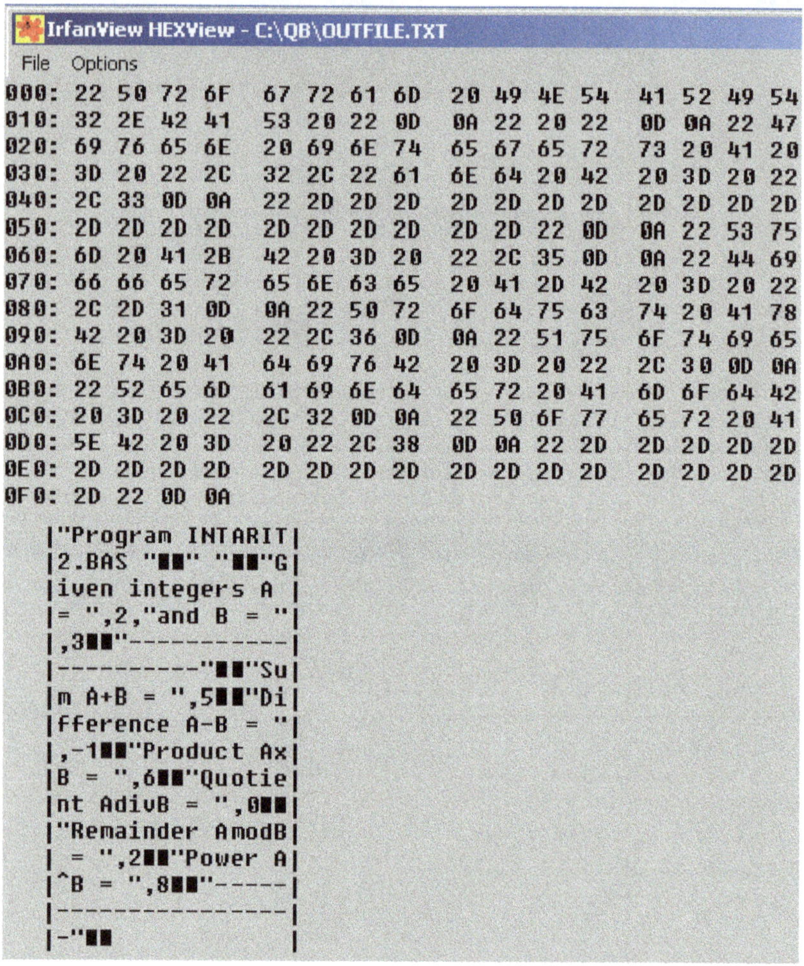

```
IrfanView HEXView - C:\QB\OUTFILE.TXT
File  Options
000: 22 50 72 6F  67 72 61 6D  20 49 4E 54  41 52 49 54
010: 32 2E 42 41  53 20 22 0D  0A 22 20 22  0D 0A 22 47
020: 69 76 65 6E  20 69 6E 74  65 67 65 72  73 20 41 20
030: 3D 20 22 2C  32 2C 22 61  6E 64 20 42  20 3D 20 22
040: 2C 33 0D 0A  22 2D 2D 2D  2D 2D 2D 2D  2D 2D 2D 2D
050: 2D 2D 2D 2D  2D 2D 2D 2D  2D 2D 22 0D  0A 22 53 75
060: 6D 20 41 2B  42 20 3D 20  22 2C 35 0D  0A 22 44 69
070: 66 66 65 72  65 6E 63 65  20 41 2D 42  20 3D 20 22
080: 2C 2D 31 0D  0A 22 50 72  6F 64 75 63  74 20 41 78
090: 42 20 3D 20  22 2C 36 0D  0A 22 51 75  6F 74 69 65
0A0: 6E 74 20 41  64 69 76 42  20 3D 20 22  2C 30 0D 0A
0B0: 22 52 65 6D  61 69 6E 64  65 72 20 41  6D 6F 64 42
0C0: 20 3D 20 22  2C 32 0D 0A  22 50 6F 77  65 72 20 41
0D0: 5E 42 20 3D  20 22 2C 38  0D 0A 22 2D  2D 2D 2D 2D
0E0: 2D 2D 2D 2D  2D 2D 2D 2D  2D 2D 2D 2D  2D 2D 2D 2D
0F0: 2D 22 0D 0A

|"Program INTARIT|
|2.BAS "■■" "■■"G|
|iven integers A |
|= ",2,"and B = "|
|,3■■"-----------|
|----------"■■"Su|
|m A+B = ",5■■"Di|
|fference A-B = "|
|,-1■■"Product Ax|
|B = ",6■■"Quotie|
|nt AdivB = ",0■■|
|"Remainder AmodB|
| = ",2■■"Power A|
|^B = ",8■■"-----|
|----------------|
|-"■■            |
```

Fig. 4.6 Program INTARIT2 output file OUTFILE.TXT as seen in hexadecimal dump. Each line of the dump shows a line count at the left, then 16 B from the file. Each byte contains two hexadecimal digits, which convert to eight binary digits. The byte-wise character correspondence is shown below the hexadecimals. Notice the two control characters carriage-return (hexadecimal 0D) and line-feed (hexadecimal 0A) which appear at the end of each printable line

```
        C = A + B
        D = A - B
        E = A * B
        F = A \ B
        G = A MOD B
        H = A ^ B
REM  ** Output to text file
        OPEN OUTFILE$ FOR OUTPUT AS #2
```

```
        WRITE #2, "Program INTARIT2.BAS"
        WRITE #2, ""
        WRITE #2, "Given integers A =", A, "and B =", B
        WRITE #2, "---------------------"
        WRITE #2, "Sum A+B =", C
        WRITE #2, "Difference A-B =", D
        WRITE #2, "Product AxB =", E
        WRITE #2, "Quotient AdivB =", F
        WRITE #2, "Remainder AmodB =", G
        WRITE #2, "Power A^B =", H
        WRITE #2, "----------------------"
        CLOSE #2
END
```

Each WRITE statement writes a line of output, including the quote-marks which enclose a string and the comma separators, and automatically inserts two control characters, carriage-return (cr, decimal 13, hexadecimal 0D) and line-feed (lf, decimal 10, hexadecimal 0A) to start a new line.

Figure 4.4a shows an input file for this program, in an ordinary text editor (Notepad). It contains integer numbers 2 and 3, with comma separator between numbers. Figure 4.4b shows the same file as a hexadecimal dump(available in more powerful editors, such as IRFANVIEW).

Figure 4.5 shows the output file as seen in NOTEPAD, and Fig. 4.6 shows the output file as seen in a hexadecimal dump.

References

1. File-Extension.com (2014) File extensions. http://www.file-extension.com/. Accessed 1 Feb 2014
2. How Stuff Works (2014) Computer memory. http://computer.howstuffworks.com/computer-memory.htm. Accessed 1 Feb 2014
3. Microsoft Developer Network (2008) Naming files paths and namespaces. http://msdn2.microsoft.com/en-us/library/aa365247.aspx. Accessed 1 Feb 2014
4. Open directory (2014) Data formats. http://www.dmoz.org/Computers/Data_Formats/. Accessed 1 Feb 2014
5. Wikipedia (2013) Millipede memory. http://en.wikipedia.org/wiki/Millipede_memory. Accessed 1 Feb 2014

Chapter 5
Alphanumeric Output

Simple alphanumeric output from a computer may be written to an ASCII file, and/or displayed on a screen, and/or printed on paper. The hardware, typography, and software for displaying output are described. An example of a program which outputs to a display is shown. The hardware, typography, and software for printing output are described. An example of a program which outputs to a printer is shown.

5.1 Flowchart

The flowchart (Fig. 5.1) is for general guidance in this chapter. Alphanumeric results from a program may be temporarily displayed on-screen or permanently written to an ASCII file for subsequent display or printout.

5.2 Display Hardware

(See Chap. 27). A display screen contains *P* px[1] in width and *Q* px in height. Various display sizes are in current use: for example Full-HD standard is $1,920 \times 1,080$ px, settable to any colour.

5.3 Display Typography

In computing, a printable character shape is called a *glyph*.[2] A *font* is a complete set of glyphs in a certain style. A *fixed-pitch font* has all glyphs of a given size occupying blocks of the same width; for example Fig. 5.2 shows glyphs in blocks of 7×12 px.

---

[1] See Sect. 19.2.

[2] From Greek $\gamma\lambda\upsilon\varphi\epsilon\iota\nu$ to carve (as in hieroglyph).

A. Parkin, *Digital Imaging Primer*, DOI 10.1007/978-3-540-85619-1_5

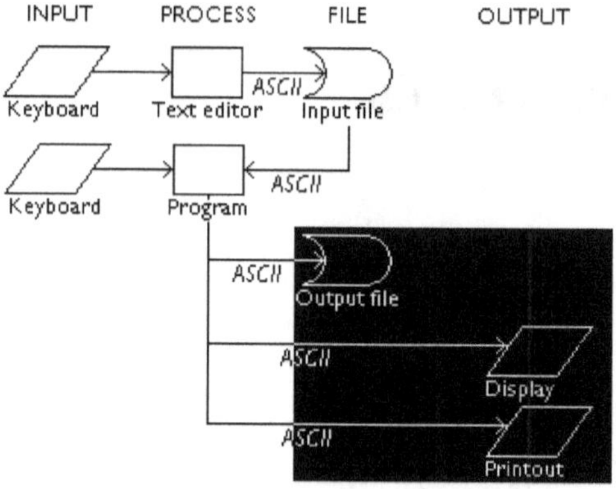

Fig. 5.1 Flowchart for number output. A program may output numeric results as an alphanumeric file, and/or a display on a screen, and/or a printout on paper

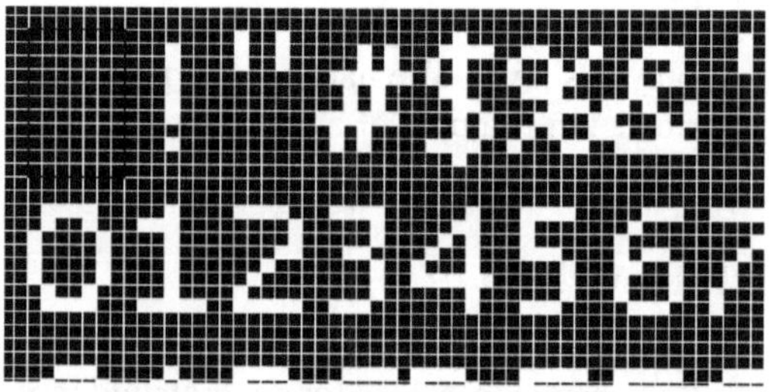

Fig. 5.2 Fixed-pitch font Lucida Console 9 pt, enlarged to show pixels. Each printable character occupies a block of 7 × 12 px. Space (hexadecimal 20), is at *top left*

Fig. 5.3 Variable-pitch font Times New Roman 9 pt, enlarged to show pixels. Each printable character occupies a block 15 px high and varying width. Space (hexadecimal 20), is at *top left*

A *variable-pitch font* has glyphs of a given size occupying blocks of different widths; for example Fig. 5.3 shows glyphs in blocks 15 px high.

To display a glyph on a screen, certain pixels of a block are set to a foreground colour (say White), and the rest are set to a background colour (say Black). A *bit-mapped font*, for example Fig. 5.4, is stored as a fixed set of glyphs in a certain block size, which are not intended to be scaled to other sizes or otherwise varied. In Windows, the storage format for a bit-mapped font is .FON.

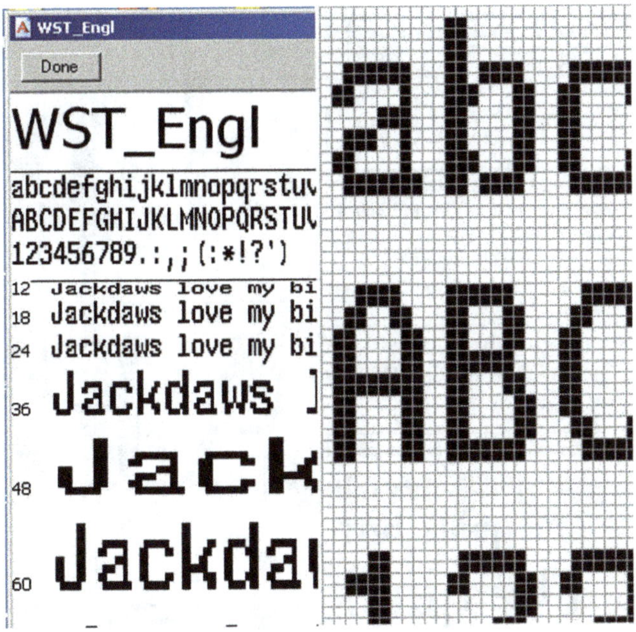

Fig. 5.4 Bit-mapped font in general-purpose text editor display. This World System Teletext font WST.FON is stored in several fixed sizes. The 18-point size is enlarged to show pixels

A *vector font*, for example Fig. 5.5, is stored as a set of parameters which can produce glyphs varying in size, weight, width, and slope, which are then *rasterized* to pixels for display. In Windows, the storage format for a vector font is TrueType Format (.TTF) or OpenType Format (.OTF).

5.4 Alphanumeric Display Software

A text editor is able to compose printable characters into words and lines, save them in an ASCII (or plain-text) file, and to recall and edit the file in a display screen. A very simple text editor, such as that incorporated in Microsoft QuickBASIC, has limited access to a few fixed-pitch bit-mapped fonts, as shown, for example in Fig. 5.6. A general-purpose text editor, such as Microsoft Notepad, allows composition, editing, and display in any (one) font which has been stored on the computer, bit-mapped or vector, fixed- or variable-pitch.

A word-processing package, such as Microsoft WORD, allows composition, editing, and display in any combination of different fonts, but saves files in a different format from ASCII.

A page-creating and page-editing system, such as TEX, or PostScript, puts all typographic aspects under the user's interactive control (see Sect. 5.9).

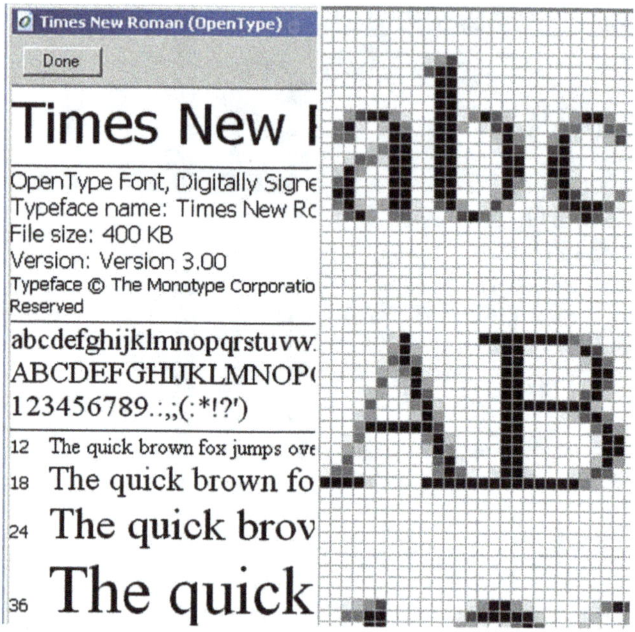

Fig. 5.5 Vector font in text editor display. This Times New Roman .OTF font can be scaled to any size. The 18-point size is enlarged to show pixels. When the glyphs are rasterized for display, 24-bit *hinting* improves the pixel rendering. When the glyphs are printed, the hinting is conveyed in the 1-bit printer-dot rendering

Fig. 5.6 Program INTARIT1. **a** Screen output. **b** Enlargement of (**a**)

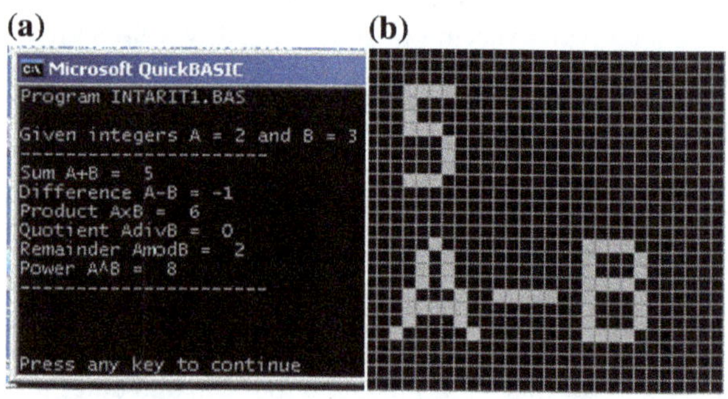

5.5 Development of Alphanumeric Display

Notable stages in displaying alphanumeric output are,

1960 Interactive display
1970 GUI graphic user interface
1978 TEX page composition
1979 METAFONT fonts

1981 IBM Codepage 437
1984 PostScript page composition
1985 WYSIWYG desk-top publishing
1985 Computer Modern fonts
1986 World System Teletext
1990 UCS universal character set
1991 TTF TrueType format
1993 PDF portable document format
1993 UTF-8 character coding
1997 OTF OpenType format

5.6 Program with Display Output

Section 3.22 shows an example of a simple program written in QuickBASIC, which asks for input to the screen then displays the results on-screen. Figure 5.6 shows the output display.

5.7 Print Hardware

Once upon a time, text was printed from metal founders' type, assembled into pages by hand compositors or by Monotype or Linotype machines. A *font*[3] was then a complete set of *sorts* in a certain style, size, weight, width, and slope, such as Times 12-point bold extended italic. This technology is now called hot-metal or letterpress printing.

In the 1960s, *photo-setting* began to replace hot-metal type-setting. Text was set by a machine which shone a light through a negative film mask containing all the characters of a font, adjusted optically to any size, to register successively on photo-sensitive paper. Photo-set galley proofs were then pasted-up to pages for printing by offset-lithography.

From the 1950s, mainframe computers could send text to a line-printer, which strikes raised glyphs through an inked ribbon onto paper, like a typewriter but with 132 characters per line. Alternatively, vector graphic output could be sent to an XY-plotter, drawing lines with a pen. For personal computers, a daisy-wheel printer also struck a raised glyph through an inked ribbon, one character at a time.

A dot-matrix printer has a matrix of small pins which can be protruded by software to form any glyph, which is then struck through an inked ribbon. The dot size is typically 200 dpi. An ink-jet printer throws minute drops of ink, in a settable glyph, onto paper. The fundamental dot size of ordinary ink-jet printers is typically 300 to

[3]From old French *fonte* melted or cast. The traditional English spelling was *fount*.

600 dpi.[4] A laser printer forms an electrostatic image of a whole page, which attracts a toner powder, which is then melted onto paper. The fundamental dot size of such printers is typically 600–2,400 dpi. Since a display screen typically has a fundamental pixel size of about 100 ppi, a a glyph can be printed at a much higher resolution than it can be displayed.

In the 1980s, *desktop publishing* (DTP) began to replace commercial photo-set publishing. This was made possible by the advent of large What-You-See-Is-What-You-Get (WYSIWYG) interactive display screens, vector fonts, and laser page-printers.

5.8 Alphanumeric Print Typography

For impact printing, a fixed font is built-in to the printer.

For dot-matrix and ink-jet printing, various bit-mapped fonts can be set by stored software. For example, Fig. 5.4 shows the standard TeleText fonts in several sizes.

For laser printing, vector fonts are stored. A vector font is created by drawing each glyph by means of straight segments and Bezier curves (quadratic or cubic splines). The glyph is then stored as a small number of parameters, in TrueType Format (TTF) or OpenType format (OTF). A single font can then be varied in size, weight, width, and slope via these parameters, to provide a font family in each style. For example, Fig. 5.5 shows the Times New Roman font in several sizes.

5.9 Alphanumeric Print Software

TeX[5] [8], followed by LaTeX[6] [2], is a page-creating and page-editing mark-up system, for printing book-quality text and images. It is based on vector graphics. So too is PostScript [1], followed by Portable Document Format (PDF)[7] [6].

The text for this book has been written in the WinShell[8] editor for the MikTex[9] version of the LaTeX implementation of the TeX typesetting system. All the figures have been created in .BMP format by the programs downloadable from the SpringerLink web-site, or by hand in PAINT, or by scanner or camera capture, and converted to Portable Network Graphics .PNG[10] [7] format by Omniformat [5] for incorporation in the text.

[4]Professional ink-jets go up to 4,800 dpi, with wide paper and high speed.

[5]Written by Donald Ervin Knuth.

[6]Written by Leslie Lamport.

[7]Written by John Warnock and others for Adobe, subsequently adopted by ISO.

[8]Written by Ingo H. de Boer.

[9]Written by one of the Christian Schenks.

[10]Written by Thomas Boutell and many others to supersede GIF, subsequently adopted by ISO.

5.10 Development of Alphanumeric Output

Notable developments in printing alphanumeric output are,

1950 Punched card or paper tape
1952 Line-printer
1959 XY plotter
1960s Photo-typesetting
1968 Electric typewriter console
1972 Daisy-wheel printer
1974 Dot-matrix printer
1975 Ink-jet printer
1978 TEX page composition
1979 METAFONT fonts
1981 IBM Codepage 437
1984 PostScript page composition
1984 Laser printer
1985 WYSIWYG desk-top publishing
1985 Computer Modern fonts
1986 World System Teletext
1990 UCS universal character set
1991 TTF TrueType format
1993 PDF portable document format
1993 UTF-8 character coding
1997 OTF OpenType format

5.11 Program with Print Output

Program INTARIT3 takes input from a previously-prepared ASCII file and sends output to an ASCII print file. The input and calculations are the same as program INTARIT2 in Chap. 4.

```
REM  **  Program INTARIT3 to do arithmetic on two
REM  **  integers, input from file, output
REM  **     to print.
REM  ** Input numbers, and filename for text output
         CLS
         PRINT ""
         PRINT "|-------------------------------------|"
         PRINT "| INTEGER ARITHMETIC 3    INTARIT3.BAS |"
         PRINT "|                                     |"
         INPUT "| Enter input file inc.TXT", INFILE$
```

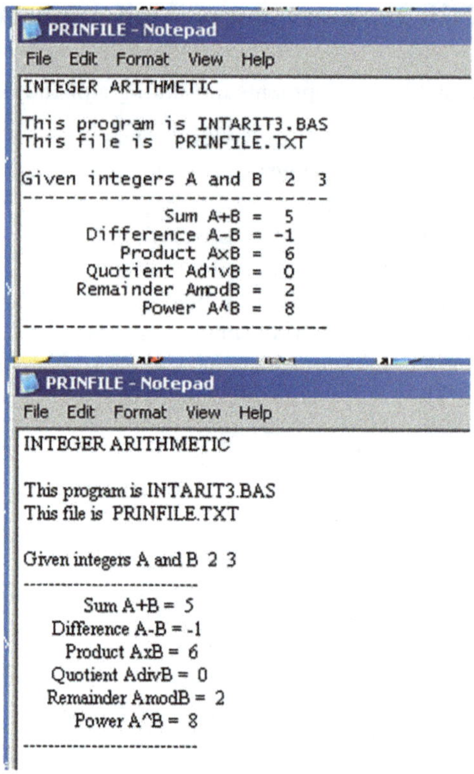

(a)

PRINFILE - Notepad

File Edit Format View Help

```
INTEGER ARITHMETIC

This program is INTARIT3.BAS
This file is  PRINFILE.TXT

Given integers A and B  2  3
----------------------------
            Sum A+B =  5
     Difference A-B = -1
        Product AxB =  6
     Quotient AdivB =  0
    Remainder AmodB =  2
          Power A^B =  8
----------------------------
```

(b)

PRINFILE - Notepad

File Edit Format View Help

INTEGER ARITHMETIC

This program is INTARIT3.BAS
This file is PRINFILE.TXT

Given integers A and B 2 3

Sum A+B = 5
Difference A-B = -1
Product AxB = 6
Quotient AdivB = 0
Remainder AmodB = 2
Power A^B = 8

Fig. 5.7 Program INTARIT3 output file PRINFILE.TXT as displayed in Notepad. **a** Using fixed-pitch Lucida Console 9 pt. **b** Using variable-pitch Times 9 pt

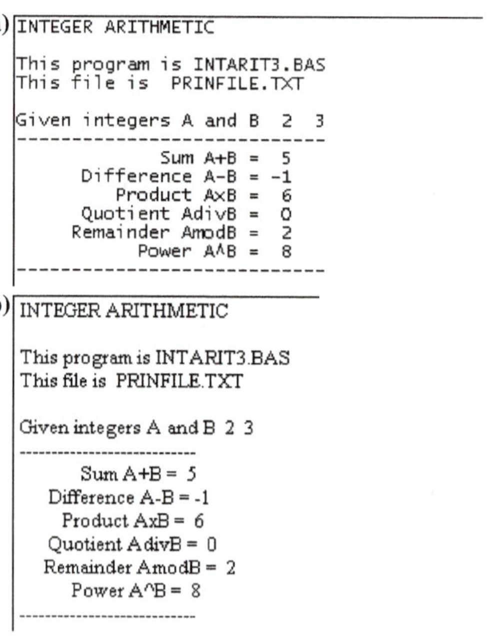

(a)

```
INTEGER ARITHMETIC

This program is INTARIT3.BAS
This file is  PRINFILE.TXT

Given integers A and B  2  3
----------------------------
            Sum A+B =  5
     Difference A-B = -1
        Product AxB =  6
     Quotient AdivB =  0
    Remainder AmodB =  2
          Power A^B =  8
----------------------------
```

(b)

INTEGER ARITHMETIC

This program is INTARIT3.BAS
This file is PRINFILE.TXT

Given integers A and B 2 3

Sum A+B = 5
Difference A-B = -1
Product AxB = 6
Quotient AdivB = 0
Remainder AmodB = 2
Power A^B = 8

Fig. 5.8 Program INTARIT3 output file PRINFILE.TXT as printed out. **a** Using fixed-pitch Lucida Console 9 pt. **b** Using variable-pitch Times 9 pt

```
          INPUT "| Enter output file inc. TXT", PRINFILE$
          PRINT "|                                     |"
          PRINT "|-------------------------------------|"
REM  **   Read text file
          OPEN INFILE$ FOR INPUT AS #1
          INPUT #1, A, B
          CLOSE #1
REM  **   Calculate
          C = A + B
          D = A - B
          E = A * B
          F = A \ B
          G = A MOD B
          H = A ^ B
          OPEN PRINFILE$ FOR OUTPUT AS #2
REM  **   Print heading
          PRINT #2, "INTEGER ARITHMETIC"
          PRINT #2, ""
          PRINT #2, "This program is INTARIT3.BAS"
          PRINT #2, "This file is", PRINFILE$
          PRINT #2, ""
REM **    Print results
          PRINT #2, USING "Given integers A and B ## ##";A;B
          PRINT #2, "----------------------------"
          PRINT #2, USING "              Sum A+B = ##"; C
          PRINT #2, USING "       Difference A-B = ##"; D
          PRINT #2, USING "          Product AxB = ##"; E
          PRINT #2, USING "       Quotient AdivB = ##"; F
          PRINT #2, USING "      Remainder AmodB = ##"; G
          PRINT #2, USING "            Power A^B = ##"; H
          PRINT #2, "----------------------------"
          CLOSE #2
END $
```

Each PRINT statement writes a line of output, like WRITE but with various formatting facilities. The formatting turns out quite differently for fixed-pitch and variable-pitch fonts. The input file INFILE.TXT is the same as for the INTARIT2 example in Chap. 4. The output file PRINFILE.TXT is shown in Fig. 5.7 as a display in Notepad, with fixed-pitch and variable-pitch fonts; in Fig. 5.8 as a printout, with fixed-pitch and variable-pitch fonts; and in Fig. 5.9 as a hexadecimal dump.

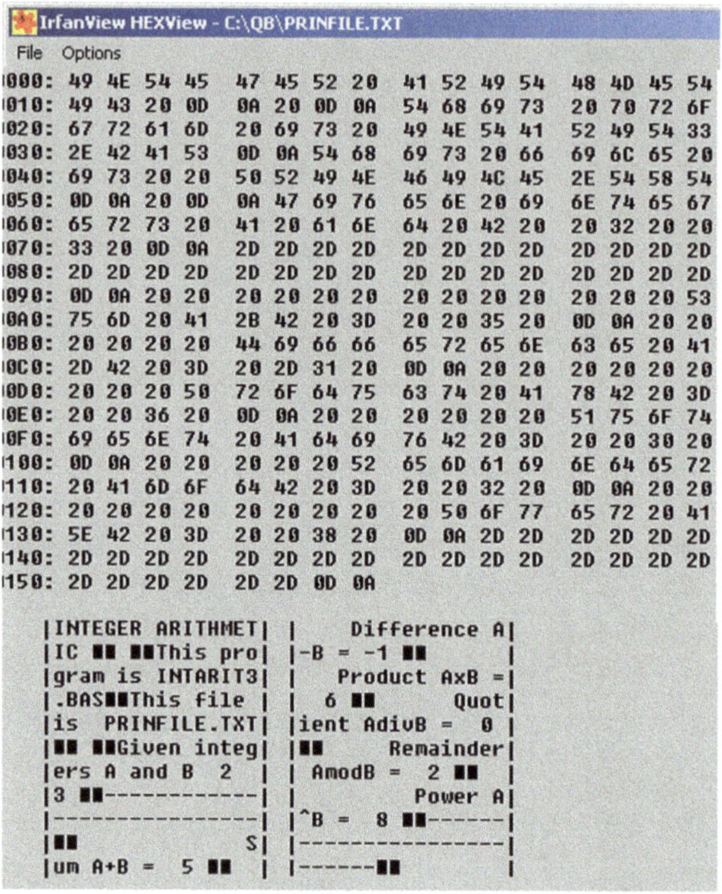

```
IrfanView HEXView - C:\QB\PRINFILE.TXT
File  Options
000: 49 4E 54 45  47 45 52 20  41 52 49 54  48 4D 45 54
010: 49 43 20 0D  0A 20 0D 0A  54 68 69 73  20 70 72 6F
020: 67 72 61 6D  20 69 73 20  49 4E 54 41  52 49 54 33
030: 2E 42 41 53  0D 0A 54 68  69 73 20 66  69 6C 65 20
040: 69 73 20 20  50 52 49 4E  46 49 4C 45  2E 54 58 54
050: 0D 0A 20 0D  0A 47 69 76  65 6E 20 69  6E 74 65 67
060: 65 72 73 20  41 20 61 6E  64 20 42 20  20 32 20 20
070: 33 20 0D 0A  2D 2D 2D 2D  2D 2D 2D 2D  2D 2D 2D 2D
080: 2D 2D 2D 2D  2D 2D 2D 2D  2D 2D 2D 2D  2D 2D 2D 2D
090: 0D 0A 20 20  20 20 20 20  20 20 20 20  20 20 20 53
0A0: 75 6D 20 41  2B 42 20 3D  20 20 35 20  0D 0A 20 20
0B0: 20 20 20 20  44 69 66 66  65 72 65 6E  63 65 20 41
0C0: 2D 42 20 3D  20 2D 31 20  0D 0A 20 20  20 20 20 20
0D0: 20 20 20 50  72 6F 64 75  63 74 20 41  78 42 20 3D
0E0: 20 20 36 20  0D 0A 20 20  20 20 20 20  51 75 6F 74
0F0: 69 65 6E 74  20 41 64 69  76 42 20 3D  20 20 30 20
100: 0D 0A 20 20  20 20 20 52  65 6D 61 69  6E 64 65 72
110: 20 41 6D 6F  64 42 20 3D  20 20 32 20  0D 0A 20 20
120: 20 20 20 20  20 20 20 20  20 50 6F 77  65 72 20 41
130: 5E 42 20 3D  20 20 38 20  0D 0A 2D 2D  2D 2D 2D 2D
140: 2D 2D 2D 2D  2D 2D 2D 2D  2D 2D 2D 2D  2D 2D 2D 2D
150: 2D 2D 2D 2D  2D 2D 0D 0A

|INTEGER ARITHMET|  |     Difference A|
|IC ■■ ■■This pro|  |-B =  -1 ■■      |
|gram is INTARIT3|  |    Product AxB =|
|.BAS■■This file |  |  6 ■■       Quot|
|is  PRINFILE.TXT|  |ient AdivB =   0 |
|■■ ■■Given integ|  |■■      Remainder|
|ers A and B   2 |  | AmodB =   2 ■■  |
|3 ■■------------|  |          Power A|
|----------------|  |^B =   8 ■■------|
|■■             S|  |-----------------|
|um A+B =   5 ■■ |  |------■■         |
```

Fig. 5.9 Program INTARIT3 output file PRINFILE.TXT as seen in hexadecimal dump. Each line of the dump shows a line count at the left, then 16 B from the file. Each byte contains two hexadecimal digits, which convert to eight binary digits. The byte-wise character correspondence is shown below the hexadecimals. Notice the two control characters carriage-return (hexadecimal 0D) and line-feed (hexadecimal 0A) which appear at the end of each printable line

References

1. Adobe Systems (1999) PostScript language reference manual 3rd edn. http://partners.adobe.com/public/developer/en/ps/PLRM.pdf. Accessed 1 Feb 2014
2. Lamport L (1994) LaTeX: A document preparation system: User's guide and reference, 2nd edn. Addison-Wesley Professional, Reading
3. LaTeX (2014) www.latex-project.org. Accessed 1 Feb 2014
4. MiKTeX (2014) www.miktex.org. Accessed 1 Feb 2014
5. Omniformat (2014) www.omniformat.com. Accessed 1 Feb 2014
6. Portable Document Format (2008) http://www.iso.org/iso/iso_catalogue/catalogue_tc/catalogue_detail.htm?csnumber=51502. Accessed 1 Feb 2014

7. Portable Network Graphics (2003) http://www.w3.org/TR/PNG/. Accessed 1 Feb 2014
8. TUG (2011) What Is TEX?. http://www.tug.org/whatis.html. Accessed 1 Feb 2014
9. WinShell (2014) www.winshell.org. Accessed 1 Feb 2014

Part II
Location in Digital Imaging

Chapter 6
Coordinate Geometry

Physical space can be represented by a formal geometric space, which can be further represented by a numeric coordinate space. Euclidean geometry is constructed from point elements of no size, with real-number coordinates, and has unlimited extent. Discrete geometry is constructed from cube elements of fixed size, with integer coordinates, and unlimited extent. Harmonic space has a harmonic scale of distances, and lies within a finite octahedron. Elliptic space has homogeneous coordinates, and lies within a finite sphere.

6.1 Geometric Space

The physical space in which we live can be represented as a *geometric space* in various ways.

Historically, geometry was developed for various purposes: for land measurement in Egypt, for philosophical understanding in Greece, for architectural practice in medieval Europe and Islam, for alternatives to Euclid in the nineteenth century, and for mathematical physics in modern times.

Logically, the idea that geometries may be classified as *groups of transformations* which preserve certain properties was introduced in 1872 by Klein[1] [3]:

- *Topological* transformations preserve only connectivity of figures.
- *Projective* transformations also preserve incidence of points lines and planes.
- *Affine* transformations also preserve lengths on parallels.
- *Euclidean* transformations also preserve lengths and angles, and the geometry allows just one parallel to a given line through a given point (the so-called parabolic metric).

[1]Felix Christian Klein (1849–1925), Professor at Erlangen, Munich, Leipzig, and Göttingen.

A. Parkin, *Digital Imaging Primer*, DOI 10.1007/978-3-540-85619-1_6

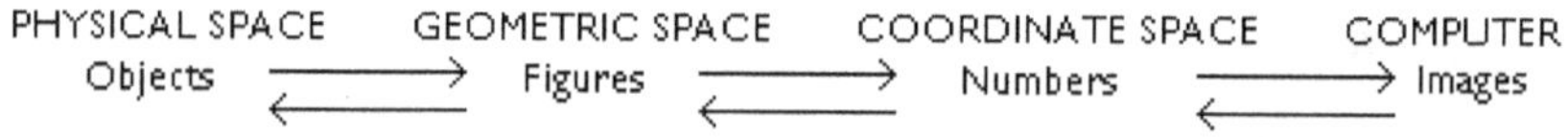

Fig. 6.1 Objects in physical space can be represented by figures in geometric space, which can in turn be represented by numbers in coordinate space, which can be processed into images by a digital computer. Conversely, images can be created by a computer, interpreted as numbers, which can be interpreted as geometric figures, which can be interpreted as physical objects

- *Riemannian* geometry has no parallel to a given line through a given point (the elliptic metric).
- *Lobatchevskian* geometry has many parallels to a given line through a given point (the hyperbolic metric).

A geometric space can be further represented by *coordinates*. Numerical coordinates for locations in a plane were introduced by Descartes[2] [1], so that every geometric figure corresponds to a set of numbers, and every geometric operation on a figure corresponds to an arithmetic operation on numbers. And since numbers and arithmetic can be represented in machine form (see Part I), we can do geometry in a computer (Fig. 6.1).

In this chapter, we construct four geometric coordinate spaces. Continuous Euclidean space is the setting for the vast body of knowledge about measurable physical objects. Discrete space is specially suited to digital imaging. Harmonic space is intimately connected with projection from space to a plane (see Chap. 8). Elliptic space is what we use for seeing (see Chap. 10).

6.2 Euclidean Coordinate Space

Euclid's[3] *Elements* [2] was the first unification of geometry as theorems under a set of explicit axioms and postulates, and incorporated results dating back to Pythagoras and possibly Egyptian and Babylonian mathematicians. Conceptually, the element of location for a Euclidean space is a *point* of no size. A line is a single extent of points, a plane a double extent and space a triple extent. If we want to make measurements, we must provide a fixed unit of length between two points, and a fixed unit of angle between two lines. All lengths and angles can be added or subdivided without limit. Segments of a line, patches of a plane, and regions of space may be empty or marked by some substance.

[2]Renée Descartes (1596–1650): born in France; lived in the Dutch Republic 1628–1649; taught at Utrecht University and elsewhere. The same idea was used, possibly as early as 1628, by Pierre de Fermat (1607–1665), lawyer in Toulouse and distinguished mathematician.

[3]Almost nothing is known about Euclid the man, beyond the possibility that he studied under Plato in Athens, and the certainty that he taught in Alexandria around 300 BC.

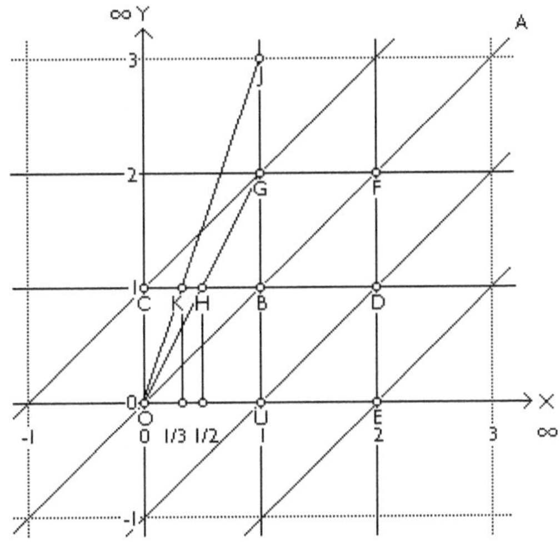

Fig. 6.2 Construction of Euclidean coordinate net in a plane. Given origin O with coordinate 0 (zero)), unit point U with coordinate 1, and unit (*right*) angle at O, throw a net of new coordinate points to cover as much of the plane as we please

Geometrically, we may assign coordinates to points on a plane by von Staudt's[4] celebrated construction of *throws* (German *werfen*) [4]. Given three *base points* and their coordinates on a line:

- *origin* point O, coordinate 0 (zero),
- *unit* point U, coordinate 1,
- *point at infinity* x on line OU, coordinate ∞[5]
- and, for this special Euclidean case, a right angle at O, value 90°.

Figure 6.2 shows the procedure:

- Construct a line Oy at 90° to Ox, with direction y.
- Construct a line OA at 45° to Ox and to Oy, with direction A.
- Join Uy (parallel to Oy), to meet OA at B.
- Join Bx (parallel to Ox), to meet Oy at C.
- Join UA (parallel to OA), to meet Bx at D.
- Join Dy (parallel to Oy), to meet Ox at E, with coordinate 2.
- Dy also meets OA at F, so join Fx (parallel to Ox) to throw a new line with coordinate 2 on Oy.
- Continue from E, making a forward throw by a parallel to OA and a parallel to Oy, to assign successive positive integer coordinates 3, 4, ... on Ox; and making an upward throw to assign successive positive integer coordinates 3, 4, ... on Oy.

---

[4]Karl Georg Christian von Staudt (1798–1867), Professor of Mathematics at Erlangen University. The construction was invented to assign rational coordinates to points in a projective plane, using incidence alone. The result, which can never be completed in practice, is often called a *net of rationality*, or a Mobius net.

[5]The ancient Greeks were very uncomfortable with the unlimited $(\alpha\pi\varepsilon\iota\rho\omega\nu)$, as they were with the irrational $(\sqrt{2})$. The modern conventions for inclusion of ∞ in arithmetic are: $n + \infty = \infty$; $n - \infty = -\infty$; $n \times \infty = \infty$; $n/\infty = 0$; $\infty/n = \infty$; $n/0 = \infty$.

- In in the opposite direction, making a backward throw to assign negative integer coordinates $-1, -2, \ldots$ on Ox and Oy.

We thus construct a net of integer points and lines as far as we please in the plane Oxy.

We divide the integer net as finely as we please by constructing points with rational coordinates:

- Join OG, to meet CB at H, with coordinate 1/2.
- Join OJ, to meet CB at K, with coordinate 1/3.
- And so on, for 1/4, 1/5, 1/6, ...
- And similarly for 2/3, 3/4, 4/5, ... and any rational coordinate.

Figure 6.3a shows the given Euclidean base points and units; Fig. 6.3b the Euclidean coordinate axis Ox with integer coordinates x; and Fig. 6.3c the Euclidean coordinate plane Oxy with integer coordinates (x, y). Between any two points further points can be inserted.

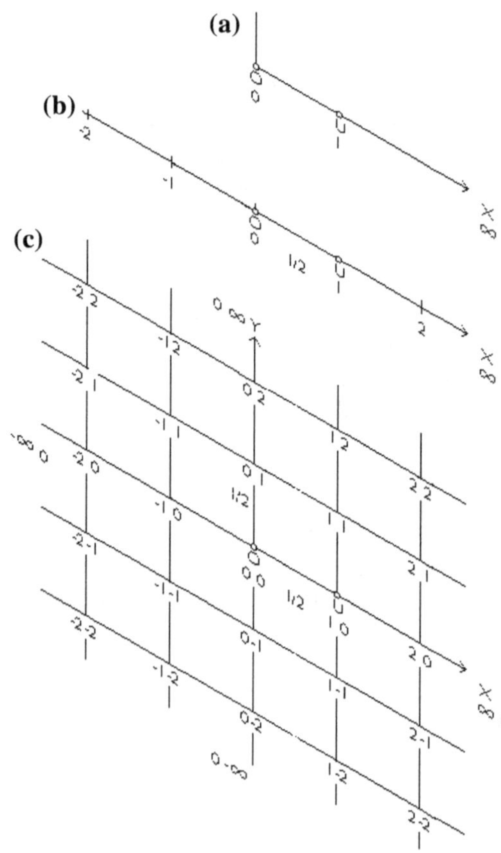

Fig. 6.3 **a** Euclidean base points. **b** Euclidean integer coordinate axis Ox. **c** Euclidean integer coordinate plane Oxy

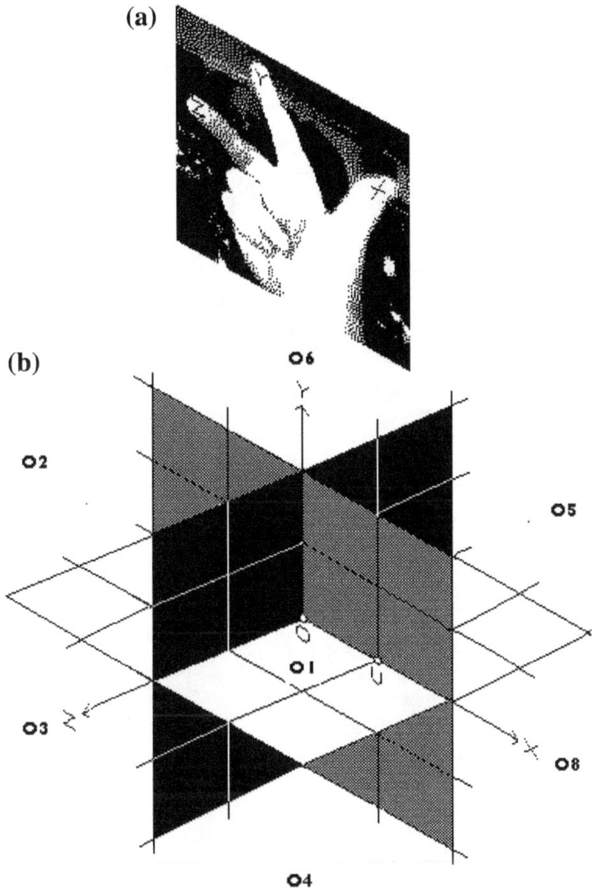

Fig. 6.4 **a** Right-handed axes. **b** Euclidean coordinate space. Each point has coordinates (x, y, z). Octants O1, O2, ..., O8 marked (O7 is hidden)

Figure 6.4 shows a z-axis added to the x- and y-axes in a *right-handed* arrangement: if the positive x-axis were rotated towards the positive y-axis, it would drive a right-handed screw in the positive direction of the z-axis. A well-known mnemonic is to extend the thumb of the right hand as x-axis, the forefinger as y-axis, and the middle finger as z-axis. If the z-axis had been assigned positive direction backwards, the axes would have been left-handed, like the fingers of the left hand. The three coordinate planes cut space into eight *octants*. Different conventions are in use for numbering the octants with right-handed and left-handed axes. Here, we shall use the numbering:

Octant 1: $(+x, +y, +z)$
Octant 2: $(-x, +y, +z)$
Octant 3: $(-x, -y, +z)$
Octant 4: $(+x, -y, +z)$
Octant 5: $(+x, +y, -z)$

Octant 6: $(-x, +y, -z)$
Octant 7: $(-x, -y, -z)$
Octant 8: $(+x, -y, -z)$.

6.3 Discrete Coordinate Space

Conceptually, the element of location for a discrete coordinate space is a cube of fixed but arbitrary size, which we shall call a *cubel* (or voxel), with one pair of opposite faces labelled x and $-x$, another pair of opposite faces labelled y and $-y$, and the third pair of opposite faces labelled z and $-z$, as shown in Fig. 6.5a. Thus the discrete element incorporates an (arbitrary) unit of length (a face width) and a unit of angle (a

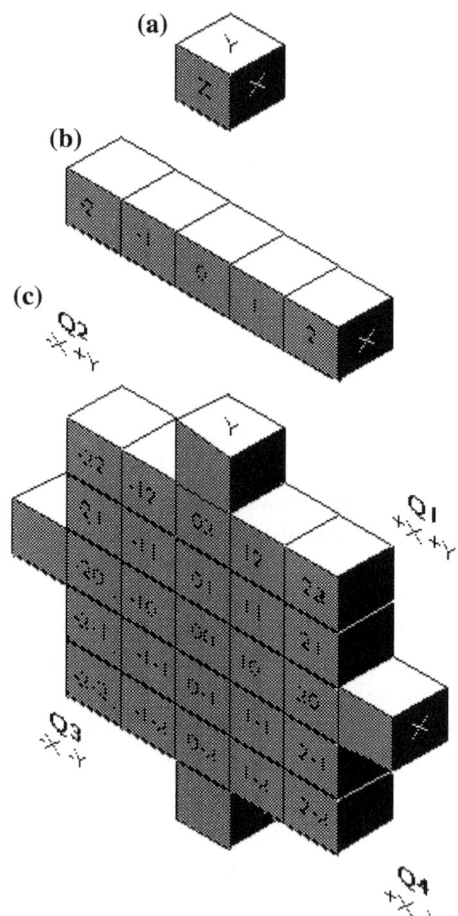

Fig. 6.5 Cube element, discrete x-axis, and discrete Oxy plane. **a** Cube element, with faces labelled. **b** x-axis, with integer coordinates continued in both directions. **c** Plane Oxy, with integer coordinates continued in all four directions. The axes define four quadrants, by the signs of the coordinates. Notice that origin O is common to all four quadrants; the positive x-axis is common to Q1 and Q4, the positive y-axis to Q1 and Q2, and so on

right angle) (whereas the Euclidean point element required units of length and angle to be supplied). All discrete lengths are additions of cubels, butted face to face, or edge to edge ($\surd 2$ face widths), or corner to corner ($\surd 3$ face widths). Subdivisions of cubels do not occur. Cubels may be empty, or marked by some substance which emits, transmits, or reflects light.

Fix an *origin* cubel O anywhere in space, and assign it coordinate 0 (zero). Fix a successor cubel abutting $-x$ face to x face of O, with coordinate 1. Fix further successors in the same way, with coordinates $2, 3, \ldots$. Fix a predecessor cubel abutting x face to $-x$ of O, with coordinate -1; and further predecessors with coordinates $-2, -3, \ldots$. We then have a *discrete line* of cubels, the x-*axis*, with single integer coordinates, as shown in Fig. 6.5b.

Fix a successor line abutting $-y$ faces to y faces of the x-axis, with coordinates (... $(-2, 1), (-1, 1), (0, 1), (1, 1), (2, 1), \ldots$; then further successor lines in the same way. And fix predecessor lines abutting y faces to $-y$ faces of the X-axis. We then have a *discrete plane* location!discrete plane of cubels, with double integer coordinates, and the Y-axis meeting the X-axis at O, as shown in Fig. 6.5c.

The x and y coordinate axes cut the discrete plane into four *quadrants*. Different conventions are in use for numbering the quadrants: here, we shall use the numbering:

Quadrant 1: $(+x, +y)$
Quadrant 2: $(-x, +y)$
Quadrant 3: $(-x, -y)$
Quadrant 4: $(+x, -y)$.

Digital images are figures in the discrete plane, and the z-faces of the cubels are *pixels*.

Fix a successor plane abutting $-z$ faces to z faces of the Oxy-plane, with coordinates $(x, y, 1)$, and further successor planes with coordinates $(x, y, 2), (x, y, 3), \ldots$. And fix predecessor planes abutting z faces to $-z$ faces of the OXY plane. We then have a *discrete space* $Oxyz$ of cubels, with triple integer coordinates, and the z-axis meeting the x-axis and the y-axis at O, as shown in Fig. 6.6.

Geometrically, Euclidean space can be converted to discrete space by replacing blocks of points by cubels, thus losing all detail within the chosen size of cubel. Numerically, Euclidean rational coordinates can be converted to discrete integer coordinates by rounding or truncating, thus losing all fractional parts.

6.4 Arithmetic Scale and Harmonic Scale

Numerically, an *arithmetic progression* is a sequence of numbers $a_1, a_2, a_3, \ldots$ with constant common difference $k = a_2 - a_1 = a_3 - a_2 \ldots$. Given two successive terms a_1 and a_2, the next term is $a_3 = a_2 + k = a_2 + (a_2 - a_1) = 2a_2 - a_1$. Given a lower neighbour a_1 and upper neighbour a_3 the arithmetic mean is $a_2 = (a_1 + a_3)/2$.

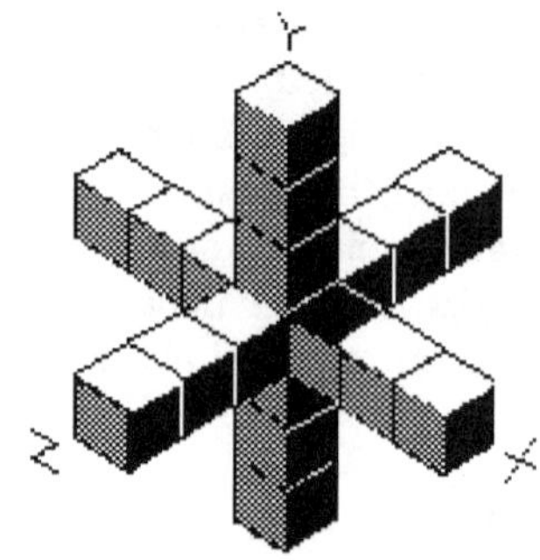

Fig. 6.6 Discrete coordinate space Oxyz, with integer coordinates continued in all six directions

Table 6.1 Arithmetic progression of integers, and reciprocal harmonic progression

a	$-\infty$	...	-4	-3	-2	-1	0	1	2	3	4	...	∞
$h = 1/a$	$-0'$	...	$-1/4'$	$-1/3'$	$-1/2'$	$-1'$	∞'	$1'$	$1/2'$	$1/3'$	$1/4'$	...	$0'$

A *harmonic progression* is a sequence of numbers $h_1,\ h_2,\ h_3,\ \ldots$ where the reciprocals of the terms are in arithmetic progression. That is, $h_1 = 1/a_1, h_2 = 1/a_2$, ... Given two successive terms h_1 and h_2, the next term is $h_3 = (h_1 \times h_2)/(2h_1 - h_2)$. Given a lower neighbour h_1 and upper neighbour h_3 the harmonic mean is $h_2 = (2 \times h_1 \times h_3)/(h_1 + h_3)$. For example, Table 6.1 shows an arithmetic progression from $-\infty$ to ∞, and the reciprocal harmonic progression from -0 to 0.

Geometrically, an *arithmetic scale* is a line marked off in steps with constant step-length k. The distances of successive steps from a fixed zero form an arithmetic progression a_i.

A *harmonic scale* is a line marked off in steps with step-lengths reciprocal to those of an arithmetic scale. The distances of successive steps from a fixed zero form a harmonic progression h_i. For example, Fig. 6.7a shows an arithmetic scale a of integers between $-\infty$ and $+\infty$, and the corresponding harmonic scale h, where $h_i = 1/a_i$ (as in Table 6.1). Notice that the infinite positive arithmetic segment between 1 and ∞ is condensed into the finite harmonic segment between $1'$ and $0'$,

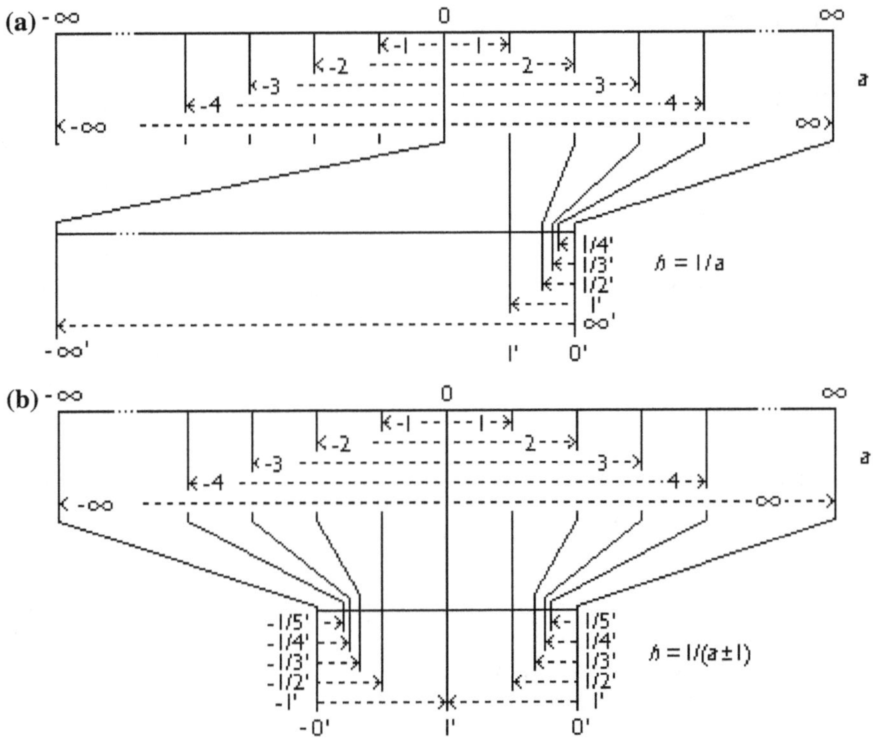

Fig. 6.7 Arithmetic and harmonic scales. **a** Where $h_i = 1/a_i$, positive arithmetic distances a between 1 and ∞ (measured from 0) compress to harmonic distances h between $1'$ and $0'$ (measured from $0'$), but arithmetic distances between 0 and 1 expand to harmonic distances between $1'$ and ∞'. **b** Where $h_i = 1/(a_i \pm 1)$, arithmetic distances a between $-\infty$ and ∞ compress to harmonic distances h between -1 and 1. The infinite harmonic segments are omitted

while the finite arithmetic segment between 0 and 1 is expanded into the infinite harmonic segment between $1'$ and ∞. The negative arithmetic scale corresponds to a separate similar harmonic scale. However, if we make $h_i = 1/(a_i \pm 1)$, (plus 1 when a_i is positive and minus 1 when a_i is negative), we get the correspondence shown in Fig. 6.7b. Now the infinite positive arithmetic segment of a between 0 and ∞ is condensed into the finite harmonic segment of h between $1'$ and $0'$, and similarly the infinite negative arithmetic segment is condensed into the finite harmonic segment between $-0'$ and $1'$.

6.5 Harmonic Coordinate Space

If we are given three base points and their coordinates as

- *origin* point O, coordinate 0 (zero),
- *unit* point U, coordinate 1,
- *finite point* x on line OU, coordinate ∞, such that length OU = length Ux,
- and a right angle at O,

and perform the construction of throws in the first quadrant, we get the result shown in Fig. 6.8 (with the Euclidean version for comparison).

x, y, and A are now finite points, and there are no parallels. The coordinates 0, 1, 2, 3, ..., ∞ form an arithmetic progression a. The construction locates the points at (primed) harmonic distances $h = 1/(a+1)$ (measured from $x = 0'$). There is a one-to-one correspondence between the Euclidean net and the harmonic net in the first quadrant. For the second quadrant, the construction must be reflected in the Oy-axis, with a discontinuity along that axis; and similarly for the third and fourth quadrants. Then every point in the unlimited Euclidean plane corresponds one-to-one with a point in the finite square x y $-x$ $-y$.

Figure 6.9a shows the given harmonic base points and units; (b) the harmonic coordinate axis Ox; and (c) the harmonic coordinate plane Oxy with integer coordinates (x, y). Between any two points further points can be inserted. Figure 6.10 shows harmonic coordinate space Oxyz.

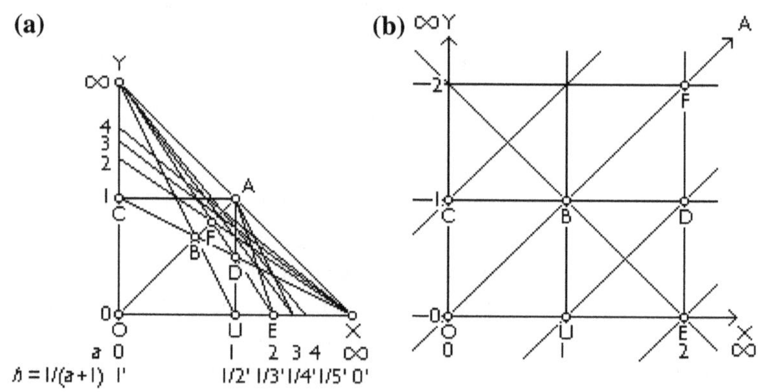

Fig. 6.8 **a** Harmonic throws on a quadrant of a plane. **b** Euclidean throws, for comparison

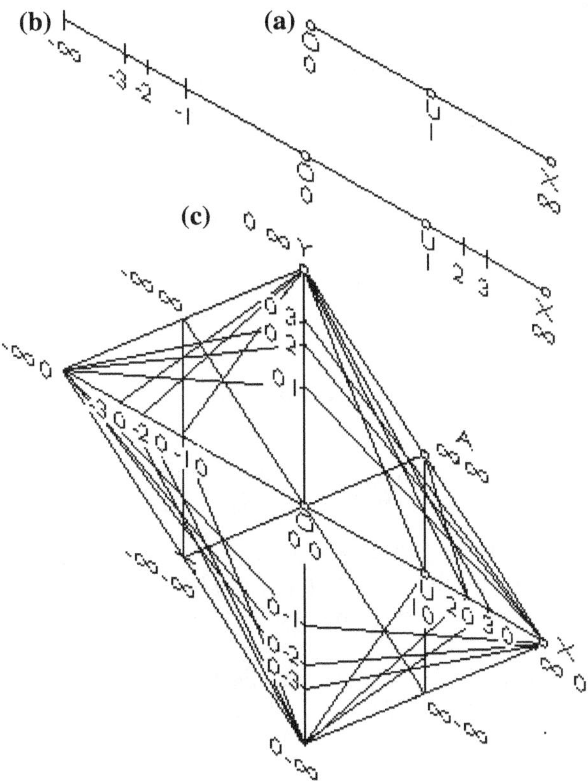

Fig. 6.9 **a** Harmonic base points. **b** Harmonic coordinate axis Ox. **c** Harmonic coordinate plane Oxy

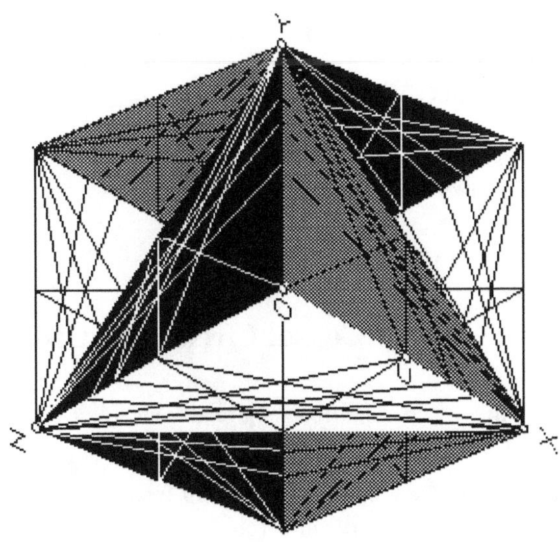

Fig. 6.10 Harmonic coordinate space. Each point has coordinates (x, y, z)

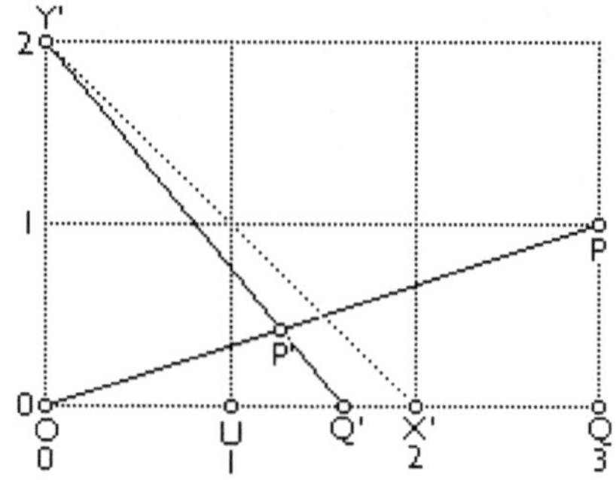

Fig. 6.11 Euclidean point P (3, 1) is on line OP with slope 1/3. Harmonic point P′ is on line Q′y′ with slope −4/3. Meet of lines OP and Q′y′ is (1.2, 0.4)

6.6 Euclidean-to-Harmonic Conversion

Figure 6.11 shows a point P with coordinates (x, y) in Euclidean space. To find the corresponding point P′ in harmonic space:

- Line OP has slope $m = y/x$.
- Slope-intercept equation of line OP is $y = mx$.
- Harmonic point Q′ has coordinates $(q', 0)$, $q' = 2 - (2/(x + 1))$.
- Line Q′y′ has slope $m' = 2/ - q'$.
- Slope-intercept equation of line Q′y′ is $y' = m'x' + 2$.
- Meet of lines OP and Q′y′ is, with coordinates (x', y').
- $x' = 2 \times (1/(m - m')$.
- $y' = m \times x'$.

For example:

- Given P = (3, 1).
- Slope of OP $m = 1/3$.
- Q′ = (3/2, 0).
- Slope of Q′y′$m' = -4/3$.
- $x' = 6/5 = 1.2$.
- $y' = 2/5 = 0.4$.

6.7 Family of Coaxal Circles and Codiskal Caps

A family of *coaxal circles* share an *axis*, their common chord with fixed length c. To construct any circle in the family:

- Let s be the length of the *sagitta* from chord to circle
- and r the radius of the circle.
- Then $r = c^2/8s + s/2$.
- Let a be the length of the *apothem* from chord to circle centre.
- Then $a = r - s$,

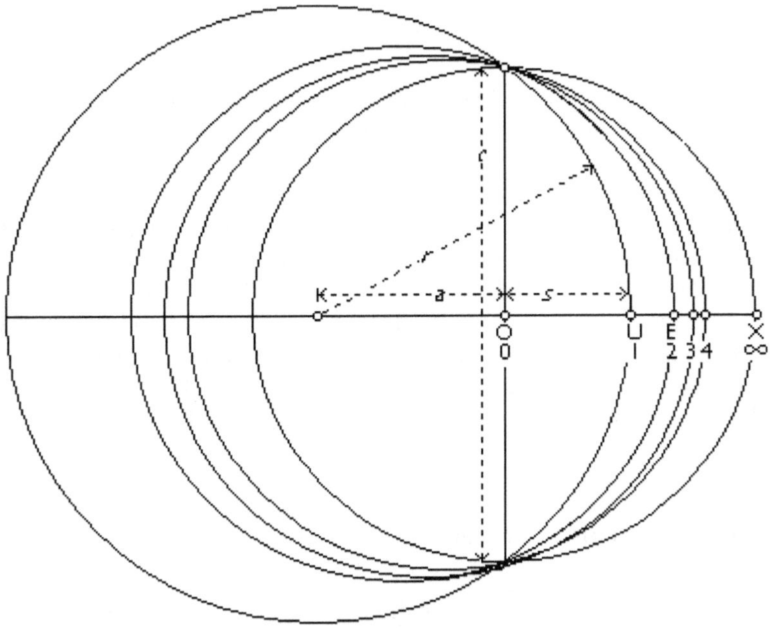

Fig. 6.12 Construction of coaxal circles. Fix chord c and set sagitta s. Find radius $r = c^2/8s + s/2$, and apothem $a = r - s$

Table 6.2 Calculation for harmonic family of coaxal circles

c	c^2	Circle n	h	s	$s/2$	$8s$	$c^2/8s$	r	a
2	4	0	1	0	0	0	∞	∞	∞
		1	1/2	1/2	1/4	4	1	5/4	3/4
		2	1/3	2/3	1/3	16/3	3/4	13/12	5/12
		3	1/4	3/4	3/8	6	2/3	25/24	7/24
		4	1/5	4/5	2/5	32/5	5/8	41/40	9/40
		...	...	...	...	...	...	...	...
		∞	0	1	1/2	8	1/2	1	0

Given chord c length 2, and sagitta s for each circle, find radius r and apothem a

as shown in Fig. 6.12. For example, given a disk with radius 1, the calculation for a harmonic family of circles is shown in Table 6.2. The circles are labelled in arithmetic progression $n = 0, 1, 2, 3, 4, ..., \infty$. The harmonic distances measured from x are $h = 1/(n + 1) = 1, 1/2, 1/3, 1/4, 1/5, ..., 0$. The sagittas measured from O are $s = 1 - h = 0, 1/2, 2/3, 3/4, 4/5, ..., 1$.

If a family of coaxal circles on a disk is rotated about the perpendicular bisector of their axis, we get a family of *codiskal spherical caps* sharing the same flat disk.

6.8 Homogeneous Coordinates

Euclidean space has two special points (at infinity) on each line, and a special line (at infinity) on each plane, and a special plane (at infinity). Harmonic space and elliptic space haves no such special elements. *Homogeneous coordinates*[6] rewrite the ordinary non-homogeneous (Cartesian) coordinates as ratios, to avoid infinite components. On a line, homogeneous $(x{:}w)$ corresponds to non-homogeneous (x/w); for example (0:1), (1:1), (2:1), ..., (1:0) correspond to 0, 1, 2, ..., ∞. On a plane (0:0:1), (1:0:1), (2:0:1), ..., (1:0:0) correspond to (0,0), (1,0), (2,0), ..., (∞,0). In space (0:0:0:1), (1:0:0:1), (2:0:0:1), ..., (1:0:0:0) correspond to (0,0,0), (1,0,0), (2,0,0), ..., (∞,0,0). Homogeneous coordinates can be multiplied by a factor without changing their significance; for example (1:2:3:4) = 5(1:2:3:4) = (5:10:15:20) = (0.25, 0.5, 0.75). Homogeneous coordinates allow all values except (0:0:0:0).

6.9 Elliptic Coordinate Space

If we are given three base points and their coordinates as

- *Origin* point O, coordinate 0 (zero).
- *Unit* point U, coordinate 1.
- *Finite point* x on line OU, coordinate ∞, such that length OU = length Ux.
- And a right angle at O,

as for harmonic space, and perform the construction of throws using coaxal circles as lines, we get the result shown in Fig. 6.13 (with the Euclidean and harmonic constructions for comparison). All lines are circular arcs. Lines through origin O are circles with radius ∞, that is, straight lines.

Figure 6.14a shows the given elliptic base points and their homogeneous coordinates; (b) the elliptic coordinate axis Ox; and (c) the elliptic coordinate plane Oxy. Similarly, in the Oxz-plane, every point has unique coordinates $(x{:}0{:}z{:}1)$ or $(x{:}0{:}z{:}0)$; and in the Oyz-plane, every point has unique coordinates $(0{:}y{:}z{:}1)$ or $(0{:}y{:}z{:}0)$. We then have the three elliptic coordinate planes, within a sphere, as shown in Fig. 6.15, where every point has unique coordinates $(x{:}y{:}z{:}1)$ or $(x{:}y{:}z{:}0)$. Between any two points further points can be inserted.

A family of spherical caps with centres on Ox, all meeting on circle yz, now provide any number of further elliptic planes on either side of coordinate plane Oyz. On each such cap there is a coordinate net of circular-arc lines. Another family with centres on Oy, all meeting on circle xz, and a third family with centres on Oz, all meeting on circle xy, provide further elliptic planes.

We then have an elliptic coordinate space Oxyz, where the points lines and planes correspond one-to-one with the Euclidean coordinate space plus its plane at infinity.

[6]Introduced in 1827 by August Ferdinand Mobius (1790–1868), Professor at Leipzig University.

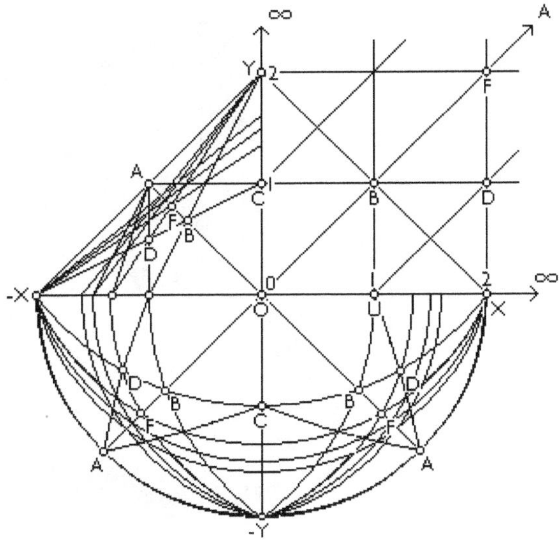

Fig. 6.13 *bottom* Elliptic throws on a disk, compared with Euclidean *top right* and harmonic *top left*. Base is origin O coordinate 0, unit point U coordinate 1, and x coordinate ∞. The construction throws any number of new coordinate points on both x-axis and y-axis, and hence covers as much of the disk as we please

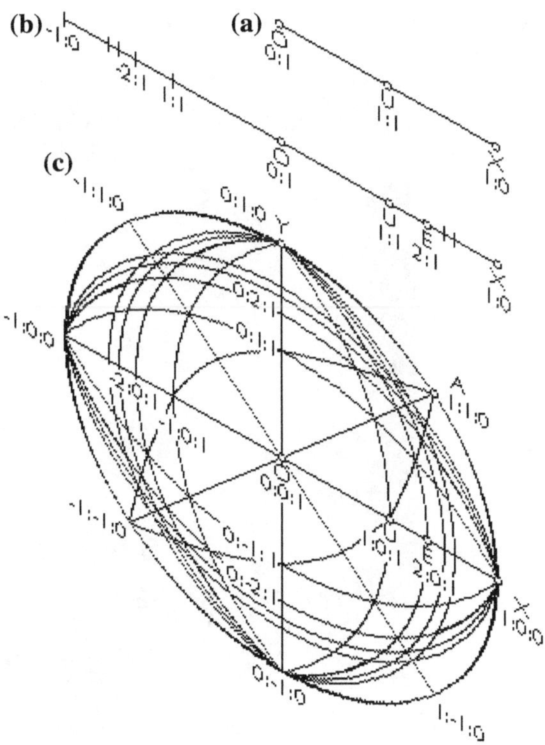

Fig. 6.14 **a** Elliptic base points (homogeneous coordinates). **b** Elliptic coordinate axis Ox. **c** Elliptic coordinate plane Oxy

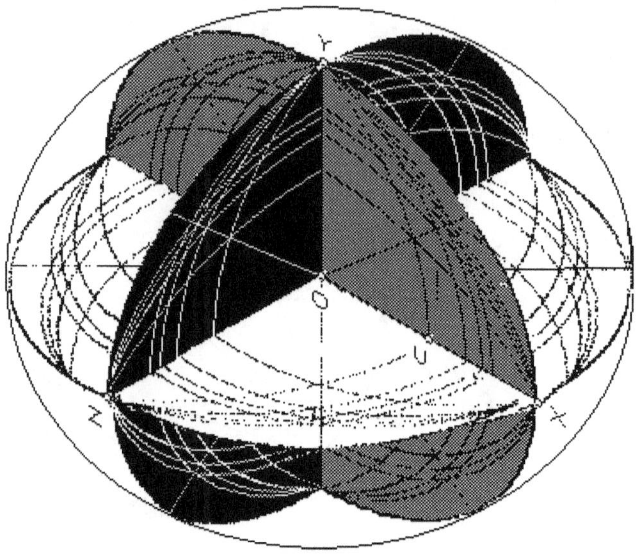

Fig. 6.15 Elliptic coordinate space. Each point has coordinates $(x{:}y{:}z{:}w)$

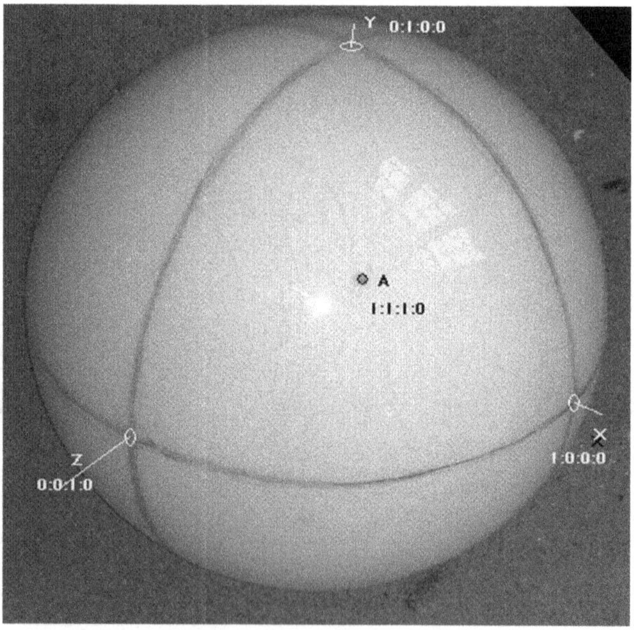

Fig. 6.16 Surface of elliptic coordinate space, with meets of coordinate planes Oxy, Oxz, Oyz, and first-quadrant mid-point A. The lines are great circles

Figure 6.16 shows the sphere enclosing elliptic space. Every line, such as xy the meet of coordinate plane Oxy with the sphere, is a great circle of the sphere. The geometry is spherical, as on the surface of the Earth (where, for example, a triangle may have three right angles). On the sphere every pair of antipodal points corresponds to a Euclidean point at infinity. If, notionally, antipodal points are identified, then every semi-great-circle of the sphere, and every line of the space, is closed.

References

1. Descartes R (1637) La Géométrie, Appendix to Discourse de la Méthode. Jan Maire, Leyden. http://www.gutenberg.org/etext/26400. Accessed 1 Feb 2014
2. Heath TL (1955) The thirteen books of Euclid's elements (trans: heath TL), vol 3. Dover, New York
3. Klein FC (1872) Review of recent researches in geometry (trans: Haskell MW 1892-3). http://arxiv.org/pdf/0807.3161v1.pdf. Accessed 1 Feb 2014
4. Staudt KGC von (1847) Geometrie der Lage, Bauer und Raspe, Nürnberg. Reprinted 2011 by Nabu Press

Chapter 7
Object Geometry

Objects in space can be measured to produce discrete coordinates specifying the object. There are many techniques, old and new, for measuring objects. A common aim is to capture the locations of a sufficient number of target points, usually as (x, y, z) coordinates in a suitable rectangular reference frame. There are several different approaches, in active development, to building a three-dimensional model of an object from the target locations, for further processing. Here we describe an approach using a space graph (or wire-frame model), and some elementary space transformations by program.

7.1 Measuring Physical Objects

We have a long-established habit of imposing a continuous Euclidean space on the world around us, and setting a convenient coordinate origin, axes, and unit length for a region containing objects of current interest (see Sect. 6.2). But physical measurement always has limited precision, according to the purpose in view and the equipment in hand [4]. For example, a metal object might be measured with a micrometer to the nearest hundredth of a millimetre, a wooden object with a ruler to the nearest millimetre, or a house with a tape to the nearest centimetre. Entirely similarly, we can set measurements for an imaginary object. So we are not locating Euclidean points (with unlimited precision), but discrete elements (with fixed finite precision). Measuring imposes a discrete space on a physical object.

7.2 Discrete Space Box

A discrete space has no inherent limits to its extent. In any practical case we use perforce a finite *space box*, a rectangular parallelepipedic region chosen to suit the purpose in hand. A *discrete space box* is specified by giving ranges for the x-, y- and

A. Parkin, *Digital Imaging Primer*, DOI 10.1007/978-3-540-85619-1_7

z-coordinates in a discrete space. For example:

$$-3 \le x \le 4, -3 \le y \le 4, -3 \le z \le 4$$

specifies a space box 8 cubels wide by 8 cubels high by 8 cubels deep, with origin near the centre of the box. However:

$$0 \le x \le 7, 0 \le y \le 7, 0 \le z \le 7$$

also specifies a space box $8 \times 8 \times 8$ cubels, but now with origin at bottom left back of the box. This lies entirely in the first octant, and has no negative coordinates. The two cases are shown in Fig. 7.1. We consider first the principles of several different techniques of measurement of objects.

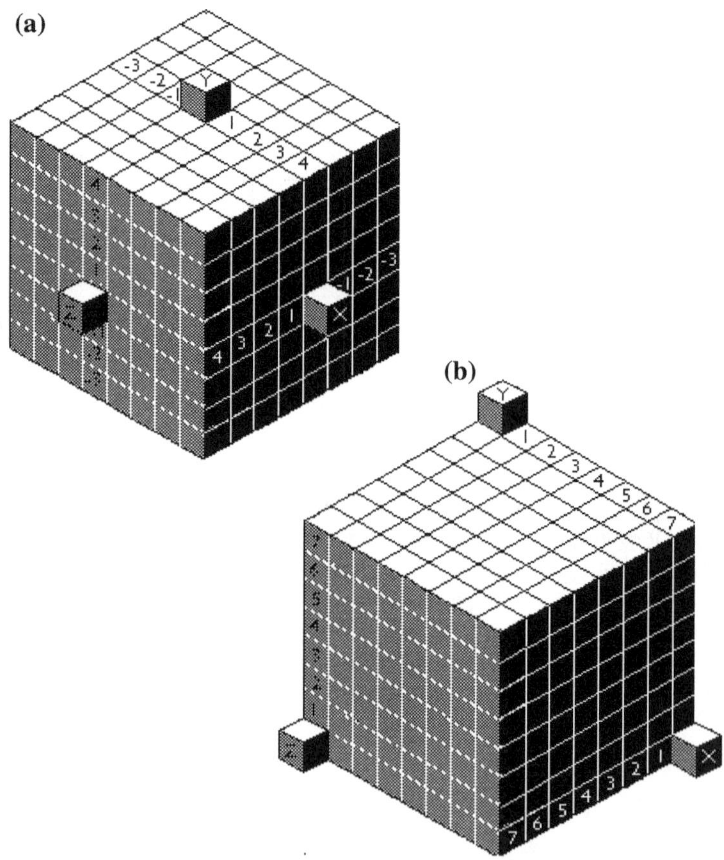

Fig. 7.1 Two discrete space boxes. **a** $8 \times 8 \times 8$ cubels, with origin near centre. **b** $8 \times 8 \times 8$ cubels, with origin at (hidden) far bottom corner

7.3 Contact Measurement

Contact measurement is suitable if the object is directly-accessible. The reference frame is usually a rectangular set of origin, axes, and planes Oxyz, fixed around the object. An engineer may use a surface plate, an angle plate, a height block, callipers, and a graduated rule. A building surveyor may use a ground mark, a spirit level, a plumb-bob, and a tape measure. Within the reference frame, the procedure is to take triples of distances (X, Y, Z) of salient corners or local maxima or minima from the reference planes, as shown in Fig. 7.2a.

A sculptor may use callipers to take triples of distances from three fixed reference points. In the late 18th century sculptors developed *pointing machines* [5], for copying a clay or wax or plaster model to stone. This had a triangle frame attachable to three fixed points on the stone block and corresponding fixed points on the model. The frame carried a pointer with a stop. For each of many salient local maxima or minima

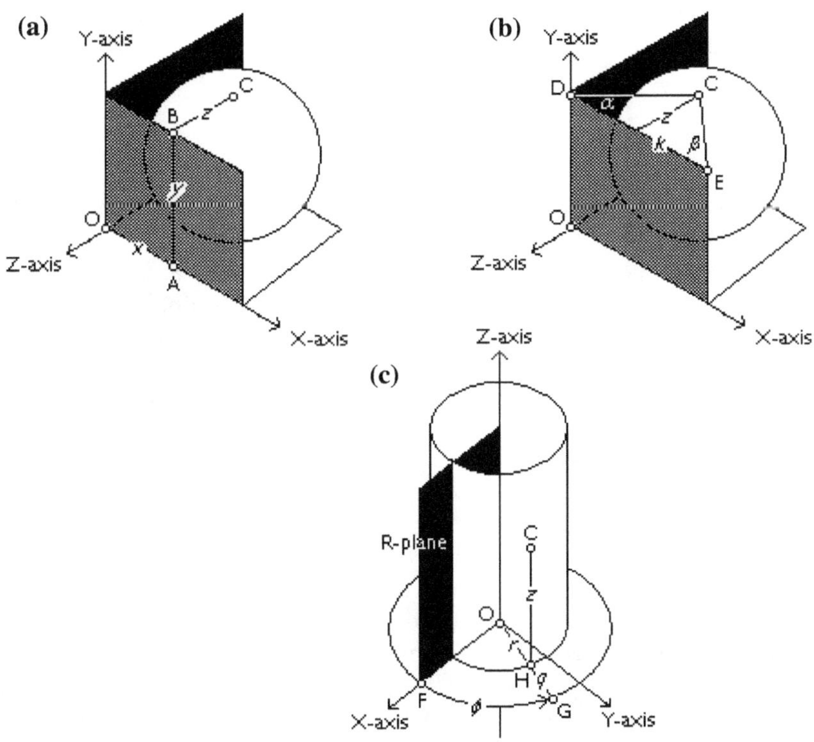

Fig. 7.2 Measuring an object. **a** In a Cartesian reference frame Oxyz, measure OA $= x$, AB $= y$, BC $= z$. **b** Triangulation from two fixed points D and E, distance k apart. Measure angle at D $= \alpha$ and angle at B $= \beta$, then distance z is $k \times \sin(\alpha) \times \sin(\beta)/(\sin(\alpha + \beta)$. **c** In a cylindrical reference frame, measure distance OF. Measure angle FG $= \phi$. Measure distance GH $= q$, then $r =$ OF $- q$. Measure distance HC $= z$. Then $x = r \times \cos(\phi)$ and $y = r \times \sin(\phi)$

the pointer was set on the model, then transferred to the block, which was drilled or carved in until the pointer stop was reached. The stone was then carved down to smooth between the salient measured points. The advantage is one measurement per point, instead of three.

A modern contact method is to use a travelling xy carriage, like a plotter, with a z-probe to touch the object at closely-spaced points, and record the coordinates in some file format. Another method uses a jointed arm, and calculates xyz-coordinates from angular measurements.

7.4 Optical Measurement

Optical measurement is suitable if the object is line-of-sight-accessible. The reference frame is again a rectangular set of origin, axes, and planes Oxyz, fixed around the object. The procedure is to measure a base-length k and the sight angles α, β of the target point from each end of the base-length, as shown in Fig. 7.2b. The perpendicular distance z of the target is then found as

$$z = k \times \sin(\alpha) \times \sin(\beta)/(\sin(\alpha + \beta)) .$$

Early land surveyors used a sighting tube and a protractor compass, and later a theodolite, to measure angles, and hence calculate (x, y, z) positions for a large number of targets.

For small objects, a modern optical method is to set a laser at D, a camera at E, and shoot a laser dot onto the object at C. The position of the dot is then calculated by trigonometry. Shooting a laser stripe rather than a spot onto the object enables a faster process.

A laser range-finder measures line-of-sight distance to target to moderate accuracy.

7.5 Global Positioning System (GPS)

A global positioning system (GPS)[1] [6] depends on multiple satellites in medium Earth orbits, which emit very precisely timed signals at gigaHerz frequencies. Four such signals are sufficient to determine the position of a target on or near the Earth's surface, currently accurate to less than three metres in general, and to a few centimetres with ground enhancements. The current GPS was developed since the 1960s by the US military, and subsequently released for public use. It is in continuous development.

[1]Invented by Roger L. Easton (1921–2014) of US Naval Research Laboratories, and many others.

7.6 Photogrammetry

If two cameras with known position take a stereo-pair of photographs, then the (X, Y, Z) coordinates of any target in the binocular overlap can be calculated (see Chaps. 11 and 12). Such photogrammetry has been used for making maps since the 19th century. Modern close-range measurement is in active development [7].

7.7 Computer Modelling of Objects

Full measurement of an object would deliver (x, y, z) coordinates for every discrete element of the object, outside and inside (as, for example, an MRI scan). The set of coordinates could be entered to a computer as a file, and this would constitute a full model of the object to that level of detail. But for most purposes something much less demanding than full measurement will do.

A *wire-frame model* is an economical way of specifying the main surface features of a solid object. For this, we measure coordinates of relatively few and coarse discrete elements (*vertices*), and record straight segments (*edges*) between adjacent vertices. We thus impose a manageable network of triangles or polygons on the object. The computer model is then a list of vertex coordinates and a list of edges referred to the vertices. A familiar example in land surveying is a triangulation between surface features, with spot heights. For built or manufactured objects there may be plans or other drawings from which coordinates of salient points may be taken. In the last resort, somebody has to do the hard work by hand, as, for example, for Newell's original teapot[2] in Fig. 7.3.

A wire-frame model may be elaborated in various ways. The straight edges between vertices may be made curved; the triangular (or polygonal) faces may be assigned colour and opacity properties; parts hidden from a given point of view may be removed [8, 9]; appearance under illumination may be rendered by ray-tracing.

Alternatively, in computer-aided design (CAD) we may be concerned to model an imagined object. We then start with vector-specified primitive shapes, which can be transformed and combined to build up the required model in highly-specialized software [1].

State-of-the-art three-dimensional scanning, modelling, processing, rendering and and motion capture is outside the scope of this book, since it depends on expensive hardware and software available only in an academic or corporate environment [2]. However, some of the simpler aspects can be enjoyably explored using only standard equipment and programming, as described below.

---

[2] Martin Edward Newell, doctoral thesis, University of Utah, 1975.

Fig. 7.3 Wire-frame model. View of model constructed from nine patches and 118 vertices

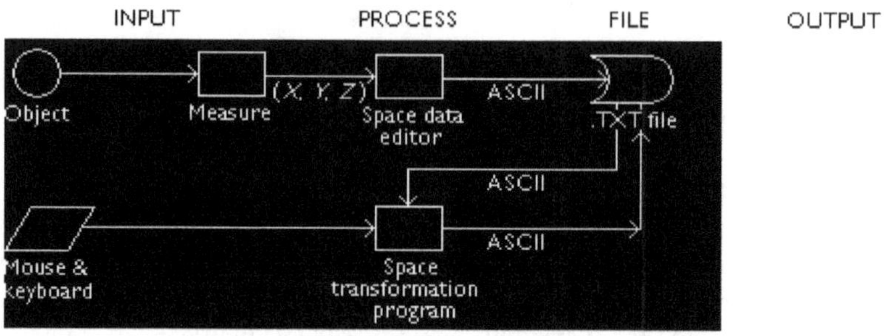

Fig. 7.4 Given object captured as a space graph or wire-frame model. Data are space coordinates (x, y, z) of vertices, and a list of edges between vertices. Data input to space-data-editor are stored in a .TXT data file. A space-transformation program can read and change the data file, thereby performing space operations on the object

7.8 Example of Computer Modelling

The flowchart (Fig. 7.4) shows a given object captured as a space graph by measuring coordinates (x, y, z) of vertices and listing edges $(V1, V2)$ between vertices. The space graph can be written as an ASCII.TXT data file by a space graph editor program, or by a text editor such as Notepad. A space-transformation program can read the data file, perform space operations on the object, and write a new data file.

7.9 Discrete Space Graph

A discrete space graph is specified numerically by

- A list of vertices $V_1, V_2, \ldots$, each as a triad of integer coordinates (x, y, z)
- A list of edges, each as a pair of vertices (V_1, V_2)
- A list of faces, each as a cycle of vertices $(V_1, V_2, V_3, \ldots)$ or of edges $(E_1, E_2, E_3, \ldots)$,

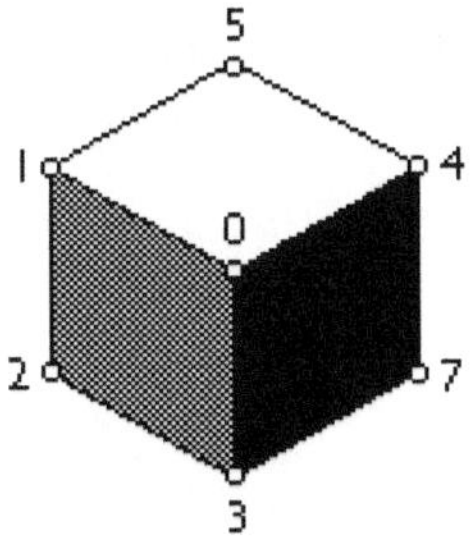

Fig. 7.5 Example of discrete space graph. The eight vertices are labelled 0, 1, 2, 3, 4, 5, 6 (*hidden*), and 7. The 12 edges are 01, 12, 23, 30, 04, 15, 26, 37, 45, 56, 67, and 74. The six faces are 0123, 0451, 1562, 2673, 3740, and 4567 $V - E + F = 8 - 12 + 6 = 2$

where an edge joins two vertices, and a face is bounded by three or more edges, hence three or more vertices. They are related by Euler's formula[3]:

$$V - E + F = 2$$

Figure 7.5 shows a simple example[4]: a cube, with $V = 8$, $E = 12$, $F = 6$, so $8 - 12 + 6 = 2$.

A simple data file structure for a space graph is

- Header

  Count of vertices
  Count of edges

- Vertex list

  Vertex identifier, X, Y, Z coordinates
  Vertex identifier, X, Y, Z coordinates
  ...

- Edge list

  Edge identifier, vertex identifier, vertex identifier
  Edge identifier, vertex identifier, vertex identifier
  ...

---

[3]Leonhard Paul Euler (1707–1783), pupil of Johann Bernoulli and collaborator with Daniel Bernoulli; Academies of St Petersburg 1727–1741, Berlin 1741–1766, St Petersburg 1766–1783; prolific founder of several branches of mathematics. He solved the problem of the seven bridges of Königsberg (is it possible to take a walk which crosses each bridge once only and ends up at the start?) by representing the map by a plane graph. The answer is no, for reasons generalized by Euler and others.

[4]The object is at this stage a numerical specification: it is shown here as an image-in-advance in isometric projection (see Sect. 8.10). The shading is added to emphasise that that a three-dimensional object is meant.

where all values are integer cubels. Faces are not needed here. Using this format, the data for the cube in Fig. 7.5, with an edge length of 65 cubels, might be

- (Header)

  (Count of vertices) 8
  (Count of edges) 12

- (Vertex list)

  0,32,32,32
  1,−32,32,32
  2,−32,−32,32
  3,32,−32,32
  4,32,32,−32
  5,−32,32,−32
  6,−32,−32,−32
  7,32,−32,−32

- (Edge list)

  0,0,1
  1,1,2
  2,2,3
  3,3,0
  4,0,4
  5,1,5
  6,2,6
  7,3,7
  8,4,5
  9,5,6
  10,6,7
  11,7,4

7.10 Space Data Editor Program

An ASCII file in this format can be composed in an ordinary text editor, such as Notepad; or in a data editor program written for the purpose, such as briefly described below (QuickBASIC source code available on the SpringerLink web-site).

```
REM  **  Program EDSGRAPH to edit space graph ASCII data
REM  **     file. User picks New or Edit. For New, user
REM  **     enters new filename and header values. For
REM  **     Edit, user enters filename to edit.
REM  **     Program reads header values and sets-up
REM  **     arrays with no data for New, old data for
```

```
REM  **     Edit, and offers change. At end, program
REM  **     (re-)writes edited file.
REM  **  User enters new or old file-name to to edit
REM  **  New file set-up: enter counts of vertices, edges
REM  **  Old file set-up: show old counts of vertices and
REM  **  edges; enter counts of additional vertices,edges
END

         SUB EDITDATA
REM  **  Subprogram EDITDATA to show data from arrays
REM  **     AVERT, AEDGE, and offer change.
END SUB
         SUB READDATA
REM  **  Subprogram READDATA to put all-zero data for
REM  **     new file, or read data from old file, into
REM  **     arrays AVERT AEDGE.
END SUB
         SUB WRITDATA
REM  **  Subprogram WRITDATA to rewrite data from
REM  **    arrays AVERT, AEDGE to ASCII file #1.
END SUB
```

For example, Fig. 7.6a shows an a space graph file for a cube, and Fig. 7.6b shows the same file after a dilation by factor 1.5.

7.11 Space Transformation

Geometrically, given any object in space, we can change its position relative to the reference frame. This can be seen in either of two ways:

- As an *active* transformation, where the reference frame OXYZ remains fixed but the object moves to a new position
- As a *passive* transformation, where the object remains fixed but the reference frame moves to a new position

The active interpretation is sometimes called an *alibi* since the object goes elsewhere; and the passive interpretation an *alias* since the same object gets other coordinate names.

Space transformations are usually grouped according to the properties which are preserved or *invariant* under the operation. Thus,

- *Rigid motions* preserve lengths and angles: reflection, translation, rotation
- *Conformal transformations* preserve angles: dilation
- *Projective transformations* preserve linearity: shearing, projection.

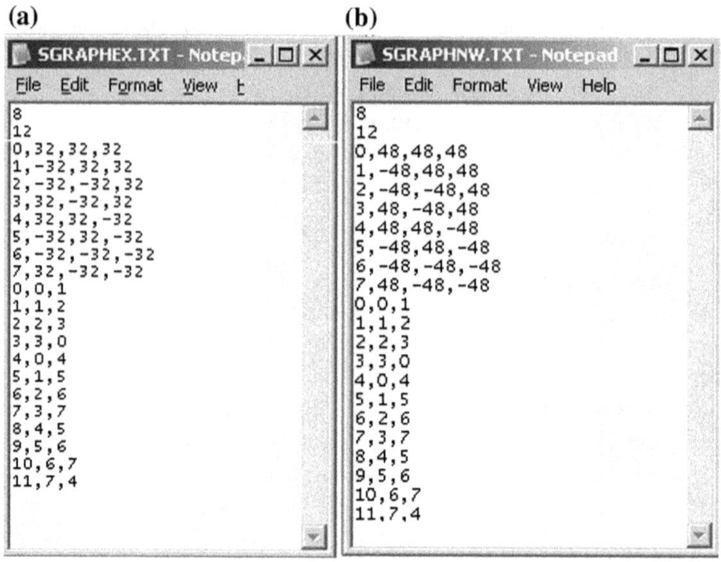

(a) SGRAPHEX.TXT - Notep

```
8
12
0,32,32,32
1,-32,32,32
2,-32,-32,32
3,32,-32,32
4,32,32,-32
5,-32,32,-32
6,-32,-32,-32
7,32,-32,-32
0,0,1
1,1,2
2,2,3
3,3,0
4,0,4
5,1,5
6,2,6
7,3,7
8,4,5
9,5,6
10,6,7
11,7,4
```

(b) SGRAPHNW.TXT - Notepad

```
8
12
0,48,48,48
1,-48,48,48
2,-48,-48,48
3,48,-48,48
4,48,48,-48
5,-48,48,-48
6,-48,-48,-48
7,48,-48,-48
0,0,1
1,1,2
2,2,3
3,3,0
4,0,4
5,1,5
6,2,6
7,3,7
8,4,5
9,5,6
10,6,7
11,7,4
```

Fig. 7.6 Space graph ASCII file for a cube. **a** As first composed. **b** After applying a dilation in all three axes by factor 1.5

Some elementary space coordinate transformations are (seen actively)

- *Reflection*, which mirrors an object in a plane
- *Translation*, which moves the object in any direction
- *Rotation*, which changes the orientation of the object
- *Dilation*, which changes the size of the object
- *Shearing*, which slants an object.

Numerically, given the location coordinates of a space object, we may change its coordinates to new values by doing arithmetic on the old values. Each transformation corresponds to the application of some function ϕ to the old values (x, y, z) of every cubel to get new values (x', y', z'):

$$(x', y', z') = \phi(x, y, z)$$

The inverse function ϕ' recovers the old values from the new

$$(x, y, z) = \phi'(x', y', z')$$

For example, if ϕ is given by:

$$(x', y', z') = (x + 8, y, z)$$

that is, add 8 to each X-coordinate: the geometrical effect is to move the object eight cubels in the positive x-direction. Then ϕ' is

$$(x', y', z') = (x - 8, y, z)$$

that is, subtract 8 from each x-coordinate: the geometrical effect is to move the object eight cubels in the negative x-direction.

For computing, the preferred method [3] of performing a transformation is to convert the given space coordinates (x, y, z) of a cubel to homogeneous form by adding a fourth coordinate $W = 1$ (see Sect. 6.8). The row vector[5] $(x, y, z, 1)$ is a 1×4 matrix, which can be multiplied by a 4×4 *transformation matrix* to produce the transformed coordinates $(x', y', z', 1)$ (see Sect. 3.15). The homogeneous coordinates are returned to Cartesian form (x', y', z') by dividing each of x', y' and z' by $W = 1$: that is, just drop the 1. For example, a matrix which multiplies every coordinate by 2 is

$$\mathbf{M} = \begin{pmatrix} 2 & 0 & 0 & 0 \\ 0 & 2 & 0 & 0 \\ 0 & 0 & 2 & 0 \\ 0 & 0 & 0 & 1 \end{pmatrix}$$

In homogeneous coordinates:

$$(x', y', z', 1) = (x, y, z, 1) \begin{pmatrix} 2 & 0 & 0 & 0 \\ 0 & 2 & 0 & 0 \\ 0 & 0 & 2 & 0 \\ 0 & 0 & 0 & 1 \end{pmatrix} = (2x, 2y, 2z, 1)\,,$$

which, by deleting the fourth coordinate 1, becomes Cartesian coordinates $(2x, 2y, 2z)$.

Each elementary transformation has a corresponding matrix. When several elementary transformations are wanted, each set of given coordinates can be multiplied serially by the several matrices; or, more efficiently, the several matrices can be multiplied together once, then the combined matrix applied once to each set of coordinates. The reason for using homogeneous coordinates and 4×4 matrices is to include translation in the same matrix scheme as the other elementary transformations. The sequence in which several transformations are applied is of great significance: matrix multiplication is not commutative.

---

[5] An alternative convention is to treat the space coordinates as a column vector, and transpose the matrices.

7.12 Reflection

Reflecting in any one or more of the space coordinate planes xy, xz, yz changes the left-to-right-handedness of an object but leaves the reflection plane, and the size and orientation of the object, unchanged.

Figure 7.7 shows an object with an asymmetric cut to make clear the reflections. The object is shown as given (unreflected), then reflected in the xy-plane, the xz-plane, the yz-plane, then in both xy- and xz-planes, then in all three planes xy, xz and xz.

To reflect an object the transformation matrix is

$$\begin{pmatrix} M & 0 & 0 & 0 \\ 0 & N & 0 & 0 \\ 0 & 0 & P & 0 \\ 0 & 0 & 0 & 1 \end{pmatrix}$$

where

- (integer) $M = -1$ reverses the x-axis, so reflects in the yz-plane,
- (integer) $N = -1$ reverses the y-axis, so reflects in the xz-plane,
- (integer) $P = -1$ reverses the z-axis, so reflects in the xy-plane,
- $M, N, P = 1$ leave unchanged (the identity matrix).

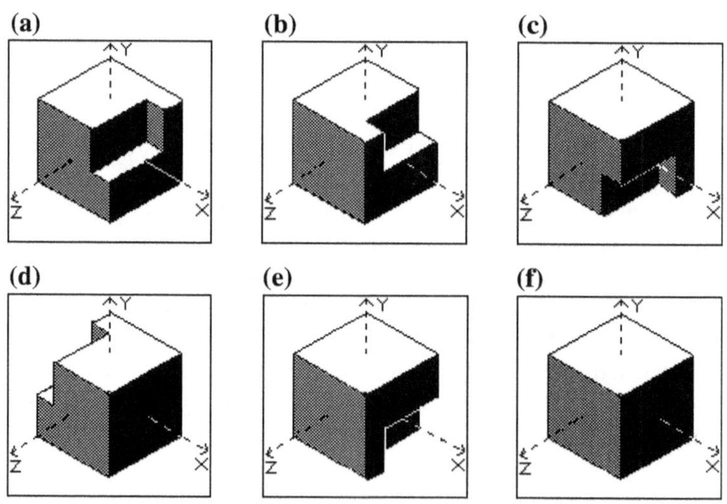

Fig. 7.7 Reflect in space. **a** Given asymmetric object (unreflected). **b** Object reflected back to front in xy (facing) plane. **c** Object reflected bottom to top in xz (horizontal) plane. **d** Object reflected left to right in yz (fore-and-aft) plane. **e** Object reflected in both xy- and xz-planes. **f** Object reflected in all three xy-, xz-, and yz-planes: the cut is now hidden

Applying the reflection matrix to the coordinates (x, y, z) of a cubel, we have

$$(x', y', z', 1) = (x, y, z, 1)\begin{pmatrix} M & 0 & 0 & 0 \\ 0 & N & 0 & 0 \\ 0 & 0 & P & 0 \\ 0 & 0 & 0 & 1 \end{pmatrix} = (\pm x, \pm y, \pm z, 1)$$

7.13 Translation

Translating an object along any one or more of the space axes x, y, z changes the location of the object in the reference frame but leaves the size and orientation unchanged.

Figure 7.8 shows a cube object as given, centred at origin; then translated along the x-axis, the y-axis, the z-axis, the x- and y-axes, and the x-, y- and z-axes. The axes of the fixed reference frame are shown in each case. The cube has vertices at ± 32 cubels symmetrically about the origin, so edge count 65 cubels.

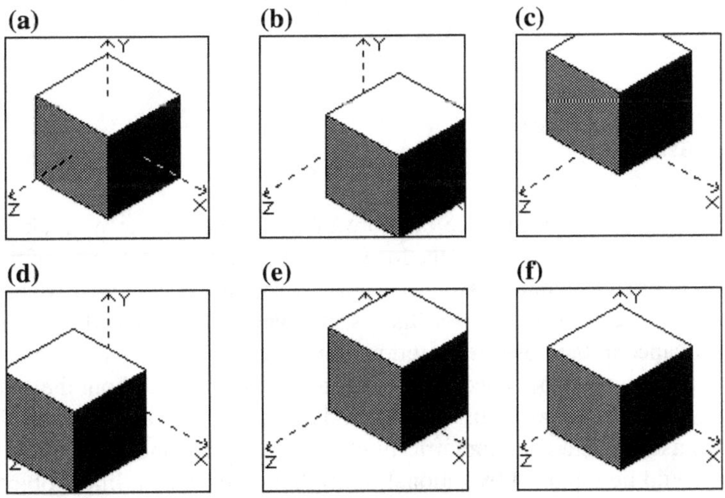

Fig. 7.8 Translate in space. **a** Cube object as given, centre at origin of reference frame, vertices ± 32 cubels. **b** Cube translated by 32 cubels along x-axis, taking centre to right of origin. **c** Cube translated by 32 cubels along y-axis, taking centre above origin. **d** Cube translated by 32 cubels along z-axis, taking centre in front of origin. **e** Cube translated by 32 cubels along x-axis, then 32 cubels along y-axis, taking centre to right of and above origin. **f** Cube translated by 32 cubels along x-axis, then 32 cubels along y-axis, then 32 cubels along z-axis, taking centre to original location of the near vertex

To translate an object in space the transformation matrix is

$$\begin{pmatrix} 1 & 0 & 0 & 0 \\ 0 & 1 & 0 & 0 \\ 0 & 0 & 1 & 0 \\ A & B & C & 1 \end{pmatrix}$$

where

- (integer) A is the magnitude of the translation along the x-axis, in cubels,
- (integer) B is the magnitude of the translation along the y-axis, in cubels,
- (integer) C is the magnitude of the translation along the z-axis, in cubels,
- $A, B, C > 0$ move positive,
- $A, B, C = 0$ leave everything unchanged (the identity),
- $A, B, C < 0$ move negative.

Applying the translation matrix to the coordinates (x, y, z) of a cubel, we have

$$(x', y', z', 1) = (x, y, z, 1) \begin{pmatrix} 1 & 0 & 0 & 0 \\ 0 & 1 & 0 & 0 \\ 0 & 0 & 1 & 0 \\ A & B & C & 1 \end{pmatrix} = (x + A, y + B, z + C, 1)$$

7.14 Rotation

Rotating an object about any one or more of the space axes x, y, z changes the orientation of the object but leaves the location of the origin and the size of the object unchanged. For a right-handed set of axes, positive rotation is clockwise when looking in the positive direction of the axis (that is, the way a right-handed screw must be turned to advance in the positive direction along the axis).

Figure 7.9 shows a cube object as given (unrotated), rotated about the x-axis, the y-axis, the z-axis, both the x- and y-axes, and all three x-, y- and z-axes.

Here we use the real trigonometric functions sin(angle) and cos(angle): in principle they could be replaced by rational multipliers. To rotate a linear object about the x-axis the transformation matrix is

$$\begin{pmatrix} 1 & 0 & 0 & 0 \\ 0 & \cos(\alpha) & \sin(\alpha) & 0 \\ 0 & -\sin(\alpha) & \cos(\alpha) & 0 \\ 0 & 0 & 0 & 1 \end{pmatrix},$$

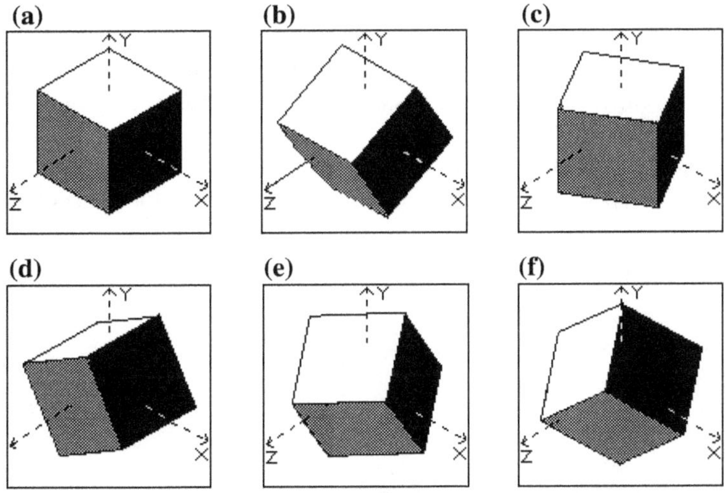

Fig. 7.9 Space rotation. **a** Given object (unrotated). **b** Object rotated by 30° about x-axis. **c** Object rotated by 30° about y-axis. **d** Object rotated by 30° about z-axis. **e** Object rotated by 30° about x-axis then by 30° about y-axis. **f** Object rotated by 30° about x-axis then by 30° about y-axis then by 30° about z-axis

where

- (real) α is the angle of rotation about the x-axis, in degrees or radians,
- $\alpha > 0$ tips edges parallel to the positive y-axis forwards, and edges parallel to the positive z-axis downwards,
- $\alpha = 0$ leaves everything unchanged (the identity),
- $\alpha < 0$ tips edges parallel to the positive y-axis backwards, and edges parallel to the positive z-axis upwards.

Applying the rotation-about-x matrix to the coordinates (x, y, z) of a cubel, we have

$$(x', y', z', 1) = (x, y, z, 1)\begin{pmatrix} 1 & 0 & 0 & 0 \\ 0 & \cos(\alpha) & \sin(\alpha) & 0 \\ 0 & -\sin(\alpha) & \cos(\alpha) & 0 \\ 0 & 0 & 0 & 1 \end{pmatrix}$$

$$= \{x, [cos(\alpha) - \sin(\alpha)]\, y, [\sin(\alpha) + \cos(\alpha)]\, z, 1\}$$

Quite similarly, to rotate a linear object about the y-axis the transformation matrix is

$$\begin{pmatrix} \cos(\beta) & 0 & -\sin(\beta) & 0 \\ 0 & 1 & 0 & 0 \\ \sin(\beta) & 0 & \cos(\beta) & 0 \\ 0 & 0 & 0 & 1 \end{pmatrix}$$

where

- (real) β is the angle of rotation about the y-axis, in degrees or radians,
- $\beta > 0$ twists edges parallel to the positive x-axis backwards, and edges parallel to the positive z-axis to the right,
- $\beta = 0$ leaves everything unchanged (the identity),
- $\beta < 0$ twists edges parallel to the the positive x-axis forwards, and edges parallel to the positive z-axis to the left.

Applying the rotation-about-y matrix to the coordinates (x, y, z) of a cubel, we have

$$(x', y', z', 1) = (x, y, z, 1) \begin{pmatrix} \cos(\beta) & 0 & -\sin(\beta) & 0 \\ 0 & 1 & 0 & 0 \\ \sin(\beta) & 0 & \cos(\beta) & 0 \\ 0 & 0 & 0 & 1 \end{pmatrix}$$

$$= (\cos(\beta) + \sin(\beta))x, y, (-\sin(\beta) + \cos(\beta))z, 1) .$$

And again, to rotate a linear object about the z-axis the transformation matrix is

$$\begin{pmatrix} \cos(\gamma) & \sin(\gamma) & 0 & 0 \\ -\sin(\gamma) & \cos(\gamma) & 0 & 0 \\ 0 & 0 & 1 & 0 \\ 0 & 0 & 0 & 1 \end{pmatrix},$$

where

- (real) γ is the angle of rotation about the z-axis, in degrees or radians,
- $\gamma > 0$ tilts edges parallel to the positive x-axis upwards, and edges parallel to the positive y-axis to the left,
- $\gamma = 0$ leaves everything unchanged (the identity),
- $\gamma < 0$ tilts edges parallel to the positive x-axis downwards, and edges parallel to the positive y-axis to the right.

Applying the rotation-about-z matrix to the coordinates (x, y, z) of a cubel, we have

Brackets in the equation were corrected; please check.

$$(x', y', z', 1) = (x, y, z, 1) \begin{pmatrix} \cos(\gamma) & \sin(\gamma) & 0 & 0 \\ -\sin(\gamma) & \cos(\gamma) & 0 & 0 \\ 0 & 0 & 1 & 0 \\ 0 & 0 & 0 & 1 \end{pmatrix}$$

$$= \{[\cos(\gamma) - \sin(\gamma)]\, x, [\sin(\gamma) + \cos(\gamma)]\, y, z, 1\} .$$

7.15 Dilation

Dilating or scaling an object along any one or more of the space axes x, y, z changes the size of the object but leaves the location and orientation unchanged.

Figure 7.10 shows a cube object as given (undilated), and dilated along the x-axis, the y-axis, the z-axis, the x- and y-axes, and the x-, y- and z-axes.

The dilation factors here are rational multipliers $N \backslash D$: first multiply by integer N, then integer-divide by D. To dilate an object the transformation matrix is

$$\begin{pmatrix} D & 0 & 0 & 0 \\ 0 & E & 0 & 0 \\ 0 & 0 & F & 0 \\ 0 & 0 & 0 & 1 \end{pmatrix}$$

where

- Rational multiplier D is the dilation factor along the x-axis,
- Rational multiplier E is the dilation factor along the y-axis,
- Rational multiplier F is the dilation factor along the z-axis,
- $D, E, F > 1$ enlarge,
- $D, E, F = 1$ leave unchanged,
- $D, E, F < 1$ diminish,
- $D, E, F = 0$ make that dimension vanish.

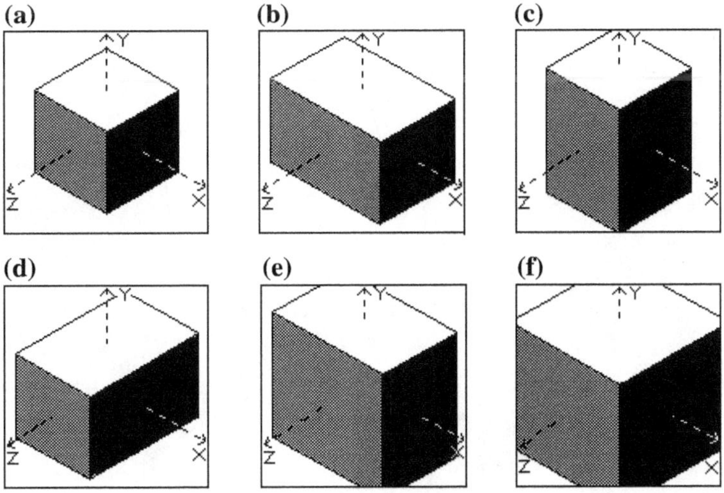

Fig. 7.10 Dilate in space. **a** Given object (undilated); say volume = 1. **b** Object dilated by rational multiplier $3\backslash 2$ = factor 1.5 along x-axis; volume = 1.5. **c** Object dilated by $3\backslash 2$ along y-axis; volume = 1.5. **d** Object dilated by $3\backslash 2$ along z-axis; volume = 1.5. **e** Object dilated by $3\backslash 2$ along both x-axis and y-axis; volume = 2.25. **f** Object dilated by $3\backslash 2$ along all three axes; volume = 3.375

Applying the dilation matrix to the coordinates (x, y, z) of a cubel, we have

$$(x', y', z', 1) = (x, y, z, 1) \begin{pmatrix} D & 0 & 0 & 0 \\ 0 & E & 0 & 0 \\ 0 & 0 & F & 0 \\ 0 & 0 & 0 & 1 \end{pmatrix} = (Dx, Ey, Fz, 1) .$$

7.16 Shearing

Shearing is harder to grasp than the other transformations. An object is *orthogonally sectioned* by any plane parallel to the xy (frontal) coordinate plane. In each sectioning plane the object has a two-dimensional *sectional shape*. Similarly, the object is orthogonally sectioned by any plane parallel to the xz (horizontal) coordinate plane, and parallel to the yz (fore-and-aft) coordinate plane. For example, Fig. 7.11a shows a cube orthogonal to the reference frame, and its sectional shape in the xy plane, a square.

Geometrically, a *shear* slants the sectional shapes of an object. For example, Fig. 7.11b shows a shear of the x-direction by a proportion k of the y-direction, which turns a square section into a rhombus.

Figure 7.12 shows examples of each of the six elementary space shears applied to a cube, with each shear factor set at 0.5.

Numerically, shear along x by ky is

$$\begin{aligned} x' &= x + ky \\ y' &= y \\ z' &= z . \end{aligned}$$

For example, if the four vertices of a square section are as given below, and the shear-factor $k = 0.5$, then shear of x by $0.5y$ produces

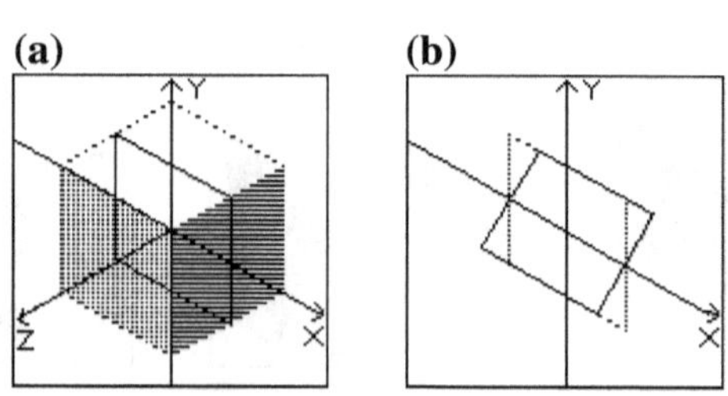

Fig. 7.11 Sectional shape and shearing. **a** A cube (*ghosted*) orthogonal to the reference frame has a square section parallel to the xy-coordinate plane. **b** A shear of x by y turns the square section into a rhombic section

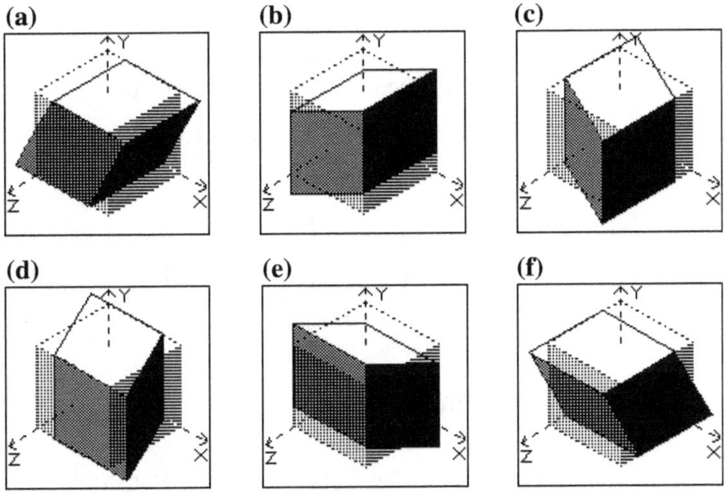

Fig. 7.12 The six elementary space shears applied to a cube. **a** Cube (*ghosted*) sheared in x by $0.5y$; changes every sectional shape parallel to frontal xy-plane; cube becomes rhombic prism on z-axis, with two faces horizontal; twin of (**b**). **b** Cube sheared in y by $0.5x$; changes every sectional shape parallel to frontal xy-plane; cube becomes rhombic prism on z-axis, with two faces vertical; twin of (**a**). **c** Cube sheared in z by $0.5x$; changes every sectional shape parallel to horizontal xz-plane; cube becomes rhombic prism on y-axis, with two faces fore-and-aft; twin of (**d**). **d** Cube sheared in x by $0.5z$; changes every sectional shape parallel to horizontal xz-plane; cube becomes rhombic prism on y-axis, with two faces frontal; twin of (**c**). **e** Cube sheared in y by $0.5z$; changes every sectional shape parallel to fore-and-aft yz-plane; cube becomes rhombic prism on x-axis, with two faces frontal; twin of (**f**). **f** Cube sheared in z by $0.5y$; changes every sectional shape parallel to fore-and-aft yz-plane; cube becomes rhombic prism on x-axis, with two faces horizontal; twin of (**e**)

- $(x, y, z) = (32, 32, 0)$ becomes $(32 + 16, 32, 0) = (48, 32, 0)$,
- $(x, y, z) = (-32, 32, 0)$ becomes $(-32 + 16, 32, 0) = (-16, 32, 0)$,
- $(x, y, z) = (-32, -32, 0)$ becomes $(-32 - 16, -32, 0) = (-48, -32, 0)$,
- $(x, y, z) = (32, -32, 0)$ becomes $(32 - 16, -32, 0) = (16, -32, 0)$.

Computationally, to shear an object the transformation matrix is

$$\begin{pmatrix} 1 & I & K & 0 \\ G & 1 & L & 0 \\ H & J & 1 & 0 \\ 0 & 0 & 0 & 1 \end{pmatrix}$$

where

- (real) G shears x by Gy,
- (real) H shears x by Hz,
- (real) I shears y by Ix,
- (real) J shears y by Jz,
- (real) K shears z by Kx,

- (real) L shears z by Ly,
- $F, G, H, I, J, K, L = 0$ leave unchanged (the identity matrix),
- $-F, -G, -H, -I, -J, -K, -L$ shear in the opposite direction.

Consider x sheared by Gy. Applying the shearing matrix to the coordinates (x, y, z) of a cubel, we have

$$(x', y', z', 1) = (x, y, z, 1) \begin{pmatrix} 1\,0\,0\,0 \\ G\,1\,0\,0 \\ 0\,0\,1\,0 \\ 0\,0\,0\,1 \end{pmatrix} = (x + Gy, y, z, 1)\,.$$

7.17 Space Transformation Program

A program to perform one or more of the elementary space transformations on any given space graph is briefly described below (QuickBASIC source code available on the SpringerLink web-site). The before and after files can be examined and edited in a data file editor such as EDSGRAPH (see Sect. 7.5), or in a text editor such as NOTEPAD.

```
REM  **  Program TRANSFOS to perform space transformations
REM  **  on any space graph. Given graph vertices and
REM  **  edges ASCII file #1, program puts vertex
REM  **  coordinates into array AVERT and edges into
REM  **  array AEDGE. User enters transformation
REM  **  constants wanted, and program sets up
REM  **  transformation matrices. Program multiplies
REM  **  given XYZ coordinates  of each vertex by matrix
REM  **  to get transformed X'Y'Z' coords in AVERT.
REM  **  Program writes new space graph to ASCII file #2.
REM  **  User enters old file-name and new file-name
         CALL RSGRDATA
         CALL INCONSTA
         CALL CALCSTRA
         CALL WSGRDATA
END

         SUB CALCSTRA
REM  **  Subprogram CALCSTRA to calculate space
REM  **  transformations of object vertices.
END SUB
         SUB INCONSTA
REM  **  Subprogram INCONSTA to enter transformation
REM  **  constants.
```

```
END SUB
         SUB RSGRDATA
REM  **  Subprogram RSGRDATA to read data from ASCII
REM  **  space graph file #1 into arrays AVERT and AEDGE.
END SUB
         SUB WSGRDATA
REM  **  Subprogram WSGRDATA to write transformed
REM  **  data from arrays AVERT and AEDGE to ASCII file #2
END SUB
```

With a program like this, the way is open to explore wire-frame modelling of all kinds of object at various scales, and to manipulate the models in various ways. There are, of course, many software packages available which include these and other transformations. Such packages are reliable, polished, and well-documented and helped. But they often impose a quite elaborate environment of their own, not compatible with other packages. And they are too expensive for the occasional user. In comparison, the programs in this book are very simple, but they do relate in a common environment, and form a flexible base for any kind of development. Programming your own tools (including fixing your own inevitable bugs) gives the best understanding and control of imaging.

For example, a regular octahedron with its eight triangular faces is a good elementary object for building wire-frame models. Any number of different octahedra in any orientation in space can be derived by the six elementary operations described above. We could write an accumulating constructor program which attaches a new octahedron to an old one, face-to-face or edge-to-edge or vertex-to-vertex. The elementary operations could then be applied to the assembled model.

References

1. AutoCAD (2013) http://www.autodesk.com/products/autodesk-autocad/overview. Accessed 1 Feb 2014
2. Blundell BG (2014) Research and writing. http://www.barrygblundell.com/default.html. Accessed 1 Feb 2014
3. Foley JA, van Dam A (1982) Fundamentals of interactive computer graphics. Addison-Wesley, Reading (IBM Systems Programming Series)
4. JCGM 200:2012 International Vocabulary of Metrology. http://www.bipm.org. Accessed 1 Feb 2014
5. Jourdain F (2002) Les Techniques de Sculpture. http://www.edu.augustins.org/pdf/second/sculp/sgen02s.pdf. Accessed 1 Feb 2014
6. US Government GPS website (2014) http://gps.gov. Accessed 1 Feb 2014
7. Luhmann T et al (2013) Close range photogrammetry and 3D imaging. Walter de Gruyter, Berlin
8. Maillot P-G (1986) Hidden-line removal. https://sites.google.com/site/patrickmaillot/english. Accessed 1 Feb 2014
9. Sutherland IE, Sproull RF, Schumacker RA (1974) Ten hidden-surface algorithms. http://design.osu.edu/carlson/history/PDFs/ten-hidden-surface.pdf. Accessed 1 Feb 2014

Chapter 8
Projection Geometry

A plane projection transforms a given figure in a plane to an image in the same plane. A plane projection has four parameters, and falls into one of 15 classes according to the incidence relations of these parameters. An example of each class is shown. In the Euclidean plane, projection preserves incidences but not lengths: an arithmetic scale projects to a harmonic scale. In the harmonic plane, contained in a square, there are 15 classes of projection corresponding to the Euclidean case; and in the elliptic plane, contained in a circle, also 15. A space projection transforms a given figure in space to an image in a plane. A space projection has four parameters, and falls into one of 15 classes according to the incidence relations of these parameters. A space projection can be constructed geometrically, or computationally by matrix multiplication. Ten classical space projections are defined, with a geometric and computational construction, and example, for each. A program to produce several classes of space projection is briefly described (full listings on Springer website). Some of the incidence classes produce relief projections, transforming a given space figure not to a plane image but to a space image. The lens projection is of special interest. All the Euclidean projections have corresponding projections in harmonic and in elliptic space.

8.1 Plane Projection

A *line graph* is a set of

- *N vertices* V_1, V_2, ...
- $(N-1)$ *edges* E_1, E_2, ...
- 1 *face* F_1,

where all the vertices lie on one line, an edge joins two vertices, and the single face is the plane containing the line. They are related by Euler's formula (see Sect. 7.9):

$$V - E + F = 2$$

A. Parkin, *Digital Imaging Primer*, DOI 10.1007/978-3-540-85619-1_8

For example, a line graph with four vertices P Q R S and three edges PQ QR RS and one face, has $4 - 3 + 1 = 2$.

A *plane graph* is a set of

- *Vertices* $V_1, V_2, \ldots,$
- *Edges* $E_1, E_2, \ldots,$
- *Faces* $F_1, F_2, \ldots,$

where all the vertices lie in one plane, an edge joins two vertices, a face is bounded by three or more edges, no two edges cross, and there are no isolated parts. They are related by Euler's formula. For example, a square has four vertices, four edges, and two faces (one inside the square and one outside), so $4 - 4 + 2 = 2$.

Projective geometry [4, 12][1] recognizes only the *incidence* and *order* of points, lines and faces, in contrast to metric geometries which also recognize length and angle. Projective constructions are done with an ungraduated straight-edge only.

A given plane-graph figure can be transformed to a plane-graph image in the same plane by a *plane projection* operation. A plane projection is specified by four parameters:

- a *centre of projection* point C,
- a *line of projection* λ,
- a pair of *corresponding points*, A in the given figure and A′ in the image, where A, A′, and C lie on the *axis of projection* AA′C.

A plane projection relates a given figure to its image by a fan of *projector* lines meeting at centre C. If C is an observer, and the projectors are lines of sight, the figure and its image look the same from C (though from outside the plane, we can also see C and the different shapes of figure and image.) Given a plane projection (C, λ, A, A′), Fig. 8.1a shows the construction for the image of a single given point, (b) the image of a given line figure, and (c) the image of a given plane figure.

There is an astonishing variety of plane projections, according to how the parameters are arranged. If we consider only the incidence and order of parameters, there are 15 classes of plane projection, as shown in Fig. 8.2, where each projection transforms the same square figure to a different image.

- In I to V, A′ is between A and C. Then as λ moves from beyond A to beyond C, in I the image is stretched longer; in II the image is just a point C; in III the image is a line; in IV and V the image is compressed shorter.
- In VI to X, A is between A′ and C. Then as λ moves from beyond A′ to beyond C, in VI the image is enlarged and compressed; in VII the image is flat; in VIII the image is just C; in IX and X the image is stretched longer.

[1] The study of projective properties was originated by Jean-Victor Poncelet (1788–1867). Served as military engineer under Napoleon, was taken prisoner in Russia 1812–1814 where he thought out the fundamentals of synthetic projective geometry, published after his return to France as Professor at Metz and at Paris. Poncelet's general term for projection was *homologie*, which later took on a more specialized meaning in topology.

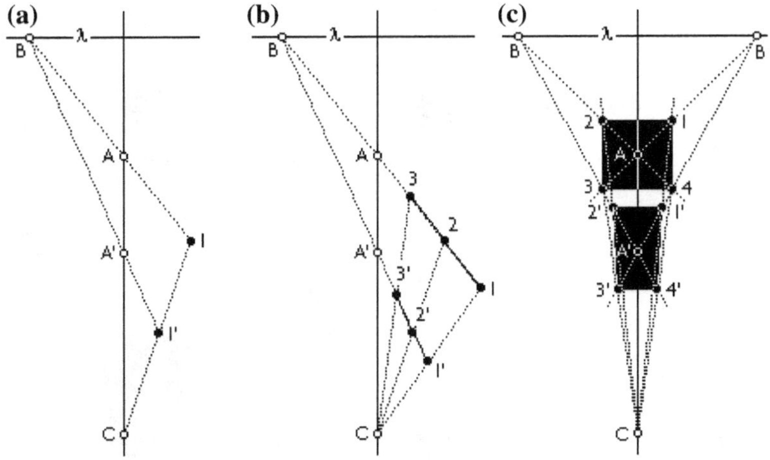

Fig. 8.1 Projective-geometric construction of plane projection, fixed by centre C, line λ, and two corresponding points A and A′ on the axis AA′C. **a** 1 is a given point. Join A1, to meet λ on B; join A′B; join C1, to meet A′B on 1′, the projected image of 1. **b** 123 is a given line figure, and 1′2′3′ is its projected image. **c** 1234 is a given plane figure, and 1′2′3′4′ is its projected image

- In XI to XV, C is between A and A′; the figure and its image are on the opposite sides of C, and the image is rotated by 180° about the axis of projection. Then as λ moves from beyond A to beyond A′, in XI the image is stretched; in XII the image is just C; in XIII the image is stretched; in XIV the image is flat; in XV the image is compressed.
- In IV IX and XIII λ is on C: the projection is then called an *elation*.

8.2 Euclidean Plane Projection

The classification in Fig. 8.2 is by incidence and order of points and lines only, without regard to lengths or angles. In the Euclidean plane, each of the these projective incidence classes can be varied by setting different lengths between the points C A and A′, and/or different angles between λ and the axis CAA′. For example, Fig. 8.3a shows λ orthogonal to CAA′, projecting a square to a trapezium; Fig. 8.3b shows λ oblique to CAA′; and Fig. 8.3c shows centre C removed to infinity.

In the Euclidean plane, projection does not preserve lengths from an object to its image, but it does preserve the *cross-ratio* of four lengths. Figure 8.4 shows given line κ with points P, Q, R, S forming an arithmetic scale (see Sect. 6.4) with equal intervals PQ, QR, RS. The distances measured positively/negatively from an arbitrary zero are p, q, r, s The cross-ratio[2] is

[2]The four distances can be permuted in six different ways to calculate the cross-ratio. We choose this form as standard.

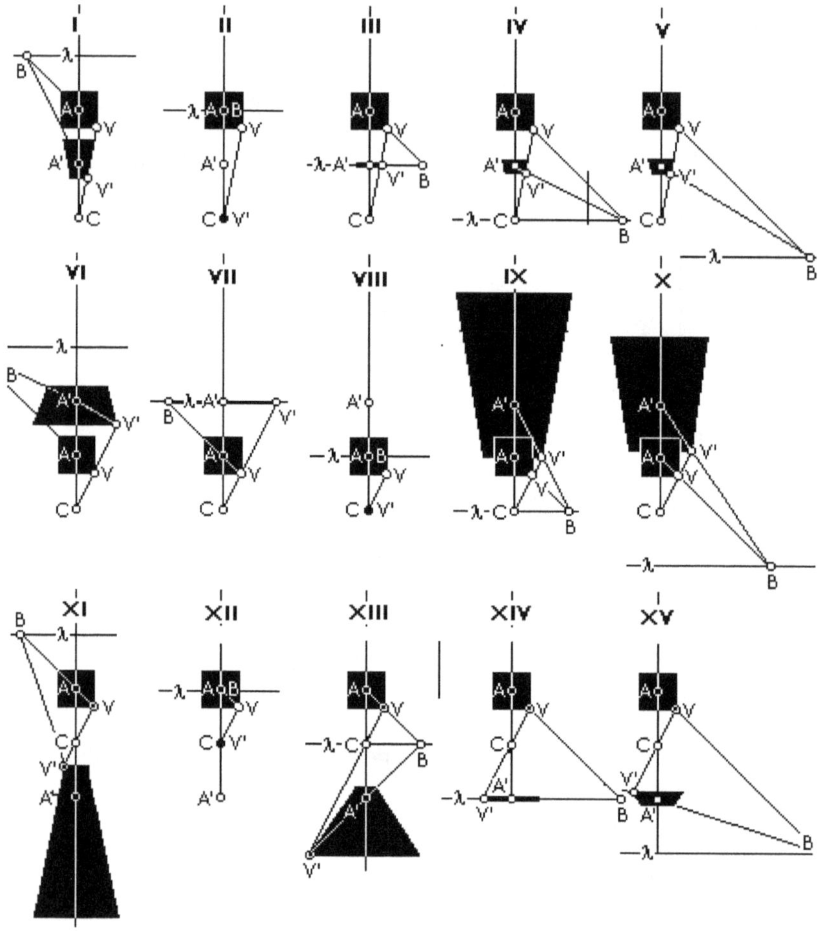

Fig. 8.2 Fifteen incidence classes of plane projection. The same square figure is projected to different images

Fig. 8.3 Different Euclidean plane projections within incidence class I. **a** λ orthogonal to axis CAA′. **b** λ oblique to axis CAA′. **c** centre C at infinity

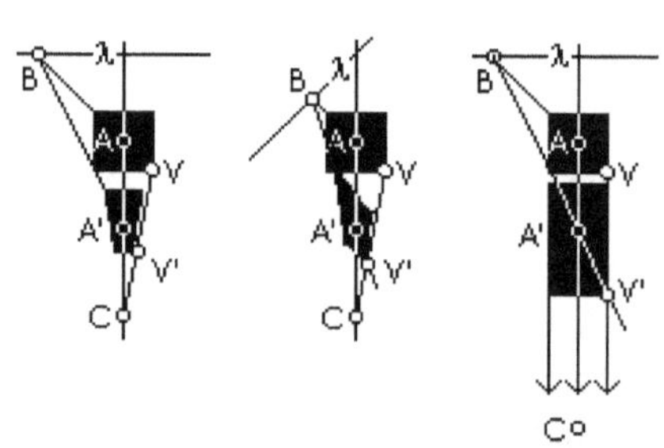

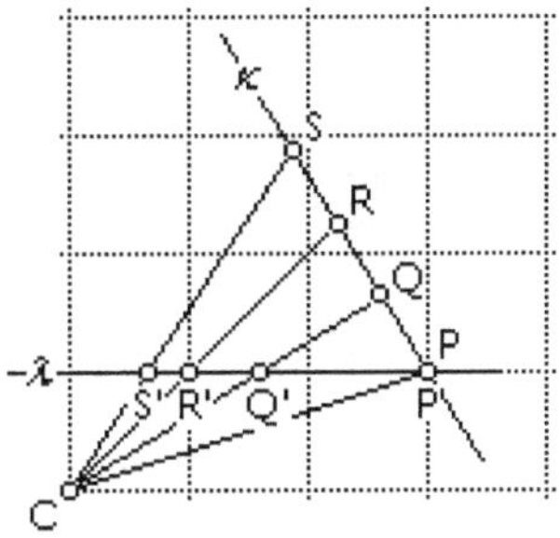

Fig. 8.4 Euclidean plane projection of arithmetic scale to harmonic scale. Given arithmetic scale PQRS on line κ, the projection from centre C to image line λ is harmonic scale P′Q′R′S′

$$\chi(\text{PQRS}) = ((r-p)(s-q))/((r-q)(s-p)) .$$

The two pairs PR and QS are said to *separate* one another on the line. For example, taking distances from P:

$$\begin{aligned}\chi(\text{PQRS}) = \chi(0123) = ((2-0)(3-1))/((2-1)(3-0))\\ ((2)(2))/((1)(3)) = 4/3 .\end{aligned}$$

Indeed, the cross-ratio of any four successive points on an arithmetic scale is 4/3.

Now draw a fan of projectors CP, CQ, CR, CS, as in Fig. 8.4. Any line not on C, such as λ, meets the projectors at image points P′, Q′, R′, S′. If PQRS is an arithmetic scale, then P′Q′R′S′ is a harmonic scale; that is to say $p' = 1/p, q' = 1/q, r' = 1/r, s' = 1/s$. The cross-ratio of P′, Q′, R′, S′ is the same as the cross-ratio of P, Q, R, S. For example, taking distances from P′ = P:

$$\begin{aligned}\chi(P'Q'R'S') &= \chi(1/0, 1/1, 1/2, 1/3)\\ &= ((1/2-\infty)(1/3-1))/((1/2-1)(1/3-\infty))\\ &= ((-\infty)(-2/3))/((-1/2)(-\infty)) = 4/3 .\end{aligned}$$

In the special case that κ is parallel to λ, the image scale P′Q′R′S′ is also an arithmetic scale.

In Fig. 8.5a, object line κ has an arithmetic scale 0, 1, 2, …, ∞, and meets image line λ at $1 = 1'$. The arithmetic scale on κ from 1 to ∞ projects to the harmonic scale on λ from $1'$ to $0'$, and the arithmetic scale from 1 to 0 projects to the harmonic scale from $1'$ to ∞'. Inversely, if the given line is λ and it is projected to κ, then the arithmetic scale on λ from 1 to ∞ projects to the harmonic scale on κ from $1'$ to $0'$, and the arithmetic scale from 1 to 0 projects to the harmonic scale from $1'$ to ∞'. In Fig. 8.5b, κ is parallel to λ, and the entire arithmetic scale projects to another arithmetic scale (in this case half the size).

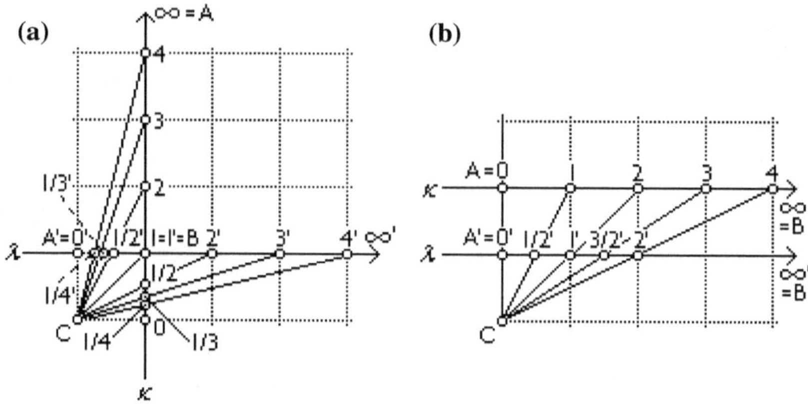

Fig. 8.5 Euclidean plane projection of arithmetic scale (class III). **a** Arithmetic scale 0, 1, 2, …, ∞ on object line κ meets line of projection λ at $1 = 1'$. Object point A is at ∞ on κ, and image point A′ is at 0′ on λ. Part arithmetic scale on κ 1, 2, 3, …, ∞ projects to part harmonic scale on λ 1′, 1/2′, 1/3′, …, 0′. **b** As (**a**) but κ parallel to λ. Object point A is 0 on κ, and image point A′ is 0′ on λ. The entire arithmetic scale on κ projects to another arithmetic scale on λ (here half the size)

8.3 Harmonic Plane Projection

The harmonic plane (see Sect. 6.5) has four quadrants, each contained within a triangle. The harmonic plane is a central projection of the Euclidean plane, quadrant by quadrant. Figure 8.6a shows the first quadrant, where

- Centre of projection C is origin O.
- Line of projection λ' is harmonic line $x'y'$.
- Pair of corresponding points are Euclidean A (at infinity) and harmonic A′, where A, A′, and O lie on the axis AA′O.
- All lines are straight.
- Euclidean distances project to harmonic distances.

Figure 8.6b shows a point P with coordinates (x, y) in Euclidean space. To find the corresponding point P′ in harmonic space:

- Line OP has slope $m = y/x$.
- Harmonic point Q′ has coordinates $(q', 0)$, $q' = 2 - (2/(x + 1))$.
- Line Q′Y′ has slope $m' = 2/ - q'$.
- Harmonic point P′ is the meet of lines OP and Q′y', with coordinates (x', y').
- $x' = 2 \times (1/(m - m'))$.
- $y' = m \times x'$.

For example:

- Given P = (3, 1).
- Slope of OP $m = 1/3$.
- Q′ = (3/2, 0).

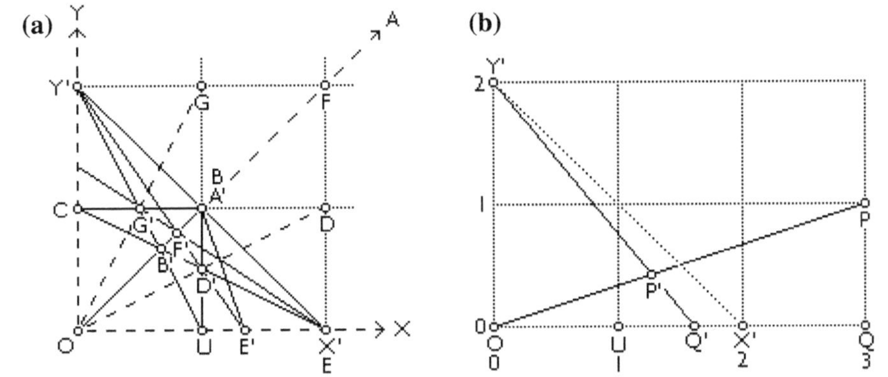

Fig. 8.6 Projection of first quadrant of Euclidean plane Oxy to harmonic plane O$x'y'$. **a** O C and U are common. A projects to A′, B to B′, D to D′, E to E′, F to F′, G to G′, x to x', and y to y'. **b** Euclidean Q projects to harmonic Q′, and P to P′

- Slope of Q′y' $m' = -4/3$.
- $x' = 6/5 = 1.2$.
- $y' = 2/5 = 0.4$.

In the harmonic plane, there are 15 classes of projection corresponding to those in the Euclidean plane.

8.4 Elliptic Plane Projection

The elliptic plane (see Sect. 6.9) is contained within a circle. All lines on origin O are straight, and all other lines are circular arcs. In the elliptic plane, there are 15 classes of projection corresponding to those in the Euclidean plane. For example, Fig. 8.7a shows the construction for image vertex V′ from given vertex V, for class VII, where

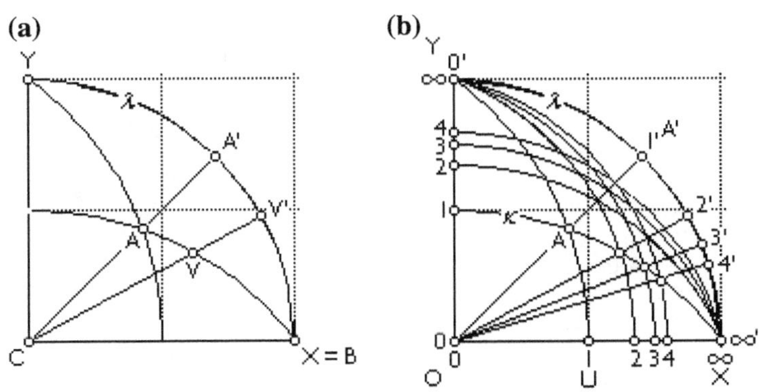

Fig. 8.7 Elliptic plane projection. **a** Projection fixed by centre C, line λ, and two corresponding points A and A′ on axis AA′C. V is a given vertex of the figure. Join AV, to meet λ on B; join A′B; join CV, to meet A′B on V′, the image of V. **b** Points $x = 0, 1, 2, \ldots, \infty$ on κ projected to 0′, 1′, 2′, …∞ on λ

- Centre of projection C is origin O.
- Line of projection λ' is elliptic line at infinity $x'y'$ (a quarter-circle).
- Object point A is at finite distance and corresponding image point A′ is on λ, on axis of projection AA′C.

Figure 8.7b shows six given points on line κ projected to image points on line λ.

8.5 Space Projection

Space projection, from an object in space to an image in space or a plane, is a straightforward extension of plane projection. An object in space can be represented by a space graph or wire-frame model (see Sect. 7.3) with

- *Vertices* V_1, V_2, ...,
- *Edges* E_1, E_2, ...,
- *Faces* F_1, F_2, ...,

where an edge joins two vertices, a face is bounded by three or more edges, no two edges cross, and there are no isolated parts. They are related by Euler's formula: for example, a cube has eight vertices, 12 edges, and six faces, so $8 - 12 + 6 = 2$.

In space, projective geometry recognizes only the incidence and order of points, lines, and planes. A given object can be transformed to an image by a *space projection* operation. A space projection is specified by four parameters:

- a *centre of projection* point C,
- a *plane of projection* π,
- a pair of *corresponding points*, A in the given object and A′ in the image, where A, A′, and C lie on the *axis* AA′C.

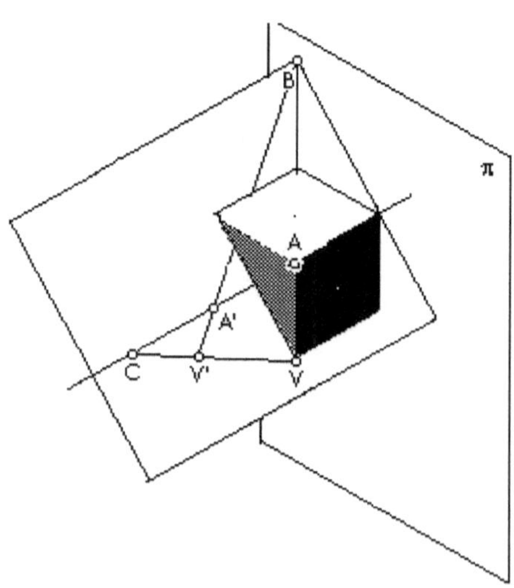

Fig. 8.8 Space projection of object to image. The projection is fixed by centre C, plane π, and two corresponding points A (*hidden here within the object*) and A′ on axis AA′C. V is a given vertex of the object. On plane VCA, join AV, to meet π on B; join A′B; join CV, to meet A′B on V′, the image of V

Geometrically, given a space projection (C, π, A, A$'$), Fig. 8.8 shows the construction for the image of a vertex of a given figure. A space projection relates a space object to its image by a bundle of projector lines all passing through a centre of projection C. If the projectors are understood as lines of sight, then the object and its image look the same when viewed from the centre C. Object and image are said to be *in perspective*.

A space projection can be done in several stages, working always in a plane. Figure 8.9 shows an example, where the (wire-frame) object is a square corridor, and the projection is one-point perspective (see item (8) in list in Sect. 8.7 below). First draw an orthographic side elevation of the object, and an orthographic plan of the object. Set out on both elevation and plan the chosen centre of projection C, plane of projection π, and corresponding points A and A$'$ (here A$'$ is on π, for a class III projection). For each vertex V, join AV to meet π at B; join A$'$B; join CV, to meet A$'$B on V$'$, the projected image of V. Transfer parallel the vertical projections from the elevation plane π to the side of the space view, and the horizontal projections from the plan plane π to the foot of the space view, and draw the space projection. There is a natural limitation of plane π to subtend no more than 90° at C. A point

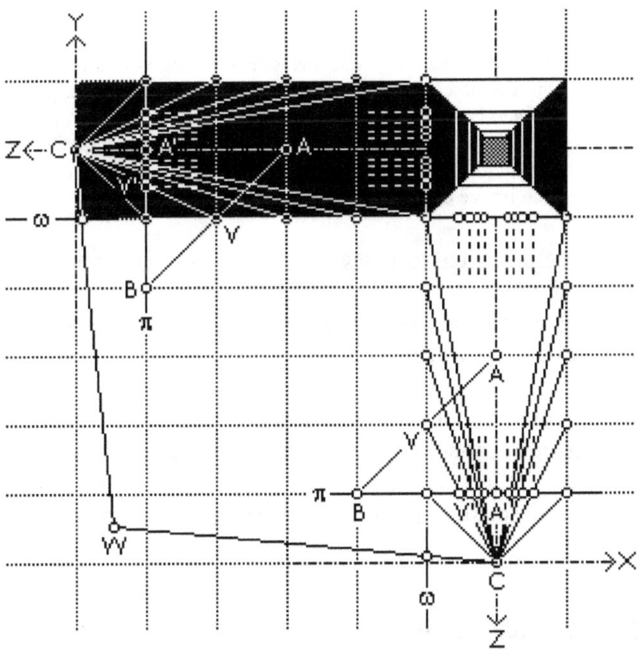

Fig. 8.9 Example of geometric construction of space projection. Draw orthographic side elevation and plan of object vertices and edges. Mark centre of projection C, plane of projection π, and corresponding pair of points A and A$'$ (here for incidence class III). For each vertex V find image V$'$. Transfer vertical and horizontal image locations to projected view, and draw image lines. For an object point such as W, project to plane ω

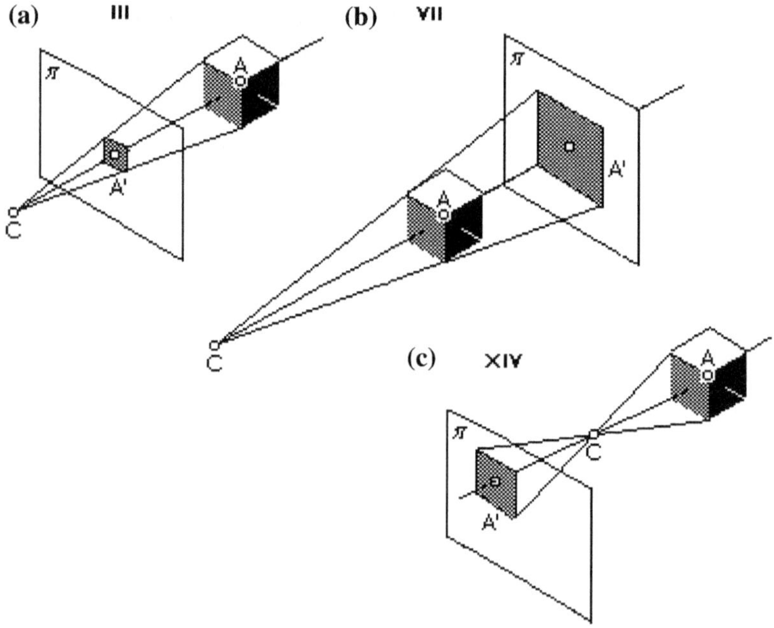

Fig. 8.10 Space projections of object to plane image. **a** Incidence class III: projection fixed by corresponding points A in the object (*centre of cube*) and A′ in the image, plane π on A′, and centre C so that A′ is between between A and C. Object projected to a nearer smaller flat image. **b** Incidence class VII: plane π on A′, and C so that A is between A′ and C. Object projected to a farther larger flat image. **c** Incidence class XIV: plane π on A′, and C between A and A′. Object projected to an inverted flat image on the opposite side of C

such as W should be projected to plane ω. Geometric construction is tedious, and needs very precise drawing.

Computationally, a space projection can be produced by a program which reads given space graph data, and calculates the projected plane image (see Sect. 8.18 below).

If we consider only the incidence and order of parameters, there are 15 classes of space projection, corresponding to the plane projections shown in Fig. 8.2. Three of these classes (III, VII, and XIV) deliver a *flat perspective* image on plane π, as shown in Fig. 8.10. These are of particular interest in digital imaging.

8.6 Euclidean Space Projection

In Euclidean space, each of the incidence classes can be varied by setting different orientations of object and plane π in space, and/or different angles between π and the axis CAA′, and/or different distances of C. For example, in incidence class III, take the origin O of the space frame as A, in or near the object to be projected, and

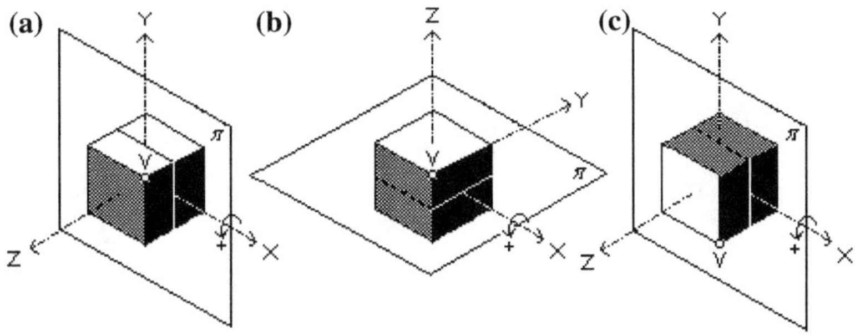

Fig. 8.11 Orientation of image plane in space. **a** Space frame OXYZ and image plane π = Oxy in initial position; cube object vertex V has xyz-coordinates (32,32,32)(say). **b** Space frame and image plane rotated by −90° about the x-axis while object remains fixed; vertex V now has xyz-coordinates (32,−32,32). **c** Object rotated by +90° about the x-axis while space frame and image plane remain fixed; vertex V now has xyz-coordinates (32,−32,32), the same as (**b**)

the image plane π as the Oxy-plane of space. We can then place the image plane in any orientation whatsoever, relative to the space object, by a rotation about one or more of the space axes, as in Sect. 7.8. Such a rotation may be understood either as a turning of the entire space frame (carrying the image plane with it) while the object remains fixed, or as an inverse turning of the object while the space frame (and the image plane) remain fixed. The resulting orientation is the same in either view. For example, Fig. 8.11a shows an object and image plane of projection in initial position, Fig. 8.11b after rotation of the frame by −90° about the x-axis, and Fig. 8.11c after rotation of the object by +90° about the x-axis.

Take the initial direction of the axis of projection AA′C as coincident with the z-axis of space, and with the same positive direction. We can then incline the axis of projection in any direction whatsoever, relative to the object, by shearing along one or more of the space axes, as in Sect. 7.10. Such a shearing may be understood either as a obliquity of the entire space frame while the object remains fixed, or as an inverse obliquity of the object while the space frame remains fixed. The resulting axial direction is the same in either view. For example, Fig. 8.12 shows a cube object, image plane, and axis of projection in initial position (Fig. 8.12a); Fig. 8.12b after shearing the x-dimension of the space frame by $1 \times z$; and Fig. 8.12c after shearing the x-dimension of the object by $-1 \times z$.

We can set the centre of projection C at any distance v = OC on the axis of projection. For v infinite, we get projectors parallel to the axis of projection; and, for the axis in standard position orthogonal to the plane of projection, an object is flattened to an image in which every width and height is unchanged, but every depth vanishes. For v finite, we get projectors diverging from the centre of projection; an object is flattened to an image in which every width and height in front of the image plane is enlarged, and every width and depth behind the image plane is diminished, proportionally to its depth. For example, Fig. 8.13a shows a cube object parallel-projected to a square image, the same size as the face of the cube; and Fig. 8.13b the same cube centrally-projected to a square image, larger than the face of the cube.

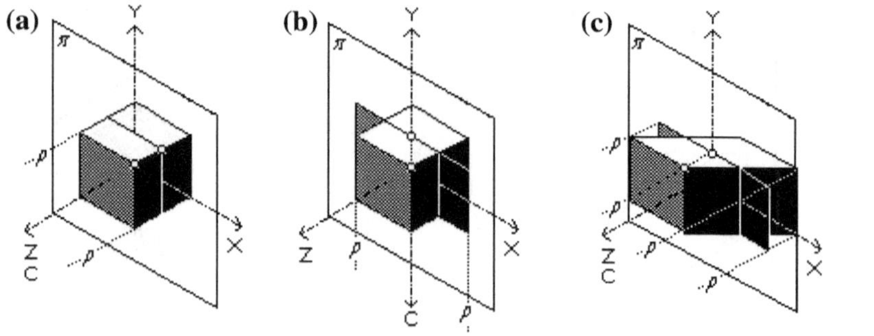

Fig. 8.12 Direction of axis of projection. **a** Space frame Oxyz, image plane Oxy, axis of projection AA′C, and parallel projectors p in initial position; marked cube vertex has xyz-coordinates (32,32,32)(say), and projects to image $x'y'$-coordinates (32,32). **b** Space frame sheared in x by 1z; axis of projection and projectors now inclined at 45° to space x-axis while object remains fixed; marked vertex now has xyz-coordinates (0,32,32) and projects to image $x'y'$-coordinates (0,32). **c** Object sheared in x by $-1z$; axis of projection and projectors remain orthogonal to image plane; marked vertex has xyz-coordinates (0,32,32) and projects to (0,32), the same as (**b**)

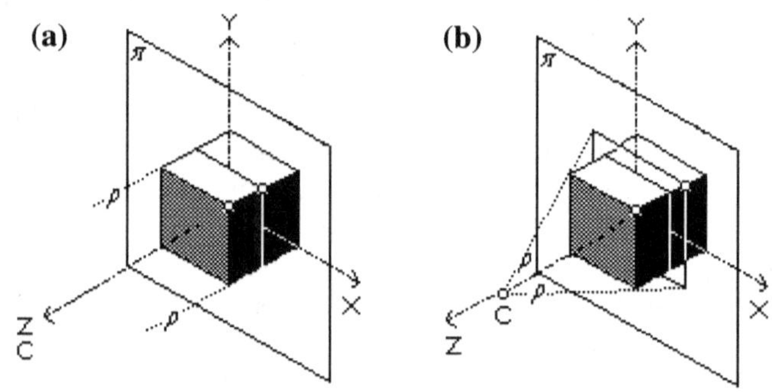

Fig. 8.13 Parallel and central projection. **a** Parallel projection: centre of projection C is infinitely-distant on the axis of projection in standard position; cube object has xyz-coordinates of vertices (±32,±32,±32)(say), and projects to image $x'y'$-coordinates (±32,±32). **b** Central projection: centre of projection C is at distance 128 on axis of projection A in standard position; cube object projects to image $x'y'$-coordinates (±43,±43)

8.7 Ten Classical Projections

Projecting the shape of a solid object, actual or imagined, from space to a flat image in a chosen plane is, and always has been, fundamental to drawing, painting, architectural design and engineering design. There is a long and interesting history of incremental development of methods of projection, from the more or less intuitive to the more or less systematic [9]. Here we shall distinguish ten classical projections from space to plane by different conditions on the Euclidean space projection parameters.

Parallel
Or axonometric,[3] where the centre of projection is infinitely-distant, so the projectors are parallel.

- *Orthogonal*,[4] where the axis of projection and the projectors are orthogonal to the plane of projection. Orthogonal projection is a Euclidean transformation, preserving parallels and lengths and angles.

(1) *Orthographic*, where the plane of projection is parallel to one of the space coordinate planes, and the axis of projection is parallel to one of the space coordinate axes,
(2) *Isometric*, where the plane of projection is inclined equally to all three space coordinate planes,
(3) *Dimetric*, where the plane of projection is inclined equally to two space coordinate planes and differently to the third,
(4) *Trimetric*, where the plane of projection is inclined differently to all three space coordinate planes.

- *Oblique*, where the axis of projection and the projectors are oblique to the plane of projection. Oblique projection is an affine transformation, preserving parallels and ratios of lengths, but not angles.[5]

(5) *Military*, where the plane of projection is parallel to the space horizontal plane, to give a true plan and two inclined elevations to same scale as plan,
(6) *Cavalier*, where the plane of projection is parallel to a space vertical plane, to give a true front elevation and inclined plan and side elevation to same scale,
(7) *Cabinet*, where the plane of projection is parallel to a space vertical plane, to give a true front elevation and inclined plan and side elevation to half scale.

Central
Or perspective, where the centre of projection is finitely-distant, so the projectors are convergent. Central projection is a projective transformation, preserving harmonic ratios of lengths and angles, but not, in general, parallels.

(8) *One-point perspective*, where the plane of projection is parallel to one of the space coordinate planes, and the axis of projection is orthogonal to that plane. Parallels to two space axes are preserved,

---

[3] *Axonometric* properly means that all three axes have the same length unit, whereas in general each axis has a different length unit.

[4] The fundamental theorem of orthogonal axonometry [5], due to Gauss, considers unit lengths on the three space axes as a *tripod*, and the image plane as the complex number plane. If the leg ends of the tripod project to complex points a, b, c and the origin projects to the null point, then $a^2 + b^2 + c^2 = 0$.

[5] The fundamental theorem of oblique axonometry [5], proposed in 1853 by K.W. Pohlke and proved in 1864 by H.A. Schwartz, says that any three distinct segments meeting at a point in the plane of projection are the oblique image of a space tripod.

(9) *Two-point perspective*, where the plane of projection is inclined to two of the space coordinate planes, and orthogonal to the third. Parallels to one space axis are preserved,

(10) *Three-point perspective*, where the plane of projection is inclined to all three space coordinate planes. No parallels are preserved.

8.8 Orthographic Projection

Geometrically, *orthographic* or Monge projection [11][6] is characterised by

- *Plane of projection*: one of the space coordinate planes xOy or xOz or yOz.
- *Axis of projection*: one of the space coordinate axes z or y or x, so meeting the plane of projection orthogonally.
- *Centre of projection*: infinitely distant.
- *Projectors*: parallel to the axis of projection.

There are six orthographic views (one for each side of each coordinate plane) as shown in Fig. 8.14:

- *Front elevation*: plane of projection = xOy-plane; axis of projection = positive z-axis.
- *Plan*: plane of projection = xOz-plane; axis of projection = positive y-axis.
- *Right-side elevation*: plane of projection = yOz-plane; axis of projection = positive x-axis.
- *Back elevation*: plane of projection = back of xOy-plane; axis of projection = negative z-axis.
- *Under-plan*: plane of projection = underside of xOz-plane; axis of projection = negative y-axis.
- *Left side elevation*: plane of projection = left side of yOz-plane; axis of projection = negative x-axis.

Monge proposed paired front elevation and plan as a flat representation of space objects. It later became customary to add a side elevation.

Geometrically, orthographic *front elevation* is shown in Fig. 8.14a. The plane of projection is $O'x'y'$, coincident with the object xy-coordinate plane, and the axis of projection is the positive z-axis, so no space rotation of the object is needed. The operations are parallel projection to plane image, and translation of image to first quadrant. A parallel-projected image will have the extent of the original object. (If a different size of image is wanted, the object may be dilated as a preliminary.)

[6]Systematic use of two orthographic projections to represent solid objects was originated about 1770 by Gaspard Monge (1746–1818), Professor at Mezieres military school 1768–1792, Minister of the Marine 1792–1793, founder and Principal of Ecole Polytechnique 1794–1809, with Bonaparte in Italy, Egypt and Syria 1796–1799, Senator of the Consulate 1800, Comte de Peluse 1808, title annulled at fall of Bonaparte.

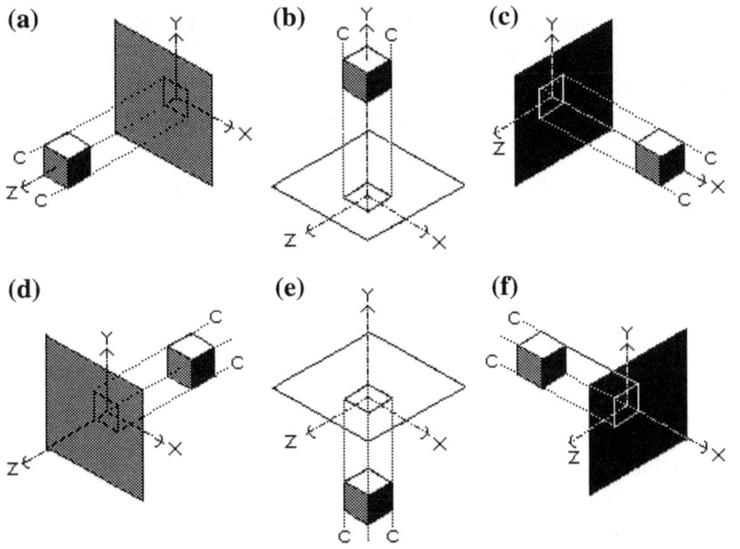

Fig. 8.14 The six orthographic views. **a** Front elevation. Plane of projection = xy-plane; axis of projection = z-axis; centre of projection C = $+\infty$. **b** Plan. Plane of projection = xz-plane; axis of projection = y-axis; centre of projection C = $+\infty$. **c** Right-side elevation. Plane of projection = yz-plane; axis of projection = x-axis; centre of projection C = $+\infty$. **d** Back elevation. Plane of projection = back of xy-plane; axis of projection = negative z-axis; centre of projection C = $-\infty$. **e** Under-plan. Plane of projection = underside of xz-plane; axis of projection = negative y-axis; centre of projection C = $-\infty$. **f** Left-side elevation. Plane of projection = left side of yz-plane; axis of projection = negative x-axis; centre of projection C = $-\infty$

A parallel-projected image will have origin at or near its centre: to put the image in the first quadrant a translation is needed.

Numerically, orthogonal parallel projection from space coordinates (x, y, z) to image coordinates (x', y') is

$$\begin{aligned} x' &= x \\ y' &= y \\ z' &= 0\,, \end{aligned}$$

that is, the x- and y-values are unchanged, and the z-value vanishes. A plane translation has the form

$$(x', y') = (x + A, y + B)\,,$$

where (x,y) are the old coordinates of a pixel in the plane, A is the count of pixels to move in the x-direction, B is the count of pixels to move in the y-direction, and (x',y') are the new coordinates.

Computationally, orthographic front elevation is found as

$$(x', y') = (x', y', 0, 1) = (x, y, z, 1)\mathbf{T}\,\mathbf{P}\,.$$

Matrix **P** parallel-projects the given object coordinates to image coordinates:

$$\mathbf{P} = \begin{pmatrix} 1\,0\,0\,0 \\ 0\,1\,0\,0 \\ 0\,0\,0\,0 \\ 0\,0\,0\,1 \end{pmatrix}$$

Matrix **T** translates image coordinates within the image plane:

$$\mathbf{T} = \begin{pmatrix} 1 & 0 & 0\,0 \\ 0 & 1 & 0\,0 \\ 0 & 0 & 0\,0 \\ A & B & 0\,1 \end{pmatrix}$$

where

- (integer) A is the magnitude of the translation along the x-axis, in px,
- (integer) B is the magnitude of the translation along the y-axis, in px,
- $A, B > 0$ move positive,
- $A, B = 0$ leave everything unchanged (the identity),
- $A, B < 0$ move negative.

Combining these two matrices, we get the orthographic front elevation matrix:

$$\begin{pmatrix} 1 & 0 & 0\,0 \\ 0 & 1 & 0\,0 \\ 0 & 0 & 0\,0 \\ A & B & 0\,1 \end{pmatrix}$$

Geometrically, orthographic *plan* is shown in Fig. 8.14b. The plane of projection is the xz-coordinate plane, and the axis of projection is the positive y-axis. To bring the image to the xy-plane we must rotate the object by 90° about the x-axis, then project the front elevation, then translate to first quadrant, and scale.

Computationally, orthographic plan elevation is found as

$$(x', y') = (x', y', 0, 1) = (x, y, z, 1)\mathbf{T}\,\mathbf{P}\,\mathbf{R}\,.$$

Matrix **R** rotates the given object about the x-axis:

$$\mathbf{R} = \begin{pmatrix} 1 & 0 & 0 & 0 \\ 0 & 0 & 1 & 0 \\ 0 & -1 & 0 & 0 \\ 0 & 0 & 0 & 1 \end{pmatrix}$$

which sends the front face underneath, the bottom face to the back, the back face to the top, and the top face to the front. The projection and translation matrices are as for front elevation.

Geometrically, orthographic *right-side elevation* is shown in Fig. 8.14c. The plane of projection is the yz-coordinate plane, and the axis of projection is the positive x-axis. To bring the image to the xy-plane we must rotate the object by $-90°$ about the y-axis, then take the front elevation.

Computationally, orthographic right-side elevation is found as

$$(x', y') = (x', y', 0, 1) = (x, y, z, 1)\mathbf{T}\,\mathbf{P}\,\mathbf{R}\,.$$

Matrix **R** rotates the given object about the y-axis:

$$\mathbf{R} = \begin{pmatrix} 1 & 0 & 0 & 0 \\ 0 & 1 & 1 & 0 \\ -1 & 0 & 0 & 0 \\ 0 & 0 & 0 & 1 \end{pmatrix}$$

which sends the front face to the left side, the right face to the front, the back face to the right side, and the left face to the back. The projection and translation matrices are as for front elevation.

Orthographic is the standard projection for engineering and architectural drawings. There are two alternative conventions for arranging the projections on a sheet:

- First-angle (used by ISO).
- Third-angle (traditionally in USA and Canada).

The three orthographic projections are the same for both presentations: only the arrangement on the sheet differs.

First-angle presentation is shown in Fig. 8.15. Monge used a vertical and a horizontal reference plane, forming four space angles numbered anticlockwise as seen from the left side. If we add a third reference plane vertical fore-and-aft, we get the octants shown. If an object is placed in the first octant, the orthographic front elevation projects back to the vertical xy-plane, the orthographic plan projects down to the horizontal xz-plane, and the orthographic right-side elevation projects left to the yz-plane. If we then hinge the plan down and the side elevation back, to lie flat with the front elevation, we have first-angle presentation.

Third-angle presentation is shown in Fig. 8.16. If an object is placed in the seventh octant, the orthographic front elevation projects forward to the vertical xy-plane, the orthographic plan projects up to the horizontal xz-plane, and the orthographic right-side elevation projects right to the yz-plane. If we then hinge the plan up and the side elevation forwards, to lie flat with the front elevation, we have third-angle presentation.

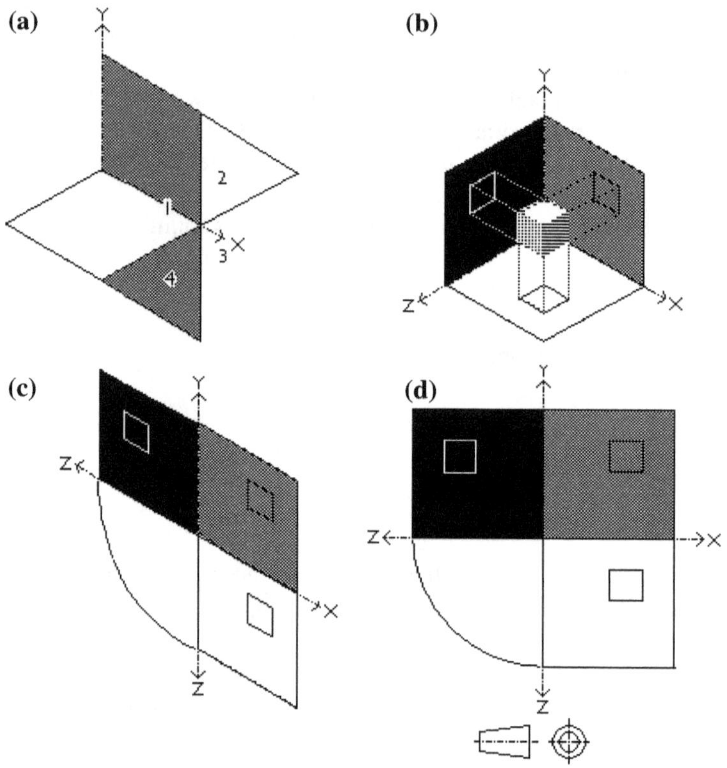

Fig. 8.15 First-angle presentation. **a** Isometric view of Monge's vertical and horizontal reference planes, and the four space angles. **b** Isometric view of object placed in octant 1, as it were transparent: front elevation projects back to xy-plane, plan projects down to xz-plane, and right-side elevation projects left to yz-plane. **c** Plan and side elevation hinged back to lie flat with front elevation. **d** First-angle orthographics: plan below front elevation (with the ISO symbol for first-angle presentation)

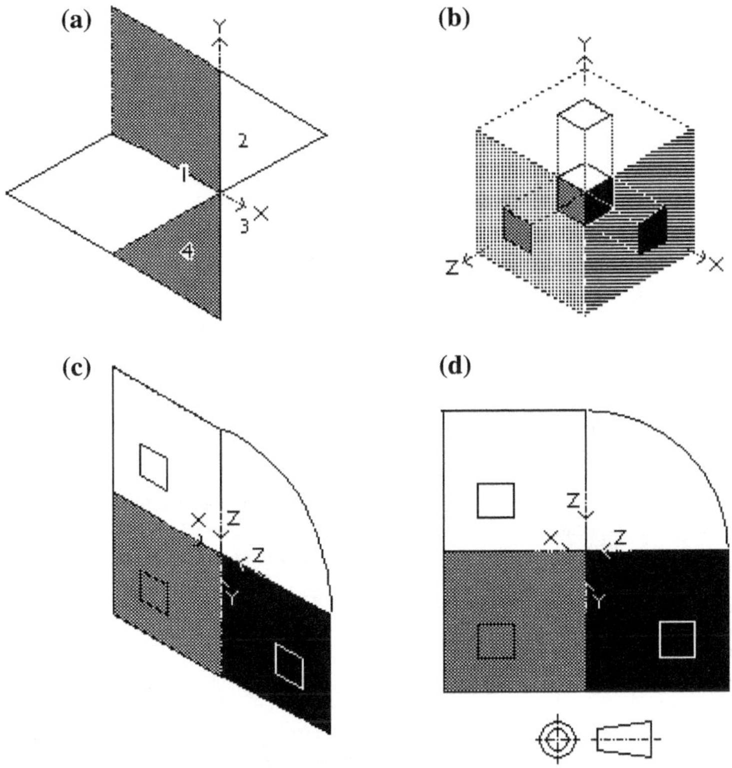

Fig. 8.16 Third-angle presentation. **a** Isometric view of Monge's vertical and horizontal reference planes, and the four space angles. **b** Isometric view of object placed in octant 7, with coordinate planes as if transparent: front elevation projects forward to xy-plane, plan projects upward to xz-plane, and right-side elevation projects rightward to yz-plane. **c** Plan and side elevation hinged forward to lie flat with front elevation. **d** Third-angle orthographics: plan above front elevation (with the ISO symbol for third-angle presentation)

8.9 Isometric Projection

Isometric projection [6, 14][7,8] is characterised by

- *Plane of projection*: equally-inclined to coordinate planes $x\mathrm{O}y$, $x\mathrm{O}z$ and $y\mathrm{O}z$.
- *Axis of projection*: orthogonal to plane of projection, so equally-inclined to coordinate axes x, y and z.

---

[7]Invented about 1820 by William Farish (1759–1837), Jacksonian Professor at Cambridge. For his lectures on mechanical engineering, Farish developed a model system of interchangeable parts, which he illustrated in isometric projection.

[8]Expounded by Thomas Sopwith (1803–1879), civil engineer and geologist, from 1834 in John Weale's superb range of technical publications. Sir Thomas Sopwith (1888–1989), aviator and yachtsman, was his grandson.

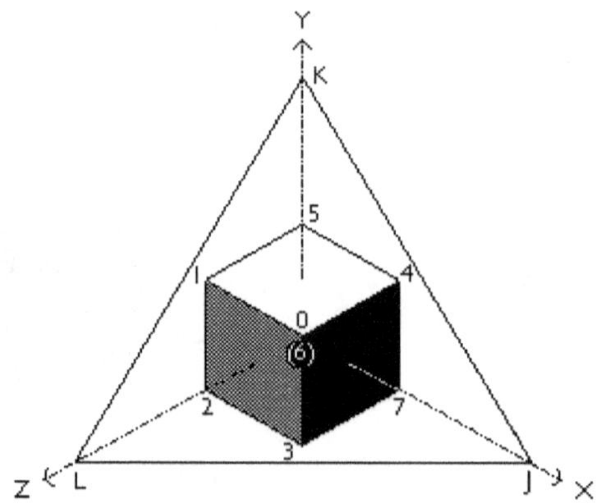

Fig. 8.17 Isometric projection. The object is a cube with centre at space origin, faces parallel to space coordinate planes, and vertices marked 0 1 2 3 4 5 6 (*hidden*) and 7. The axis of projection is the body diagonal 06 of the cube, orthogonal to the paper. The plane of projection is the paper, which meets the space axes x y z at J K L, and meets the space coordinate planes xy xz yz along JK JL KL. JKL is an equilateral triangle, and the profile of the cube 123745 is a regular hexagon. Lengths along the projected axes 0J 0K 0L are 0.8165 of true space lengths

- *Centre of projection*: infinitely distant.
- *Projectors*: parallel to axis of projection.

Figure 8.17 shows a typical isometric projection of a cube (as already used for various illustrations). The plane of projection is the paper, equally-inclined to the space axes which it meets at J K and L. The axis of projection (orthogonal to the paper) coincides with the body diagonal of the cube from vertex 0 to vertex 6. The image shows three square faces of the cube as rhombs, with angles 60° and 120°. The profile of the cube is a regular hexagon. Lengths along the projected axes 0J 0K 0L are all scaled alike by $(\sqrt{2}/\sqrt{3} = 1.4142/1.7321 = 0.8165$. A circle projects to an ellipse, with major axis equal to the diameter of the circle.

Instead of inclining the plane of projection to the space reference frame, we get the same result if we fix the xy-plane as plane of projection and incline the object. The required inclination of the object is rotation about the y-axis by $-45°$ and rotation about the x-axis by 35.2664°. The rotation about the y-axis brings edge 03 orthogonal to the z-axis, and the rotation about the x-axis inclines edge 03 to the xy-image plane. Figure 8.18 shows the orthographic projections of a cube thus inclined: the front elevation is indeed the isometric view.

There are eight isometric views: one for each octant of the coordinate planes, as listed in Table 8.1 and shown in Fig. 8.19.

Computationally, isometric projection is implemented as

$$(x', y') = (x', y', 0, 1) = (x, y, z, 1)\mathbf{T}\,\mathbf{P}\,\mathbf{R2}\,\mathbf{R1}\,.$$

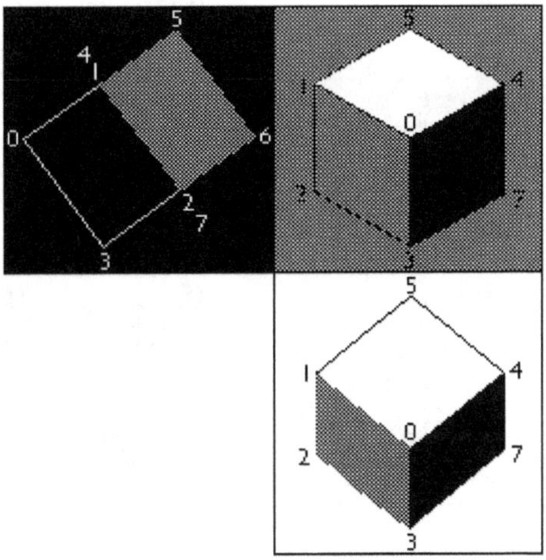

Fig. 8.18 Orthographic projections of cube in isometric position. The cube has been rotated about the y-axis by $-45°$, and rotated about the x-axis by $35.2664°$, to bring the body diagonal 06 into coincidence with the z-axis. The orthographic front elevation is then the isometric projection

Table 8.1 The eight isometric views

Octant	Central vertex	Profile vertices	Hidden vertex	Rot.y	Rot.x
1	0	123745	6	−45	35.2664
2	1	230456	7	45	35.2664
3	2	015673	4	45	−35.2664
4	3	012674	5	−45	−35.2664
5	4	037651	2	−135	35.2664
6	5	047621	3	135	35.2664
7	6	154732	0	135	−35.2664
8	7	032654	1	−135	−35.2664

Vertices numbered as in Fig. 8.17. For each octant, one vertex is central in the image; of the other seven vertices, like Pleiades, six form a regular hexagonal profile and one vertex is hidden. The rotations are in degrees

For the octant-1 view, matrix **R1** rotates the given object about the y-axis:

$$\mathbf{R1} = \begin{pmatrix} \cos(-45) & 0 & -\sin(-45) & 0 \\ 0 & 1 & 0 & 0 \\ \sin(-45) & 0 & \cos(-45) & 0 \\ 0 & 0 & 0 & 1 \end{pmatrix}$$

Matrix **R2** rotates the given object about the x-axis:

$$\mathbf{R2} = \begin{pmatrix} 1 & 0 & 0 & 0 \\ 0 & \cos(35.664) & \sin(35.2664) & 0 \\ 0 & -\sin(35.2664) & \cos(35.2664) & 0 \\ 0 & 0 & 0 & 1 \end{pmatrix}$$

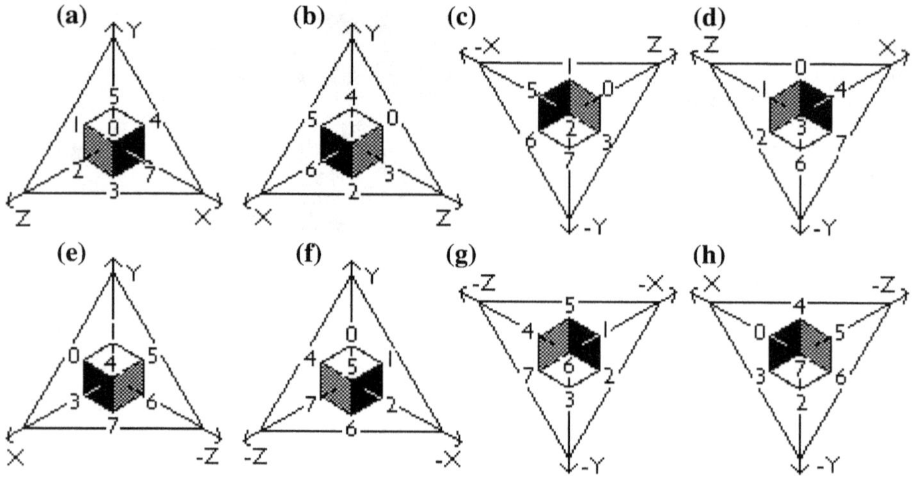

Fig. 8.19 The eight isometric views. **a** Octant 1. Object rotated about y-axis by $-45°$, and about x-axis by $35.2664°$. **b** Octant 2. Object rotated about y-axis by $45°$, and about x-axis by $35.2664°$. **c** Octant 3. Object rotated about y-axis by $45°$, and about x-axis by $-35.2664°$. **d** Octant 4. Object rotated about y-axis by $-45°$, and about x-axis by $-35.2664°$. **e** Octant 5. Object rotated about y-axis by $135°$, and about x-axis by $35.2664°$. **f** Octant 5. Object rotated about y-axis by $-45°$, and about x-axis by $35.2664°$. **g** Octant 6. Object rotated about y-axis by $135°$, and about x-axis by $35.2664°$. **h** Octant 7. Object rotated about y-axis by $135°$, and about x-axis by $-35.2664°$. **i** Octant 8. Object rotated about y-axis by $-135°$, and about x-axis by $-35.2664°$

which may be combined as one isometric rotation matrix:

$$\begin{pmatrix} 0.7071 & -0.4082 & 0 & 0 \\ 0 & 0.8165 & 0 & 0 \\ -0.7071 & -0.4082 & 0 & 0 \\ 0 & 0 & 0 & 1 \end{pmatrix}$$

The projection and translation matrices are as for orthographic front elevation. For the octant 2 to 8 views, the rotations are as shown in Table 8.1.

8.10 Dimetric Projection

Dimetric projection is a partial generalisation of isometric, where lengths along two projected axes are scaled equally, and lengths along the third axis are scaled differently. It is characterised by

- *Plane of projection*: equally-inclined to two coordinate planes, differently-inclined to the third coordinate plane.

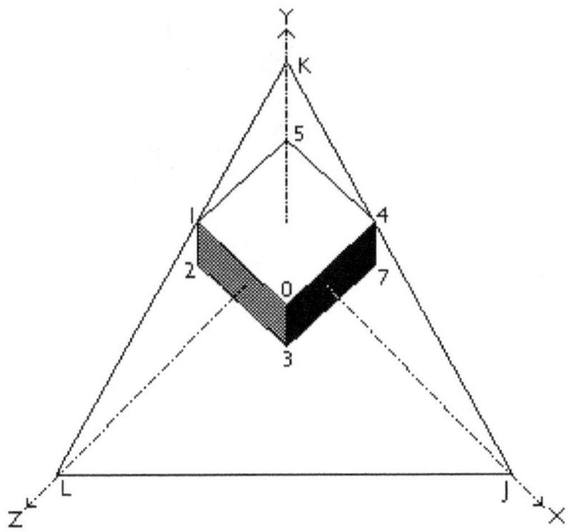

Fig. 8.20 Dimetric projection. The object is a cube with centre at space origin, faces parallel to space coordinate planes, and vertices marked 0 1 2 3 4 5 6 (*hidden*) and 7. The axis of projection is orthogonal to the paper, and goes through the centre of the cube. The plane of projection is the paper, which meets the space axes x y z at J K L, and meets the space coordinate planes xy xz yz along JK JL KL. JKL is an isosceles triangle. Lengths along the projected axes 0J and 0K are scaled alike, but lengths along 0L are scaled differently

- *Axis of projection*: orthogonal to plane of projection, so equally-inclined to two coordinate axes, differently-inclined to the third coordinate axis.
- *Centre of projection*: infinitely distant.
- *Projectors*: parallel to axis of projection.

Figure 8.20 shows a typical dimetric projection of a cube. The plane of projection is the paper, equally-inclined to space axes x and z, which it meets at J and L, but differently-inclined to space axis y, which it meets at K. The axis of projection (orthogonal to the paper) goes through the centre of the cube. The image shows the top, front and right-side faces of the cube (like the isometric), but now the top face 0154 is a rhomb, and the front and side faces are two different parallelograms. The profile of the cube is a hexagon with pairs of opposite sides parallel. Lengths along the projected axes 0J and 0L are scaled alike, but lengths along the third projected axis OK are scaled differently.

Instead of inclining the plane of projection to the space reference frame, we get the same result if we fix the xy-plane as plane of projection and incline the object. For the example shown, the object is rotated about the y-axis by $-45°$ (the same as the isometric rotation) and rotated about the x-axis by $70°$ (double the isometric rotation). Figure 8.21 shows the orthographic projections of a cube thus inclined: the front elevation is indeed the dimetric view.

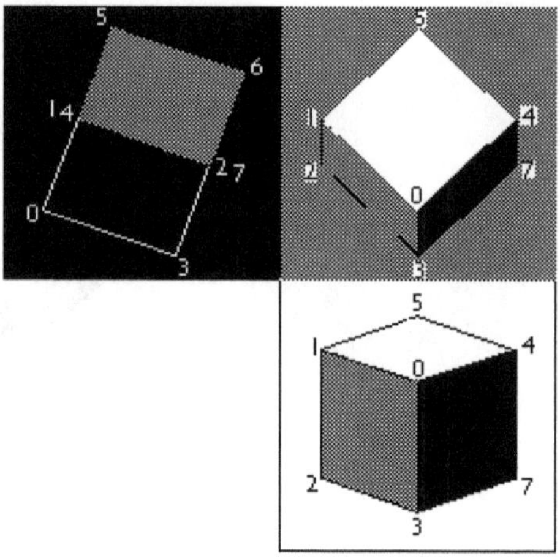

Fig. 8.21 Orthographic projections of cube in a dimetric position. The cube has been rotated about the y-axis by $-45°$, and rotated about the x-axis by $70°$, to bring the top face towards the xy-plane. The orthographic front elevation (on the xy-plane) is then the dimetric projection

Computationally, a dimetric projection is just an isometric projection where the second rotation is varied from the isometric value

$$(x', y') = (x', y', 0, 1) = (x, y, z, 1)\mathbf{T}\,\mathbf{P}\,\mathbf{R2}\,\mathbf{R1}\,.$$

For the octant-1 view the rotation matrices are

$$\mathbf{R1} = \begin{pmatrix} \cos(-45) & 0 & -\sin(-45) & 0 \\ 0 & 1 & 0 & 0 \\ \sin(-45) & 0 & \cos(-45) & 0 \\ 0 & 0 & 0 & 1 \end{pmatrix}$$

$$\mathbf{R2} = \begin{pmatrix} 1 & 0 & 0 & 0 \\ 0 & \cos(\beta) & \sin(\beta) & 0 \\ 0 & -\sin(\beta) & \cos(\beta) & 0 \\ 0 & 0 & 0 & 1 \end{pmatrix}$$

where

- $\beta = 0$ shows top face of cube 0154 as a line.
- $0 < \beta < 35.2664°$ shows top face as narrow rhomb.
- $\beta = 35.3664°$ shows isometric.
- $35.2664 < \beta < 90°$ shows top face as wide rhomb.
- $\beta = 90°$ shows top face as a square.

The projection and translation matrices are as for orthographic front elevation. A family of dimetric views is got from an isometric by varying the degree of the second rotation, as shown in Fig. 8.22.

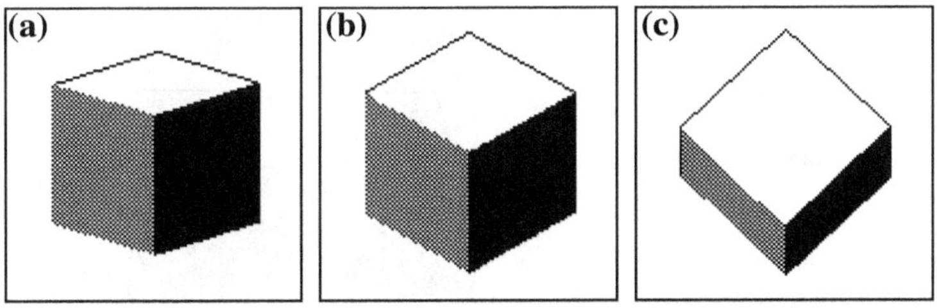

Fig. 8.22 Family of dimetric views from isometric. **a** Shallow dimetric: cube rotated by −45° about y-axis and 18° about x-axis, then projected as orthographic front elevation. **b** Isometric: cube rotated by −45° about y-axis and 35.2664° about x-axis, then projected as orthographic front elevation. **c** Steep dimetric: cube rotated by −45° about y-axis and 70° about x-axis, then projected as orthographic front elevation

8.11 Trimetric Projection

Trimetric projection is a further generalisation of isometric, where lengths along each projected axis are scaled differently. It is characterised by:

- *Plane of projection*: differently-inclined to all three coordinate planes,
- *Axis of projection*: orthogonal to plane of projection, so differently-inclined to all three coordinate axes,
- *Centre of projection*: infinitely distant,
- *Projectors*: parallel to axis of projection.

Figure 8.23 shows a typical trimetric projection of a cube. The plane of projection is the paper, differently-inclined to space axes x, y and z, which it meets at J K and L.

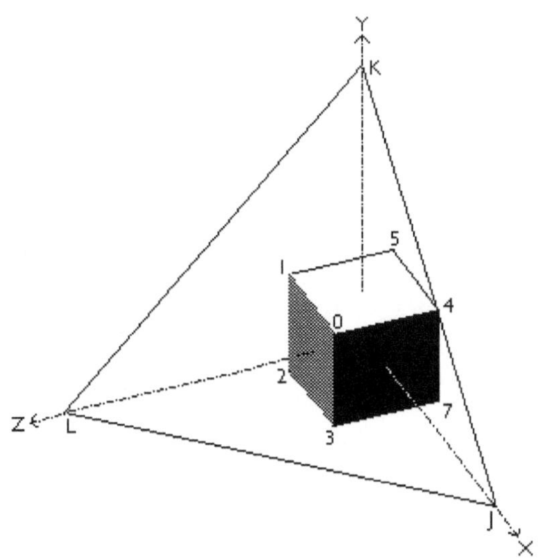

Fig. 8.23 Trimetric projection. The object is a cube with centre at space origin, faces parallel to space coordinate planes, and vertices marked 0 1 2 3 4 5 6 (*hidden*) and 7. The axis of projection is orthogonal to the paper, and goes through the centre of the cube. The plane of projection is the paper, which meets the space axes x y z at J K L, and meets the space coordinate planes xy xz yz along JK JL KL. JKL is a scalene triangle. Lengths along the projected axes 0J OK and 0 L are scaled differently

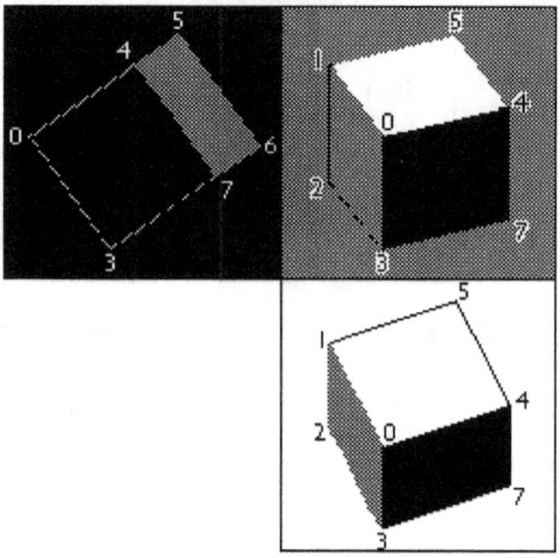

Fig. 8.24 Orthographic projections of cube in a trimetric position. The cube has been rotated about the y-axis by $-67.5°$, and rotated about the x-axis by $18°$, to bring the right-side face towards the xy-plane. The orthographic front elevation (on the xy-plane) is then the trimetric projection

The axis of projection (orthogonal to the paper) goes through the centre of the cube. The image shows the top, front and right-side faces of the cube (like the isometric), but now the faces are three different parallelograms. The profile of the cube is a hexagon with pairs of opposite sides parallel. Lengths along the projected axes 0J OK and 0L are scaled differently.

Instead of inclining the plane of projection to the space reference frame, we get the same result if we fix the xy-plane as plane of projection and incline the object. For the example shown, the object is rotated about the y-axis by $-67.5°$ and rotated about the x-axis by $18°$. Figure 8.24 shows the orthographic projections of a cube thus inclined: the front elevation is indeed the trimetric view.

Computationally, a trimetric projection is like an isometric projection where the two rotations are varied from the isometric value:

$$(x', y') = (x', y', 0, 1) = (x, y, z, 1)\mathbf{T}\,\mathbf{P}\,\mathbf{R2}\,\mathbf{R1}\,.$$

For the example shown the rotation matrices are

$$\mathbf{R1} = \begin{pmatrix} \cos(-67.5) & 0 & -\sin(-67.5) & 0 \\ 0 & 1 & 0 & 0 \\ \sin(-67.5) & 0 & \cos(-67.5) & 0 \\ 0 & 0 & 0 & 1 \end{pmatrix}$$

$$\mathbf{R2} = \begin{pmatrix} 1 & 0 & 0 & 0 \\ 0 & \cos(18) & \sin(18) & 0 \\ 0 & -\sin(18) & \cos(18) & 0 \\ 0 & 0 & 0 & 1 \end{pmatrix}$$

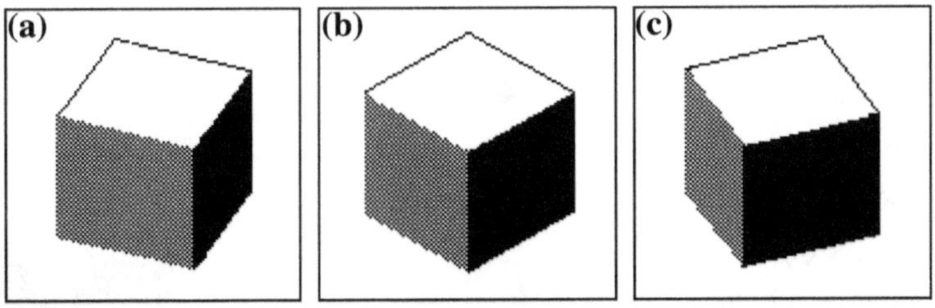

Fig. 8.25 Family of trimetric views. **a** Isometric: cube rotated by −45° about y-axis and 35.2664° about x-axis, then projected as orthographic front elevation. **b** Shallow dimetric: cube rotated by −45° about y-axis and 18° about x-axis, then projected as orthographic front elevation. **b** Steep dimetric: cube rotated by −45° about y-axis and 70° about x-axis, then projected as orthographic front elevation

The projection and translation matrices are as for orthographic front elevation. A family of trimetric views is got by varying the value of one of the rotations, as shown in Fig. 8.25.

8.12 Military Projection

Geometrically, *military projection*[9] is characterised by

- *Plane of projection*: horizontal space coordinate plane xz,
- *Axis of projection*: oblique to plane of projection,
- *Centre of projection*: infinitely distant,
- *Projectors*: parallel to axis of projection.

Figure 8.26 shows a military projection of a cube. The plane of projection is the paper, parallel to the ground. The axis of projection is inclined at 45° to the plane of projection. The image shows the top face as square, and the front and right faces as rhombs, with angles 45° and 135°. Lengths along all three projected axes are scaled alike. A horizontal circle projects to a circle, and a vertical circle to an ellipse.

Instead of inclining the axis of projection to the plane of projection, we get the same result if we fix the XY-plane as plane of projection, and rotate then shear the object. A rotation of the object about the Y-axis by −45° brings edge 03 round to the front

[9]In common use since the seventeenth century for drawing fortifications and street plans. It is easy to construct by first drawing a true plan, then rotating it by a suitable angle, usually 45° or 30°, then drawing verticals at the corners of the plan, and adding two elevations to the same scale as the plan. Systematically developed for drawings of historical architecture by Auguste Choisy (1841–1909), Professor of Architecture at l'Ecole Normale de Ponts et Chausees 1877–1901, notably in his Histoire de l'Architecture, 1899, where he writes "Le lecteur a sous ses yeux, a la fois, le plan, l'exterieur de l'edefice, sa coupe et ses dispositions interieures." Adopted in the 1920s as the projection of choice for architectural presentation by Van Doesberg, Le Corbusier, Gropius, and many others. Often loosely called axonometric projection, but more correctly called oblique parallel.

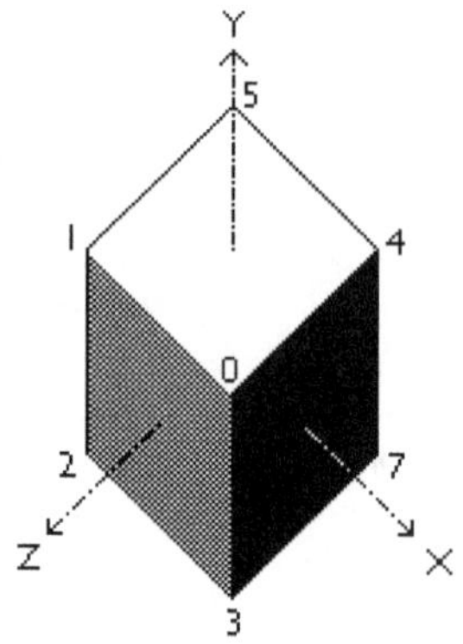

Fig. 8.26 Military projection. The object is a cube with centre at space origin, faces parallel to space coordinate planes, and vertices marked 0 1 2 3 4 5 (6 *hidden*) and 7. The plane of projection is the XZ coordinate plane, parallel to the ground, and the axis of projection is inclined at 45° to it. Lengths along the projected x-, y- and z-axes are to the same scale

orthogonal to the Z-axis. Then a shear along the Y-axis by -1Z tips horizontal faces down at the front and up at the back, to leave them inclined at 45° to the horizontal. Figure 8.27 shows the orthographic projections of a cube thus rotated and sheared: the front elevation is indeed the military projection.

Computationally, military projection is implemented as

$$(x', y') = (x', y', 0, 1) = (x, y, z, 1)\mathbf{T\,P\,S\,R}\,.$$

For the example shown, matrix $\mathbf{R}$ rotates the given object about the y-axis by $-45°$:

$$\mathbf{R} = \begin{pmatrix} \cos(-45) & 0 & -\sin(-45) & 0 \\ 0 & 1 & 0 & 0 \\ \sin(-45) & 0 & \cos(-45) & 0 \\ 0 & 0 & 0 & 1 \end{pmatrix}$$

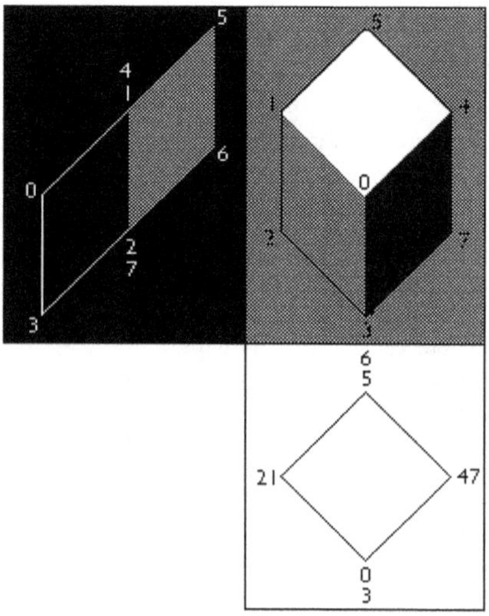

Fig. 8.27 Orthographic projections of cube rotated and sheared. The cube has been rotated about the y-axis by $-45°$, then sheared along the y-axis by $-1z$, to incline the top and bottom faces. The orthographic front elevation is then the military projection

Matrix **S** shears the object along the y-axis by $-1z$:

$$\mathbf{S} = \begin{pmatrix} 1 & 0 & 0 & 0 \\ 0 & 1 & 0 & 0 \\ -1 & 0 & 1 & 0 \\ 0 & 0 & 0 & 1 \end{pmatrix}$$

The projection and translation matrices are as for orthographic front elevation.

8.13 Cavalier Projection

Geometrically, *cavalier projection*[10] is characterised by

- *Plane of projection*: frontal space coordinate plane xy,
- *Axis of projection*: oblique to plane of projection,
- *Centre of projection*: infinitely distant,
- *Projectors*: parallel to axis of projection.

Figure 8.28 shows a cavalier projection of a cube. The plane of projection is parallel to the front face. The axis of projection is inclined at 45° to the plane of projection. The image shows the front face as square, and the top and right faces as rhombs, with angles 45° and 135°. Lengths along all three projected axes are scaled alike. A circle in the XY plane projects to a circle, and a circle in the XZ or YZ plane to an ellipse.

Instead of inclining the axis of projection to the plane of projection, we get the same result if we shear the object along both the x-axis and the y-axis by $-0.7071z$. The two shears incline the right and top faces by 45° to the front face. Figure 8.29 shows the orthographic projections of a cube thus double-sheared: the front elevation is indeed the cavalier projection.

Computationally, cavalier projection is implemented as

$$(x', y') = (x', y', 0, 1) = (x, y, z, 1)\,\mathbf{TPS}\,.$$

[10]In common use in Chinese and Japanese landscapes and interiors since early times, and in Europe as an easily-constructed alternative to perspective. First draw a true front elevation, then draw obliques at a suitable angle, usually 45° or 30°, at the corners of the elevation, and add top and side elevations to the same scale as the front elevation. It is like military projection, but with true front elevation instead of true plan. A *cavalier* is a platform in a fortification, some ten feet higher than the surrounding works, used as an observation or gunnery post; so a cavalier projection approximates to a view from a cavalier. The supposed connection with the mathematician Bonaventura Cavalieri (1598–1647) of Bologna is nonsense. Sometimes called Chinese perspective.

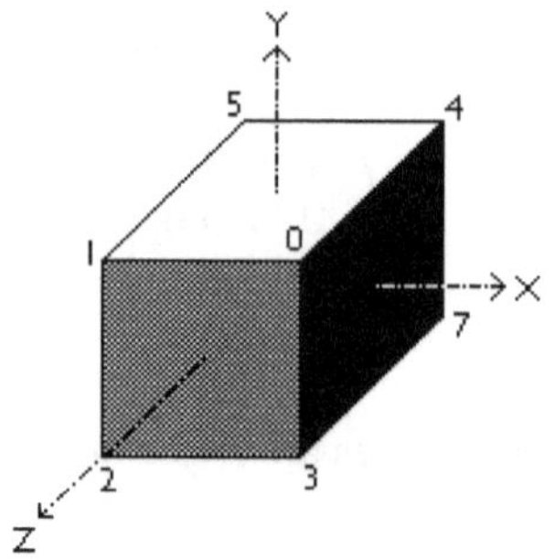

Fig. 8.28 Cavalier projection. The object is a cube with centre at space origin, faces parallel to space coordinate planes, and vertices marked 0 1 2 3 4 5 (6 *hidden*) and 7. The plane of projection is the xy-coordinate plane, parallel to face 0123, and the axis of projection is inclined at 45° to it. Lengths along the projected x-, y- and z-axes are to the same scale

Fig. 8.29 Orthographic projections of cube double-sheared. The cube has been sheared along the x-axis by $-0.7071z$ and along the y-axis by $-0.7071z$, to incline the top and right faces. The orthographic front elevation is then the cavalier projection

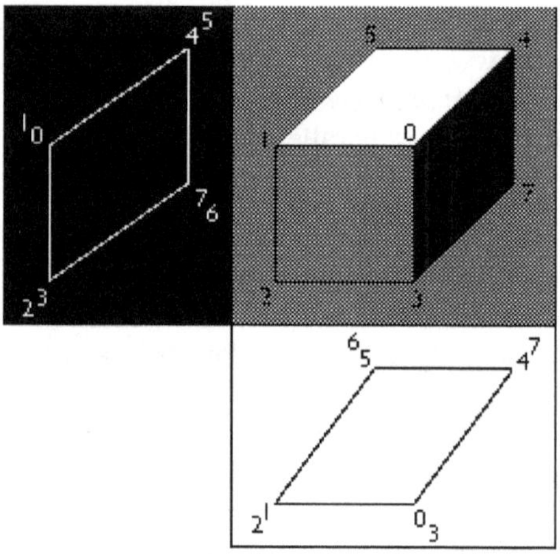

For the example shown, matrix **S** shears the object along the x-axis by $-0.7071z$ and along the y-axis by $-0.7071z$:

$$\mathbf{S} = \begin{pmatrix} 1 & 0 & 0\;0 \\ 0 & 1 & 0\;0 \\ -0.7071 & -0.7071 & 1\;0 \\ 0 & 0 & 0\;1 \end{pmatrix}$$

The projection and translation matrices are as for orthographic front elevation.

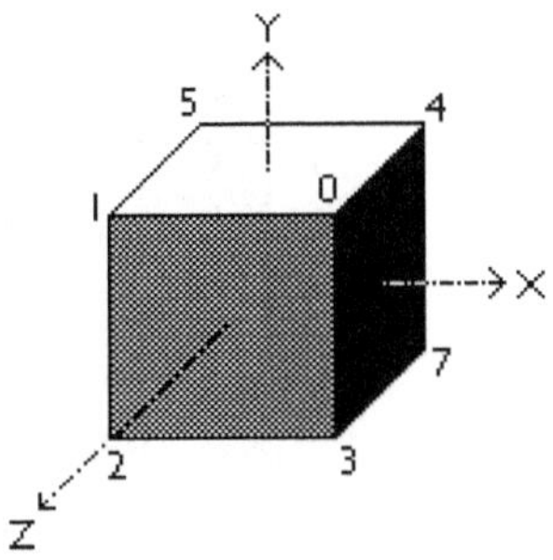

Fig. 8.30 Cabinet projection. The object is a cube with centre at space origin, faces parallel to space coordinate planes, and vertices marked 0 1 2 3 4 5 (6 *hidden*) and 7. The plane of projection is the xy-coordinate plane, parallel to face 0123, and the axis of projection is inclined at 45° to it. Lengths along the projected z-axis are half the scale of lengths along the projected x- and y-axes

8.14 Cabinet Projection

Geometrically, *cabinet projection*[11] is like cavalier projection except that the scale of the top and side elevation is half that of the front elevation, to give a somewhat more natural impression of the depth. It is characterised by

- *Plane of projection*: frontal space coordinate plane xy,
- *Axis of projection*: oblique to plane of projection,
- *Centre of projection*: infinitely distant,
- *Projectors*: parallel to axis of projection.

Figure 8.30 shows a cabinet projection of a cube. The plane of projection is parallel to the front face. The axis of projection is inclined at 45° to the plane of projection. The image shows the front face as square, and the top and right faces as rhombs, with angles 45° and 135°. Lengths along the projected z-axis are half the scale of lengths along the projected x- and y-axes.

Instead of inclining the axis of projection to the plane of projection, we get the same result if we shear the object along both the x-axis and the y-axis by $-0.3536z$. The two shears incline the right and top faces by 45° to the front face. Figure 8.31 shows the orthographic projections of a cube thus double-sheared: the front elevation is indeed the cabinet projection.

Computationally, cabinet projection is implemented as

$$(x', y') = (x', y', 0, 1) = (x, y, z, 1)\,\mathbf{TPS}\,.$$

For the example shown, matrix $\mathbf{S}$ shears the object along the x-axis by $-0.3536z$ and along the y-axis by $-0.3536z$:

[11] Possibly named from the older sense of cabinet as a room, or from the newer sense as a piece of furniture.

Fig. 8.31 Orthographic projections of cube double-sheared. The cube has been sheared along the x-axis by $-0.3536z$ and along the y-axis by $-0.3536z$, to incline the top and right faces. The orthographic front elevation is then the cabinet projection

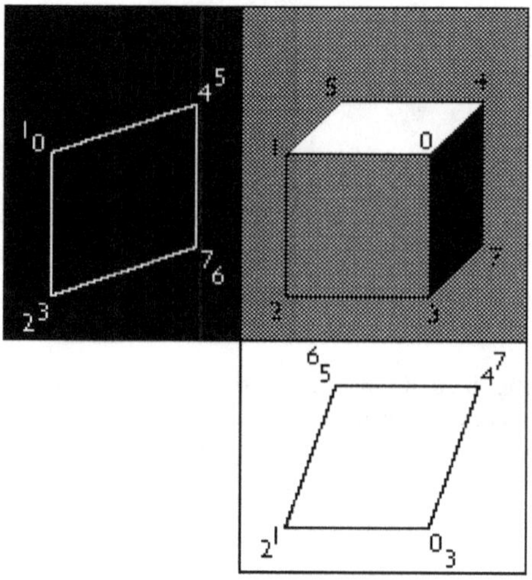

$$\mathbf{S} = \begin{pmatrix} 1 & 0 & 0 & 0 \\ 0 & 1 & 0 & 0 \\ -0.3536 & -0.3536 & 1 & 0 \\ 0 & 0 & 0 & 1 \end{pmatrix}$$

The projection and translation matrices are as for orthographic front elevation.

8.15 One-Point Perspective

One-point perspective projection[12] is characterised by

- *Plane of projection*: space coordinate plane xy,
- *Axis of projection*: space coordinate axis z, so meeting the plane of projection orthogonally,
- *Centre of projection*: finitely distant at V units from origin,
- *Projectors*: diverging from the centre of projection.

[12] Filippo Brunelleschi (1377–1446), architect of the dome of Florence cathedral and other profoundly original buildings, demonstrated (according to Vasari) about 1415 an astoundingly accurate perspective picture of the Baptistery of Florence. The painting had a small hole at the vanishing point; the viewer looked through the hole from the back, either at the actual Baptistery or at a mirror reflecting the painting. However, his precise method is not known and the picture is lost. His friends, such as Donatello, were evidently in possession of the method in the 1420s. A construction for the one-point perspective image of given objects on a squared pavement was first published [1] by Leon Battista Alberti (1404–1472), Florentine architect, painter, poet and scholar: see Sect. 13.5.

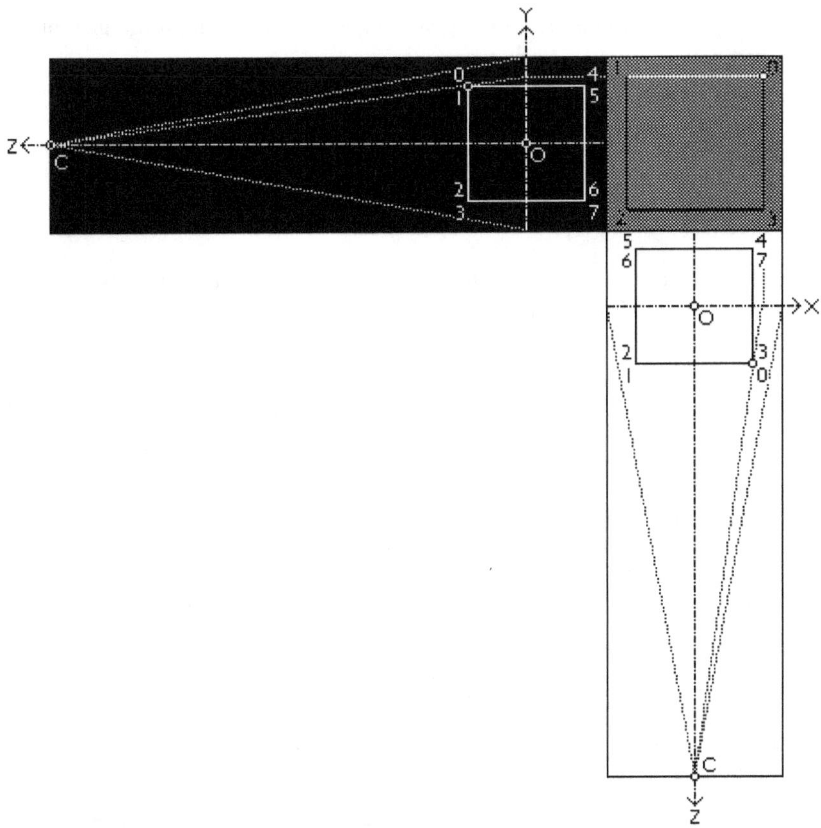

Fig. 8.32 One-point perspective projection of cube. The cube is centred at space origin O, with side length 64 and with faces parallel to the space coordinate planes. Distance from origin O to centre of projection C is $V = 263$. **a** Orthographic right-side elevation, for geometric construction. **b** One-point perspective image, from geometric construction or numeric calculation. **c** Orthographic plan, for geometric construction. Object vertex 0(32,32,32) projects to image vertex 0(64,64)

Object arithmetic scales parallel to the x-axis or the y-axis remain arithmetic scales parallel to the x-axis or the y-axis in the image. Object arithmetic scales parallel to the z-axis become harmonic scales converging to a single vanishing point in the image.

Geometrically, a one-point perspective is constructed from an orthographic side elevation and plan of the object, as shown in Fig. 8.32.

Computationally, one-point perspective is calculated directly from the space coordinates of the object as

$$(x', y') = (x', y', 0, 1) = (x, y, z, 1)\, \mathbf{T}\, \mathbf{C}\,.$$

Matrix **C** projects, from centre C, the given object coordinates (without rotation) to image coordinates:

$$\mathbf{C} = \begin{pmatrix} V & 0 & 0 & 0 \\ 0 & V & 0 & 0 \\ 0 & 0 & 0 & -1 \\ 0 & 0 & 0 & V \end{pmatrix}$$

where V is the distance OC. Notice that the product of the homogeneous coordinates and the central projection matrix, exceptionally, has W-coordinate equal to $(V - z)$; to reduce this to 1, the other coordinates must be divided by $(V - z)$.

Matrix **T** translates image coordinates to the first quadrant of the image plane, as in the preceding cases. The result is the same as Fig. 8.32b.

8.16 Two-Point Perspective

Two-point perspective projection has two forms. The first is characterised by

- *Plane of projection*: rotated about the space x-axis by some angle, but not the y-axis or the z-axis,
- *Axis of projection*: orthogonal to the plane of projection,
- *Centre of projection*: finitely distant at V spacels from origin,
- *Projectors*: diverging from the centre of projection.

Instead of rotating the plane of projection about the x-axis by $\alpha°$, we get the same result if we fix the xy-plane as plane of projection and rotate the object about the x-axis by $-\alpha°$. Object arithmetic scales parallel to the x-axis remain arithmetic scales parallel to the x-axis in the image. Object arithmetic scales parallel to the y-axis become harmonic scales converging to one vanishing point in the image; and object arithmetic scales parallel to the z-axis become harmonic scales converging to another vanishing point in the image.

Geometrically, a two-point perspective is constructed from an orthographic side elevation and plan of the (rotated) object, as shown in Fig. 8.33.

Computationally, a two-point perspective is calculated directly from the space coordinates of the object as

$$(x', y') = (x', y', 0, 1) = (x, y, z, 1)\,\mathbf{T}\,\mathbf{C}\,\mathbf{R1}\,.$$

Matrix **R1** rotates the given object about the x-axis by $\alpha°$:

$$\mathbf{R1} = \begin{pmatrix} 1 & 0 & 0 & 0 \\ 0 & \cos(\alpha) & \sin(\alpha) & 0 \\ 0 & -\sin(\alpha) & \cos(\alpha) & 0 \\ 0 & 0 & 0 & 1 \end{pmatrix}$$

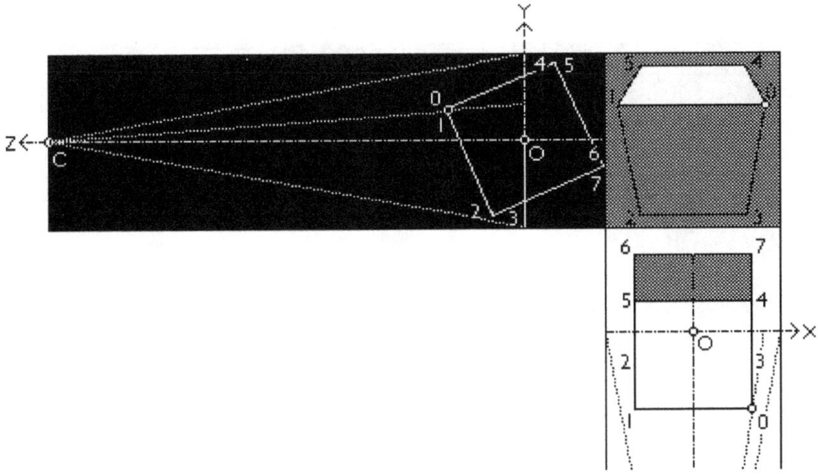

Fig. 8.33 Two-point perspective projection of cube, by x-rotation. The cube is centred at space origin O, with side length 64, and rotated about the x-axis by 22.5°. Distance from origin O to centre of projection C is $V = 263$. **a** Orthographic right-side elevation rotated about x-axis, for geometric construction. **b** Two-point perspective image, from geometric construction or numeric calculation. **c** Orthographic plan rotated about x-axis, for geometric construction. Object vertex 0(32,−32,32) projects to image vertex 0(44,−57)

Matrix **C** projects, from centre C, the given rotated object coordinates to image coordinates:

$$\mathbf{C} = \begin{pmatrix} V & 0 & 0 & 0 \\ 0 & V & 0 & 0 \\ 0 & 0 & 0 & -1 \\ 0 & 0 & 0 & V \end{pmatrix}$$

where V is the distance OC. Matrix **T** translates image coordinates to first quadrant, as in the preceding cases. The result is the same as Fig. 8.33b.

The second form of two-point perspective projection is characterised by

- *Plane of projection*: rotated about the space y-axis by some angle, but not the x-axis or the z-axis,
- *Axis of projection*: orthogonal to the plane of projection,
- *Centre of projection*: finitely distant at V spacels from origin,
- *Projectors*: diverging from the centre of projection.

Instead of rotating the plane of projection about the y-axis by $\beta°$, we get the same result if we fix the xy-plane as plane of projection and rotate the object about the y-axis by $-\beta°$. Object arithmetic scales parallel to the y-axis remain arithmetic scales parallel to the y-axis in the image. Object arithmetic scales parallel to the x-axis become harmonic scales converging to one vanishing point in the image; and object arithmetic scales parallel to the Z-axis become harmonic scales converging to another vanishing point in the image.

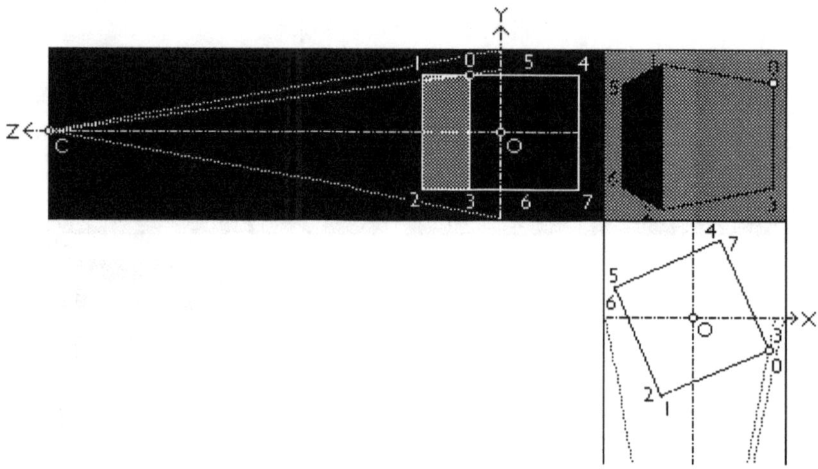

Fig. 8.34 Two-point perspective projection of cube, by y-rotation. The cube is centred at space origin O, with side length 64, and rotated about the y-axis by 22.5°. Distance from origin O to centre of projection C is $V = 263$. **a** Orthographic right-side elevation rotated about y-axis, for geometric construction. **b** Two-point perspective image, from geometric construction or numeric calculation. **c** Orthographic plan rotated about y-axis, for geometric construction. Object vertex 0(32,32,32) projects to image vertex 0(57,43)

Geometrically, this form of two-point perspective is constructed like the first form from an orthographic side elevation and plan of the (rotated) object, as shown in Fig. 8.34.

Computationally, this form of two-point perspective is calculated directly from the space coordinates of the object as

$$(x', y') = (x', y', 0, 1) = (x, y, z, 1)\, \mathbf{T}\,\mathbf{C}\,\mathbf{R2}\,.$$

Matrix **R2** rotates the given object about the y-axis by $\beta°$:

$$\mathbf{R2} = \begin{pmatrix} \cos(\beta) & 0 & -\sin(\beta) & 0 \\ 0 & 1 & 0 & 0 \\ \sin(\beta) & 0 & \cos(\beta) & 0 \\ 0 & 0 & 0 & 1 \end{pmatrix}$$

Matrix **C** projects, from centre C, the given rotated object coordinates to image coordinates:

$$\mathbf{C} = \begin{pmatrix} V & 0 & 0 & 0 \\ 0 & V & 0 & 0 \\ 0 & 0 & 0 & -1 \\ 0 & 0 & 0 & V \end{pmatrix}$$

where V is the distance OC. Matrix **T** translates image coordinates to first quadrant, as in the preceding cases. The result is the same as Fig. 8.34b.

8.17 Three-Point Perspective

Three-point perspective projection is characterized by

- *Plane of projection*: rotated about the space x-axis by some angle, and about the y-axis by some angle, but not the z-axis,
- *Axis of projection*: orthogonal to the plane of projection,
- *Centre of projection*: finitely distant at V length units from origin,
- *Projectors*: diverging from the centre of projection.

Instead of rotating the plane of projection about the x-axis by $\alpha°$ and about the y-axis by $\beta°$, we get the same result if we fix the xy-plane as plane of projection and rotate the object about the x-axis by $-\alpha°$ and about the y-axis by $-\beta°$. Object arithmetic scales parallel to the x-axis become harmonic scales converging to one vanishing point in the image; object arithmetic scales parallel to the y-axis become harmonic scales converging to another vanishing point in the image; and object arithmetic scales parallel to the z-axis become harmonic scales converging to a third vanishing point in the image.

Geometrically, a three-point perspective is constructed from an orthographic side elevation and plan of the (rotated) object, as shown in Fig. 8.35.

Computationally, a three-point perspective is calculated directly from the space coordinates of the object as

$$(x', y') = (x', y', 0, 1) = (x, y, z, 1)\,\mathbf{T\,C\,R2\,R1}\,.$$

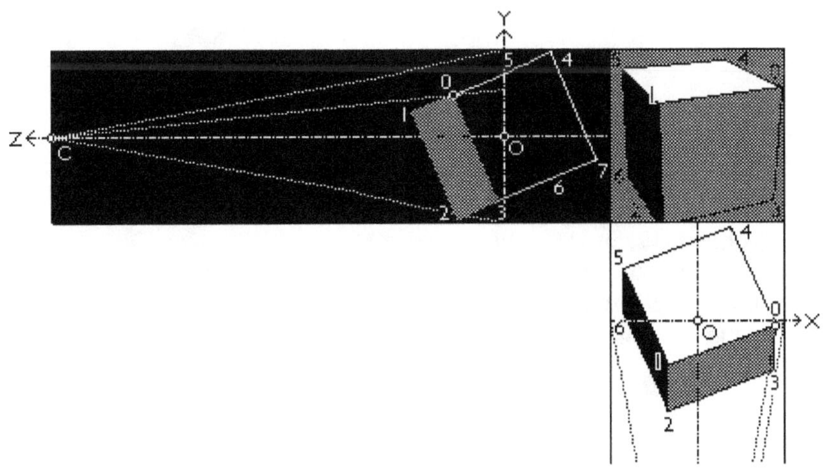

Fig. 8.35 Three-point perspective projection of cube. The cube is centered at space origin O, with side length 64, rotated about the x-axis by 22.5° and about the y-axis by 22.5°. Distance from origin O to centre of projection C is $V = 263$. **a** Orthographic right-side elevation rotated, for geometric construction. **b** Three-point perspective image, from geometric construction or numeric calculation. **c** Orthographic plan rotated, for geometric construction. Object vertex 0(32,−32,32) projects to image vertex 0(45,−39)

Matrix **R1** rotates the given object about the x-axis by $\alpha°$:

$$\mathbf{R1} = \begin{pmatrix} 1 & 0 & 0 & 0 \\ 0 & \cos(\alpha) & \sin(\alpha) & 0 \\ 0 & -\sin(\alpha) & \cos(\alpha) & 0 \\ 0 & 0 & 0 & 1 \end{pmatrix}$$

Matrix **R2** rotates the given object about the y-axis by $\beta°$:

$$\mathbf{R2} = \begin{pmatrix} \cos(\beta) & 0 & -\sin(\beta) & 0 \\ 0 & 1 & 0 & 0 \\ \sin(\beta) & 0 & \cos(\beta) & 0 \\ 0 & 0 & 0 & 1 \end{pmatrix}$$

Matrix **C** projects, from centre C, the given rotated object coordinates to image coordinates:

$$\mathbf{C} = \begin{pmatrix} V & 0 & 0 & 0 \\ 0 & V & 0 & 0 \\ 0 & 0 & 0 & -1 \\ 0 & 0 & 0 & V \end{pmatrix}$$

where V is the distance OC. Matrix **T** translates image coordinates to first quadrant, as in the preceding cases. The result is the same as Fig. 8.35b.

8.18 Space Projection Program

A projection can be implemented by a program, as shown in Fig. 8.36, which does the following:

- Read a .TXT input file of data specifying a space graph, with coordinates (x, y, z) of vertices, and identifiers of edge ends.
- Accept user input specifying any required space transformations of the object.
- Set up the appropriate transformation and projection matrices.
- Multiply the coordinates of each vertex by the matrices to get image coordinates (x, y).

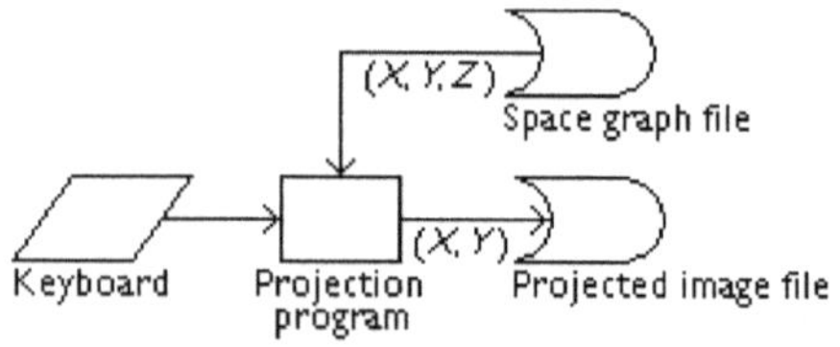

Fig. 8.36 Projection by program. Space coordinates (x, y, z) of object are given in a numeric file. Program calculates plane coordinates (x, y) of image, outputting a .BMP file

- Calculate the edge segments between the image vertices.
- Write a .BMP output file containing the image (for display or printout by a GUI editor).

A program may be written to produce just one type of projection, or more than one type at user choice. For example, the program below (QuickBASIC source code available on the SpringerLink web-site) reads a file specifying any space graph (or wire-frame model) (see Sect. 7.4). It lets the user do elementary space transformations (see Sect. 7.6), then pick one of several projections to the plane. It writes a 1-bit image file.

```
REM  **  Program PROJECTI to perform space transformation
REM  **  on given space graph and projection to plane
REM  **  image. User enters file-name.TXT #1 containing
REM  **  coordinates of object vertices and list of edges.
REM  **  User enters file-name.BMP #2 to contain output
REM  **  image. Program reads file #1 and puts vertex and
REM  **  edge data into arrays AVERT and AEDGE. User enters
REM  **  space transformations wanted; program sets up 4x4
REM  **  matrices MROT for rotation, MTRA for translation,
REM  **  MSCA for scaling, MSHE for shearing, MREF for
REM  **  reflection. Program multiplies XYZ coordinates of
REM  **  each vertex in AVERT by matrices to get transformed
REM  **  X'Y'Z' coords in AVERT. User picks orthographic or
REM  **  isometric or perspective projection. Program sets
REM  **  up matrix MORT or MISO or MPER and multiplies to
REM  **  GET X"Y" coords in AVERT. Then for each edge in
REM  **  AEDGE program calculates Bresenham segment into
REM  **  AIMAGE. Program  writes AIMAGE to .BMP file #2
REM  **  Main program
REM  **  User enters file-names for input and output
         CALL RSGRDATA
         CALL INCONST
         CALL CALCSTRA
REM  **  Ask orthogonal or isometric or perspective
REM  **  projection
            CALL CALCORTH
            CALL CALCISOM
            CALL CALCPERS
         CALL CALIMAG
         CALL WBMPHEAD
         CALL WBMPDATA
         CALL REVBHEAD
END
```

```
            SUB CALCISOM
REM  **     Subprogram to calculate isometric coords X Y, to
REM  **     show top, front, and right side.
END SUB
            SUB CALCORTH
REM  **     Subprogram to calculate orthographic coords X Y,
REM  **     to show front elev or plan or right side elev.
END SUB
            SUB CALCPERS
REM  **     Subprogram to calculate perspective image
REM  **     coords X Y.
END SUB
            SUB CALCSTRA
REM  **     Subprogram to calculate space transformations of
REM  **     object vertices.
END SUB
            SUB CALIMAG
REM  **     Subprogram to set vertices in array AIMAGE, and
REM  **     calculate Bresenham line for each edge in array
REM  **     AEDGE, end1 and end2 pointing to AVERT(V), and
REM  **     set line pixels in array AIMAGE.
END SUB
            SUB INCONST
REM  **     Subprogram to enter constants for required dilation
REM         rotation, translation, shearing, and reflection.
END SUB
            SUB PUTBGR (INTE%)
REM  **     Puts single greyscale intensity INTE% into
REM  **     file #2 as 3 bytes.
END SUB
            SUB PUTBYTE (INTE%)
REM  **     Subprogram to put 1 byte in file #2
END SUB
            SUB PUTI (INTE)
REM  **     Puts integer value passed into file #2 as 2
REM  **     bytes and increments pointer.
END SUB
            SUB PUTL (INTL&)
REM  **     Puts long integer into file #2 as 4 bytes .
END SUB
            SUB REVBHEAD
REM  **     Revises header of .BMP file #2 with values in bytes
END SUB
            SUB RSGRDATA
REM  **     Subprogram RSGRDATA to read data from ASCII
```

```
REM  **  space graph file #1 into arrays AVERT and AEDGE.
END SUB
         SUB WBMPDATA
REM  **  Subprogram to write data from array AIMAGE to
REM  **  .BMP file #2. Values 0 or 1 are converted as
REM  **  groups of 8 to byte. AX% must finish on a
REM  **  long (4-byte) boundary.
END SUB
         SUB WBMPHEAD
REM  **  Subprogram to write header and colour-table
REM  **  of output .BMP file #2.
END SUB
```

The projected images shown above were done by this program, with shading added.

8.19 Relief Projection

Whereas a projection belonging to space incidence class III, IV, or VII (see Fig. 8.2 above) delivers a *flat* image, a projection in class I, IV, V, VI, IX, X, XI, XIII, or XV delivers a *relief* image. Figure 8.37 shows an example in class VI.

Projecting the shape of a solid object, actual or imagined, from space to a relief image is, and always has been, important in sculpture. Large stone sculptures in ancient Assyria, Egypt, Greece, Rome, and medieval Europe are usually reliefs. Relief perspective was used by Donatello, Ghiberti, and other Florentine sculptors in stone and metal soon after Alberti's publication of a construction for flat perspective (see Sect. 13.5). The Italian masters distinguished:

- *alto-rilievo*, where more than half the object stands out, and is undercut,
- *mezzo-rilievo*, where less than half the object stands out, with no undercuts,

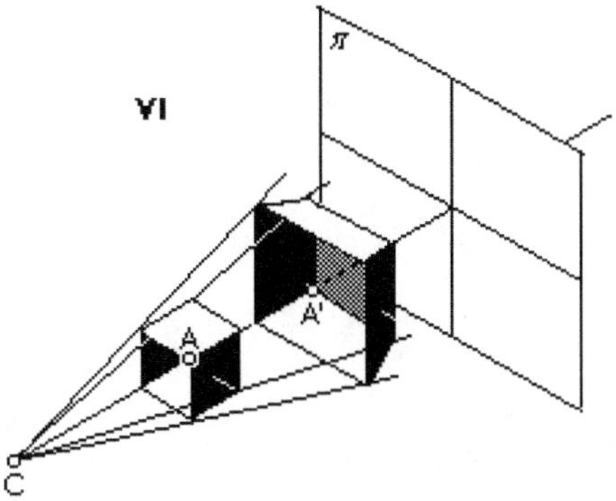

Fig. 8.37 Relief projection. Incidence class VI is fixed by object point A, image point A′, centre C so that A is between A′ and C, and plane π beyond A′. Object is projected to a relief image farther and larger than the object

- *basso-rilievo*, where the relief is low,
- *rilievo-schiacciato*, where the relief is only just discernible, as in some of Donatello's virtuoso pieces.

The first publication of specifically relief-perspective methods appears to have been by Johann Adam Breysig (1766–1831) [3], painter and stage designer for the theatre at Königsberg. It was further developed by Noël-Germain Poudra, mathematician and scholar at Paris [13]. It is described with interesting comments by Helmholtz [8].

Geometrically, a relief perspective may be constructed quite like a flat perspective from an orthographic side elevation and plan of the object, as shown in Fig. 8.38. When image point A′ lies on plane of projection π the image is flat. As A′ moves towards A the relief rises. A simple technique is to make a series of flat faces which are sections of the object parallel to the plane of projection π, then assemble them to form a relief perspective like scenery flats on a theatre stage. A topographic relief map can be made similarly by cutting out contour levels and spacing them up from a base plane. Low relief model soldiers, called "flats" [2], were produced in huge numbers in Germany in the 19th century. They were 30 cm high and 2 mm thick, cast in pewter from slate moulds, then painted.

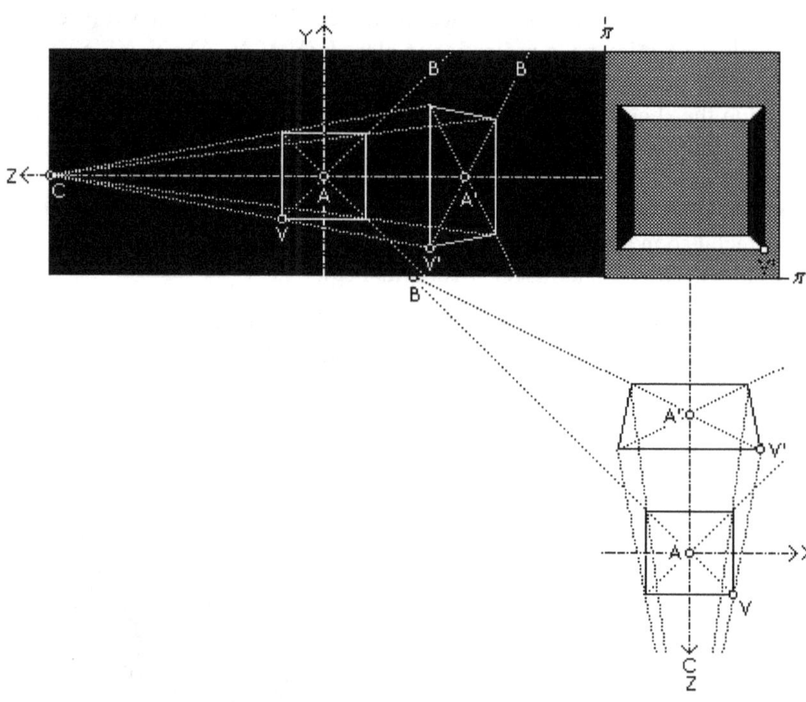

Fig. 8.38 Construction of relief projection. Object centre is A at space origin O, with faces parallel to the space coordinate planes. Projection centre C is on the positive z-axis, image point A′ on the negative z-axis, and plane of projection π beyond A′ **a** Orthographic right-side elevation. **b** Relief image faces parallel to π. **c** Orthographic plan

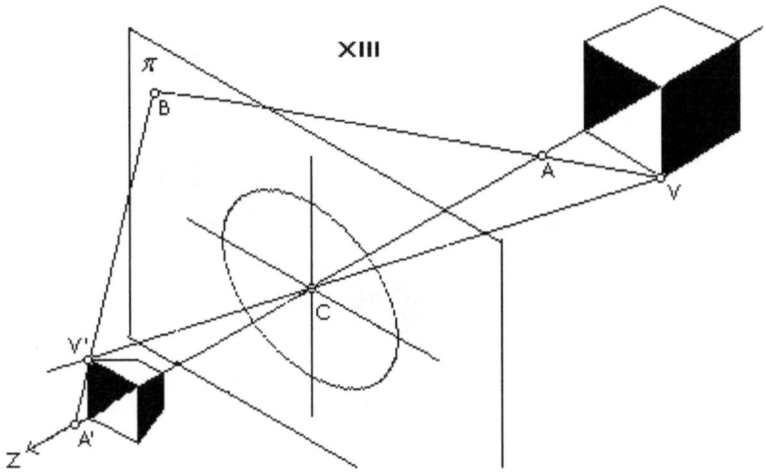

Fig. 8.39 Lens projection, class XIII. When plane of projection π is on centre C (an elation), and corresponding points A and A′ are at equal distances on opposite sides of C, the image is an inverted relief, the same as that produced by a converging lens at centre C

When A and A′ are on opposite sides of C (classes XI, XIII, and XV), we get an *inverted relief.* The elation (class XIII) with π on C, A on one side of π and A′ at equal distance on the other side of π, as shown in Fig. 8.39 gives an inverted relief which is of great importance, since it is the image produced by a converging lens (see Sect. 9.4). We may call this the *lens projection.*

8.20 Harmonic Space Projection

Just as the harmonic plane is a central projection of the Euclidean plane, quadrant by quadrant (see Sect. 8.4), so too is harmonic space a central projection of Euclidean space, octant by octant. And just as a harmonic plane projection is limited to a segment xy of the image line λ, so too is a harmonic space projection limited to a triangle xyz of the image plane π. Figure 8.40a shows a section through a harmonic space projection of class VII, where the centre C is at space origin O, the image plane π is a bounding plane of harmonic space, and corresponding points A and A′ are on the perpendicular to the image plane.

To project all of harmonic space from the coordinate origin O, eight image planes are needed, forming an octahedron. Figure 8.40b shows centre C at harmonic space origin, image plane $\pi_5 = xy - z$ in octant 5, $\pi_6 = -xy - z$ in octant 6, (hidden $\pi_7 = -x - y - z$ in octant 7), and $\pi_8 = x - y - z$ in octant 8. Compare this to Euclidean space, where six image planes are needed needed, in the directions z, y, x, $-z$, $-y$, and $-x$.

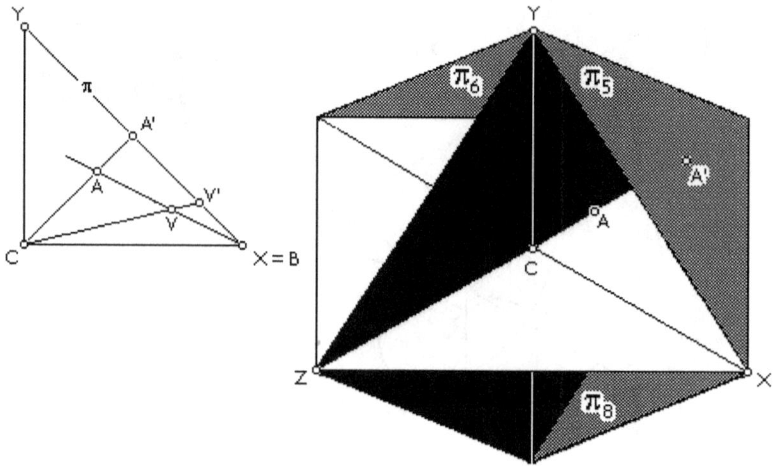

Fig. 8.40 Harmonic space projection, incidence class VII

Relief perspectives can be made in harmonic space in the same way as in Euclidean space (see Sect. 8.20), by setting image point A′ between image plane π and corresponding point A.

8.21 Elliptic Space Projection

Just as the elliptic plane corresponds one-to-one with the Euclidean plane, quadrant by quadrant (see Sect. 8.5), so too does elliptic space correspond one-to-one with Euclidean space, octant by octant. But elliptic space is not a central projection of Euclidean space, because elliptic lines not passing through centre C are circular arcs. However, an elliptic space projection from C has an image sphere σ instead of an image plane π. Figure 8.41a shows a section through an elliptic space projection of class VII, where the centre C is at space origin O, the image sphere σ is the bounding sphere of elliptic space, and corresponding points A and A′ are on the normal to the image sphere.

To project all of elliptic space from the coordinate origin O, only the image sphere is needed, without any discontinuity. Figure 8.41b shows centre C at elliptic space origin, and half of the image sphere σ. Compare this to Euclidean space, where six image planes are needed needed, and harmonic space, where eight image planes are needed. Elliptic space is a good model for vision.

Relief perspectives can be made in elliptic space in the same way as in Euclidean space (see Sect. 8.19) or harmonic space, by setting image point A′ between image sphere σ and corresponding point A.

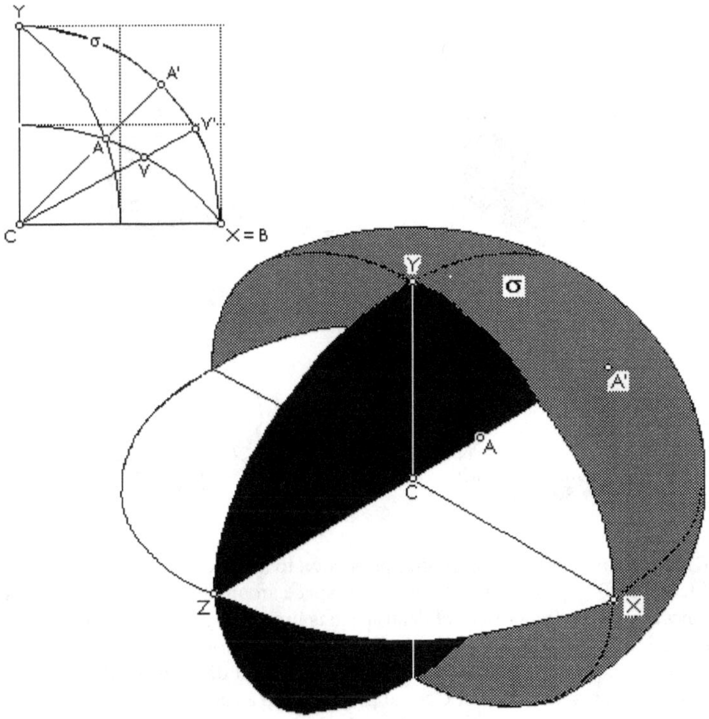

Fig. 8.41 Elliptic space projection, incidence class VII

8.22 Diorama and Panorama

Figure 8.42 shows how central projection can be extended to wide regions of space around the viewpoint. In Fig. 8.42a, we make an ordinary perspective projection from the viewpoint at origin of the axes, looking along the positive z-axis. Manuals of perspective drawing usually recommend restricting the view to 30° either side of the z-axis. If we want to include more than 45° either side we must turn to look along the x-axis positive or negative. The perspective image is then on three sides of a cube. And similarly for the other directions. All of space surrounding the viewpoint projects to a perspective image on the six faces of a cube.

In Fig. 8.42b, the projection is to a cylinder instead of four of the planes, and limited to a band 45° above and below the horizon. This is the principle of the *diorama* (across-view). Very large full-cylinder painted dioramas were built as public entertainments in the 19th century [7]. Smaller relief-perspective dioramas of 90° or less were built as displays in natural history and other museums. Some cinema systems are built to project onto a part-cylinder screen. Some cameras are made to take very wide-angle images, which are usually called panoramic. There is ongoing

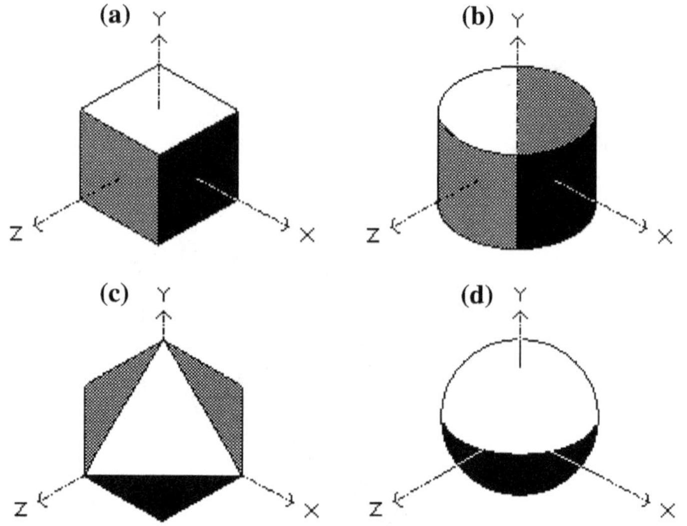

Fig. 8.42 Diorama and panorama. **a** Central projection to a plane in Euclidean space is limited to a square subtending 90° in all six directions. All of space around the viewpoint projects to a cube. **b** Central projection to a cylinder in Euclidean space is limited to a band subtending 90° vertically. The space around the viewpoint in this band projects to a cylindrical diorama. **c** Central projection to a plane in harmonic space is limited to a triangle in all eight directions. All of space around the viewpoint projects to an octahedron. **d** Central projection to a sphere in elliptic space is unlimited. All of space around the viewpoint projects to a spherical panorama

development of techniques for stitching together multiple images to make even wider-angle images [15].

In Fig. 8.42c, the projection is in harmonic space, contained in an octahedron. There appear to be no actual examples of this.

In Fig. 8.42d the projection is in elliptic space, contained in a sphere. This is the principle of the *panorama* (everything-view). A planetarium has an overhead hemispherical panorama. For another hemispherical example see [10].

References

1. Alberti LB (1435) De Pictura. English edition: Alberti LB (1970) (translator: Spencer JR) Yale UP, New Haven. http://www.noteaccess.com/Texts/Alberti/. Accessed 1 Feb 2014
2. Baden A (2009) Flats. http://www.geocities.com/alexanderbaden/contents1.html. Accessed 1 Feb 2014
3. Breysig, JA (1798) Versuch einer Erlauterung der Reliefperspektive. Magdeburg
4. Coxeter HSM (2003) Projective geometry. Springer, New York
5. Dorrie H (1965) 100 Elementary theorems. Dover, New York
6. Farish W (1822) On isometrical perspective. Camb Philol Trans 1
7. Gernsheim H, Gernsheim A (1956) LJM Daguerre: the history of the Diorama and the Daguerreotype. Secker and Warburg, London

8. Helmholtz H von (1967) Handbuch der Physiologischen Optik. Voss, Leipzig, vols 3. pp 1856–1867. English translation of 1910 edition (1925) Optical Society of America, republished (1962) Dover, New York
9. Lawrence S (2003) History of descriptive geometry in England. In: Proceedings of 1st conference construction history, Madrid. http://gilbert.aq.upm.es/sedhc/biblioteca_digital/Congresos/CIHC1/CIHC1_121.pdf. Accessed 1 Feb 2014
10. Lehar S (2014) Bubble World Diorama. http://cns-alumni.bu.edu/slehar/webstuff/bubblemodel/diorama.html. Accessed 1 Feb 2014
11. Monge G (1799) Géometrie Descriptive. Baudouin, Paris
12. Poncelet J-V (1822) Traité des Proprietes Projectives des Figures: ouvrage utile à qui s'occupent des applications de la géométrie descriptive et d'opérations géométriques sur le terrain, vol 2, 2nd edn. Gauthier-Villars, Paris, pp 1865–1866
13. Poudra N-G (1860) Traité de Perspective en Relief. Correard, Paris. Reprinted (2012) Hachette Livre Bnf, USA. Plates. http://gallica.bnf.fr/ark:/12148/bpt6k6215114w. Accessed 1 Feb 2014
14. Sopwith T (1838) A treatise on isometrical drawing. 2nd edn. J Weale, London. https://openlibrary.org/works/OL13132045W. Accessed 1 Feb 2014
15. Szeliski R (2005) Image alignment and stitching. http://www.caam.rice.edu/~zhang/caam699/p-files/Im-Align2005.pdf. Accessed 1 Feb 2014

Chapter 9
Lens Geometry

A converging lens is a device for projecting an object to an image, and is therefore of fundamental interest in digital imaging, as a simple glass disk in a camera, and as a complex organic arrangement in a viewing eye. The characteristics of a lens are defined. The locations of object and image relative to the lens can be found by geometric ray-tracing or by calculation. Examples are given for a real image, on the opposite side of a lens from an object. The real image is an inverted relief projection in the incidence classification of Chap. 8. It can be cast on a surface, in which case the depth of field becomes significant. Further examples are given for a virtual image, on the same side of a lens as an object. Two or more lenses can be compounded for increased power. The simple geometric treatment of lenses ignores the diffractive nature of light; when this is recognized a limit on the resolving power of a lens emerges.

9.1 Converging Lens

A *converging lens* is an analogue (continuous) device which accepts light emitted or reflected from objects in a cone of space in front of it, and delivers a real image of the objects in a cone of space behind it, or a virtual image in a cone of space in front of it. The image is a relief projection (see Sect. 8.19) of the objects.

Physically, a simple converging lens is a transparent disk with at least one curved surface. It is usually of glass, refractive index around 1.5, operating in air, refractive index 1.

Geometrically, a lens has a plane through its circumference, and an axis normal to its plane at its centre C. The *curvature* of each surface is usually spherical, and is measured reciprocally as $1/r$ where r is the radius of the sphere: the smaller the radius the greater the curvature.

Figure 9.1a shows a *plano-convex* lens, with one curved surface and one flat surface. The curved surface has centre K, radius r, and curvature $1/r$. The flat surface has centre at infinity, radius infinite, and curvature $1/\infty = 0$. Figure 9.1b shows a

A. Parkin, *Digital Imaging Primer*, DOI 10.1007/978-3-540-85619-1_9

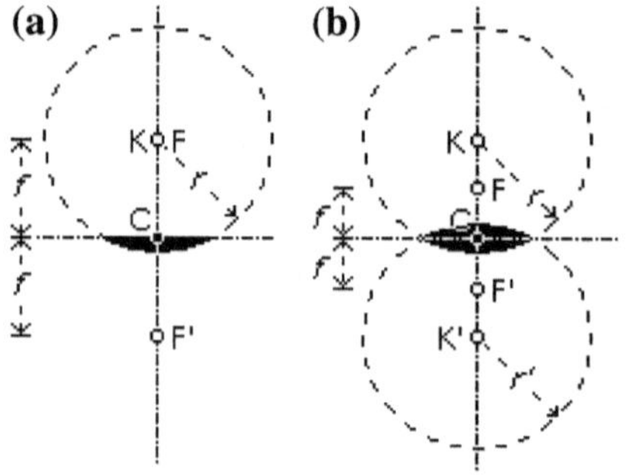

Fig. 9.1 Converging lens, in plan section. **a** Plano-convex lens has one spherically-curved surface with centre K and radius r, curvature $1/r$. The other surface is flat, curvature 0. **b** Bi-convex lens has two spherically-curved surfaces, one with centre K, radius r, curvature $1/r$, and one with centre K′, radius r', curvature $1/r'$

Table 9.1 Converging lens curvatures, dioptric powers, and focal lengths

Form	r	r'	$1/r$	$1/r'$	DP	f
Plano-convex	2	∞	0.5	0	0.5	2
Plano-convex	1	∞	1	0	1	1
Plano-convex	0.5	∞	2	0	2	0.5
Bi-convex	2	1	0.5	1	1.5	0.67
Bi-convex	1	1	1	1	2	0.5
Bi-convex	0.5	1	2	1	3	0.33

Curvatures r and r' in metres, dioptric powers dpt in dioptres, focal lengths f in metres

bi-convex lens, with two curved surfaces. One surface has centre K, radius r, and curvature $1/r$; the other has centre K′, radius r', and curvature $1/r'$. A lens has two *focal points* F and F′ on the axis, each at *focal length* f from the centre C.

Optically, the strength or *dioptric power* DP of a lens is equal to the sum of the curvatures of both surfaces. If the radii are in metres, the dioptric power is in *dioptres*[1]:

$$DP = 1/r + 1/r' \text{ dioptres}\,.$$

Clearly, the focal length f is the reciprocal of the dioptric power:

$$f = 1/DP = 1/(1/r + 1/r') \text{ metres}\,.$$

For example, Table 9.1 shows some curvatures, dioptric powers, and focal lengths.

Light travels more slowly in glass than in air. Figure 9.2a shows a source at focal point F emitting light as an expanding spherical wave. The wave meets the lens,

[1]The term was introduced in its modern sense by Felix Monoyer in 1872. The SI equivalent is inverse metre, m^{-1}.

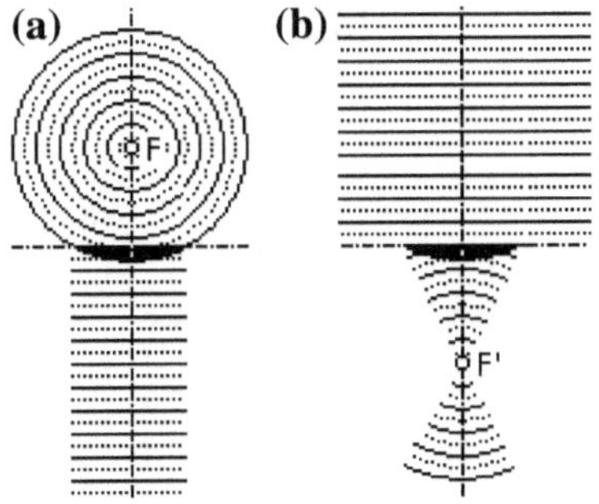

Fig. 9.2 Light refraction by lens, in plan section. **a** Source at focal point F emits light as an expanding spherical wave, refracted by lens to emerge as a plane wave. **b** Light from source at great distance arrives as a nearly plane wave, refracted by lens to emerge as a spherical wave contracting to focal point F′

slows differentially, and emerges as a plane wave. Conversely, Fig. 9.2b shows light emitted by a source at great distance arriving as a plane wave. The wave meets the lens, slows differentially, and emerges as a spherical wave contracting to focal point F′ (and expanding again beyond F′). The lens acts the same in either direction.

A simple theory of lenses is based on three idealizations:

- *Geometrical ray-tracing*, which represents light emitted spherically from a source by a bundle of straight-line *rays* normal to the wave-fronts.
- Limitation to *paraxial* rays, which make small angles with the axis of the lens.
- Limitation to a *thin lens*, where refraction of rays is deemed to occur at the plane of the lens.

Geometrical ray-tracing ignores the diffraction which occurs with spherical waves (see Sect. 9.8 below). The paraxial limitation allows differences between angular widths and linear widths to be ignored. A thin lens ignores the complications of thick, compound, and non-spherical lenses with several refracting surfaces.

9.2 Real Image

A lens forms a *real image* of an object by converging rays of light. A real image is one which can be thrown onto a physical surface, as distinct from a virtual image, which cannot.

Figure 9.3a shows a geometric construction for locating a real image by tracing rays. In plan section, C is the centre of a lens with focal points F and F′, and A is a point on an object. A ray which arrives at the lens parallel to axis CF is refracted through the back focal point F′. A ray which arrives at the lens via front focal point F is refracted to emerge parallel to the lens axis, and meets the ray through F′ at image point A′. A third ray which goes through C emerges unrefracted to A′. Any two of these rays determine an image point, with the third as a check.

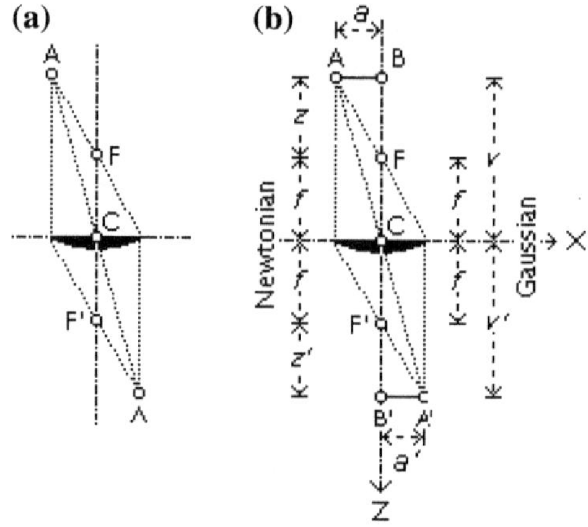

Fig. 9.3 Construction of real image by ray-tracing. **a** Lens with centre C and focal points F and F′, and object point A. Ray from A incident parallel to lens axis is deviated to emerge through back focal point F′. Ray from A incident through centre C is undeviated. Ray from A incident through front focal point F is deviated to emerge parallel to lens axis. All three rays meet at A′ the image of A. **b** Size variables, Newtonian at *left*, Gaussian *at right*

Figure 9.3b shows two alternative conventions for measuring distances in thin lens optics. On the left, *Newtonian* distances are

- Focal length $f = \mathrm{CF} = \mathrm{CF}'$,
- Object distance $z = \mathrm{FA}$,
- Image distance $z' = \mathrm{F'A'}$.

They are related by the *Newtonian lens equations*:

$$zz' = f^2$$
$$\text{lateral magnification } m = f/z = z'/f = a'/a\,,$$

where a is object width AB, and a' is image width $A'B'$. On the right, *Gaussian* distances are

- Focal length $f = \mathrm{CF} = \mathrm{CF}'$,
- Object distance $v = \mathrm{CA}$,
- Image distance $v' = \mathrm{CA}'$.

They are related by the *Gaussian lens equations*:

$$1/v + 1/v' = 1/f$$
$$\text{lateral magnification } m = v'/v = a'/a\,,$$

where a is object width AB, and a' is image width A′B′.

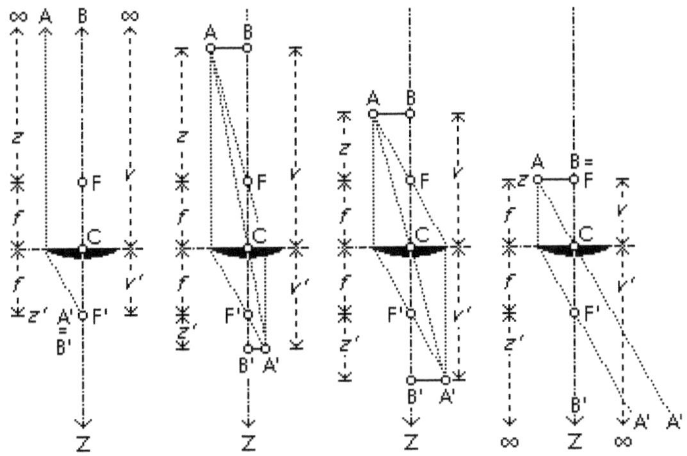

Fig. 9.4 Location of real image. **a** Object at infinite distance CB; image at back focal point F′, diminished to nothing. **b** Object AB at three times the focal length f; image A′B′ inverted, and half-size. **c** Object AB at twice the focal length; image A′B′ inverted, and same-size. **d** Object AB at front focal point F; image at infinite distance and infinite size

Sign conventions for optical distances can be set in more than one way. Here we shall treat all distances as positive except where explicitly stated.

Figure 9.4 shows the real image formed by a converging lens as the object distance CB (z or v) decreases from infinity to focal length f. We see that,

(a) when the object is at very great distance from the lens, the image is at back focal point F′, diminished to very small,
(b) when the object is greater than $2f$ from the lens, the image is inverted and diminished,
(c) when the object is exactly $2f$ from the lens, the image is inverted and same-size,
(d) when the object is exactly f from the lens, the image becomes infinite.

9.3 Gaussian Nomogram

A convenient rough guide to the location of real images is provided by a *nomogram* [1],[2] as shown in Fig. 9.5. This relates dioptric power of lens (and focal length f), object distance v, and image distance v', in Gaussian measure. The horizontal axis is calibrated as an arithmetic progression of dioptres dpt from 0 to 25, and a harmonic progression of object distances $v = 1000/\text{dpt}$ in millimetres. The vertical axis is

[2]Graphical method for approximate solution of equations, invented by Philibert Maurice d'Ocagne (1862–1938), Professor at Ecole Polytechnique, Paris.

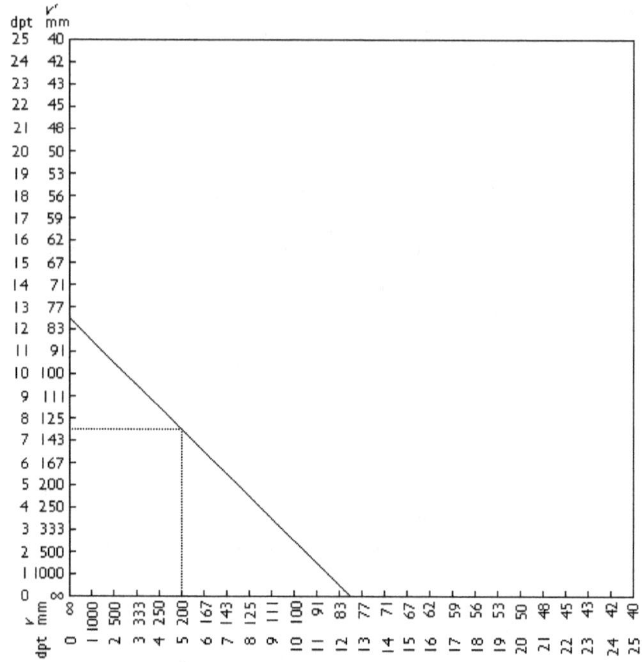

Fig. 9.5 Nomogram relating dioptric power of lens, focal length, object distance, and image distance. Say a lens has power 12.5 dpt, focal length $f = 80\,\text{mm}$. Draw the diagonal line through 80 mm on both axes. Then, given object distance $v = 200\,\text{mm}$, read off image distance $v' = 133\,\text{mm}$. Or, given object distance $v = 133\,\text{mm}$, read off image distance $v' = 200\,\text{mm}$

calibrated likewise as dioptres and image distances $v' = 1000/\text{dpt}$ in millimetres. Given the power dpt of a lens (or the focal length 1000/dpt), draw a diagonal line through dpt on both axes. Then given an object distance v mm on the horizontal axis, the real image distance v' mm can be read off on the vertical axis; or given a real image distance v', the object distance v can be read off. For example, given lens power 12.5 dpt (so $f = 80\,\text{mm}$), and object distance $v = 200\,\text{mm}$. Then

$$
\begin{aligned}
1000/v + 1000/v' &= 1000/f \\
1000/200 + 1000/v' &= 1000/80 \\
1000/v' &= 12.5 - 5 = 7.5 \\
v' &= 133\,\text{mm}
\end{aligned}
$$

and the lateral magnification is

$$
m = 133/200 = 0.67\,.
$$

9.4 Lens Image as Space Projection

Another way to locate a real image is as a space projection. Recalling from Sect. 8.1 the projection class XIII, where

- Centre C is the centre of the lens.
- Plane of projection π is the plane of the lens, an elation.
- Corresponding points G and G′ are on opposite sides of π at equal distances GC $= \mathrm{G'C} = 2f$.

The construction is shown in Fig. 9.6, compared to the ray-trace. If the object distance FB $= z$ is measured as a multiple of focal distance CF $= f$, and the image distance F′B′ $= z'$ as a multiple of $f' = f$, we have the harmonic relation:

$$z'/f' = 1/(z/f) = f/z\,.$$

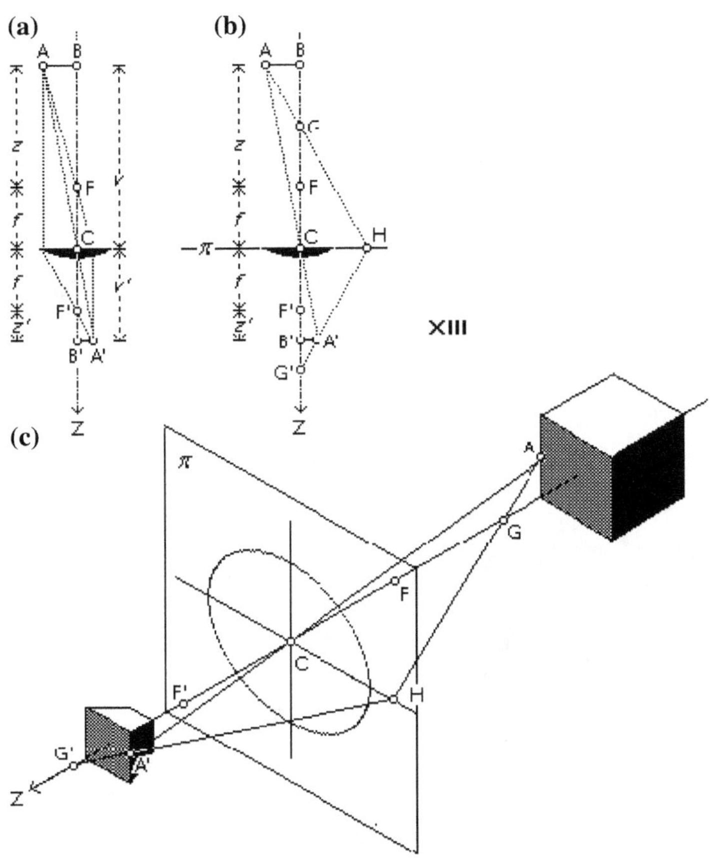

Fig. 9.6 Lens image is a projection class XIII. **a** Ray-trace for lens. **b** Construction for projection where plane of projection π is on centre C, and corresponding points G and G′ are on opposite sides of π at equal distances GC $= \mathrm{G'C} = 2f$. Then $z/f = f'/z'$. **c** Space view of (**b**)

9.5 Depth of Field

Projection is a geometric operation which fixes a geometric image point for every object point. But a lens is a physical device which collects a diverging cone of rays from every object location and deviates it to a converging cone of rays whose vertex is the focus. A screen placed at the focus will display a very small spot (but not a geometric point, because of lens aberrations, and diffraction). The image made of these focussed spots is the best possible. If the screen is moved to cut a cone before or after it reaches its focus, it displays a disk, larger with increasing distance from the focus. The image made of these disks is blurred. Figure 9.7 shows three cases in plan section and in space. Object A_1 has image focussed at A_1', and displays a blurred image on a screen placed beyond its focus. Object A_2 has image focussed at

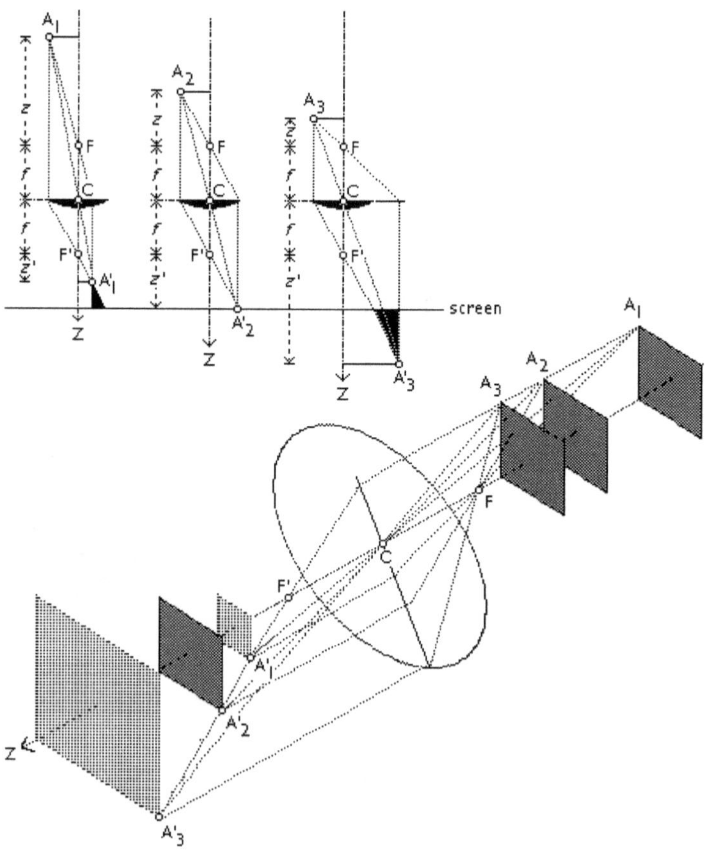

Fig. 9.7 Depth of field. Object spot A_1 is focussed to image spot A_1' and displays a disk on a screen placed beyond the focus. Object spot A_2 is focussed to image spot A_2' and is sharp on a screen placed at the focus. Object spot A_3 would be focussed to image spot A_3' but displays a disk on a screen placed before the focus

A'_2, and displays a sharp image on a screen placed at its focus. Object A_3 has image focussed at A'_3, and displays a blurred image on a screen placed before its focus. As a screen is moved from A'_1 through A'_2 to A'_3 the image begins blurred, becomes sharp, then blurred again.

Sharpness of image falls off gradually either side of the focus. We can set near and far limits to some acceptable sharpness by fixing a *circle of confusion* c, the diameter of the disk where the image is deemed to pass from sharp to blurred. Then, for a lens, where

- Lens diameter $= d$
- Lens focal length $= f$
- Lens f-number $= N = f/d$
- Circle of confusion diameter $= c$
- Object Newtonian distance $= z$
- Image Newtonian distance $= z'$

calculate object near and far depth of field as

$$z_{\mathrm{N}} = (f \times z)/(f - (N \times c))$$
$$z_{\mathrm{F}} = (f \times z)/(f + (N \times c))$$

and image near and far depth of field as

$$z'_{\mathrm{N}} = (f' \times z')/(f' - (N \times c))$$
$$z'_{\mathrm{F}} = (f' \times z')/(f' + (N \times c))\,.$$

For example, Fig. 9.8 shows, in arbitrary length units:

- Lens diameter $d = 4$.
- Lens focal length $f = 4$.
- Lens f-number $N = 1$.
- Circle of confusion diameter $c = 0.3$ (set very large for illustration).
- Object distance $z = 8$.
- Image distance $z' = 2$.

Calculate object near and far depth of field as

$$z_{\mathrm{N}} = (4 \times 8)/(4 - (1 \times 0.3)) = 8.65$$
$$z_{\mathrm{F}} = (4 \times 8)/(4 + (1 \times 0.3)) = 7.44$$

and image near and far depth of field as

$$z'_{\mathrm{N}} = (4 \times 2)/(4 - (1 \times 0.3)) = 2.16$$
$$z'_{\mathrm{F}} = (4 \times 2)/(4 + (1 \times 0.3)) = 1.86\,.$$

Object distance z is the harmonic mean between object near depth of field z_N and far depth of field z_F (see Sect. 6.5). For example, if $z = 8$ and $z_N = 8.65$ and $z_F = 7.44$, then $(2 \times 8.65 \times 7.44)/(8.65 + 7.44) = 8$. Similarly image distance z' is the harmonic mean between image near depth of field z'_N and far depth of field z'_F. Evidently, the object far field is greater than the object near field, but the image far field is less than the image near field.

It is sometimes useful to calculate a *hyper-focal distance h*, the object distance where the far depth of field is at infinity. That is to say, when $z = h$ the images of all objects out to infinity are deemed sharp.

$$h = f^2/(N \times c).$$

For example, in Fig. 9.8 $h = 4^2/(1 \times 0.3) = 53.33$.

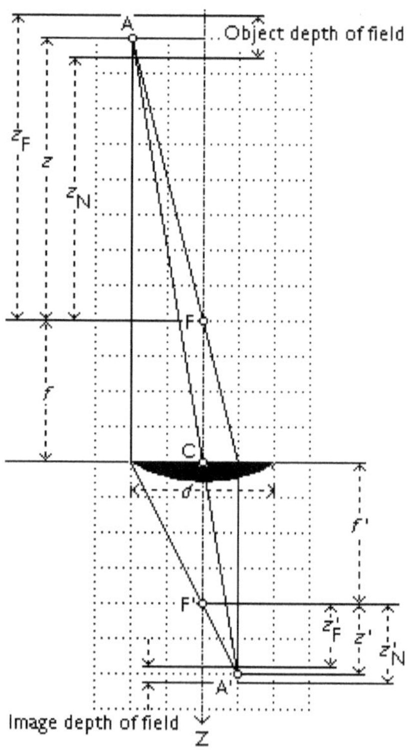

Fig. 9.8 Depth of field example as calculated in the text. For clarity, the horizontal scale has been set implausibly large. In practice the circle of confusion would be set much smaller, and the vertical dimensions would be much larger than the horizontal dimensions

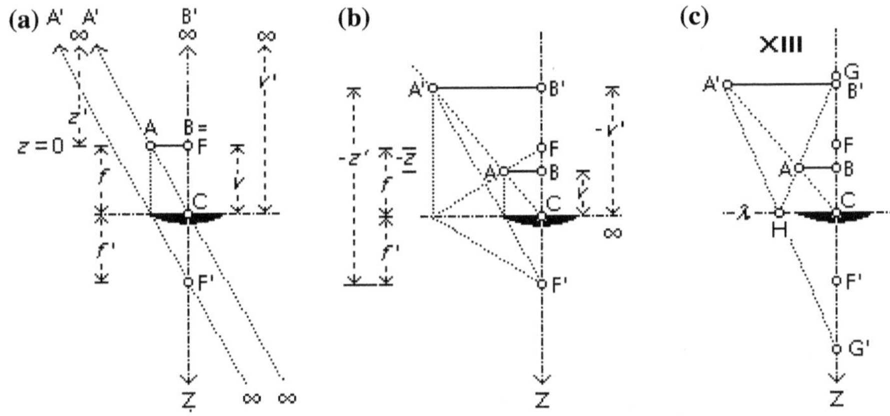

Fig. 9.9 Formation of virtual image. **a** Ray-trace for object AB at focal distance f, virtual image A$'$B$'$ infinity. **b** Ray-trace for object AB at less than focal distance f, virtual image A$'$B$'$ upright and magnified in front of lens. **c** Projection class XIII construction for virtual image

9.6 Virtual Image

When an object is closer to a converging lens than the focal distance f, the lens forms a *virtual* image. A virtual image is on the same side of the lens as the object, and upright; but it is formed by diverging rays, and so cannot be thrown onto a screen. Figure 9.9a shows object AB at front focal point F; the lens forms a real image at $-\infty$ and a virtual image at $+\infty$. In Newtonian distances, as z approaches 0 and z' approaches $\infty z \times z'$ approaches f^2. In Gaussian distances, when $1/v = 1/f$ and $1/v' = 0$, $1/v + 1/v' = 1/f$. Figure 9.9b shows object AB closer than F; the virtual image A$'$B$'$ is upright, in front of the lens, and larger than AB. A ray which arrives at the lens parallel to axis CF is refracted through the back focal point F$'$. A ray which arrives at the lens from front focal point F is refracted to emerge parallel to the lens axis, and meets the ray through F$'$ at image point A$'$. A third ray which goes through C emerges unrefracted to A$'$. In Newtonian distances, $-z$ is negative, measured from F to B, and $-z'$ is negative, measured from F$'$ to B$'$, so $-z \times -z' = f^2$. In Gaussian distances, v is positive, measured from C to B, and $-v'$ is negative, measured from B$'$ to C, so $1/v - 1/v' = 1/f$.

Figure 9.9c shows the construction for a virtual image when the lens is regarded as a projection in class XIII.

9.7 Compound Converging Lens

Two or more converging lenses may be placed on a common axis to make a *compound converging lens*. Lens 1, nearest the object, forms a first image, real or virtual, which is the object for lens 2 to form a second image, and so on if more lenses are compounded.

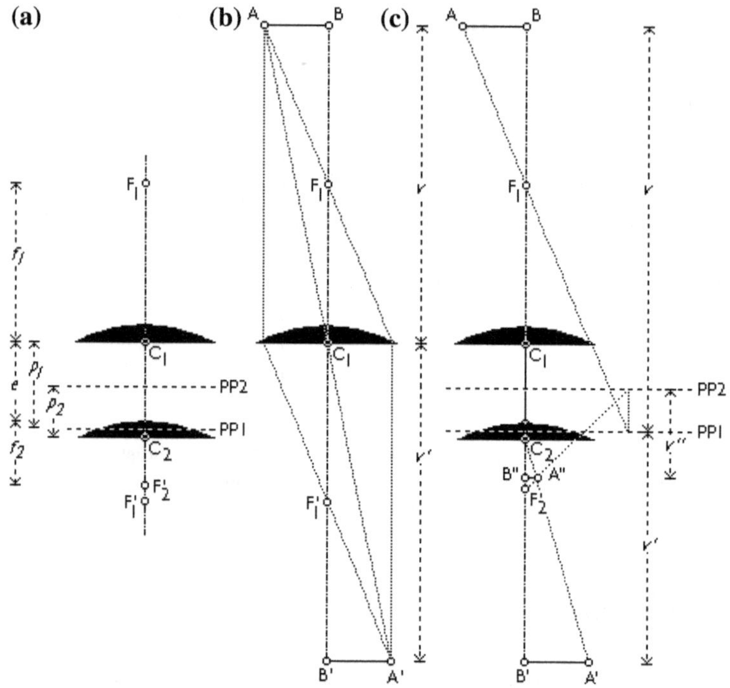

Fig. 9.10 Compound converging lens. **a** Lens 1 has centre C_1 and focal length f_1; lens 2 has centre C_2 and focal length f_2. The distance between the two lenses is e. Principal plane 1 PP1 is distance p_1 from lens 1, and principal plane 2 PP2 is distance p_2 from lens 2. **b** Lens 1 forms image A′B′ from object AB. Single magnification is v'/v **c** Lens 2 forms image A″B″ from object A′B′. Compound magnification is v''/v

Figure 9.10a shows lens 1 with centre C_1 and focal length f_1, and lens 2 with centre C_2 and focal length f_2. The distance between lens 1 and lens 2 is $e = C_1C_2$. *Principal plane 1* PP1 is the reference plane for object-side measurements, and is parallel to the lens planes at distance p_1 behind lens 1. *Principal plane 2* PP2 is the reference plane for image-side measurements, and is parallel to the lens planes at distance p_2 in front of lens 2. Given f_1, f_2, and e, we may find:

- Focal length of compound $f = (f_1 \times f_2)/(f_1 + f_2 - e)$.
- Distance of principal plane 1 $p_1 = (f \times e)/f_2$.
- Distance of principal plane 2 $p_2 = (f \times e)/f_1$.

Figure 9.10b shows rays for image A′B′ formed from object AB by lens 1 alone. In a compound, a ray incident on lens 1 is deemed to refract at PP1, then continue from the corresponding point on PP2 to emerge from lens 2. Figure 9.10c shows rays for image A″B″ formed from object A′B′ by lens 2.

For example, Table 9.2 shows the calculation for a compound of two given converging lenses, with different focal lengths, at varying distance apart.

Table 9.2 Compound of two given lenses with varying distance apart

dpt_1	dpt_2	f_1	f_2	e	f	p_1	p_2	DP
5.55	8.33	0.180	0.120	0	0.072	0	0	13.9
5.55	8.33	0.180	0.120	0.010	0.074	0.006	0.004	13.4
5.55	8.33	0.180	0.120	0.020	0.077	0.013	0.008	13.0
5.55	8.33	0.180	0.120	0.030	0.080	0.020	0.013	12.5
5.55	8.33	0.180	0.120	0.040	0.083	0.028	0.018	12.0
5.55	8.33	0.180	0.120	0.050	0.086	0.036	0.024	11.6

dpt_1 and dpt_2 are the powers of lens 1 and lens 2, in dioptres. $f_1 = 1/dpt_1$ and $f_2 = 1/dpt_2$ are the focal lengths of lens 1 and lens 2, in metres. e is the distance between lens 1 and lens 2, in metres. $f = (f_1 \times f_2)/(f_1 + f_2 - e)$ is the focal length of the compound, in metres. $p_1 = (f \times e)/f_2$ is the distance from lens 1 to principal plane 1, and $p_2 = (fe)/f_1$ from lens 2 to principal plane 2, in metres. DP is the dioptric power of the compound

9.8 Diffraction Limit of Lens

In geometrical optics the image of the edge of an object is idealized as a perfectly sharp profile, and the the image of an arbitrarily small object is idealized as an arbitrarily small point. But in fact light diffracts at an edge, so that the image shows bright and dark interference fringes parallel with the edge, getting rapidly fainter with distance from the edge. And a point object (such as an illuminated pinhole, or a star in the night sky) is focussed by a (supposed perfect) lens, not to a point but to an *Airy disk*[3]: a bright disk surrounded by a series of fainter and fainter rings, as shown in Fig. 9.11a. The size of the disk is inverse to the diameter of the entrance pupil of the lens, and varies with the wavelength of the light from the object. Rayleigh[4] established the empirical formula

$$\sigma = 1.22\lambda/d\,,$$

where d is the diameter of the entrance pupil, λ is the wavelength of the light, and σ is the angle subtended by the centre of the bright disk to first minimum, in radians.[5] The *Rayleigh criterion* is that the smallest angular separation of two point sources that can just be distinguished apart is σ: that is, the centre of one disk lies on the first dark ring of the other disk, as shown in Fig. 9.11b.

Table 9.3 shows how the angular diffraction limit decreases as the diameter of a lens increases, for yellow-green light to which the eye is most sensitive.

[3] Sir George Biddell Airy (1801–1892); Lucasian Professor, Cambridge University 1826–1828; Plumian Professor 1828–1835; Astronomer Royal 1835–1881; KCB 1872.

[4] John William Strutt (1842–1919), 3rd Baron Rayleigh 1873; Cavendish Professor, Cambridge University 1879–1884; Nobel prize 1904; President, Royal Society 1905–1908; many discoveries in optics (including why the sky is blue) and in acoustics.

[5] One radian $= 360/2\pi \approx 57.2958°$. One degree $= 2\pi/360 \approx 0.01745$ radians.

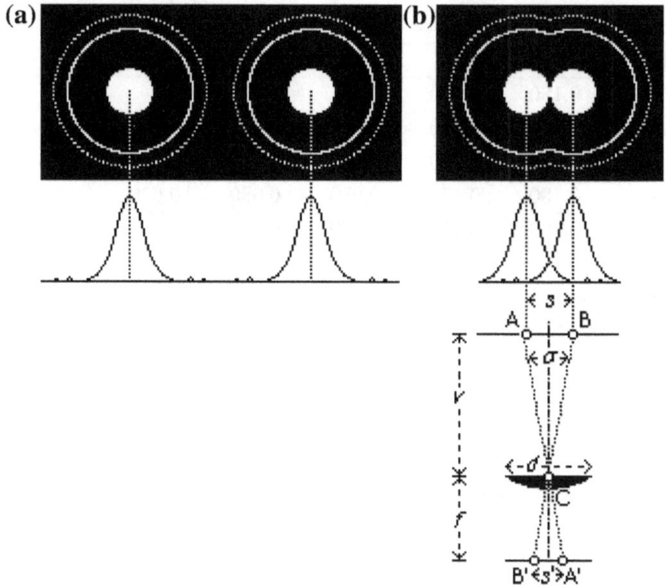

Fig. 9.11 Diffraction limit of lens. **a** Airy disks of two point objects, and their intensity profiles. **b** The angular diffraction limit σ of a lens is when the centre of one disk lies on the first minimum of the other disk. The linear diffraction limit s' is when two images A′ and B′ of point objects A and B are separated by $s' = 1.22 \times \lambda \times N$, where $N = f/d$ is the f-number of the lens

Table 9.3 Angular diffraction limit of lens

d mm	1.22λ mm	σ rad	σ deg	σ arc-min
1	0.000671	0.000671	0.0384	2.307
10	0.000671	0.0000671	0.00384	0.231
100	0.000671	0.00000671	0.000384	0.023
1000	0.000671	0.000000671	0.0000384	0.002

d is diameter of lens; λ is wavelength of yellow-green light = 550 nm; σ is angular diffraction limit in radians, degrees and arc-minutes

Consider now two object points A and B with angular separation σ and linear separation s, at object distance v from a lens with diameter d and focal length f, as in Fig. 9.11b. A lens with focal length f and (entrance pupil) diameter d is said to have *f-number* $N = f/d$. When the object distance v is much greater than the linear separation s, the angular separation σ is approximately d/v radians. In the focussed image A′B′ the linear separation s' is

$$s' = 1.22 \times \lambda \times N\,,$$

where λ is the wavelength of the light, $N = f/d$ is the f-number of the lens, f is the focal length of the lens, and d is the diameter of the lens. For example, a camera lens might have

$$\begin{aligned}
d &= 17.8\,\text{mm} \\
f &= 50\,\text{mm} \\
N &= 50/17.8 = 2.8 \\
\lambda &= 550\,\text{nm (yellow-green)} \\
s' &= 1.22 \times 550 \times 2.8 = 1879\,\text{nm} \approx 0.002\,\text{mm}.
\end{aligned}$$

That is to say, the diffraction limit of the lens is 0.002 mm (though only a very good lens will get close to the limit). For a given wavelength of light, the diffraction limit of a lens varies inversely with the f-number N. So, for example:

$$\begin{aligned}
d &= 35.7\,\text{mm} \\
f &= 50\,\text{mm} \\
N &= 50/35.7 = 1.4 \\
\lambda &= 550\,\text{nm (yellow-green)} \\
s' &= 1.22 \times 550 \times 1.4 = 939\,\text{nm} \approx 0.001\,\text{mm}.
\end{aligned}$$

Reference

1. Ocagne PMd' (1899) Traité de Nomographie. Gauthier-Villars, Paris

Chapter 10
Viewing Geometry

The human eye is an extremely sensitive organ for projecting a continually-changing internal image of an external scene, and transforming the image to neural signals. The eye incorporates an adjustable converging lens which forms an inverted relief image on the photo-sensitive retina. The reduced eye is a simplified model for geometry and calculation. The resolving power of a normal eye can be calculated, and the acuity of an actual eye can be measured by a Snellen test. The eye has a diffraction limit which varies somewhat with pupil diameter. A glass lens can correct certain defects of the eye. Simple and compound magnifiers are in common use.

10.1 Flowchart

The flowchart (Fig. 10.1) is for general guidance in this chapter. A scene in space has three coordinates of location (x, y, z), relative to a mobile viewer. Light E incident on the scene is reflected in the direction of the viewer with varying intensity distribution (b, g, r) in the short, middle, and long wavebands of the visible spectrum. An eye takes continually-changing discrete samples of location and intensity, which it delivers to the viewer as visual sensations S. The viewer uses the plethora of visual sensations, often combined with sensations of touch, hearing, taste, and smell, to maintain a stable perception of the objects in the scene. (For binocular viewing, see Chap. 12. For perception of Space, see Chap. 14. For perception of Brightness, Hue, and Saturation see Chap. 17).

10.2 Physiology of Eye

A human eye is an approximate sphere with radius about 12 mm, containing the functional parts shown in Fig. 10.2a:

A. Parkin, *Digital Imaging Primer*, DOI 10.1007/978-3-540-85619-1_10

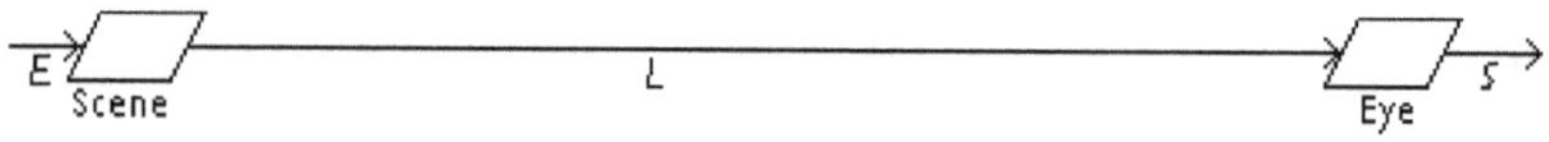

Fig. 10.1 Viewing a scene. Incident light E falls on a scene in the view of an eye. The scene reflects, per point location, differential intensities of light L. The eye takes discrete samples of location, relative to the viewer's frame of reference, and of intensity in the short-, middle-, and long-wavelength bands of the visible spectrum, and outputs a continual stream of sensation S to the viewer. The viewer maintains a stable perception of the scene

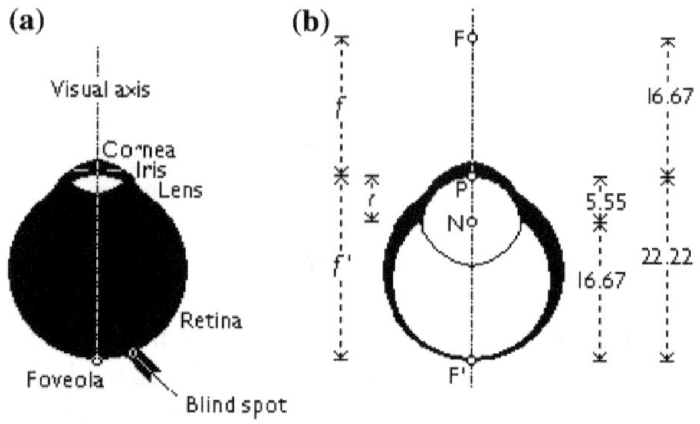

Fig. 10.2 Plan section of left eye and reduced eye. **a** Functional parts. **b** Reduced eye model superimposed on actual eye. F is front focal point, P is principal point, N is nodal point, and F′ is back focal point of reduced eye. f is front focal length, r is radius of curvature, and f' is back focal length of reduced eye

- Cornea
- Iris
- Lens
- Retina
- Foveola.

The *cornea* is a transparent part of the outer coat, with radius of curvature about 6 mm. It covers about a sixth of the eyeball, and is the only part of the eye directly open to the surrounding air. The *iris* is a coloured ring behind the cornea, with an open central *pupil*. The diameter of the pupil is changed by involuntary muscles from 8 mm when the retina is receiving dim light, to 2 mm or even 1 mm when the retina is receiving bright light. The *lens* is a transparent biconvex body surrounded by muscles which change its curvature. When the muscles are relaxed the lens curvature is at a minimum, and distant objects are brought to focus on the retina; when the muscles are fully tensed the lens curvature is at a maximum, and near objects are brought to focus. The *retina* is a multi-layered inner surface of receptor cells covering well over half of the interior of the eye. Each receptor cell registers the light falling on it as a binary neural signal, cycling 0 1 0 1 … with increasing rapidity as the intensity of light increases. The *foveola* contains only *cone* receptor cells, which are particularly sensitive to bright light and to colour. The diameter of the foveola is about 0.4 mm, subtending a visual angle of about 1.5°.[1] The foveola is surrounded by the *fovea centralis*, which contains many cone receptors and some *rod* receptors which are particularly sensitive to dim light. The diameter of the fovea is about 1.2 mm, or 4.5°. The rest of the retina contains mostly rod receptors. The whole retina contains an estimated 7 million cones and 120 million rods.

The eye is maintained in shape by the transparent liquids in front of and behind the lens. The eyeball can be voluntarily rotated in its orbit, about a centre of rotation, to direct the visual axis anywhere in a cone of at least 160° without moving the head. The *visual axis* through the centres of cornea, pupil, and lens, meets the retina at the foveola.

10.3 Geometrical Optics of Eye

The geometrical optics of the eye are complex, because refraction occurs at the corneal surface and at the front and back surfaces of the lens, varying with accommodation. A full description uses three pairs of cardinal points: two focal points, two principal points, and two nodal points. A *reduced eye* is a simplification which is sufficiently accurate for many purposes. Figure 10.2b shows Emsley's [1] 60-dioptre reduced eye, relative to the actual eye. The reduced eye is a transparent body with the refractive index of water, 1.33. It has a single spherical refracting surface at a single principal point P, with centre at a single nodal point N and radius of curvature r. There are two focal points, F and F′. Notice that unlike a simple converging lens, where the rays emerge to form an image in air, in the eye the rays remain in the denser medium so that the back focal length f' is greater than the front focal length f. The reduced eye has:

- Refractive index $n' = 1.33$.
- Front focal length FP $f = 16.67\,\text{mm}$.
- Radius of curvature NP $r = 5.55\,\text{mm}$.
- Back focal length PF′ $f' = 22.22\,\text{mm}$.

The dioptric power DP (see Sect. 9.1) of such a body is (refractive index of air)/(front focal length) $= n/f = 1/0.01667\,\text{m} = 60\,\text{dpt}$; or (refractive index of body)/(back focal length) $= n'/f' = 1.33/0.02222\,\text{m} = 60\,\text{dpt}$.

The actual eye varies its focal length by varying the radius of curvature r combining cornea and lens. When the ciliary muscles of the lens are fully relaxed the curvature is at its flattest, and an object at the *far point* is focussed on the retina. The far point is customarily set at $v = 6000\,\text{mm}$. The radius of curvature is then

$$r = (n' - n)/(n/v + n'/v') = (1.336 - 1)/(1/6000 + 1.336/24) = 6\,\text{mm}\,.$$

[1] The little-finger nail at arm's length subtends roughly 1°; the thumb joint 2°; and the flat back of hand-with-thumb 10°.

The focal lengths of the actual eye for the far point are

$$f = (n/(n' - n))r = (1/(1.336 - 1)) \times 6 = 17.8\,\text{mm}$$
$$f' = (n'/(n' - n))r = (1.336/(1.336 - 1)) \times 6 = 23.86\,\text{mm}\,.$$

That is to say, the back focus F′ is very nearly on the retina.

When the eye is fully accommodated the curvature is at its greatest, and an object at the *near point* is focussed on the retina. The near point is customarily set at $v = 250\,\text{mm}$. The radius of curvature is then

$$r = (n' - n)/(n/v + n'/v') = (1.33 - 1)/(1/250 + 1.336/24) = 5.63\,\text{mm}\,.$$

The focal lengths of the actual eye for the near point are

$$f = (n/(n' - n))r = (1/(1.336 - 1)) \times 5.63 = 16.76\,\text{mm}$$
$$f' = (n'/(n' - n))r = (1.336/(1.336 - 1)) \times 5.63 = 22.38\,\text{mm}\,.$$

10.4 Size Variables of Eye

In the eye, the lens projects a continuous inverted image of the object (as in Sect. 9.4) onto the retina. Figure 10.3 shows object ABCDE forming an image A′B′C′D′E′ on the retina of the reduced eye. In Gaussian optics a ray arriving at the front nodal point at a given angle leaves the back nodal point at the same angle. In the reduced eye the two nodal points become one, N, at the centre of curvature of the single refracting surface. So N is the centre of projection from object to image (and also, in reverse, from image to object). The retina which receives the continuous image is discrete in structure, composed of very many cones and rods very densely packed more or less hexagonally, as shown much regularized in Fig. 10.3b. In a central projection the sample count S and angular widths α and ω subtended are the same for object and image, but the linear widths a, a', w, w' and distances v, v' are different for object and image. For small angles, the difference between curvilinear and linear widths is negligible.

Table 10.1 shows the size variables of the eye. The fundamental variables are a' the diameter of a cone sampling element, v' the distance of the retina from N, and d the diameter of the pupil. If we further set the object distance v and object width w, we can calculate all the rest.

For example, Table 10.2 shows values calculated for the reduced eye, in millimetres and inches.

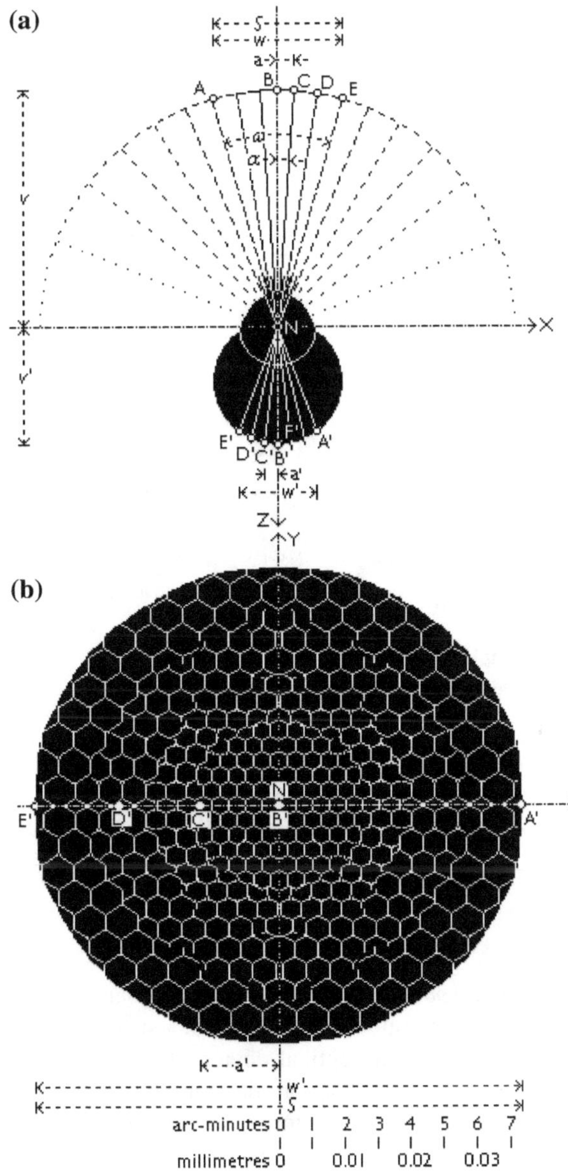

Fig. 10.3 Size variables of eye. Coordinate origin is at nodal point N; x-axis is positive to the right; y-axis is positive upwards; z-axis is positive to back of head (*right-handed set*). **a** Plan view. v is object distance; v' image distance; S count of samples; w object width; w' image width; a object sample width; a' image sample width; α angular sample width; ω angular object and image width. **b** Back elevation of (**a**) showing retinal cones

Table 10.1 Size variables for the eye, as shown in Fig. 10.3

Object			
Object sample width	a	$= 1/F_S$	Length units
Object sampling freq.	F_S	$= S/w$	Samples per unit len.
Object resolution lim.	F_{LP}	$= F_S/2$	Line-pairs per unit len.
Object distance*	v	$= v'/m$	Length units
Object width*	w	$= w'/m$	Length units
Common			
Pupil diameter*	d	$= f/n$	Length units
(Back) focal length	f	$= v'$	Length units
F-number	F/n	$n = f/d = f/d$	
Width/distance const.	k	$= w/v$	
Magnification im./ob.	m	$= v'/v$	
Count samples in width	S	$= \omega/\alpha$	Samples
Angular sample width	α	$= \arctan(a'/v')$	Angle units
Angular sampling freq.	Φ_S	$= 1/\alpha$	Samples per unit angle
Angular resolution lim.	Φ_{LP}	$= \Phi_S/2$	Line-pairs per unit angle
Angular width	ω	$= 2 \times \arctan(w/2v)$	Angle units
Image			
Image sample width*	a'	$= 1/F'_S$	Length units
Image sampling frequency	F'_S	$= S/w'$	Samples per unit len.
Image resolution limit	F'_{LP}	$= F'_S/2$	Line-pairs per unit len.
Image distance*	v'	$= m \times v$	Length units
Image width	w'	$= k \times v$	Length units

Given values for the asterisked variables, the rest can be calculated

10.5 Resolving Power of Eye

The *resolving power* or *resolution limit* of a sampling device, taking regularly-spaced samples of a continuous object, is the finest detail which can be registered accurately by the device. For an optical device, the usual measure of resolution limit is the highest-frequency of object grating which can be reliably distinguished from a uniform grey. Reliable means more than 50 % correct recognitions in repeated tests with the grating in various orientations.

A grating has alternating Black and White lines, as at BC, CD, ... in Fig. 10.3, so the finest detail is a line-pair, that is one Black line and one White line, BC and CD. At the resolution limit of the eye, the images of successive line-pairs fall on successive pairs of retinal cones, B′C′ and D′E′. Clearly, the resolution limit is reached at half the sampling frequency.[2] So for a perfect reduced eye viewing a grating the resolution

[2]Half the sampling frequency is often called the *Nyquist frequency* for Harry Theodor Nyquist (1889–1956), a founder of communication theory, at ATT, later Bell, Telephone Laboratories.

Table 10.2 Sizes calculated for the reduced eye

Object		mm	in	
Sample width	a	0.028	0.0011	Length units
Sampling freq.	F_S	36	910	Samples per unit length
Resolution limit	F_{LP}	18	455	Line-pairs per unit length
Object distance*	v	250	10	Length units
Object width*	w	200	8	Length units
Common				
Pupil diameter*	d	4	0.16	Length units
(Back) focal length	f	22.22	0.875	Length units
F-number	F/n	F/5.5	F/5.5	
Width/distance constant	k	0.80	0.80	
Magnification image/object	m	0.0895	0.0895	
Count of samples in width	S	6812	6812	Samples
Angular sample width	α	0.00645	0.00640	Degrees
Angular sampling freq.	Φ_S	156	156	Samples per degree
Angular resolution limit	Φ_{LP}	78	78	Line-pairs per degree
Angular width	ω	43.80	43.80	Degrees
Image				
Sample width*	a'	0.0025	0.0001	Length units
Sampling frequency	F'_S	400	10160	Samples per unit length
Resolution limit	F'_{LP}	200	5080	Line-pairs per unit length
Image distance*	v'	22.22	0.875	Length units
Image width	w'	17.90	0.705	Length units

The asterisked variables are given, and the rest are calculated

limit is, in angular measure:

$$\Phi_{LP} = \Phi_S/2 = 156/2 = 78 \text{ lppd}$$

and in linear measure on the grating at near distance 250 mm is

$$F_{LP} = F_S/2 \ = 36/2 = 18 \text{ line-pairs per millimetre}$$

But actual eyes are more complex physiologically than the reduced eye, and have variations and defects, congenital or age-related, so that actual resolving power differs person to person, and between left and right eyes in one person. To find the actual resolution limit for a particular eye we can do the following test:

- In a GUI construct a grating with, say, 60 line-pairs, each Black line and each White line 1 px wide, that is 2 px per line-pair, total width $w = 120$ px.
- Print the grating with line width $a = 1$ mm (see Sect. 23.7).

- Set the grating at various distances from the eye, and note the threshold distance v mm at which the grating is just resolved.
- Angular width of line $\alpha = \arctan(a/v) = \arctan(1/v)°$.
- Angular width of line-pair $= 2 \times \alpha°$.
- Angular resolution limit $\Phi_{\text{LP}} = 1/(2 \times \alpha)$ lppd.
- Sample width on grating at threshold distance $a = v \times \tan(\alpha)$ mm.
- Linear resolution limit on grating $F_{\text{LP}} = 1/2 \times a$ lppmm.

For example, if the threshold distance $v = 3000$ mm, then the angular width of a line $\alpha = \arctan(1/3000) = \arctan(0.000333) = 0.0191°$. The angular width of a line-pair $= 2 \times \alpha = 0.0382°$, and the angular resolution limit $\Phi_{\text{LP}} = 1/0.0382 = 26$ lppd. The corresponding linear sample width measured on the grating $a = 6000 \times \tan(0.0191) = 2$ mm, and the linear resolution limit on the grating $F_{\text{LP}} = 1/(2\times2) = 0.25$ lppmm.

10.6 Snellen Acuity

In ophthalmology, resolution or *separation acuity* is measured as the smallest letters which can be read on a *Snellen chart*.[3] Figure 10.4a shows a typical Snellen chart, reduced to about quarter-size.

Snellen proposed that an eye is normal, needing no correction, if it can resolve angular sample width $\alpha =$ one arc-minute, that is angular sampling frequency $\Phi_{\text{S}} = 60$ spd $=$ angular resolution limit $\Phi_{\text{LP}} = 30$ lppd. He designed letters with strokes to subtend $\alpha =$ one arc-minute when viewed from different distances, from 60 m down to 3 m. In Fig. 10.4b we see that $\tan(\alpha) = a/v$. So when $\alpha = 1/60°$ and $v = 6000$ mm, stroke width $a = v \times \tan(\alpha) = 6000 \times \tan(1/60) = 6000 \times 0.000291 = 1.74$ mm. Table 10.3 shows the calculated stroke widths a for viewing distances v. Each letter occupies a square of 5×5 stroke-widths a, as shown in Fig. 10.5. To make a chart for viewing distance 6 m (or 20 ft), arrange a row of $v = 6000$ mm, $a = 1.74$ mm letters, and mark it 6/6 (or 20/20). Then arrange a row of $v = 7500$ mm, $a = 2.18$ mm letters, and mark it 6/7.5 (or 20/25). And so on, up to 6/60 (or 20/200), and down to 6/3 (or 20/10).

A normal eye can read row 6/6 at 6 m, and has Snellen acuity 6/6. A somewhat defective eye can read row 6/7.5 at 6 m, but not row 6/6, so has Snellen acuity 6/7.5. An eye which can read only the top row (with spectacles), subtending one arc-minute at 60 m, has Snellen acuity 6/60, which counts as legally blind. The best eyes can exceed Snellen acuity 6/3, approaching the calculated limit of a geometrically-perfect reduced eye, Snellen acuity 6/2.6.

Table 10.4 shows the size variables of the normal eye, Snellen 6/6. Here the fundamental values are Φ_S the one-minute angular sampling frequency, v' the distance of the retina from N, and d the diameter of the pupil. If we further set the object

[3] Hermann Snellen (1834–1908), Professor of Ophthalmology, University of Utrecht, and pupil of Franciscus Cornelis Donders (1818–1889), originator of acuity measurement.

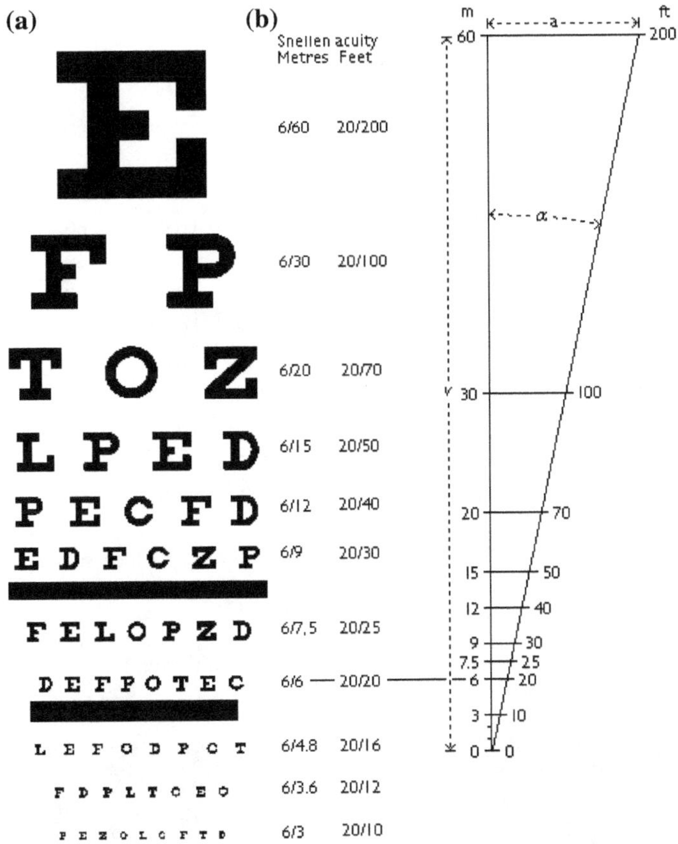

Fig. 10.4 **a** Snellen chart, shown here about quarter-size. A standard chart has the top E 87 mm square, and is set at distance 6 m (or 20 feet). **b** Construction of chart. a is width of letter stroke, in millimetres; v is nominal viewing distance, in metres (or feet), at which a subtends one arc-minute

distance v and object width w, all the rest are calculated. So for a Snellen 6/6 normal eye viewing a grating at near distance 250 mm, the resolution limit is, in angular measure:

$$\Phi_{\mathrm{LP}} = 30\,\mathrm{lppd}$$

and, in linear measure on the grating, is

$$F_{\mathrm{LP}} = 6.5\,\mathrm{lppmm}\ (=166\,\mathrm{lppi})\,.$$

And for a Snellen 6/3 very good eye the resolution limit is, in angular measure:

$$\Phi_{\mathrm{LP}} = 30\,\mathrm{lppd}$$

Table 10.3 Calculation of Snellen chart

$SnAc$ m	$SnAc$ ft	v mm	a mm	F_S spmm	F_{LP} lppmm	α deg.	Φ_S spd	Φ_{LP} lppd
6/60	20/200	60000	17.4	0.057	0.029	0.017	60	30
6/30	20/100	30000	8.73	0.114	0.057	0.017	60	30
6/20	20/70	20000	5.82	0.172	0.086	0.017	60	30
6/15	20/50	15000	4.36	0.229	0.115	0.017	60	30
6/12	20/40	12000	3.49	0.286	0.143	0.017	60	30
6/9	20/30	9000	2.62	0.382	0.191	0.017	60	30
6/7.5	20/25	7500	2.18	0.458	0.229	0.017	60	30
6/6	20/20	6000	1.74	0.573	0.286	0.017	60	30
6/4.8	20/16	4800	1.40	0.716	0.358	0.017	60	30
6/3.6	20/12	3600	1.05	0.955	0.477	0.017	60	30
6/3	20/10	3000	0.87	1.145	0.573	0.017	60	30
6/2.6	20/7.7	2308	0.67	1.489	0.745	0.017	60	30

$SnAc$ m is Snellen acuity in metres; $SnAc$ ft is Snellen acuity in feet; v is nominal viewing distance in millimetres; $a = v \times \tan(1/60)$ is linear width of stroke or gap which subtends 1 arcmin at distance v, in mm; $F_S = 1/a$ is sampling frequency, in samples per mm; $\alpha = \arctan(a/v)$ is angular sample width, in samples per degree; $F_{LP} = F_S/2$ is linear resolution limit, in line-pairs per mm; $\Phi_S = 1/\alpha$ is angular sampling frequency, in samples per degree; and $\Phi_{LP} = \Phi_S/2$ is angular resolution limit, in line-pairs per degree. The separated bottom line is the Snellen acuity calculated for a geometrically perfect eye

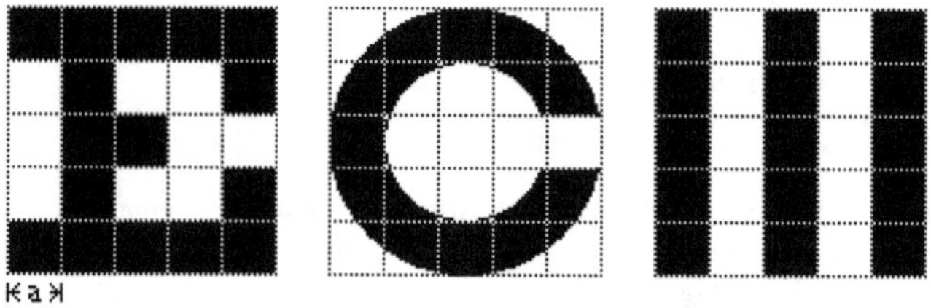

Fig. 10.5 Snellen E, Landolt C, and grating. For a given viewing distance v, a small square (*dotted*) width a subtends 1 arc-minute at the eye, a letter subtends 5 arc-minutes, and a line-pair subtends 2 arc-minutes

and, in linear measure on the grating, is

$$F_{LP} = 13.1\,\text{lppmm}\ (=334\,\text{lppi})\,.$$

A Snellen chart can also be made using a *Landolt C*[4]: a ring with outer diameter five units, inner diameter three units, and a gap one unit wide, as shown in Fig. 10.5, presented with the gap in four or eight different orientations. A person has to detect

[4]Edmund Landolt (1846–1926), private clinic, Paris. Worked with Helmholtz, Donders, and Snellen. Introduced ring test 1888, now the preferred standard.

Table 10.4 Sizes calculated for normal eye, Snellen acuity 6/6, $\Phi_S = 60$ spd, and for very good eye, Snellen acuity 6/3, $\Phi_S = 120$ spd

		6/6	6/6	6/3	6/3	
Object		mm	in	mm	in	
Sample width	a	0.073	0.0029	0.038	0.0015	lu
Sampling freq.	F_S	13.1	333	26.3	667	spul
Resolution limit	F_{LP}	6.5	166	13.1	334	lppul
Object distance*	v	250	10	250	10	Lu
Object width*	w	200	8	200	8	Lu
Common						
Pupil diameter*	d	4	0.16	4	0.16	lu
(Back) focal length	f	22.38	0.88	22.38	0.88	lu
F-number	F/n	F/5.6	F/5.6	F/5.6	F/5.6	
Width/distance constant	k	0.80	0.80	0.80	0.80	
Magnification image/object	m	0.0895	0.895	0.0895	0.0895	
Count of samples in width	S	2616	2616	5232	5232	Samples
Angular sample width	α	0.0167	0.0167	0.0083	0.0083	Degrees
Angular sampling freq.*	Φ_S	60	60	120	120	spd
Angular resolution limit	Φ_{LP}	30	30	60	60	lppd
Angular width	ω	43.60	43.60	43.60	43.60	Degrees
Image						
Sample width	a'	0.0068	0.0003	0.0043	0.0002	lu
Sampling frequency	F'_S	147	3734	233	5907	spul
Resolution limit	F'_{LP}	73.5	1868	116	2953	lppul
Image distance*	v'	22.38	0.88	22.38	0.88	lu
Image width	w'	17.90	0.705	17.90	0.705	lu

The asterisked variables are given, and the rest are calculated (compare Tables 10.1 and 10.2)

the position of the gap. The ring avoids the variation in recognisability of Snellen letters, and is usable for people with non-Latin alphabets, and for illiterates.

Or again, a chart can be made with gratings. In Table 10.3, the Snellen row with nominal viewing distance v has line width $a = v \times \tan(1/60) = v \times 0.000291$ mm. So a grating with this line-width measures the same acuity as the Snellen row. For example, the 6/60 top row has a grating with $a = 60,000 \times 0.000291 = 17.45$ mm; the 6/6 normal row has a grating with $a = 6000 \times 0.000291 = 1.74$ mm; and the 6/3 best row has a grating with $a = 3000 \times 0.000291 = 0.87$ mm.

Figure 10.6 shows a E. Snellen, a C. Landolt, and a grating as imaged on the retinal cones at normal one-arc-minute and very good 0.5-arc-minute acuities.

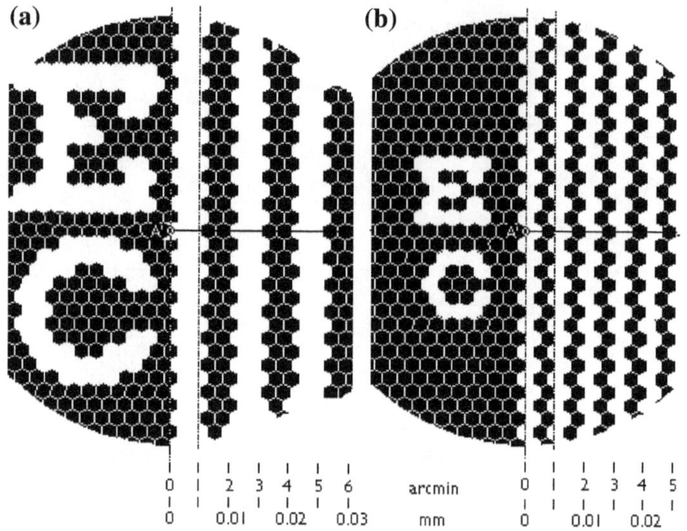

Fig. 10.6 Standard and very good acuity. **a** Image of Snellen E and Landolt C for acuity 6/6 m = 20/20 ft, and grating with frequency $F_{\mathrm{LP}} = 30$ lppd, projected on the foveal cones. **b** Image of E. Snellen and C. Landolt, acuity 6/3 m = 20/10 ft, and grating with frequency $F_{\mathrm{LP}} = 60$ lppd, projected on the foveal cones

10.7 Vernier Acuity

Besides separation acuity, which is measurable by looking at a grating, the eye has remarkable *Vernier acuity*.[5] This is measured as the angular width of the smallest detectable offset between two segments of a line. For example, Fig. 10.7 shows a Vernier acuity of 0.25 arc minute, or half a cone diameter. Under good conditions, the actual Vernier limit of the eye is about 0.13 arc minute. It is thought that this is achieved by processing combinations of cone signals.

10.8 Diffraction Limit of Eye

When diffraction is taken into account, any lens has a diffraction limit of resolution, as described in Sect. 9.8, and shown again in Fig. 10.8a. Table 10.5 shows values for a supposed perfect eye lens, with light of wavelength 550 nm (to which the eye is most

[5]Pierre Vernier (1580–1637), mathematician, surveyor and engineer in government service in Hapsburg Burgundy. Invented secondary scale for subdividing primary scale of measuring instruments, based on alignment of two lines.

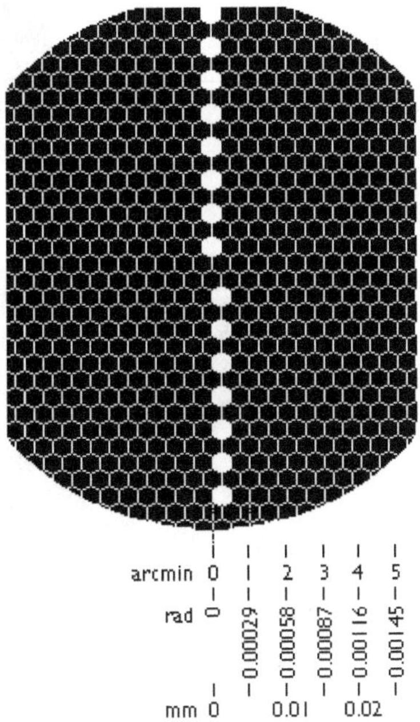

Fig. 10.7 Vernier acuity. The angular offset of the line segments is 0.25 arc-minute, or half a cone diameter. A good eye can do better than this

sensitive), at pupil diameters $d = 1, 2, 4, 6$, and 8 mm. For very bright light, when the pupil diameter is 2 mm or less, the resolution of the eye is diffraction-limited. When the pupil diameter is 3 mm or more, the resolution is cone-limited, as shown in Fig. 10.8b.

10.9 Corrective Converging Lens

The standard near point of the eye is usually taken to be 250 mm or ten inches. The fully-accommodated eye of a young child will focus objects at a near point of perhaps 200 mm or eight inches. With increasing age, the ciliary muscles become weaker and the ligaments slacker, so that the near point becomes gradually farther away. Reading-glasses with converging lenses can be used to correct such presbyopic eyes, as shown in Fig. 10.9. In Fig. 10.9a object AB is at the large distance v at which the fully-accommodated eye forms a real image $A'B'$ on the retina. In Fig. 10.9b object AB is at the standard near distance $v = 250$ mm: the eye cannot accommodate any more, so the image is formed behind the retina, and looks blurred. In Fig. 10.9c a converging lens is placed in front of the eye, so the focal length of the lens-plus-eye is shortened, and the real image again forms on the retina in sharp focus.

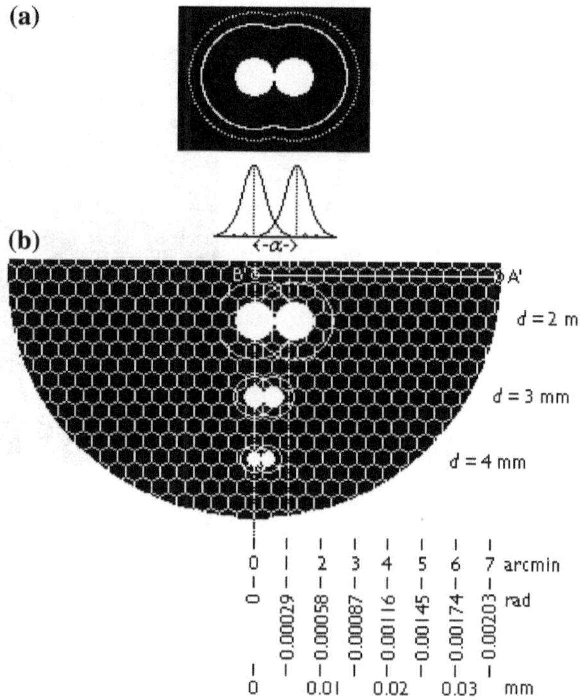

Fig. 10.8 Diffraction limit of eye. **a** Airy disks of two point objects, with their intensity profiles. The angular radius from centre to first minimum is α. Two point objects are just resolvable when the separation is α. **b** Point images on the retina when the pupil has diameter $d = 2$, 3, and 4 mm

Table 10.5 Airy disk in image of point object at various pupil diameters

d mm	α rad	α min
1	0.00067	2.31
2	0.00033	1.16
3	0.00023	0.77
4	0.00017	0.58
5	0.00014	0.46
6	0.00011	0.39
7	0.000097	0.33
8	0.000085	0.29

d mm is pupil diameter in mm; α is angular radius of central disk on retina in radians and in arc-minutes

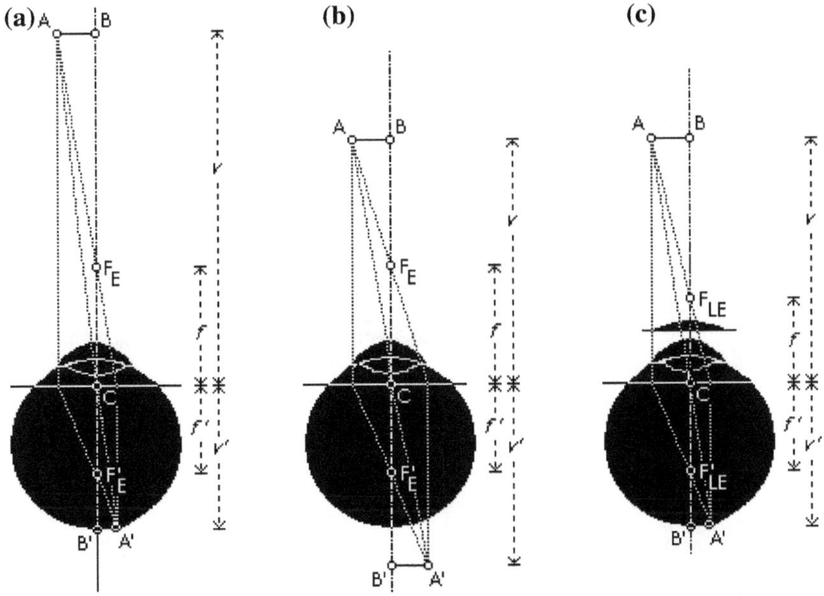

Fig. 10.9 Corrective converging lens. **a** Presbyopic eye fully-accommodated has front focal point F_E, and forms a sharp image on retina when object AB is at distance v, say 500 mm (not to scale). **b** When object AB is brought to distance $v = 250$ mm, image is formed behind retina, hence blurred. **c** Converging lens brings front focal point of lens-plus-eye combination F_{LE} closer, and so again forms sharp image on retina

10.10 Virtual Image

When an object AB is at distance v equal to or less than the focal length of a converging lens f, the lens forms a *virtual* image. A virtual image is on the same side of the lens as the object, and upright; but it is formed by non-converging rays, and so cannot be thrown onto a screen. Figure 10.10 shows ray-traces for two cases:

(a) When object distance v = focallength f, image distance v' is infinite (rays coming from the focal point emerge parallel to the axis).
(b) When object distance v is less than focal length f, image distance v' is greater than focal length.

Given any two of v, v', and f, we may calculate the third from the Gaussian lens equation:

$$1/v + 1/v' = 1/f \text{length units.}$$

For example:

(a) Given $v = f = 133$ mm, $1/v' = 1/f - 1/v = 1/133 - 1/133 = 0$; hence $v' = 1/0 = \infty$.

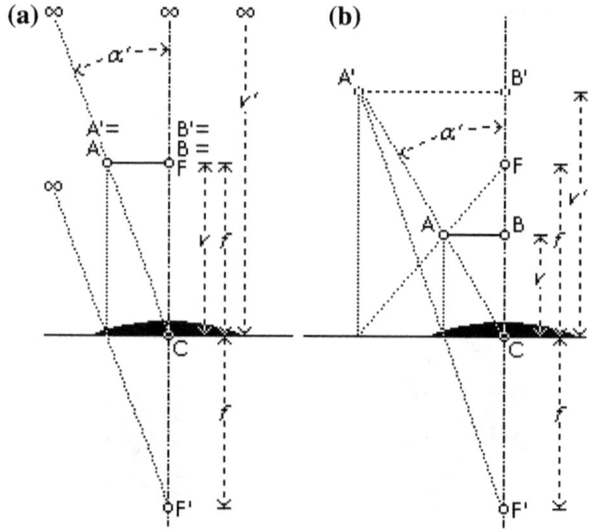

Fig. 10.10 Formation of virtual image. **a** When object AB is at front focal point of lens F, virtual image A′B′ is upright at infinity. **b** When object AB is closer than F, virtual image A′B′ is upright beyond F

(b) Given $v = 100\,\text{mm}$ and $f = 133\,\text{mm}$, $1/v' = 1/f - 1/v = 1/133 - 1/100 = 0.0075 - 0.01 = -0.0025$; hence $v' = 1/-0.0025 = -400\,\text{mm}$.

Virtual image distance turns out negative because it is on the same side of the lens as the object.

The linear width of a virtual image is not easily measured. However, the *angular width* α' of a virtual image A′B′ is that subtended at lens centre C:

$$\alpha' = \arctan(w/v)\ \text{degrees}\,,$$

where w is the linear width of the object AB and v is the object distance. For example, given $w = 1$ mm in Fig. 10.10, we can find the angular width of the image α':

(a) $\alpha' = \arctan(1/133) = 0.43°$.
(b) $\alpha' = \arctan(1/100) = 0.57°$.

10.11 Simple Magnifier

A converging lens placed in front of the eye acts as a simple magnifier. There are two ways to use a simple magnifier. The first way, shown in Fig. 10.11b, is to set the object AB at distance v = focal length of lens f. The virtual image A′B′ is then at distance v' = infinity, and is viewed from about distance f by the fully-relaxed eye. The *angular magnification m* is the ratio of the angular width α' of the virtual image (with the lens), to the angular width α of the object at the standard near distance 250 mm (without the lens). We have:

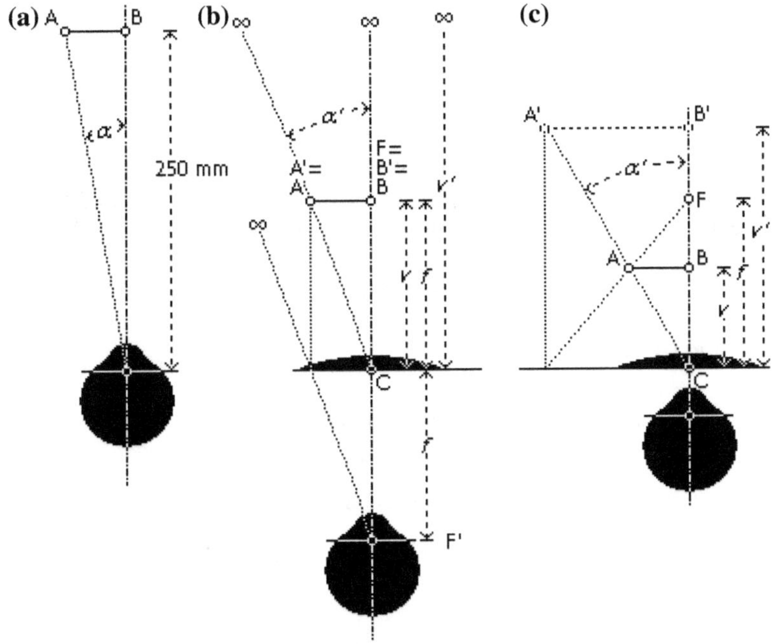

Fig. 10.11 Simple magnifier. **a** Reference angular width α of object AB at standard near distance 250 mm viewed by fully-accommodated eye. **b** Object AB at $v = f$, image A′B′ at $v' =$ infinity, viewed by fully-relaxed eye. **c** Object AB at v less than f, image A′B′ at v' greater than f, viewed by eye close to lens

$$\begin{aligned}\alpha &= \arctan(w/250)^\circ \\ \alpha' &= \arctan(w/f)^\circ \\ m &= \alpha'/\alpha = 250/f \approx DP/4\,,\end{aligned}$$

where w is the width of the object in mm, f is the focal length of the lens in mm, and DP is the dioptric power of the lens in dioptres. For example, given object width $w = 1$ mm, and converging lens focal length $f = 133$ mm, $DP = 7.5$ dpt:

$$\begin{aligned}\alpha &= \arctan(1/250) = \arctan(0.004) = 0.23^\circ \\ \alpha' &= \arctan(1/133) = \arctan(0.0075) = 0.43^\circ \\ m &= 0.43/0.23 = 1.87 = 250/133 \approx DP/4 = 7.5/4\,.\end{aligned}$$

The second way, shown in Fig. 10.11c, is to set the object AB at distance v less than f, so that virtual image A′B′ is formed at standard near distance $v' = 250$ mm. The image is viewed by the accommodated eye close to the lens. We then have

$$\alpha = \arctan(w/250)^\circ$$
$$\alpha' = \arctan(w/v')^\circ$$
$$m = \alpha'/\alpha = (250/f) + 1 \approx (DP/4) + 1\,,$$

where w is the width of the object in mm, v' is the image distance, f is the focal length of the lens in mm, and DP is the dioptric power of the lens in dioptres. For example, given object width $w = 1$ mm, and converging lens focal length $f = 133$ mm:

$$1/v = 1/250 + 1/133 = 0.004 + 0.0075 = 0.0115\,;$$

hence $v = 1/0.0115 = 86.8$ mm,

$$\alpha = \arctan(1/250) = \arctan(0.004) = 0.23^\circ$$
$$\alpha' = \arctan(1/86.8) = \arctan(0.0115) = 0.66^\circ$$
$$m = 0.66/0.23 = 2.87 = (250/133) + 1 \approx (DP/4) + 1\,.$$

10.12 Compound Magnifier

A compound converging lens can be used as a powerful magnifier, or *loupe*, close to the eye. The compound lens has a focal length f found as in Sect. 9.7:

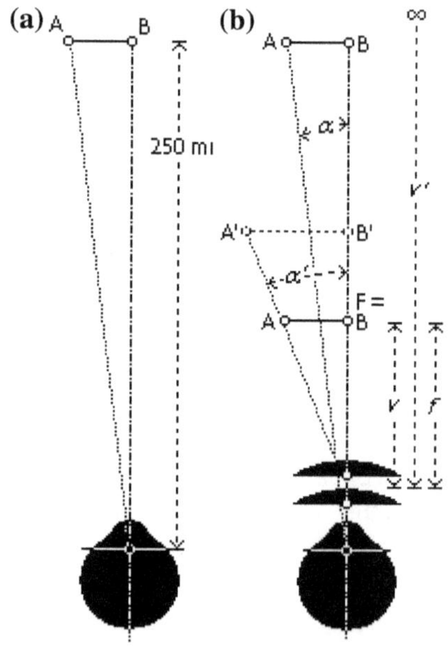

Fig. 10.12 Compound magnifier. **a** Reference angular width α of object AB at conventional near distance $v = 250$ mm. **b** Compound lens has focal length f. Its object AB is at distance v, and its virtual image A′B′ is at $v' =$ infinity. The angular magnification is α'/α

- Focal length of compound $f = (f_1 f_2)/(f_1 + f_2 - d)$,
- Distance of principal plane 1 $p_1 = (fd)/f_2$
- Distance of principal plane 2 $p_2 = (fd)/f_1$,

where d is the distance between the lenses. Figure 10.12b shows object AB set at distance $v =$ focal length f; then virtual image A′B′ is at distance $v' =$ infinity (that is, rays from any point at object distance emerge parallel to the axis).

For example:

- Given lens 1 focal length $f_1 = 180\,\text{mm} = 5.55\,\text{dpt}$,
- and given lens 2 focal length $f_2 = 120\,\text{mm} = 8.33\,\text{dpt}$,
- and given distance between lenses $d = 20\,\text{mm}$.
- Find compound focal length $f = (f_1 f_2)/(f_1 + f_2 - d) = (180 \times 120)/(180 + 120 - 20) = 77\,\text{mm} = 12.96\,\text{dpt}$.
- Set object AB at distance $v = f = 77\,\text{mm}$.
- Find virtual image distance v' from $1/v + 1/v' = 1/f$, that is $1/v' = 1/83 - 1/83$, so $v' =$ infinity.
- Find angular magnification $m = \alpha'/\alpha = 250/f = 250/77 = 3.25 \approx DP/4 = 12.96/4$.

Reference

1. Emsley HH (1950) Visual Optics. Hatton Press, London

Chapter 11
Stereo Projection

Paired eyes capture a pair of slightly-different internal images of a fixated object. We can make a stereo-pair of external images by projecting an object via a pair of centres which are laterally-separated like the eyes. A stereo-pair can be constructed by geometry, or by calculation, or by paired cameras, or even out of nothing by random dots. Examples of each method are given.

11.1 Stereoscopy

The principle of stereoscopy, discovered by Wheatstone[1] before 1833 (as noted in [5]) and demonstrated in 1838 [7], is to present to the two eyes two slightly different flat images of one and the same solid object, invoking binocular depth perception (see Chap. 12). The theory and practice of stereoscopy was notably developed by Brewster[2] [2].

A *stereo-pair* of images, left and right, is a pair of projections from an object to a plane, using two centres laterally separated by the inter-pupillary distance of the eyes. A stereo-pair can be constructed by geometry, or computed from numerical data, or assembled from two camera images, or even by manipulating a random-dot image. A stereo-pair can be viewed with or without various devices, as described in Chap. 12.

[1] Sir Charles Wheatstone (1802–1875), Professor of Experimental Philosophy at King's College, London.

[2] Sir David Brewster (1781–1868), Principal of Edinburgh University, many discoveries in optics, inventor of the kaleidoscope in 1819 and the de-centred-lens stereoscope in 1849, prolific and combative writer.

A. Parkin, *Digital Imaging Primer*, DOI 10.1007/978-3-540-85619-1_11

11.2 Stereo Projection by Geometry

Given an orthographic plan and elevation of an object, we can construct a stereo-pair by two central projections class III in the scheme of Sect. 8.2, as shown in Figs. 11.1 and 11.2. Geometrically, to construct a stereo-pair first fix reasonable values for three variables:

- *Separation s* of centres C_L and C_R,
- *Image distance* v' from centres,
- (Single) *Image width* w'.

Separation s is usually set as the inter-pupillary distance of the eyes, say 62 mm (varying from person to person between 60 and 64 mm). A stereo-pair where s is much greater than 64 mm exaggerates depths when viewed with a suitable device. Image distance v' is usually set as the near distance for reading, about 250 mm. Stereoscopic effects are not very sensitive to small differences in these two settings. A stereo-pair for direct viewing should have single image width w' not greater than separation s. For viewing with a common Brewster stereoscope the standard image width is 75 mm.

Five more variables are

- *Object distance* v from centres,
- (Single) *object field width* w,
- *Magnification* m of image relative to object,
- *Binocular overlap width* b,
- *Angle of view* ϕ,

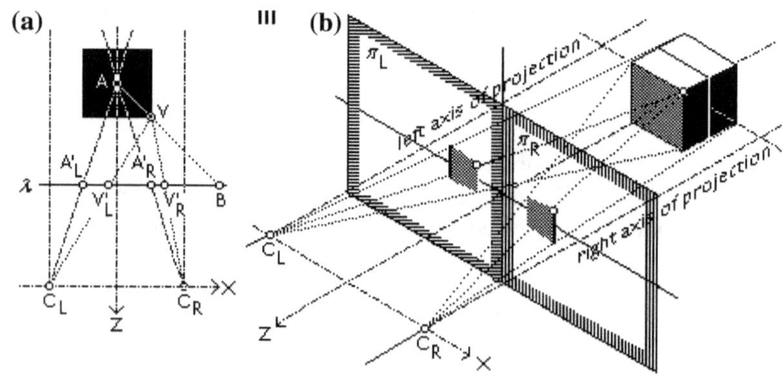

Fig. 11.1 Construction and general arrangement of stereo-pair, projection class III. **a** Plane projective-geometric construction of two images from given figure. Projection fixed by two centres C_L and C_R, line λ, and corresponding points A and A'_L, and A and A_R. V is a given vertex of the figure. Join AV, to meet λ on B; join A′B; join C_LV to meet A′B on V'_L, the *left image* of V; join C_RV to meet A′B on V'_R, the *right image* of V. **b** Space projection of stereo-pair π_L and π_R from given object

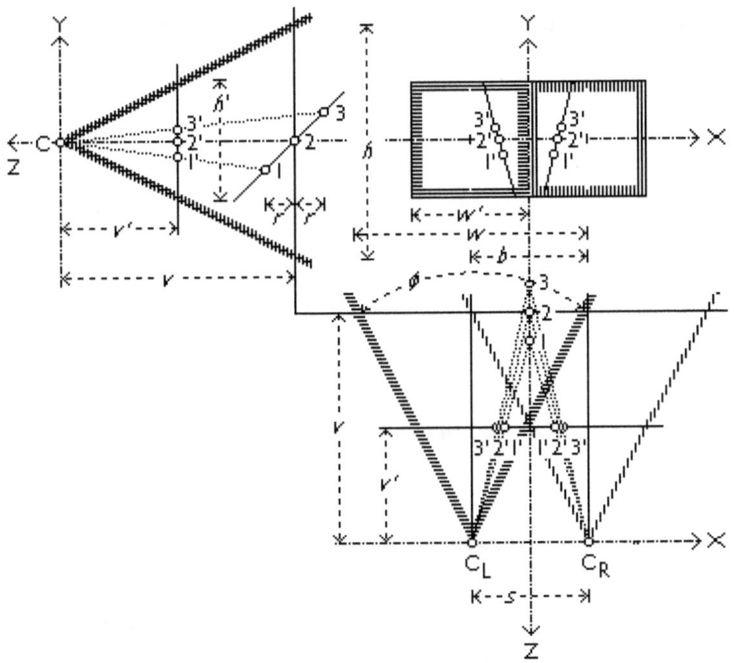

Fig. 11.2 Stereo projection by geometry. **a** Orthographic right-side elevation in yz-plane. **b** Orthographic front elevation in xy-plane. **c** Orthographic plan in xz-plane. Objects are three blobs 1 2 3, at distance v. Left and right central projections are from centres C_L and C_R to stereo-pair of images on image-plane at distance v'

related as

$$\begin{aligned}
m &= v'/v = w'/w \\
w &= v \times 2 \times \tan(\phi/2) \\
b &= w - s \\
\phi &= 2 \times \arctan((w/2)/v) = 2 \times \arctan((w'/2)/v')\,.
\end{aligned}$$

So, given s, v' and w', we may set any one of v, w, m, b or ϕ, and calculate the remaining four. For example, suppose we set object distance $v = 500$ mm. Then

$$\begin{aligned}
m &= v'/v = 250/500 = 0.25 \\
w &= w'/m = 62/0.25 = 248\,\text{mm} \\
b &= w - s = 248 - 62 = 188\,\text{mm} \\
\phi &= 2 \times \arctan((w/2)/v) = 2 \times \arctan(124/500) = 28^\circ\,.
\end{aligned}$$

Or suppose we set binocular overlap width $b = 150\,\text{mm}$. Then

$$\begin{aligned} w &= b + s = 150 + 62 = 212\,\text{mm} \\ m &= w'/w = 62/212 = 0.29 \\ v &= v'/m = 250/0.29 = 855\,\text{mm} \\ \phi &= 2 \times \arctan((w/2)/v) = 2 \times \arctan(106/855) = 14^\circ\,. \end{aligned}$$

Image height h' can be set to whatever we please; we then have image aspect ratio w'/h' and object field height h, where $w/h = w'/h'$.

An alternative geometric method is to place the object first on the left axis of projection, rotated about its y-axis by angle $+\rho$, then on the right axis of projection rotated by angle $-\rho$, where $\rho = \arctan(s/2v)$. For example, if $s = 62\,\text{mm}$ and $v = 500\,\text{mm}$, $\rho = 3.5^\circ$. The axes of projections from C_L and C_R remain parallel.

11.3 Stereoscopic Difference

Helmholtz defines [3] *stereoscopic difference* as the distance d between left and right images of the same object in a stereo-pair. One way to calculate this is shown in Fig. 11.3. In similar triangles V C_L C_R and V V'_L V'_R we have

$$s/v = d/(v - v')$$

where s is the separation of centres C_L and C_R, v is the distance of object V, v' is the distance between centre and image, and d is the distance between images V'_L and V'_R. Hence

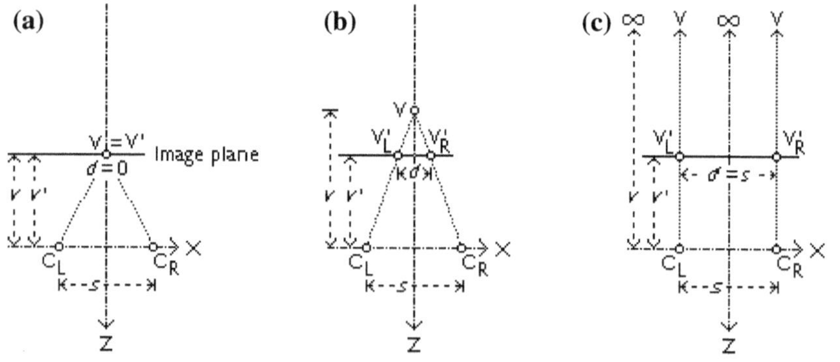

Fig. 11.3 Stereoscopic difference. Centres of projection C_L C_R are separated by distance s. Object V is distance v from centres. Images V_L and V_R are distance v' from centres. Stereoscopic difference is d. **a** Object V on image-plane: $v = v'$ and $d = 0$. **b** $v > v'$ and $0 < d < s$. **c** Object V at ∞, and $d = s$

Table 11.1 Stereoscopic difference

s (mm)	v' (mm)	v (mm)	d (mm)
62	250	250	0
62	250	500	31
62	250	750	41.3
62	250	1000	46.5
62	250	∞	62

Given constant separation s and image distance v', and varying object distance v, calculate stereoscopic difference $d = s \times ((v - v')/v)$

$$
\begin{aligned}
d &= s \times (v - v')/v \\
s &= v \times d/(v - v') \\
v' &= v - ((d \times v)/s)) \\
v &= -(s \times v')/(d - s)\,.
\end{aligned}
$$

All objects at the same distance v have the same stereoscopic difference d. When an object is itself on the image plane, object distance v equals image distance v', and stereoscopic difference $d = 0$. As the object recedes, stereoscopic difference increases. As the object approaches infinity, stereoscopic difference approaches separation s. For example, Table 11.1 shows the stereoscopic differences for five object distances.

11.4 Stereo Projection by Computation

Computationally, the geometric construction can be matched by using a perspective projection program (as described in Sect. 8.18) twice, once for the left image and once for the right image. Figure 11.4 shows a processing flow for a computed stereo-pair. To compute a stereo-pair, a procedure is

- First create a data file, which specifies a space graph (wire-frame object) as vertex-coordinates and edges (as in Sect. 7.4).
- Calculate any required rotations of object about x-axis and/or y-axis so that the z-axis is the desired direction of view (as in Sect. 7.8).
- Set separation s, image distance v', and image width w' (as in Sect. 11.2 above).
- Set any one of object distance v, object width w, magnification m, binocular overlap b, or angle of view ϕ, and calculate the other four (as in Sect. 11.2 above).
- Run program for left image, to read object data, rotate object, translate rotated object by $+s/2$ along x-axis, project perspective image, convert image to first quadrant, and output image as a bitmap file.

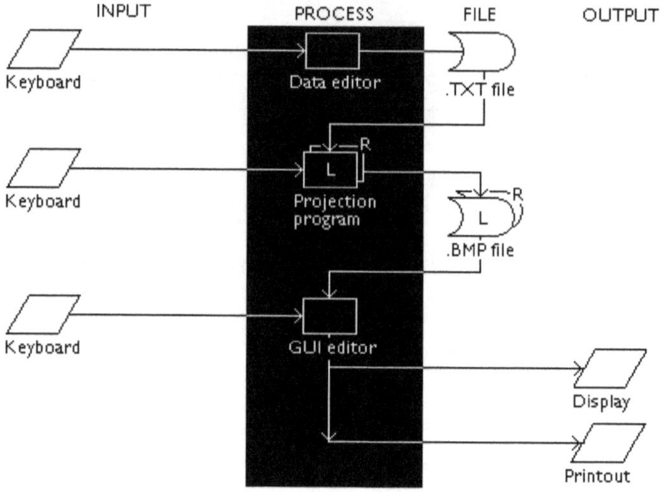

Fig. 11.4 Processing flow for computed stereo-pair. User writes a data file specifying a space graph (*wire-frame*) object in suitable format. Projection program runs twice, once for left image and once for right image. At each run, program reads data file; accepts input of stereo projection values; does perspective projection; and writes image to file in .BMP format. GUI editor combines images side by side, for display or printout

- Run program for right image, to read object data, rotate object, translate rotated object by $-s/2$ along x-axis, project perspective image, convert image to first quadrant, and output image as a bitmap file.
- Transfer both images to an image editor and abut right image to left image, to display the stereo-pair for uncrossed viewing, or abut left image to right image for crossed viewing.

The measurements here are made in (integer) image pixels at the pitch intended for viewing the images.[3]

For example, consider an object of seven vertices 1, 2, ..., 7 and five edges, shown in Fig. 11.5. The object is specified in the following data file SGSQ.TXT, where the measurements are in px, pitch 3.58 ppmm:

```
8
5
0,0,0,0
1,0,-83,42
2,0,0,0
3,0,111,-42
4,111,111,0
5,-111,111,0
```

[3]To find the pixel-pitch of an editor screen, display a line (say) 100 px long; if this measures n mm on the screen, the pitch is $n/100$ ppmm. Or, for a printer, print and measure a 100-px line.

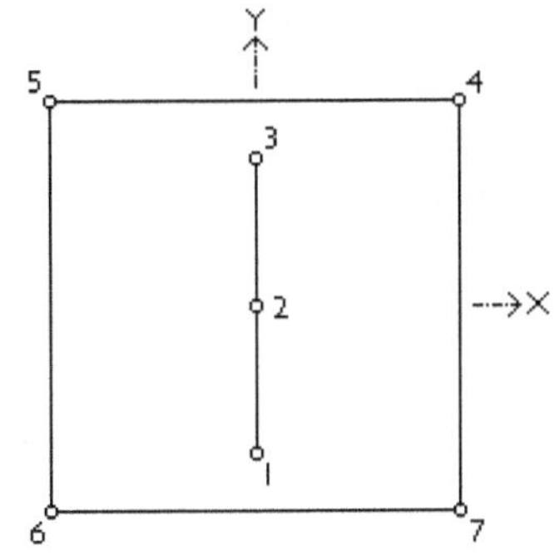

Fig. 11.5 Space graph of object. The square 4567 and blob 2 are on the xy-plane. Blob 1 is out in front (positive z), and blob 3 is out behind (negative z)

```
6,-111,-111,0
7,111,-111,0
0,1,3
1,4,5
2,5,6
3,6,7
4,7,4
```

Set values for a computed stereo-pair like in Sect. 11.2, with object distance $v =$ 500 mm, and pixels = 3.58× mm:

- Separation $s = 62\,\text{mm} = 222\,\text{px}$,
- Image distance $v' = 250\,\text{mm} = 895\,\text{px}$,
- Image width $w' = s = 62\,\text{mm} = 222\,\text{px}$,
- Object distance $v = 500\,\text{mm} = 1790\,\text{px}$,
- Magnification $m = v'/v = 250/500 = 0.5$,
- Object width $w = w'/m = 62/0.5 = 124\,\text{mm} = 444\,\text{px}$,
- Binocular overlap $b = w - s = 124 - 62 = 62\,\text{mm} = 222\,\text{px}$,
- $\phi = 2 \times \arctan((w/2)/v) = 2 \times \arctan(124/500) = 28°$.

The settings for program PROJECTI used in these examples are shown in Table 11.2. Figure 11.6 shows two stereo-pairs computed by program PROJECTI (see Sect. 8.18), from data file SGSQ.TXT. When viewed crossed (see Chap. 12), the square and the middle blob establish the reference depth, the lower blob stands out in front, and the upper blob falls back behind. Figure 11.7 shows the same with some 4-bit shading added.

For another example, a cube object is specified in the following data file SGCU.TXT.

```
8
12
0,111,111,111
1,-111,111,111
2,-111,-111,111
```

Table 11.2 Settings for program PROJECTI used in examples Figs. 11.6, 11.7 and 11.8

Parameter	Fig. 11.6a	Fig. 11.6b	Fig. 11.8	Fig. 11.9a	Fig. 11.9b	Fig. 11.9c
Data file	SGSQ	SGSQ	SGCU	SGCU	SGCU	SGCU
Scaling	N	N	N	0.5	0.5	0.5
Rotation X deg.	N	22.5	22.5	0	0	0
Rotation Y deg.	N	22.5	22.5	±3.5	±3.5	−41.5, − 48.5
Translation px	±111	±111	±111	N	N	N
Shear about Y	N	N	N	N	N	1Z
Orth/Iso/Pers	P	P	P	O	I	O
Object dist.px	1790	1790	1790	–	–	–
Image dist.px	895	895	895	–	–	–

```
3,111,-111,111
4,111,111,-111
5,-111,111,-111
6,-111,-111,-111
7,111,-111,-111
0,0,1
1,1,2
2,2,3
3,3,0
4,0,4
5,1,5
6,2,6
7,3,7
8,4,5
9,5,6
10,6,7
11,7,4
```

Figure 11.8 shows a stereo-pair computed by program PROJECTI from data file SGCU.TXT.

11.5 Parallel Stereo Projection

The usual way to compute a stereo-pair is by two perspective projections with laterally-separated centres. However, two parallel projections with laterally-separated centres will also produce a stereo-pair, as if the object is seen from a very great distance—an extreme "tele-photo effect".

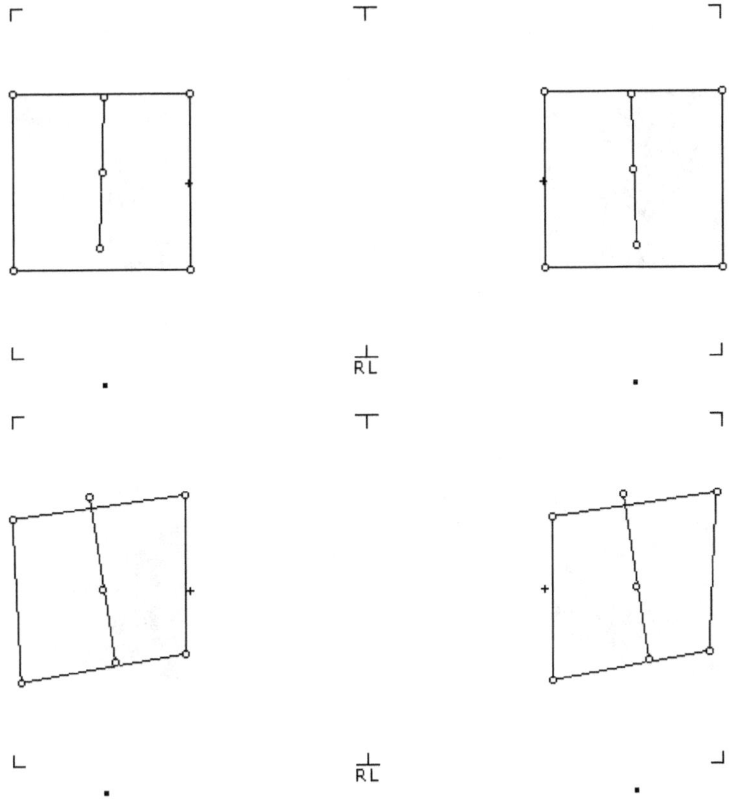

Fig. 11.6 Stereo-pairs computed by program PROJECTI from data file SGSQ.TXT. Image marked R is intended for the right eye, and placed on the left for crossed viewing. Cross the eyes so that the two dots below the pair appear as three: one sharp between two fuzzy. **a** Object not rotated. **b** Object rotated by 22.5° about x and 22.5° about y. The small crosses within the pair are actually $s = 62.4$ mm apart: this printed reproduction is slightly reduced

For example, Fig. 11.9 shows stereo-pairs using (a) orthographic, (b) isometric, and (c) military projections as described in Chap. 8. The object is the same cube as in the previous examples. In a parallel projection, the magnification m is necessarily 1, that is, same-size, so there is no perspective diminution. Here the program has first diminished the object to half-size, to get an image comparable with the perspective examples. Instead of translating one object right and left, here two identical objects are placed on the parallel left and right axes, turned by $\pm \arctan(62/750) = 4.7°$. In viewing, the small crosses merge as one.

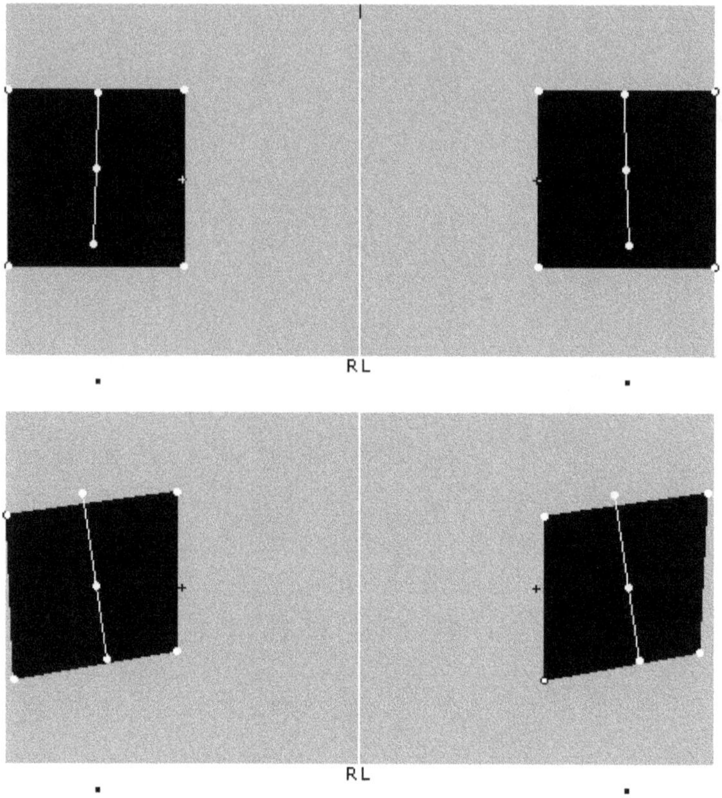

Fig. 11.7 As Fig. 11.6 in 4-bit mode

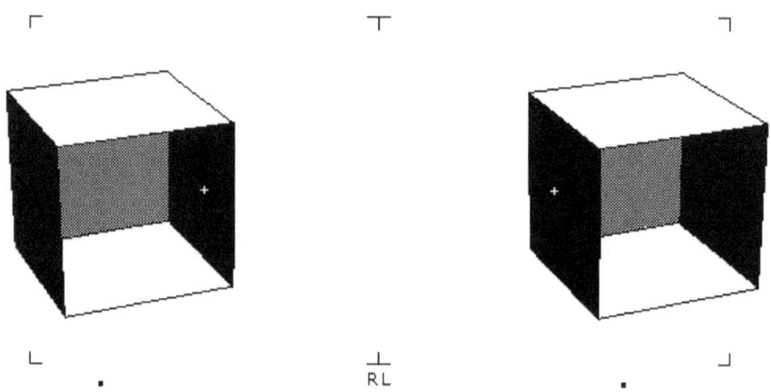

Fig. 11.8 Stereo-pair computed by program PROJECTI from data file SGCU.TXT, for crossed viewing. Object rotated by 22.5° about x and 22.5° about y. *Small crosses* are $s = 62\,\text{mm} = 234\,\text{px}$ apart. Shading added, and reproduction on page slightly reduced

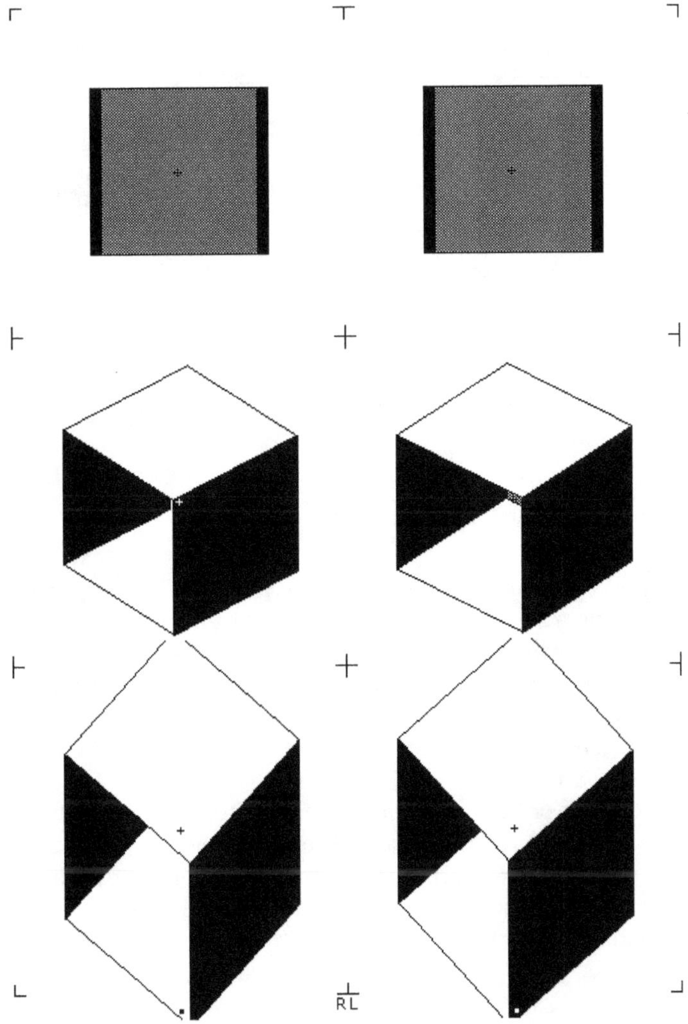

Fig. 11.9 Parallel-projected stereo-pairs for crossed viewing. Object is the same as for Fig. 11.7, reduced to 1/2 size and rotated about y-axis by 3.5° for *left image*, and −3.5° for *right image*. **a** Orthographic front elevation. **b** Isometric projection. **c** Military, or parallel oblique projection

11.6 Magnified Stereo Projection

If the image plane is set beyond the object we have a projection of class VII in the scheme of Sect. 8.1, and the stereo geometry is as shown in Fig. 11.10. The images are then magnified relative to the object, the image for the left centre is to the right of the image for the right centre, for crossed viewing (see Chap. 12).

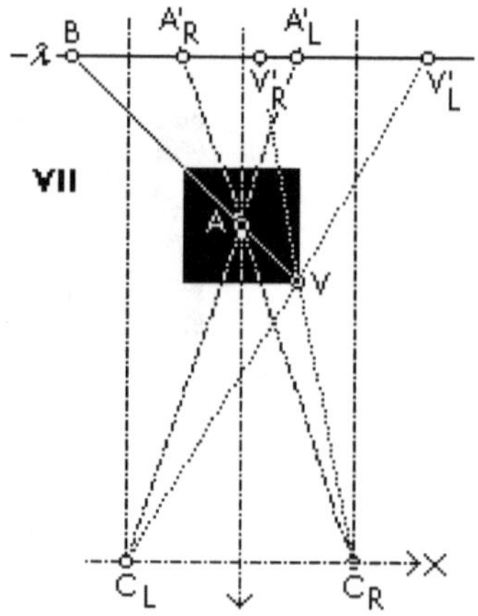

Fig. 11.10 Stereo projection class VII. The stereo-pair is magnified and is for crossed viewing

11.7 Stereo Projection by Camera

If a physical object is available, the obvious way to get a stereo-pair of images is by camera.[4] A camera stereo-pair is just two shots of the same object, taken from two camera positions which are laterally separated. For a static object, the shots can be successive. For a moving object, two cameras are needed (or one camera with a double-periscope device which gives two views at once). Figure 11.11 shows the construction and general arrangement for stereo projection by camera, class XIII in the scheme of Sect. 8.1.

Figure 11.12 shows orthographic plan, side, and front elevation (third-angle presentation) for a camera stereo-pair. The two centres of projection C_L and C_R are the centres of the camera lenses. Each image is projected, rotated by a half-turn about its lens axis, to the camera sensor. Notice that, for object 1 closest to the centres, the images $1'$ are farthest apart in the stereo-pair, whereas in Fig. 11.2 they are closest together in the stereo-pair.

11.8 Camera Stereoscopic Difference

In a camera stereo-pair, stereoscopic difference is the distance *d* between left and right images of the same object. It is like the direct case in Sect. 11.3, but now the images are inverted and the other side of the lenses. One way to calculate the difference is shown in Fig. 11.13. In similar triangles V C_L C_R and V V'_L V'_R we have

[4]Wheatstone's first stereo-pairs were line drawings. William Henry Fox Talbot invented the Talbotype negative/positive photographic process in 1839, and Wheatstone got Beard and Collen to make the first photographic stereo-pair (a portrait of Charles Babbage) in 1841.

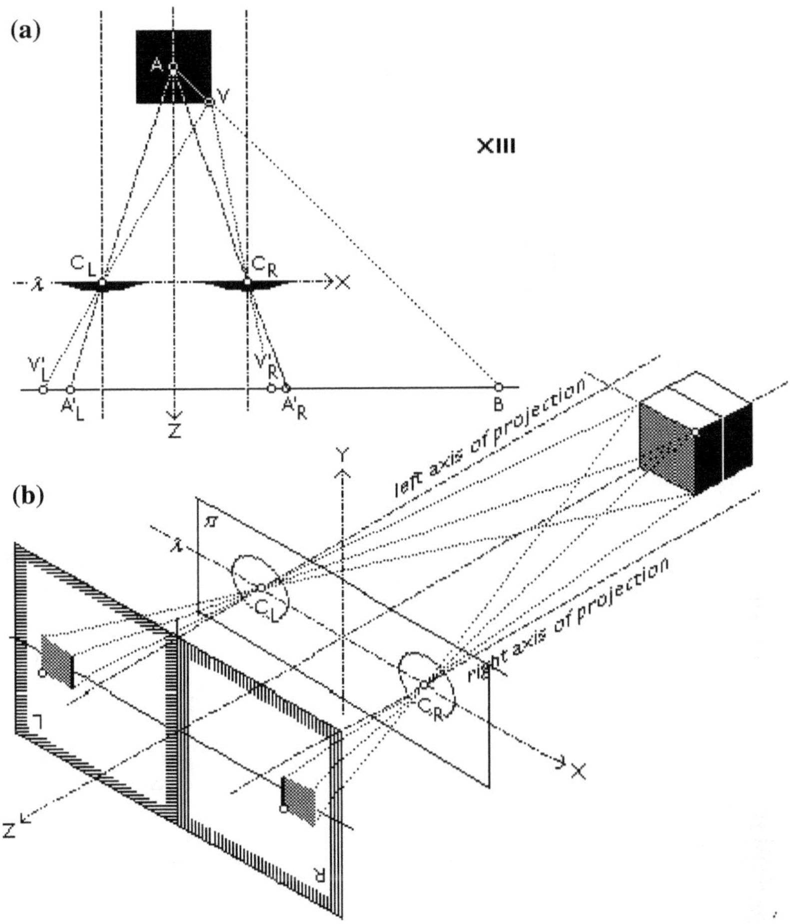

Fig. 11.11 Construction and general arrangement of camera stereo-pair, projection class XIII. **a** Plane projective-geometric construction of two images from given figure. Projection fixed by two centres $\mathrm{C_L}$ and $\mathrm{C_R}$, line λ, and corresponding points A and $\mathrm{A'_L}$, and A and $\mathrm{A_R}$. V is a given vertex of the figure. Join AV, to meet λ on B; join A′B; join $\mathrm{C_L V}$ to meet A′B on $\mathrm{V'_L}$, the *left image* of V; join $\mathrm{C_R V}$ to meet A′B on $\mathrm{V'_R}$, the *right image* of V. **b** Space projection of stereo-pair from given object. Each stereo image is rotated by 180° about its lens axis

$$s/(v - v') = d/v\,,$$

where s is the separation of centres $\mathrm{C_L}$ and $\mathrm{C_R}$, v is the distance of object V, d is the distance between images $\mathrm{V_L}$ and $\mathrm{V_R}$, and v' is the distance of images $\mathrm{V_L}$ and $\mathrm{V_R}$. Hence

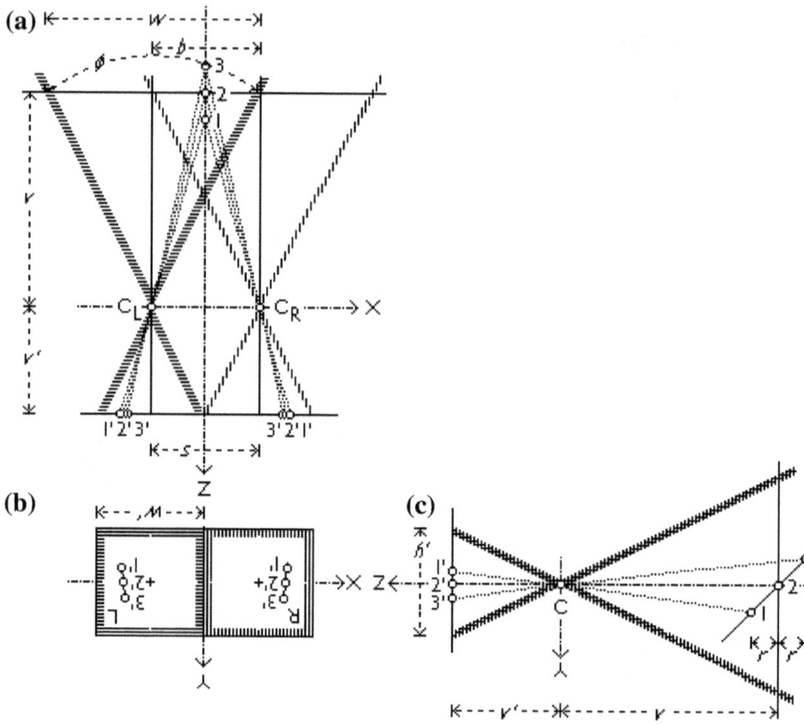

Fig. 11.12 Stereo camera geometry (third-angle presentation). **a** Orthographic plan: with objects 1 2 3, camera lenses C_L and C_R, separation s, and images $1'$ $2'$ $3'$. Left field and right fields of view (differently-edged) have a binocular overlap b. **b** Orthographic front elevation: the stereo-pair, left and right, with paired images $1'$ $2'$, $3'$. **c** Orthographic side elevation: with objects 1 2 3, camera lenses C, and images $1'$ $2'$ $3'$. For clarity, the sensor is shown here much farther from the lens than it should be

$$
\begin{aligned}
d &= (v \times s)/(v + v') \\
s &= (d \times v)/(v + v') \\
v &= ((d \times v) - (v + v'))/s \\
v' &= ((d \times v)/s) - v\,.
\end{aligned}
$$

When $v = 2v'$, stereoscopic difference $d = 2s$. As the object recedes, stereoscopic difference decreases. In the limit when the object approaches infinity, stereoscopic difference approaches separation s. All objects at the same distance v have the same stereoscopic difference d. For example, Table 11.3 shows the camera stereoscopic differences for five object distances.

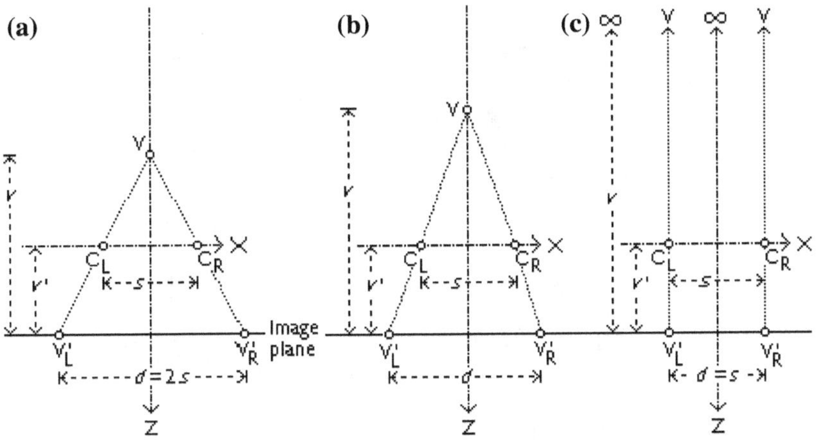

Fig. 11.13 Camera stereoscopic difference. Centres of lenses C_L C_R are separated by distance s. Object distance is v, image distance is v', and stereoscopic difference is d. **a** Object distance $v = 2v'$, and $d = 2s$. **b** $v > 2v'$, and $2s < d < s$. **c** Object V at ∞, and $d = s$

Table 11.3 Camera stereoscopic difference

s (mm)	v' (mm)	v (mm)	d (mm)
62	250	250	31
62	250	500	41.3
62	250	750	46.5
62	250	1000	49.6
62	250	∞	62

Given constant separation s and image distance v', and varying object distance v, calculate stereoscopic difference $d = (v \times s)/(v + v')$

11.9 Stereo-pair from Native Camera Images

When using a given camera and a given display to make a stereo-pair we may consider the camera-and-display as a single combined device, as shown in Fig. 11.14. A given camera takes an image with

- Camera pixel count in width P px,
- Camera angular width of field $\phi°$.

A given monitor screen displays an image with

- Display pixel frequency F'_d px per unit length.

The camera-and-display combination then has

- Native image width $w'_n = P/F'_d$ unit lengths,
- Native image distance $v'_n = w'_n/(2 \times \tan(\phi/2))$ unit lengths.

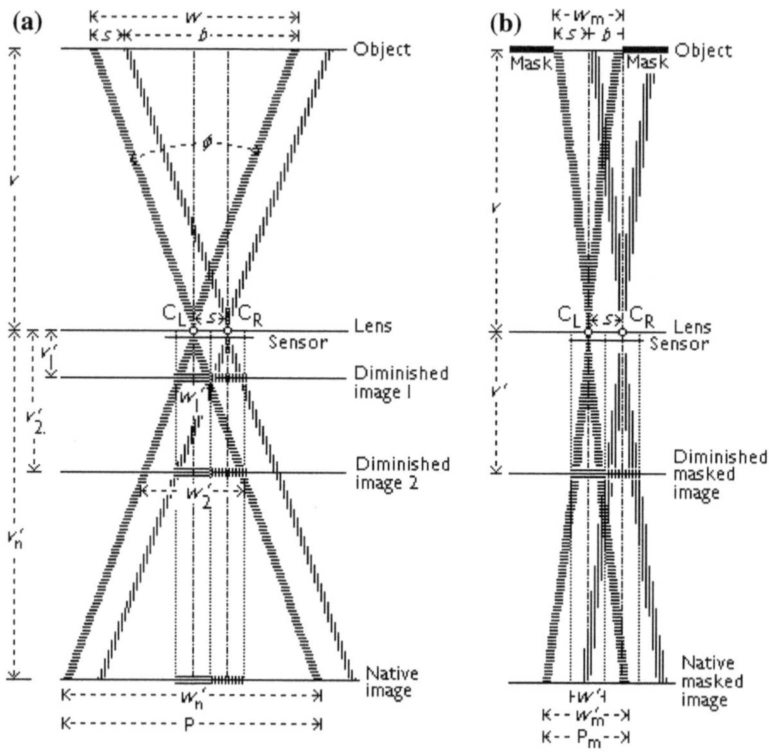

Fig. 11.14 Combined camera-and-display (or camera-and-printer) geometry. *Top half* is camera input; *bottom half* is display (or printer) output. **a** Given two laterally-separated native images. **b** Masked set-up on a photographic bench

These values are constants for a given camera-and-display combination. The native image width w'_n is what we always get from the combination of this camera with this display (which may often be wider than can be shown without panning the display). For example, if a camera has pixel count $P = 2304$ px and a display has frequency $F'_d = 3.58$ ppmm, then the native displayed image width is $w'_n = 2304/3.58 = 644$ mm. And if the camera has angular width $\phi = 40°$, then the native image distance is $v'_n = 644/(2 \times \tan(20)) = 644/0.728 = 885$ mm.

Similarly, a given printer outputs an image with

- Printer pixel frequency F'_p px per unit length.

The camera-and-printer combination then has

- Native image width $w'_n = P/F'_p$ unit lengths,
- Native image distance $v'_n = w'_n/(2 \times \tan(\phi/2))$ unit lengths.

These values are constants for a given camera-and-printer combination. The native image width w'_{n} is what we get from the combination of this camera with this printer (which may often be wider than can be printed without tiling the printout to more than one sheet). For example, if a camera has pixel count $P = 2304\,\mathrm{px}$ and a printer outputs at frequency $F'_{\mathrm{p}} = 3.78\,\mathrm{ppmm}$, then the native printed image width is $w'_{\mathrm{n}} = 2304/3.78 = 609\,\mathrm{mm}$. And if the camera has angular width $\phi = 40°$, then the native printed image distance is $v'_{\mathrm{n}} = 609/(2 \times \tan(20)) = 609/0.728 = 837\,\mathrm{mm}$.

Suppose we are given two camera images of a scene, taken from positions s (say 62) mm apart. To make a stereo-pair:

- Display the left image in an editor with known frequency F'_{d} (say 3.58) ppmm, note the pixel width P (say 2304) px, hence get the native image width $w'_{\mathrm{n}} = P/F'_{\mathrm{d}} = 2304/3.58 = 644\,\mathrm{mm}$:
- Required stereo image width $w'_1 = s = 62\,\mathrm{mm}$.
- Diminish the native image by magnification $m = s/w'_{\mathrm{n}} = 62/644 = 0.10$.
- Do the same for the right image.
- Abut right image to left for a stereo-pair to view uncrossed, or left image to right to view crossed,

as shown in Fig. 11.15a. The stereo images are diminished from 2304 to 230 px, so lose detail. The correct stereo viewing distance is $v'_1 = m \times v'_{\mathrm{n}} = 0.10 \times 644 = 64\,\mathrm{mm}$: too close for comfort. If viewed at the normal near distance of 250 mm the effect is a so-called "wide-angle" view. Alternatively:

- Diminish the native images by $m = 250/w'_{\mathrm{n}} = 250/644 = 0.39$,
- Crop diminished images to width $s = 62\,\mathrm{mm} = 62 \times 3.58 = 222\,\mathrm{px}$ around the middle of each diminished image,
- Abut the cropped images,

as shown in Fig. 11.15b. The correct stereo viewing distance is $v'_2 = m \times v'_{\mathrm{n}} = 0.39 \times 644 = 251\,\mathrm{mm}$. Or again:

- Crop undiminished images to width $s = 62\,\mathrm{mm} = 62 \times 3.58 = 222\,\mathrm{px}$ around the middle of each undiminished native image,
- Abut the cropped images,

to get a stereo-pair as shown in Fig. 11.15c. This keeps the full detail of the images, but loses extent. The correct viewing distance $v'_{\mathrm{n}} = 644\,\mathrm{mm}$: rather far. If viewed at the normal near distance of 250 mm the effect is a so-called "telephoto" view.

To get a good stereo-pair, use a fixed jig which ensures that the camera positions are level and parallel and about 62 mm apart. When the images are cropped, any two corresponding pixels in the images can be chosen as middles of the stereo-pair frames.

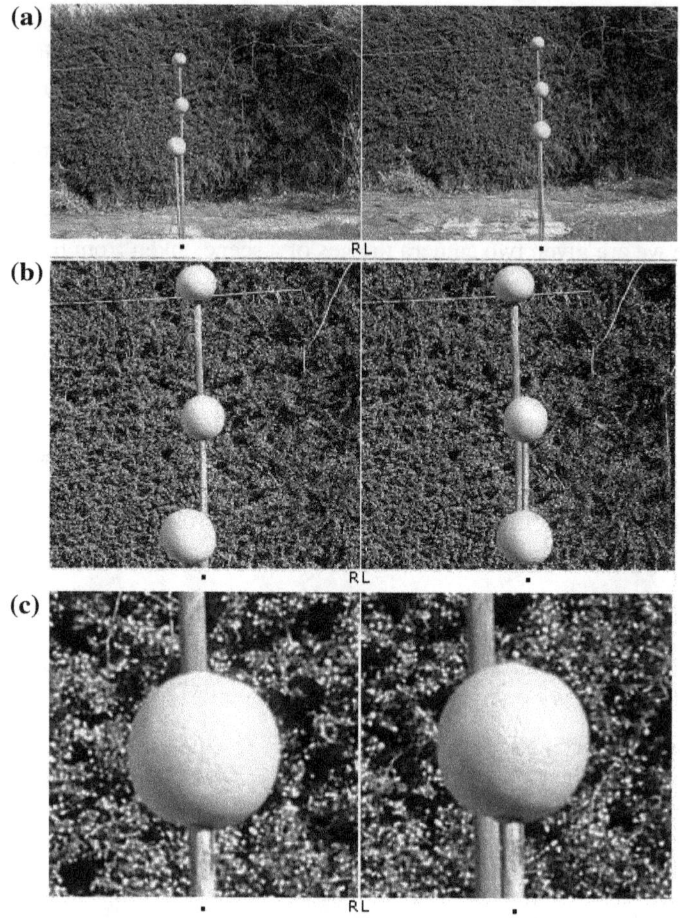

Fig. 11.15 Examples of stereo-pairs from native camera images, for crossed viewing. **a** Whole native images diminished by $m = 0.10$, to 62 mm wide. **b** Native images diminished by $m = 0.29$, then cropped to 62 mm. **c** Undiminished native images cropped to 62 mm. Reproduction on page slightly reduced

A digital camera image is usually given as a .JPG file, which can be reduced to a 24-bit .BMP colour file, or an 8-bit grey-scale file, or a 4-bit file, or a 1-bit file (see Appendix B). Stereo effects survive any such intensity reductions, as shown in Fig. 11.16.

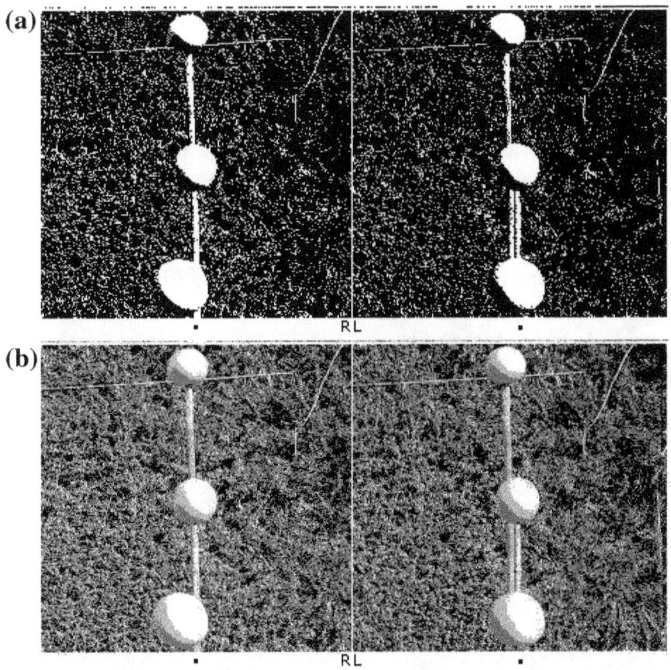

Fig. 11.16 Here Fig. 11.15b is reduced to **a** 1-bit mode and **b** 8-bit grey-scale

11.10 Stereo-pair from Masked Camera Images

Where an object is fairly small and easily-accessible, a well-controlled camera stereo-pair can be made on a photographic bench. First fix reasonable values for three variables:

- *Separation* s of centres C_L and C_R, say 62 mm,
- *Image distance* v' from centres, say 250 mm,
- (Single) *image width* w', say 62 mm for direct viewing (or 75 mm for viewing with a common Brewster-Holmes stereoscope).

Then

- Taking the bench as the xz-plane, mark z-axis.
- Mark left and right camera axes parallel to z-axis, at distances $s/2 = 31$ mm from z-axis.
- Place object on z-axis.
- Measure width of object (binocular overlap required) b, say 62 mm.
- Cut a mask with width $w = b + s = 62 + 62 = 124$ mm, and any suitable height, say also 124 mm.
- Place mask accurately centred on, and orthogonal to, left camera axis, around object.

- Place camera lens on, and orthogonal to, left axis, at such distance v from object that right edge of mask is just visible at right of object in camera view. Mark object distance v.
- Shoot left image.
- Move mask and camera to right axis, and shoot right image.
- Transfer both images to an image editor (such as MS PAINT), with given display frequency F_d = say 3.58 ppmm.
- Note the pixel width of the mask in each displayed image P'_m = say 647 px.
- Calculate displayed mask width $w'_m = P'_m/F'_d = 647/3.58 = 181$ mm.
- Calculate required magnification $m = w'/w'_m = 62/181 = 0.34$.
- Diminish each displayed image by m.
- For a stereo-pair to view uncrossed, abut right diminished image to left; for crossed abut left image to right.

Figure 11.17a shows a stereo-pair of masked camera images using the measurements listed above, and Fig. 11.17b a stereo-pair computed using the same measurements. We see that where precision is important, computation from a given wire-frame object is accurate to within 1 px, which is practically impossible to match in a camera

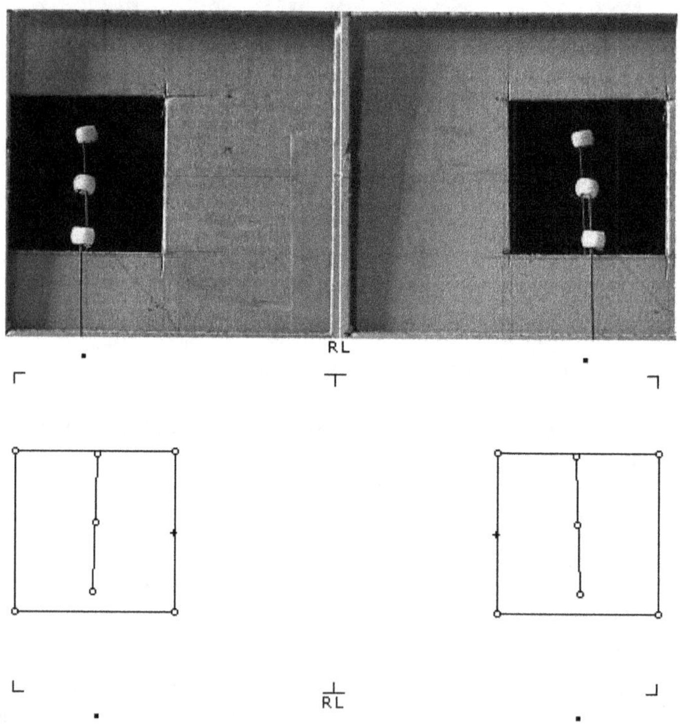

Fig. 11.17 **a** Stereo-pair from masked camera images, for crossed viewing. **b** Stereo-pair computed by program, for crossed viewing

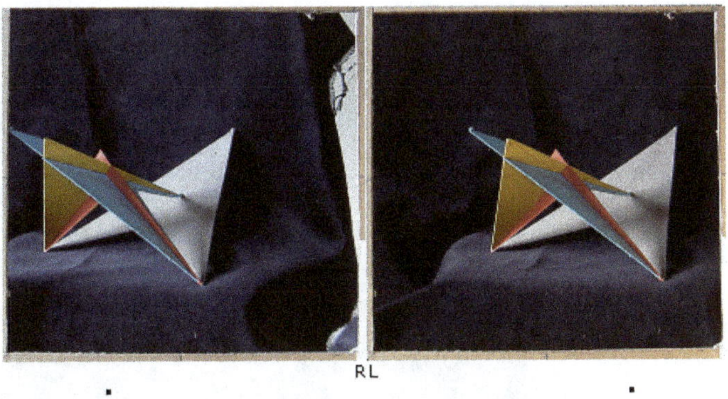

Fig. 11.18 Another stereo-pair from masked camera images, for crossed viewing. The discrepancy in the backgrounds at the right is outside the binocular overlap, so does not affect the stereo effect. The object is four complete quadrangles from a projective geometry

set-up. On the other hand, a natural scene is easily masked for a camera, but practically impossible to supply as a wire-frame object.

For another example, Fig. 11.18 shows a masked stereo-pair where the binocular overlap $b = 150$ mm, mask width $w = 212$, and object distance $v = 500$ mm.

- Measured width of object (binocular overlap required) $b = 150$ mm.
- Cut a mask with width $w = b + s = 150 + 62 = 212$ mm, and height also 212 mm.
- Place mask accurately centred on, and orthogonal to, left camera axis, around object.
- Place camera lens on, and orthogonal to, left axis, at such distance v from object that right edge of mask is just visible at right of object in camera view. Mark object distance v.
- Shoot left image.
- Move mask and camera to right axis, and shoot right image.
- Transfer both images to an image editor (such as MS PAINT), with given display frequency $F_{\mathrm{d}} =$ say 3.58 ppmm.
- Note the pixel width of the mask in each displayed image $P'_{\mathrm{m}} = 1060$ px.
- Calculate displayed mask width $w'_{\mathrm{m}} = P'_{\mathrm{m}}/F'_{\mathrm{d}} = 1060/3.58 = 296$ mm.
- Calculate required magnification $m = w'/w'_{\mathrm{m}} = 62/296 = 0.21$.
- Diminish each displayed image by m.
- For a stereo-pair to view uncrossed, abut right diminished image to left; for crossed abut left image to right.

Figure 11.19 shows a stereo-pair of the kind made by camera in huge numbers in the late 19th century. The standard width of each image is 75 mm, intended for uncrossed viewing with a Brewster-Holmes de-centred-lens stereoscope.

Fig. 11.19 Typical camera stereo-pair, image-width 75 mm for uncrossed viewing with a Brewster-Holmes stereoscope. The object is Sir Charles Wheatstone and family

11.11 Random-Dot Stereo-pair

Another way to construct a stereo-pair is to start with two identical random images (such as those produced by the program described in Sect. 20.6), then to shift a region of one of the images by a small distance to the left or right. To ordinary inspection, no depth is apparent; but when viewed with crossed or uncrossed visual axes the shifted region stands out, or drops back, in depth.

For example, Fig. 11.20a shows a pair of 24 × 24-px images. The R-image is the original. In the L-image an 8 × 8-px region has been shifted by 1 px to the right, by careful manipulation in a GUI image editor. If the pair are viewed with visual axes crossed (left eye directed at the L-image and right eye directed at the R-image, so that the two dots below appear as three), then a 7 × 8-px region stands out in front.

Fig. 11.20 Random-dot stereo-pair. **a** R–L: in the L-image, an 8 × 8-px region has been shifted to the right by 1 px. In crossed viewing, the region stands out in front of the surround; in uncrossed viewing, the region sinks back. **b** L–R: L-image and R-image of (**a**) swapped left to right. In crossed viewing, the region sinks back from the surround; in uncrossed viewing, the region stands out in front

If the visual axes are uncrossed towards parallel, so that the two dots again appear as three, then a 9 × 8-px region falls back behind. (The uncrossed view is best seen with a Brewster-Holmes stereoscope about 250 mm from the images.) In Fig. 11.20b, the L-image is the original, and the R-image has the shifted regions.

Figure 11.21 shows how the stereo effect arises. Figure 11.21a is a plan of Fig. 11.20b, with the random-dot stereo-pair L–R on line λ, and with a patch of the L-image shifted to the right. In uncrossed viewing, the left eye sees more of the surround to the left of the patch, and less to the right. The right eye sees less of the surround to the left of the patch, and more to the right. This looks the same as Fig. 11.21b, the projection of a patch which actually is in front of the surround.

For another example, Fig. 11.22 shows three nested square regions shifted 1, 2, and 3 px, producing four depths. The stereo effect is entirely dependent on horizontal shifts: vertical shifts have no effect.

Random-dot stereo-pairs of this kind were used by Julesz[5] to demonstrate that depth perception must arise after the left and right retinal nerve impulses are combined at the optic chiasma, as if viewed by the single central eye of a Cyclops, at the Cyclopean point K midway between the left and right eyes [4].

The stereo effect here may be compared with the relief projection described in Sect. 8.19, and with the auto-stereo effect described in Chap. 13.

[5]Bela Julesz (1928–2003), Bell Telephone Laboratories 1956–1989, and Rutgers University 1989–2003.

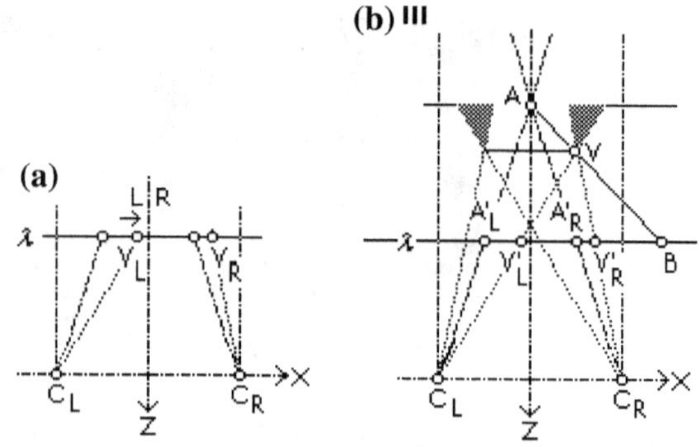

Fig. 11.21 Random-dot stereo-pair looks the same as a projected stereo-pair. **a** Random-dot images L–R lie on line λ, one image seen (uncrossed) by eye C_L and one by eye C_R. In the image *on the left*, a patch containing vertex V is shifted to the right, to reveal part of the original image *on the left* and hide another part *on the right*. **b** (Class III) projection fixed by two centres C_L and C_R, line λ, and corresponding points A and A'_L, and A and A_R. Vertex V is in front of the surround. Join AV, to meet λ on B; join A′B; join $C_L V$ to meet A′B on V'_L, the left-eye image of V; join $C_R V$ to meet A′B on V'_R, the right-eye image of V. The left eye sees more of the surround on the left, and less on the right; and the right eye oppositely

11.12 Anaglyph Stereo-pair

An *anaglyph* is made by filtering a left-eye image, drawn or photographed, to show only reds, and a right-eye image to show only greens and blues. The images are then printed or displayed or projected as one frame. Viewing this frame through a cyan filter shows the red image as black/grey/white, whereas viewing through a red filter shows the green/blue image as black/grey/white. So a viewer wearing complementary cyan and red glasses receives the appropriate stereoscopic effect. Wilhelm Rollmann published the method in Leipzig in 1852, for drawings, and Louis Ducos du Hauron patented a photographic method in Paris in 1891. Anaglyphic methods have been used for audiences in cinemas and TV, and sometimes for individual viewing in books on solid geometry and similar subjects.

11.13 Polarized Stereo-pair

Instead of separating projected images, movie or still, by complementary wavebands as in anaglyphs, a rather better effect is got by opposite polarization of the projector lights, and then viewing through left-right-reversed polarizing filter glasses. This became a practical method with the availability of Edwin Land's Polaroid sheets from 1929.

Fig. 11.22 Random-dot stereo-pair with four depths. **a** R–L: the L-image is the original; in the R-image, three nested square regions have been shifted to the right by 1, 2, and 3 px. In crossed viewing, the nested squares sink back from the surround. **b** R–L: the R-image is the original; in the L-image, three nested square regions have been shifted to the right by 1, 2, and 3 px. In crossed viewing, the nested squares stand out from the surround. In uncrossed viewing, the depths are reversed

11.14 Lenticular Printing

A lenticular sheet is made from transparent plastic with closely-spaced cylindrical lenses impressed on the front surface (typically 40/in). When laminated to a printout, with the cylinders vertical, it deviates the light reflected from the printout. When viewed monocularly from a certain angle, the viewer sees a set of 1/40-in stripes of the printout, and from another angle another set of stripes. If a stereo-pair of images are combined in alternating vertical stripes (see Sects. 23.22 and 39.3), and the viewing angles coincide with the interocular distance, we have a stereo effect without glasses or other intervening device. A stereo-pair printed at 720 ppi will have alternating stripes 9 px wide. The effect can be extended to display (say) three stereo pairs taken from different viewpoints, combined in 3-px stripes.

Lenticular printing [1, 6] is often done for joke postcards and other gimmicks, and sometimes for more elaborate wall-mounted or back-lit displays.

References

1. Alpaslan Z, Sawchuk AA (2004) 3D Interaction with autostereoscopic displays. http://www.zahiralpaslan.com/research/3DIASD_ZYAlpaslan_web.pdf. Accessed 1 Feb 2014
2. Brewster D (1856) The Stereoscope: its history, theory, and construction. John Murray, London. https://archive.org/details/stereoscopeitsh01brewgoog. Accessed 1 Feb 2014
3. Helmholtz H von (1967) Handbuch der Physiologischen Optik. Voss, Leipzig, 3 vols 1856–1867. English translation of 1910 edition (1925) Optical Society of America, republished (1962) Dover, New York
4. Julesz B (1971) Foundations of cyclopean perception. Chicago University Press, Chicago
5. Mayo H (1833) Outlines of human physiology, 3rd edn. Burgess and Hill, London. http://catalog.hathitrust.org/Record/006104741. Accessed 1 Feb 2014
6. Visual Creative Graphics Innovations (2012). http://www.vicgi.com/index.html. Accessed 1 Feb 2014
7. Wheatstone C (1838) Contributions to the physiology of vision, on some remarkable, and hitherto unobserved, phenomena of binocular vision. Phil Trans Roy Soc 128:371–394. http://www.stereoscopy.com/library/wheatstone-paper1838.html. Accessed 1 Feb 2014

Chapter 12
Stereo Viewing

Paired eyes give an immediate perception of depth in surrounding space, with great acuity. Viewing a stereo-pair of images gives a similar perception. Seven modes of viewing a stereo-pair, with or without special devices, are described.

12.1 Flowchart

The flowchart (Fig. 12.1) is for general guidance in this chapter. A stereo-pair of digital images, created as in Chap. 11, can be viewed with or without a viewing device by the two eyes. A scene in space is viewed by continually-changing paired samples of location and intensity. The two cases deliver visual sensations S which are similar.

12.2 Binocular Geometry

In binocular vision, the eyes work as a coupled pair, darting continually from one fixation to another. Figure 12.2 shows one fixation, with the two visual axes converging on an object point P. Each eye projects, via centre C, an image of the objects around P in sharp focus around fovea F. The reference frame for the object field is the horizontal circle through P and the two centres C, and the vertical circle through P. The reference frame for each image field is the projected horizontal arc and vertical arc through fovea F.

Figure 12.3 shows the binocular field in orthographic elevations and plan (first angle arrangement). The extreme width of the combined object fields is about 200°, but the width of the binocular overlap is about 120°.

Physiologically, the nerve fibres from the two eyes go to the *optic chiasma*, whence those from the left halves of both retinas go to the left half of the brain, and those from the right halves of both retinas to the right half of the brain (as correctly surmised

A. Parkin, *Digital Imaging Primer*, DOI 10.1007/978-3-540-85619-1_12

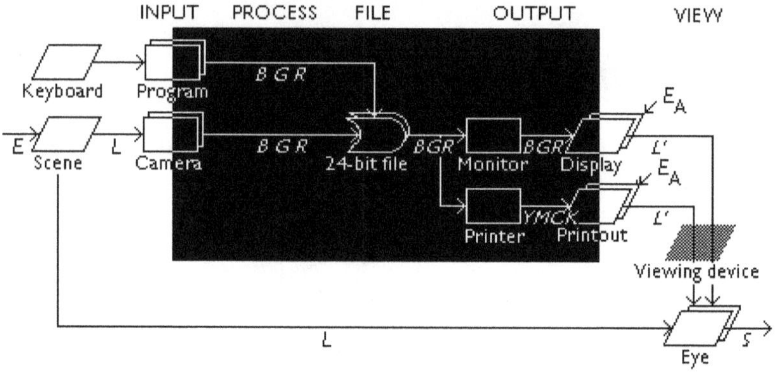

Fig. 12.1 Stereo viewing. A stereo-pair of digital images can be made by two runs of a program, or two shots of a camera, stored as two files, and displayed or printed as a pair, to be viewed with or without a device by the two eyes. The resulting sensation S is like the ordinary view by the two eyes of a scene in space

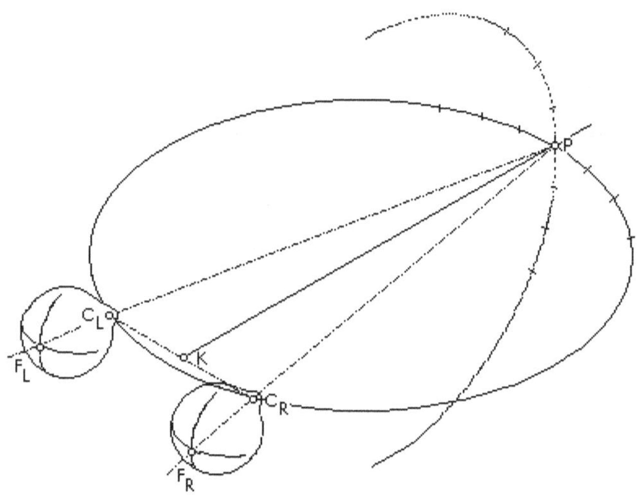

Fig. 12.2 Binocular geometry. Left eye has optical centre C_L, fovea F_L, and visual axis C_L F_L; and right eye similarly. In any given fixation, the visual axes converge on object point P, which is imaged at the two foveae on the two retinas. Object positions are referred to the horizontal circle through C_L, C_R, and P, and the vertical circle through P. Image positions are referred to the horizontal and vertical arcs through F. K is the Cyclopean point midway between the eyes

by Newton). Neurologically, the two slightly-different images from the left and right retinas are then combined as one perceptual field, as if seen from the single eye of a Cyclops at the Cyclopean point K midway between the eyes (see Sect. 11.11).

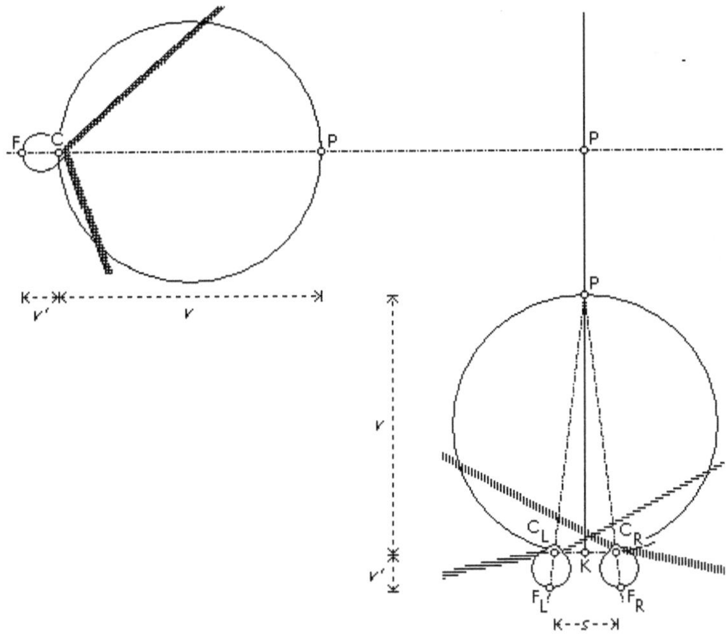

Fig. 12.3 Binocular field. **a** Side elevation, field limited by brow and cheek. **b** Front elevation. **c** Plan, fields limited by nose. Centres C_L and C_R have separation s. Object P, at distance v from centres, is imaged at F_L and F_R, at distance v' from centres

12.3 Binocular Fusion

Geometrically, a *Vieth-Muller circle*[1] goes through the optical centres of the eyes C_L and C_R, and the point of fixation P. Figure 12.4 shows that, for a given symmetrical fixation on object point P, the two images of P are focussed on the foveae, which are *corresponding points* of the left and right retinas. Any other object point on the V–M circle, such as 4 or 5, subtends the same *angle of parallax* ϑ at the eyes, so its two images also fall on corresponding points of the retinas.

A *horopter*[2] is a line, surface, or volume in object space in which objects appear to have the same position in the image fields of both eyes, and so are seen cyclopically as single or *fused*. The *geometric horizontal horopter* is the Vieth-Muller circle for a given fixation. The *geometric vertical horopter* is the vertical through the fixation P.

But it is known experimentally that the *empirical horizontal horopter* lies outside the Vieth-Muller circle (the Hering-Hillebrand deviation). When fixation is at the near point, say 250 mm, the empirical horopter is close to the V–M circle; at fixation distance 6 m (abathic distance), the horopter is close to the frontal plane; and at

[1]Proposed by Franciscus Aguilonius in 1613, now named for G.U.A. Vieth (1818) and J.P. Muller (1840).

[2]From $o\rho\iota o$ boundary or limit, and $o\pi\tau\eta\rho$ one who looks.

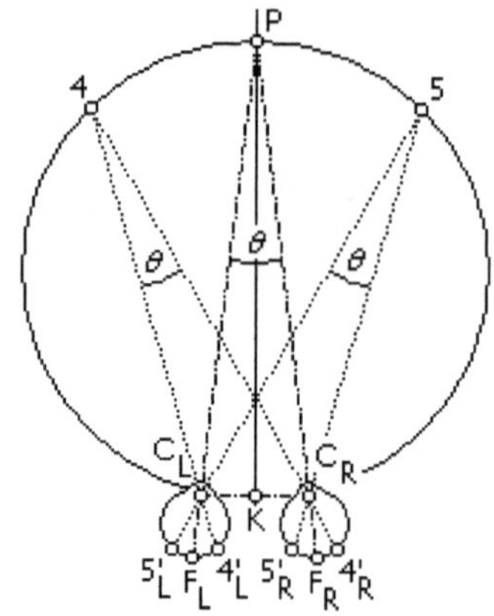

Fig. 12.4 Corresponding points. For a given fixation on object P, the images of P project to the foveae, which are the principal corresponding points of the retinas. The Vieth-Muller circle goes through P and centres C_L and C_R. An object on the circle, such as *4* or *5*, subtends at the eyes the same angle ϑ as P, so their images also project to corresponding points of the retinas

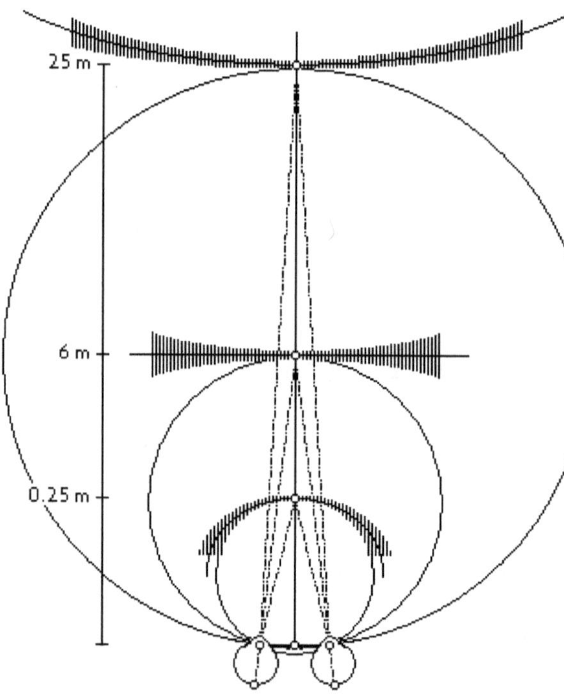

Fig. 12.5 Vieth-Muller circle, empirical horizontal horopter, and Panum's fusional area for three different fixations (not to scale). When fixation distance is about 6 m, the horopter is close to the frontal plane

fixations greater than 6 m the horopter curves away from the frontal plane, as shown in Fig. 12.5.

When the eyes fix on an object at P, the pair of images of P fall on corresponding points of the two retinas, and *fuse* to a single perception of P as if seen from the Cyclopean point K. And the paired images of any object within *Panum's fusional area* [5][3] also fuse, though they project to non-corresponding points. The fusional

[3]Peter Ludvig Panum (1820–1885), Professor of Physiology at Kiel, then at Copenhagen, University.

area lies around the horopter, quite narrow at the fixation point, and wider with increasing lateral distance, as shown in Fig. 12.5.

12.4 Binocular Disparity

When the eyes fix on an object at P, the paired images of objects outside the fusional area project to non-corresponding retinal points, and are not fused but seen double (dioptically). There is a retinal *disparity* between the pair of images. Figure 12.6 shows three objects: 1, 2 and 3. The eyes are fixated on object 2, so left image $2'_L$ falls on left fovea F_L and right image $2'_R$ falls on right fovea F_R. Object 1 is closer than the fusional area, so its images fall temporally of both the foveae, that is to the left of the left fovea and to the right of the right fovea: this is called convergent or *crossed disparity*. Object 3 is farther than the fusional area, so its images fall nasally

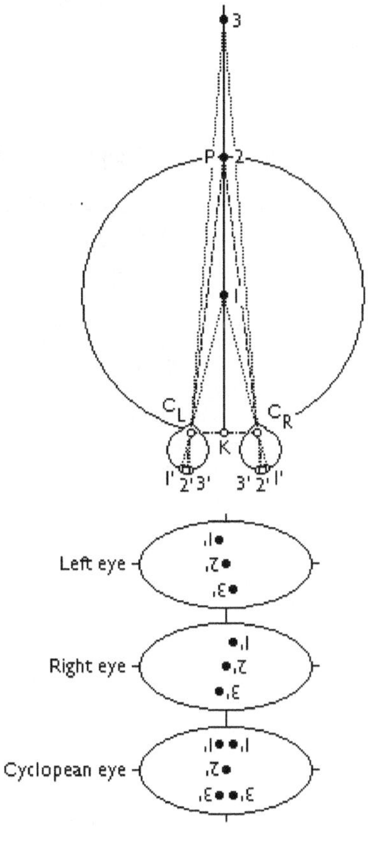

Fig. 12.6 Binocular disparity. When the eyes fixate object 2, images 2′ are projected to the foveae, which are corresponding points, so 2 is perceived single. In the same fixation, object 1 projects images 1′ to non-corresponding points with crossed disparity, so 1 is perceived double. In the same fixation, object 3 projects images 3′ to non-corresponding points with uncrossed disparity, so 3 is also perceived double. The ellipses below show the inverted retinal images of the left eye alone, the right eye alone, and the Cyclopean eye at K

of both the foveae, that is to the right of the left fovea and to the left of the right fovea: this is called divergent or *uncrossed disparity*. The crossed disparity for closer objects is greater than the uncrossed disparity for farther objects.

12.5 Depth Acuity

Depth perception is very acute at the near point, about 250 mm from the eyes, and falls off with increasing distance. Figure 12.7 shows the left eye with centre of curvature C, the Cyclopean point K midway between the eyes, object 1 which is fixated on the visual axis, object 2 which is farther than 1, and the retinal images $1'$ and $2'$. We have the distances

- Half-separation of eyes $s/2$ length units (say 31 mm),
- Object 1 distance v_1 length units,
- Object 2 distance v_2 length units,
- Threshold distance of depth discrimination $t = v_2 - v_1$ length units,

and the angles

- Half-angle of parallax for object $1 \vartheta_1 = \arctan(v_1/(s/2))^\circ$,
- Half-angle of parallax for object $2 \vartheta_2 = \arctan(v_2/(s/2)^\circ$,
- Threshold angle of depth discrimination $\tau = \vartheta_2 - \vartheta_1^\circ$.

It is known experimentally [3] that the minimum angle of retinal disparity is $\tau = 5$ arc-s (=0.083 arc-min $= 0.0014^\circ$) for normal observers, and for the best observers is $\tau = 2.5$ arc-s (=0.042 arc-min $= 0.0007^\circ$). This is a remarkable finding,

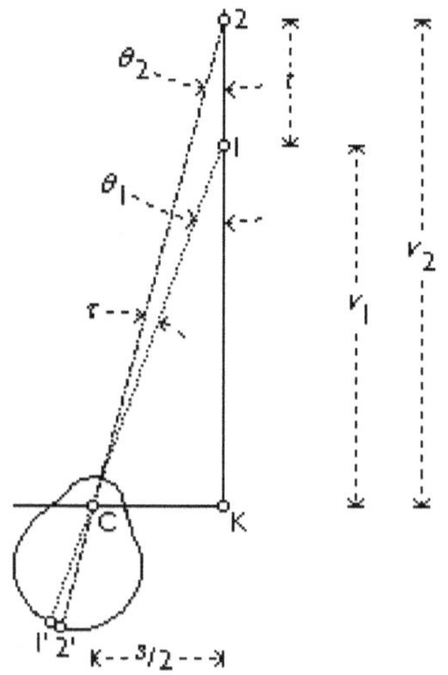

Fig. 12.7 Depth acuity. Left eye, centre C, is at distance $s/2$ from Cyclopean point K. Eye fixates object 1 at distance v_1 and half-angle of parallax ϑ_1. Object 2 is at distance v_2 and half-angle ϑ_2. Threshold distance of depth discrimination is t and threshold angle is τ

since the angle subtended by adjacent cones in the fovea is about 28 arc-s (=0.47 arc-min = 0.0078°), so the minimum disparity cannot be registered in a single retinal image. It is thought that Cyclopean processing in the brain, using the two disparate retinal images, produces depth perception; and that this is not fully-developed in the infant until 10 months [6].

Geometrically, object 2 will be just-perceptibly separated in depth from object 1 when

$$\vartheta_2 = \vartheta_1 + \tau$$

that is, for the best case

$$\text{arc-tan}(v_2/(s/2)) = \text{arc-tan}(250/31) + 0.0007^\circ$$

So we may calculate

- Set $v_1 = 250\,\text{mm}$.
- Then $\vartheta_1 = \text{arc-tan}\ (250/31) = 82.9314^\circ$
- and $\vartheta_2 = \vartheta_1 + 0.0007 = 82.9321^\circ$,
- hence $v_2 = 31 \times \tan(82.9321) = 250.025\,\text{mm}$
- and $t = v_2 - v_1 = 250.025 - 250 = 0.025\,\text{mm}$.

When the eye fixates object 1 at the the near point, 250 mm, object 2 will just be seen in relief if its depth is 0.025 mm from 1. Table 12.1 shows the depth acuity thus calculated for increasing object distances, assuming inter-pupillary half-distance $s/2 = 31\,\text{mm}$. Even at 1,000 m, a depth of some 700 m can be detected binocularly. Binocular vision is a very highly-evolved process, partly physiological and partly psychological, partly innate and partly learned. The paired eyes continually dart from one fixation to another, sequentially gathering information which the brain integrates to a stable perception of the visual field. The conscious person effortlessly distinguishes between movements of objects and movements of the eyes.

Table 12.1 Depth acuity thresholds

v_1	v_2	t
250 mm	250.025 mm	0.025 mm
500 mm	500.102 mm	0.102 mm
1 m	1.00041 m	0.41 mm
10 m	10.0409 m	41 mm
100 m	104.245 m	4.2 m
1000 m	1687.03 m	687 m

v_1 is the distance of the fixated reference object 1; v_2 is the distance of the just-deeper object 2; $t = v_2 - v_1$ is the just-detectable depth

12.6 Viewing Stereo-pair of Images

To view a stereo-pair, the left and right images must be presented separately to the two eyes. There are several different ways to do this, with or without a specially-made *stereoscope*. The history of stereoscopic devices is reviewed early by Helmholtz [2] and later by Judge [4]. Here we describe the viewing methods according to the optical principles used.

12.7 Direct Viewing, Uncrossed

Figure 12.8a shows direct viewing of a stereo-pair with uncrossed visual axes, described by Helmholtz as follows:

> This can be done without any apparatus by placing the two pictures side by side, one on the right and the other on the left, with the corresponding points in the two views at about the same distance apart as the observer's inter-pupillary distance. Then with the visual axes of the two eyes parallel, the observer will see both pictures with both eyes and will obtain the stereoscopic effect. It is true that in this case he will see with each eye not only the picture intended for that eye, but alongside of it the picture intended for the other eye as well. Accordingly, when the proper position of the eyes has been obtained, the observer will see apparently three images side by side, the two outer ones being viewed with one eye alone (the right one by the left eye, and the left one by the right eye) and not making any 'plastic' impression, whereas the one in the middle, which is viewed by both eyes together, will produce such an impression.
>
> In this mode of experiment the presence of the three images creates a disturbance; and, besides, the accommodation has to be changed for near objects without converging the eyes at all, which is contrary to the way we are accustomed to using our eyes. And so it takes some practice to learn how to get stereoscopic vision without any other external aid. Incidentally, the visual illusion obtained in this fashion is just as perfect as it is when the instrument is used which is about to be described. ...
>
> The sole object of the so-called *stereoscope* for viewing stereoscopic pictures is to facilitate and maintain the proper adjustment of the observer's eyes and to get rid of the disturbance due to the secondary images. It has no special advantage so far as the actual production of the visual illusion is concerned.

12.8 Direct Viewing, Crossed

Figure 12.8b shows direct viewing with crossed visual axes, described by Helmholtz as follows:

> Instead of directing the visual axes of the two eyes to a very remote point so that they are practically parallel, they may be converged on a nearer point, and the two images made to fuse, by turning the right eye toward the one on the left and the left eye toward the one on the

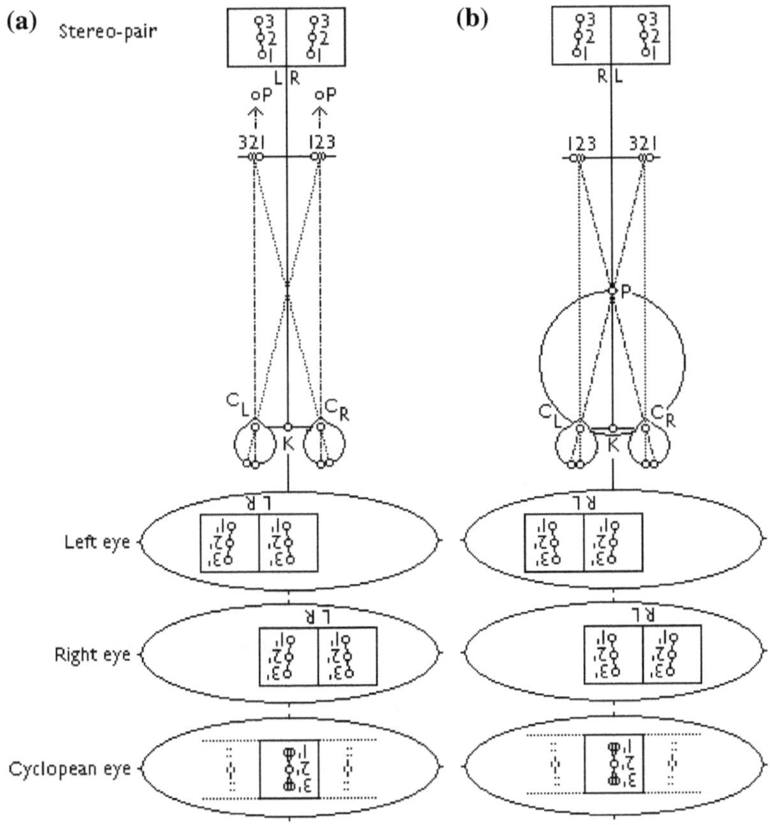

Fig. 12.8 Direct viewing of stereo-pair. **a** Uncrossed viewing: elevation of stereo-pair, plan of stereo-pair viewed with uncrossed visual axes, retinal images of left eye, right eye, and Cyclopean eye. **b** Crossed viewing: as (**a**), but with crossed visual axes

> right, so that the lines of fixation of the two eyes intersect at a point between the observer and the plane of the pictures. In this case the adjustment of the eyes is the same as if they were both gazing at this intermediate point, the stereoscopic object appearing to be situated there, that is, nearer the eyes than the pictures actually are. But, of course, in this experiment the picture intended for the right eye must be placed on the left-hand side, and that for the left eye on the right-hand side; otherwise the stereoscopic parallax will be negative, and the relief reversed.

Helmholtz remarks that "owing to the difficulty in getting the proper accommodation, it is easier for a far-sighted person to fuse the images when the visual axes are crossed, whereas in the case of a near-sighted person it is easier when the visual axes are not crossed."

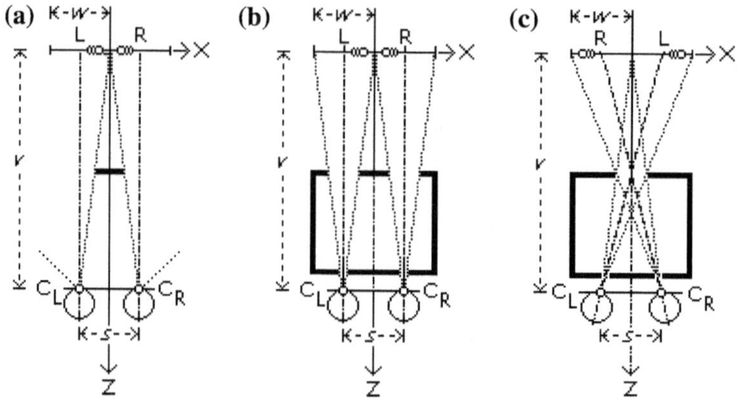

Fig. 12.9 Viewing with masks. **a** Single mask. Typical sizes: $s = 60$ mm, $v = 250$ mm, $w = 60$ mm. **b** Double mask, uncrossed visual axes. Typical sizes: $s = 60$ mm, $v = 250$ mm, $w = 60$ mm. **c** Double mask, crossed visual axes. Typical sizes: $s = 60$ mm, $v = 250$ mm, $w \geq 60$ mm

12.9 Viewing with Masks

A simple improvement in direct viewing is to block the two outside un-fused images by a suitably-placed mask. Figure 12.9a shows viewing with a single mask.

Somewhat better control is given by two masks, one with small square openings near the eyes and one with larger square openings farther away. These masks may be made as the front and back of a box, or as the ends of square tubes. Figure 12.9b shows uncrossed viewing with a double mask.

Figure 12.9c shows crossed viewing with a double mask. This was indeed the principle of Elliot's very early stereoscope.[4] An Elliot stereoscope can be constructed for stereo-pairs of virtually any size.

12.10 Viewing with Plane Mirrors

If two plane mirrors are set at 45° across the parallel visual axes, reflecting outwards, then the left and right images of a stereo-pair can be placed, separately, facing inwards, at equal distances v. We can thus produce a stereo-pair with any convenient image distance v, and corresponding image width w larger than s. Figure 12.10a shows such an arrangement. This was the principle of Wheatstone's original stereoscope made by Murray and Heath, and demonstrated in 1838 (see Sect. 11.1). This had peep-holes for better location of the eyes, and a screw mechanism for adjusting both viewing

[4]James Elliot was an Edinburgh mathematics teacher who showed his stereoscope in 1839, with drawn pictures, and described it in a letter to the Philosophical Magazine in 1852. Brewster (1856) claimed that Elliot's invention dated from 1834 and had priority over Wheatstone's, but Elliot himself made no such claim.

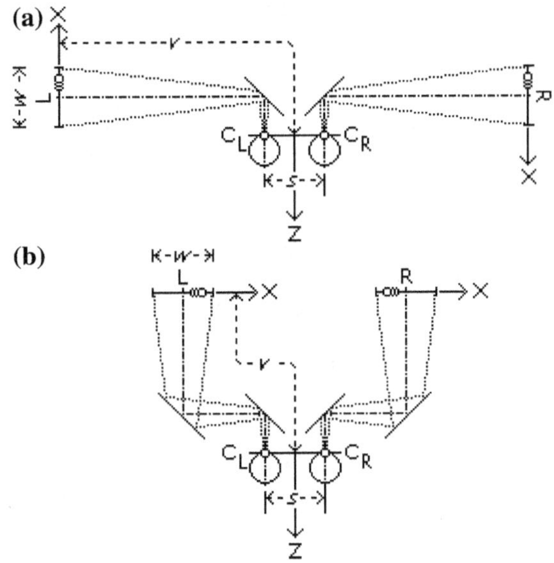

Fig. 12.10 Viewing with plane mirrors. **a** With two mirrors. Typical sizes: $s = 62\,\text{mm}$, $v \geq 250\,\text{mm}$, $w \geq 62\,\text{mm}$. **b** With four mirrors. Typical sizes: $s = 62\,\text{mm}$, $v \geq 250\,\text{mm}$, $w \geq 62\,\text{mm}$

distances together; the mirrors were 4-in squares, and the image widths could be a foot or more. Each mirror perverts its image left/right, so printing reads backwards.

Figure 12.10b shows an improved reflecting stereoscope, with four mirrors. The two reflections of each image restore the original handedness, so printing reads correctly.

12.11 Viewing with Plane Prisms

The mirrors in Sect. 12.10 can be replaced by prisms, with advantage, since prisms lose less light and do not tarnish.

The critical angle of a typical air-glass surface is about 42°, so a triangular prism with angles $90/45/45°$ totally-reflects a ray incident normal to a short face, deviating the ray by 90°, as shown in Fig. 12.11a. A parallelogrammic prism with angles $45/135°$ makes two such reflections to restore the original direction, as shown in Fig. 12.11b. A thin prism with apical angle $\alpha°$ refracts an incident ray twice, at the two surfaces. The minimum angle of deviation $\delta°$ of the emergent ray is

$$\delta = (n - 1)\alpha\,,$$

where n is the refractive index of the glass of the prism, typically 1.5–1.6, and the minimum deviation is when the entry face is at $\alpha/2°$ to the incident ray. So, for example, a prism with angle 8° will deviate a ray by 4°, as shown in Fig. 12.11c.

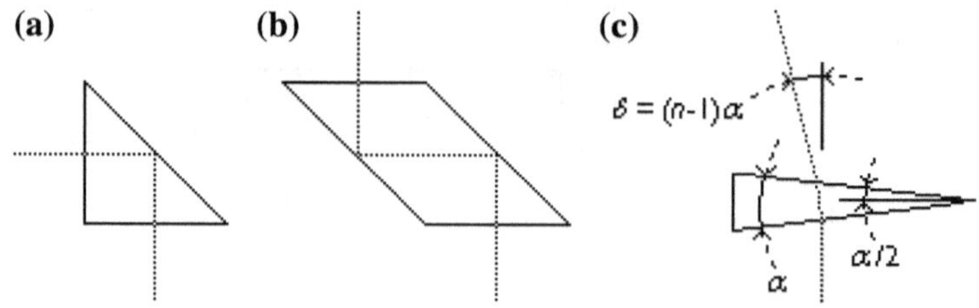

Fig. 12.11 Prisms. **a** 90/45/45-degree prism totally-reflects an incident ray with a deviation of 90°. **b** 45/135-degree prism totally-reflects an incident ray twice, emerging parallel to the incident ray. **c** Thin prism with apical angle $\alpha°$ deviates an incident ray by $\delta = (n-1)\alpha°$

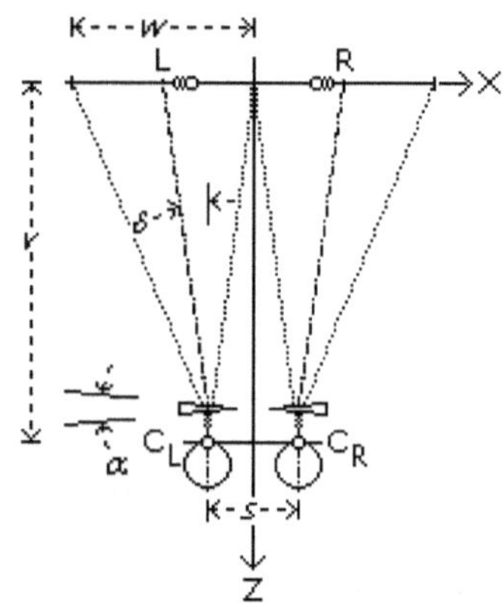

Fig. 12.12 Viewing with plane prisms. Each prism has apical angle $\alpha°$, and the visual axes are each deviated outwards by $\delta°$. The stereo-pair images may then be wider than s. Typical sizes: $s = 60$ mm, $v = 250$ mm, $\delta = 4°$, $w = 77$ mm

If two thin prisms, each with apical angle $\alpha°$, are placed across the parallel visual axes, bases outwards, the visual axes are each deviated outwards by $\delta°$. Figure 12.12 shows such an arrangement. The stereo-pair images may then have a maximum width $w = 2(s + (v \tan(\delta)))$. The prisms may be set in a spectacle frame, to make a neat stereoscope.

12.12 Viewing with De-centred Lenses

If a convex lens is cut across into two halves, and the halves are set on the visual axes, cut edges outwards, we get a prismatic deviation and also a magnification. Figure 12.13 shows the principle of the lenticular stereoscope invented by Brewster in 1849 and constructed by instrument-makers in Paris and London [1]. This was far more convenient than Wheatstone's mirror stereoscope, and earned great popularity, particularly in the hand-held version designed by Holmes.[5]

[5]Oliver Wendell Holmes Sr, American physician and writer.

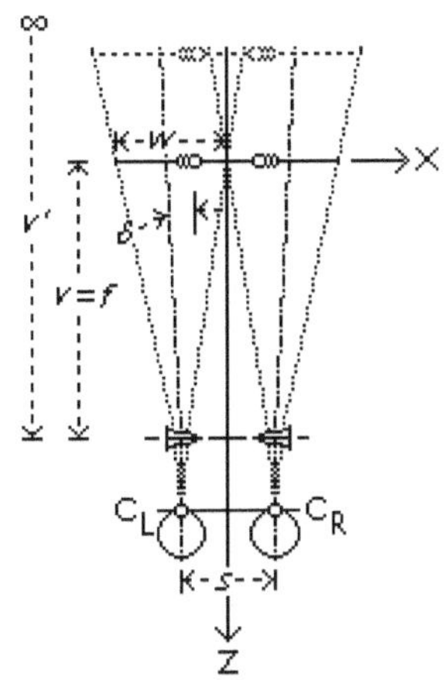

Fig. 12.13 Viewing with de-centred lenses. Each half-lens has focal length f. When the stereo-pair is set at distance f, each eye sees a sharply-focussed virtual image at v' = infinity. The half-lenses give a magnification, and also act as prisms, so parallel visual axes are deviated outwards by $\delta°$

The stereo-pair is the object, placed at distance v = focal distance f of the lenses. We then have a binocular simple magnifier (see Sect. 10.11). The lens equation is

$$1/v + 1/v' = 1/f$$

where v is the object distance, v' is the virtual image distance, and f is the focal distance. From this, since $v = f$, v' = infinity.

The angular magnification μ is the ratio of the angular width of the virtual image (with the lens), to the angular width of the object at the conventional near point, 250 mm (without the lens), which reduces to

$$\mu = 250/f \approx DP/4\,,$$

where f is the focal length in millimetres, and DP is the dioptric power in dioptres.

For example, a typical Brewster-Holmes lenticular stereoscope, as mass-produced in the late 19th century, might have,

- Lens focal length $f = 180$ mm, hence $DP = 1/0.180 = 5.55$ dpt,
- Angular magnification $\mu = 250/180 = 1.39 \approx 5.55/4$,
- Separation of eyes $s = 62$ mm,
- Prismatic deviation $\delta = 4°$,
- Width of each image of stereo-pair $w = s + (f \times \tan(4)) = 62 + (180 \times 0.07) = 75$ mm.

Figure 11.19 shows a stereo-pair for such a viewer.

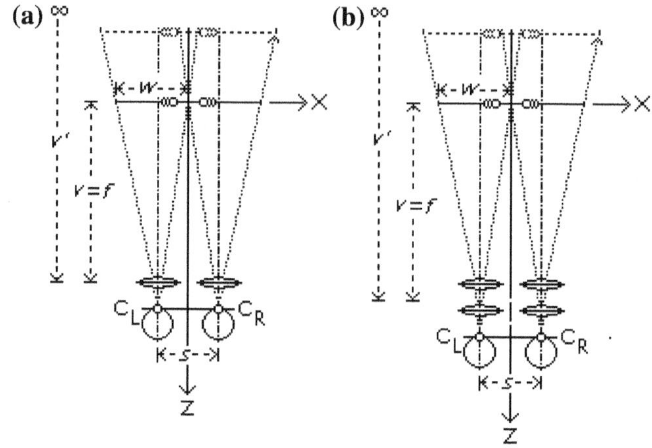

Fig. 12.14 Viewing with centred lenses. **a** Simple lenses. Each lens has focal length f. When the stereo-pair is set at distance f, each eye sees a sharply-focussed virtual image at $v' =$ infinity, with magnification. **b** Compound lenses. Each compound lens has focal length f. When the stereo-pair is set at distance f, each eye sees a sharply-focussed virtual image at $v' =$ infinity, with magnification

12.13 Viewing with Centred Lenses

Figure 12.14a shows a stereoscope using two converging lenses, each with focal length f, centred on the parallel visual axes, s apart. A stereo-pair, placed at distance $v = f$, is the object for these lenses, and the virtual images are in sharp focus, erect, and magnified. The width w of each picture of the stereo-pair is limited to the separation s of the eyes. The angular magnification is $\mu = 250/f$ (where f is in mm). For example, a typical stereoscope of this type might have

- Lens focal length $f = 120\,\text{mm}$, hence $DP = 1/0.120 = 8.33\,\text{dpt}$,
- Distance of stereo-pair $v = 120\,\text{mm}$,
- Angular magnification $\mu = 250/120 = 2.1 \approx 8.33/4$,
- Separation of eyes $s = 62\,\text{mm}$,
- Width of each image of stereo-pair $w = s = 62\,\text{mm}$

Figure 12.14b shows a stereoscope using two compound lenses, each with focal length f, centred on the parallel visual axes, s apart. A stereo-pair, placed at distance $v = f$, is the object for these lenses, and the virtual images are in sharp focus, erect, and magnified. The width w of each picture of the stereo-pair is limited to the separation s of the eyes. The angular magnification is $\mu = 250/f$ (where f is in mm). Helmholtz recommends [2] a stereoscope of this type, with

- Lens 1 focal length $f_1 = 180\,\text{mm}$, hence $DP_1 = 1/0.180 = 5.55\,\text{dpt}$,
- Lens 2 focal length $f_2 = 120\,\text{mm}$, hence $DP_2 = 1/0.120 = 8.33\,\text{dpt}$,
- Distance between lens 1and lens 2 $d = 22.5\,\text{mm}$,
- Focal length of compound $f = (f_1 f_2)/(f_1 + f_2 - d) = (180 \times 120)/(180 + 120 - 22.5) = 78\,\text{mm}$, hence $DP = 1/0.078 = 12.8\,\text{dpt}$,

- Distance of stereo-pair $v = f = 78\,\text{mm}$,
- Angular magnification $\mu = 250/78 = 3.2 \approx DP/4 = 12.8/4$,
- Separation of eyes $s = 62\,\text{mm}$,
- Width of each image of stereo-pair $w = s = 62\,\text{mm}$.

The tubes holding the lenses are adjustable laterally for eye-separation, and longitudinally for focus. Helmholtz remarks that "the irregularities of refraction in prismatic lenses are reduced as much as possible. A special advantage of this type of instrument is that it enables us to use higher magnifications than can be obtained with ordinary [Brewster] stereoscopes and therefore to obtain a still more natural impression. But it must be mentioned that such additional magnification is hardly ever feasible except with glass transparencies."

Objects seen through this instrument "will appear not only much larger and much farther but also more substantial than they do in an ordinary form of stereoscope, which almost always requires too much convergence and so makes the objects appear in low relief. Another very great advantage in this improved apparatus is the complete absence of fatigue and smarting of the eyes which are so apt to occur otherwise."

12.14 State of the Art

A digital stereo-pair of images may be

- Printed as a pair at high resolution, and viewed by any of the classical methods described above,
- Displayed as a pair on a computer screen at lower resolution, and viewed by a classical method,
- Projected to a single frame by red/cyan anaglyph separation (see Sect. 11.12) or by polarized-light separation (see Sect. 11.13), and viewed with glasses,
- Printed combined as vertical stripes, and viewed by lenticular sheet (see Sect. 11.14).

References

1. Brewster D (1856) The stereoscope. John Murray, London
2. Helmholtz H von (1962) Handbuch der Physiologischen Optik. Voss, Leipzig, 3 vols 1856–1867. English translation of 1910 edition (1925) Optical Society of America, republished (1962) Dover, New York
3. Howard HJ (1919) A test for the judgment of distance. Am J Ophthal 2:654
4. Judge AW (1935) Stereoscopic photography, 2nd edn. Chapman and Hall, London. https://archive.org/details/stereoscopicphot029899mbp. Accessed 1 Feb 2014
5. Panum PL (1858) Physiologiske Untersuchungen uber das Sehen mit Zwei Augen. Schwessche Buchhandlung, Kiel
6. Tyler CW (1982) In: Duane's foundations of clinical ophthalmology, Chap. 24. Lippincott, Philadelphia

Chapter 13
Auto-stereo Imaging

The wallpaper effect is the false fusion of laterally-separated features in a single image. It can produce a vivid perception of depth. Examples are given of auto-stereo images constructed by hand, by a perspective construction, and by random-dot computation. A general procedure for converting a depth-map to an auto-stereogram is described, with examples.

13.1 False Fusion

Besides the binocular fusion of two corresponding images (see Sect. 12.3), we may also observe *false fusion* of two similar but disparate images. Figure 13.1a shows the eyes fixating distant point P and two fingers L and R held up on the visual axes. The visual axes are uncrossed relative to the fingers, so an image L′ of the left finger falls on the left fovea, and an image R′ of the right finger falls on the right fovea. These two images correspond, so fuse, and are seen as a single virtual finger at P. However, another image R′ of the right finger falls temporally of the left fovea, and another image L′ of the left finger falls temporally of the right fovea. These two images are disparate, and are seen to the right and left of the fused image. The rather odd perception is of one clear opaque finger in the middle, at distance P (so appearing bigger), and one fuzzy transparent finger either side, at correct distance.

Figure 13.1b shows the eyes fixating near point P and two fingers L and R again held up on the visual axes. The visual axes are now crossed relative to the fingers, so an image L′ of the left finger falls on the right fovea, and an image R′ of the right finger falls on the left fovea. These two images correspond, so fuse, and are seen as a single virtual finger at P. However, another image R′ of the right finger falls nasally of the right fovea, and another image L′ of the left finger falls nasally of the left fovea. These two images are disparate, and are seen to the left and right of the fused image. The perception is now of one finger in the middle, at distance P (so appearing smaller), and one fuzzy finger either side, at correct distance.

A. Parkin, *Digital Imaging Primer*, DOI 10.1007/978-3-540-85619-1_13

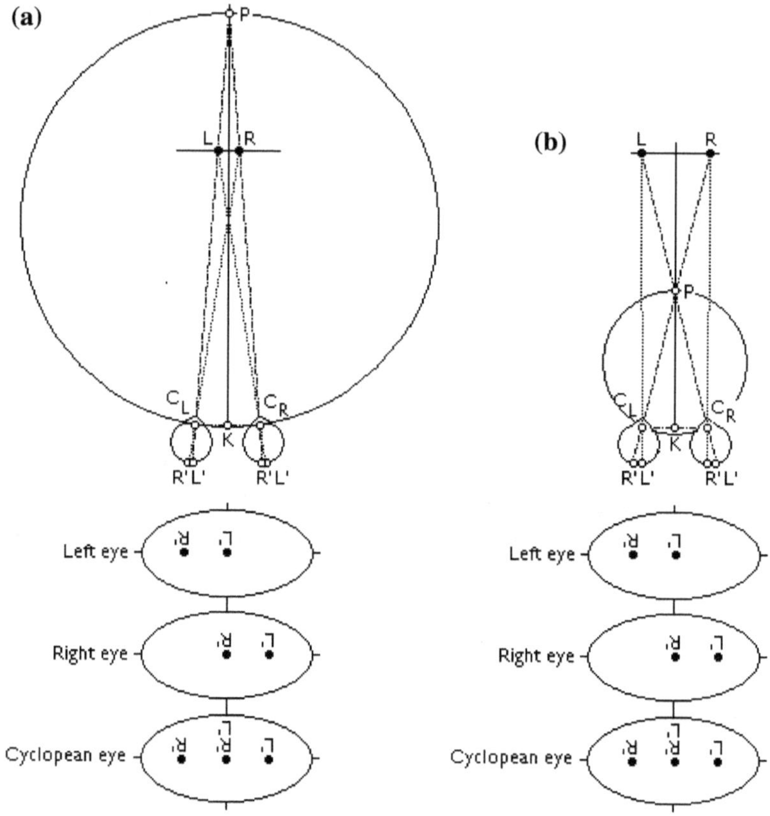

Fig. 13.1 False fusion of two similar objects. **a** Uncrossed viewing: two similar objects L and R are viewed with fixation P farther than the objects. The ellipses show the inverted images for the left eye alone, right eye alone, and both eyes together. There is a false-fused Cyclopean virtual image at distance P. **b** Crossed viewing: two similar objects L and R are viewed with fixation P closer than the objects. There is a false-fused Cyclopean virtual image at distance P

13.2 Wallpaper Effect

Brewster remarked [3] that a pattern with a horizontal repeat, such as a wallpaper, when viewed with crossed visual axes, appears to lie in a plane closer than its actual distance. He wrote:

> If we, therefore, look at a papered wall without pictures, or doors, or windows, or even at a considerable portion of a wall, at the distance of three feet and unite two of the figures, two flowers, for example—at the distance of twelve inches from each other horizontally, the whole wall or visible portion of it will appear covered with flowers as before but as each flower is now composed of two flowers united at the point of convergence of the optic axes, the whole papered wall with all its flowers will be seen suspended in the air at the distance of six inches from the observer! At first the observer does not decide upon the distance of the suspended wall from himself. It generally advances slowly to its new position, and when

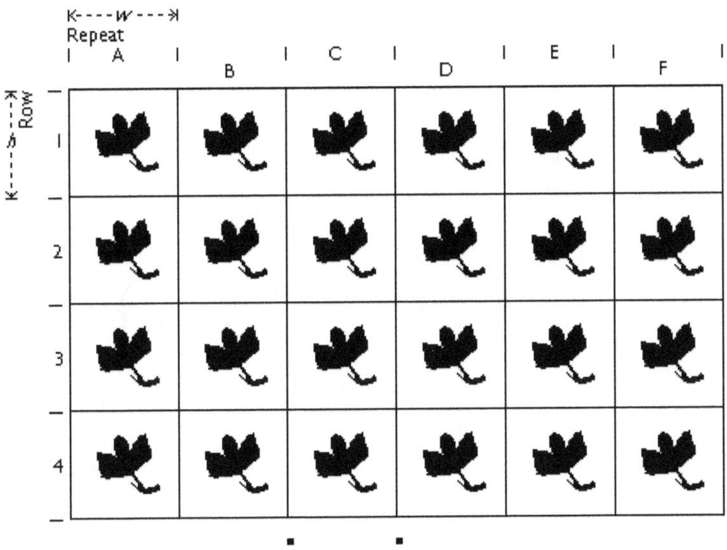

Fig. 13.2 Wallpaper effect. The repeating pattern unit is a square of 64 × 64 px, containing a leaf. A row contains six repeats A, B,..., F, and there are four rows 1, 2,..., 4

> it has taken its place it has a very singular character. The surface of it seems slightly curved. It has a silvery transparent aspect. It is more beautiful than the real paper, which is no longer seen, and it moves with the slightest motion of the head. If the observer, who is now three feet from the wall, retires from it, the suspended wall of flowers will follow him, moving farther and farther from the real wall, and also, but very slightly farther and farther from the observer. When he stands still, he may stretch out his hand and place it on the other side of the suspended wall, and even hold a candle on the other side of it to satisfy himself that the ghost of the wall stands between the candle and himself.

Figure 13.2 is an example of the wallpaper effect. First look at the pattern so that the two guide dots below are distinct. Then cross the visual axes to converge closer, so that the two dots appear as three: a virtual pattern appears quite clear, closer than the page, and somewhat diminished. Next uncross the visual axes as if for a distant object, so that the two dots again appear as three: the pattern now appears farther than the page, and somewhat enlarged. The effect is also got with over-crossed axes (four apparent guide dots).

13.3 Extended False Fusion

The wallpaper effect is an extended case of false fusion. Figure 13.3 shows, in plan, a laterally-repeating pattern at distance v, with three repeats A B and C. In Fig. 13.3a, the visual axes are uncrossed, to converge beyond the pattern at distance v'. At a certain adjustment of the eyes, the image of repeat A in the left eye and the image of

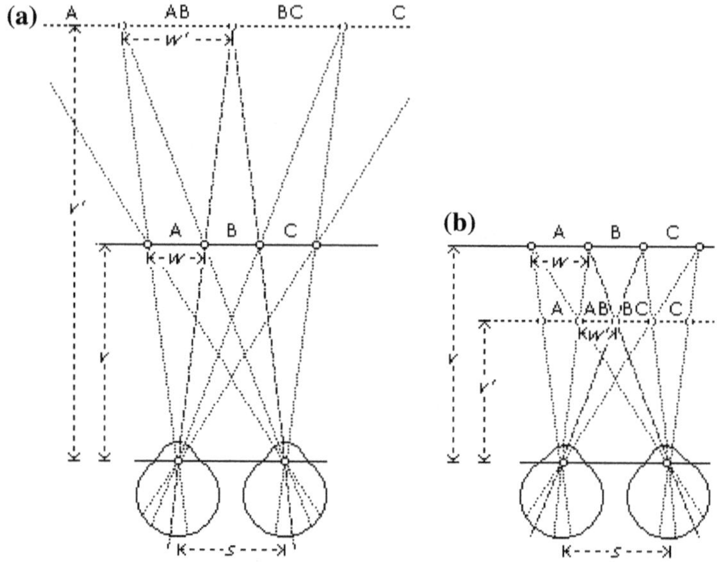

Fig. 13.3 False fusion, uncrossed and crossed viewing. **a** Uncrossed viewing. A laterally-repeating pattern is at distance v, and has three repeats A B C. The visual axes converge beyond the pattern, at distance v'. The left retinal image of repeat A and the right retinal image of repeat B fall on corresponding points, and so fuse to appear as object AB at distance v'; similarly B and C appear as BC. Apparent width w' of AB and BC is greater than actual width w of A B and C. **b** Crossed viewing. The wallpaper pattern is again at distance v, with three repeats A B C. The visual axes converge closer than the pattern, at distance v'. The right retinal image of repeat A and the left retinal image of repeat B fall on corresponding points, and so fuse to appear as object AB at distance v'; similarly B and C appear as BC. Apparent width w' of AB and BC is less than actual width w of A B and C

repeat B in the right eye fall on corresponding retinal points, and so fuse to appear as a single virtual object AB at distance v'. Similarly B fuses with C to appear as BC. So we get two clear enlarged virtual repeats, apparently farther than the actual pattern, and a rather fuzzy A from the left eye alone and C from the right eye alone. In Fig. 13.3b, the visual axes are crossed, to converge closer than the pattern at distance v'. At a certain adjustment of the eyes, the image of repeat A in the right eye and the image of repeat B in the left eye fall on corresponding retinal points, and so fuse to appear as a single virtual object AB at distance v'. Similarly B fuses with C to appear as BC. So we get two clear diminished virtual repeats, closer than the pattern, and a rather fuzzy A from the right eye alone and C from the left eye alone.

If we know the actual pattern distance v and the repeat width w and the separation of the eyes s, we may calculate the apparent distance v'. For uncrossed viewing:

$$v' = (s \times v)/(s - w)$$

or for crossed viewing:

$$v' = (s \times v)/(s + w)\,.$$

For example, uncrossed viewing with $v = 250\,\text{mm}$, repeat width $w = 18\,\text{mm}$, and separation of eyes $s = 62\,\text{mm}$, will give apparent distance $v' = (62 \times 250)/(62 - 18) = 352\,\text{mm}$. Crossed viewing will give $v' = (62 \times 250)/(62 + 18) = 194\,\text{mm}$. Brewster's figures are accurate: if viewing distance $v = 36$ in, repeat width $w = 12$ in, and separation of eyes $s = 2.4$ in, then apparent distance for crossed viewing $v' = (2.4 \times 36)/(2.4 + 12) = 6$ in.

13.4 Auto-stereo

False fusion may be manipulated to make an *auto-stereogram*, that is, a single image which appears at varying depths when viewed with crossed or uncrossed visual axes. Since the apparent depth v' depends on the repeat width w, we may set different repeat widths for different rows, so the rows go to different apparent depths. Figure 13.4 shows row 1 with repeat width $w = 52\,\text{px}$, row 2 with 56 px, row 3 with 60 px, and row 4 with 64 px. Viewed uncrossed, narrowest row 1 appears closest, and widest row 4 farthest. Viewed crossed, narrowest row 1 appears farthest, and widest row 4 closest. In both viewings, the closer rows appear less high and diminished, while the farther rows appear higher and enlarged. When viewed crossed or uncrossed, the

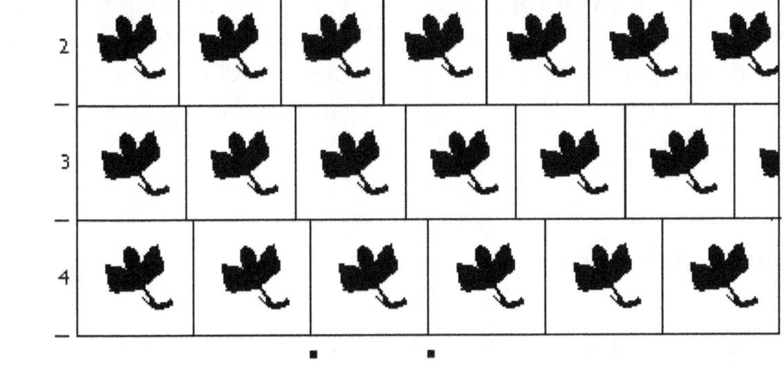

Fig. 13.4 Rows at different depths. There are four rows 1 to 4, each 64 px high. Row 1 has repeat width $w = 52\,\text{px}$, row 2 56 px, row 3 60 px, and row 4 64 px. Viewed uncrossed, row 1 appears closest, and row 4 farthest. Viewed crossed, row 1 appears farthest, and row 4 closest

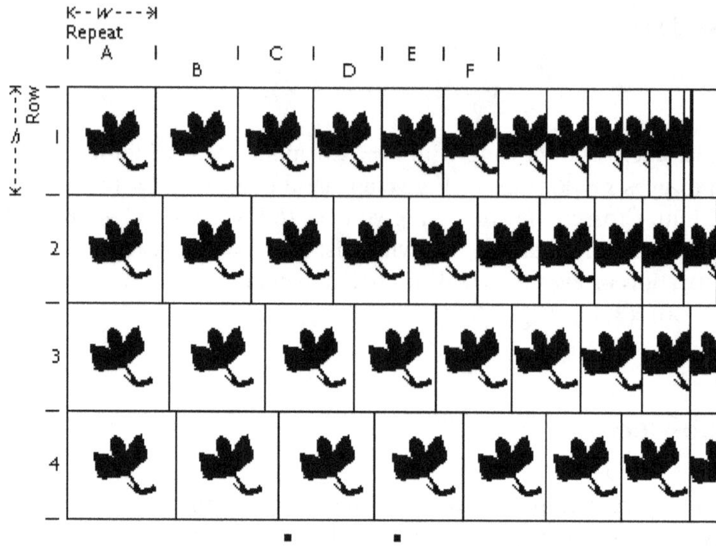

Fig. 13.5 Different depths within row. There are four rows 1 to 4, each 64 px high. Row 1 has successive repeat widths $w = 52, 48, 44, ..., 1$ px. Row 2 has 56, 52, 48, ... px. Row 3 has 60, 56, 52, ... px. Row 4 has 64, 60, 56, ... px. Viewed uncrossed, bottom left units appear farthest, and top right closest. Viewed crossed, bottom left units appear closest, and top right farthest

letters should appear as A AB BC CD DE EF F: five fusions, with un-fused singles left and right. It is also possible to over-cross, so that the two guide dots become four, to see A B AC BD CE DF E F: 4 jumped fusions and 4 un-fused.

Within any row, repeat A has width w, which establishes an apparent depth v'. If repeat B also has width w, fusion AB will appear at depth v'. But if repeat B has width decreased to $w - j$ AB will appear at greater depth (uncrossed) or lesser depth (crossed). Conversely, if repeat B has width increased to $w + j$ AB will appear at lesser depth (uncrossed) or greater depth (crossed). Figure 13.5 shows row 1 with repeat widths w decreasing from 52 px to 1 px; row 2 decreasing from 56 px; row 3 decreasing from 60 px; and row 4 decreasing from 64 px. Viewed uncrossed, the fused units at bottom left appear farthest, and at top right closest. Viewed crossed, the bottom left units appear closest, and top right farthest. In both viewings, the closer rows appear diminished, and the farther rows enlarged.

13.5 Perspective Auto-stereo

In 1435 Alberti described [1] (without illustrations) a geometric construction for a 1-point perspective image of a pavement of square tiles, to form the receding ground plan of any scene (see Sect. 8.15 for an equivalent geometric construction and a numeric calculation). Figure 13.6 shows Alberti's construction:

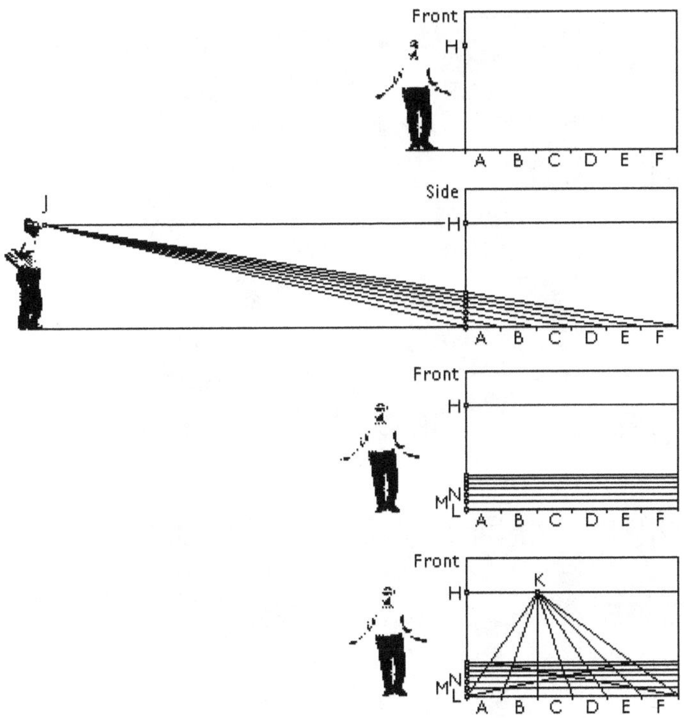

Fig. 13.6 Alberti's construction of 1-point perspective of a square pavement. The steps are described in the text

(a) Draw the required image boundary. On the bottom edge mark any integer number of equal widths A, B, C, … . On the left edge mark the eye-height H of a (scaled) man standing on the bottom edge of the image.
(b) Draw the horizon through H. On the horizon extended to the left mark the viewpoint J at the required viewing distance. Join J and each width mark, to meet the left edge at L, M, N, …
(c) Draw horizontals on M, N,…
(d) On the horizon, mark the required vanishing point K, and join K and each width mark. The pavement can be extended by drawing diagonals.

Buildings or other objects take their vertical scale from the horizontal scale at the appropriate depth on the pavement. A man with his feet at any depth will have his eye-level on the horizon.

Figure 13.7a shows a perspective image of a pavement of 6 × 6 tiles, a back wall of 6 × 4 tiles, and a ceiling of 6 × 6 tiles, drawn by Alberti's construction. It is interesting that such an image will appear in auto-stereo depth when viewed with crossed visual axes, or in a curious reverse perspective depth when viewed uncrossed. Any repeated object can be placed in this space. For example, Fig. 13.7b shows four repeats of a man standing on the third row of the pavement. Viewed with crossed axes, five repeats appear in depth, the outer two fuzzy.

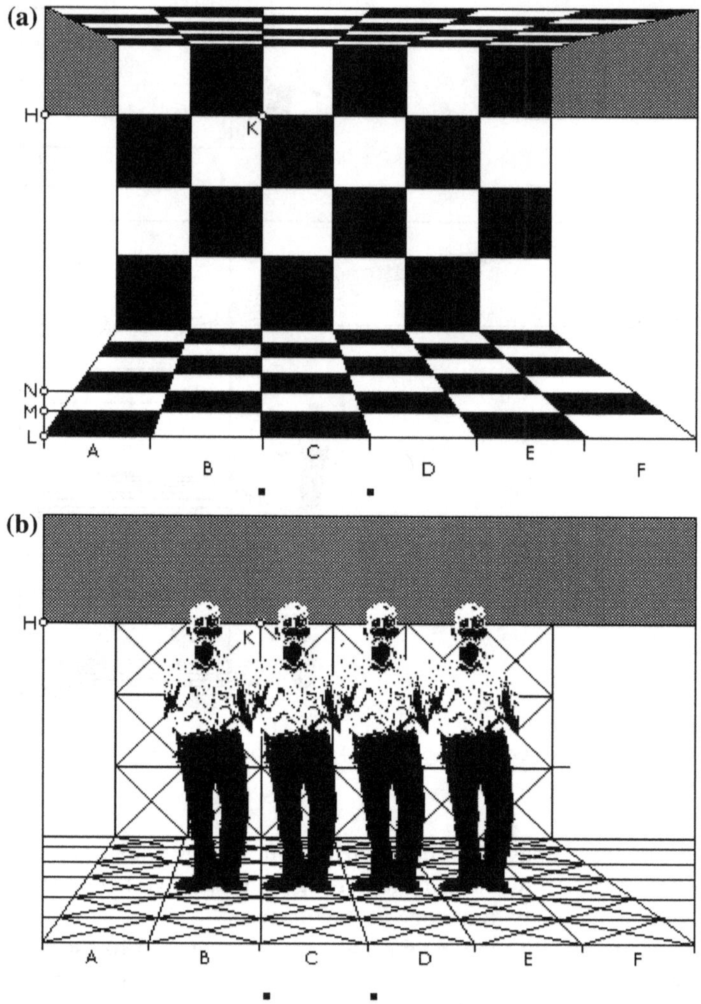

Fig. 13.7 Perspective auto-stereo. **a** Each row of the pavement has a perspective-diminished height and width, with vanishing point K on horizon H, and likewise the ceiling. The back wall units are square, at the depth of the far edge of the pavement. **b** The four repeats of a man stand on the third row from the front of the pavement; when viewed with crossed axes, five repeats appear in depth

13.6 Random-dot Auto-stereo

A *random-dot auto-stereogram* has a laterally-repeating pattern unit composed of small dots, randomly set as black or white (or any range of intensities). Figure 13.8 shows an example where the bottom row has a random-dot unit 16 dots wide, the next row up has the unit curtailed to 15 dots wide, and so on to the top row where the unit is curtailed to one dot wide. The auto-stereo effect is a recession of depth

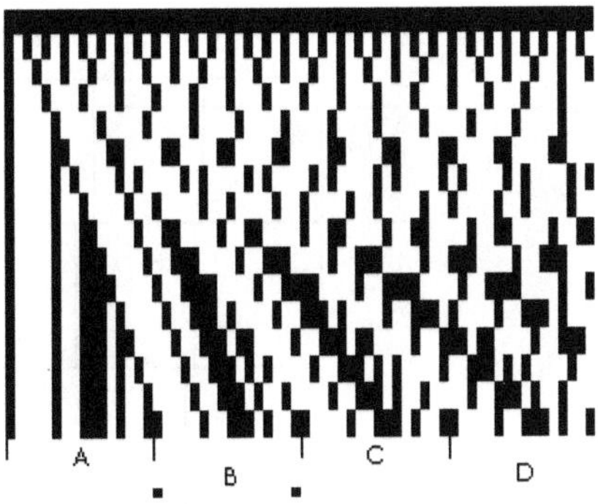

Fig. 13.8 Random-dot auto-stereogram. Bottom row has random-dot unit 16 dots wide, next row up 15 dots, ..., to row 1 dot wide. Compare Fig. 13.7

like Fig. 13.7, but not a perspective. In a perspective, a far object is imaged smaller than a same-sized near object, whereas in an auto-stereogram a far object is imaged the same size as a near object, and consequently looks bigger.

Figure 13.9 is like the pattern in Fig. 13.5, but now made of random-dot units. It is like a random-dot stereo-pair (see Sect. 11.11) in that depth is not present in ordinary viewing, but only in crossed or uncrossed binocular viewing. It is unlike a random-dot stereo-pair in that there is only a single image, depending on false fusion for depth. Simple random-dot images of this kind can be constructed quite easily in an image editor.

13.7 Computed Random-dot Auto-stereo

A *computed random-dot auto-stereogram*, often called a single-image random-dot stereogram (SIRDS), is an ingenious exploitation of false fusion, invented by Tyler and Clarke[1] in 1990 [7]. Clarke's original auto-stereogram programs were written in BASIC and run on an Apple II computer. The simple look-back algorithm has subsequently been improved by others [5, 6]. Computed random-dot auto-stereograms have had considerable commercial success as hidden-image puzzles, both in the original form and in various elaborations [2, 4].

To compute a random-dot auto-stereogram there are two stages. In stage one, the user creates a *depth-map*; that is, an image where the depth of each pixel is set as the colour index I. A 1-bit depth-map has just two depths: $I_2 = 0$ Black or 1 White. A 4-bit depth-map has 16 depths: 0 Black $\leq I_{16} \geq$ 15 White. An 8-bit depth-map has 256 depths: 0 Black $\leq I_{256} \geq$ 255 White. The user also specifies a *base-width* for the repetition interval of the lowest depth, and a *step-width* for a jump to the next depth. In stage two, a program reads the depth-map row by row, replaces the specified depths

[1] Christopher W. Tyler (1933-) researcher at Bell Laboratories 1974–1975 (under Julesz), researcher at Smith-Kettlewell Institute since 1975, Professor at City University, London since 2013. Maureen B. Clarke, programmer at Smith-Kettlewell Institute.

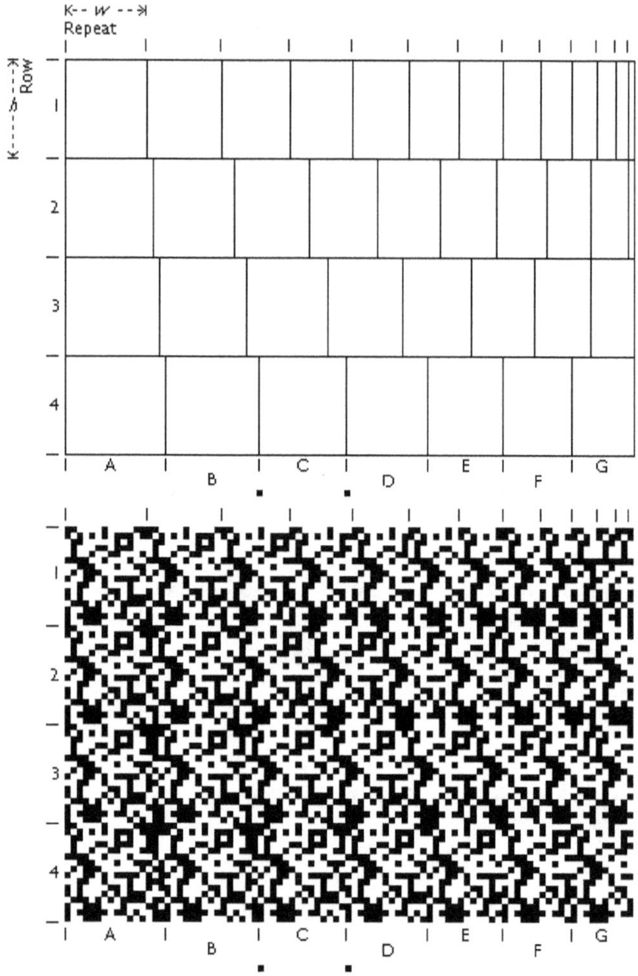

Fig. 13.9 Random-dot auto-stereogram. There are four rows 1 to 4, each 16 dots high. Row 1 has successive repeat widths $w = 13, 12, 11, \ldots, 1$ dots. Row 2 has 14, 13, 12, ...dots. Row 3 has 15, 14, 13, ...dots. Row 4 has 16, 15, 14, ...dots. Viewed uncrossed, *bottom left* units appear farthest, and *top right* closest. Viewed crossed, bottom *left units* appear closest, and *top right* farthest. Compare Fig. 13.5

by different-width runs of random-intensity pixels, and finally outputs the required auto-stereogram.

The row-by-row processing is illustrated in Fig. 13.10. Say the chosen base-width = 3 px and the step-width = 1 px. Figure 13.10a shows a 1-bit depth-map image, 32×24 px, with row 11 marked (row 0 is at the bottom). Figure 13.10b shows the depth-map image file for row 11 (4 B). The program reads bytes from the depth-map and unpacks the bytes to bits in a depth vector, Fig. 13.10c. The program sets up a random vector, Fig. 13.10d, of pixels. Processing depth vector (**c**) pixel-by pixel:

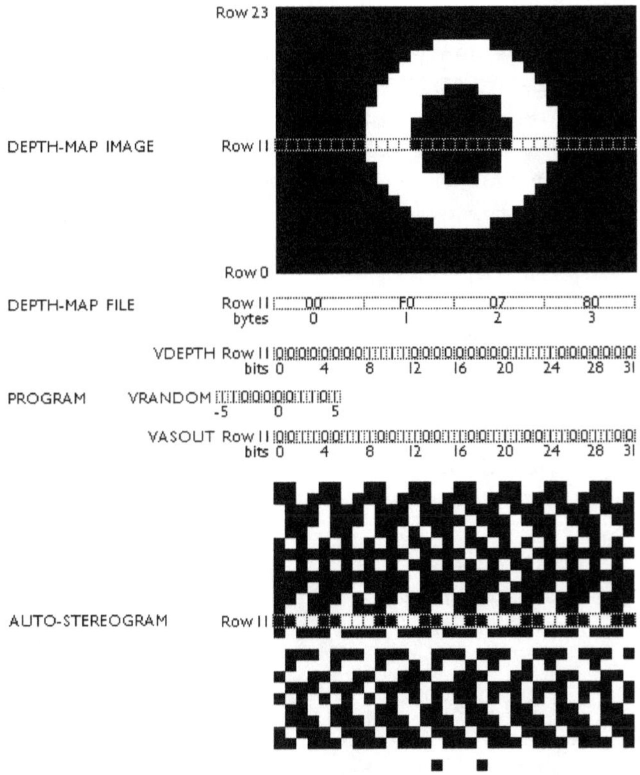

Fig. 13.10 Processing for one row of a computed auto-stereogram. **a** 1-bit depth-map image, with row 11 marked. **b** Row 11 in the file for this depth-map. **c** Program depth vector for row 11. **d** Program random vector. **e** Program output vector. **f** Completed auto-stereo image

- Starting depth = 0, so current repetition width = 4.
- For pixels 0–3, copy from random vector to output vector.
- For pixels 4–7, repeat copy from random vector to output vector.
- At pixel 8, depth changes to 1, so current repetition width becomes 5, so start a new run of copying from random vector to output vector.
- At pixel 12, depth reverts to 0, so start a new run of width 4.
- At pixel 21, depth changes to 1, so start a new run of width 5.
- At pixel 25, depth reverts to 0, so start a new run of width 4.

When the completed image is viewed with crossed visual axes, the ring stands out in front of the surround.

A program which produces an auto-stereogram from a given 4-bit depth-map is briefly described below (QuickBASIC source code available on the SpringerLink web-site). Most of the program deals with input and output. The central conversion of depths to repetition widths is done in the sub-program CALCAS.

```
REM   **  Program ASTEREO4.BAS to make an auto-stereogram
REM   **     from a 4-bit depth-map. Depth-map is 4-bit BMP
REM   **     file, width a multiple of 32 pixels.User enters
REM   **     filename of depth-map, filename of auto-stereo-
REM   **     gram output, base-width of repeat, and step-
REM   **     width of change of depth. Program processes
REM   **     each row of depth-map to row of auto-stereogram.
REM
REM   **     Main program
REM          User enters file-names, base_width, and step-width
             CALL R4W1HE
             FOR YROW% = 0 TO MAXY%
                CALL CRERAN
                CALL RDEPT4
                CALL CALCAS
                CALL WASOUT

             NEXT YROW%
             CALL REVBHEAD
END


            SUB CALCAS
REM   **    Sub-program to calculate auto-stereogram row
REM         VASOUT from depth-map row VDEPTH and random vector
REM         VRANDOM.
REM   **    Process first pixel in row
REM   **    Process rest of pixels in row
END SUB
            SUB CRERAN
REM   **    Subprogram to create a sequence of pixels in
REM   **    random vector VRANDOM, each set as intensity
REM   **    0 or 1 at random.
END SUB
            SUB R4W1HE
REM   **    Subprogram R4W1HEAD to read header of 4-bit
REM   **       .BMP input file #1, and preliminary-write
REM   **       header and colour-table of 1-bit output file
REM   **       #2 (to be revised later).
END SUB
            SUB RDEPT4
REM   **    Subprogram to read one row of 4-bit depth-map
REM   **       file #1 and load into depth vector VDEPTH.
REM   **       The data are 0 to 15.
END SUB
```

```
            SUB REVBHEAD
REM   **    Revises header of file #2 with values in bytes.
END SUB
            SUB WASOUT
REM   **    Subprogram WASOUT writes data from auto-stereogram
REM   **    vector VASOUT to 1-bit .BMP file #2. Take VASOUT
REM   **    values 0 or 1 in groups of 8 and convert to byte.
REM   **    Finish row on a long (4-byte) boundary.
END SUB
```

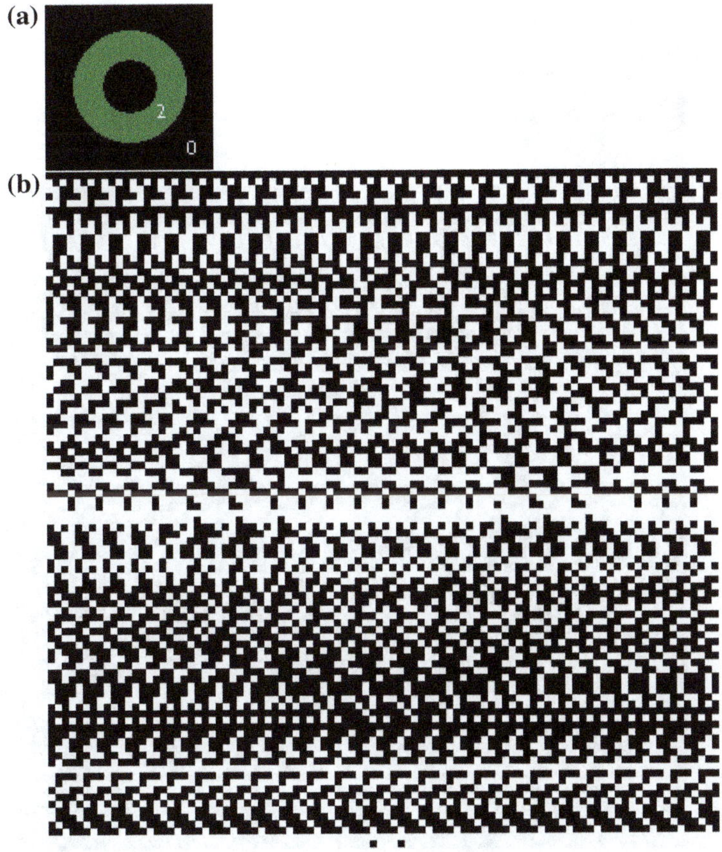

Fig. 13.11 Auto-stereogram with two depths. **a** Depth-map 96 × 96 px, with two depths coded 0 and 2, base-width 3 px, step-width 1 px. **b** Auto-stereogram (enlarged ×4): the ring should stand out in front of the surround. The pair of guide dots should appear as three for crossed viewing. If four dots appear, you are over-crossing and multiple rings are produced

13.8 4-bit Random-dot Auto-stereogram Examples

Figure 13.11 shows a 4-bit depth-map (a) which uses only two of the 16 possible coding colours, and the auto-stereogram (b) produced by the program described above. The image has 96 × 96 px; the base-width is 3 px; and the step-width is 1 px. The pair of guide dots are base-width + 1 apart.

Figure 13.12 shows a 4-bit depth-map (a) which uses only four depths, and the auto-stereogram (b). The image has 96 × 96 px; the base-width is 3 px; and the step-width is 1 px. The pair of guide dots are base-width + 1 apart. Compare this image with the projected relief in Sect. 8.20, and with the stereo-pair in Sect. 11.11.

(a) (b)

Fig. 13.12 Auto-stereogram with four depths. **a** Depth-map 96 × 96 px, with four depths coded 0, 1, 2, and 3, base-width 3 px, step-width 1 px. **b** Auto-stereogram (enlarged ×4): three nested squares should stand out from the surround. The pair of guide dots should appear as three for crossed viewing. If four dots appear, you are over-crossing and multiple squares are produced

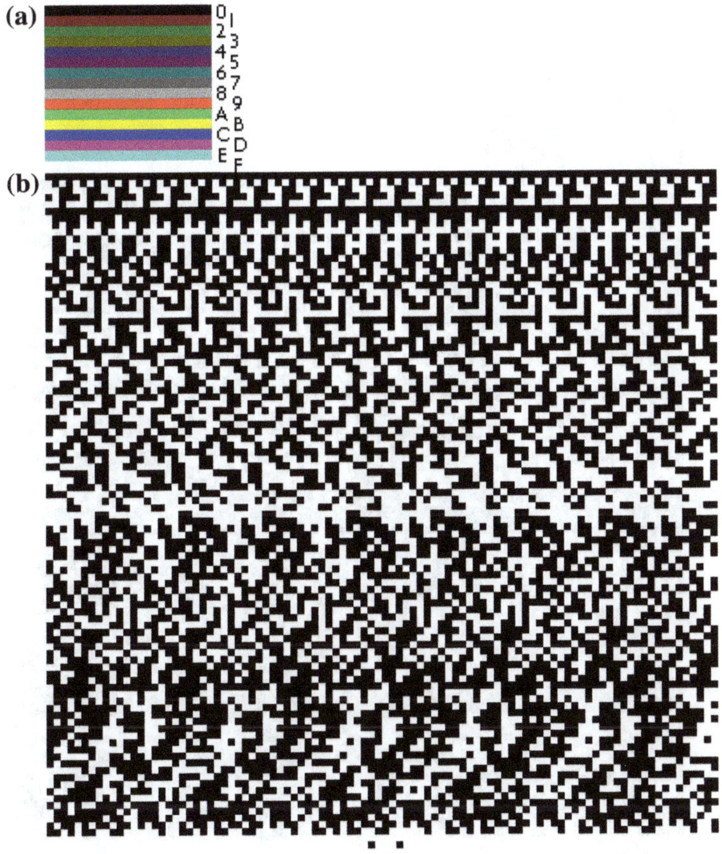

Fig. 13.13 Auto-stereogram with 16 depths. **a** Depth-map 96 × 96 px, with 16 depths coded 0 to F, base-width 3 px, step-width 1 px. **b** Auto-stereogram (enlarged ×4): 16 rows should stand out. The pair of guide dots should appear as three for crossed viewing. If four dots appear, you are over-crossing and the depths are muddled

Figure 13.13 shows a 4-bit depth-map (a) which uses all 16 depths, and the auto-stereogram (b). The image has 96 × 96 px; the base-width is 3 px; and the step-width is 1 px. The pair of guide dots are base-width + 1 apart. Compare this depth-map-driven image with the manually-constructed image in Fig. 13.8.

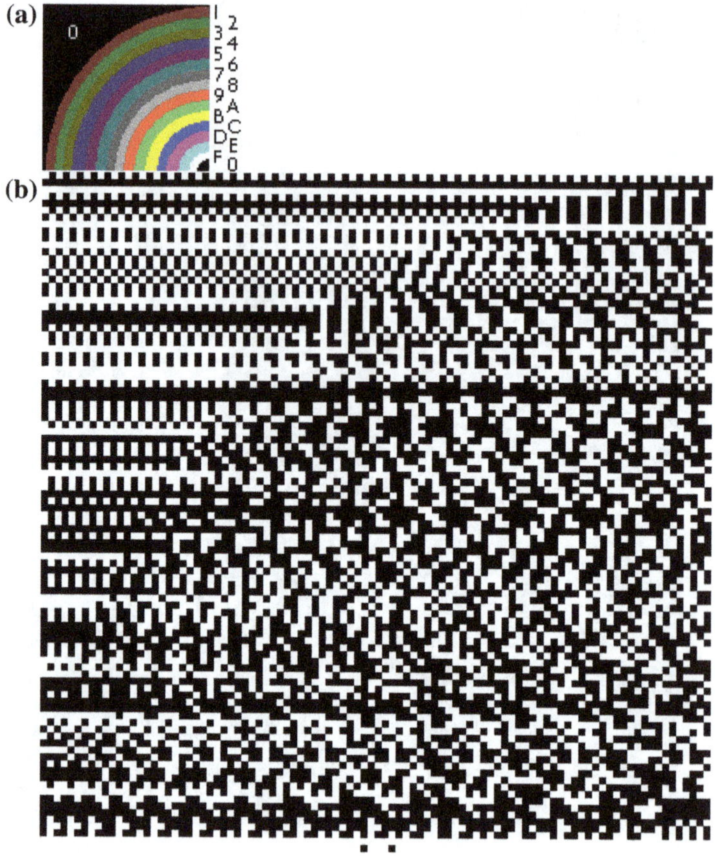

Fig. 13.14 Auto-stereogram with 16 depths. **a** Depth-map 96 × 96 px, with 16 depths coded 0 to F, base-width 3 px, step-width 1 px. **b** Auto-stereogram (enlarged ×4): one octant of a sphere should stand out from the surround. The pair of guide dots should appear as three for crossed viewing. If four dots appear, you are over-crossing and the depths are muddled

An auto-stereogram contains only flat patches at various apparent depths. But if these are suitably arranged we can get a layered approximation to a curved surface. Figure 13.14 shows a depth-map (a) which uses all 16 depths, and the auto-stereogram (b) which approximates an octant of a sphere. The image has 96 × 96 px; the base-width is 3 px; and the step-width is 1 px.

A depth-map can be originated by scanning a flat object in 1-bit silhouette, then changing figure and ground to 4-bit colour codes. Figure 13.15 shows such a depth-map (a) where the scan of a hand has been given two depths. In the auto-stereogram (b) the silhouette of the hand stands out. The image has 128 × 160 px; the base-width is 4 px; and the step-width is 1 px.

(a) (b)

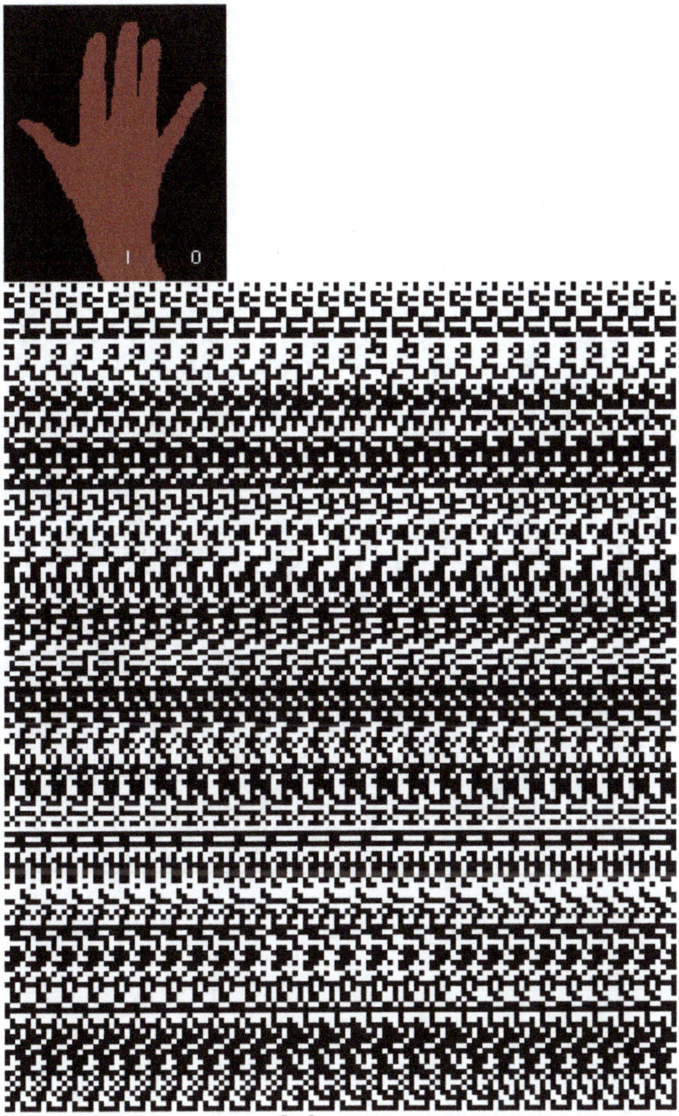

Fig. 13.15 Auto-stereogram from scanner. **a** Depth-map 128 × 160 px, with depths coded 0 and 1, base-width 4 px, step-width 1 px. **b** Auto-stereogram (enlarged ×3): the silhouette of the hand should stand out from the surround

A depth-map can also be originated by taking a camera image, reducing it to 4-bit colour (see Sect. 28.8), then changing the colours to depth-codes in suitable patches. All light-created detail must be obliterated, otherwise the result is hopeless confusion. Figure 13.16 shows (a) a camera image reduced to 4-bit colour; (b) the

Fig. 13.16 Auto-stereogram from camera. **a** Camera image 224 × 150 px, reduced to 4-bit colour. **b** Depth-map, with depths coded 0 to 8. **c** Auto-stereogram, base-width 17 px, step-width 1 px (enlarged ×2): the dinghy should stand out from the several depths of water and background

depth-map with changed colour-codes; and (c) the resulting auto-stereogram. The image has 224 × 150 px; the base-width is 17 px; and the step-width is 1 px. It must be said that this is usually a tedious task, and the result is usually unrewarding. Good results can sometimes be got from photographs of buildings or rooms.

References

1. Alberti LB (1435) De Pictura. English edition: Alberti LB (1970) (translation: Spencer JR) Yale UP, New Haven. http://www.noteaccess.com/Texts/Alberti/. Accessed 1 Feb 2014
2. Baccei TJ, Salitsky R (1994) Random dot stereogram and method for making same. US Patent 5,371,672. 6 Dec 1994
3. Brewster D (1844) On the knowledge of distance given by binocular vision. Trans Roy Soc Edinb 15:663–674
4. Magic Eye Inc (1994) Magic Eye: 3D hidden treasure. Andrews and McNeel, Kansas City
5. Steer WA (1995) Stereograms: technical details. http://www.techmind.org/stereo/stech.html. Accessed 1 Feb 2014
6. Thimbleby H, Witten IH, Inglis S (1994) Displaying 3D images: algorithm for single-image random-dot stereograms. IEEE Comput 27:38-48. http://cs.swan.ac.uk/~csharold/sirds/ieee3d.pdf. Accessed 1 Feb 2014
7. Tyler CW, Clarke MB (1990) The autostereogram: stereoscopic displays and applications. Proc SPIE 1256:182-197. Previously available at: http://www.ski.org/CWTyler_lab/CWTyler/TylerPDFs/TylerClarkeAutosterSPIE1979.pdf. But see the unhappy story at: https://sites.google.com/a/eidactics.com/eidactics/home/s-k

Chapter 14
Perception of Space

We routinely impose different coordinate spaces on the physical space around us, for different purposes. For general investigation of objects (actual or imagined), Euclidean space is standard, where lengths and angles remain unchanged when moved anywhere in space. For measurement of objects, the instruments in use impose a discrete coordinate space. For representing solid objects as flat images, by geometry or by calculation or by a lens, we project into an octant of harmonic space, where lengths get smaller with distance from the centre of projection. And for viewing, an elliptic space is natural, where lengths get smaller and eventually disappear in a sphere around the current centre.

14.1 Choice of Coordinate Space

For general investigation of objects (actual or imagined) in the physical space around us, we use Euclidean coordinate space as standard. There is no universally-privileged location. A frame of reference OXYZ is defined for current purposes by three base points O U x and a right angle (see Sect. 6.2). An object can be placed anywhere and moved anywhere without changing its shape.

For measuring objects, we use perforce a discrete coordinate space, where the measuring instruments impose a finite limit of precision, and the coordinate values are integers (see Chap. 7). To convert Euclidean coordinates to discrete coordinates, all that is needed is to truncate, or round, real numbers to integers. Measuring is a "touch-and-go" regime, where a measuring rod and an angle protractor can be carried anywhere and matched against an object throughout space.

For representing space objects as images on a plane, we project from Euclidean space to harmonic space. The projection can be done by geometric construction, or by numeric calculation of coordinates (see Chap. 8), or optically by a lens (see Chap. 9). There is a currently-privileged location, the centre of projection. Image

A. Parkin, *Digital Imaging Primer*, DOI 10.1007/978-3-540-85619-1_14

lengths get steadily smaller with object distance from the centre, and angles change correspondingly. If we send an assistant with a measuring rod out from the centre, the assistant and the rod get steadily smaller and eventually disappear, though his reports of measurements continue unchanged. Projection to the harmonic plane is limited to at most one octant of space. Projection is a "see-and-turn" regime.

For viewing (see Chaps. 10 and 12), elliptic space is natural. There is a currently-privileged location, the Cyclopean centre of the viewer. Objects in space are projected to continually-changing retinal images, and retinal images are combined to a stable perception of the whole scene. A different viewer has a different reference frame, in which near objects look big and far objects look small, eventually vanishing. For a given viewpoint, elliptic coordinates of objects are the same as Euclidean coordinates, but elliptic space represents the entire visible world in a sphere of arbitrary size.

A very good imaging procedure would be to project objects surrounding a given centre, not to flat images in harmonic space, but to spherical-cap images in elliptic space. This cannot be done with ordinary displaying and printing technology (but may perhaps be developed with stereo movie-goggles and very elaborate software).

14.2 Phenomenology of Perceived Space

The core procedures of digital imaging, from computer input to output, are physically well-defined, and the initial stages of viewing, from intensity of light to neural signals, are quite well-understood. But the category leap from neural signals to perceptions in consciousness is a deep mystery [2].

Phenomenologically, my perceived sensation of Space corresponds to a geometrical elliptic space (see Sect. 6.9 and [1]), where my eyes are at origin O, and an incoming ray of light is a straight line on O, as shown in Fig. 14.1a. If I stand on an empty plain and look ahead my gaze is centred along O $-z$ and extends peripherally almost to a hemisphere bounded by a vertical circle x y $-x$ $-y$, as shown in Fig. 14.1b. If I sweep my gaze right round I see the horizon as a straight line, which is evidently a circle $-z$ $-x$ z x at great distance. I see the ground beneath my feet at P as a flat surface, which is evidently a shallow dish curving up in all directions to meet the horizon. If I stand between railway lines either side of P, I see the rails as straight lines which are evidently shallow curves meeting on the horizon at $-z$. If I turn round I see the railway lines again meeting the horizon at z. If there is another pair of railway lines crossing my pair at Q, I see them meeting at $-x$ and at x. If there is a layer of cloud above, I see it as flat everywhere, evidently a bowl which goes down all round to meet the horizon. If the sky is unclouded I see a blue hemisphere at infinite distance, the top half of the bounding sphere of my elliptic visual space. I see sizes diminishing in harmonic progression with increasing distance.

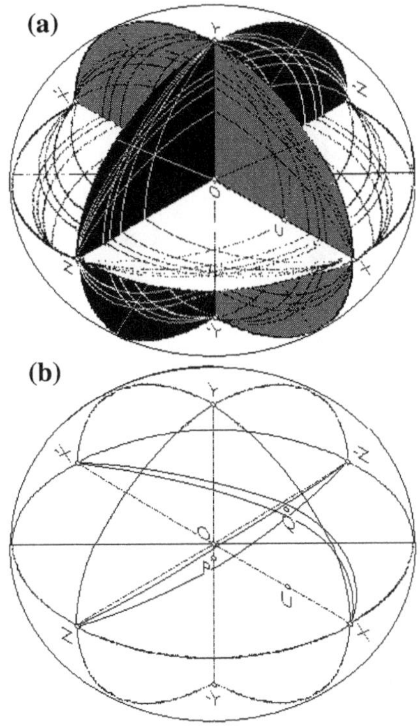

Fig. 14.1 Perception of space. **a** Elliptic space $Oxyz$. **b** Perceptual space. Viewer's eyes are at origin O, and viewer's feet are on the ground at P. Horizon is circle $-z$ $-x$ z x. Railway lines are curves $-z$ P z and $-x$ Q x. Sky is hemisphere y $-x$ $-z$ x z

For small parts of perceptual space, I am accustomed to use a correspondence with a geometrical Euclidean space, with centre at O. And for making pictures, I am accustomed to projecting the contents of a small viewing cone to a Euclidean plane.

References

1. Lehar S (2004) The boundaries of human knowledge: a phenomenological epistemology. http://cns-alumni.bu.edu/~slehar/webstuff/book2/Boundaries.html. Accessed 1 Feb 2014
2. Smythies JR (1994) The walls of Plato's cave. Avebury Press, Aldershot

Part III
Intensity in Digital Imaging

Chapter 15
Photometry

In an image, difference of location is marked by difference of intensity of light, emitted or reflected. Photometry measures intensity of light as seen by a standard eye, from utter darkness to dangerous brightness, without regard to colour. This Chapter describes the photometry of light sources and of surfaces, with examples.

15.1 Electro-magnetic Radiation

Electro-magnetic radiation (EMR) is a stream of accelerated electric charges (such as the photons emitted by the sun), and/or (in an irreducible duality) a self-propagating wave (such as the coupled electric and magnetic fields emitted by a wire conducting an alternating current). The *power* of a source of EMR is the rate at which it emits energy, measurable in Watts = Joules per second [1]. Constant energy spreads over an ever-increasing spherical wave-front, so the *intensity* at distance r metres from a point source is proportional to $1/r^2$, measurable in Watts per m^2. In free space, velocity is 299,792,458 m/s (the value is exact because the metre has been defined thus in SI since 1983). In any other medium, velocity is less. Frequency is measurable in Hertz, and wavelength in metres. The energy of an elementary particle is measurable in electron-volts: it increases with frequency, so decreases with wavelength.

$$velocity = frequency \times wavelength$$

EMR varies over a very large range of wavelengths, from pico-metres (gamma-ray), through nano-metres (X-ray, ultra-violet, visible light), millimetres (infra-red), centimetres (microwave), to kilometres (radio).

A. Parkin, *Digital Imaging Primer*, DOI 10.1007/978-3-540-85619-1_15

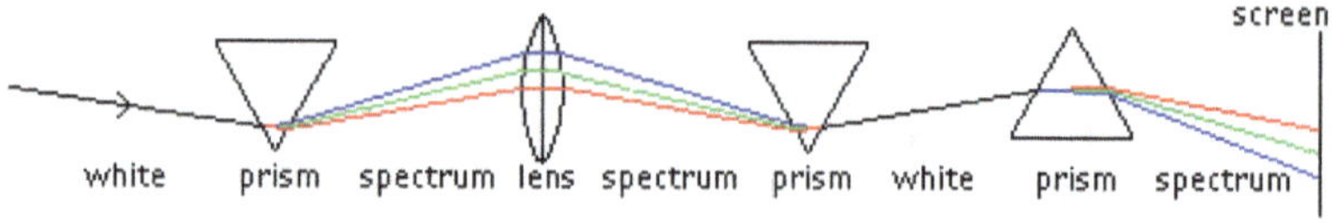

Fig. 15.1 Dispersion and re-combination of light. Collimated white light is dispersed by a prism to a spectrum at a converging lens. The lens converges the spectrum to another prism, which combines the wavelengths to white light. The white light is again dispersed to a spectrum on a screen

15.2 Radiometry

In physics, *radiometry* [5] is the measurement, by instruments, of EMR in the ultra-violet, visible-light, and infra-red bands of the electro-magnetic spectrum.

When EMR hits matter it is variously reflected, absorbed, or transmitted. At infra-red and longer wavelengths the energy of photons is dissipated as heat. At visible and shorter wavelengths the energy is sufficient to change the chemical bonds of molecules, either transiently as in retinal pigments[1] or chlorophyll,[2] or with permanent damage to DNA and other biological molecules.

At the boundary of a transparent medium, such as glass, a beam of radiation undergoes *dispersion*, whereby shorter wavelengths are slowed more than longer wavelengths. Newton demonstrated about 1670 that a prism spreads a collimated beam of white light (such as that arriving from the sun) into a spectrum, with red deviated least and violet deviated most; and a spectrum can be re-combined into white light, as shown in Fig. 15.1.

A slit placed in the spectrum isolates a narrow band of wavelengths. The power per unit area at various wavelengths can be measured with some form of bolometer[3] which detects changes in electrical properties of a metal exposed to EMR. The *spectral power distribution* (SPD) of a source can then be plotted as power against wavelength, as shown in Fig. 15.2a.

An ideal *equal-energy* white light would have power equal at all wavelengths. Sunlight has a mild peak at around 500 nm, and is somewhat lower in short wavelengths than in long. Light from a blue sky but not direct sunlight is high in short wavelengths and steadily declines to longer wavelengths. A tungsten lamp is low in short wavelengths and steadily increases to longer. Fluorescent sources have various power distributions according to the mixture of phosphors used in their manufacture.

An opaque surface has a characteristic SPD of reflection. Figure 15.2b, for example, shows SPDs of reflection of white light by white paper, grass, violets, and black ink.

---

[1]Discovered 1958 by George Wald (1906–1997), Harvard University.

[2]Discovered 1937 and 1960 by Robert Hill (1899–1991), Agricultural Research Council.

[3]Invented 1879 by Samuel Pierpoint Langley (1834–1906), American astronomer and pioneer of flight. A bolometer is an instrument of remarkable sensitivity: Langley detected the heat from a cow at a quarter of a mile. Named from $\beta o \lambda \eta$ sunbeam or ray.

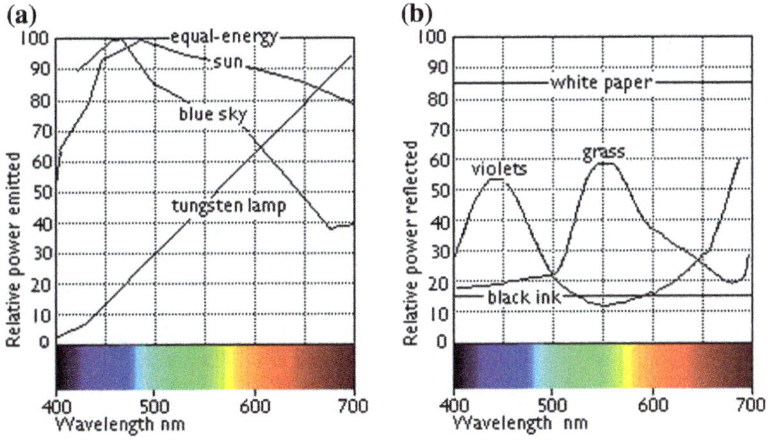

Fig. 15.2 Spectral power distributions. **a** Relative power of several light sources at different wavelengths. **b** Relative reflection of white light by paper, grass, violets, and black ink

15.3 Photometry

Visible light is EMR which can be detected by the human eye, that is with wavelengths from about 380 nm to about 750 nm. It is emitted by various sources, either natural, such as the sun, or artificial, such as lamps and computer screens. Light is reflected variously by surfaces of objects, such as the moon, landscapes, people, and computer printouts.

Photometry [2, 7] is the measurement, by eye or by instruments which match the sensitivity of the eye, of energy in visible light. It is like radiometry, but restricted to the visible spectrum, and adjusted to the varying sensitivity of the eye across this spectrum.

The sensitivity of a standard observer when viewing light at different spectral wavelengths is specified in the CIE V(λ) photopic and the V′(λ) scotopic curves[4] shown in Fig. 15.3. The photopic curve plots a standard observer's sensation of brightness against wavelength when the eye is adapted to day vision (see Sect. 10.2), for luminance levels from 1 to 10^6 cd/m^2. The scotopic curve is when the eye is adapted to night vision, for luminance levels 10^{-2}–10^{-6} cd/m^2. In mesopic vision from 10^{-2} to 1 cd/m^2 both are active. Photopic vision is dominated by cones in the fovea, which are most sensitive to wavelengths around 555 nm and gradually less sensitive to wavelengths on either side. Scotopic vision is dominated by rods in the periphery of the retina, which are most sensitive to wavelengths around 505 nm. Thus if a red object and a blue object have the same brightness in day vision, when dusk comes

[4] Averaged from many experiments and issued by La Commission Internationale de l'Eclairage (CIE) in 1924 and 1951 [2].

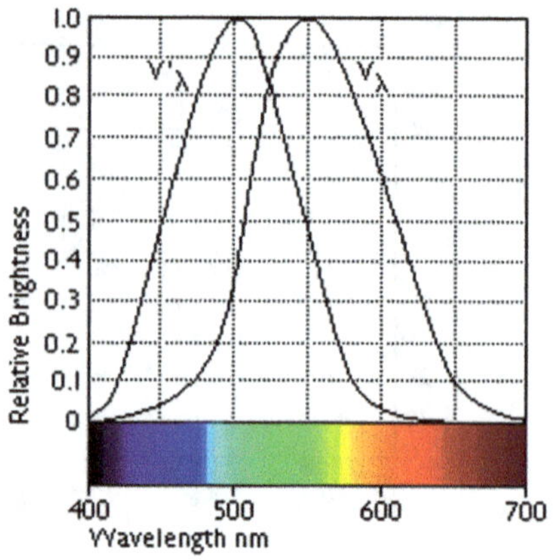

Fig. 15.3 CIE V(λ) photopic curve and V′(λ) scotopic curve: relative sensation of brightness received by a standard observer across the visible spectrum in day vision and in night vision

on and the eye gradually adapts to night vision the blue will appear brighter and the red darker.[5]

Photometrically, given a light of known spectral power distribution, the brightness sensed by the observer is found as

$$B = k\Sigma(P_\lambda \times V_\lambda)\,,$$

where B is brightness, P_λ is the power at wavelength λ, V_λ (or V'_λ) is the CIE standard value at wavelength λ, and k is a constant.

15.4 Radian and Steradian

In the plane, angles are measured in degrees or in radians. A right angle measures 90° or π radians. Figure 15.4a shows a circle with radius r, and a sector with angle ϑ cutting off circumferential length c on the circle. The radian measure of the angle is

$$\vartheta = c/r \text{ radians}\,.$$

Clearly, one radian is the measure when $c = r$, independent of the size of the circle. Since the complete circumferential length of a circle is $2\pi r$, degree and radian measure are related as

$$\text{degrees} = \text{radians} \times (180/\pi)$$
$$\text{radians} = \text{degrees} \times (\pi/180)\,.$$

[5]First described in 1825 by Jan Evangelista Purkyně (1787–1869) Professor of Physiology at Prague University.

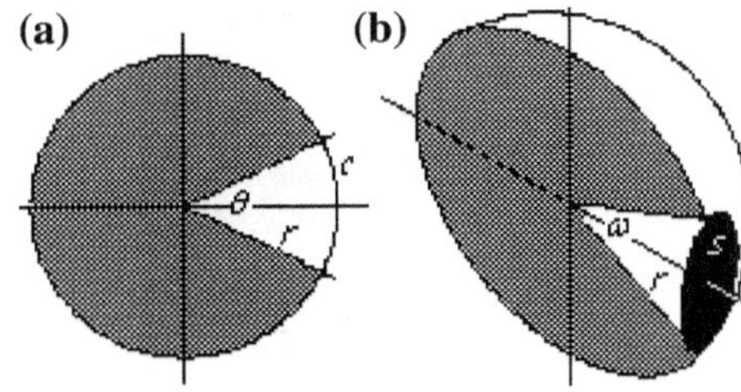

Fig. 15.4 Radian and steradian. **a** Radian measure of plane angle $\theta = c/r$. **b** Steradian measure of solid angle $\omega = s/r^2$

A sector with angle approximately 57.30° contains one radian and subtends $1/2\pi \approx 1/6.28 \approx 15.91\,\%$ of the circumference of the circle.

In space, angles are measured in steradians (or sometimes square degrees). Figure 15.4b shows a sphere cut in half, with radius r, and a cone with solid angle ω cutting off surface area s on the sphere. The steradian measure of the solid angle is

$$\omega = s/r^2 \text{ steradians}.$$

Clearly, one steradian is the measure when $s = r^2$, independent of the size of the sphere. A right circular cone with vertex at centre of the sphere and vertex angle approximately 65.54° contains one steradian. Since the complete surface of a sphere has area $4\pi r^2$, one steradian subtends $1/(4\pi) \approx 1/12.57 \approx 7.96\,\%$ of the surface of the sphere.

15.5 Point Source

In photometry, an idealized *point source* radiates light evenly in all directions into space. A sphere around a point source, with radius r m, has central solid angle $\omega = 4\pi$ steradians and surface area $s = 4\pi r^2\ \text{m}^2$. The *luminous flux* F produced by a point source is the light flowing through a spherical surface centred on the source. The SI unit of flux is one *lumen* (lm), the light flow in solid angle $\omega = 1$ steradian, from a point source with the intensity I of one standard candle.[6] Intensity, flux, and solid angle are related as

$$\begin{aligned} I \text{ cd} &= F \text{ lm}/\omega \text{ sr} \\ F \text{ lm} &= I \text{ cd} \times \omega \text{ sr} \\ \omega \text{ sr} &= F \text{ lm}/I \text{ cd}. \end{aligned}$$

[6] The SI unit of *luminous intensity* I is one *candela* (cd), the luminous intensity, in a given direction, of a point source emitting monochromatic radiation of frequency 540 × 1012 Hz, with radiant intensity in that direction 1/683 watt per steradian, adjusted by the CIE V(λ) standard observer. Where photometric values need to be distinguished from radiometric, the SI convention is to subscript them as I_v, F_v, …(for vision), or I_e, F_e, …(for energy). Here we assume that values I, F, …without subscript are photometric I_v, F_v, ….

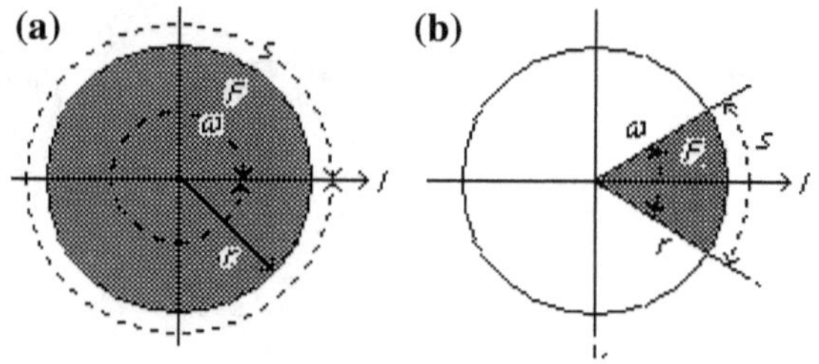

Fig. 15.5 Point source. **a** Point source radiates flux F lm in solid angle $\omega = 4\pi$ sr through surface area $s = 4\pi r^2 \text{m}^2$ of sphere radius rm. Intensity $I = F/4\pi$ cd is the same in every direction. **b** Same source radiates flux $F = I \times \omega$ lm in solid angle $\omega = 1$ sr through surface area $s = 1\,\text{m}^2$ of sphere

Table 15.1 Typical point sources

Source	F lm	I cd
230 v 11-W compact fluorescent lamp	600	48
230 v 60-W incandescent lamp	700	56
230 v 100-W incandescent lamp	1400	111

Figure 15.5a shows, in section, a point source radiating total flux F lm in solid angle $\omega = 4\pi$ sr, that is in all directions through surface area $s = 4\pi r^2\,\text{m}^2$ of a sphere with radius $r = 1$ m. The intensity I cd $= F$ lm/4π sr is the same in every direction. Figure 15.5b shows the same point source radiating flux F lm $= I$ cd $\times\omega$ sr in solid angle $\omega = 1$ sr through surface area $s = 1\text{m}^2$ of a sphere, that is in a cone with base area $s = 1\,\text{m}^2$. The intensity in the direction of the axis of the cone is I cd. For example, given total flux $F = 600$ lm, intensity $I = 600/12.57 = 47.75$ cd.

In practice, the primary standard candela is maintained by national laboratories [3, 4, 6], and disseminated by secondary standard lamps. Luminous flux of a lamp is usually measured by the manufacturer using a goniometer to take readings in many directions, or an Ulbricht[7] sphere which integrates the light radiated in all directions. In commerce, lamps are now labelled with lumens of luminous flux as well as watts of radiant power. Incandescent lamps deliver some 8–18 lm/w; compact fluorescent lamps (CFL) 40–60 lm/w. An actual lamp can be treated as a point source (less the dead cone for the base fitting) if the action distances are more than five times the greatest dimension of the lamp. Table 15.1 lists some typical point sources.

[7]Friedrich Richard Ulbricht (1849–1923), Professor at Dresden Technical University and later President of Royal Saxon Railways.

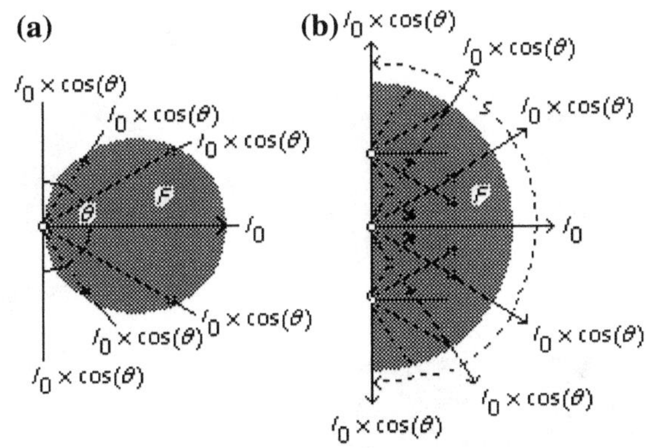

Fig. 15.6 Perfectly-diffusing (Lambertian) extended source. **a** Each point in a Lambertian source radiates intensity I_0 cd in the direction normal to the surface, and intensity $I_\vartheta = I_0 \times \cos(\vartheta)$cd in a direction at angle ϑ sr to the normal. **b** Aggregate flux radiated by Lambertian source is F lm $= I_0 \times 2\pi$. Intensity normal to surface is $I_0 = F/2\pi$; intensity at angle ϑ is $I_0 \times \cos\vartheta$

15.6 Extended Source

An *extended source* is a surface which is emitting or reflecting or transmitting light. For example, a display screen is an extended source emitting light; a printout is an extended source reflecting incident light. A perfectly-diffusing, or *Lambertian*[8] source is like a collection of half-point sources, each radiating into a half-space. Lambert's cosine law for emission says that, for one point in a diffusing surface, the intensity I_ϑ in a direction at angle ϑ to the surface normal is $I_0 \times \cos(\vartheta)$. In Fig. 15.6a the lengths of the arrows in the sphere are proportional to the intensities, greatest I_0 on the normal, and falling off to zero as ϑ goes to $\pi/2$ radians $= 90°$:

$$I_\vartheta = I_0 \times \cos(\vartheta) \text{ cd},$$

where I_0 is the intensity along the normal to the surface, and ϑ is the angle to the normal. Aggregating half-points over the whole diffusing surface, the flux radiates through a hemispherical surface area $s = 2\pi r^2$, with intensity according to direction, as shown in Fig. 15.6b. Intensity, flux, and solid angle are related as

$$\begin{aligned} I_0 \text{ cd} &= F \text{ lm}/2\pi \text{ sr} \\ F \text{ lm} &= I_0 \text{ cd} \times 2\pi \text{ sr} \\ 2\pi \text{ sr} &= F \text{ lm}/I_0 \text{ cd}. \end{aligned}$$

[8]Johann Heinrich Lambert (1728–1777), Professor at Prussian Academy of Sciences, Berlin.

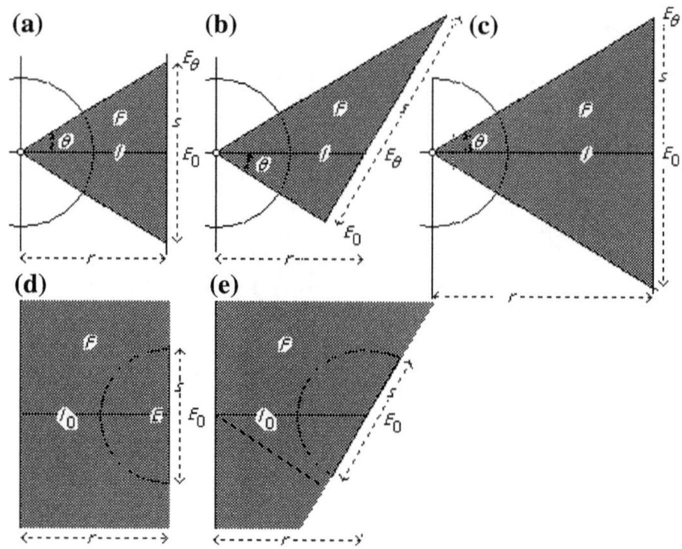

Fig. 15.7 Light arriving at a surface. **a** Point source with intensity I cd radiates flux F lm to illuminate surface area sm^2 at distance rm. Illuminance $E_\vartheta = (I \times \cos(\vartheta))/r^2$ lx. **b** Illuminance E varies directly with cosine of angle of incidence. **c** Illuminance varies inversely with distance r. **d** Lambertian extended source with intensity I cd radiates flux F lm to illuminate surface area sm^2 at distance rm. Illuminance $E = F/s$ lx. **e** Illuminance is the same for inclined surface

For example, given flux $F = 300$ lm, intensity along normal $I_0 = 300/6.28 = 47.75$ cd; at 30° $I_{30} = 47.75 \times 0.87 = 41.35$ cd; at 60° $I_{60} = 47.75 \times 0.50 = 23.87$ cd; at 90° $I_{90} = 47.75 \times 0 = 0$ cd.

15.7 Light Arriving at a Surface

Illuminance E is the flux density (that is, the flux per unit area) arriving at a surface from one or more sources. The SI unit of illuminance is one *lux* (lx),[9] that is one lumen of flux F incident on one square metre of surface area s.

For a point source, Fig. 15.7a shows, in section, intensity I cd radiating flux F lm which illuminates a surface area s m^2 at distance r m from the source. If the illuminated surface s is spherical, the illuminance is

$$E \text{ lx} = F \text{ lm}/r^2 \text{ m}^2 ,$$

[9] An old unit of illuminance is one *foot-candle* (fc), that is one lumen of flux F incident on one square foot of surface area s. 1 fc = 10.764 lx.

where r is distance from source, F is flux on s (which is $I \times \omega$), and E is illuminance. For example, if $I = 47.75\,\text{cd}$, $\omega = 1\,\text{sr}$, and $r = 2\,\text{m}$, then $E = 47.75\,\text{lm}/4\,\text{m}^2 = 11.94\,\text{lx}$. If the illuminated surface is flat, the illuminance is

$$E_\vartheta \text{ lx} = (I \text{ cd} \times \omega \times \cos(\vartheta) \text{ lm})/r^2 \text{ m}^2 ,$$

where r is the distance of the illuminated surface, ω is the solid angle subtended by s, ϑ is the angle between the normal to the illuminated surface and the direction of the point source, I is the intensity, and E_ϑ is the illuminance at a point where the direction of the source makes angle ϑ with the surface normal. For example, $E_0 = (47.75\,\text{cd} \times 1 \times 1)/4 = 11.94\,\text{lx}$; while $E_{30} = (47.75\,\text{cd} \times 1 \times 0.87)/4 = 10.39$. Figure 15.7b shows the illuminated surface inclined, and (c) shows the illuminated surface at greater distance, spreading flux over a greater area.

For a Lambertian extended source. Figure 15.7d shows, in section, the source radiating flux F lm which illuminates surface area $s\,\text{m}^2$, arriving through a hemisphere (area $2\pi r^2$). We have

$$E \text{ lx} = F \text{ lm}/s \text{ m}^2 ,$$

where s is area illuminated, F is flux, and E is illuminance. For example, $E = 300\,\text{lm}/25.13\,\text{m}^2 = 11.94\,\text{lx}$. Figure 15.7e shows that the flux arrives through the same hemisphere when the surface is inclined to the source.

In practice, two illuminances can be compared in a simple visual photometer, such as that devised by Bunsen[10] in 1844. Two sources are placed at opposite ends of a bench marked in millimetres, and between them a movable head holding a sheet of paper containing a spot of grease which is translucent. The sources illuminate opposite sides of the paper, and mirrors allow an observer to compare the two sides. The head is moved this way or that until the grease spot disappears: the illuminances are then equal. The relative intensities of the sources can be calculated from the distances. If one of the sources is a known standard, then a measure of the other source is found.

Alternatively, illuminance can be measured by a photographic light-meter laid on the surface, with a hemispherical diffuser to collect the incident light. A purpose-made lux-meter is a similar photoelectric detector, calibrated in lx. Table 15.2 lists some typical illuminances from various sources, and Table 15.3 lists some recommended illuminances.

---

[10] Robert Wilhelm Bunsen (1811–1899), Professor of Chemistry at Heidelberg University.

Table 15.2 Typical illuminances

	E lx
Full moonlight on Earth	0.27
Wax candle at 1 foot	10
Dimly-lit room	15–50
Living room	50–200
Brightly-lit office	300–500
TV studio	1000
Overcast sky light	1000
Clear sky light	10,000–25,000
Direct sunlight	32,000–130,000

Table 15.3 Recommended illuminances

Source	E lx
Public spaces	30
Reading	100
High-contrast large-size working	300
High-contrast small-size working	500
Low-contrast large-size working	500
Low-contrast small-size working	1000

15.8 Light Leaving a Surface

Emittance M is the flux density leaving a surface. The SI unit of emittance is the same as the SI unit of illuminance, one *lux* (lx), that is one lumen of flux F emitted from $1\,\mathrm{m}^2$ of surface area s.

Figure 15.8a shows a Lambertian reflecting surface (such as a sheet of paper), with area $s\,\mathrm{m}^2$, illuminated by a point source. The *reflection factor* $0 \le rf \le 1$ of the sheet is the ratio between the illuminance E arriving and the emittance M leaving

$$
\begin{aligned}
rf &= M\ \mathrm{lx})/E\ \mathrm{lx} \\
M\ \mathrm{lx} &= rf \times E\ \mathrm{lx}\,.
\end{aligned}
$$

For example, if a point source with total flux $F = 600\,\mathrm{lm}$ illuminates an A4 sheet of paper, area $s = 0.062\,\mathrm{m}^2$, at distance 1 m; then the flux on the sheet is $F = 0.062/4\pi \times 600 = 2.96\,\mathrm{lm}$, and the illuminance is $E = 2.96/0.062 = 47.75\,\mathrm{lx}$. If the reflection factor of the sheet is $rf = 0.80$, then the emittance $M = 0.80 \times 47.75 = 38.20\,\mathrm{lx}$.

Figure 15.8b shows a Lambertian self-luminous surface (such as a display screen) with area $s\,\mathrm{m}^2$. The emittance M is the flux F emitted, averaged across the emitting area s:

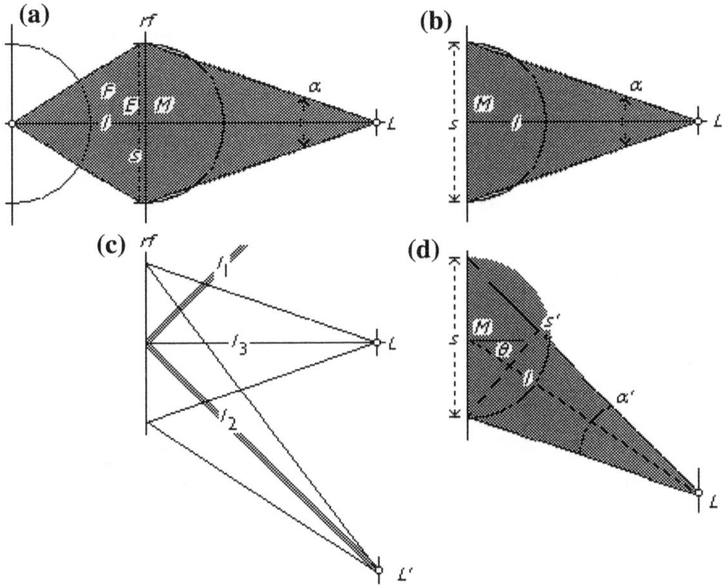

Fig. 15.8 Light leaving a surface. **a** Lambertian reflecting surface with reflection factor *rf*, illuminated by point source and measured from same side. **b** Lambertian self-luminous surface. **c** Specular reflecting surface, illuminated by small source with intensity I_1. **d** Lambertian self-luminous surface viewed from angle ϑ

$$M \text{ lx} = F \text{ lm})/s \text{ m}^2 .$$

The *luminance L* of an extended source, reflecting or self-luminous, is the flux density radiated per solid viewing angle. The SI unit of luminance is one *candela per square metre* (cd/m^2),[11] that is one lumen of flux F from 1 m^2 of source area a, collected in one steradian of solid angle of view α.

$$L \text{ cd/m}^2 = M \text{ lx}/\pi \text{ sr} .$$

For a Lambertian reflecting surface (such as a matt sheet of paper), luminance depends on illuminance E and reflection factor rf. For a surface with varying rf (such as a 1-bit image printed on paper), luminance is averaged across the area s selected. For a specular (glossy) reflecting surface (such as a mirror), luminance also depends on angle of view. Figure 15.8c shows a specular surface illuminated by a small source with intensity I_1 cd. The luminance L' viewed from a direction equal and opposite to the source gets high intensity $I_2 = I_1 \times rf$ cd; but the luminance L viewed from the surface normal gets low intensity $I_3 = 0$ cd.

For a uniform Lambertian self-luminous surface (such as a White CRT display screen), luminance depends on emittance M only. For example, if the screen is A4

[11] One candela per square metre is also called one *nit* (Latin *nitere* to shine or gleam).

Table 15.4 Typical luminances [4]

Source	L cd/m^2
Starlit sky	5×10^{-4}
Black computer screen	1×10^{-2}
White computer screen	1×10^{2}
Full moon from Earth	3×10^{3}
Clear blue sky	4×10^{3}
Spermaceti candle flame	1×10^{4}
230 v 60 W opal lamp	1.2×10^{5}
Noon sun from Earth	1.6×10^{9}
Lightning flash	8×10^{10}

size ($s = 0.297 \times 0.210\,\text{m} = 0.062\,\text{m}^2$) viewed at distance 0.250 m, so viewing angle $\alpha = 1$ sr, and the emittance is registered as $M = 314\,\text{lm/m}^2$, then $L = 314/\pi = 100\,\text{cd/m}^2$. Or if the viewing solid angle is $\alpha = 0.5$ sr at distance 0.125 m, covering area $s = 0.031\,\text{m}^2$, and the emittance is again registered as $M = 314\,\text{lm}$, then again $L = 314/\pi = 100\,\text{cd/m}^2$. If luminance is measured in a direction at angle ϑ from the normal to the source, the intensity I' is decreased by $\cos(\vartheta)$, but the area s' viewed is increased by the same factor, as shown in Fig. 15.8d.

For a partly-Lambertian surface (such as a White LCD display screen), luminance falls off as the direction of view departs from the surface normal.

In practice, two luminances can be compared in a simple visual device which brings, by mirrors or prisms, a target luminance against a known standard luminance as two halves of the observer's field of view. Calibrated shutters or filters acting on the two luminances allow the observer to match the luminances, and hence calculate the luminance of the target.

More conveniently, luminance can be measured by using a a photoelectric detector with lenses and an aperture to collect the emittance M from a small area s of the extended source. Table 15.4 lists some typical luminances of extended sources.

15.9 Photometry Examples

Table 15.5 lists the relations between photometric units occurring in digital imaging, followed by some examples.

Figure 15.9 shows a pixel of an object recorded in a scanner. For each pixel, area $s\,\text{m}^2$, the light source is an extended one with luminance $L_1\,\text{cd/m}^2$ measurable by photographic light meter. Each pixel receives illuminance $E = \pi \times L_1\,\text{lx}$ at very small distance d m. Each pixel reflects emittance $M = rf \times E\,\text{lx}$ according to its reflection factor. The sensor records luminance $L_2 = M/\pi\,\text{cd/m}^2$. For example, say $L_1 = 2{,}000\,\text{cd/m}^2$, $E = 6{,}283\,\text{lx}$, say $rf = 0.80$, $M = 5{,}026\,\text{lx}$, and $L_2 = 1{,}600\,\text{cd/m}^2$.

Table 15.5 Photometric units in digital imaging

Name	Value	Unit	Relation
Solid angle	ω	sr	s / r^2
Surface area	s	m^2	$\omega \times r^2$
Distance	r	m	$\sqrt{(s / \omega)}$
Intensity	I	cd	F / ω
Flux	F	lm	$I \times \omega$
Illuminance	E	lx	F / s
Reflection factor	rf		M / E
Emittance	M	lx	$rf \times E$
Luminance	L	cd/m^2	M / π

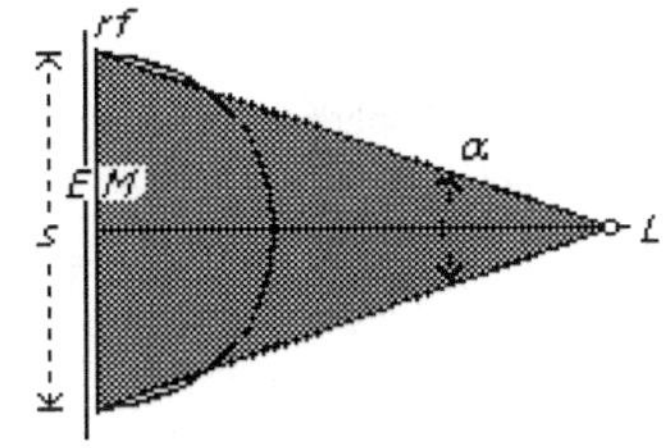

Fig. 15.9 Scanner. Internal extended source sheds illuminance E on object pixel which has area s and reflection factor rf, so emittance is M. Sensor records luminance L in angle α

Figure 15.10 shows a pixel of an object, lit by outdoor sky, recorded in a camera. The extended source has luminance L_1 cd/m^2 measurable by photographic light meter. Each pixel receives illuminance $E = \pi \times L_1$ lx, and reflects emittance $M = rf \times E$ lx according to its reflection factor. The sensor records luminance $L_2 = M/\pi$ cd/m^2. For example, given cloudy sky $L_1 = 2{,}000$ cd/m^2, $E = 6{,}283$ lx, say $rf = 0.80$, $M = 5{,}026$ lx, and $L_2 = 1{,}600$ cd/m^2.

Figure 15.11 shows a pixel of an image displayed on a monitor screen. The luminance of an all-White screen is L_1 cd/m^2 measurable by photographic light meter. A White pixel emits illuminance $E = \pi \times L_1$ lx, and a pixel with relative factor rf emits emittance $M = rf \times E$ lx. The sensor records luminance $L_2 = M/\pi$ cd/m^2. For example, given White screen $L_1 = 100$ cd/m^2, $E = 314$ lx, say $rf = 0.80$, $M = 251$ lx, and pixel 0.80 $L_2 = 80$ cd/m^2.

Figure 15.12 shows an A4 sheet of paper, area $s = 0.0624\,\text{m}^2$, lit by a point source with flux F at distance d m. The source intensity $I = F/4\pi$. The solid angle ω subtended by the sheet is equivalent to s/d^2, and the flux in this solid angle is $F = I \times \omega$. The illuminance received by the sheet is $E = F/s$. If the reflection factor of the sheet is rf, the emittance $M = rf \times E$ lx. The luminance of the sheet is $L = M/\pi$ cd/m^2. For example, given total source flux $F = 600$ lm and source distance $d = 1$ m, $I = 47.75$ cd, solid angle $\omega = 0.0624$ sr, flux in this angle $F_\omega = 0.0624$ lm, illuminance on sheet $E = 47.75$ lx. Say reflection factor of sheet is $rf = 0.80$, emittance from sheet $M = 38.2$ lx, and luminance of sheet $L = 12.16$ cd/m^2.

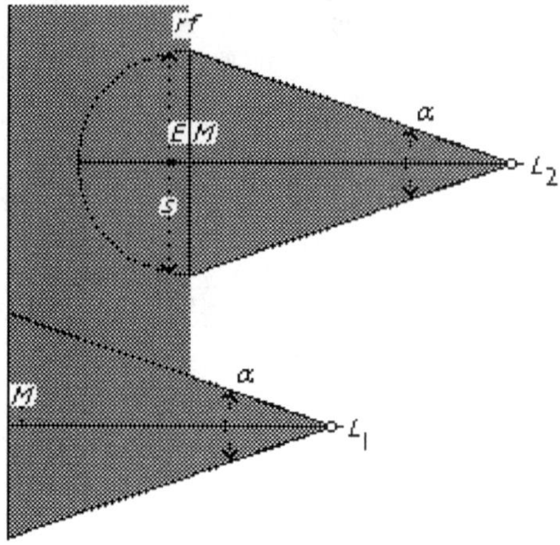

Fig. 15.10 Camera. Sky is extended source with luminance L_1. Sky sheds illuminance E on object pixel which has area s and reflection factor rf, so emittance is M. Sensor records luminance L_2 in angle α

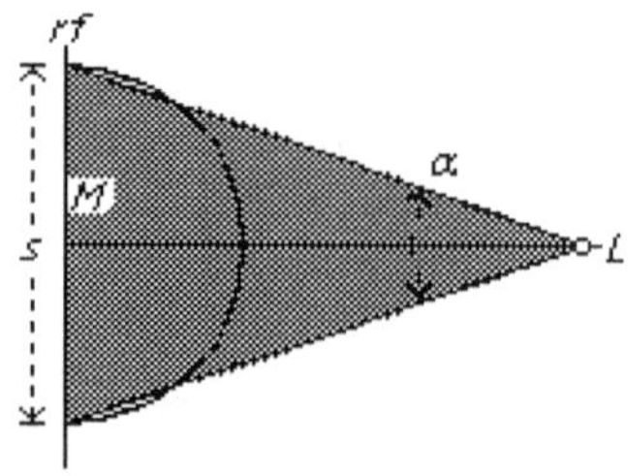

Fig. 15.11 Display. Self-luminous pixel has emittance M, measurable as luminance Lcd/m^2 in angle α

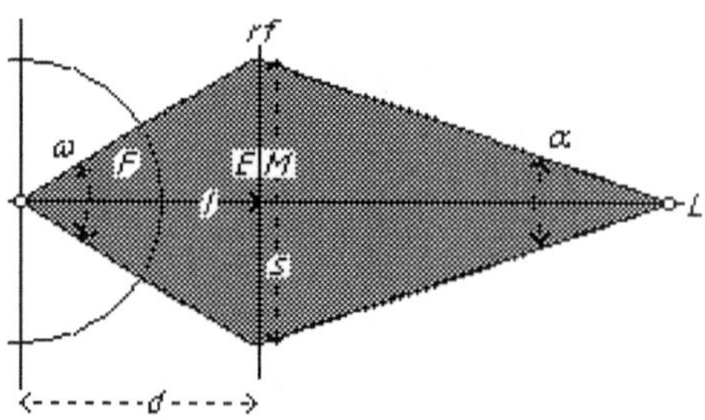

Fig. 15.12 Printout. Point source radiates flux F at distance d over A4 sheet, area s and reflection factor rf. Illuminance E becomes emittance M, measurable as luminance L in angle $z\alpha$

If we wanted the printout sheet in the last example to present the same luminance as an all-White monitor screen displaying luminance $L = 100\,\text{cd/m2}$, we could calculate as follows. Let the luminance of the printout sheet be $L = 100\,\text{cd/m}^2$. Emittance $M = \pi L = 314\,\text{lx}$; illuminance $E = M/rf = 392\,\text{lx}$; flux arriving at and leaving sheet $F = E \times s = 24.5\,\text{lm}$; intensity of point source at distance $d = 1\,\text{m}$ has to be $I = F/\omega = 393\,\text{cd}$; so total flux of point source is $I \times 4\pi = 4934\,\text{lm}$.

References

1. BIPM (2006) The international system of units, 8th edn. http://www.bipm.org/en/si/si_brochure/. Accessed 1 Feb 2014
2. CIE S 010/E:2004 Photometry: the CIE system of physical photometry. http://www.techstreet.com/cie/products/1210115. Accessed 1 Feb 2014
3. National Institute of Standards and Technology (2014). http://nist.gov. Accessed 1 Feb 2014
4. National Physical Laboratory (2014) Light. http://www.npl.co.uk/educate-explore/factsheets/light/. Accessed 1 Feb 2014
5. Palmer J (1999) Radiometry and photometry FAQ. http://fp.optics.arizona.edu/Palmer/rpfaq/rpfaq.htm. Accessed 1 Feb 2014
6. Physikalisch-Technische Bundesanstalt (2014). http://www.ptb.de. Accessed 1 Feb 2014
7. Walsh JWT (1926) Photometry, 1st edn. Constable, London. https://archive.org/details/Photometry. 3rd edn reprinted 1965, Dover, New York

Chapter 16
Colorimetry

Colorimetry extends the measurement of light to distinguish hue and saturation as well as photometric brightness, always in relation to a standard eye. The general-purpose CIE colour spaces are defined. The sRGB colour space is specifically designed for computer displays, and the YMCK colour space for printing. Examples and conversions between colour spaces are given.

16.1 Colorimetry

Colorimetry [11] is, like photometry (see Chap. 15), the measurement, by eye or by instruments which match the sensitivity of the eye, of energy in visible light. But whereas photometry is achromatic, taking only the aggregate of power on a grey-scale, colorimetry distinguishes different spectral power distributions in a colour space.

Various systems of colorimetry or colour spaces have been put forward by Newton, Goethe, Grassmann, Helmholtz, Maxwell, Chevreul, Rood, and many others. For reflecting surfaces, Munsell[1] published an atlas of colour chips in 1905, improved by the Optical Society of America in 1929, and still in quite widespread use [12]. Ostwald[2] devised a similar system in 1916, which was taken up by European artists and schools. The Pantone swatch of 15 base pigments and about 1,000 spot colours is a notably successful commercial product for the print and paint trades [12]. However, the international standards for colour specification are the CIE colour spaces.

[1] Albert Henry Munsell (1858–1918), artist and teacher at Massachusetts College of Art and Design.

[2] Friedrich Wilhelm Ostwald (1853–1932), Professor of Physical Chemistry, Leipzig University.

A. Parkin, *Digital Imaging Primer*, DOI 10.1007/978-3-540-85619-1_16

16.2 CIE XYZ Colour Space

Experiments[3] have long since demonstrated that any three light sources A, B, C can be combined to match almost any given target source T, provided that none of A, B, C can be matched by a combination of other two. In such experiments, the subject sees T in one half of a field, and a combination of A, B, C in the other half of the field. By varying the intensity t of the target T and the intensities a, b, c of the sources A, B, C, the two halves can usually be made to match:

$$t\mathrm{T} \equiv a\mathrm{A} + b\mathrm{B} + c\mathrm{C}\,.$$

But there are always some targets T which cannot be matched, no matter how the sources A, B, C are chosen. In these cases a match can always be made by adding some intensity of one, or two, of the sources to the target. Then, for example

$$t\mathrm{T} + a\mathrm{A} \equiv b\mathrm{B} + c\mathrm{C}$$

that is

$$t\mathrm{T} \equiv -a\mathrm{A} + b\mathrm{B} + c\mathrm{C}\,.$$

In the 1920s careful tests[4] on many subjects established matches across the entire spectrum for normal colour vision.

Figure 16.1a shows the intensities b, g, r of monochromatic primaries $\mathrm{B} = 460\,\mathrm{nm}$, $\mathrm{G} = 530\,\mathrm{nm}$, and $\mathrm{R} = 650\,\mathrm{nm}$, which match monochromatic targets across the spectrum. The intensity units are adjusted so that $b + g + r = 1$ at each wavelength. Notice that all the matches have some negative values, except for the primaries B, G, R.

Based on these results, in 1931 La Commission Internationale de l'Éclairage (CIE) defined a system of colour specification [1, 11] which (with a few extensions) has been the international standard ever since. In order to remove negative coordinates, three ideal primaries X, Y, and Z are chosen, and the intensities of these ideal primaries which combine to match the spectral colours are calculated. These calculated *distribution coefficients* $\bar{x}$, $\bar{y}$, $\bar{z}$, are listed in the standard for each 5 nm across the spectrum, and shown in Fig. 16.1b. Notice that the $\bar{y}$ curve is in fact the CIE V(λ) curve for luminance (see Sect. 15.3), and that the areas under the $\bar{x}$ and $\bar{z}$ curves have been made equal to the area under the $\bar{y}$ curve.

To measure a given target colour T in the CIE system, some form of colorimeter is needed [9, 11]. A colorimeter must provide three actual primary sources R, G, B close to

[3] By Thomas Young around 1800; James Clerk Maxwell around 1850 onwards; Hermann Helmholtz around 1850 onwards.

[4] By William David Wright at Imperial College, London and independently John Guild at the National Physical Laboratory, Teddington.

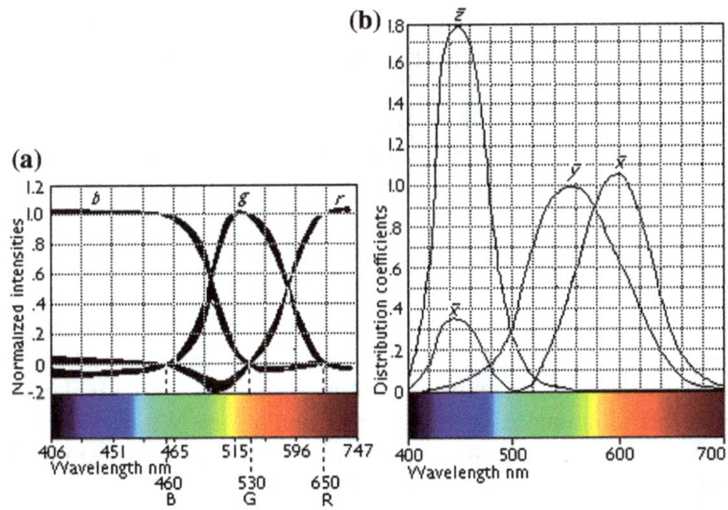

Fig. 16.1 Matches of spectral targets by experimental and ideal primaries. **a** Normalized intensities (b, g, r) of experimental monochromatic primaries B = 460 nm, G = 530 nm, R = 650 nm which match monochromatic targets across the spectrum (including negative values). **b** Relative intensities $\bar{x}$, $\bar{y}$, $\bar{z}$ of ideal primaries X Y Z calculated to match monochromatic targets across the spectrum (adjusted to all-positive values)

- R wavelength 700 nm,
- G wavelength 546 nm,
- B wavelength 436 nm,

and provide ways to

- Vary the intensities of the primaries.
- Mix the three intensities.
- Possibly mix one, or two, of the primaries with a target colour.
- Compare the target (or target mixture) with the primary mixture, until matched.

For a self-luminous target (such as a display screen), the simplest visual colorimeter might use three colour filters to isolate three primaries, sliding shutters to vary their intensities, and a mirror box to mix the intensities. Then mirrors or totally-reflecting prisms bring the target and primary mixture into two halves of an observer's field of view. The observer varies the intensities of the primaries (and possibly the intensity of the target) until the two halves match in colour and brightness.

A photo-electric colorimeter would use three photo-electric cells with filters which match the long-, medium- and short-wavelength sensitivities of the eye as in Fig. 16.1. When the target is presented the sensor voltages are proportional to the R, G, B values wanted, and the colorimeter can be constructed to give direct readings.

For more precise measurement, a spectrometer can be used. This might isolate monochromatic primaries by placing slits in a spectrum spread by a prism or a diffraction grating, and mixing these monochromatic primaries to match a target.

Given the actual (R, G, B) values which match the target, the ideal *tristimulus values* (X, Y, Z) which would match the target are calculated as

$$(X, Y, Z) = 1/0.17697 \begin{pmatrix} 0.49 & 0.31 & 0.20 \\ 0.17697 & 0.81740 & 0.01063 \\ 0.00 & 0.01 & 0.99 \end{pmatrix} (R, G, B)$$
$$= ((0.49R + 0.31G + 0.20B), (0.17697R + 0.81740G + 0.01063B), (0.00R + 0.01G + 0.99B)).$$

For a reflecting target (such as a printout) the spectral power distribution of the illuminating source must be taken into account. An ideal *equal-energy* source E would have an SPD which is completely flat across the spectrum. An ideal perfectly-diffusing surface illuminated by this source would look White, by definition. A warmer source, such as a tungsten lamp, has an SPD sloping up from the blue end to the red end of the spectrum: a perfect diffuser then looks a warmer White, and all reflecting surfaces look warmer. A cooler source, such as daylight, has an SPD sloping down from the blue end to the red end of the spectrum: a perfect diffuser then looks a cooler White, and all reflecting surfaces look cooler. The CIE 1931 standard specified three practical sources, called illuminant A (tungsten lamp), B (sunlight), and C (overcast daylight), and later replaced B and C by a daylight series D and a fluorescent series F. Table 16.1 lists the A D and E illuminants.

For a reflecting target, a reflection colorimeter is used. A common design has a fixed light source, three colour filters, and a photo-electric cell attached to a galvanometer. Say the light source is illuminant A tungsten with tristimulus values $X = 1.0985$, $Y = 1.0000$, $Z = 0.3558$, which are in ratio 100:91:32. First put a red filter in front of the light source, reflect the light from a standard white surface (magnesium carbonate), and adjust the galvanometer reading to 100. Next replace the white surface by the target surface and get the X reading. Then repeat with a green filter, adjusting the white reading to 91, to get Y from T. Finally repeat with a blue filter, adjusting the white reading to 32, to get Z from T.

Table 16.1 CIE illuminants

Illuminant	Example	X	Y	Z	Temp. (K)
A	Tungsten lamp	1.0985	1.0000	0.3558	2860
D50	Horizon daylight	0.9642	1.0000	0.8249	5000
D55	Mid-morn daylight	0.9568	1.0000	0.9215	5500
D65	Noon daylight	0.9505	1.0000	1.0888	6500
D75	North skylight	0.9497	1.0000	1.2264	7500
E	Equal-energy	1.0000	1.0000	1.0000	5450

The tristimulus values (X, Y, Z) are the relative intensities of the ideal primaries X, Y, Z which match the source. The colour temperature is that of a comparable Planckian black-body radiator, in degrees Kelvin

Table 16.2 CIE actual primaries R G B

	Wavelength (nm)	X	Y	Z	Sum	x	y	z
R	700	0.0114	0.0041	0.0000	0.0155	0.7347	0.2653	0.0000
G	546	0.3706	0.9709	0.0119	1.3534	0.2738	0.7174	0.0088
B	436	0.3275	0.0175	1.6231	1.9683	0.1664	0.0089	0.8246

The tristimulus values (X, Y, Z) are the relative intensities of the ideal primaries X, Y, Z which match the actual CIE primaries R, G, B. The chromaticity coordinates (x, y, z) are the relative intensities adjusted so that $x + y + z = 1$

Chromaticity coordinates (x, y, z) are normalized from tristimulus values as

$$x = X/(X + Y + Z)$$
$$y = Y/(X + Y + Z)$$
$$z = Z/(X + Y + Z),$$

so that $x + y + z = 1$. Chromaticity coordinates are listed in the standard. Table 16.2 shows the tristimulus values and chromaticity coordinates of the three actual primaries R, G, B, of the CIE system.

16.3 CIE xyY Colour Space

Figure 16.2a shows a two-dimensional *rg chromaticity chart* for the three-dimensional data in Fig. 16.1a. It has experimental primary B as origin, experimental primary R as unit point on the horizontal axis, and experimental primary G as unit point on the vertical axis. Any colour-match specified by (b, g, r) plots as chromaticity coordinates (r, g), with $b = 1 - (r + g)$. The spectrum is seen as a horseshoe curve which lies outside the primary triangle BGR (because of the negative values in the matches), everywhere except for the three vertices B, G, and R.

Figure 16.2b shows the CIE xy chromaticity chart. It has ideal primary Z as origin, ideal primary X as unit point on the horizontal axis, and ideal primary Y as unit point on the vertical axis. Any colour-match specified by (x, y, z) plots as chromaticity coordinates (x, y), with $z = 1 - (x + y)$. The spectrum is now seen as a horseshoe curve which lies on or inside the ideal primary triangle ZXY everywhere. The blue and red ends of the spectral curve are joined by a straight line (the *alchyne*) representing the non-spectral purples, which are mixtures of spectrally-separated blues and reds. The *gamut*[5] of all visible colours lies on or within the closed spectral curve of the chromaticity chart. The additive mixture of any two colours lies on the straight line between them, and the additive mixture of any three colours lies on or within the

[5]In medieval music, *gamma ut* was the lowest note of the scale. Later, *gamut* came to mean the entire musical scale, and later still the whole of any spread.

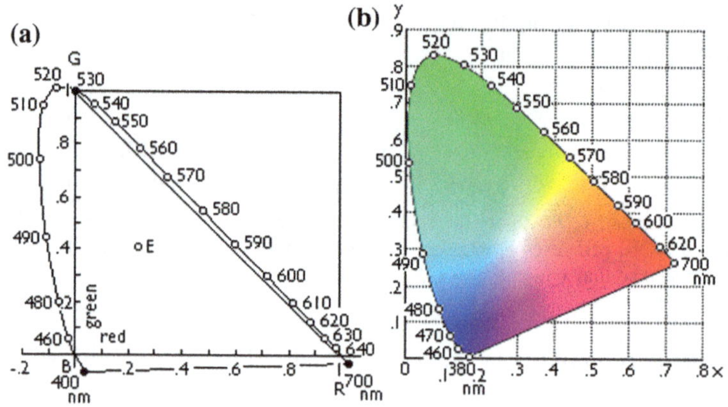

Fig. 16.2 Chromaticity charts for experimental and ideal primaries. **a** Chromaticity chart for experimental primaries B G R. The spectral curve lies outside the primary triangle everywhere except the three vertices. **b** Chromaticity chart for ideal primaries X Y Z. The spectral curve lies on or inside the primary triangle everywhere. E is an ideal equal-energy white

triangle between them.[6] Since the spectral curve is convex, it is clear that any triangle formed by three spectral primaries must exclude some visible colours, which can be matched only by mixing some of one or two primaries with the target.

In the CIE 1931 system a colour can be specified in any of three ways:

- Tristimulus values X, Y, Z,
- Chromaticity coordinates x, y, and luminance Y,
- Dominant wavelength λ nm and purity p.

Tristimulus values are the intensities of the ideal primaries which would mix to match the target colour.

Chromaticity coordinates are the abscissa and ordinate of the target plotted on the chromaticity chart; and luminance is the directly-measured tristimulus Y.

Dominant wavelength corresponds to the sensation of Hue, and is found on the chromaticity chart by drawing a straight line from the appropriate white point W through target T to meet the spectral curve at wavelength λ nm. (If T is a non-spectral purple, the dominant wavelength is taken as the *complementary* wavelength where WT does meet the spectral curve.) Purity corresponds to the sensation of Saturation, and is found as the ratio of lengths WT/Wλ. For example, Fig. 16.3 shows four colours T1, T2, T3, T4 plotted on the chromaticity chart.

[6] Additive colour mixing can be represented as weights and distances in a barycentric system, as first formulated in 1861 by Hermann Gunther Grassmann (1809–1877), schoolteacher in Stettin, recognized as a philologist, but not as a mathematician until after his death. Maxwell made great use of such a system, and a barycentric additive mixing diagram is often called a *Maxwell triangle*.

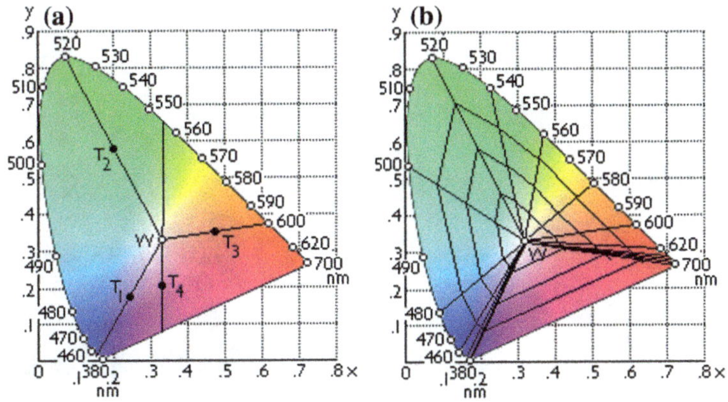

Fig. 16.3 CIE dominant wavelength and purity. **a** T1 is a mixture of blue and white, dominant wavelength $\lambda = 445\,\text{nm}$ and purity $p = 50\,\%$. T2 is a mixture of green and white, dominant wavelength $\lambda = 520\,\text{nm}$ and purity $p = 50\,\%$. T3 is a mixture of red and white, dominant wavelength $\lambda = 600\,\text{nm}$ and purity $p = 50\,\%$. T4 is a mixture of purple and white, complementary dominant wavelength $\lambda = 555\,\text{nm}$ and purity $p = 50\,\%$. **b** Dominant wavelength λ nm and purity $p\,\%$ as a coordinate scheme

16.4 CIE LAB Colour Space

In 1976 the CIE issued a standard for L*a*b* (or LAB) colour space [1].[7] *CIE LAB* is a colour space convertible to and from the CIE XYZ colour space which aims at *perceptual uniformity*, that is to say equal Euclidean distances in the colour space represent equal differences in sensation of Brightness, Hue, and Saturation (which is not the case for CIE XYZ). CIE LAB has three dimensions: L*, a*, and b*. $0 \leq L* \leq 100$ specifies lightness; $a* < 0$ greenness, $a* > 0$ redness; $b* < 0$ blueness, $b* > 0$ yellowness. The space is a cube containing all visible colours, and the LAB chromaticity chart is a section of this cube at a given L*-level. Figure 16.4 shows the chromaticity chart with the hexagonal gamut corresponding to the RGB triangle in Fig. 16.2. In CIE LAB, mixtures do not lie on straight lines.

16.5 sRGB Colour Space

In 1999 the International Electro-technical Commission (IEC) issued a standard for sRGB colour space [5]. *sRGB* is a colour space which is native to a standard digital-imaging system, and which is a sub-space of CIE colour space. That is to say, sRGB colours are specified as 24-bit (R, G, B) triples where $0 \leq R \leq 255$, $0 \leq G \leq 255$, $0 \leq B \leq 255$; and these triples can be converted to, or from, CIE values (X, Y, Z). sRGB is explained by its originators in [10].

[7] Based on work by E.Q. Adams in 1942 and R.S. Hunter in 1948.

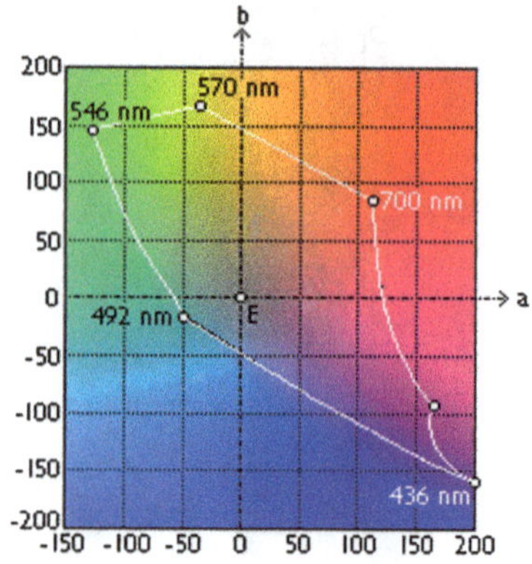

Fig. 16.4 CIE L*a*b* chromaticity chart. The gamut of CIE primaries (R = 700 nm, G = 546 nm, B = 436 nm) and secondaries is marked. White point is equal-energy E. (Compare with Fig. 16.2)

The sRGB primaries are the ITU set used for almost all computer and television display screens [8], as shown in Table 16.3 with the corresponding CIE xyY and CIE LAB values. Figure 16.5 shows the sRGB primaries and secondaries plotted on the CIE xyY and CIE LAB chromaticity charts.

The sRGB standard recommends (for computational reasons) an encoding γ of $1/2.4 = 0.4167$, and a decoding γ of 2.4, which give effectively the same result as 0.4545 and 2.2. The sRGB standard also recommends viewing environments as

- Luminance 80 cd/m^2 (typical of displays).
- Illuminant white D65 (typical of displays).
- Image surround reflectance 20 %.
- Encoding ambient illuminance 64 lx.
- Encoding ambient white D50 (slightly warmer than D65).
- Encoding viewing flare 1 % (contrast reduction due to ambient light).
- Display ambient illuminance 200 lx (typical for ordinary viewing).
- Display ambient white D50 (typical for ordinary viewing).
- Display viewing flare 5 % (contrast reduction due to ambient light).

Table 16.3 sRGB primaries and white

Name		sRGB			CIExyY			CIELAB	
	R	*G*	*B*	*x*	*y*	*Y*	*L*∗	*a*∗	*b*∗
Red	255	0	0	0.6401	0.3300	0.2126	53.24	80.10	67.22
Green	0	255	0	0.3000	0.6000	0.7152	87.74	−86.18	83.18
Blue	0	0	255	0.1500	0.0600	0.0722	32.31	79.19	−107.86
White D65	255	255	255	0.3127	0.3290	1.0000	100.00	0.00	−0.01

(R, G, B) are the 24-bit digital colour values; (x, y, Y) are the CIE chromaticity coordinates and luminance; and $(L*, a*, b*)$ are the CIE LAB luminance and chromaticity coordinates

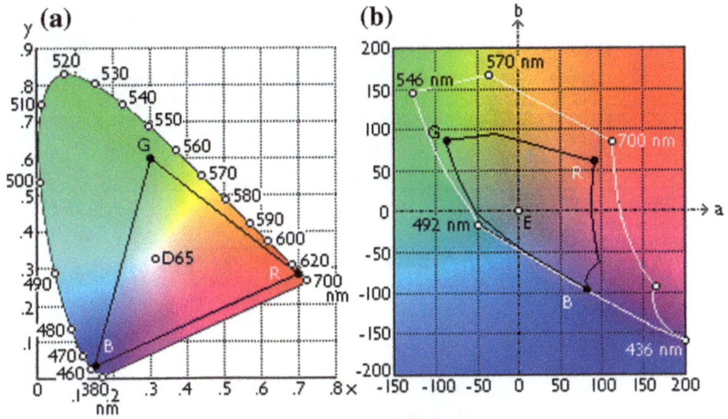

Fig. 16.5 sRGB gamut. **a** In the CIE xyY chart, the sRGB primaries are vertices of a triangle, and the secondaries lie on the three sides of the triangle. **b** In the CIE LAB chart, the sRGB primaries and secondaries are vertices of a hexagon

16.6 CMYK Colour Space

Whereas for displaying an image three primary sources R G B add colours to a black screen (per pixel), for printing the same image four primary inks C M Y K subtract colours from a white reflecting paper (per pixel). An *ideal subtractive colour space* would have only three inks, C M Y, exactly inverse to the additive primaries R G B. That is to say, an ideal C ink would have dominant wavelength complementary to R and the same purity as R, and similarly M and G, and Y and B. Figure 16.6a shows the ideal spectral power distributions.

However, available printing inks are far from the ideal in several respects:

- The C M Y inks are not exact complementaries to the sRGB primaries.
- The pass levels of the inks are lower than they should be, and the stop levels are higher.
- The pass bands of the inks overlap.
- Printout colour varies with the precise white and surface properties of the substrate.
- Ink colours vary from one manufacturer to another.

A great improvement is obtained by using an additional black ink K, to replace solid CMY and proportional parts of other combinations of inks. The conversion from RGB to CMYK with 100 % removal of under-colour is then

$$
\begin{aligned}
K &= \mathrm{MIN}(100 - R_{\mathrm{pc}}, 100 - G_{\mathrm{pc}}, 100 - B_{\mathrm{pc}})\\
C &= 100 - R_{\mathrm{pc}} - K\\
M &= 100 - G_{\mathrm{pc}} - K\\
Y &= 100 - B_{\mathrm{pc}} - K\,,
\end{aligned}
$$

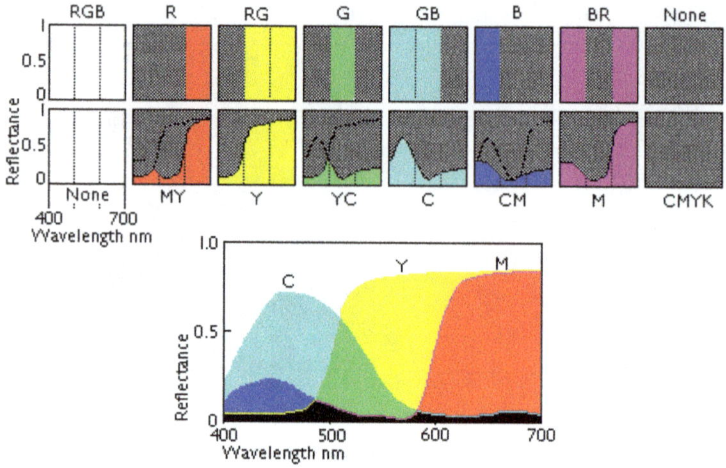

Fig. 16.6 Spectral power distributions for printing inks. **a** SPDs for ideal CMY inks. **b** SPDs for typical CMYK inks. **c** SPDs for ISO 2846 offset-lithography inks. The cyan ink is complementary to red 630 nm; the magenta ink is complementary to green 570 nm, and the yellow ink is complementary to blue 430 nm

where $R_{pc} = 100R/255$, $G_{pc} = 100G/255$, $B_{pc} = 100B/255$, and C, M, Y, K are percentages in the range 0–100. Figure 16.6b shows spectral power distributions for a typical set of CMYK ink-jet printer inks. Figure 16.6c shows SPDs for the standard commercial offset-lithography inks [7], to which desktop printer inks might aspire.

In an ordinary desk-top system the default mapping from RGB values in the image file to the CMYK values printed out is fixed by the software supplied with the printer by the manufacturer. This default mapping is reliably repeatable, and may be acceptable as a fact of digital life. A simple procedure to find the default mapping is

- Construct in an image editor eight patches: black (0, 0, 0), red (255, 0, 0), yellow (255, 255, 0), green (0, 255, 0), cyan (0, 255, 255), blue (0, 0, 255), magenta (255, 0, 255), and white (255, 255,255).
- Print the patches.
- Scan the printout, and display the scan.
- Note the new R G B values of the display of the scan.
- Calculate the CIE xyY coordinates from the new R G B values.
- Plot the CMYK colour space on the CIE chromaticity chart.

Such a "round-trip" procedure of course conflates the particularities of the printer and the scanner, but it is quite precise and repeatable. Figure 16.7 shows a typical ink-jet YMCK gamut plotted within the sRGB gamut, in the CIE xyY and LAB chromaticity charts. It is evident that the purest (most saturated) colours in sRGB are *out-of-gamut* for CMYK, so a printout of a given image is perforce duller than its display.

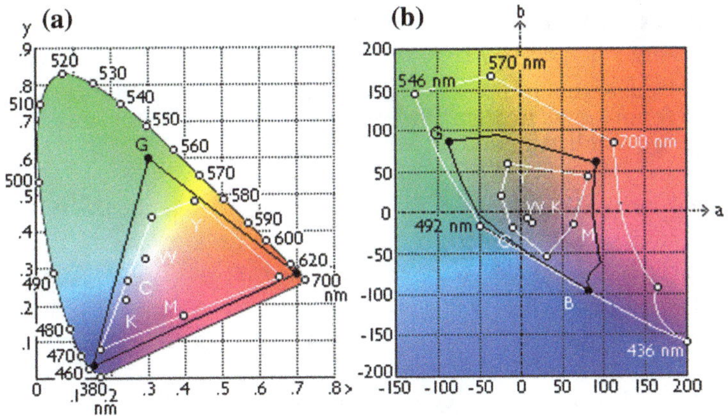

Fig. 16.7 CMYK gamut within sRGB gamut. **a** In the CIE xyY chart, the sRGB gamut is marked by the triangle RGB, and the YMCK gamut by the hexagon y yC C CM M My. **b** In the CIE LAB chart, both gamuts are hexagonal

On the other hand, we sometimes have a larger context, where we need to juggle the input colour spaces of different cameras and scanners, the display spaces of different monitors, or the printout spaces of different printers. For example, we may want to send digital images to a commercial printer who uses the ISO 2846 offset-lithography inks, or to a high-quality photographic printer who uses CMY dyes. We then need some kind of colour management system to take control of the mapping from one colour space to another. Current colour management systems use device profiles and heavy proprietary software.

The International Color Consortium (ICC) issues and manages a standard format for colour profiles for input and output devices [4]. A *colour profile* is a data set which enables conversion between the actual colour space of a device and a standard *profile colour space* (PCS). ICC defines its PCS as CIE XYZ with white point D50 and recommended viewing conditions. An input profile converts input device colour values to PCS values, and an output profile converts PCS values to output device values. The conversions are by interpolation in a look-up table. Where a given sRGB colour is out-of gamut for YMCK, the ICC scheme allows choice of *rendering intent*: to clip to the nearest CMYK, or to reduce all the RGB values to fit the CMYK space.

A printer profile is constructed backwards by first printing the CMYK gamut as a grid of samples, then measuring the spectral power distribution of each patch by colorimeter, then calculating the corresponding PCS values. This arduous task is usually done by manufacturers for their printers, inks, and various substrates; or by companies offering a service over the internet.

16.7 Numerical Conversion Between Colour Spaces

The various standards define numerical conversions from one colour space to another [2, 3]. These formulas are implemented in the pair of QuickBASIC programs briefly described below, and listed in full in Vol. 2.

```
REM  **  Program SRGB2ETC to convert given sRGB
REM  **     values to CIEXYZ, CIExyY, CIELAB, and
REM  **     CMYK. User enters filename for output, and
REM  **     sRGB values to convert. Program does
REM  **     calculations and prints results.
REM  **  Print headings
REM  **  Enter sRGB values to convert
         CALL RGB2CMYK
         CALL RGB2CIES
         CALL PRINTOUT
END PROGRAM
         SUB RGB2CMYK
REM  **  Subprogram RGB2CMYK to convert RGB to CMYK
         SUB RGB2CIES
REM  **  Subprogram RGB2CIES to convert RGB to XYZ
REM  **     and xyY and Lab
REM  **     Normalize RGB to range 0 to 1
REM  **     Apply decoding gamma
REM  **     Adapt to D65
REM  **     Calculate x y Y chromaticity coordinates
REM  **     Calculate L a b values
         SUB PRINTOUT
REM  **  Subprogram PRINTOUT to print results
```

```
REM  **  Program ETC2SRGB to convert given CIEXYZ
REM  **     or CIExyY or CIELAB or YMCK values to
REM  **     sRGB. User enters filename for output, and
REM  **     values to convert. Program does
REM  **     calculations and prints results.
REM  **  Print headings
REM  **  Enter values to convert
         IF INTYPE = 4 THEN CALL CMYK2RGB
         IF INTYPE = 3 THEN CALL LAB2RGB
         IF INTYPE = 2 THEN CALL XYY2RGB
         IF INTYPE = 1 THEN CALL XYZ2RGB
         CALL PRINTOUT
END PROGRAM
```

```
        SUB CMYK2RGB
REM  ** Subprogram CMYK to convert given CMYK values
REM  **    to RGB.
        SUB LAB2RGB
REM  ** Subprogram LAB2RGB to convert given CIE L*a*b*
REM  **    values to RGB
REM  **    Convert L*a*b* to XYZ
REM  **    Convert XYZ to RGB
        SUB XYY2RGB
REM  ** Subprogram XYY2RGB to convert given CIE xyY
REM  **    values to RGB
REM  **    Convert xyY to XYZ
REM  **    Convert XYZ to RGB
        SUB XYZ2RGB
REM  ** Subprogram XYZ2RGB to convert given CIE XYZ
REM  **    values to RGB
        SUB PRINTOUT
REM  ** Subprogram PRINTOUT to print results
```

References

1. CIE (2014) Colorimetry series. http://www.cie.co.at/Publications/Standards. Accessed 1 Feb 2014
2. Hoffmann G (2003) CieLab color space. http://docs-hoffmann.de/cielab03022003.pdf. Accessed 1 Feb 2014
3. Hoffmann G (2013) CIE XYZ color space PDF. http://maintenance5.com/Gernot-Hoffmann-CIE-Color-Space-pdf-e3875.html. Accessed 1 Feb 2014
4. International color consortium. http://www.color.org
5. IEC 61966-2-1 (1999) sRGB colour space. http://webstore.iec.ch/. Accessed 1 Feb 2014
6. ISO Standard 15076-1:2010 (2010) Image technology color management—architecture, profile format, and data structure. http://www.iso.org/iso/catalogue_detail.htm?csnumber=40317. Accessed 1 Feb 2014
7. ISO Standard 2846-1:2006 (2006) Graphic technology: colour and transparency of printing ink sets for four-colour printing—Part 1. http://www.iso.org/iso. Accessed 1 Feb 2014
8. ITU-R Recommendation BT 709–5 (2008) Parameter values for the HDTV standards for program and international exchange. http://www.itu.int/rec/R-REC-BT.709/en. Accessed 1 Feb 2014
9. Padgham CA, Saunders E (1975) The perception of light and colour. Bell, London
10. Stokes M, Anderson M, Chandrasekar S, Motta R (1996) A standard default color space for the internet—sRGB. http://www.w3.org/Graphics/Color/sRGB. Accessed 1 Feb 2014
11. Wright WD (1973) The measurement of colour, 4th edn. Adam Hilger, London
12. x-rite (2014). http://www.xrite.com. Accessed 1 Feb 2014

Chapter 17
Perception of Brightness, Hue, and Saturation

Psycho-physics relates measurable physical stimuli to observable perceptual responses. There is a long and disputatious history of brightness perception, much of which is irrelevant to the normal perception of digital images. The transfer function of the eye from intensity input to perceptual output appears to be linear, for both additive and subtractive gradients. Perception of colour depends on three types of receptor in the retina. Perception of features in a scene or an image depends on retinal structures specialised for that function. Perception of the current scene or an image depends on continual combination of fixations.

17.1 Perception of Brightness

Psycho-physically, *brightness* is the sensation delivered to consciousness corresponding to physiological activity in the eye and brain, which in turn corresponds to physical intensity of light emitted or reflected by a surface in view, measurable by photometry (see Chap. 15).

17.2 Brightness Threshold

Consider a psycho-physical experiment which presents a viewer with an utterly dark field of view, luminance L_0, and within it a patch of variable luminance L_1. When $L_1 = L_0$ the viewer of course sees nothing, just the "dark light" or noise caused by spontaneous quantum events in the retina. When L_1 is increased somewhat, the viewer sometimes sees the patch, and sometimes not. The *lower sensory threshold t* is taken as the L_1 which is seen in 50 % of trials.

A. Parkin, *Digital Imaging Primer*, DOI 10.1007/978-3-540-85619-1_17

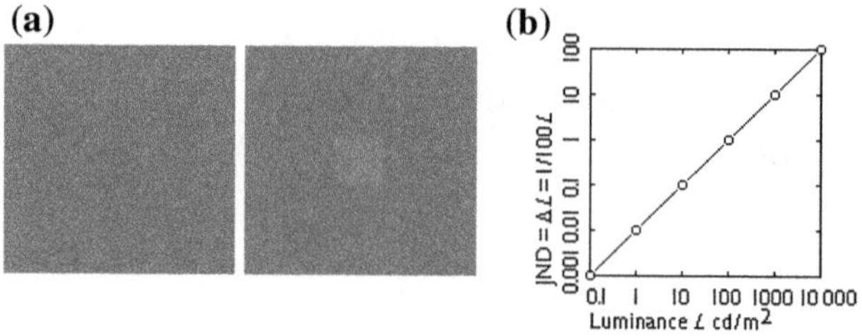

Fig. 17.1 Weber's Law. **a** Luminance of inner patch L_2 = luminance of surround L_1: viewer never sees inner patch. **b** Luminance $L_2 > L_1$: viewer sees inner patch. A just noticeable difference (JND) is when subject sees inner patch in 50 % of trials. Magnitude of JND is $\triangle L = L_2 - L_1$. **c** Log-log plot of JND $= 1/100\triangle L$ against L

17.3 Brightness Difference

A similar experiment presents a viewer with a field of fixed luminance L_1, and within it a patch of variable luminance L_2, as shown in Fig. 17.1b. A *just noticeable difference* (JND) is the luminance increment $\triangle L = L_2 - L_1$ which is seen in 50 % of trials.

Weber[1] proposed[2] that the increment of luminance $\triangle L$ needed for a JND is in constant ratio k to the initial luminance L:

$$\triangle L/L = k\,.$$

This is now known as *Weber's Law*, and the constant k as the *Weber ratio.*

It follows from Weber's Law that the increments of luminance $\triangle L$ needed for successive JNDs, from threshold upwards, are in *geometrical progression*, with common ratio $r = 1 + k$. Under careful experimental conditions, the constant k for JNDs of brightness sensation is found to be from 1/60 to 1/200, depending on the ambient light level and the viewer's eyesight and experience. Under typical conditions for viewing a computer screen, it is usually accepted that $k = 1/100$. Figure 17.1c is a log-log plot of increment $\triangle L$ against base luminance L: a straight line with slope 1 in 100.

A numerical measure often used where L_2 is the luminance of a small figure and L_1 the luminance of a large uniform ground is the *Weber contrast* C_W:

$$C_W = (L_2 - L_1)/L_1\,.$$

[1] Ernst Heinrich Weber (1795–1878), Professor of Anatomy and Physiology, University of Leipzig; a founder of psycho-physics.

[2] Weber's investigations in the 1840 s followed Bouguer's observation in the 1760s, confirmed by Lambert, that a doubling of number of candles is perceived as a constant increase of illumination.

17.4 Fechner's Law

Fechner,[3] generalized Weber's Law, making two further assumptions: that JNDs are subjectively equal, and that JNDs are additive across a given sensory range (brightness, loudness, weight ...). So then the sensory difference $\triangle S$ (of any magnitude) is proportional to the difference of physical luminance $\triangle L$ divided by the (lower) luminance L:

$$\triangle S = c\,(\triangle L/L) = k\,,$$

where c is a constant of proportionality. He then treated $\triangle S$ and $\triangle L$ as infinitesimals, and integrated to get

$$S = c\,(\log L) + a\,,$$

that is: sensation S is proportional to the logarithm of the physical luminance L, plus a constant of integration a. To find a in terms of L-values, let L_0 be the sensory threshold, at which $S = 0$; then

$$0 = c\,(\log L_0) + a\,,$$

hence

$$a = -c\,(\log L_0)\,,$$

so we can write:

$$S = c\,(\log L) - c\,(\log L_0) = c\,(\log L - \log L_0)\,,$$

hence *Fechner's Law*[4]

$$S = c\,(\log(L/L_0))\,.$$

17.5 Brightness Transfer Function of Eye

In communications engineering, the *transfer function* of a device relates its output to its input:

---

[3]Gustav Theodor Fechner (1801–1887), Professor of Physics, University of Leipzig.

[4]Now often called the *Weber-Fechner Law*, though Weber's Law relates two physical variables L and $\triangle L$ in a perceptual task, while Fechner's Law relates a perceptual variable S to a physical variable L.

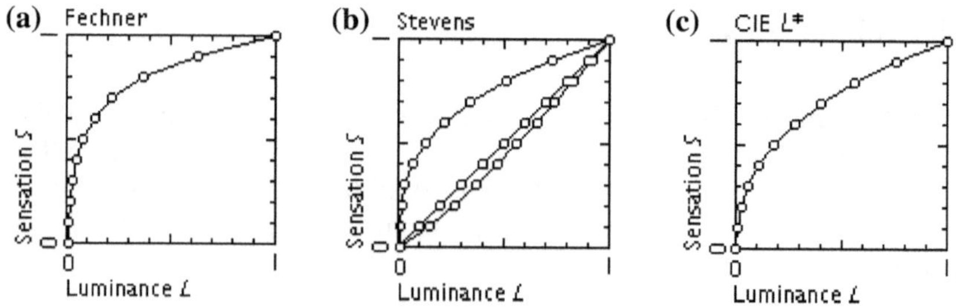

Fig. 17.2 Comparison of proposed transfer functions between luminance L and sensation S. **a** Fechner's Law. **b** Stevens' Power Law, where the uppermost curve has $\alpha = 0.33$, the middle curve $\alpha = 1$, and the lowest curve $\alpha = 1.2$. **c** CIE Lightness

$$\text{output} = \phi\,(\text{input})\,.$$

If the designated input range and output range are both normalized between 0 and 1, the transfer function can be plotted with input as abscissa and output as ordinate: such a plot is often called a *tone reproduction curve* (TRC). If we consider the eye as a device, the transfer function for brightness will be from normalized luminance L input to normalized brightness S output (as in Fig. 17.2).

We now calculate the brightness transfer function proposed by Fechner. Fechner's Law applies to luminances from a non-zero threshold upwards (since zero has no logarithm). A typical LCD display has a normalized range of luminance from Black $= L_0 = 0.007$ to White $= L_1 = 1$. So, setting $S_1 = 1$, we have

$$\begin{aligned} 1 &= c\,(\log(1/0.007)) = c(2.155) \\ c &= 1/2.155 = 0.4641 \\ S &= 0.4641(\log(L/L_0)\,. \end{aligned}$$

It follows from Fechner's Law that a geometrical progression of luminance L is perceived as an arithmetical progression of sensation S. In a geometrical progression, the *geometric mean* between any two terms A and B is $C = \sqrt{}(A \times B)$. To construct a geometrical progression of $N + 1$ luminances $L_0, L_1, L_2, \ldots, L_N$, given $L_0 > 0$ and L_N:

- Insert $N - 1$ geometric means between L_0 and L_N, with common ratio r.
- L_N is equal to $L_0 \times r^N$.
- Therefore $r^N = L_N/L_0$
- and $r = (L_N/L_0)^{1/N}$.
- We then have $L_1 = rL_0, L_2 = rL_1, L_3 = rL_2, \ldots, L_N = rL_{N-1}$.

In this case we have

- Insert 9 geometric means of luminance L_N between L_0 LCD Black 0.007 and L_1 White 1, with common ratio r.
- $1 = 0.007 \times r^{10}$.

Table 17.1 Ten equal steps of brightness, as proposed by Fechner, Stevens, and CIE

S	Fech.		St.33		St1.0		St1.2		CIE $L*$	I_{256}
	L	I_{256}	L	I_{256}	L	I_{256}	L	I_{256}	L	
0	0.007	27	0.007	27	0.007	27	0.007	27	0.002	17
0.1	0.011	33	0.008	28	0.106	92	0.153	109	0.011	33
0.2	0.019	42	0.015	38	0.205	124	0.267	140	0.030	52
0.3	0.031	53	0.033	54	0.304	149	0.371	162	0.062	72
0.4	0.051	66	0.049	76	0.404	169	0.470	181	0.112	94
0.5	0.083	82	0.129	100	0.503	187	0.564	197	0.184	118
0.6	0.137	103	0.218	128	0.602	203	0.656	210	0.281	143
0.7	0.226	130	0.344	157	0.702	217	0.745	223	0.407	170
0.8	0.371	162	0.512	188	0.801	231	0.832	234	0.567	197
0.9	0.609	203	0.729	221	0.901	243	0.917	245	0.763	225
1	1	255	1	255	1	255	1	255	1	255

S is relative magnitude of brightness sensation. For each proposal, L is relative physical luminance on a typical LCD display, and $I_{256} = (L^{0.4545}) \times 255$ is gamma-encoded intensity index as held on an 8-bit image file

- $r^{10} = 1/0.007$.
- $r = (1/0.007)^{1/10} = 1.6424$.

So we calculate 11 relative luminances which should correspond to equal-step sensations of brightness according to Fechner's Law, as shown in Table 17.1.

For example, brightness $S_{0.1}$ has luminance $L_{0.1} = 1.6424 \times 0.007 = 0.011$. Figure 17.2a shows this Fechner series as a transfer function, plotting sensation S against luminance L. Figure 17.3a shows this Fechner series as a display of 11 patches. For each patch, the 8-bit file intensity index $I_{256} = L^{0.4545} \times 255$. For example, brightness $S_{0.5}$ has index $I_{256} = 0.083^{0.4545} \times 255 = 0.3226 \times 255 = 82$.

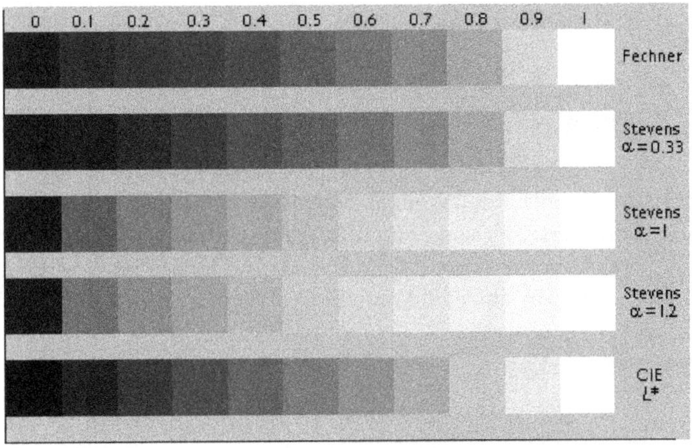

Fig. 17.3 Comparison of proposed gradients. **a** Fechner's Law. **b** Stevens' Power Law with $\alpha = 0.33$. **c** Stevens' Power Law with $\alpha = 1$. **d** Stevens' Power Law with $\alpha = 1.2$. **e** CIE Lightness

17.6 Stevens' Power Law

Fechner's Law was accepted as the model for all sensory modalities for nearly a century. From the 1950 s onwards Stevens[5] made extensive new studies of sensory discrimination, with many innovations of method. He proposed that sensation magnitude S grows as a power function of physical luminance L:

$$S = c(L - L_0)^{\alpha} ,$$

hence

$$L = (S/c)^{1/\alpha} + L_0 ,$$

where L is luminance, S is sensation, c is a constant of proportionality, α is an exponent which varies from one sensory modality to another and L_0 is luminance threshold corresponding to $S = 0$. Stevens gives $\alpha = 0.5$ for a point source, $\alpha = 0.33$ for a 5° emitting source viewed in the dark, and $\alpha = 1.2$ for reflecting grey papers.

So, setting $\alpha = 0.33$, $L_0 = 0.007$, and $S = 1$, we have

$$\begin{aligned} 1 &= c\,(1 - 0.007)^{0.33} \\ c &= 1.0023 \\ L &= (S/1.0023)^{3.03} + 0.007 \end{aligned}$$

and similarly for $\alpha = 1$ and $\alpha = 1.2$. We may then calculate luminances which should correspond to equal-step sensations of brightness according to Stevens' Power Law, as shown in Table 17.1, plotted as transfer functions in Fig. 17.2b, and displayed in Fig. 17.3b–d. When $\alpha = 0.33$, the curve is quite similar to Fechner's. When $\alpha = 1$, $L = S + L_0$, and the curve is a straight line. When $\alpha = 1.2$, the curve bends upwards slightly.

17.7 CIE Lightness

In 1976 CIE issued a colorimetric standard called L*u*v* (or CIE LUV), intended for use in television, printing and computer graphics. This maps the 1931 standard variables (x, y, z) to new variables $(L*, u*, v*)$, of which *Lightness* $L*$ is intended as the sensory correlate of physical luminance Y. The definition is

$$L* = 116\,(Y/Y_n)^{1/3} - 16 ,$$

[5]Stanley Smith Stevens (1906–1973), Professor of Psycho-physics at Harvard University (and the first anywhere).

where $L*$ is perceived lightness, Y is luminance of sample, and Y_n is the luminance of reference White, usually normalized to 100. If we write S for $L*$, and L for the luminance of the sample relative to White = 1, we have

$$S = 116 \times (L^{1/3}) - 16\,,$$

hence

$$L = (100 \times S + 16)/116)^3$$

where $0 \leq L \leq 1$ and $0 \leq S \leq 1$. We may then calculate luminances which should correspond to equal-step sensations of lightness according to the CIE formula, as shown in Table 17.1, plotted as a transfer function in Fig. 17.2c, and displayed in Fig. 17.3e. The curve is quite similar to the Fechner and to the Stevens 0.33 curve.

17.8 The Brightness Transfer Function is Linear

In Fig. 17.3 the background is just slightly paler than a 0.5 luminance (intensity 192 instead of 186). Gradient (c), Stevens with $\alpha = 1$, is the one with its middle step matching the background. We conclude that the transfer function for brightness is simply

$$S = L\,,$$

where both S and L are normalized between 0 and 1.

The widespread assumption in descriptions of digital imaging that brightness sensation is logarithmically related to physical luminance is erroneous. The reason for the error seems to lie in a failure to appreciate the importance of viewing conditions, as pointed out by Hoffmann [1][6]:

"All this cannot be applied to images, because the Weber Law is a result of variable adaptation (sitting in a dark room and observing two large patches). For imaging, the adaptation is more or less fixed, the Weber Law is not valid. ... The resolution for dark greys is not generally better than for light greys."

That is to say, a laboratory psycho-physical experiment is a long series of careful judgements of single differences, singly presented, where the viewer adapts to the surround level at each stage, little by little. Viewing a complete grey-scale, or any image, requires simultaneous or a very short series of judgements of multiple differences, all presented together, where the viewer has but a single overall adaptation to the ambient illuminance.

It is well-known that visual perception is very good at adapting to current viewing conditions in a complex and involuntary way, to provide various constancies. For

[6] Gernot Hoffmann (*1944), Formerly Professor at Fachhochschule OOW, Emden.

example, constancy of shape means that a circular plate looks circular even though the retinal image is elliptical. Constancy of size means that a person 20 m away looks the same size as a person 10 m away even though the retinal image is half the length. Constancy of brightness means that a sheet of white paper still looks white in dim light, and a black book still looks black in strong light, even though the luminance of the paper is much less than the luminance of the book. We may speculate that something of this kind happens in viewing a grey-scale image containing a great mixture of greys between Black and White. It seems that we instantly and involuntarily identify the average luminance of the whole image as mid-grey, or normalized luminance $L = 0.5$, and judge the lighter greys as departing from the average towards 1 White, and the darker greys towards 0 Black, in a simple linear manner. That is, we see the multiple mixed luminances of an image pretty much as they physically are. The transfer function of the eye from luminance to Brightness is linear.

17.9 Subtractive Brightness Gradient

When a gradient is displayed on a screen, the physical variable is additive light, going from screen minimum (Black) to screen maximum (White). All the proposed transfer functions except the linear one predict small physical differences between dark-grey JNDs and large physical differences between pale-grey JNDs. But when a gradient is printed as a half-tone, the physical variable is subtractive ink coverage, going from uncovered paper (White) to solid ink (Black). Then the prediction is small physical differences between pale-grey JNDs and large between dark-grey JNDs, contradicting the additive case (as seen in Stevens' $\alpha = 1.2$ curve for grey papers). But if the transfer function is linear, all physical differences are equal for JNDs, the additive and subtractive cases are symmetrical, and there is no contradiction.

To make a subtractive gradient of greys corresponding to an additive gradient of greys, the relative intensity of light reflected (from a White paper) must be equal to the relative intensity of light emitted (from a display screen). That is to say, the coverage of Black ink K must be $(1 - L)$ where L is the additive luminance (normalized between 0 and 1). Any half-toning method (see Chap. 32) will do just this. Figure 17.4a shows a subtractive series of greys with a linear gradient (Stevens' Power Law with $\alpha = 1$). This has been half-toned by hand with with counted-pixel coverages of ink to give the normalized reflectances L as marked. Figure 17.4b is a corresponding additive series, and Fig. 17.4c is a scanned standard step-wedge (see Sects. 22.13 and 22.14).

It is not easy to compare an additive display to a subtractive printout. However, a very good measure of the luminance of a printed grey is by direct matching of that part of the printout to a patch of the step-wedge, under the same incident light. The relative luminance is

$$L = 1/(10^{\mathrm{d}}),$$

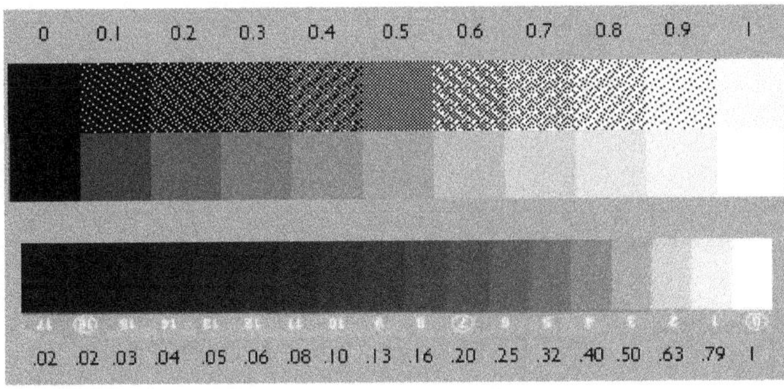

Fig. 17.4 Comparison of subtractive equal-step grey-scale, additive equal-step grey-scale, and step-wedge. **a** Subtractive linear series, with normalized reflectances marked. **b** Additive linear series. **c** Standard step-wedge, with normalized reflectances marked

where d is the density marked on the step-wedge (see Sect. 22.13). For example, say step-wedge patch 3 matches the middle patch of the subtractive grey series. Step-wedge patch 3 has density $d = 0.3$, $10^{0.3} = 1.9953$, and $1/1.9953 = 0.5$. So the middle patch of the grey series has luminance (relative reflectance) 0.5.

On the other hand, the luminance of a displayed grey can be measured by using the eye-dropper in MS PAINT, then Colors—Edit Colors—Define Custom Colors to find the R = G = B encoded intensity index I_{256}. The relative luminance is

$$L = (I_{256}/255)^{2.2}\,.$$

For example, say the middle patch of the additive grey series has index $I_{256} = 186$. $186/255 = 0.7294$, and $0.7294^{2.2} = 0.5 = L$.

17.10 Perception of Colour

Physiologically, there are three types of cone receptors in the normal human retina,[7] responding to light of long L, medium M, and short S wavelengths. In the fovea the proportional counts are 63 % L, 31 % M, and 6 % S on average, with considerable variation from person to person. The central foveola contains nearly equal counts

[7] The evolution of colour vision perhaps began with detectors of ultra-violet radiation (to avoid it) and of infra-red/heat (to catch, or escape from, the emitter). Retinal cones of one or more types developed along various evolutionary paths: humans and bees have three types, most amphibians reptiles and birds four, some butterflies six, and mantis shrimps 12. Some animal groups have lost previously-evolved cone types: dogs, cats, and cows have only two (to detect blues and greens but not reds), and marine mammals only one. Old-world primates have apparently re-evolved to three types of cone (possibly to detect the reds of ripe fruits). In humans there are genetic and gender variations: some 8 % of men and 0.4 % of women in the UK lack one or two of the cone types, and a few women are believed to have four cone types.

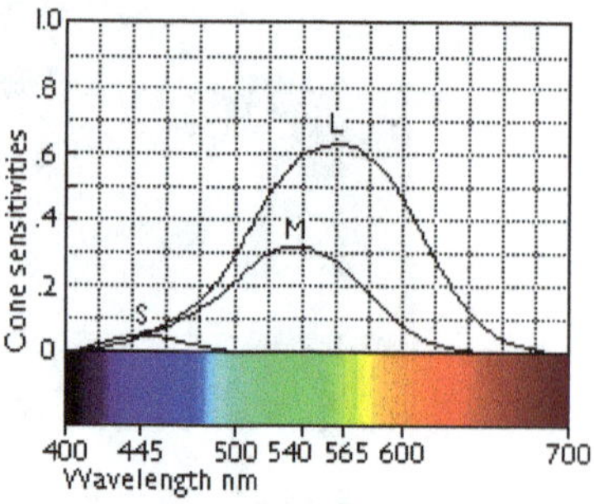

Fig. 17.5 Cone fundamentals. Relative sensitivity of L M and S cones for 10° field, weighted by population of cones. 1.0 is total cumulative response of all three cones

of L and M cones, while the S cones are mostly towards the edge of the fovea. The spectral sensitivity curves of the three types are similar in shape, with peaks at 565 nm for L cones, 540 nm for M cones, and 445 nm for S cones.

Figure 17.5 [4] shows the sensitivity curves for a 10° field, weighted by proportion of each cone type in the retina; that is to say, the retinal response across the spectrum. Evidently, for wavelengths above 525 nm or so, the colour information comes from the difference between the L and M responses; while below 525 nm it comes from the difference between the S and the combined L and M responses.

Neuro-physiologically, nerve fibres from the cones and rods leave each retina at the *blind spot* where there are consequently no receptors. Nerves from the left side of both retinas are combined at the optic chiasma; and similarly the right. The left-side information is processed in the left lateral geniculate body deep in the brain, and passed to the left visual cortex; and similarly the right.

It is thought that the brain organizes the three types of cone responses and the rod responses into a red-green opposing pair, a blue-yellow opposing pair, and an overall brightness signal, as proposed by Hering.[8] It is interesting to note that in colour television the camera produces three signals R, G, B, which are converted to a luminance L and two colour differences RL and BL for transmission, then reconstituted as R, G, B at the receiver. The transmission bandwidth for L, RL, BL is only about half that required for R, G, B [3].

17.11 Perception of Hue

Psycho-physically, *hue* is the sensation delivered to consciousness corresponding to physiological activity in the eye and brain, which in turn corresponds to the wavelength distribution of light emitted or reflected by a surface in view, measurable by colorimetry (see Chap. 16).

[8]Karl Ewald Konstantin Hering (1834–1918), Professor of Physiology at Prague University (succeeding Purkyně), then at Leipzig. Controverted Helmholtz on colour vision: it is now accepted that Helmholtz and Hering were both partially correct.

Hue extremes are CIE dominant wavelength about 400 nm, hue violet, and about 700 nm, hue red. The extremes have inherently low brightness, and the middle range high brightness, (see the CIE V(λ) curves, Sect. 15.3).

Hue difference has a JND of about 2 nm in the yellow-greens, and much greater towards the extremes. Wright found about 150 JNDs of hue in the spectrum, by plus those in non-spectral mixtures [5]. The Farnsworth Munsell 100 Hue Test is based on discriminating 100 hues [6].

17.12 Perception of Saturation

Psycho-physically, *saturation* is the sensation delivered to consciousness corresponding to physiological activity in the eye and brain, which in turn corresponds to the spectral power distribution of light emitted or reflected by a surface in view, measurable by colorimetry (see Chap. 16).

Saturation extremes are CIE purity zero for a completely unsaturated neutral, to purity 1 for a monochromatic spectral colour. High saturations, particularly of blues and reds, have inherently low brightness.

It seems likely that saturation difference JNDs are around 10 in the yellows, 50 in the reds, and 60 in the blue-violets.

17.13 Perception of Feature

It is known that the retina has structures which detect features within the projected image. For example, in the ganglion layer of the retina, *ganglion cells* connect to a group of cones to form contrast-detectors, as shown in Fig. 17.6. A centre-on group fires rapidly when the centre is brighter than the surround, and a centre-off group the reverse. A contrast-detector has a lower threshold of difference of luminances.

Combinations of such cells produce neural signals for corners, edges, regions, and other features in an image.

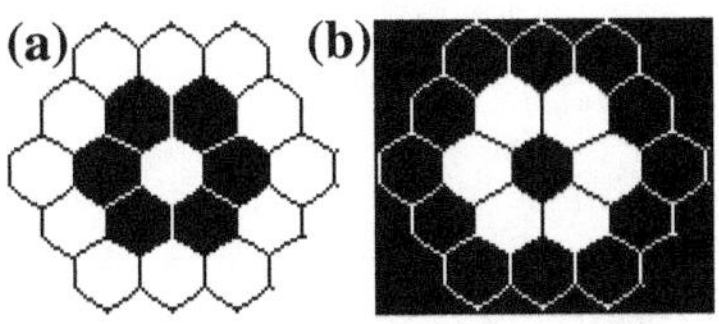

Fig. 17.6 Ganglion cells. **a** Centre-on cell fires when centre cone is receiving high intensity and surrounding cones are receiving low intensity. **b** Centre-off cell fires when centre cone is receiving low intensity and surrounding cones are receiving high intensity

17.14 Perception of Scene

My eyes dart about in ever-changing brief binocular fixations on features of interest in the current scene. At each fixation my retinas get paired images with a small sharp foveal region surrounded by a large fuzzy peripheral region. The retinal images are projected distributions of intensities of light. The intensities in the images go as binary-coded neural signals to my brain. I build up a conscious exploratory attention-led steady perception of solid objects located in a space which is centred *here*, at my current Cyclopean point. If I move, I perceive the same objects from a different centre. If an object moves, I perceive that it has moved, not I. The physical activities are throughout discrete, digital, and stepped. The psychological results are continuous, analogue, and smooth.

Clearly, visual perception is astonishingly complex, and it is little-understood. Perhaps the best account to date has been Marr's[9] three-stage computational model [2], where the first stage is a primal two-dimensional sketch of the view, the second is a "2 1/2"-dimensional sketch adding texture and other detail, and the third is a three-dimensional perceptual representation of the view.

References

1. Hoffmann G (2003) The gamma question. ftp://ftp.ekspress.ee/incoming/reklaam/Prepress/Doc/Gamma.pdf. Accessed 1 Feb 2014
2. Marr DC (1982) Vision: a computational investigation into the human representation and processing of visual information. WH Freeman, New York. Re-issued 2010 MIT Press, Boston. http://uploading.com/files/e1159a61/0262514621Vision.rar/. Accessed 1 Feb 2014
3. Padgham CA, Saunders E (1975) The perception of light and colour bell. Bell and Sons, London
4. Stockman A, Sharpe LT (2000) 10-degree cone fundamentals. http://www.cvrl.org/cones.htm. Accessed 1 Feb 2014
5. Wright WD (1946) Researches in normal and defective colour vision. Henry Kimpton, London
6. x-rite (2013) Farnsworth Munsell 100 Hue Test. http://www.xrite.com/online-color-test-challenge. Accessed 1 Feb 2014

[9]David Courtnay Marr (1945–1980), Professor of Psychology, Massachusetts Institute of Technology.

Part IV
Complexity in Digital Imaging

Chapter 18
Complexity

Complex systems may be separated into disorganized and organized. Disorganized complex systems are best studied statistically; organized by computer modelling. Digital images may be treated as disorganized systems by adapting statistical information theory, and by using elementary classical statistical methods. Statistical calculations are defined, with 8-bit examples. Digital images may often be usefully analyzed by histogram; but their normality of distribution or otherwise is usually of little interest. Exploration of organized complex systems has been very closely connected with the development of computing, and particularly with digital imaging. Several lines of current investigation are indicated.

18.1 Complex Systems

Weaver [15] distinguished *disorganized complexity* from *organized complexity*.

Disorganized complexity is seen in a system which has very many components acting independently, such as molecules in a gas. Tracing the behaviour of the system as a whole is hopelessly long and difficult. But a statistical approach may find and measure quite simple and stable patterns of probability which emerge from low-level random events. Many natural complex systems were studied in this way in the 19th century and onwards.

Organized complexity is seen in a system which has very many components interacting. That is to say, each component has simple rules of behaviour which are affected by, and in turn affect, other components. Natural examples of such systems are the growth of plants, the development of patterns in animal coats, and the differentiation of stem cells to specialized cells. Fruitful methods here are mathematical and computer models, now in active development. Artificial examples are cellular automata, and evolutionary computer programs, which are already in computable form.

A. Parkin, *Digital Imaging Primer*, DOI 10.1007/978-3-540-85619-1_18

Digital images may be treated as disorganized complex systems, using the methods of statistical information theory. Or they may be treated as organized complex systems, using the methods of computer programming.

18.2 Statistical Information Theory

Shannon developed a fundamental theory of communication based on a statistical definition of information [11, 12]. His model was an information *source* with a repertoire of distinct symbols, emitting a *message* as a time-sequence of symbols, into a *channel*, and thence to a *receiver*. He developed measures for the *entropy* of a source, the *capacity* of a channel, the limiting rate of information transfer, and the best coding for channels with noise.

In statistical information theory the measure H of the entropy of a source is

$$H = -(\Sigma(i)1 \text{ to } k)p(i) \times \log_2(p(i)),$$

where k is the count of symbols which can be emitted, and $p(i)$ is the probability of each symbol being emitted. The logarithm of a fraction is negative, so the minus signs make the entropy positive. Shannon used this as a measure of the *uncertainty* which is removed by knowing the value of one symbol. Entropy is zero if all the probabilities p are either 0 or 1. It is maximum if all the probabilities p are equal. Where entropy is low, various compression codings are possible.

A digital image may be regarded as an information source with a repertoire of intensities, and a scan as a location-sequence of intensities emitted into a visual channel, and thence to a binocular receiver.

The entropy of an image is the average number of bits needed to store or transmit 1 px. For a 1-bit image the entropy is

$$H = -p \times log_2(p) - q \times \log_2(q),$$

where p is the probability of a pixel being Black, and $q = (1 - p)$ is the probability of a pixel being White. Entropy is zero if both the probabilities are either 0 or 1. It is maximum if both the probabilities are equal. For example, a 1-bit image where all pixels are Black has entropy zero; and where half the pixels are Black has entropy 1 bit. For an 8-bit image the entropy is

$$H = -(\Sigma(i)1 \text{ to } 256)p(i) \times \log_2(p(i)),$$

where 256 different intensities can be emitted, and $p(i)$ is the probability of each being emitted. Where entropy is low, various compression codings are possible (such as ZIP and JPEG).

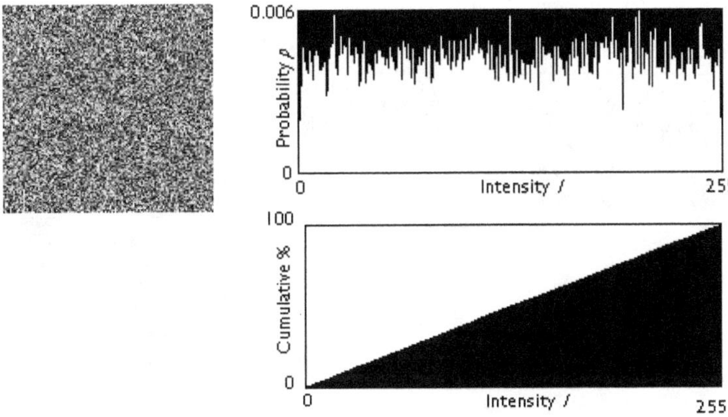

FREQUENCY TABLE

I	Freq.	Prob.	Cu.Freq.	Cu.%
0	30	.002	30	0
1	49	.003	79	0
2	71	.004	150	1
3	63	.004	213	1
4	58	.004	271	2
5	57	.003	328	2
6	69	.004	397	2
7	63	.004	460	3
...	...	...	...	...
248	64	.004	15995	98
249	65	.004	16060	98
250	57	.003	16117	98
251	46	.003	16163	99
252	62	.004	16225	99
253	60	.004	16285	99
254	69	.004	16354	100
255	31	.002	16385	100

STATISTICS OF IMAGE

Image bit-depth	8
Image width, px	128
Image height, px	128
Image total px	16384
Mean	128
Mode	206
Lowest intensity	0
Highest intensit	255
Variance	5373
Standard deviatn	73
Skew	-0
Kurtosis	-1
Lower quartile	63
Median	126
Upper quartile	190
Inten.complexity	8
Locat.complexity	14
Total complexity	112
Entropy	7.985

Fig. 18.1 8-bit random image. **a** Image 128 × 128 px, with 256 intensities randomly set by program. **b** Histogram. **c** Ogee plot. **d** Frequency table. **e** Statistics

18.3 Classical Statistics of Digital Images

Graphically, a digital image is a collection of (usually) square pixels in a (naturally) rectangular window, each pixel showing an intensity of light emitted or reflected. A viewer perceives groups, patterns, and structures of pixels, which have all kinds of meanings.

Numerically, a digital image is represented by an array of integer triples (x, y; I), where x and y are location coordinates, and I is the intensity code. A computer program can pick out groups, patterns, and structures (which may or may not be the same as a viewer's perceptions), which have all kinds of mathematical properties.

Statistically, a digital image is a *finite population* of pixels, with a *population count* of N px. An individual pixel of the population is identified by its location (x, y). The population has one discrete *variate*, intensity I. A large part of classical

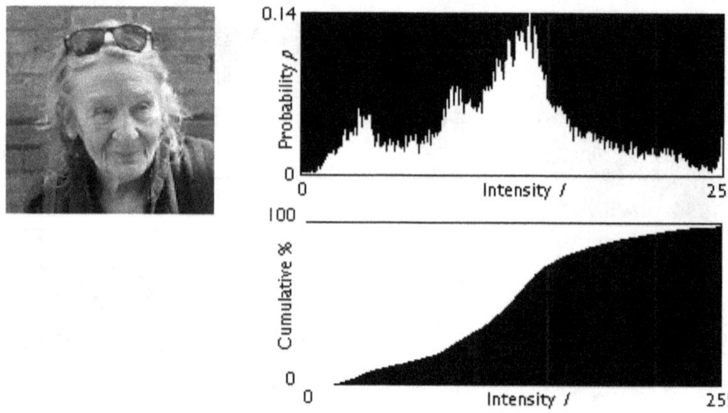

FREQUENCY TABLE

I	Freq.	Prob.	Cu.Freq.	Cu.%
0	1	.000	1	0
1	0	.000	1	0
2	0	.000	1	0
3	1	.000	2	0
4	2	.000	4	0
5	2	.000	6	0
6	3	.000	9	0
7	2	.000	11	0
...	...	...	...	...
248	6	.000	16255	99
249	4	.000	16259	99
250	11	.001	16270	99
251	7	.000	16277	99
252	11	.001	16288	99
253	19	.001	16307	100
254	25	.002	16332	100
255	53	.003	16385	100

STATISTICS OF IMAGE

Image bit-depth	8
Image width, px	128
Image height, px	128
Image total px	16384
Mean	122
Mode	138
Lowest intensity	0
Highest intensit	255
Variance	2610
Standard deviatn	51
Skew	0
Kurtosis	-0
Lower quartile	91
Median	126
Upper quartile	149
Inten.complexity	8
Locat.complexity	14
Total complexity	112
Entropy	7.585

Fig. 18.2 8-bit camera image. **a** Image 128 × 128 px, with 256 intensities set by camera. **b** Histogram. **c** Ogee plot. **d** Frequency table. **e** Statistics

statistics is concerned with questions of sampling from a large population. In digital imaging, on the other hand, we have the simpler situation of complete enumeration (or census) of the population.

We illustrate with 8-bit grey-scale images as examples. An 8-bit digital image (see Part VII) has pixels showing one of 256 intensities, so $0 \leq I \leq 255$. Figure 18.1a shows an example of a random image (generated by program GENRAN8: see Sect. 38.2). Figure 18.2a shows an example of a camera image.

18.4 8-bit Frequency Table

In a given 8-bit image, we can count the frequency F_0, F_1, F_2, ..., F_{255} of each intensity I_{256} over the whole population of pixels. The sum of these frequencies $N = F_0 + F_1 + F_2 + \cdots + F_{255}$ is the count of all pixels in this image, = (width

Table 18.1 Frequency table for an 8-bit image

I	Freq	Prob	CuF	CuPC
0	F_0	p_0	CuF_0	$CuPC_0$
1	F_1	p_1	CuF_1	$CuPC_1$
2	F_2	p_2	CuF_2	$CuPC_2$
...	...	...	...	...
255	F_{255}	p_{255}	CuF_{255}	$CuPC_{255}$
	N	1		

I is intensity of pixel; Freq is frequency; Prob is probability; CuF is cumulative frequency; and CuPC is cumulative percent. F_{I} is the frequency count of intensity I, summing to N; p_{I} is the probability $p_I = F_{\text{I}}/N$, summing to 1

in px) × (height in px). The (empirical) probability of each intensity is $p_0 = F_0/N$, $p_1 = F_1/N$, $p_2 = F_2/N$,..., $p_{255} = F_{255}/N$. In making a frequency table, we can add a cumulative frequency count CuF and a cumulative percent count $CuPC$ for each value of the variate I. We then have a form like Table 18.1.

For example, Fig. 18.1d shows the start and finish of the frequency table for random image Fig. 18.1a. Figure 18.2d shows the start and finish of the frequency table for camera image Fig. 18.2a.

18.5 8-bit Histogram

Having made a frequency table, we can construct a *histogram*. The name was introduced by Pearson[1] in 1891. In lectures he used a bar graph to group events spread out in time (hence a time-diagram or *historic-diagram*), by plotting frequency (or probability) against value of variate. The histogram has become a fundamental tool of descriptive statistics, to give an immediate graphical summary of the variety in a population.

For example, Fig. 18.1b shows a histogram for random image Fig. 18.1a. Figure 18.2b shows a histogram for camera image Fig. 18.2a. Here the abscissa is intensity I, marked 0–255; the ordinate is probability p, that is, frequency normalized in the range 0–1. There are 256 bars, one for each intensity, each with a bar height proportional to the frequency. The proportionality may be chosen however we please: here we take the maximum value (the mode) as 100 and reduce the others to percentage of mode.

[1] Karl Pearson, (1857–1936) Professor of Eugenics (later Genetics), London University, 1911–1933; friend and follower of Galton.

18.6 8-bit Ogee Plot

We can also construct an *ogee*[2] plot of cumulative percentage against intensity. Here the abscissa is the same 256 intensities $0 \leq I \leq 255$, and the ordinate is cumulative percent from 0 to 100, as calculated in the frequency table.

For example, Fig. 18.1c shows an ogee plot for random image Figs. 18.1a; and 18.2c for camera image Fig. 18.2a.

18.7 Histograms in Digital Imaging

Histograms of the frequencies of intensities are quite often useful when planning or judging or comparing intensity transformations, in 8-bit (see Chap. 39) or 24-bit (see Chap. 47) images.

18.8 Normal Distribution

Classical statistics is typically concerned with estimating the distributions of single variates (height, weight, income, ...) in large populations (UK men 18–65, UK women over 65, ...). In such studies a histogram of the frequencies of occurrence of (suitably grouped) values of the variate shows a bell-shaped curve profile, where most occurrences have values near the middle of the scale, falling off in frequency to lower and higher values in both directions. This is taken as a fact of nature, and called a *normal distribution* of a variate. On the other hand, in digital imaging each image is an independent population, and there is no special significance in a normal or other distribution of intensities.

How a normal distribution arises from a large number of independent random events is demonstrated in Fig. 18.3, which shows the essentials of Galton's Quincunx.[3] The machine is a vertical board with pins placed in alternating rows (here six rows). A ball dropped on a pin has a 0.5 probability of continuing on a left path and a 0.5 probability of a right path. So after the top pin there are $2^1 = 2$ paths with expectation 1/2 and 1/2. After the next row of pins there are $2^2 = 4$ paths, but two merge to give expectation 1/4, 2/4, and 1/4. That is to say, in the long run about twice as many balls would go through the middle gap as through each of the outer gaps. After the next row of pins there are $2^3 = 8$ paths, but merging gives expectation 1/8, 3/8, 3/8, and 1/8. After the sixth row of pins there are $2^6 = 64$ paths, but merging gives expectation 1/64, 6/64, 15/64, 20/64, 15/64, 6/64, and 1/64. The balls

[2] also called an *ogive*, *OG*, or by gentlemen a *cyma reversa*.

[3] Sir Francis Galton, F.R.S. (1822–1811), was a polymath [2] who followed Quetelet in developing statistical methods, including correlation and regression analyses. The quincunx was built in 1894. It is named for the five-spot pattern as seen on a die.

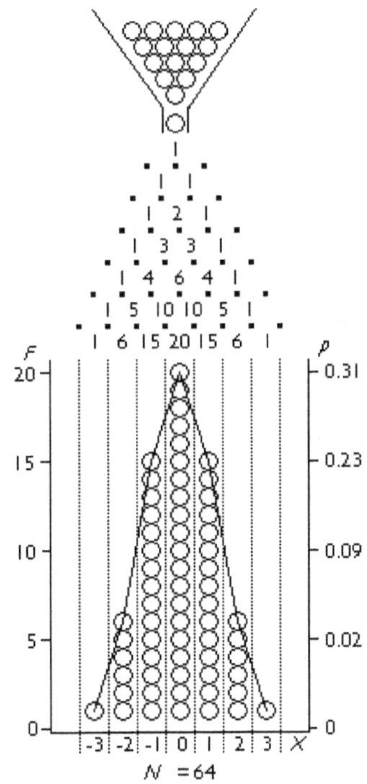

Fig. 18.3 Galton's Quincunx. 64 balls drop from a hopper onto alternating rows of pins. At each pin a ball has equal chances of going left or right. The *numbers* show the expected count of balls to pass through each gap. The balls stack up in bins, and the tops of the stacks make a curved profile. The *curve* is a visualization of the probability distribution. N is the count of balls dropped; X are the collecting bin labels; F is the frequency count of balls in bins; $p = F/N$ is the corresponding probability

are collected in (here seven) bins to form a histogram, and the tops of the bars show a curved profile. The curved profile is a visual record of the probability distribution for this run of the machine. A bigger quincunx makes a better curve, approximating a bell-shaped normal distribution.

18.9 Standard Normal Distribution

The normal distribution has been formalized mathematically as the *Standard Gaussian*[4] *Normal* (SGN) probability density function (for continuous values of a variate) or the SGN probability mass function (for discrete values) Fig. 18.4a shows the curve. Pearson's original aim was to summarize a large set of biological data by four measures applied to this curve, which he called the mean, variance, skewness, and kurtosis.

- The *mean* μ is a single value, which best represents the whole population—the average.
- The *mode* is the most frequent value in the population.

[4]Johann Carl Friedrich Gauss (1777–1855), mathematician, physicist, and Professor of Astronomy, University of Göttingen. Discoveries in probability preceded by De Moivre 1738 and followed by Laplace 1810.

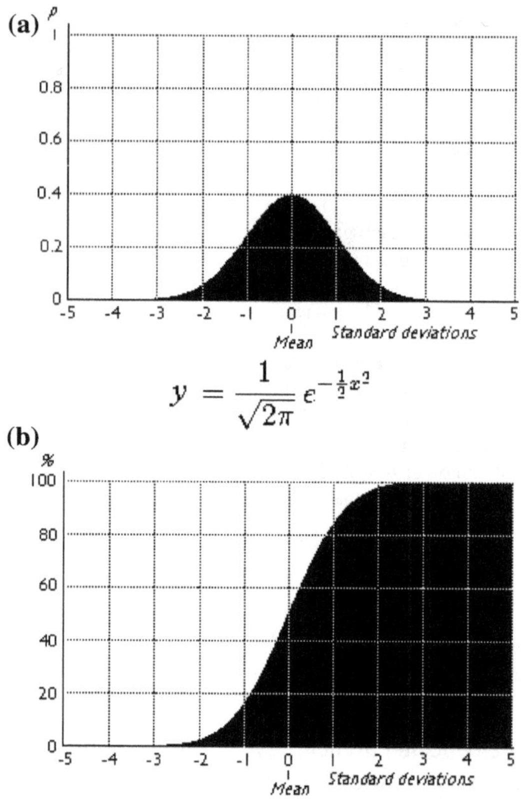

Fig. 18.4 Standard Gaussian Normal distribution. **a** Standard curve, where the abscissa has the variate values arranged symmetrically about the mean, and the ordinate is the probability for each value. The formula shown plots the curve with mean $\mu = 0$ and standard deviation $\sigma = 1$. The mode is the same as the mean, which has probability $1/\sqrt{2\pi} = 0.399$.
b Cumulative curve, where the ordinate is the percentage of observations lying below that value

- The *variance* σ^2, or dispersion, summarizes how far the values occurring in a population deviate from the mean.
- The *standard deviation* σ is the square root of the variance, which is a more useful measure of deviation from the mean.
- *Skewness* is the asymmetry of a distribution about its mean. A curve where the peak is displaced towards low values and shows a longer tail of high values has positive skewness. A curve where the peak is displaced towards high values and shows a longer tail of low values has negative skewness.
- *Kurtosis* is the convexity of a distribution curve, compared with a Standard Gaussian Normal curve. A flatter or platy-kurtic curve has kurtosis less than 0; and a sharper or lepto-kurtic curve greater than 0.

The Standard Gaussian Normal curve is symmetrical about its mean value $\mu = 0$, and the mean is the same as the mode. The variance $\sigma^2 = 1$, and so the standard deviation $\sigma = 1$. The skewness is 0 and the kurtosis is 0.

Figure 18.4b shows the SGN distribution plotted as a cumulative curve. The abscissa is the variate values as before, but here the ordinate is the percentage (probability $\times 100$) of observations lying below that value. The curve starts horizontal, rises with increasing slope to an inflection, then with decreasing slope to finish horizontal.

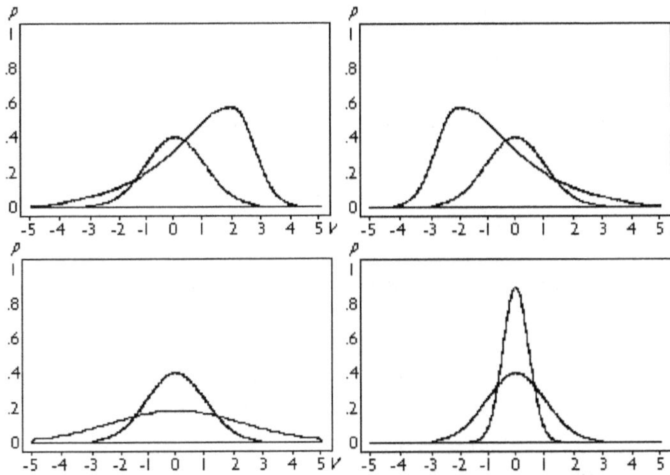

Fig. 18.5 Departures from the Standard Gaussian Normal distribution. **a** Negatively-skewed curve, compared with standard. Peak displaced *to the right*, tail longer *to the left*. **b** Positively-skewed curve, compared with standard. Peak displaced *to the left*, tail longer *to the right*. **c** Curve with negative kurtosis, flatter than standard. **d** Curve with positive kurtosis, sharper than standard

18.10 Departures from Standard Normal Distribution

Figure 18.5 shows how actual distributions may differ from the standard normal. In Fig. 18.5a the curve has negative skewness with a small tail to the right and a big tail to the left. In Fig. 18.5b the curve has positive skewness (for example, income distribution shows positive skewness, with most people earning little and a few earning much). In Fig. 18.5c the curve is platy-kurtic, flatter than the standard, indicating that the observations are fairly evenly-spread across the range. In Fig. 18.5d the curve is lepto-kurtic, sharper than the standard, indicating that the observations are mainly close to the mean, with rather few outliers.

18.11 8-bit Statistical Calculations

$$\text{Mean} = ((0 * F_0) + (1 * F_1) + (2 * F_2) + \cdots + (255 * F_{255})/N\,,$$

where F_0 is the 0-count, F_1 the 1-count, …, and N the total population count.

$$\text{Variance} = \text{SUM}\,((I - M)^2)/N\,,$$

where N is the population count, M is the mean, and the sum is taken over all the I from 0 to 255.

$$\text{Standard deviation} = \sqrt{\text{variance}}$$
$$\text{Skewness} = \text{SUM}\,((I - M)^3)/(N \times SDEV^3)\,,$$

where N is the population count, M is the mean, and the sum is taken over all the I from 0 to 255.

$$\text{Kurtosis} = (\text{SUM}\,((I - M)^4)/(N \times SDEV^4)) - 3\,,$$

where N is the population count, M is the mean, and the sum is taken over all the I from 0 to 255.

If the frequencies are converted to percentages, then cumulated as in Sect. 18.4 above, they can be grouped as four *quartiles*, ten *deciles*, or any other division of the percentages. The lower quartile intensity is that which first cumulates to 25 %; the median to 50 %, and the upper quartile to 75 %.

For an 8-bit image:

- Intensity complexity = 8 bits. The intensity complexity of an image is the number of binary digits needed to specify the intensity of any one pixel. It is a logarithmic form of count-intensity-levels. It is used as a classification of images throughout this book.
- Location complexity = $\log_2(N)$ bits, where N = (width X in pixels) × (height Y in pixels).[5] The location complexity is the number of binary digits needed to locate any one pixel of an image. It is a logarithmic form of image-size-in-pixels, and is irrespective of actual displayed or printed size-in-inches or size-in-millimetres.
- Total complexity = (location complexity) × (intensity complexity) bits. The total complexity measures the resources needed to store an image. It is a logarithmic form of file-size-in-bytes.

See Sect. 39.26 for program DOSTATS8 which calculates statistics for an 8-bit image, as shown above.

18.12 Organized Complex Systems

Von Neumann[6] and Ulam[7] devised algorithms for a self-replicator [10] in the 1940s. This had a 29-state constructor and an instruction tape; the instruction tape contained instructions to change itself, as well as to change the constructor. The work was done by brain-power and pencil and paper.

---

[5] $\log_2(N) = \log_{10}(N)/\log_{10}(2)$; or for computation $\log_2(N) = \ln(N)/\ln(2)$.

[6] John von Neumann (1903–1957), Los Alamos National Laboratory; Princeton University.

[7] Stanislaw Marcin Ulam (1909–1980), Los Alamos National Laboratory; Colorado University.

Turing[8] devoted his last two years to a possible explanation of how patterns can grow from initially homogeneous states in biological cell growth (thus apparently decreasing entropy in the system) [14]. His proposal was that chemical reaction-diffusion processes could account for cell differentiation, patterns in animal coats and sea-shells, phyllotaxis in plants, and other phenomena. The work was done by brain-power and pencil and paper.

Thom[9] invented in the 1960 s the geometrical theory of catastrophes [13], in which small changes in a complex system lead to major changes of equilibrium.

Lorenz[10] originated chaos theory in the 1970s [4], studying complex systems such as weather patterns in which small differences in initial conditions of deterministic systems lead to diverging behaviours which cannot be predicted.

Conway[11] invented the Game of Life in the 1970s [3], in which a small initial pattern of cells in a regular two-dimensional grid grows into often surprising complexity under remarkably simple rules. At each time-step the rules determine the next state of each cell according to its current state and the states of its neighbours. The work was done by brain-power and a GO board.

Mandelbrot[12] developed a theory of fractals in the 1970s [7]. A fractal is a geometric shape which is self-similar at any scale, in a space with fractal dimensions. He and many others implemented the theory in iterative computer programs to produce spectacular images.

Wolfram[13] began a careful study of one-dimensional elementary cellular automata [16] in the 1980s. Such an automaton has one of 256 rules (of which 88 are distinct), and at each time-step the rule determines the next state of a cell according its current state and the states of its two neighbours. Wolfram proposed (controversially) that such *simple programs* constitute a fundamental tool for all science. The work was done by brain-power and computer.

In all of this early work the often surprising evolution of an organized complex system was a major theme.

18.13 Cellular Automata

Current cellular automata are usually implemented as ways of creating digital images, where

- there is a regular grid of cells, arranged on a torus which is cut and flattened, so that it becomes an array with left-right and top-bottom wrapping,

[8]Alan Mathison Turing (1912–1954), Manchester University.

[9]René Frédŕic Thom (1923–2002), Strasbourg University.

[10]Edward Norton Lorenz (1917–2008), Massachusetts Institute of Technology.

[11]John Horton Conway (*1937), Princeton University.

[12]Benoit B. Mandelbrot (1924–2010), IBM; Yale University.

[13]Stephen Wolfram (*1959), Wolfram Research.

- a state is defined for a cell, say 0 Black or 1 White,
- a neighbourhood is defined for a cell, usually the four edge-butting or the eight edge-or-vertex-butting cells,
- a transition rule is defined for a cell, as a function of current state and neighbourhood state,
- an initial configuration is defined, say all cells in state 0 except for a small set of seed cells in state 1,
- a clock emits time-steps, when every cell in the configuration changes according to the transition rule.

When such an automaton is set going, it generates a dynamic digital image.

Investigation of different initial state/rule configurations shows that

- some die as a permanent uniform all-0 image,
- some become cancerous as a permanent all-1 image,
- some survive as a permanent static image,
- some survive as a permanently-repeating cycle of two or more images,
- some travel forever as a moving image,
- some become (technically) chaotic as unpredictable divergences of behaviour,
- a tiny error in a transition may lead to a mutation which may be advantageous or destructive.

18.14 Complexity Programming

Computer programs which produce dynamic or static digital images are in active development in many fields.

LIFE and its variations have a community of enthusiasts, with many free and commercial programs available [6]. A LIFE program written in Microsoft Quick-BASIC is briefly described below, and available for download at the SpringerLink web-site.

```
REM   **   Program to play LIFE. User picks starter
REM   **   configuration and continuation options.
REM   **   Program sets starter configuration and
REM   **   continuation options. Program continues
REM   **   to calculate new configurations until
REM   **   stopped.
REM   **   Main program
REM   **   Starter options and steps to pause
REM   **   Show array A and continuation options
AGAIN:     CLS
REM **     Calculate neighbours in array A and set
REM   **   array B
REM   **   Copy array B to array A
```

```
REM   **   Intruder option
           GOTO AGAIN
FINISH: END
```

For example, Fig. 18.6 shows output from this program. Writing and changing such a program (say, adding colour to display an archaeological record of when a pixel last changed state) is a good introduction to cellular automata.

In morphogenesis, a BASIC program modelling biological pattern formation is at [9]. Superb examples of this approach are in [8]. A model of cell differentiation which creates a French flag is described in [5]. Figure 18.7 shows four stages in the process.

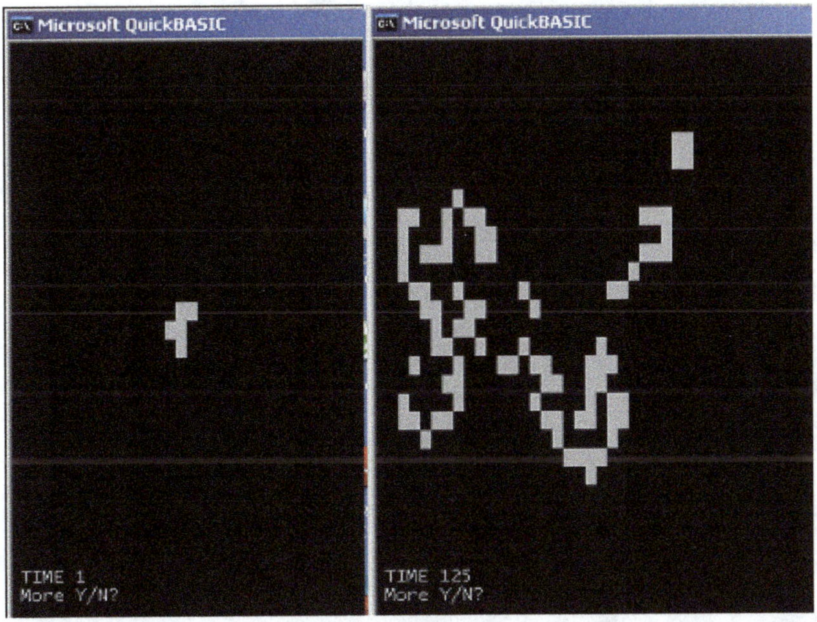

Fig. 18.6 Example of LIFE program output. **a** Screen shows initial configuration (an R-pentomino) at time-step T = 1. **b** On-going evolution at T = 125

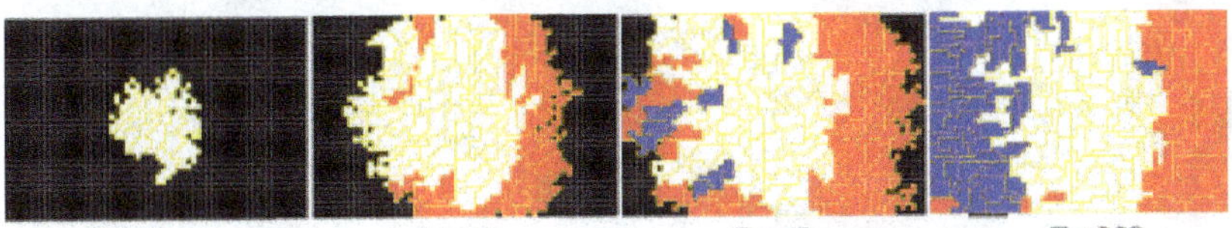

T = 15 T = 35 T = 45 T = 200

Fig. 18.7 Example of morphogenesis. An undifferentiated cell with a 60 × 40-cell French flag as target is exposed to a chemical reaction-diffusion process. As the cell develops chemical gradients create differentiation. See [5]

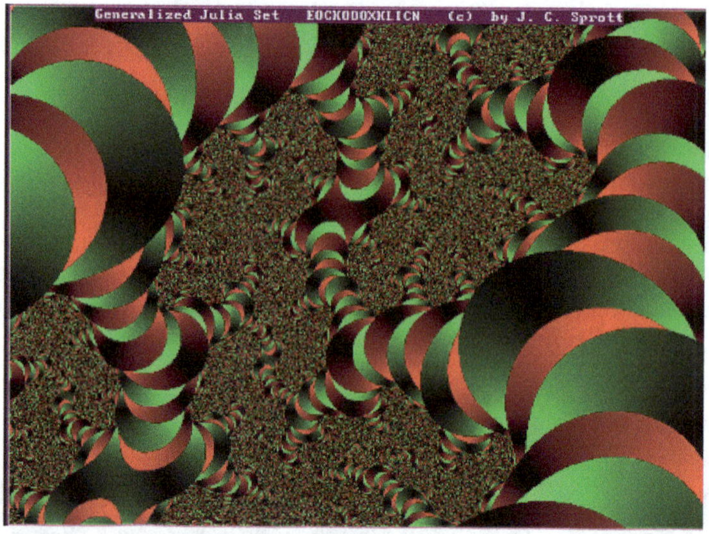

Fig. 18.8 Example of fractal, taken from a plethora in [1]

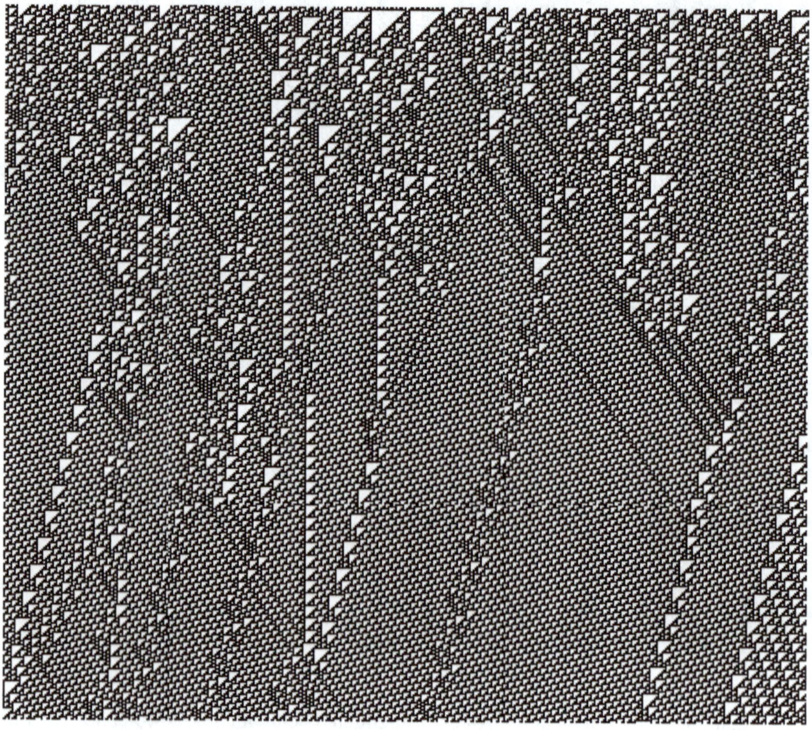

Fig. 18.9 Example of pattern generated by a 1-dimensional cellular automaton under Wolfram Rule 110, taken from [16]

Chaos and fractals have also become popular fields, with many free and commercial programs available [1]. Figure 18.8 shows a typical example.

In cellular automation, Wolfram Research is a notable centre for research and publication [17]. Figure 18.9 shows an example of a pattern generated by a 1-dimensional cellular automaton under Rule 110.

References

1. Chaos and Fractal directory (2014). http://www.dmoz.org/Science/Math/Chaos_and_Fractals/Software/. Accessed 1 Feb 2014
2. Galton web-site (2014). http://www.galton.org. Accessed 1 Feb 2014
3. Gardner M (1970) John Conway's new solitaire game "life". Sci Am 223:120–123
4. Gleik J (1988) Chaos. Penguin, New York
5. Knabe JF, Nehaniv CL, Schilstra MJ (2008) Evolution and morphogenesis of differentiated multicellular organisms. http://panmental.de/papers/FlagPottsGRNALife11.pdfpage0001.html. Accessed 1 Feb 2014
6. LIFE directory (2013). http://www.dmoz.org/Computers/Artificial_Life/Cellular_Automata/. Accessed 1 Feb 2014
7. Mandelbrot B (1982) The fractal geometry of nature. WH Freeman, New York
8. Meinhardt H 2003 The Algorithmic beauty of sea shells, 3rd edn. Springer, Heidelberg
9. Meinhardt H (2014) A simple program to simulate biological pattern formation. http://www.eb.tuebingen.mpg.de/research/emeriti/hans-meinhardt/biuprog.html. Accessed 1 Feb 2014
10. Neumann J von, Burks AW (1966) Theory of self-reproducing automata. http://web.archive.org/web/20080328110016/, http://www.walenz.org/vonNeumann/. Accessed 1 Feb 2014
11. Shannon CE (1949) Communication in the presence of noise. Proc IRE 37: 10–21. http://nms.csail.mit.edu/spinal/shannonpaper.pdf. Accessed 1 Feb 2014
12. Shannon CE, Weaver W (1949) A mathematical theory of communication. Illinois University Press, Urbana
13. Thom R (1989) Structural stability and morphogenesis. Addison-Wesley, Reading, MA
14. Turing AM (1952) The chemical basis of morphogenesis. Phil Trans Roy Soc B 237(641):37–72. doi:10.1098/rstb.1952.0012
15. Weaver W (1948) Science and complexity. Am Sci 36:536-544. http://people.physics.anu.edu.au/~tas110/Teaching/Lectures/L1/Material/WEAVER1947.pdf. Accessed 1 Feb 2014
16. Wolfram S (2002) A new kind of science. http://www.wolframscience.com/nksonline/toc.html. Accessed 1 Feb 2014
17. Wolfram Research web-site (2014). http://wolfram.com/. Accessed 1 Feb 2014

Part V
1-bit Imaging

Chapter 19
1-bit Image

This book covers bit-mapped, not vector, images. The element of location in a bit-mapped image is a pixel, with integer coordinates in a finite window. The element of intensity is a tone with integer value from 0 to 255. A 1-bit image contains just two tones, 0 Black and 255 White, so the intensity of each pixel can be specified by one binary digit, that is one bit. Examples are shown of typical 1-bit images created by program, graphic user interface, scanner, and camera.

19.1 Bit-mapped Graphics and Vector Graphics

In computer graphics there are two radically different traditions. *Bit-mapped* or raster or "paint" graphics are based on the discrete grid structures of scanners, cameras, display screens and printers. *Vector* or "draw" graphics are based on continuous primitive forms: points, lines, and curves, geometrically assembled into shapes. They can be scaled without loss of definition, and are preferred for architectural and for type-face design; but have to be rasterized for display or printing.

This book deals with bit-mapped graphics.

19.2 Image Location

In a bit-mapped image the element of location is a *pixel*: a square[1] of indeterminate size. Pixels abut left and right, top and bottom, to form a *discrete plane*, as shown in Fig. 19.1.

[1]Rectangular pixels occur in some contexts, and require careful treatment.

A. Parkin, *Digital Imaging Primer*, DOI 10.1007/978-3-540-85619-1_19

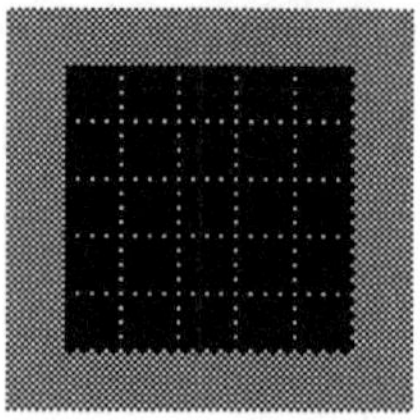

Fig. 19.1 A discrete plane of pixels, extending in all directions without limit

19.3 Location Coordinates

We choose any pixel in the plane as *origin*, and assign to it the integer 0, as its *location coordinate*. We then assign the positive integers 1, 2, ... to successive pixels to the right of origin, and the negative integers −1, −2, ... to successive pixels to the left of origin. We now have a *discrete Cartesian x-axis* (see Sect. 6.3) as shown in Fig. 19.2, where every pixel in the axis has a unique integer coordinate x, and every integer x corresponds to a unique pixel in the axis. Each pixel in the axis abuts just two neighbours, and no further pixels can be inserted between neighbours. Clearly, there is a 1–1 correspondence between the pixels of the axis and the elements of an integer vector $\mathbf{V}$ (see Sect. 3.10):

$$\mathbf{V} = \big(\ldots\ V_{-2}\ V_{-1}\ V_0\ V_1\ V_2\ \ldots\big)$$

where the pixel with coordinate J corresponds to the vector element with subscript J.

Starting from origin 0, we assign the positive integer coordinates 1, 2, ... to successive pixels upwards from origin, and the negative integer coordinates −1, −2, ... to successive pixels downwards from origin. We now have a *discrete Cartesian y-axis* as shown in Fig. 19.3, orthogonal to the x-axis at origin, where again every pixel in the

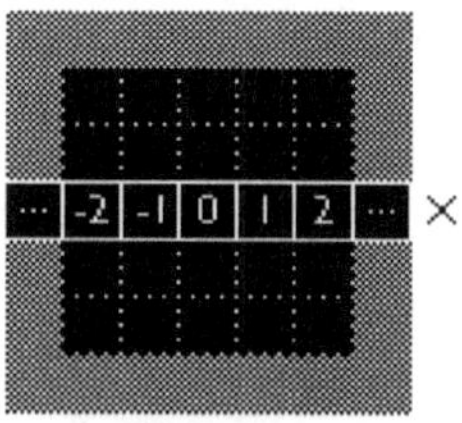

Fig. 19.2 Discrete Cartesian x-axis, with integer coordinates extending to right and left without limit

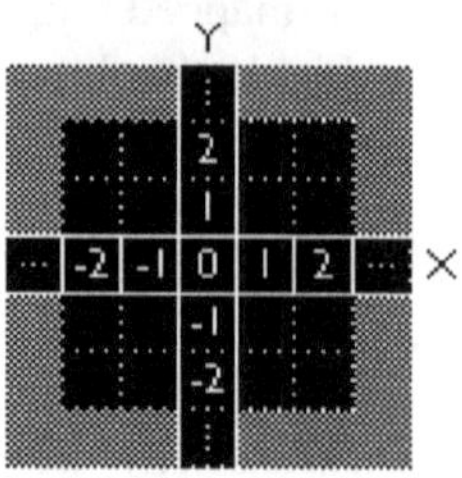

Fig. 19.3 Discrete Cartesian x-axis and y-axis, with integer coordinates extending right and left, and *up* and *down*, without limit

axis has a unique integer coordinate y, and every integer y corresponds to a unique pixel in the axis. Each pixel in the axis abuts just two neighbours, and no further pixels can be inserted between neighbours. Clearly, there is a 1–1 correspondence between the pixels of the axis and the elements of an integer vector.

19.4 Discrete Cartesian Plane

A *discrete Cartesian plane*, as shown in Fig. 19.4 is the geometric product of two discrete Cartesian axes meeting orthogonally at a common origin pixel. Each pixel in the plane belongs to one column, which has an X-coordinate, and to one row, which has a y-coordinate. So each pixel has a unique pair of integer coordinates (x, y), and each pair of integers (x, y) corresponds to a unique pixel in the plane. Each pixel in the plane abuts just four neighbours, and no further pixels can be inserted between neighbours. Clearly, there is a 1–1 correspondence between the pixels of the plane and the elements of an integer matrix **A** (see Sect. 31.1):

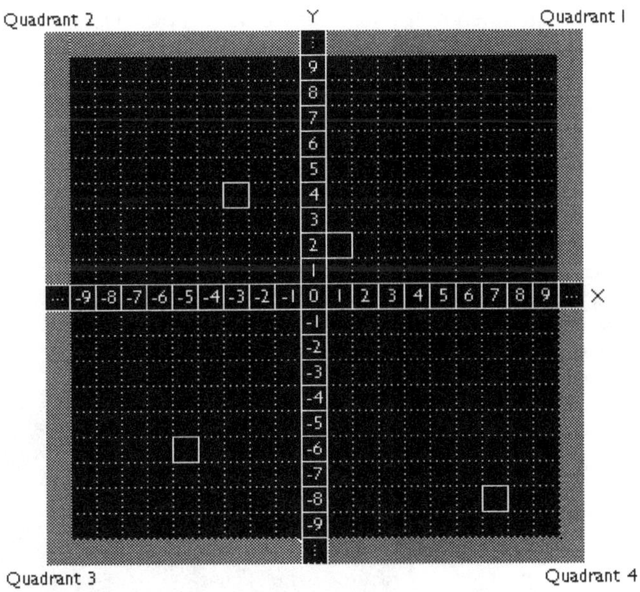

Fig. 19.4 Discrete Cartesian plane, unlimited to right, left, upward, and downward. Marked pixels have coordinates (1, 2) in first quadrant, (−3, 4) in second quadrant, (−5, −6) in third quadrant, and (7, −8) in fourth quadrant

$$\mathbf{A} = \begin{pmatrix} \cdots & \cdots & \cdots & \cdots & \cdots & \cdots & \cdots \\ \cdots & A_{-22} & A_{-12} & A_{02} & A_{12} & A_{22} & \cdots \\ \cdots & A_{-21} & A_{-11} & A_{01} & A_{11} & A_{21} & \cdots \\ \cdots & A_{-20} & A_{-10} & A_{00} & A_{10} & A_{20} & \cdots \\ \cdots & A_{-2-1} & A_{-1-1} & A_{0-1} & A_{1-1} & A_{2-1} & \cdots \\ \cdots & A_{-2-2} & A_{-1-2} & A_{0-2} & A_{1-2} & A_{2-2} & \cdots \\ \cdots & \cdots & \cdots & \cdots & \cdots & \cdots & \cdots \end{pmatrix}$$

where the pixel with x-coordinate J and y-coordinate K corresponds to the matrix element with subscript J, K.

By convention, the *first quadrant* is that with $(+x, +y)$ coordinates, the *second quadrant* is that with $(-x, +y)$ coordinates, the *third quadrant* is that with $(-x, -y)$ coordinates, and the *fourth quadrant* is that with $(+x, -y)$ coordinates.

A discrete Cartesian plane is a sub-space of discrete coordinate space.

19.5 Discrete Cartesian Window

A discrete plane has no inherent limits. In any practical case we use perforce a finite *window*, a rectangular region chosen to suit the purpose in hand. A *discrete Cartesian window* is specified by giving a range for the x-coordinate (the column) and for the y-coordinate (the row). For example,

$$-4 \leq x \leq 4, -4 \leq y \leq 4$$

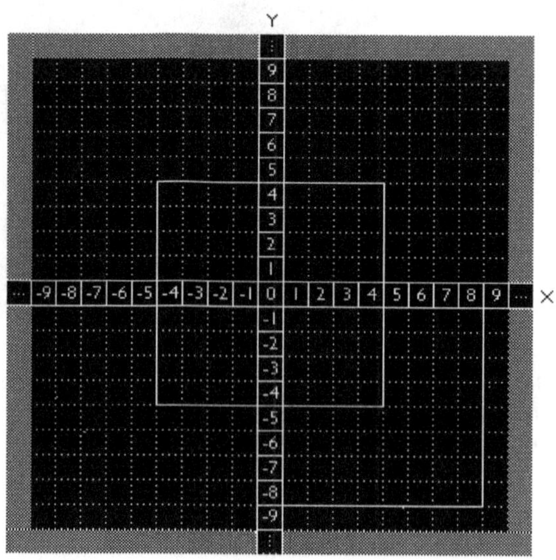

Fig. 19.5 Two discrete Cartesian windows, each 9×9 px. The central window has its origin at centre, and admits zero, positive, and negative coordinates. The fourth-quadrant window has its origin at top left, and admits zero and positive x-coordinates and zero and negative y-coordinates

specifies a window 9 px wide and 9 px high, with origin at the centre of the window. However,

$$0 \leq x \leq 8, 0 \leq y \leq -8$$

also specifies a window 9 px wide and 9 px high, but now with origin at top left of the window. This lies entirely in the fourth quadrant. The two cases are shown in Fig. 19.5.

Notice that the usual GUI image editor window uses x-coordinates positive to the right and y-coordinates positive downwards, so the origin is at the top left of the window, inverting the first quadrant.

19.6 Image Intensity

In digital imaging, we define the element of intensity as a *tone*: a reproducible level of light. Internally, a typical current computer systems offers 256 tones, numbered 0 to 255, in each of three channels, labelled B, G and R. In an image, every pixel has a pixel intensity specified as an integer triple (B, G, R), where $0 \leq \mathrm{B} \leq 255$, $0 \leq \mathrm{G} \leq 255$, and $0 \leq \mathrm{R} \leq 255$. Black is specified as (0, 0, 0), White as (255, 255, 255), and a series of greys as (B = G = R).

Externally, the luminance L, measured in candelas per square metre (cd/m^2) (see Chap. 15), of the internal tones depends on the properties and settings of the input and output devices in use (see Chap. 29).

19.7 Bit-depth 1

A fundamental classification of digital images is by *bit-depth*. Two intensities, say Black (0, 0, 0) and White (255, 255, 255), can be specified by an economical *intensity index* I_2 taking a 1-bit value 0 or 1. We then say that the discrete plane has *bit-depth 1*, or that it is the *1-bit plane*. The intensity index I_2 refers to a two-entry *colour-table* in a 1-bit .BMP file (see Chap. 26). Each of the two index values points to an intensity triple (B, G, R). The colour-table may be set up in any way we please by using a file editor such as EDBMP1.BAS described in Chap. 26. The default setting, here called 1BW, is shown in Table 19.1.

Table 19.1 1-bit default colour-table 1 BW

Index I_2	B	G	R	Name
0	0	0	0	Black
1	255	255	255	White

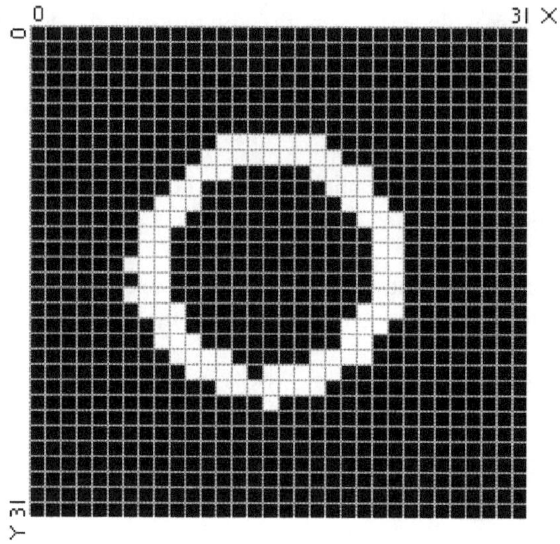

Fig. 19.6 Example of 1-bit image, 32 × 32 px, enlarged ×8 in an image editor. Notice that the display origin is at top left, and the coordinates begin at 0

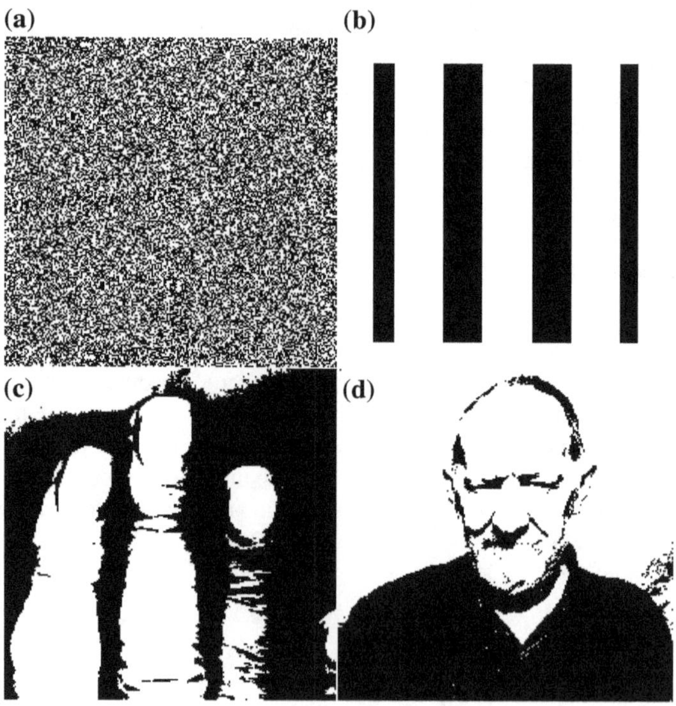

Fig. 19.7 Examples of 1-bit images with colour-table 1BW. **a** Image generated by 1-bit program: 222 × 222 px randomly set as Black or White. **b** Image created by PAINT in 1-bit mode. **c** Fingers scanned in 1-bit mode. **d** Scene photographed by 24-bit camera then reduced to 1-bit

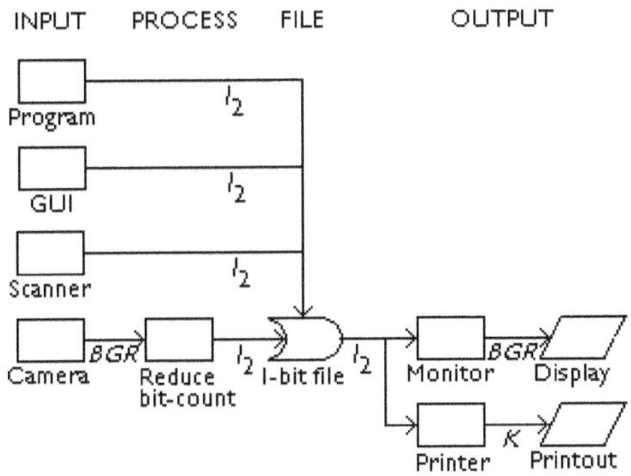

Fig. 19.8 Flowchart for 1-bit image. A 1-bit image may be input alpha-numerically from keyboard via a program, or graphically from mouse/keyboard via a GUI editor, or from a flat object via a scanner, or from a space scene via a camera followed by bit-count reduction. The image is stored in a 1-bit file. It may then be displayed on a screen or printed on paper

19.8 1-bit Digital Image

Graphically, a *1-bit digital image* is a set of $x \times y$ px, each pixel showing one of two intensities, say (0,0,0) or (255, 255, 255). Numerically, it is a set of integer triples:

$$(x, y; I_2)\,,$$

where (for a first-quadrant image) $0 \leq x \leq x_{\max}$, $0 \leq y \leq y_{\max}$, $I_2 = 0$ or 1.

For example, Fig. 19.6 shows a 1-bit digital image with 32 × 32 px, as it might appear enlarged in a GUI image editor display (see Chap. 21). Figure 19.7 shows examples of 1-bit digital images input by program, GUI image editor, scanner, and camera.

19.9 Flowchart

The flowchart Fig. 19.8 is for general guidance in this Part. To make a 1-bit image, we can use a program (see Chap. 20) or a graphic user interface (GUI) (see Chap. 21) or a scanner in 1-bit mode (see Chap. 22), or a camera in 24-bit mode (see Sect. 46.6) and a bit-count reduction (see Appendix B). The image can be stored in a 1-bit .BMP file (see Chap. 26), then displayed (see Chap. 27) or printed (see Chap. 28) for viewing (see Chap. 29). Statistics for a 1-bit image can be calculated (see Chap. 18).

Chapter 20
1-bit Program Input

One way to create a digital image is by writing a program. The program examples in this book are written in the Microsoft QuickBASIC language, which is widely understood and easily learned. The examples illustrate principles, which can be reworked in more powerful languages. To create an image by program, a user inputs any data required, the program sets up an integer array representing the pixels of the image, performs some algorithm to set the array elements to various values, and writes out the array serially to an image file (which can be passed elsewhere for display or printout or further processing). An example program creates a random distribution of Black and White pixels, with no intentional features. Algorithms are given for constructing some elementary features: line, circle, sinusoid, and Bézier curve. A plane graph is any collection of lines. A program to prepare and edit plane graph data, and a program to create the plane graph, are shown. A measure of resolution for a programmed image is given.

20.1 Image Creation by Program

The flowchart in Fig. 20.1 is for general guidance in this chapter. The programming examples in this book are written in Microsoft QuickBASIC (see Sect. 32.1). They are intended to illustrate principles in an elementary way. Mike Kamermans makes the point very well [8]: "This is working code, and you can trace everything I do. Regardless of whether you *want* to, I consider it important that you can. And the kicker is this code is in the public domain. I waive all rights to it. Do with it what you want, to me this is trivial code and it would be ridiculous to claim it as anything other than 'I just wrote this to demonstrate how things work'."

Each quite short main program calls sub-programs which do the detailed work, including reading and writing 1-, 4-, 8-, and 24-bit bitmap files. Full program code is on the SpringerLink web-site. The program comments are used in the book text as brief descriptions of what the program does. Programmers may enjoy re-writing in more powerful languages and re-combining for various practical purposes.

A. Parkin, *Digital Imaging Primer*, DOI 10.1007/978-3-540-85619-1_20

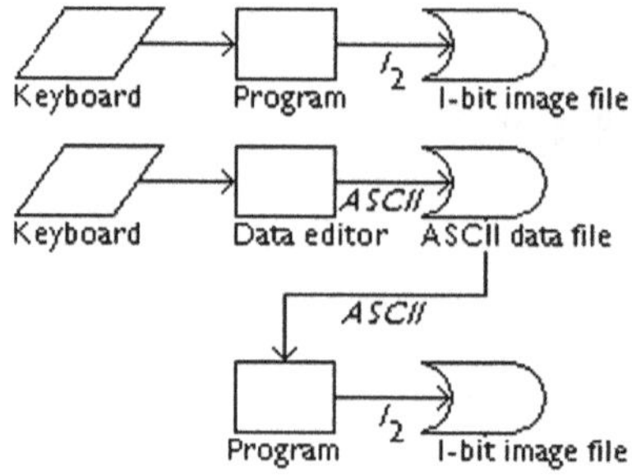

Fig. 20.1 Image creation by program. A program may create a 1-bit image under direct keyboard control, and store it as a bitmap file. A program may read data, prepared by a data editor or other source, from an ASCII or other suitable file, create a 1-bit image, and store it in a bitmap file where each pixel has intensity index I_2

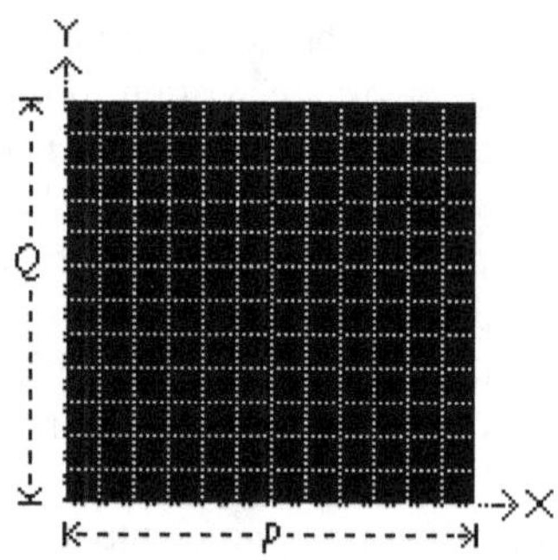

Fig. 20.2 Image array in a program. Coordinate origin is bottom left, x-axis to the right, y-axis upwards (first-quadrant). P is count of pixels in width; Q is count of pixels in height

The general form of a 1-bit image-creating program is,

- user inputs file-names and values of variables,
- program reads input files if any,
- program sets up an integer array of $P \times Q$ elements, to represent the pixels of the image, each element set initially to (say) value 0 to represent Black,
- program performs an algorithm to re-set certain elements to value 1, to represent White,
- program writes out the array serially to a 1-bit image file.

Figure 20.2 shows the image array in a program. The location coordinates (x, y), where $0 \leq x \leq P, 0 \leq y \leq Q$, correspond to the coordinates of a discrete Cartesian window in the first quadrant (see Sect. 19.2). The array is serialized to a file row-by-row from the bottom row upwards, that is (0,0), (1,0), (2,0), …, (0,1), (1,1), (2,1), …, …, (P, Q).[1] In a program the pixels are of indeterminate size, so there are no lengths, only pixel-counts.

[1]Notice that the pixel coordinates in a GUI screen invert the Y-axis, like reading text (see Sect. 21.2).

20.2 Random Process

In nature, a *random*[2] process is one which repeats an event with N possible outcomes, and there is no known way to predict which of the N will occur at each step. But in a large number of repetitions there may be a stable relative frequency of each of the N outcomes. If we count the relative frequencies, and normalize them in the range 0–1, we have the *empirical probabilities* $p_1, p_2, \ldots, p_N$ of the N outcomes. For example, tossing a coin has $N = 2$, $p_H = p_T = 1/2 = 0.5$. Rolling a die has $N = 6$, $p_1 = p_2 = p_3 = p_4 = p_5 = p_6 = 1/6 = 0.1667$. Drawing a card has $N = 52$, $p_{each} = 1/52 = 0.0192$.

In mathematics, a *discrete random variable* R takes each of a given set of N distinct values $r_1, r_2, \ldots, r_N$ with given probabilities $p_1, p_2, \ldots, p_N$, where the probabilities sum to 1. The case where there are just two values, 0 and 1, so $q = 1 - p$, is called a *Bernoulli distribution*.[3]

20.3 1-bit Random Image

If we want to make a random image, it is usually impractically tedious to do it by hand: we need to write a program which incorporates a random process. In computing, a *pseudo-random number generator*[4] produces a series of numbers which can be taken as random for most ordinary purposes. The numbers are generated by a determinate arithmetical formula from a chosen *seed*: if we know the formula the numbers are predictable and the series is repeatable; if not, not. In QuickBASIC, the statement RANDOMIZE generates a succession of (7-digit) random numbers *RND* between 0 and 1, starting from a settable seed.

Suppose we want a pixel to have intensity White with given probability p. We write a program which scans sequentially through all the elements of an image array, generates a new random value *RND* for each element, and compares *RND* with p. If $p >= RND$ the program sets intensity White, else set intensity Black.

---

[2]The root meaning is from the old French *randir* to gallop headlong without aim or care. Johnson's dictionary cites Shakespeare and Milton. Venn developed the idea of a random walk, using the digits of π in 1866.

[3]Jakob Bernoulli (1654–1705), Professor of Mathematics at Basel University 1687–1705. His major work on probability, *Ars Conjectandi*, was published posthumously in 1715. The family produced several generations of distinguished mathematicians.

[4]invented in 1946 by John von Neumann and Stanislaw Ulam for calculations for the hydrogen bomb, and subsequently the foundation for Monte Carlo simulations in many fields.

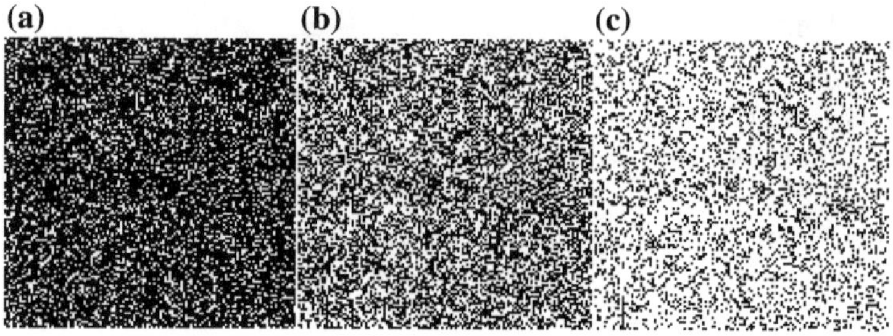

Fig. 20.3 Three 1-bit uniform random images. **a** Probability p of White is 0.25, probability $(1-p)$ of Black is 0.75. **b** $p = 0.50$, $(1-p) = 0.50$. **c** $p = 0.75$, $(1-p) = 0.25$

20.4 1-bit Uniform Random Image

In a *uniform* random distribution a given probability p applies to all the pixels alike. The result is an all-over texture or ground, which is perfectly meaningless. (But even here the incorrigible human tendency to seek meaning discerns incipient features, like faces in the fire, or weasels in clouds.) For example, Fig. 20.3 shows three uniform random images, each 128×128 px, produced by the program GENRAND1 described below.

20.5 Program to Generate Random Images

A program to generate a uniform 1-bit probability distribution or a gradient with any of four profiles is described briefly below, with source code at the SpringerLink web-site.

```
REM  **  Program GENRAND1.BAS to generate 1-bit
REM       random distribution, uniform or linear
REM       or exponential or normal or radial. User
REM       enters file-name for output .BMP file,
REM       required image  width and height, and
REM       type of distribution  wanted. Program
REM       writes 1-bit .BMP file header and colour
REM       -table, then generates distribution
REM       into array AIMAGE. Program  writes image
REM       from AIMAGE to output file and revises
REM       header.
REM  **  Enter file-name for output, width, height, and
REM       type of distribution wanted
REM  **  Re-set width to a multiple of 32
```

```
REM  **  Main program to generate distribution
REM       into array AIMAGE.
REM  **  Generate distribution.
REM       Uniform: probability of White is same everywhere.
REM       Linear: probability of White is distance from
REM        left edge.
REM       Exponential: probability of White is distance
REM        squared.
REM       Normal: Right half of normal curve, width 3
REM        standard dev.
REM       Radial: probability of White is distance from origin
         CALL WRIAIMA
         CALL REVHEA1
FINIS: END

         SUB CREHEA1
REM  **  Subprogram to write header and colour-table of
REM  **  1-bit .BMP file #1.
END SUB
         SUB PUTBYTE (ASCII%)
REM  **  Subprogram to put 1 byte in file #1
END SUB
         SUB PUTI (INTE%)
REM  **  Sub-program PUTI to put integer value in binary
REM  **     file at pointer position (set outside) and
REM  **     increment pointer.
END SUB
         SUB PUTL (INTL&)
REM  **  Sub-program PUTL to put long integer into
REM  **     binary file as 4 bytes
END SUB
         SUB REVHEA1
REM  **  Sub-program REVHEA1 to revise header of 1-bit
REM  **     .BMP file #1 with values in bytes.
END SUB
         SUB WRIAIMA
REM  **  Sub-program WRIAIMA to write image from
REM  **     array AIMAGE to .BMP file #1. Row must have
REM  **     multiple of 4 bytes = 32 bits. Pack each
REM  **     sequence of 8 bits as 1 byte and write
REM  **     to output file.
END SUB
```

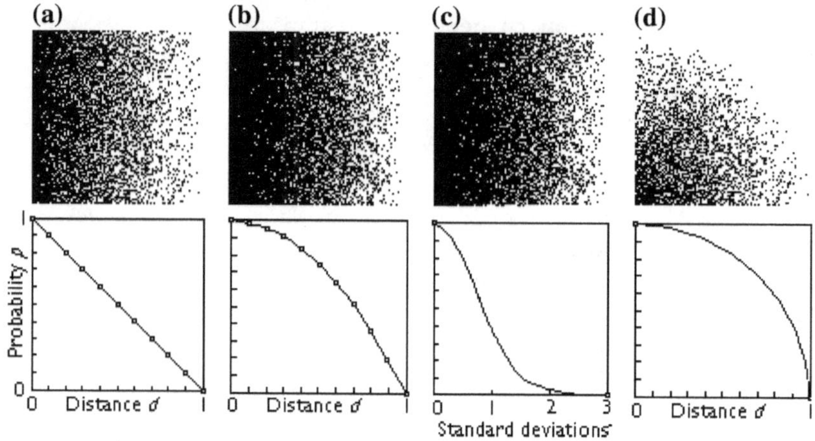

Fig. 20.4 Four 1-bit gradient random images. **a** Linear gradient has straight-line profile. **b** Exponential gradient has exponential profile. **c** Normal gradient has half Gaussian Normal profile. **d** Radial gradient has quadrant profile

20.6 1-bit Gradient Random Image

In a *gradient* random distribution the probability p varies with some function of distance d from (say) the left-hand edge of the image. Figure 20.4 shows some examples. In a linear gradient p is directly proportional to d; in an exponential gradient p is proportional to a power of d. In a normal gradient, p is proportional to standard-deviations-distance from left-hand edge (to give a profile of half a Gaussian Normal curve). A radial gradient varies p with distance from a point, say the bottom left corner of the image.

From here, the way is open to explore more elaborate conditions on random distributions, and to impose random processes on images created by other means.

20.7 1-bit Elementary Features

Meaning comes by introducing structures into an image. Let us call a group of pixels with a definite structure of intensities a *feature* in an image. A graphic user interface image editor, such as MS PAINT (see Chap. 21) includes an efficient set of programs to generate features, wrapped for convenient use. The examples below show the principles underlying such programs.

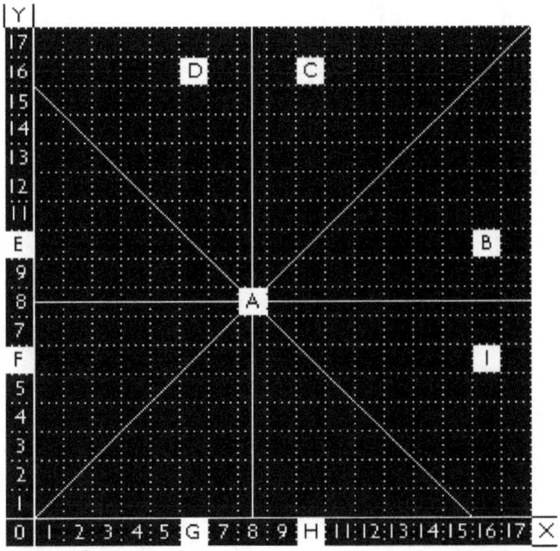

Fig. 20.5 Discrete line. Given two end-pixels, find the pixels for the line segment between them. There are eight possible cases. Lines AC, AD, AG, and AH are steep; lines AD, AE, AF, and AG are backwards; lines AF, AG, AH, and AI are downwards. All eight cases can be reduced to line AB, which is not steep, not backwards, and not downwards

20.8 Discrete Line

Given the coordinates of two ends, (x_0, y_0) and (x_1, y_1) we can construct a *discrete line* feature by finding the coordinates (x, y) of all the pixels between the ends. We can use Bresenham's[5] celebrated line algorithm [4]. There are eight possible cases as shown in Fig. 20.5.

The algorithm first reduces all cases to one. In pseudo-code:

```
If ABS(Y_1 - Y_0) > ABS(X_1 -  X_0) then the line is STEEP,
      so swap X_0,Y_0 and swap X_1,Y_1
If X_0 > X_1 then the line is BACKWARDS, so swap X_0,X_1
      and swap Y_0,Y_1
```

[5] Jack E. Bresenham, (*1937), IBM 1960–1987, Winthrop University 1987–1997. In 2001 he wrote: "I was working in the computation lab at IBM's San Jose development lab. A Calcomp plotter had been attached to an IBM 1401 via the 1407 typewriter console. [The algorithm] was in production use by summer 1962, possibly a month or so earlier. Programs in those days were freely exchanged among corporations so Calcomp (Jim Newland and Calvin Hefte) had copies. When I returned to Stanford in Fall 1962, I put a copy in the Stanford comp center library. A description of the line drawing routine was accepted for presentation at the 1963 ACM national convention in Denver, Colorado. It was a year in which no proceedings were published, only the agenda of speakers and topics in an issue of Communications of the ACM. A person from the IBM Systems Journal asked me after I made my presentation if they could publish the paper. I happily agreed, and they printed it in 1965." (www.nist.gov/dads/HTML/bresenham.html).

```
Set integer constants:
     If Y_0 <= Y_1 then the line is DOWNWARDS, so set
          YSTEP = -1 else set YSTEP = 1
     X-difference  DX = X_1 - X_0
     Y-difference  DY = Y_1 - Y_0
```

Figure 20.6 shows the principle of Bresenham's line algorithm, using real variables *s*, *t*, *tdec*, *tinc*:

```
Slope s = DY/DX
Errorterm t = 0.5 (the initial distance of the real line
     from the top of the current pixel row)
Errorterm decrement tdec = s
Errorterm increment tinc = 1
```

Now start an iteration at $x = x_0$, $y = y_0$, and continue by stepping $x = x + 1$ until $x = x_1$. Errorterm *t* starts positive, and accumulates negative decrements *tdec* until it reaches 0, which causes a step in *y* and re-sets *t* to a positive value:

```
If STEEP then plot (Y,X) else plot (X,Y)
Re-set t = t - tdec
If t <= 0 then
     Y = Y + 1 (to plot next pixel on next row up)
     t = t + tinc (to re-set errorterm for next row up)
(repeat)
```

For example, given A = (8, 8) and B = (16, 10), Table 20.1 shows the successive values of the variables in the algorithm to plot the line AB in Fig. 20.6.

But in a digital computer integer arithmetic is faster than real arithmetic, and is exact. The real variables (lower-case) can be replaced by integer variables (upper-case), by multiplying by $2 \triangle x$, as follows:

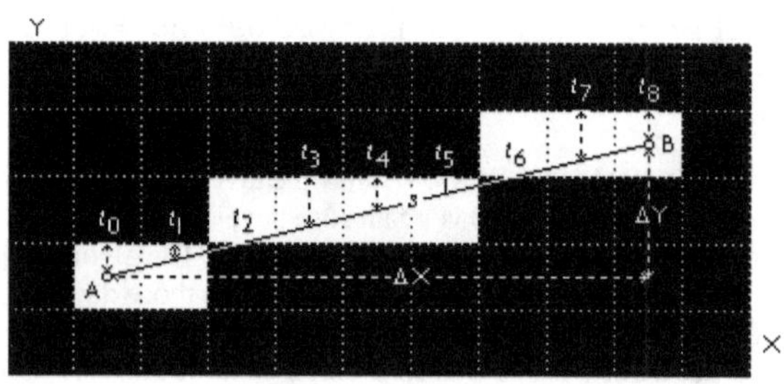

Fig. 20.6 Principle of Bresenham's line algorithm, using real variables. Starting at A, we step the *x*-coordinate by 1, and test to find whether the *y*-coordinate should be increased by 1 or not (it should not). Then at the new pixel, we again step the *x*-coordinate by 1 and again test to find whether the *y*-coordinate should be increased (it should). And so on until we reach end B

Table 20.1 Successive values for line AB in Fig. 20.6

Δx	*tdec*	*tinc*	x	y	t	(Test $t \leq 0$)	y	t
8	0.25	1	8	8	0.25	(False)	8	0.25
			9	8	0	(True)	9	1
			10	9	0.75	(False)	9	0.75
			11	9	0.50	(False)	9	0.50
			12	9	0.25	(False)	9	0.25
			13	9	0	(True)	10	1
			14	10	0.75	(False)	10	0.75
			15	10	0.50	(False)	10	0.50
			16	10	0.25	(False)	10	0.25

The given ends are $A = (8, 8)$ and $B = (16, 10)$. Columns 1, 2 and 3 are the calculated constants. Columns 4 and 5 are the pixels plotted. Column 6 is the error term t after subtraction of *tdec*. Column 7 is the result of the test. Columns 8 and 9 are the values carried forward to the next iteration

```
Initial value of error term T = 0.5 * 2
DX = DX
Decrement TDEC = s * 2 * DX
     = DY/DX * 2 * DX
     = 2 * DY
Increment TINC = 1 * 2 * DX
     = 2 * DX
```

A program which produces any number of lines from given end-pairs is briefly described below, with source code at the SpringerLink web-site:

```
REM  **  Program MKLINE to make lines between given ends
REM  **     in a 1-bit image 128 x 128 pixels. User
REM  **     enters filename for output, and coords of
REM  **     any number of end-pairs. Program tests and
REM  **     swaps for"steep" and "backwards", then does
REM  **     Bresenham line algorithm to set line pixels
REM  **     in image array. Finally program writes image
REM  **     array as 1-bit .BMP file.
REM  **  Main program
REM  **  Enter file-name for output
REM  **  Initialize image array AIMAGE to all-black
REM  **  Input coordinates of any number of end-pairs,
REM  **  do Bresenham line algorithm  and set pixels in
REM  **  image array AIMAGE.
REM  **  If STEEP then swap X0,Y0 and swap X1,Y1
REM  **  If BACKWARDS then swap X0,X1 and swap Y0,Y1
```

```
REM   **   Initialize iteration variables
REM   **      Iterate to set edge pixels white
REM   **      More or done
           CALL WBMPHEAD
           CALL WBMPDATA
           CALL REVBHEAD
END

           SUB PUTI (INTE%)
REM   **   Puts integer value passed into binary file as 2
REM   **   bytes and increments pointer
END SUB
           SUB PUTL (INTL&)
REM   **   Puts long integer into binary file #2 as 4 bytes
END SUB
           SUB REVBHEAD
REM   **   Revises header of file #2 with values in bytes.
END SUB
           SUB WBMPDATA
REM   **   Subprogram to write data from array AIMAGE
REM   **   to .BMP file #2 with row-length 128.
REM   **   1-bit locations are converted as groups of
REM   **   8 to byte.
END SUB
           SUB WBMPHEAD
REM   **   Subprogram to write header and colour-table of
REM   **   1-bit .BMP file #2, 128 X 128 pixels.
END SUB
```

Figure 20.7 shows an example of the output of the all-integer algorithm, and Table 20.2 shows the successive values of the variables.

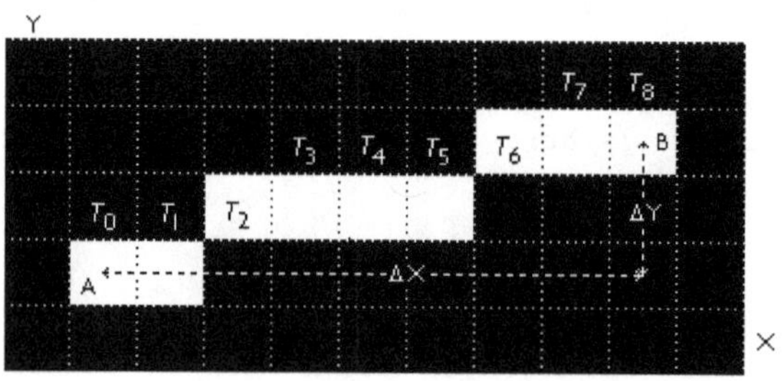

Fig. 20.7 Example of Bresenham's all-integer algorithm. Starting at A, we step the x-coordinate by 1, and test to find whether the y-coordinate should be increased by 1 or not (it should not). Then at the new pixel, we again step the x-coordinate by 1 and again test to find whether the Y-coordinate should be increased (it should). And so on until we reach end B

Table 20.2 Successive values for line AB in Fig. 20.7

Δx	*TDEC*	*TINC*	x	y	T	(Test $T \leq 0$)	y	T
8	4	16	8	8	4	(False)	8	4
			9	8	0	(True)	9	16
			10	9	12	(False)	9	12
			11	9	8	(False)	9	8
			12	9	4	(False)	9	4
			13	9	0	(True)	10	16
			14	10	12	(False)	10	12
			15	10	8	(False)	10	8
			16	10	4	(False)	10	4

The given ends are $A = (8, 8)$ and $B = (16, 10)$. Columns 1, 2 and 3 are the calculated constants. Columns 4 and 5 are the pixels plotted. Column 6 is the error-term T after subtraction of *TDEC*. Column 7 is the result of the test. Columns 8 and 9 are the values carried forward to the next iteration

20.9 Discrete Length

The real coordinate plane has an arbitrary unit of length U. The length of a line segment is measured by a real coefficient, which multiplies or divides the unit. The line segment between two points with coordinates $(x1, y1)$ and $(x2, y2)$ is the hypotenuse of the right-angled triangle with sides $\Delta x = x1 - x2$ and $\Delta y = y1 - y2$, that is

$$h = \surd(\Delta x^2 + \Delta y^2)\text{units}\,.$$

The discrete coordinate plane has a natural unit of length: the width of a pixel. Pixels may either abut side to side, or touch corner to corner. Figure 20.8 shows line segments AB and AC composed of only abutting pixels; AD composed of only touching pixels; and AE composed of both abutting and touching pixels.

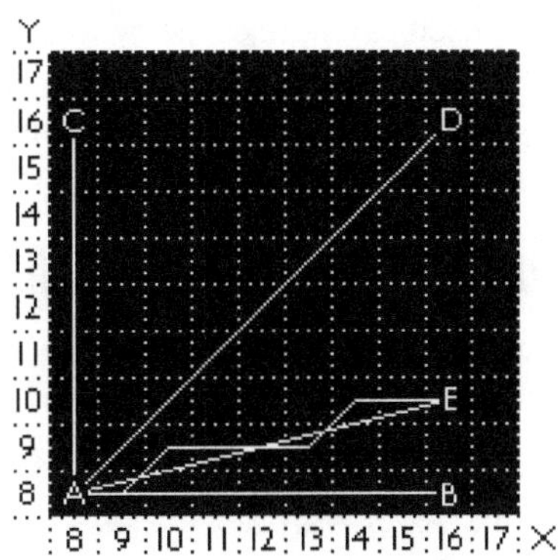

Fig. 20.8 Discrete length. Line segments AB and AC contain only abutting pixels; AD only touching pixels; AE both abutting and touching pixels

Table 20.3 Examples of discrete length

Segment	Δx	Δy	H	h
AB	8	0	8	8
	8	1	8	8.0622
AE	8	2	8	8.2462
	8	3	9	8.5440
	8	4	9	8.9443
	8	5	10	9.4340
	8	6	10	10.000
	8	7	10	10.6301
AD	8	8	11	11.3137
AC	0	8	8	8

The (unavailable) real length h is shown for comparison

Line segment AB has discrete length Δx, and AC has discrete length Δy. AD has discrete length $\sqrt{2} \Delta X$, where $\sqrt{2}$ is approximated by the rational multiplier:

$$\sqrt{2} \approx 10{,}000 \setminus 7071$$

To find the discrete length of any other line segment, such as AE, we take the greater of Δx and Δy, and add $\sqrt{2} - 1$ for each occurrence of the lesser of Δx and Δy:

$$H = \max(\Delta x, \Delta y) + \min(\Delta x, \Delta y)(\sqrt{2} - 1)\,\text{px}\,.$$

This is approximated in integer arithmetic by the rational multiplier:

$$\sqrt{2} - 1 \approx 2929 \setminus 7071\,.$$

So,

$$H = \max(\Delta x, \Delta y) + 2929 \times \min(\Delta x, \Delta y) \setminus 7071\,\text{px}\,.$$

The discrete length of the hypotenuse (zig-zag) is larger than the real length (straight) by less than 8 %.

Table 20.3 shows examples of discrete lengths of line segments.

20.10 Discrete Plane Graph

A *discrete plane graph* (see Sect. 8.1) is a set of:

- *Vertices* $V_1, V_2, \ldots,$
- *Edges* $E_1, E_2, \ldots,$
- *Faces* $F_1, F_2, \ldots,$

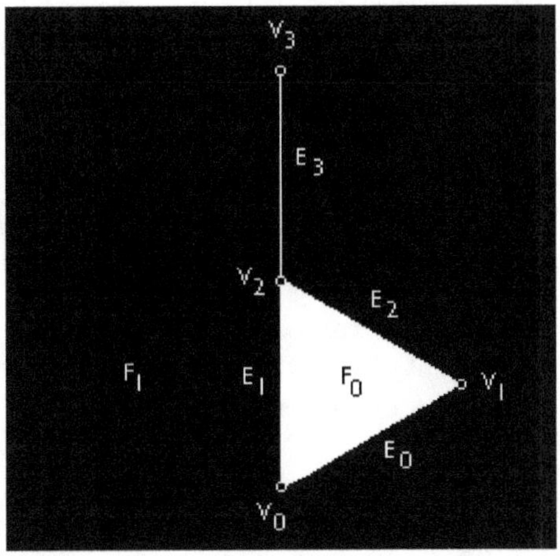

Fig. 20.9 Example of discrete plane graph. $V - E + F = 4 - 4 + 2 = 2$

where an edge joins two vertices, a face is bounded by three or more edges, no two edges cross, and there are no isolated parts. They are related by Euler's formula:

$$V - E + F = 2\,.$$

Figure 20.9 shows a simple example, with four vertices, four edges, and two faces.

A discrete plane graph can be fully specified numerically by

- a list of vertices V_1, V_2, ..., each a pair of integer coordinates (x, y),
- a list of edges, each a pair of vertices (V_1, V_2),
- a list of faces, each a cycle of vertices $(V_1, V_2, V_3, \ldots)$ or of edges $(E_1, E_2, E_3, \ldots)$

in a file with a suitable structure. A simple data file structure is (all values integer)

- Header

  Count of vertices
  Count of edges
  Count of faces

- Vertex list

  Vertex identifier, x, y coordinates
  Vertex identifier, x, y coordinates
  ...

- Edge list

  Edge identifier, vertex identifier, vertex identifier
  Edge identifier, vertex identifier, vertex identifier
  ...

- Face list

  Face identifier, vertex identifier, vertex identifier, vertex identifier
  Face identifier, vertex identifier, vertex identifier, vertex identifier
  ...

Using this format, the data for the graph in Fig. 21.3 would be

- (Header)

  (Count of vertices) 4
  (Count of edges) 4
  (Count of faces) 2

- (Vertex list)

  0,127,226
  1,211,178
  2,127,129
  3,127,30

- (Edge list)

  0,0,1
  1,0,2
  2,1,2
  3,2,3

- (Face list)

  0,0,1,2 .

Notice that the coordinate origin used here is at top left. It may often be sufficient to specify only the vertex coordinates and the edge-list of a plane graph, omitting the faces: in that case Euler's formula does not apply.

An ASCII file in this format can be composed in an ordinary text editor, such as NOTEPAD; or in a data editor written for the purpose, as follows.

20.11 Data Editor Program

The program does the following:

```
REM  **  Program EDPGRAPH to edit plane graph ASCII data
REM  **     file. User picks New or Edit. For New, user
REM  **     enters new filename and header values. For
REM  **     Edit, user enters filename to edit.
REM  **     Program reads header values and sets-up
REM  **     arrays with no data for New, old data for
```

```
REM  **     Edit, and offers change. At end, program
REM  **     (re-)writes edited file.
REM
REM  **  Enter filename to edit
REM  **  New file set-up: enter counts of vertices edges
REM  **  and faces.
REM  **  Old file set-up: show old counts of vertices edges
REM  **  and faces, enter counts of additional vertices
REM  **  and edges
REM  **  Call sub-programs
         CALL READDATA
         CALL EDITDATA
         CALL WRITDATA
END PROGRAM

     SUB READDATA
REM  **  Subprogram READDATA to put all-zero data for
REM  **     new file, or read data from old file, into
REM  **     arrays AVERT AEDGE AFACE.
     SUB EDITDATA
REM  **  Subprogram EDITDATA to show data from arrays
REM  **     AVERT, AEDGE, AFACE, and offer change.
     SUB WRITDATA
REM  **  Subprogram WRITDATA to rewrite data from
REM  **    arrays AVERT, AEDGE, AFACE to ASCII file #1
```

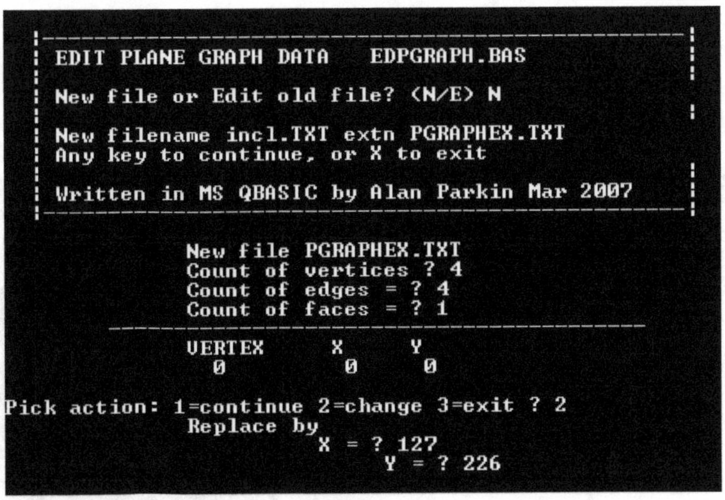

Fig. 20.10 Data editor screen

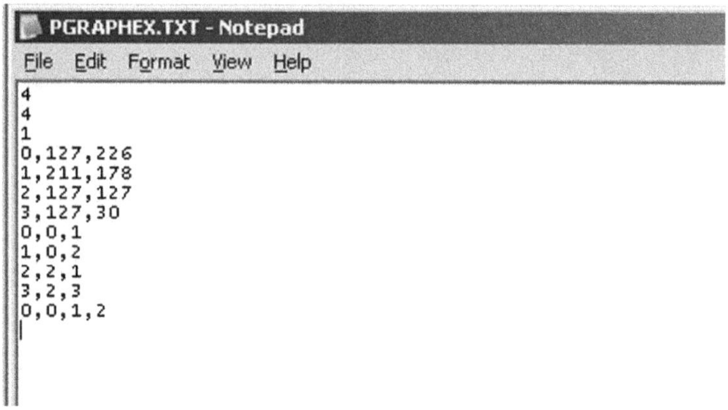

Fig. 20.11 File output from data editor, as seen in NOTEPAD

The source code is at the SpringerLink web-site. Figure 20.10 shows the starting screen of this program. Figure 20.11 shows an output file of this program.

20.12 Plane Graph Program

We can write a program which will read an ASCII data file specifying the vertices and edges of a plane graph, and make a bitmap file of the graph, which can be displayed in a GUI graphic editor. Such a program is MKPGRAPH, briefly described below and with source code at the SpringerLink web-site. For example, Fig. 20.12 shows the plane graph specified by the data file in Fig. 20.11, as made by program MKPGRAPH, and displayed on an ordinary image editor screen, with coordinate origin at top left. The program has been set to turn out a Black ground with White figure.

Fig. 20.12 Plane graph displayed in PAINT screen. The data file is that shown in Fig. 20.11

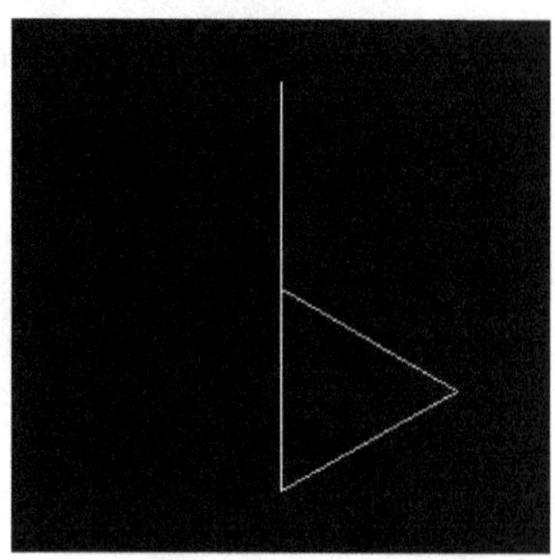

```
REM  **  Program MKPGRAPH to make plane graph from
REM  **     ASCII data in a 256 x 256-pixel window.
REM  **     User enters .TXT filename for input
REM  **     and .BMP filename for output. Program
REM  **     reads input file, sets-up arrays, puts
REM  **     data into arrays. For each edge, program
REM  **     tests and swaps for "steep" and
REM  **    "backwards", then does Bresenham line
REM  **     algorithm to set edge pixels in image
REM  **     array. Program writes image array as
REM  **     1-bit .BMP file.
REM  **  Main program
REM  **  Enter input and output file-names
         INPUT #1, CVERT%
         INPUT #1, CEDGE%
         INPUT #1, CFACE%
         CALL READDATA
         CALL MAKEIMAG
         CALL WBMPHEAD
         CALL WBMPDATA
         CALL REVBHEAD
         CLOSE #2
END

         SUB MAKEIMAG
REM  **  Sub-program MAKEIMAG to do Bresenham line algorithm
REM  **  for each edge, and set edge pixels in image array.
END SUB
         SUB READDATA
REM  **  Sub-program READDATA to read vertex and edge
REM  **     data from input file.
END SUB
         SUB REVBHEAD
REM  **  Revises header of file #2 with values in bytes.
END SUB
         SUB WBMPDATA
REM  **  Subprogram to write data from array AIMAGE
REM  **  to .BMP file #2 with row-length 128.
REM  **  1-bit locations are converted as groups of
REM  **  8 to byte.
END SUB
         SUB WBMPHEAD
REM  **  Subprogram to write header and colour-table of
REM  **  1-bit .BMP file #2, 128 X 128 pixels.
END SUB
```

20.13 Discrete Circle

Given the coordinates of the centre A $= (X_A, Y_A)$ and the length of the radius R, we can find the coordinates (x, y) of all the pixels of a discrete circle, by using Bresenham's [5] circle algorithm (apparently based on previous work by Pitteway [9]). The principle is shown in Fig. 20.13.

- Place, temporarily, the centre-pixel A at origin (0, 0), and the first pixel B of the circle at $(R, 0)$.
- Observe that only the pixels in the first octant of the circle need to be calculated, since by symmetry we then have the pixels in the other seven octants.
- Set $x = R$, $y = 0$ (that is, pixel B).
- The next pixel in the circle must be either C $= (x, y + 1)$ or D $= (x - 1, y + 1)$.
- The (real) distance AC of C from the centre is

  $$c = \sqrt{(X^2 + (Y + 1)^2)}\,,$$

  and the (real) error of C from the (real) circle is

  $$tc = c - R\,.$$

  The (integer) squared distance AC^2 is

  $$C = x^2 + (y + 1)^2$$

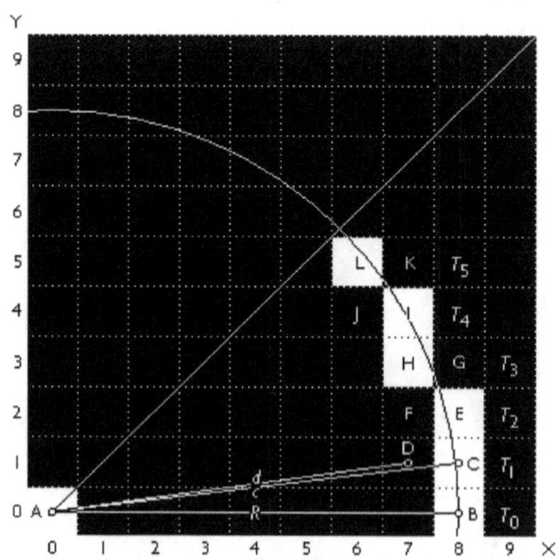

Fig. 20.13 Principle of Bresenham's circle algorithm. Given centre A and radius R, the algorithm finds the pixels for the discrete circle

and the (integer) squared error is

$$TC = C - R^2 .$$

- On the other hand, the (real) distance AD of D from the centre is

$$d = \sqrt{((x-1)^2 + (y+1)^2)}$$

and the (real) error of D from the (real) circle is

$$td = d - R .$$

The (integer) squared distance AD^2 is

$$D = (x-1)^2 + (y+1)^2$$

and the (integer) squared error is

$$TD = D - R^2 .$$

- Now set the error term T

$$T = TC + TD$$

- If $T < 0$, (as here) then TC is the lesser error, so take C as the next pixel, and increase y to $y+1$; or if $T \geq 0$, then TD is the lesser error, so take D as the next pixel, decrease x to $x-1$, and increase y to $y+1$.
- Continue the process from C to select E, H, I, and L, when the first octant, hence the whole circle, is complete.

From the above, we have

$$T = x^2 + (y+1)^2 - R^2 + (x-1)^2 + (y+1)^2 - R^2$$

which, multiplied out, becomes

$$T = 2x^2 - 2x + 2y^2 + 4y + 3 - 2R^2$$

Put the initial values $x = R$, $y = 0$, to get

$$T_0 = 2R^2 - 2R + 3 - 2R^2$$

that is

$$T_0 = 3 - 2R$$

We also find that, when $T_i < 0$

$$T_{i+1} = T_i + 4Y + 6$$

or when $T_i \geq 0$

$$T_{i+1} = T_i + 4(y - x) + 10$$

For example, Table 20.4 shows successive values calculated for Fig. 20.13.

Program MKCIRCLE is described briefly below, with source code at the Springer-Link web-site.

```
REM  **  Given centre (XA%, YA%) and radius RAD%
REM  **  Initialize array AIMAGE to all-black
REM  **  Initialize variables
REM  **  Plot all eight octants
REM  **  Test and re-set T% and X%
REM  **  Increase Y%
```

Figure 20.14 shows an example of circles plotted by this algorithm.

Table 20.4 Successive values for circle shown in Fig. 20.13

Step	Pixel	x	y	T_{out}	T_{in}	T	$(T < 0)$
0	B	8	0	1	−14	−13 (True)	
1	C	8	1	4	−11	−7 (True)	
2	E	8	2	9	−6	3 (False)	
3	H	7	3	1	−12	−11 (True)	
4	I	7	4	10	−3	7 (False)	
5	L	6	5	8	−3	5 (False)	

T_{out} is for the outer pixel (as, here, C, E, G, I, K); T_{in} is for the inner pixel (as, here, D, F,H, J, L)

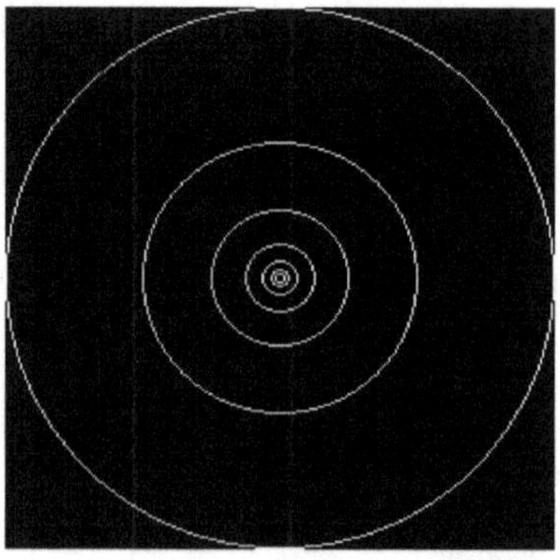

Fig. 20.14 Discrete circles plotted by Bresenham's algorithm. The given centre is (128,128) and the radii are 2, 4, 8, 16, 32, 64, and 128

20.14 Discrete Arc-length

In the real coordinate plane a circle with radius r has diameter d

$$d = 2\,r \text{ units}\,.$$

The *arc-length* s of the entire circle is

$$s = \pi\, d = \pi\, 2\,r = 3.1415\ldots 2\,r \text{ units}\,,$$

which is an irrational number, tailing off into fuzziness. The arc-length q of a quadrant is therefore

$$q = (\pi/2)\,r = 1.5707\ldots r \text{ units}\,.$$

But in the discrete coordinate plane (Fig. 20.15) a circle with integer radius R has a centre pixel, and diameter D

$$D = 2R + 1 \text{ px}\,,$$

which is an odd integer. A circle with even-integer diameter $D = 2R$ has no centre pixel. The *pixel count* P of a quadrant constructed by the discrete circle algorithm is the integer

$$P = (((\sqrt{}2)D) - 1) \setminus 2 \text{ px}\,,$$

which is, in integer arithmetic

$$P = (((10000\,D) \setminus\ 7071) - 1) \setminus 2 \text{ px}$$

(since the rational multiplier $10{,}000 \setminus 7{,}071 \approx \sqrt{}2$).

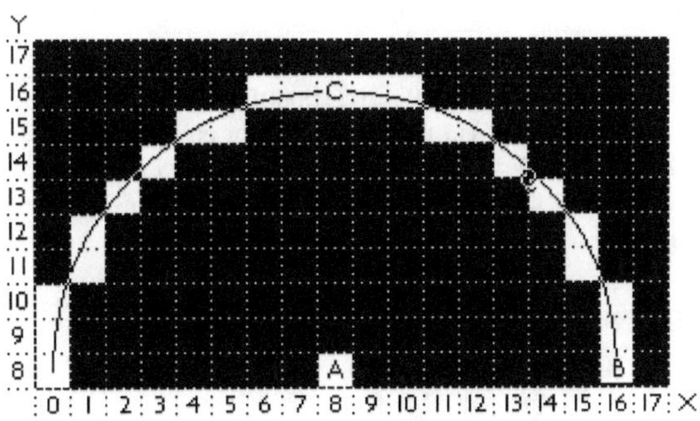

Fig. 20.15 Discrete arc-length. Circle with radius AB = $R = 8$ px has diameter $D = 2R + 1 = 17$ px. Quadrant BC has pixel count $P = 11$ px, and has arc-length $Q = 13$ px

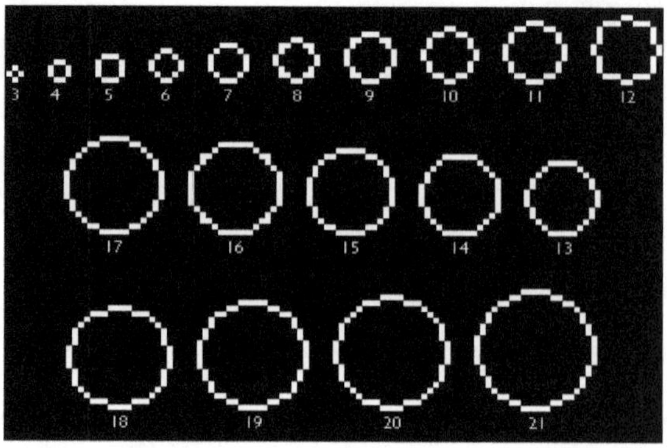

Fig. 20.16 Discrete circles with diameters from 3 to 21 px. Odd diameters have a central pixel; even diameters have no central pixel. The pixel counts P and quadrantal arc-lengths Q are calculated in Table 20.5

Table 20.5 Diameter, pixel count, and quadrantal arc-length of discrete circles shown in Fig. 20.16

D	P	Q	q
3	1	2	2.3562
4	2	3	3.1416
5	3	3	3.9270
6	3	4	4.7124
7	4	5	5.4978
8	5	6	6.2832
9	5	7	7.0686
10	6	7	7.8540
11	7	8	8.6394
12	7*	9	9.4248
13	8	10	10.2102
14	9	10	10.9956
15	10	11	11.7810
16	10	12	12.5664
17	11	13	13.3518
18	12	14	14.1372
19	12*	14	14.9226
20	13	15	15.7080
21	14	16	16.4934

D is diameter in pixels; $P = (((10000D) \setminus 7071) - 1) \setminus 2$ px; $Q = (11005D) \setminus 14012$ px. The real arc-length $q = (\pi/4)d$ units is shown for comparison. The two starred numbers are wrong, reason unknown: for $D = 12$, P should be 8, but 7.98 is truncated to 7; and for $D = 19$, P should be 13, but 12.92 is truncated to 12

The *discrete arc-length* S of the entire circle is, from the real plane

$$S = \pi D \, \text{px}$$

and the discrete arc-length Q of a quadrant is

$$Q = (\pi/4) \, D \, \text{px}$$

which is, in integer arithmetic

$$Q = (11,005 \, D) \setminus 14{,}012 \, \text{px}$$

(since the rational multiplier $11{,}005 \setminus 14{,}012 \approx \pi/4$).

For example, Fig. 20.16 shows discrete circles with diameters from 3 to 21 px, and Table 20.5 shows the calculated pixel counts P and quadrantal arc-lengths Q.

20.15 Discrete Sinusoidal Curve

A *discrete sinusoidal curve* can be derived directly from a discrete circle. Figure 20.17 shows the principle.

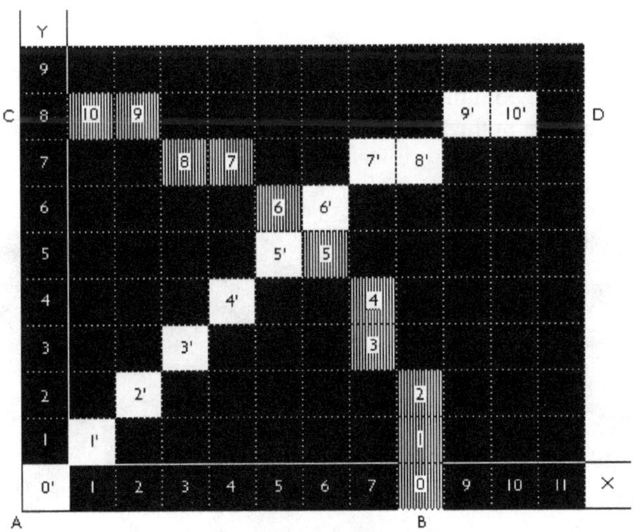

Fig. 20.17 Principle of generating discrete sine curve. The generating circle has centre pixel A at origin (0, 0), and radius AB px on the x-axis at (8, 0). The circle quadrant BC has a pixel count of $P = 11$ px, numbered 0 to 10. The curve quadrant AD also has a pixel count of $P = 11$ px, numbered $0'$–$10'$. For each circle pixel $0 \leq N \leq 10$ with coordinates (x, y), curve pixel $0' \leq N' \leq 10'$ has coordinates $(x' = N, y' = y)$

- Let the *generating circle* have centre pixel A at origin (0, 0), and radius AB = R.
- The pixel count for the quadrant BC is $P = (((10{,}000(2R+1)\backslash 7{,}071) - 1)\backslash 2$ px (as shown in the previous section).
- Number these P pixels $0 \le N \le (P-1)$.
- For each circle pixel N with coordinates (X, Y), mark a curve pixel N' with coordinates (N, Y).
- We then have the first quadrant AD of a *sine curve.*
- Continuing round the circle, number pixels $P \le N \le (2P-1)$, and mark curve pixels N' in the second quadrant.
- Continuing, number pixels $2P \le N \le (3P-1)$, and mark curve pixels N' in the third quadrant.
- Continuing, number pixels $3P \le N \le (4P-1)$, and mark curve pixels N' in the fourth quadrant.

Figure 20.18 shows the symmetries of a sine curve. There are two axes of reflection, orthogonal to the x-axis at P and $3P$; and two poles of rotation by a half-turn, on the x-axis at 0 and $2P$. These symmetries determine eight octants of the curve. If we have the coordinates of a pixel in the first octant (from 0 to $P/2$), then we can find also the coordinates for

Octant 4: by reflection in the axis at P
Octant 5: by rotation about the pole at $2P$
Octant 8: by reflection in the axis at $3P$

and if we have the coordinates of a pixel in the second octant (from $P/2$ to P), then we can find also the coordinates for:

Octant 3: by reflection in the axis at P
Octant 6: by rotation about the pole at $2P$
Octant 7: by reflection in the axis at $3P$

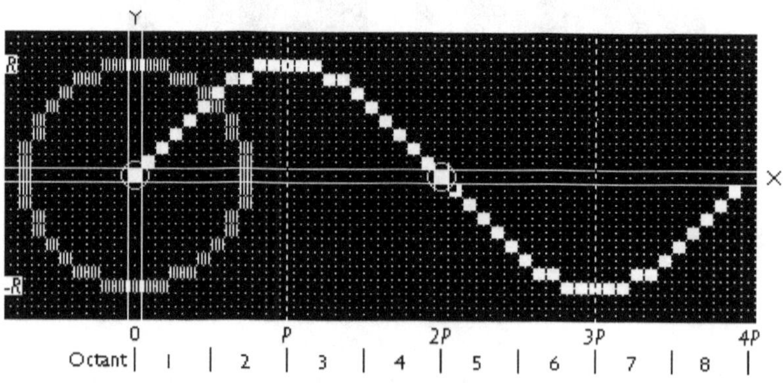

Fig. 20.18 Discrete sine curve and its symmetries. The curve has wavelength $4P$ and amplitude R. There are reflection axes at $x = P$ and $3P$, and 180-degree rotation poles at 0 and $2P$. Octant 1 reflects to octant 4; octant 2 reflects to octant 3; and octants 1–4 rotate to octants 5–8

Now, Bresenham's circle algorithm (Sect. 51.1) finds the coordinates of the pixels of the first octant of a generating circle, and these determine the coordinates of the pixels of all eight octants of the corresponding sine curve. A sine curve algorithm is thus (see the SpringerLink web-site for full listing):

```
REM  **  Program MKSINCUR to make sine curves, with .BMP
REM  **     file output. User enters filename for output,
REM  **     and any number of amplitude values. Program
REM  **     does Bresenham circle algorithm to find
REM  **     pixels of first octant of circle, then sets
REM  **     all eight octants of sine curve in image
REM  **     array. Program writes image array as 1-bit
REM  **     .BMP file.
REM  **  Enter output filename
REM  **  Initialize image array AIMAGE to all-black
         CALL MAKESINE
         CALL WBMPHEAD
         CALL WSICODAT
         CALL REVBHEAD
END

         SUB MAKESINE
REM  **  Sub-program MAKESINE to accept input of amplitude
REM  **     of sine curve which starts at (0,0). Program
REM  **     does Bresenham circle algorithm, and sets
REM  **     pixels of curve in image array.
REM  **  Set start and input amplitude
REM  **  Initialize
REM  **  Plot all eight octants of sine curve
REM  **  Test and re-set T% and X%
REM  **  Increase Y% and XX%
REM  **  More or done
END SUB
        SUB PUTI (INTE%)
REM  **  Puts integer value passed into binary file
REM  **  as 2 bytes and increments pointe.r
END SUB
         SUB PUTL (INTL&)
REM  **  Puts long integer into binary file #2 as 4 bytes
END SUB
         SUB REVBHEAD
REM  **  Revises header of file #2 with values in bytes.
REM  **  Next update bfSize length of file in bytes, as long
REM  **  Next update biSizeImage length of image data in
```

```
REM    **   bytes, as long.
END SUB
            SUB WBMPHEAD
REM    **   Subprogram to write header and colour-table of
REM    **   output .BMP file #2.
END SUB
            SUB WSICODAT
REM    **   Subprogram WSICODAT write data from array AIMAGE
REM    **      (which has width 0 to 128, height -63 to 64)
REM    **      to .BMP file #2. 1-bit locations are converted
REM    **      as groups of 8 to byte, to finish row on a
REM    **      long (4-byte) boundary.
END SUB
```

Figure 20.19 shows examples of sine curves plotted by this algorithm.

A *discrete cosine curve* has the same shape as a sine curve, but starts at a maximum, descends to a minimum, and rises again to a maximum. Geometrically, a cosine curve is a sine curve with the leftmost quadrant removed and replaced as the rightmost. Numerically, a cosine curve with amplitude R can be calculated from a circle with radius R, as shown in Fig. 20.20.

Figure 20.21 shows the symmetries of a cosine curve. There are two axes of reflection, orthogonal to the x-axis at 0 and $2P$; and two poles of rotation by a half-turn, on the x-axis at P and $3P$. These symmetries determine eight octants of the curve. If we have the coordinates of a pixel in the first octant (from 0 to $P/2$), then we can find also the coordinates for

Octant 4: by rotation about the pole at P
Octant 5: by reflection in the axis at $2P$
Octant 8: by rotation about the pole at $3P$

and if we have the coordinates of a pixel in the second octant (from $P/2$ to P), then we can find also the coordinates for

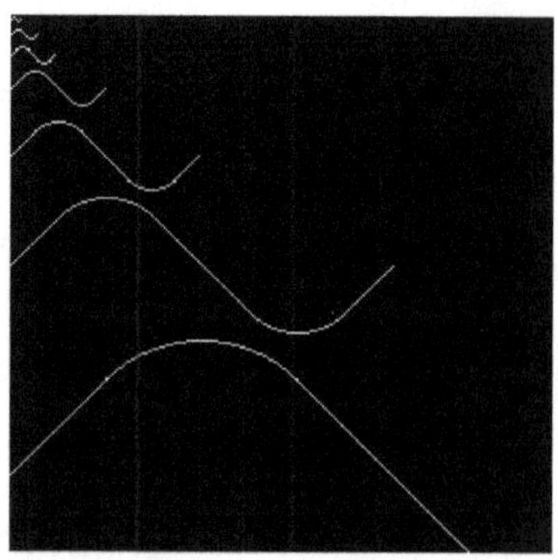

Fig. 20.19 Discrete sine curves plotted by an adaptation of Bresenham's algorithm. The amplitudes and start coordinates are 1,(0,126); 2,(0,119); 4,(0,109); 8,(0,93); 16,(0,61); 32,(0,9); and 64,(0,−91)

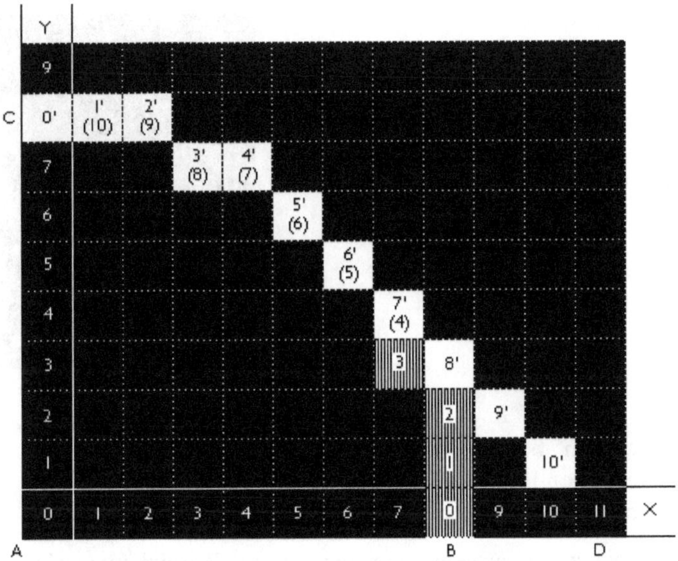

Fig. 20.20 Principle of generating discrete cosine curve. The generating circle has centre pixel A at origin (0, 0), and radius AB = R px on the x-axis at (8, 0). The circle quadrant BC has a pixel count of $P = 11$ px, numbered 0–10. The curve quadrant CD also has a pixel count of $P = 11$ px, numbered $0'$–$10'$. For each circle pixel $0 \leq N \leq 10$ with coordinates (x, y), curve pixel $0' \leq N' \leq 10'$ has coordinates $(x' = N, y' = x)$

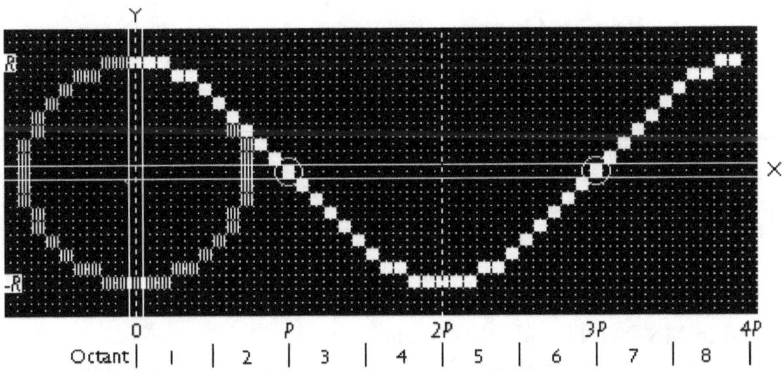

Fig. 20.21 Discrete cosine curve and its symmetries. The curve has wavelength $4P$ and amplitude R. There are reflection axes at $X = 0$ and $2P$, and 180-degree rotation poles at P and $3P$. Octant 1 rotates to octant 4; octant 2 rotates to octant 3; and octants 1–4 reflect to octants 5–8

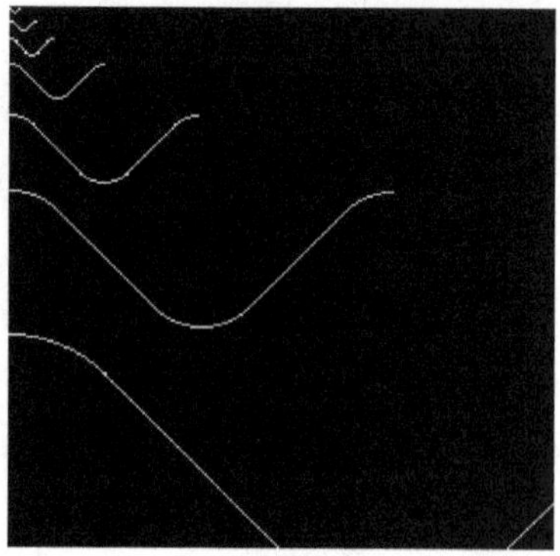

Fig. 20.22 Discrete cosine curves plotted by an adaptation of Bresenham's algorithm. The amplitudes and start coordinates are 1,(0,126); 2,(0,119); 4,(0,109); 8,(0,93); 16,(0,61); 32,(0,9); and 64,(0,−91)

Octant 3: by rotation about the pole at P
Octant 7: by reflection in the axis $2P$
Octant 6: by rotation about the pole at $3P$.

To make a discrete cosine curve we can again use the Bresenham circle algorithm to find the coordinates of the pixels in the first octant of the circle, as for a sine curve, but with a different plotting part. Figure 20.22 shows examples of cosine curves plotted by this algorithm.

20.16 Discrete Bézier Curve

A mechanical *spline* (from splint/splinter), as used for railway curves, ship lofting, and aircraft design, is a flexible strip, anchored at a few points, which takes up the curve of least strain. A mathematical spline is a polynomial curve weighting function which delivers a smooth curve through or near given fixed *control points*.[6] Splines are in general space curves with N controls. Here we consider plane Bézier cubic curves [8], with four controls: two *anchors* A and D, and two *handles* B and C. Figure 20.23 shows some examples. Convenient implementations of Bézier curves are included in MS PAINT and other image editors.

The controls form a polygon ABCD. The curve lies within the convex hull of the control polygon. Anchors A and D are the ends of the curve. If A B C D are in a straight line, the curve is just the segment AD. As B and C move independently away from the line they pull the curve in one direction or another. Figure 20.24 shows how the curve changes with the placing of the controls.

[6]The mathematics originated with Sergei Natanovich Bernstein [1]. Practical computer applications were developed (with a good deal of commercial secrecy) by Paul de Casteljau [6] at Citroen, Carl de Boor [3] at General Motors, James Ferguson [7] at Boeing, Malcolm Sabin [10] at British Aircraft Corporation, and Pierre Bézier [2] at Renault.

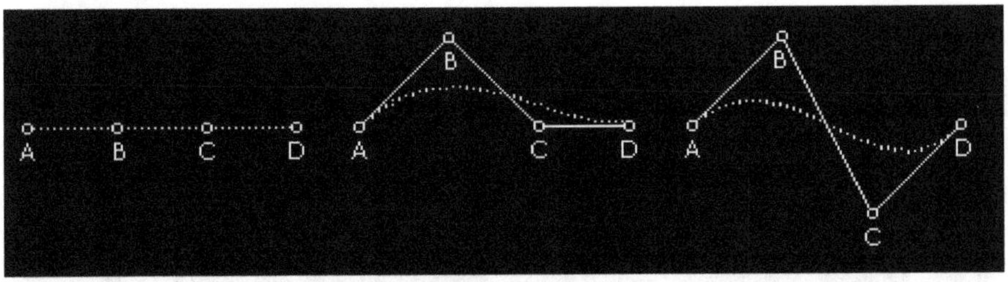

Fig. 20.23 Examples of plane Bézier cubic curves. Anchors A and D are the ends of the curve. As handles B and C move away from the line AD they pull the curve in that direction

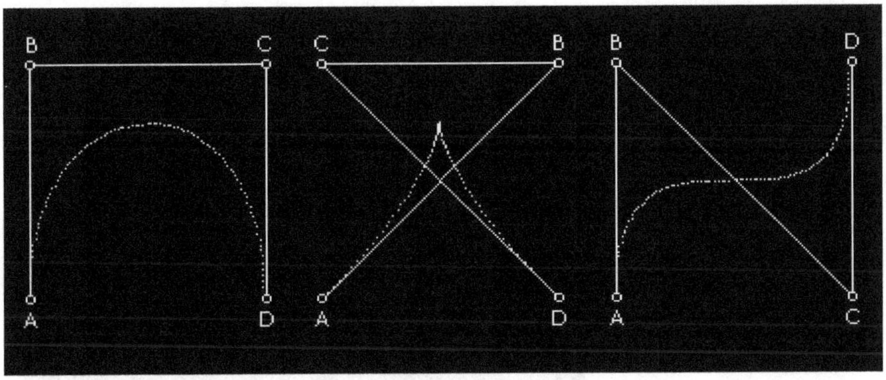

Fig. 20.24 Bézier curves with controls differently placed at the four corners of a square

The tangent to the curve at A is the polygon edge AB, and the tangent at D is the edge CD, so separate curves can be joined smoothly by aligning the polygon edges. Figure 20.25 shows curves thus aligned.

A discrete Bézier curve can be constructed by calculating the coordinates of its pixels from a polynomial weighting function. However, a remarkable property of a Bézier curve is its recursive decomposition into two sub-curves by de Casteljau's geometric algorithm. Figure 20.26 shows the principle:

```
Given curve ABCD
Fix point AB so that the distances A to AB and AB to B
     are in ratio m:(1-m)
Fix point BC so that the distances B to BC and BC to C
     are also in ratio m:(1-m)
Join AB and BC
Fix point CD so that the distances C to CD and CD to D
     are also in ratio m:(1-m)
```

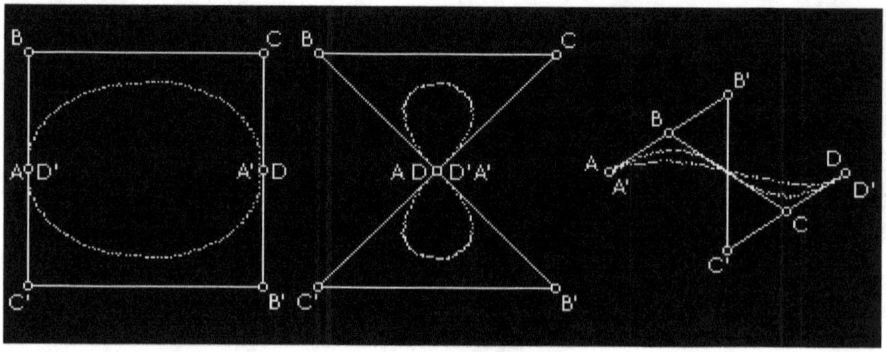

Fig. 20.25 Pairs of Bézier curves with tangents at A and D aligned to give a smooth join. In the middle example A and D coincide, so the curve is closed

Fig. 20.26 Principle of decomposing a Bézier curve into two sub-curves. The curve A ABCD D with controls A B C D separates into two sub-curves: A to ABCD with controls A AB ABC ABCD, and ABCD to D with controls ABCD BCD CD D

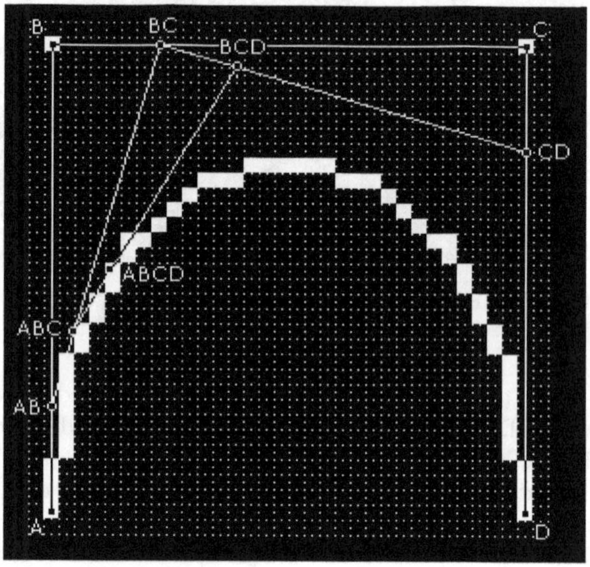


```
Join BC and CD
Fix point ABC so that the distances AB to ABC and ABC
     to BC are in ratio m:(1-m)
Fix point BCD so that the distances BC to BCD and BCD
     to CD are also in ratio m:(1-m)
Join ABC and BCD
Fix point ABCD so that the distances ABC to ABCD and ABCD
     to BCD are in ratio m:(1-m)
Now point ABCD lies on the curve
```

```
Moreover, the sub-curve between A and ABCD has the controls
     A AB ABC ABCD; and the sub-curve between ABCD and D
     has the controls ABCD BCD CD D
```

Clearly, any number of points ABCD on the curve can be marked by stepping m from 0 to 1.

A program for plotting a Bézier curve is briefly described below (see the Springer-Link web-site for source code). In this program

- M steps through a series of values,
- for each M-value, sub-program DECAST calculates the coordinates of a pixel on the curve,
- DECAST calls sub-program INTERP, which makes a linear interpolation in the current segment.

```
REM  **  Program MKBEZIER to make Bezier curve in
REM  **     window 128 x 128 pixels. User enters
REM  **     filename for output, coordinates of
REM  **     four control points A B C D, and count
REM  **     N of points to be calculated on curve.
REM  **     Program does de Casteljau algorithm, puts
REM  **     controls and curve points in image array,
REM  **     and writes image array as 1-bit .BMP file.
REM  **  Enter output filename
REM  **  Initialize image array AIMAGE to all-black
REM  **  Enter coords for control points,
REM  **     and count of points on curve.
REM  **  Set control pixels in image array
REM  **  Find pixels on curve and set in image array
         FOR M% = 0 TO N%
            CALL DECAST(BEZX%, BEZY%, M%, N%)
            AIMAGE(BEZX%, BEZY%) = 1
         NEXT M%
REM  **  Do output to 1-bit .BMP file
         CALL WBMPHEAD
         CALL WBMPDATA
         CALL REVBHEAD
END

         SUB DECAST (KMX%, KMY%, M%, N%)
REM  **  Sub-program to find coordinates (KMX%, KMY%)
REM  **     of each of M% pixels of Bezier curve
         CALL INTERP(ABX%, ABY%, AX%, AY%, BX%, BY%, M%,
  N%)
         CALL INTERP(BCX%, BCY%, BX%, BY%, CX%, CY%, M%,
```

```
 N%)
          CALL INTERP(CDX%, CDY%, CX%, CY%, DX%, DY%, M%,
 N%)
          CALL INTERP(ABBCX%, ABBCY%, ABX%, ABY%, BCX%,
BCY%, M%, N%)
          CALL INTERP(BCCDX%, BCCDY%, BCX%, BCY%, CDX%,
CDY%, M%, N%)
          CALL INTERP(KMX%, KMY%, ABBCX%, ABBCY%, BCCDX%,
BCCDY%, M%, N%)
END SUB

          SUB INTERP (PMX%, PMY%, END1X%, END1Y%, END2X%,
END2Y%, M%, N%)
REM  **   Sub_program INTERP to interpolate the Mth
REM  **     pixel of N between A and B
END SUB
          SUB PUTI (INTE%)
REM  **   Puts integer value passed into binary file
REM  **   as 2 bytes and increments pointer.
END SUB
          SUB PUTL (INTL&)
REM  **   Puts long integer into binary file #2 as 4 bytes
END SUB
          SUB REVBHEAD
REM  **   Revises header of .BMP file #2 with values in bytes.
END SUB
          SUB WBMPDATA
REM  **   Subprogram to write data from array AIMAGE
REM  **   to .BMP file #2 with row-length 128.
REM  **   1-bit locations are converted as groups of
REM  **   8 to byte.
END SUB
          SUB WBMPHEAD
REM  **   Subprogram to write header and colour-table of
REM  **   1-bit .BMP file #2, 128 X 128 pixels.
END SUB
```

Figure 20.27 shows three examples generated by program MKBEZIER (compare Fig. 20.24).

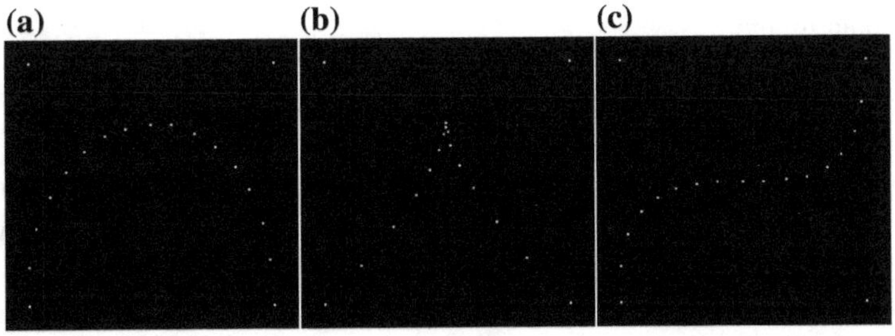

Fig. 20.27 Examples of Bézier curves generated by program MKBEZIER. **a** Control points A(10,10), B(10,117), C(117,117), D(117,10). **b** Control points A(10,10), B(117,117), C(10,117), D(117,10). **c** Control points A(10,10), B(10,117), C(117,10), D(117,117). Each curve has 16 interpolations

20.17 Programmed Image Resolution

The *resolution limit* of an image is the finest detail which can be discerned in it. The usual measure of resolution of an image is the highest frequency of line-pairs which are discernible. A *line-pair* is a Black bar and a White bar of equal widths. In a digital image the smallest width of a bar is 1 px, so the smallest width of a line-pair is 2 px. In a displayed or printed image we can measure the frequency as line-pairs per unit length. But in a program a pixel has no inherent size: we can only measure a relative frequency as line-pairs per image width

$$F_{\mathrm{LP}} = P \,/\, 2 \text{ line-pairs per image width}\,,$$

where F_{LP} is highest frequency of line-pairs, and P is the count of pixels in the width of the image. The reciprocal of the limit is a reasonable measure of resolution of

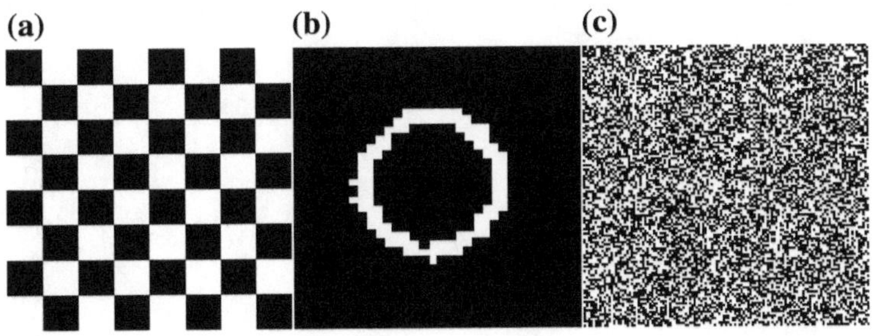

Fig. 20.28 Programmed image resolution. **a** Image width $W = 8$; highest frequency of line-pairs $F_{\mathrm{LP}} = 4$; resolution of location 1/4. **b** Image width $W = 32$; highest frequency of line-pairs $F_{\mathrm{LP}} = 16$; resolution of location 1/16. **c** Image width $W = 128$; highest frequency of line-pairs $F_{\mathrm{LP}} = 64$; resolution of location 1/64

location in an image

$$1\ /\ F_{\mathrm{LP}}\,.$$

The count of pixels P in a programmed image is ultimately limited by the size of memory available to the program, which sets an upper limit of relative resolution for an image created by a program.

For example, Fig. 20.28 shows (a) a 1-bit image where width $W = 8$, so the highest frequency of line-pairs $F_{\mathrm{LP}} = 8/2 = 4$, so resolution of location $= 1/4$. In (b), $W = 32$; $F_{\mathrm{LP}} = 16$; resolution $= 1/16$. In (c), $W = 128$; $F_{\mathrm{LP}} = 64$; resolution $= 1/64$.

References

1. Bernstein SN (1912) Démonstration du Théoreme de Weierstrass fondee sur le calcul des probabilites. Comm Soc Math Kharkov 13:1–2
2. Bézier PE (1970) Emploi des Machines a Commande Numerique. Masson, Paris
3. Boor Cde (1962) Bicubic spline interpolation. J Math Phys 41:212–218
4. Bresenham JE (1965) Algorithm for computer control of a digital plotter. IBM Syst J 4(1):25–30
5. Bresenham JE (1977) A linear algorithm for incremental digital display of circular Arcs. CACM 20(2):100–106
6. Casteljau PF (1959) Outillage Méthodes Calcul. Technical Report, André Citroën Automobiles SA, Paris
7. Ferguson JC (1964) Multivariate curve interpolation. JACM II:221–228
8. Kamermans M (2014) A Primer on Bézier Curves. http://pomax.github.io/bezierinfo/
9. Pitteway MLV (1967) Algorithm for drawing ellipses or hyperbolae with a digital plotter. Comput J 10(3):282–289
10. Sabin MA (1968) Conditions for continuity of surface normals between two adjacent parametric patches. Technical Report, British Aircraft Corporation

Chapter 21
1-bit Graphic Input

Another way to create a digital image is through a graphic user interface (GUI). A user can set up a window on a display screen containing the required image pixels, and set the pixels individually or in groups to different intensities: in the 1-bit case Black or White. In this book the Microsoft PAINT GUI is used throughout. A GUI offers several ready-made elementary features such as line, circle, and curve. Simple geometric constructions can be done in a GUI. Text can be added within an image, or an image can be set within text.

21.1 Image Creation by Graphic User Interface

The flowchart (Fig. 21.1) is for general guidance in this chapter.

21.2 Graphic User Interface

Current personal computers use a *graphic user interface*(GUI[1]) [1–3] for graphic input and graphic processes. A GUI is a combination of various *widgets*:

- a display screen which, can be dynamically divided into *windows*,
- *menu lists* of optional actions,
- a *pointer*, which can be moved across the screen by a *mouse* or other manual device,
- a *click button* for the pointer to select a window or select an action from a menu.

---

[1]The GUI idea was originated around 1962 by a team at the Stanford Research Institute and later at the Palo Alto Research Center led by Douglas Carl Engelbart (1925–2013). Ivan Edward Sutherland's Ph.D. thesis in 1963 (supervised by Prof. Shannon and guided by Prof. Minsky and Prof. Coons) was accompanied by a highly-specialized working system.

A. Parkin, *Digital Imaging Primer*, DOI 10.1007/978-3-540-85619-1_21

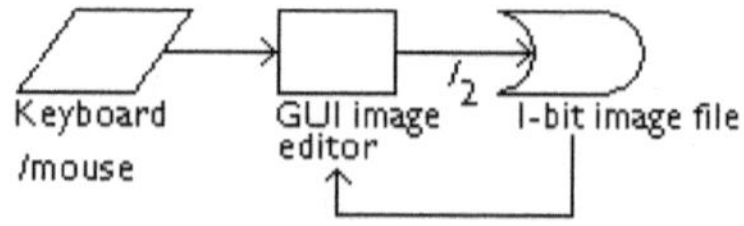

Fig. 21.1 Image creation by GUI. A user may create a 1-bit image via a GUI image editor, with interactive display, and store it as a bitmap file. A user may also display and change a 1-bit image already stored as a bitmap file

The examples shown here were made with the Microsoft PAINT image editor, supplied with the Microsoft Windows operating system. There are many similar free and proprietary image editors [4].

Figure 21.2 shows a PAINT GUI screen. The working window is initially White, and in 1-bit mode allows marks in Black (like pen on paper). The working window, or any sub-window, can be reversed to White marks on Black. The pixel coordinates (x,y) of the working widow have origin at top left, x increasing to the right, and y increasing downwards (like reading text), as shown in in Fig. 21.3. Notice that screen coordinates are inverted relative to the program array described in Sect. 20.1. There are optional (x,y) rules top and left, optional grid lines, and a coordinate indicator which displays the current position of the pointer in the working window. The working window can be enlarged by 100 % steps to 800 % to address single pixels with accuracy. The working window can be set to any size permitted by computer memory. It can be panned across or scrolled up and down to cover an area bigger than the current display screen.

21.3 Graphic Input by GUI Image Editor

A GUI image editor, such as PAINT, wraps a large set of programs for graphic input and processing in a well-arranged and easily-learned form. It provides nearly everything that we want for everyday graphic input. In 1-bit mode we can set up and display an image array of $P \times Q$ px, interactively assign intensity 0 or 1 to the pixels, and write the array sequentially to a 1-bit .BMP image file.

Elementary commands in a typical editor, such as PAINT, are,

- Mark a pixel,
- Draw a line between two end-pixels,
- Fill a face,
- Erase a pixel,
- Draw a box between two diagonal vertices (optionally square),
- Draw an ellipse within a box (optionally circular),
- Draw a curve on four control vertices,
- Enter text from keyboard.

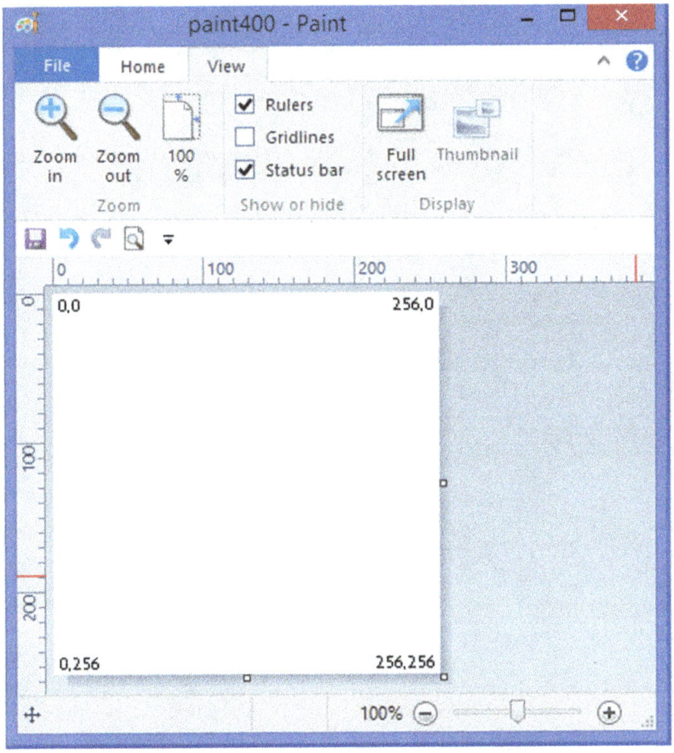

Fig. 21.2 Microsoft PAINT GUI screen. The white input window here is 256 × 256px, with origin (0, 0) at top left, and admits only non-negative pixel coordinates. The screen is set in 1-bit (monochrome) mode, so admits only Black or White pixels

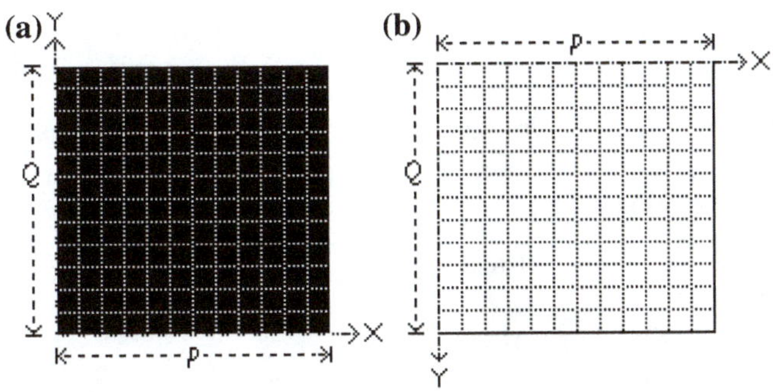

Fig. 21.3 **a** Program array: coordinate origin is bottom left, x-axis to the right, y-axis upwards (first-quadrant). P is count of pixels in width; Q is count of pixels in height. **b** GUI image editor window: coordinate origin is *top left*, x-axis *to the right*, y-axis downwards (*inverted first-quadrant*). P is count of pixels in width; Q is count of pixels in height

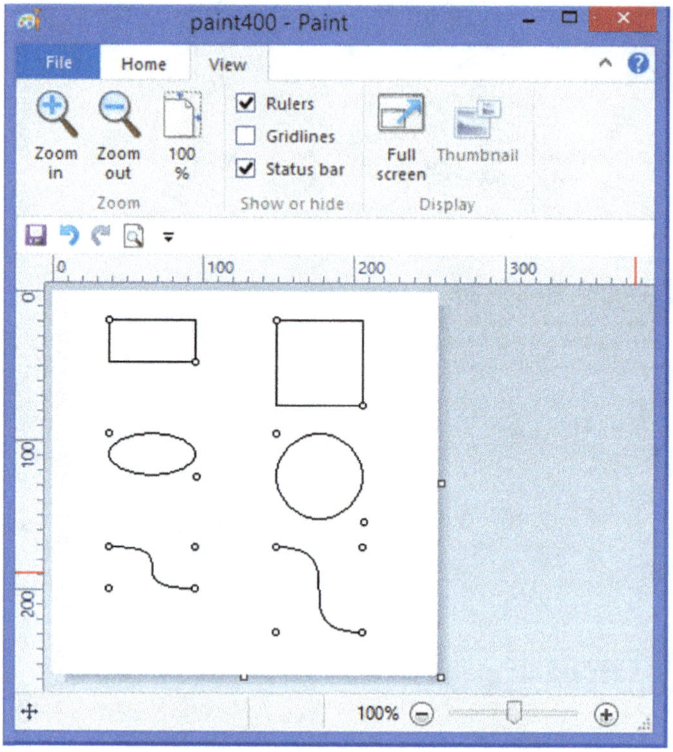

Fig. 21.4 PAINT screen with boxes, ellipses, and curves. **a** Rectangular box on two diagonal vertices. **b** Square box on two diagonal vertices. **c** Ellipse within rectangular box set by two diagonal vertices. **d** Circle within square box set by two diagonal vertices. **e** Curve set by two diagonal anchors (top left, bottom right), and two diagonal handles (top right, bottom left). **f** Another curve set by two diagonal anchors, and two diagonal handles

Figure 21.4 shows boxes, ellipses, and curves marked on the PAINT screen. Further inputs, such as text, use the keyboard.

Elementary operations are,

- Flip (reflect) horizontal or vertical,
- Stretch (dilate),
- Skew (shear) horizontal or vertical,
- Rotate by 90° increments,
- Set transparent or opaque.

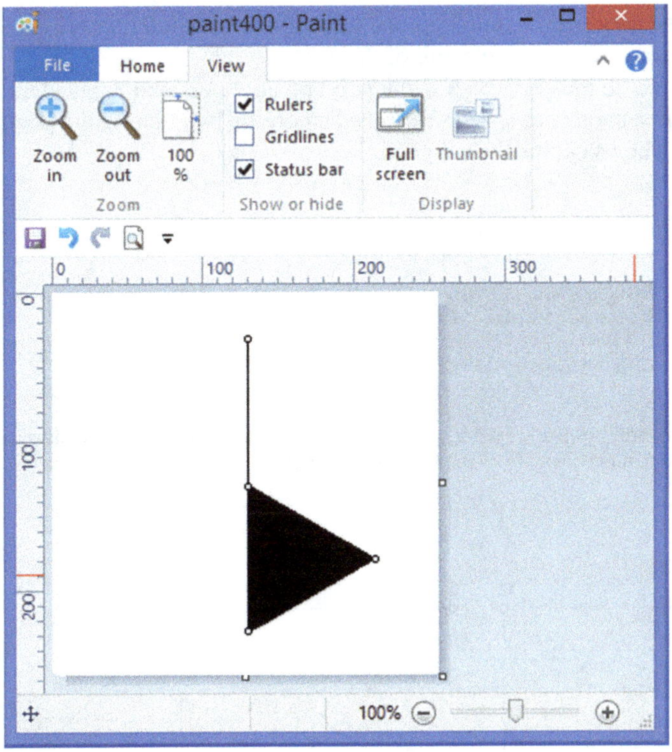

Fig. 21.5 Discrete plane graph in PAINT screen. The origin is at top left, x increasing to the right and y increasing downwards. The marks are black on a white ground

21.4 Plane Graph by GUI Editor

Any discrete plane graph (see Sect. 20.10) can be created or modified by the first four commands. For example, Fig. 21.5 shows a graph with four vertices, four edges, and two faces.

21.5 Curve by GUI Editor

The curves in PAINT are plane Bézier cubics (see Sect. 20.16). They have four controls: two *anchors* A and D, and two *handles* B and C. Figure 21.6 shows some examples.

The controls form a polygon ABCD. The curve lies within the convex hull of the control polygon. Anchors A and D are the ends of the curve. If A B C D are in a straight line, the curve is just the segment AD. As B and C move independently away from the line they pull the curve in one direction or another. Figure 21.7 shows how the curve changes with the placing of the controls.

The tangent to the curve at A is the polygon edge AB, and the tangent at D is the edge CD, so separate curves can be joined smoothly by aligning the polygon edges. Figure 21.8 shows curves thus aligned.

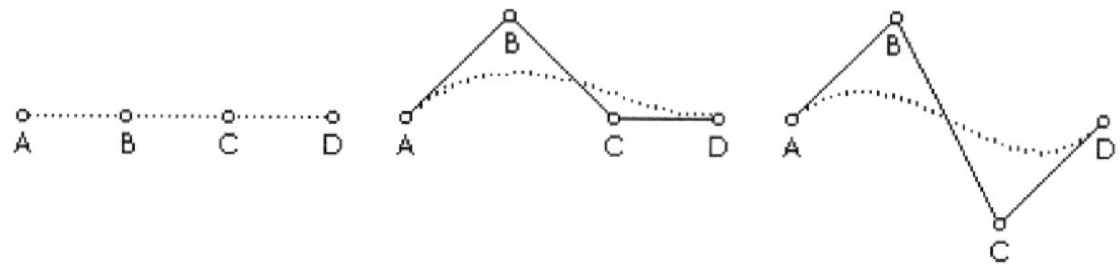

Fig. 21.6 Examples of plane Bézier cubic curves. Anchors A and D are the ends of the curve. As handles B and C move away from the line AD they pull the curve in that direction

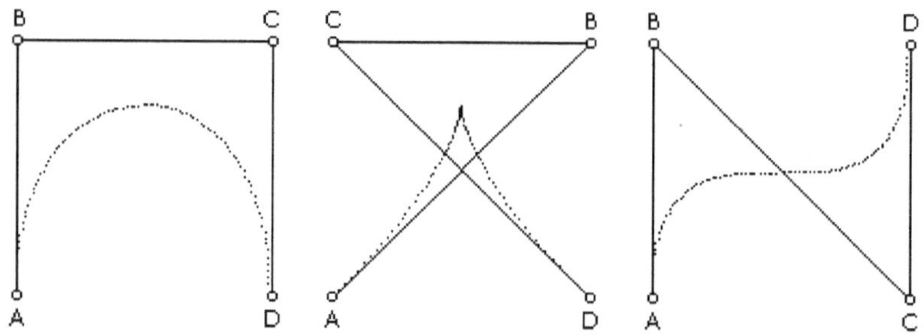

Fig. 21.7 Bézier curves with controls differently placed at the four corners of a square

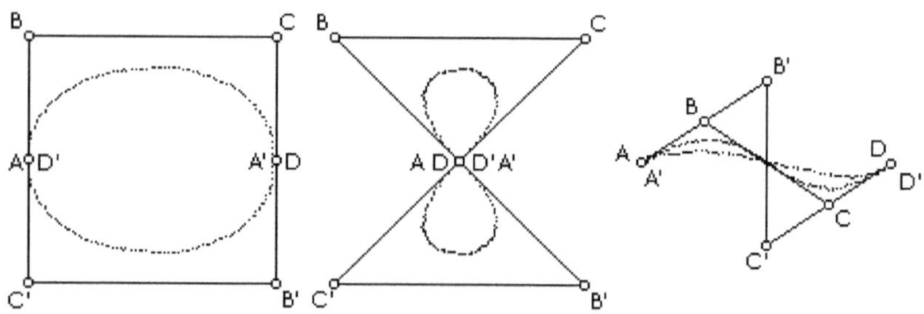

Fig. 21.8 Pairs of Bézier curves with tangents at A and D aligned to give a smooth join. In the middle example A and D coincide, so the curve is closed

21.6 Construction by GUI Editor

A great variety of constructions ab initio is possible by combined use of GUI elements and transformations. For example, Fig. 21.9 shows a construction for a hexagonal grid:

- Required hexagon width 32 px.
- Set out 90°, 60°, 30° triangle 17 px wide and 10 px high (since $\tan 30° = 0.5773$, and 17×0.5773 rounds to 10).
- Copy, paste, and flip triangle to make hexagon.
- Erase unwanted pixels.
- Copy and paste hexagon.

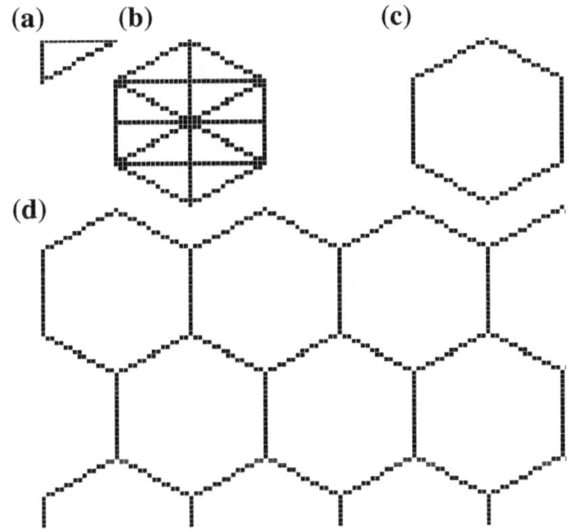

Fig. 21.9 GUI construction for hexagonal grid. **a** Elementary 90°, 60°, 30° triangle. **b** Hexagon made from triangles. **c** Unwanted pixels erased. **d** Grid copied and pasted from hexagon

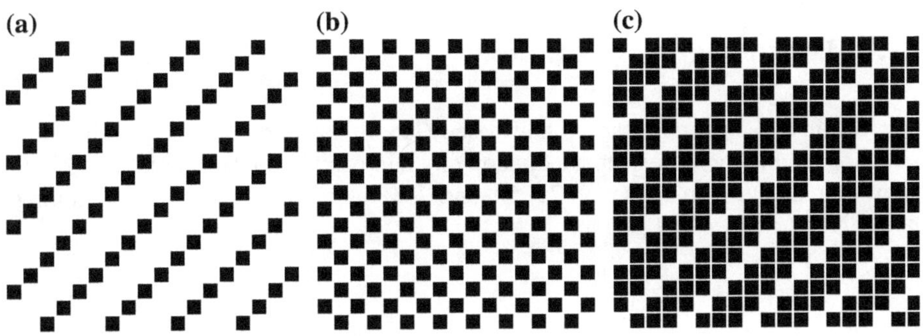

Fig. 21.10 Example of three GUI half-tones. **a** Black coverage 25 %. **b** Black coverage 50 %. **c** Black coverage 75 %

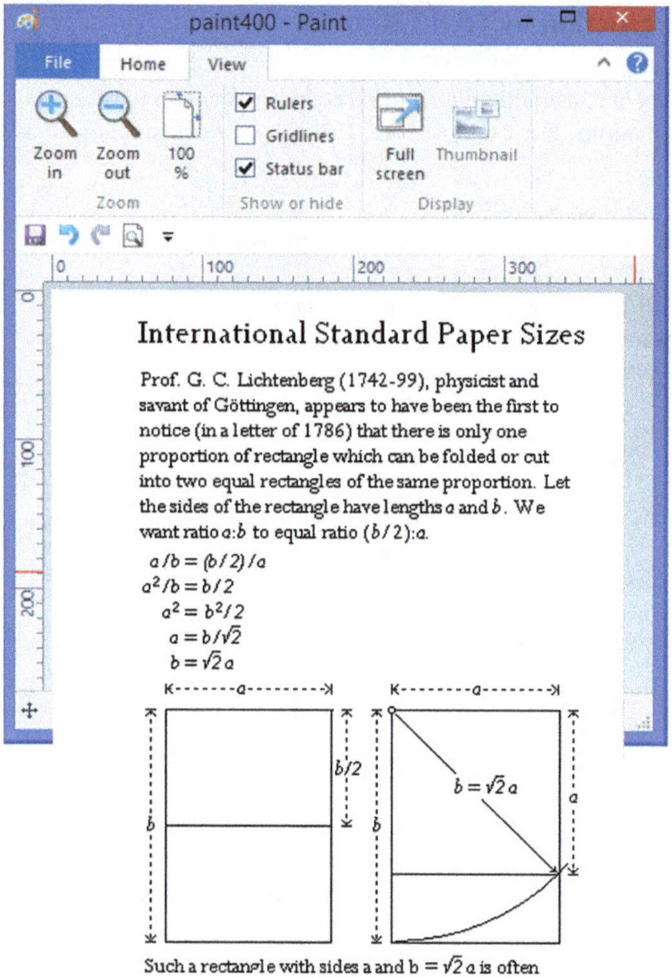

Fig. 21.11 PAINT image editor creating page image containing both text and sub-image

21.7 Half-toning by GUI Editor

Regions in a 1-bit image can be half-toned (laboriously) in a GUI editor by constructing a small unit then repeatedly copying and pasting. For example, Fig. 21.10 shows three square images with 25, 50, and 75 % black coverage. (For half-toning by program, see Sect. 39.16.)

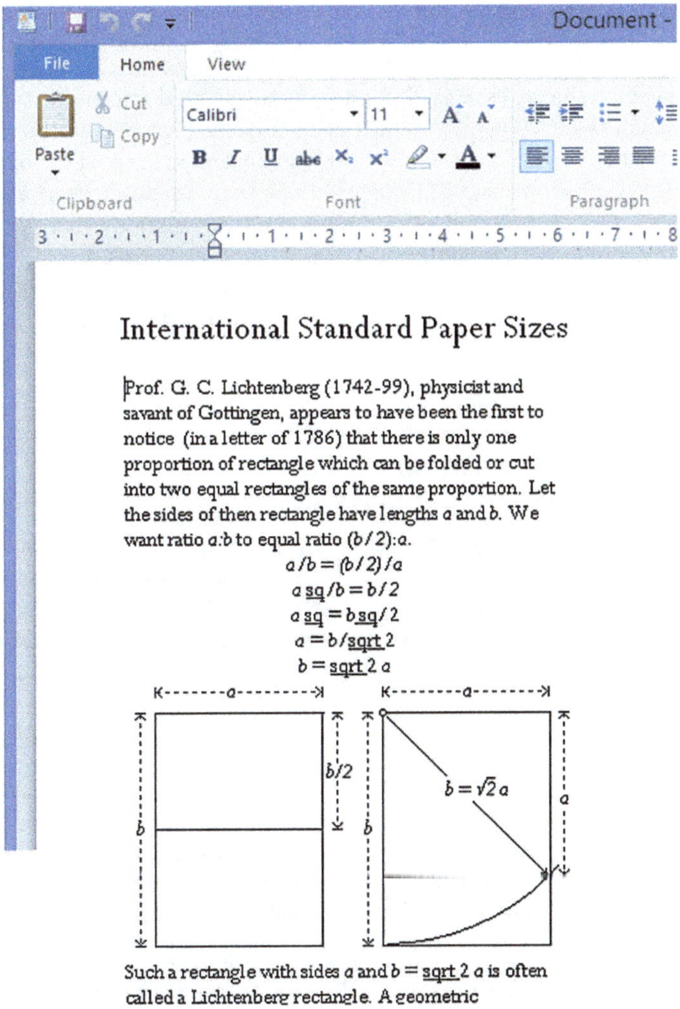

Fig. 21.12 WORDPAD text editor creating page image containing both text and sub-image

21.8 Text Within Image

In a GUI image editor we can compose a document containing both text and sub-images as a series of pages, each stored as a bitmap image. For example, Fig. 21.11 shows the PAINT image editor creating such a page. The text can be in any available fount and size. The whole page is open to pixel-wise manipulation. The resulting image can be displayed or printed using the GUI image editor.

21.9 Image Within Text

Bitmap images produced by a GUI editor can be inserted into many types of text documents, such as WORDPAD, WORD, and similar text-processors. First create the bitmap image, then copy and paste into the document text in the position required. For example, Fig. 21.12 shows the WORDPAD text editor creating a page. WORDPAD is supplied with Windows, and offers quite good text options (but not superiors or square-root symbols), outputting a file in Rich Text Format (.RTF). The more elaborate text-processors offer better text options, and output files in various formats. Simple ASCII text-editors such as NOTEPAD usually do not allow insertion of bitmap images.

21.10 GUI Image Resolution

The relative resolution of an image created by a GUI image editor is measurable in the same way as for a program (see Sect. 20.17). The upper limit is set by the physical size of memory in the computer.

However, when an image is displayed or printed it then has an actual resolution limit (see Chaps. 27 and 28).

References

1. Engelbart DC (1962) Augmenting human intellect: a conceptual framework. http://www.dougengelbart.org/pubs/augmenting-human-intellect.html. Accessed 1 Feb 2014
2. Reimer J (2005) A history of the GUI. http://arstechnica.com/articles/paedia/gui.ars. Accessed 1 Feb 2014
3. Sutherland IE (1963) SketchPad: a man-machine graphical communication interface. http://www.cl.cam.ac.uk/techreports/UCAM-CL-TR-574.pdf. Accessed 1 Feb 2014
4. Wikipedia (2014) Comparison of raster graphics editors. http://en.wikipedia.org/wiki/Comparison_of_raster_graphics_editors

Chapter 22
1-bit Scanner Input

A digital image can be created by taking input from a scanner. A scanner has a light source which illuminates a flat or nearly-flat object, which reflects differential intensities to a sensor, which registers intensities pixel-by-pixel, for input to a computer. In the 1-bit case, the intensities are Black or White. The size variables of a scanner are defined. A measure of resolving power is given, and the acuity of a scanner is compared with that of the eye. The transfer function of a device between its input and output is defined, and the power-law gamma function of input and output devices is explained. The intensity variables of a scanner are defined, including optical density as used in a standard photographic step-wedge. Examples of 1-bit scanning are shown.

22.1 Flowchart

The flowchart Fig. 22.1 is for general guidance in this chapter.

22.2 1-bit Scanner

A scanner [2] is a device which accepts a more-or-less flat object, illuminates it, senses the light reflected, and delivers a digital image file. A scanner operating in 1-bit mode delivers a 1-bit image file. The functional parts of a scanner are

- Object window,
- Light source,
- Sensor,
- Scanning mechanism,
- Analogue-to-digital converter (ADC),
- Thresholder.

A. Parkin, *Digital Imaging Primer*, DOI 10.1007/978-3-540-85619-1_22

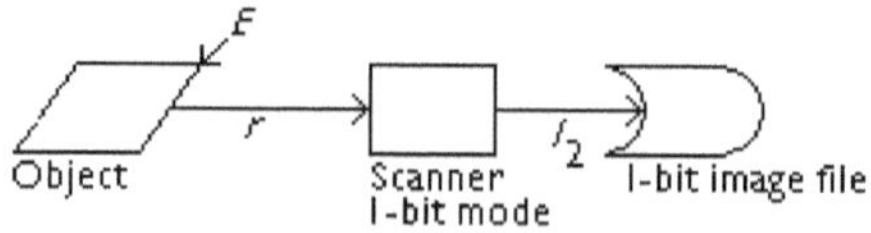

Fig. 22.1 Flowchart for 1-bit scanner input. A scanner provides a fixed illuminance E over a flat object, and receives per-sample reflectance r. When operating in 1-bit mode the scanner delivers an image file of discrete pixels with Black or White intensities I_2

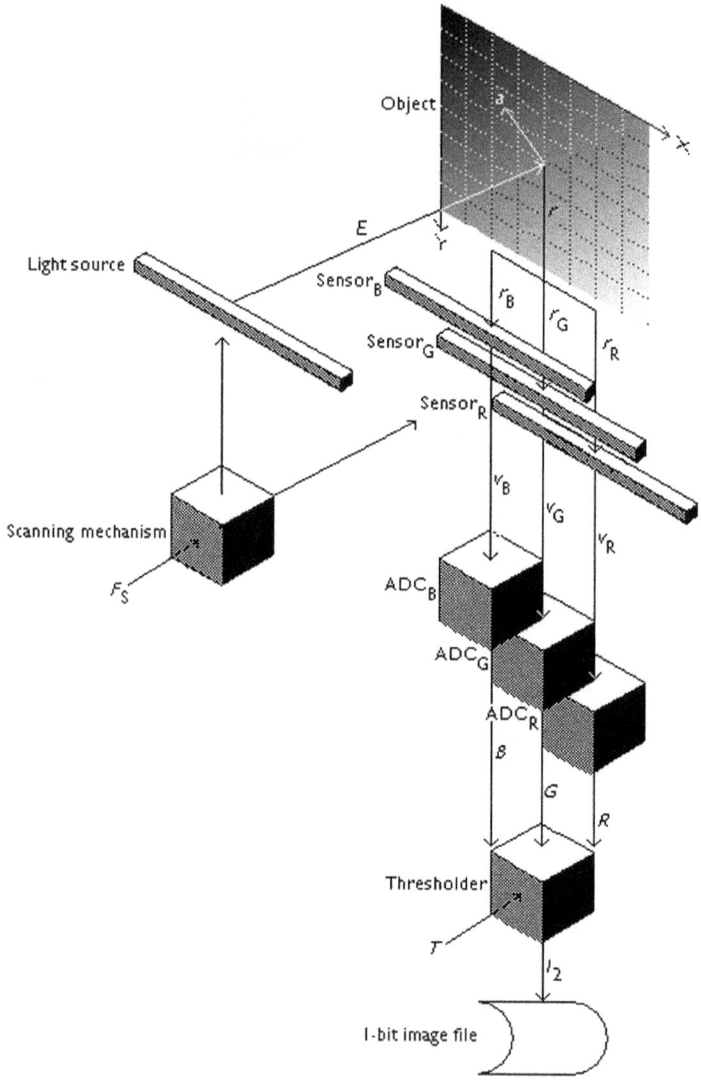

Fig. 22.2 1-bit scan, from object to image file

22.3 Object Window

In a typical flat-bed scanner the object window (or platen) is a glass sheet about A4 size. If an object is truly flat, such as a sheet of paper, it is placed face down in the window, with a White lid closed on top of it. If an object is only approximately flat, such as a hand, it is laid in the window with the lid open for a Black background or closed for a White background. Figure 22.2 shows an object as seen by the scanner (that is, looking through the window at the front of the object), with location coordinate origin at top left, X positive to the right and Y positive downwards.

In a sheet-fed scanner the physical arrangements are different, but functionally similar.

22.4 Light Source

In a typical flat-bed scanner, the light source has the form of a horizontal strip, as shown in Fig. 22.2. The intensity of light arriving at the object is measured as *illuminance* E lux (see Sect. 15.7); but since this is constant in a scanner, we may normalize it as $E = 1$. At each point location (x, y) some fraction a of the incident light is absorbed, and some fraction r is reflected (by an opaque object), or transmitted (by a transparent object such as a photographic film negative). The intensity of light reflected or transmitted is measured as *luminance* L cd/m^2 (see Sect. 15.8); which we may here normalize as fraction r between 0 and 1.

22.5 Sensor

The sensor is usually a photo-electric *charge-coupled device* (CCD).[1] A CCD has a large number of small *buckets*, each of which senses the luminance L in front of it, and outputs a voltage v proportional to the luminance. The location precision of a CCD is fixed by the size of the buckets: in a general-purpose scanner it is typically 1/600 in; for scanning photographs 1/2400 in; and for special work up to 1/5400 in. The intensity sensitivity or *dynamic range* of a CCD lies between a floor and a ceiling. The floor is the lowest intensity at which light input can be distinguished from the inherent quantum noise of the device: it is well below the lower limit of the eye (hence the importance of CCD sensors for astronomy). The ceiling is the highest intensity of light at which the device functions correctly: beyond this buckets become over-saturated and blooming occurs. In a scanner the internal light source is adjusted to match the ceiling of the CCD.

[1] Invented in 1969 by Willard Sterling Boyle (1924–2011) and George Elwood Smith (born 1930) at Bell Laboratories, New Jersey, awarded Nobel prize for Physics 2009.

A colour scanner has three sensor *channels*: the Blue channel has a filter which passes a band of short wavelengths 400–500 nm, the Green passes 500–600 nm, and the Red passes 600–700 nm. Each channel may record at 24 bpp or 36 bpp or 48 bpp.

22.6 Scanning Mechanism

The scanning mechanism operates by stepping the light source and the sensor channels down the object window by y-increments. At each y-increment the sensors take samples of luminance by x-increments.

A typical scanner has a settable *sampling frequency* F_S. The fundamental sampling frequency, say 600 spi, can be reduced by combining samples, down to say 30 spi, or interpolating up to say 1200 spi. Thus the object is sensed as a sequence of per-pixel voltage samples (v_B, v_G, v_R), as shown in Fig. 22.2.

22.7 Analogue-to-Digital Converter (ADC)

An 8-bit ADC is a device which converts a real-valued (analogue) input voltage $0 \leq v \leq 1$ to an integer (digital) output $0 \leq I \leq 255$. A colour scanner has one such ADC in each channel, to digitize Blue, Green and Red intensities.

An ADC has two voltage inputs: a fixed reference voltage k and a variable sample voltage $0 \leq v \leq k$. The operation of an ADC is equivalent to the following algorithm:

```
comment K is the input reference voltage
comment V is the input sample voltage
comment B is the output byte
real K, V
integer C, D
binary array B(1:8)
begin
   for C = 1,1,8 do
   begin
      D = 2^C
      K = K/D
      if V >= K then
      begin
         V = V-K
         B(C) = 1
      else
```

Table 22.1 Example of analogue-to-digital conversion by algorithm

C=	1	2	3	4	5	6	7	8
D=	2	4	8	16	32	64	128	256
k=	2.5000	1.2500	0.6250	0.3125	0.1560	0.0780	0.0390	0.0195
v=	1.1490	1.1490	0.5240	0.2115	0.0555	0.0555	0.0165	0.0165
B(C)=	1	0	1	1	1	0	1	0

Input $k = 5.0000$, input $v = 3.6490$. Output is 10111010 binary = BA hexadecimal = 186 decimal

```
            B(C) = 0
        end
    end
end
```

Table 22.1 shows an example.

22.8 Thresholder

When operating in 24-bit colour mode, a scanner delivers pixels $(x, y; \mathrm{B}, \mathrm{G}, \mathrm{R})$, each with a triple of integer intensities, where $0 \leq \mathrm{B} \leq 255$, $0 \leq \mathrm{G} \leq 255$, $0 \leq \mathrm{R} \leq 255$ (see Chap. 46).

When operating in 8-bit grey-scale mode, a scanner applies weights[2] to the sensor (b, g, r) intensities, and delivers for each pixel a single integer *intensity index*:

$$I_{256} = 0.114\mathrm{B} + 0.587\mathrm{G} + 0.299\mathrm{R}\,,$$

where $0 \leq I_{256} \leq 255$. The grey-scale index points to an 8-bit colour-table 8GS in an 8-bit file (see Chap. 37).

When operating in 1-bit Black and White mode, a scanner further applies a settable *threshold* $0 \leq T \leq 255$ to the 8-bit intensities I (that is, B or G or R), to produce a binary index:

$$I_2 = 0 \quad \text{if} \quad I \leq T$$
$$I_2 = 1 \quad \text{if} \quad I > T\,.$$

Thus when the threshold is low, most of the middle greys will go to White; and when the threshold is high, most of the middle greys will go to Black. The binary index points to a 1-bit colour-table in a 1-bit file (see Chap. 26).

[2]The weights given here are widely used in digital imaging. They are taken from [1], and are based on the NTSC CRT phosphors of 1953. See [3].

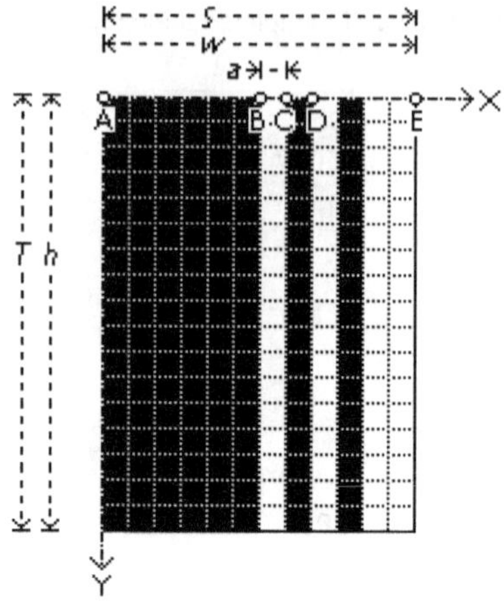

Fig. 22.3 Scanner size variables. A scanner imposes an integer sampling grid on a continuous object. Coordinate origin is at top left; x-axis positive to the right; y-axis positive downwards. w is width of object and h height. S is count of samples in width of object and T is count of samples in height. a is sample width

Table 22.2 Size variables for scanner, as shown in Fig. 22.3

Object			
Object sample width	a	$= 1/F_S$	Length units
Object sampling freq. *	F_S		Samples per unit length
Object resolution lim.	F_{LP}	$= F_S/2$	Line-pairs per unit length
Object height *	h		Length units
Count samples in width	S	$= w \times F_S$	Samples
Count samples in height	T	$= h \times F_S$	Samples
Object width *	w		Length units

The asterisked variables are settable, then the rest can be calculated

22.9 Scanner Size Variables

In a flat-bed scanner the object is placed face-down on the scanner window. Figure 22.3 shows the size variables[3] as seen by the scanner: that is, front elevation of object and back elevation of window. A scanner has a settable sampling frequency F_S in software which combines or interpolates the fundamental bucket samples. Table 22.2 shows the relations of the size variables.

For example, given object width $w = 8$ in and setting sampling frequency $F_S = 300$ spi, sample width is $a = 1/300 = 0.0033$ in, and count of samples is $S = 8 \times 300 = 2400$ samples. Or, given object width $w = 8$ in and required count of samples $S = 1024$, sample width $a = w/S = 8/1024 = 0.0078$ in, so set sampling frequency $F_S = 1/0.0078 = 128$ spi.

[3]Scanner sizes are commonly in inch units. Metres = inches $\times 0.0254$; inches = metres $\times 39.37$; millimetres = inches $\times 25.4$; inches = millimetres $\times 0.0394$.

22.10 Resolving Power of Scanner

The resolving power, or resolution limit, of a sampling device, taking regularly-spaced samples of a continuous object, is the finest detail which can be registered accurately by the device. For an optical device, the usual measure of resolution limit is the highest-frequency of object grating which can be reliably distinguished from a uniform grey. Reliable means more than 50 % correct recognitions in repeated tests with the grating in various orientations.

A grating has alternating Black and White lines, as at BC, CD, ... in Fig. 22.3, so the finest detail is a line-pair, that is one Black line and one White line, BC and CD. At the resolution limit, the images of successive line-pairs fall on successive pairs of sensor elements. Clearly, the resolution limit is reached at half the sampling frequency. So we have the resolution limit of a scanner in linear measure on the grating

$$F_{\mathrm{LP}} = F_{\mathrm{S}}/2\,\mathrm{lppul}\,.$$

For example, if sampling frequency F_{S} is 300 spi, the resolution limit is $F_{\mathrm{LP}} = 300/2 = 150$ lppi.

A simple and convenient tool for measuring resolution is a *line-pair test strip*. This is a carefully-produced succession of line-pairs with spacing varying from about 0.2 lppmm (or 5 lppi) to about 3 lppmm (or 76 lpi). Figure 22.4a shows a typical test strip, with spacings marked in line-pairs per mm, scanned in 8-bit grey-scale mode

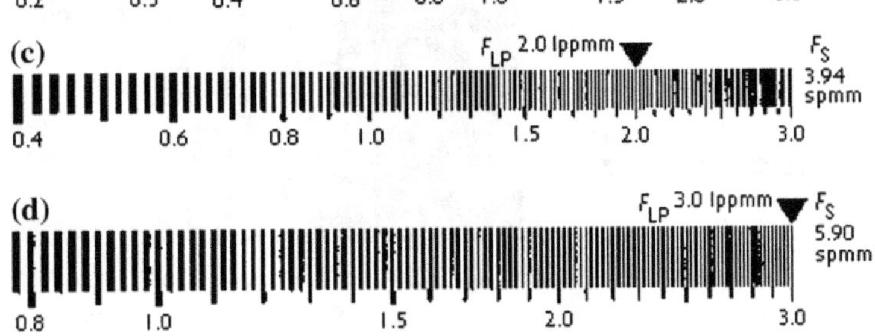

Fig. 22.4 Line-pair test strip scanned at different sampling frequencies. **a** Scan in 8-bit grey-scale mode, sampling frequency $F_S = 3.27$ spmm (= 83 spi), resolution limit $F_{\mathrm{LP}} = 1.6$ lppmm. **b** Scan in 1-bit mode at = 2.95 spmm (= 75 spi), resolution limit 1.5 lppmm. **c** Scan at 3.94 spmm (= 100 spi), resolution limit 2.0 lppmm. **d** Scan at 5.90 spmm (= 150 spi), resolution limit 3.0 lppmm

at sampling frequency $F_S = 3.27\,\text{spmm} = 83\,\text{spi}$, so the resolution limit $F_{LP} = 3.27/2 = 1.6\,\text{lppmm}$, as marked. In Fig. 22.4b, the strip is scanned in 1-bit mode at $F_S = 2.95\,\text{spmm} = 75\,\text{spi}$, so $F_{LP} = 2.95/2 = 1.5\,\text{lppmm}$. Line-pairs with wider spacing occupy more than two samples, and line-pairs with closer spacing occupy less than two samples. In Fig. 22.4c, $F_S = 3.94\,\text{spmm}$, so $F_{LP} = 3.94/2 = 2.0\,\text{lppmm}$. In Fig. 22.4d, $F_S = 5.90\,\text{spmm}$, so $F_{LP} = 5.90/2 = 3.0\,\text{lppmm}$.

22.11 Acuity of Scanner

It may be interesting to compare the acuity of a scanner with that achieved by the eye (see Sect. 10.6). Figure 22.5a shows in plan a very good Snellen 6/3 eye viewing an object grating at distance $v = 10\,\text{in}$, and in elevation the hexagonally-packed retinal cones projected onto the object grating. Figures 22.5b–e show the square samples of

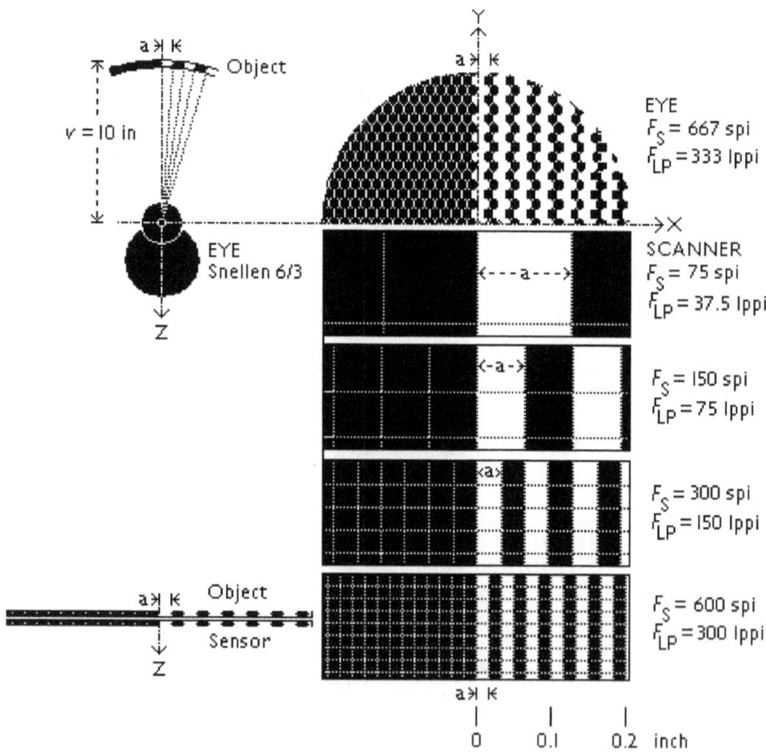

Fig. 22.5 Acuity of eye and scanner. **a** Snellen 6/3 eye viewing object at distance $v = 10\,\text{in}$ projects sample width $a = 0.0015\,\text{in}$ onto object. **b** Scanner operating at sampling frequency $F_S = 75\,\text{spi}$ has sample width $a = 1/75 = 0.0133\,\text{in}$. **c** At $F_S = 150\,\text{spi}$, $a = 1/150 = 0.0067\,\text{in}$. **d** At $F_S = 300\,\text{spi}$, $a = 1/300 = 0.0033\,\text{in}$. **e** At $F_S = 600\,\text{spi}$, $a = 1/600 = 0.00167\,\text{in}$

Table 22.3 Acuity of eye, and of scanner operated at four different sampling frequencies, as shown in Fig. 22.5

Variable	6/3 eye	(b)	(c)	(d)	(e)	
Sample width $a = 1/F_S$	0.0015	0.013	0.0067	0.0033	0.0017	inch
Sampling freq. * F_S	667	75	150	300	600	spi
Resolution limit $F_{LP} = F_S/2$	334	37.5	75	150	300	lppi
Count smpls in width $S = w/a$	5333	615	1194	2424	4800	samples
Object distance * v	10	0	0	0	0	inches
Object width * w	8	8	8	8		inches

Asterisked values are given, and the rest calculated

a scanner set at four different sampling frequencies. The calculated values are shown in Table 22.3. We see that the acuity of a (contact) scan at 600 spi is nearly as high as the best eye viewing at near distance.

22.12 Transfer Function

In communications engineering, the *transfer function* of a device relates its output to its input:

$$\text{output} = \phi(\text{input}) .$$

If the designated input range and output range are both normalized between 0 and 1, the transfer function can be plotted with input as abscissa and output as ordinate. When the device is processing an intensity value, such a plot is often called a *tone reproduction curve* (TRC).

A display monitor[4] has a built-in power-law transfer function of the form

$$L = v^{\gamma} ,$$

where v is the input voltage (normalized[5] between 0 and 1), γ is an exponent applied to the input, and L is the output luminance (normalized between 0 and 1). The

[4] In a CRT this is imposed by the physics of the electron guns. In an LCD, which works differently, the same function is imposed for compatibility with the once-dominant CRTs.

[5] Given a number N in the range N_{min} to N_{max}, it is normalized in the range 0 to 1 as

$$N_{\text{norm}} = (N - N_{\text{min}})/(N_{\text{max}} - N_{\text{min}}) .$$

Given a number N_{norm} normalized in the range 0 to 1, it is de-normalized in the range N_{min} to N_{max} as

$$N = (N_{\text{norm}} \times (N_{\text{max}} - N_{\text{min}})) + N_{\text{min}}$$

.

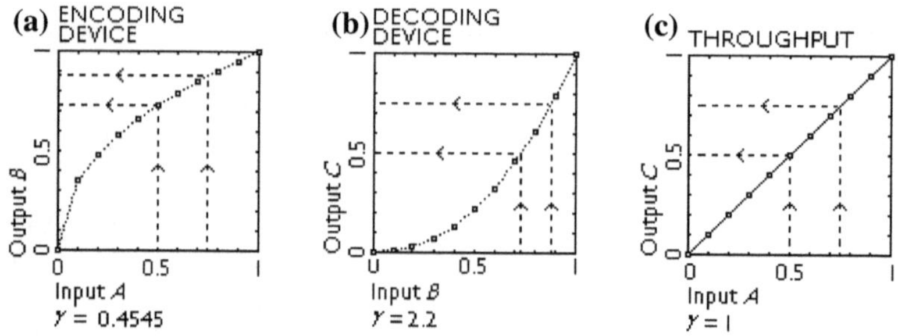

Fig. 22.6 Tone reproduction curves, $\gamma = 2.2$. **a** Normalized input A encoded to normalized output $B = A^{0.4545}$. **b** Input B decoded to output $C = B^{2.2}$. **c** When encoding is followed by decoding, final output C = initial input A

Windows operating system uses a standard display (or *decoding*) $\gamma = 2.2$ (possibly different in other environments, and adjustable in the operating software).

To calculate a TRC with power γ,

- given input value A and range A_{min} to A_{max},
- $A_{norm} = (A - A_{min})/A_{max}$,
- A^{γ}_{norm} = antilog(log(A_{norm}) $\times \gamma$),
- $B_{norm} = A^{\gamma}_{norm}$,
- given output range B_{min} to B_{max},
- output value $B = B_{min} + (B_{norm} \times (B_{max} - B_{min})$.

To counteract display gamma it is standard practice for scanners, cameras, and programs, to apply an *encoding* gamma with value $1/\gamma = 1/2.2 = 0.4545$. Figure 22.6a shows a TRC for an encoding device, Fig. 22.6b a TRC for a decoding device, and Fig. 22.6c the linear result when Fig. 22.6a is followed by Fig. 22.6b. For example, if a scanner registers mid-grey as 0.5 (normalized between 0 Black and 1 White), it will encode it as $0.5^{0.4545} = 0.73$, and a display will decode it as $0.73^{2.2} = 0.5$; that is, as registered. The tone-code passed from scanner to display for mid-grey will be, not $0.5 \times 255 = 128$, but $0.73 \times 255 = 186$.

Clearly, however, a TRC has no effect on the extremes 0 and 1 of the normalized range, but only on intermediate values. In a 1-bit image all intermediate intensities are thresholded to either 0 or 1, which are passed by a scanner as $I_2 = 0$ or 1.

22.13 Scanner Intensity Variables

In 1-bit scanning we have, per sample, the following intensity variables:

- fractional reflectance r from object,
- voltage v from sensor,
- encoded 8-bit intensity B G R from ADC,
- 1-bit intensity index I_2 from thresholder to file.

A sensor receives analogue reflectance r and outputs voltage v proportional to r. An ADC receives voltage v, calculates $i_{norm} = v_{norm}^{0.4545}$, and outputs 8-bit integer tone-code between 0 and 255 $I = i_{norm} \times 255$. A thresholder with threshold T receives I and outputs $I_2 = 0$ if $I \leq T$, or $I_2 = 1$ if $I > T$.

22.14 Optical Density

When constant illuminance $E = 1$ falls on a sample of an object, a fraction $0 \leq a \leq 1$ is absorbed and a fraction $0 \leq r \leq 1 = (1 - a)$ is reflected. *Optical density* $0 \leq d$ is a standard measure of absorption of incident light, defined as the logarithm, to base 10, of the ratio of illuminance $E = 1$ to reflectance r:

$$d = \log_{10}(1/(1 - a)) = \log_{10}(1/r)\,,$$

where d is optical density, a is fractional absorptance, and r is fractional reflectance. Since optical density is a logarithmic measure, it takes positive real values from 0 upwards. Practical values rarely exceed 1.7. We have

$$\begin{aligned} 10^d &= 1/r \\ E &= 1 = r \times 10^d \\ r &= 1/10^d\,. \end{aligned}$$

22.15 Step-wedge

A simple and convenient tool for measuring scanner performance as to intensity is a standard photographic *step-wedge* test card. A typical test card is shown in Fig. 22.7a. The step-wedge has (say) 18 steps at equal intervals of optical density, interval $d = 0.1$. The steps are marked 0 (White, $d = 0$) to 17 (Black, $d = 1.7$). Optical density is a logarithmic measure of absorptance, so $d = 0$ means that none of the incident light is absorbed, and all is reflected.[6] $d = 1$ means that 90 % is absorbed and 10 % reflected; $d = 2$ means that 99 % is absorbed and 1 % reflected.[7] The density markings on a step-wedge are thus the opposite of the digital convention 0 Black and 1 White.

[6]In practice a perfect reflecting surface is never achieved: the best laboratory White surface (thick magnesium oxide newly-deposited on metal) reflects about 98 % of incident light (density 0.009), and ordinary White paper about 85 % (density 0.07). In practice, optical density is taken as relative to the current good-enough White.

[7]In practice a perfect absorbing surface is never achieved: the best laboratory Black surface (Black velvet) reflects about 0.0025 % of incident light (density 2.6), and ordinary rough Black paper about 15 % (density 0.82), or 8 % (density 1.1) if glossy.

Fig. 22.7 Scanner step-wedge test. **a** 8-bit grey-scale scan of test card step-wedge. d is optical density (0 for density 0.0, 1 for density 0.1, 2 for density 0.2, ..., 17 for density 1.7). I is calculated encoded tone-code. **b** 1-bit scan of (**a**) with threshold $T = 192$. **c** 1-bit scan with threshold $T = 128$. **d** 1-bit scan with threshold $T = 64$

Table 22.4 shows the calculated values for a perfect step-wedge. The conversions are

$$
\begin{aligned}
r &= 1/10^d \\
d &= \log_{10}(1/r) \\
i &= r^{0.4545} \\
r &= i^{2.2} \\
I &= 255 \times i \\
i &= I/255 .
\end{aligned}
$$

22.16 Scanner Step-wedge Test

A scanner in 1-bit mode converts greys to White or Black according to the threshold setting. In the perfect case, we know (see Table 22.4) the reflectance r of each patch of a standard step-wedge, hence the voltage v output by the sensor, hence the 8-bit intensity I output by the ADC, hence the threshold T which should separate each patch from its neighbours. To measure the accuracy of a 1-bit scanner in capturing intensity,

Table 22.4 Calculated values for perfect step-wedge

d	r	i_{norm}	I
0	1	1	255
0.1	0.79	0.90	229
0.2	0.63	0.81	207
0.3	0.50	0.73	186
0.4	0.40	0.66	168
0.5	0.32	0.60	152
0.6	0.25	0.53	136
0.7	0.20	0.48	123
0.8	0.16	0.43	111
0.9	0.13	0.40	101
1	0.10	0.35	89
1.1	0.08	0.32	81
1.2	0.06	0.28	73
1.3	0.05	0.26	65
1.4	0.04	0.23	59
1.5	0.03	0.20	52
1.6	0.02	0.17	43
1.7	0.02	0.17	43

The columns are: Optical density d of step-wedge patch; reflectance $r = 1/10^d$; normalized intensity $i_{norm} = r^{0.4545}$; encoded 8-bit intensity $I = i_{norm} \times 255$

- scan a step-wedge at a series of threshold settings,
- compare each scan with the actual step-wedge,
- note how closely the actual scans agree with the calculated values in Table 22.4.

For example, Fig. 22.7a shows an 8-bit scan of a test card step-wedge, marked with the calculated encoded tone-codes I. Figure 22.7b is a 1-bit scan of Fig. 22.7a with threshold $T = 192$: it correctly separates $d = 0.2$ from 0.3. In Fig. 22.7c the threshold $T = 128$: it should separate $d = 0.6$ from 0.7 but actually separates 0.5 from 0.6. In Fig. 22.7d the threshold $T = 64$: it should equivocate on $d = 1.3$ but actually separates 1.0 from 1.1. A perfect intensity scan is hard to achieve, because of tiny variations in the step-wedge, the scanner, and the scanning conditions.

22.17 1-bit Scan of Text

When an scanned image is subsequently displayed on a screen (see Chap. 27), one scan sample corresponds to one display pixel. If the scan has been made at sampling frequency F_S samples per declared unit, and the display has a fixed frequency of F_P px per unit, the image will be displayed with a *magnification/diminution* m:

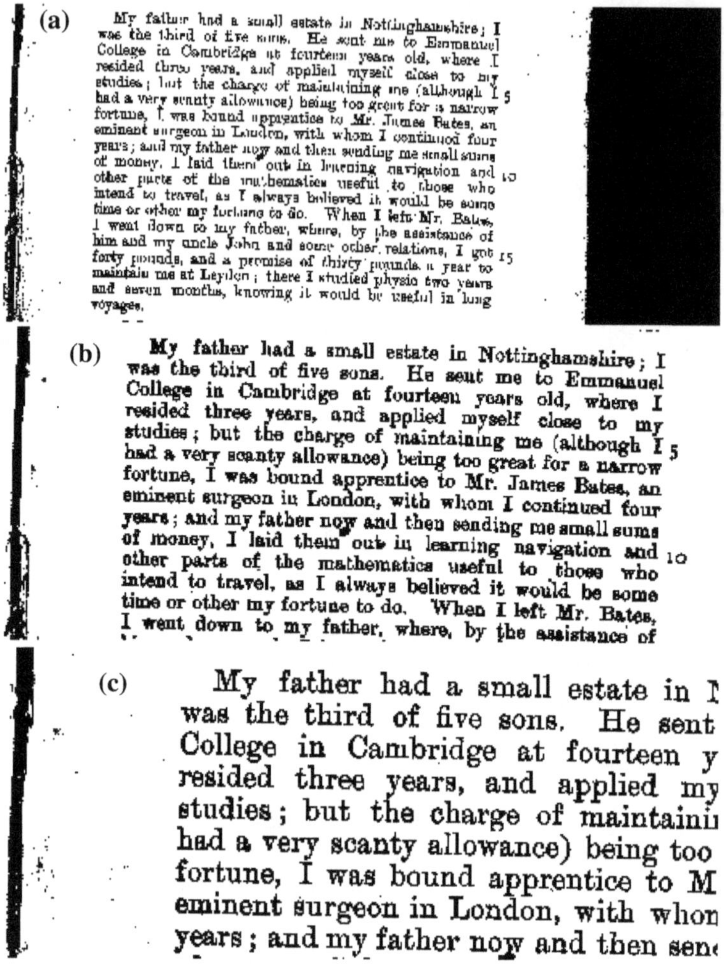

Fig. 22.8 1-bit scans of text. **a** Scan sampling frequency $F_S = 75$ spi; printout pixel frequency $F_P = 96$ ppi; magnification $m = 75/96 = 0.78$. **b** $F_S = 100$ spi; $F_P = 96$ ppi; $m = 100/96 = 1.04$. **c** $F_S = 150$ spi; $F_P = 96$ ppi; $m = 150/96 = 1.56$

$$m = F_S / F_P\,.$$

Similarly, when an scanned image is printed out (see Chap. 28), one scan sample corresponds to one print pixel. If the scan has been made at sampling frequency F_S samples per unit, and the printer is set at F_P px per unit, the image will be printed with a magnification/diminution m:

$$m = F_S / F_P\,.$$

For 1-bit scans of ordinary book text we usually want to display or print at about same-size, that is F_S should be set approximately equal to F_P. For example, Fig. 22.8 shows three 1-bit scans of the same text, each 446 × 200 px, scanned at sampling frequencies $F_S = 75$, 100, and 150 spi, with threshold $T = 128$, and printed at pixel frequency $F_P = 96$ ppi. The magnifications are approximately three-quarters, one, and one-and-a-half. In practice, it is usual to scan text at 300 spi and print at 300 ppi.

22.18 1-bit Scan of Grating

If a continuous-tone grating is scanned in 1-bit mode at a sampling frequency F_S which is the same as the grating frequency, the phase of sampling is significant. Figure 22.9 shows two extreme cases. On the left, the continuous grating Fig. 22.9a is scanned with sampling in exact anti-phase with the grating and thresholded to 1-bit, to produce an all-Black (or all-White) image Fig. 22.9e. On the right, the sampling is in exact phase with the grating, to produce an exact image. In practice, there are always small departures from such exactitude.

If a grating is scanned at a sampling frequency F_S which is close to the grating pixel frequency F_P, the scan shows a lower-frequency interference pattern (see Moiré effect in Sect. 23.18). For example, Fig. 22.10a shows a grating of equally-spaced Black and White bars each 1 px wide, with pixel count $P = 400$ px and pixel frequency $F_P = 81$ ppi. In Fig. 22.10b, Fig. 22.10a has been scanned at the slightly lower sampling frequency $F_S = 79$ spi. The sample count is $S = P \times F_S/F_P = 400 \times 79/81 = 390$. The scan has nine fewer pixels than the object, so there are nine interference bands.

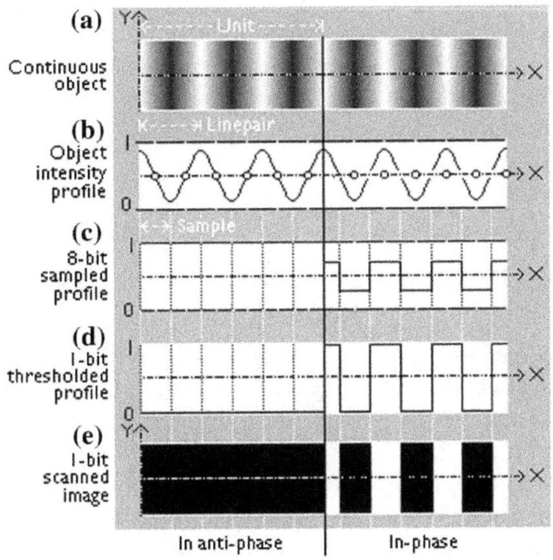

Fig. 22.9 Continuous grating and 1-bit scan. **a** Object grating with continuous grey-scale intensities. **b** Intensity profile of (**a**). **c** Intensity profile of 8-bit sampling of (**a**), in exact anti-phase on the left and exact phase on the right. **d** Intensity profile of (**c**) after thresholding to 1-bit. **e** Resulting 1-bit image of (**a**)

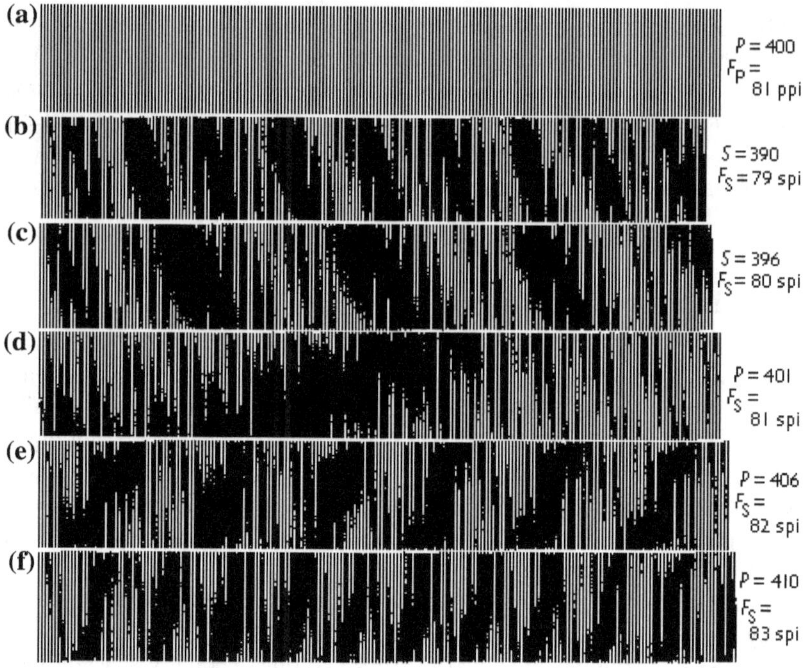

Fig. 22.10 Object grating and five scans. **a** Object grating with pixel count $P = 400\,\text{px}$, pixel frequency $F_P = 81\,\text{ppi}$. **b** Scan of (**a**) at sampling frequency $F_S = 79\,\text{spi}$, showing 9-pixel deficit. **c** Scan at sampling frequency $F_S = 80\,\text{spi}$, showing 4-pixel deficit. **d** Scan at sampling frequency $F_S = 81\,\text{spi}$, showing 1-pixel excess. **e** Scan at sampling frequency $F_S = 82\,\text{spi}$, showing 6-pixel excess. **f** Scan at sampling frequency $F_S = 83\,\text{spi}$, showing 10-pixel excess

In Fig. 22.10c, $F_S = 80\,\text{spi}$, and there are 4 interference bands. In Fig. 22.10d, $F_S = 81\,\text{spi}$, but there is a phase difference of 1 px, so the scan has one more pixel than the object and there is one interference band. In Fig. 22.10e, $F_S = 82\,\text{spi}$, so there are six interference bands. And in Fig. 22.10f, $F_S = 83\,\text{spi}$, so there are ten interference bands.

22.19 1-bit Scan of Half-tone

A half-tone image is composed of varying-sized dots on a square grid. The grid is in effect two gratings at right angles, with a two-dimensional frequency from about 80 dpi (for a newsprint image) to about 300 dpi (for a glossy magazine image). If a printed half-tone is scanned with a sampling frequency close to the half-tone dot frequency, Moiré interference occurs in two dimensions, as for gratings in one dimension described above. The usual recommendation to avoid such interference is to scan at a frequency at least twice the half-tone frequency, which gives a magnification of

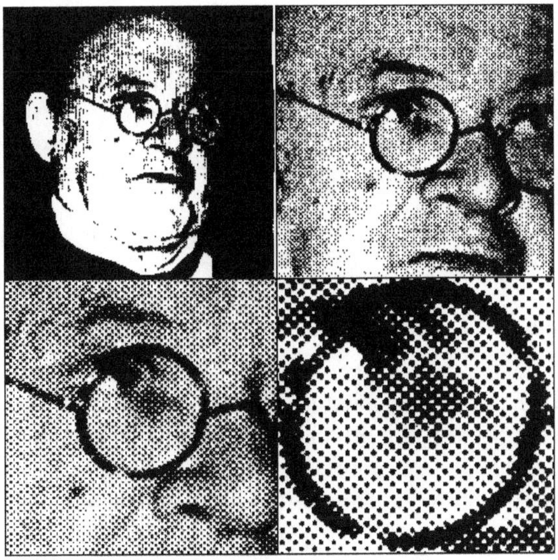

Fig. 22.11 1-bit scans of 80 dpi half-tone. **a** Scan sampling frequency $F_S = 100$ spi; printout pixel frequency $F_P = 96$ ppi; magnification $m = 100/96 = 1.04$. **b** $F_S = 200$ spi; $F_P = 96$ ppi; $m = 200/96 = 2.08$. **c** $F_S = 300$ spi; $F_P = 96$ ppi; $m = 300/96 = 3.12$. **d** $F_S = 600$ spi; $F_P = 96$ ppi; $m = 600/96 = 6.25$

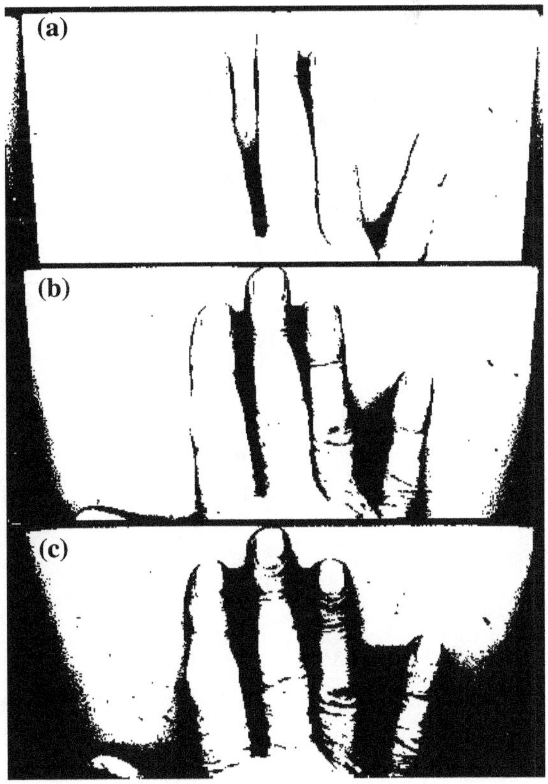

Fig. 22.12 1-bit scans of object. **a** Scan sample frequency $F_S = 50$ spi; printout pixel frequency $F_P = 96$ ppi; magnification $m = 100/96 = 0.52$. Threshold $T = 16$. **b** As (**a**), but $T = 24$. **c** As (**a**), but $T = 32$

at least 2. For example, Fig. 22.11 shows four scans of the same newspaper half-tone image, which has dot frequency 80 dpi. Each scan is 223 × 223 px, scanned at sampling frequencies $F_S = 100, 200, 300$, and 600 spi, with threshold $T = 128$, and all printed at pixel frequency F_P about 96 ppi. The magnifications are about 1, 2, 3, and 6. In Fig. 22.11a the original half-tone dots are replaced by the interference pattern (which may sometimes be quite pleasing). In Fig. 22.11b–d the original half-tone dots are resolved.

22.20 1-bit Scan of Object

If the object scanned is not flat but has relief, the scanner light-source will create chiaroscuro. A 1-bit scan will then give markedly different results according to where the threshold T is set. Figure 22.12 shows three scans of fingers, with threshold set at 16, 24, or 32. The setting of the threshold is very delicate.

References

1. ITU-R Recommendation BT 709-5 (2008) Parameter values for the HDTV standards for program and international exchange. http://www.itu.int/rec/R-REC-BT.709/en. Accessed 1 Feb 2014
2. Open Directory (2014) Scanners. http://www.dmoz.org/Computers/Hardware/Peripherals/Scanners/. Accessed 1 Feb 2014
3. Poynton CA (2009) Color FAQ. http://www.poynton.com/ColorFAQ.html. Accessed 1 Feb 2014

Chapter 23
1-bit Location Process

In digital imaging, computer processes operate on locations or intensities or both. A discrete location transformation changes all the locations of an image in some systematic way. A location transformation can be done geometrically, or numerically, or computationally by matrix arithmetic. Six elementary 1-bit location transformations are defined, with examples and with a program. Symmetry groups are defined, and examples of the rotation, single-translation, and double-translation groups are shown. Gratings are defined, and examples of superimposition of gratings are show. The locations of two images can be combined by program. The count of locations in an image can be reduced (and pixel size increased) by program.

23.1 Flowchart

The flowchart (Fig. 23.1) is for general guidance in this chapter.

23.2 Discrete Transformation

A *discrete transformation* is an operation on a discrete image which changes the entire image in some systematic way. If we have a 1-bit image in a file (produced by program, interactive GUI, scanner or camera), we may apply,

- a transformation of image locations (this chapter),
- a transformation of image intensities (Chap. 24),
- a Fourier transformation of locations to frequencies and back again (Chap. 25).

These are plane transformations of plane coordinates (x, y). For space transformations of space coordinates (x, y, z) see Chap. 7. For projections from space coordinates to plane coordinates see Chap. 8.

A. Parkin, *Digital Imaging Primer*, DOI 10.1007/978-3-540-85619-1_23

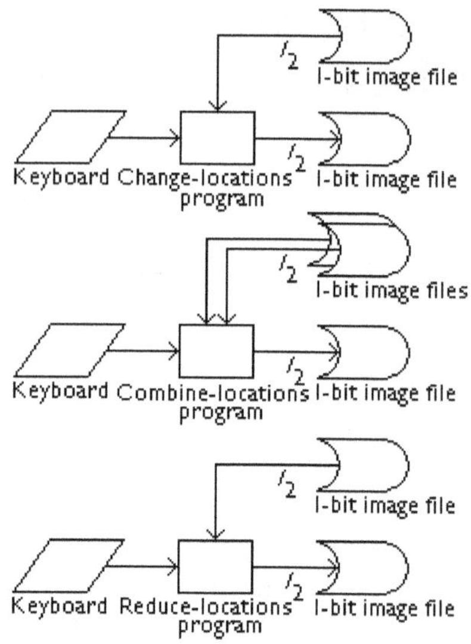

Fig. 23.1 1-bit location process. A change-locations program reads a given 1-bit image file, changes the location coordinates, and writes a new image file. A combine-locations program reads two 1-bit files and writes a new combined-image file. A reduce-location program reads a given 1-bit image file and writes a new file with fewer pixels

23.3 1-bit Location Transformation

Geometrically, a transformation of image locations is a quasi-physical operation on an image, such as turning it over on its back (the same as reflecting it in a fixed mirror-line), or moving it to a new place, or rotating it about some fixed pole. (see also space location transformation, which operates on solid objects rather than flat images, Sect. 7.5). Location transformations are usually grouped according to the geometric properties which are *invariant*, that is, preserved under the operation. Thus,

- *rigid motions* preserve lengths and angles: reflection, translation, rotation,
- *conformal transformations* preserve angles: dilation, inversion,
- *projective transformations* preserve linearity: shearing, projection.

Numerically, a location transformation changes the values of the location coordinates (x, y) of every pixel in a plane image to new values (x', y') by applying some function-ϕ:

$$(x', y') = \phi(x, y) .$$

Computationally, we use 3×3 integer matrices (see Sect. 3.11). To do this, we write the Cartesian coordinates of each pixel (x, y) as homogeneous coordinates $(x, y, 1)$, then multiply by a transformation matrix, and finally ignore the third coordinate 1 to get the transformed coordinates (x', y'). The general form of the calculation is

$$(x', y', 1) = (x, y, 1)\begin{pmatrix} A & B & C \\ D & E & F \\ G & H & I \end{pmatrix}$$
$$= ((xA + yD + G), (xB + yE + H), (xC + yF + I)) .$$

The reason for introducing homogeneous coordinates is to allow efficient and uniform treatment of all the transformations, including translation. We define below:

- *reflection* of an image about a mirror-line,
- *translation* of an image along a direction,
- *rotation* of an image about a pole,
- *dilation* of an image about an origin,
- *shearing* of an image along an axis,
- *inversion* of an image about a circle.

A GUI image editor, such as PAINT, provides a convenient wrapper for some of the transformations. Typical names for transformations in an editor are

- *flip* image about a mirror-line (Reflection),
- *copy and Paste* image to new location (Translation),
- *rotate* image about a pole,
- *stretch/Shrink* image to smaller or larger size (Dilation),
- *skew* image along an axis (Shearing).

23.4 Reflection

Reflection moves each pixel of an image from one side of a mirror-line to the other side. Pixels on the mirror-line remain fixed. The *sense* of the image (left/right, clockwise/anticlockwise) is reversed.

Figure 23.2a shows a working window and an image within it; Fig. 23.2b shows the image reflected in a mirror-line through the origin at 0°, that is the x-axis itself; and Fig. 23.2c shows it reflected in a mirror-line through the origin at 30°. In this and the following figures, the working window is taken as a first-quadrant frame, with origin at bottom left, x increasing from 0 to maximum to the right, and y increasing from 0 to maximum upwards. The ground is taken as intensity 0 Black, and the figure as intensity 1 White.

Numerically, an explicit equation which reflects an image in a mirror-line through the origin and inclined at $\alpha°$ to the x-axis is

$$(x', y') = (\cos(\alpha)x + \sin(\alpha)x\,, \quad \sin(\alpha)y - \cos(\alpha)y)\,.$$

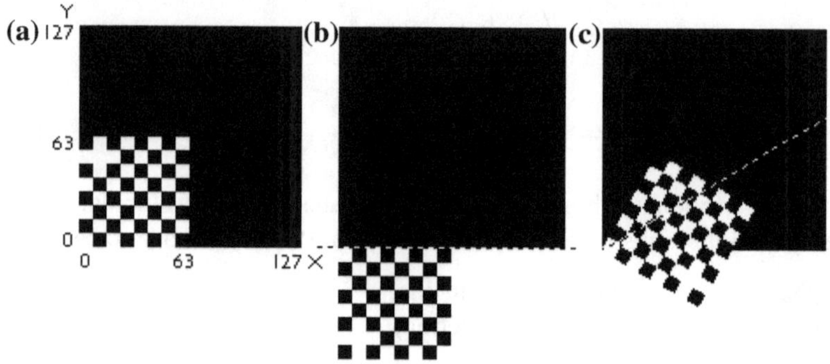

Fig. 23.2 Reflection. **a** 1-bit image of 64 × 64 px displayed in a 128 × 128 first-quadrant working window. **b** Image reflected in mirror-line (*dashed*) through the origin at 0°; the reflection is wholly outside the window. **c** Image reflected in mirror-line (*dashed*) through the origin at 30°; the reflection is partly outside the window

For example, to reflect an image in the x-axis, $\alpha = 0$, so

$$(x', y') = (\cos(0)x + \sin(0)x\,, \quad \sin(0)y - \cos(0)y)$$
$$(x', y') = (x, -y)\,.$$

The reflection operates from the origin of the (first-quadrant) reference frame, so the whole image in the first quadrant is carried into the fourth quadrant outside the frame.

Computationally, reflection in a mirror-line through the origin at $\alpha°$ can be achieved by multiplying by a 3 × 3 reflection matrix:

$$(x', y', 1) = (x, y, 1)\begin{pmatrix} \cos(2\alpha) & \sin(2\alpha) & 0 \\ \sin(2\alpha) & -\cos(2\alpha) & 0 \\ 0 & 0 & 1 \end{pmatrix}.$$

For example, to reflect an image in a mirror-line through the origin at $\alpha = 0°$, as shown in Fig. 23.2b:

$$(x', y', 1) = (x, y, 1)\begin{pmatrix} 1 & 0 & 0 \\ 0 & -1 & 0 \\ 0 & 0 & 1 \end{pmatrix} = (x, -y, 1)\,.$$

Or, to reflect an image in a mirror-line through the origin at $\alpha = 30°$, as shown in Fig. 23.2c:

$$(x', y', 1) = (x, y, 1)\begin{pmatrix} 0.5 & 0.867 & 0 \\ 0.867 & -0.5 & 0 \\ 0 & 0 & 1 \end{pmatrix}$$

as shown in Fig. 23.2c.

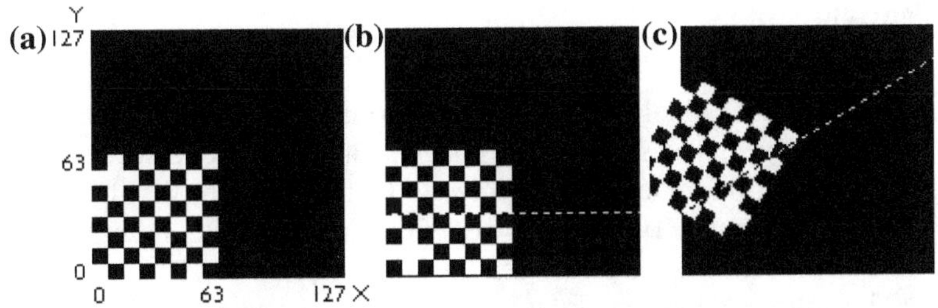

Fig. 23.3 Reflection. **a** 1-bit image of 64 × 64 px displayed in a 128 × 128 first-quadrant working window. **b** Image reflected in mirror-line (*dashed*) through (0, 32) at 0°. **c** Image reflected in mirror-line (*dashed*) through (0, 32) at 30°; reflection is partly outside the window

Again, Fig. 23.3a shows a working window and an image within it; Fig. 23.3b shows the image reflected in a mirror-line through (0, 32) at 0°, that is parallel to the x-axis; and Fig. 23.3c shows it reflected in a mirror-line through (0, 32) at 30°. To reflect an image in a mirror-line through $(0, H)$ we have

$$(x', y', 1) = (x, y - H, 1) \begin{pmatrix} \cos(2\alpha) & \sin(2\alpha) & 0 \\ \sin(2\alpha) & -\cos(2\alpha) & 0 \\ 0 & H & 1 \end{pmatrix} .$$

For example, to reflect an image in a mirror-line through (0, 32) at $\alpha = 0°$, as shown in Fig. 23.3b:

$$(x', y', 1) = (x, y - 32, 1) \begin{pmatrix} 1 & 0 & 0 \\ 0 & -1 & 0 \\ 0 & 32 & 1 \end{pmatrix} .$$

Or, to reflect an image in a mirror-line through (0, 32) at $\alpha = 30°$, as shown in Fig. 23.3c:

$$(x', y', 1) = (x, y - 32, 1) \begin{pmatrix} 0.5 & 0.867 & 0 \\ 0.867 & -0.5 & 0 \\ 0 & 32 & 1 \end{pmatrix} .$$

Graphically, a GUI editor reflects by selecting a sub-window, and *flipping* it horizontally (in x) and/or vertically (in y). The reflections are about the centre-lines of the selected sub-window. Notice that the usual GUI editor window has origin at top left, x increasing from 0 to maximum to the right, and y increasing from 0 to maximum downwards; and the ground is usually intensity 1 White with figure 0 Black.

23.5 Translation

Translation moves each pixel of an image by a certain distance in a certain direction. No pixel remains fixed. The sense of the image is preserved. Two successive reflections in the same mirror-line give the *identity* transformation; that is, leaving the original image unchanged. Two successive reflections in parallel mirror-lines give a translation by twice the distance between the mirror-lines.

Figure 23.4a shows a first-quadrant working window and a 1-bit image within it; Fig. 23.4b shows the image translated to the right; and Fig. 23.4c shows the image translated to the right and upwards.

Numerically, an explicit equation which translates an image by G px parallel to the x-axis and H px parallel to the y-axis is

$$(x', Yy) = (x + G, y + H).$$

When $G < 0$, the image moves to the left in the frame; when $G > 0$, the image moves to the right in the frame. When $H < 0$, the image moves down in the frame; when $H > 0$, the image moves up in the frame. When $G = 0$ and $H = 0$ the image is unchanged.

Computationally, an equation for translation is

$$(x', y', 1) = (x, y, 1) \begin{pmatrix} 1 & 0 & 0 \\ 0 & 1 & 0 \\ G & H & 1 \end{pmatrix}.$$

For example, to translate an image in the x-direction by $G = 32$ px, as shown in Fig. 23.4b:

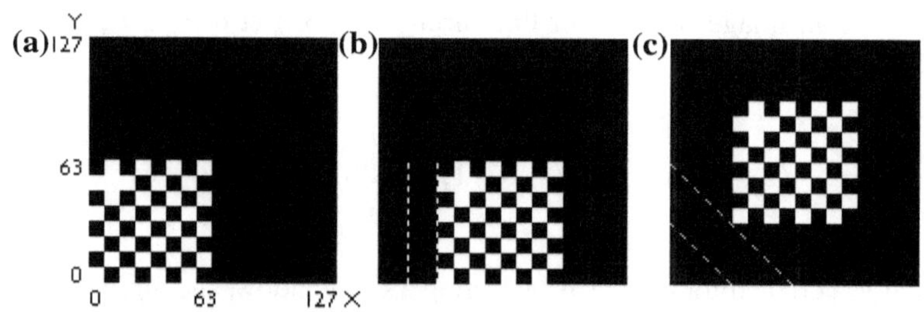

Fig. 23.4 Translation. **a** 1-bit image of 64 × 64 px displayed in a 128 × 128 working window. **b** Image translated by 32 px rightwards in the same frame. This is the same as two successive reflections in mirror-lines (*dashed*) 16 px apart. **c** Image translated by 32 px rightwards and 32 px upwards in the same frame. This is the same as two successive reflections in mirror-lines (*dashed*) 22 px apart

$$(x', y', 1) = (x, y, 1) \begin{pmatrix} 1 & 0 & 0 \\ 0 & 1 & 0 \\ 32 & 0 & 1 \end{pmatrix} = (x + 32, y, 1) \,.$$

Or, to translate an image in the x-direction by $G = 32$ px and in the y-direction by $H = 32$ px, as shown in Fig. 23.4(c):

$$(x', y', 1) = (x, y, 1) \begin{pmatrix} 1 & 0 & 0 \\ 0 & 1 & 0 \\ 32 & 32 & 1 \end{pmatrix} = (x + 32, y + 32, 1) \,.$$

Graphically, a GUI editor translates by selecting a sub-window, copying it, and pasting it to a new location (or by dragging and releasing).

23.6 Rotation

Rotation moves each pixel of an image through a certain angular arc of a circle whose centre is the pole of the rotation. The pole remains fixed. The sense of the image is preserved. Angles are conventionally measured anticlockwise from the X-axis. Two successive reflections in mirror-lines which meet give a rotation by twice the angle between the mirror-lines. The meet of the mirror-lines is the pole of the rotation.

Figure 23.5a shows a first-quadrant working window and a 1-bit image within it. Figure 23.5b shows the image rotated about the origin (0, 0) by 30°; and Fig. 23.5c shows the image rotated about (0, 32) by 30°.

Numerically, an explicit equation which rotates an image about the origin (0, 0) by $\alpha°$ is

$$(x', y') = ((\cos(\alpha) - \sin(\alpha))x, \sin(\alpha)x + \cos(\alpha)y) \,,$$

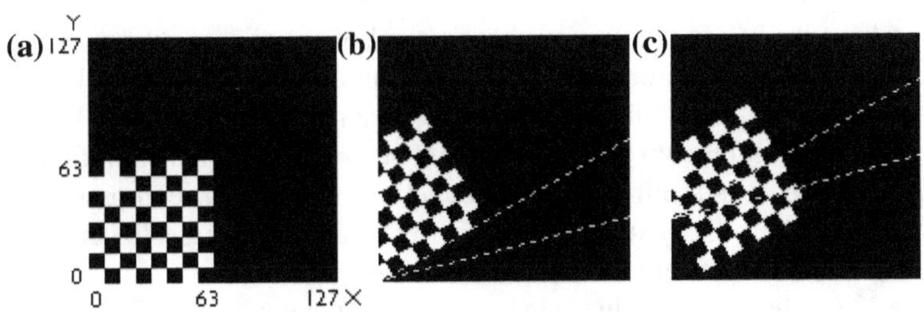

Fig. 23.5 Rotation. **a** 1-bit image of 64 × 64 px displayed in a 128 × 128 working window. **b** Image rotated about origin (0, 0) by 30° in the same window. This is the same as two successive reflections in mirror-lines (*dashed*) 15° apart. **c** Image rotated about (0, 32) by 30°. This is the same as two successive reflections in mirror-lines (*dashed*) 15° apart

where α is angle measured from the positive x-axis. When $\alpha < 0$, the image is rotated clockwise; when $\alpha > 1$, the image is rotated anticlockwise. To rotate about any pole (G, H), first translate (G, H) to (0, 0), then rotate about origin, then translate (0, 0) back to (G, H)

$$(x', y') = ((\cos(\alpha) - \sin(\alpha))(x - G), (\sin(\alpha)x + \cos(\alpha))(y - H)) .$$

Computationally, an equation for rotation about (G, H) by $\alpha°$ is

$$(x', y', 1) = (x - G, y - H, 1) \begin{pmatrix} \cos(\alpha) & -\sin(\alpha) & 0 \\ \sin(\alpha) & \cos(\alpha) & 0 \\ G & H & 1 \end{pmatrix} .$$

For example, to rotate about the origin by 30°, as shown in Fig. 23.5b:

$$(x', y', 1) = (x, y, 1) \begin{pmatrix} 0.867 & -0.5 & 0 \\ 0.5 & 0.867 & 0 \\ 0 & 0 & 1 \end{pmatrix} .$$

Or, to rotate about (0, 32) by 30°, as shown in Fig. 23.5c:

$$(x', y', 1) = (x - 0, y - 32, 1) \begin{pmatrix} 0.867 & -0.5 & 0 \\ 0.5 & 0.867 & 0 \\ 0 & 32 & 1 \end{pmatrix} .$$

Graphically, a GUI editor rotates by selecting a sub-window, and setting 90, 180 or 270° about the centre of the sub-window. Some editors allow any number of degrees.

23.7 Dilation

Dilation about the origin (0, 0) moves each pixel of an image in the x-direction by a certain proportion of its original x-distance, and in the y-direction by a certain proportion of its original y-distance. The origin remains fixed. If the dilation factor is positive, it is a magnification or enlargement or stretch or expansion. If the dilation factor is negative, it is a diminution or reduction or shrink or compression. If the dilation is equal in x and y it is a *similarity*: the sense and angles of the image are preserved, but all lengths are increased or decreased.

Figure 23.6a shows a working window and an image within it; Fig. 23.6b shows the image dilated by a factor of 0.5 (that is, shrunk to half size) in the x-direction; and Fig. 23.6c shows the image dilated by a factor of 2 (that is stretched to twice size) in both x- and y-directions.

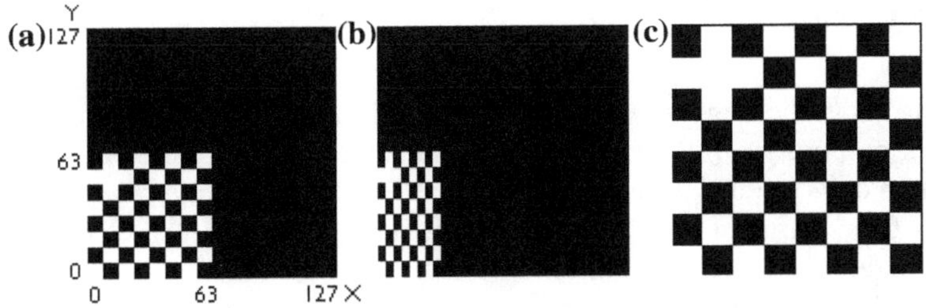

Fig. 23.6 Dilation. **a** 1-bit image of 64×64 px displayed in a 128×128 working window. **b** Image dilated by 0.5 in x-direction. **c** Image dilated by 2 in both x- and y-directions

Numerically, an equation which dilates an image by factor a in the x-direction is

$$(x', y') = ((a \times x), y)$$

and by factor e in the y-direction is

$$(x', y') = (x, (e \times y)) .$$

When $a < e$, the image is compressed left to right; when $a > e$, the image is expanded bottom to top. When $a = e = 1$, the image is unchanged.

Computationally, an equation for dilation about the origin (0, 0) by a factor a in the x-direction and by a factor e in the y-direction:

$$(x', y', 1) = (x, y, 1) \begin{pmatrix} a & 0 & 0 \\ 0 & e & 0 \\ 0 & 0 & 1 \end{pmatrix} .$$

For example, to dilate about the origin by $a = 0.5$, as shown in Fig. 23.6b:

$$(x', y', 1) = (x, y, 1) \begin{pmatrix} 0.5 & 0 & 0 \\ 0 & 1 & 0 \\ 0 & 0 & 1 \end{pmatrix} .$$

Or, to to dilate about the origin by $a = 2$ and by $e = 2$, as shown in Fig. 23.6c:

$$(x', y', 1) = (x, y, 1) \begin{pmatrix} 2 & 0 & 0 \\ 0 & 2 & 0 \\ 0 & 0 & 1 \end{pmatrix}$$

Graphically, a GUI editor dilates by selecting a sub-window, and *shrinking/stretching* by a percentage input from the keyboard. Shrinking omits pixels, so may give some possibly unexpected changes to the image. Stretching repeats pixels, so may take part of the image outside the working window. (see also Sect. 23.21 below.)

23.8 Shearing

Shearing about the origin (0,0) moves each pixel of an image in the x-direction by a certain proportion of its original y-distance, and in the y-direction by a certain proportion of its original x-distance. The effect is a slanting of the image. The origin remains fixed. Straight lines remain straight lines, and the sense of the image is preserved, but not lengths or angles.

Figure 23.7a shows a working window and an image within it; Fig. 23.7b shows the image sheared by a factor of $0.5y$ in the x-direction; and Fig. 23.7c shows the image sheared by a factor of $0.5y$ in the x-direction and by a factor of $0.5x$ in the y-direction.

Numerically, an equation which shears an image by factor dY in the x-direction and by factor bx in the y-direction is

$$(x', y') = (x + (d * y), y + (b * x)) .$$

Computationally, an equation for shearing by factor dy in the x-direction and by factor bx in the y-direction is

$$(x', y', 1) = (x, y, 1) \begin{pmatrix} 1 & b & 0 \\ d & 1 & 0 \\ 0 & 0 & 1 \end{pmatrix} .$$

For example, to shear by $d = 0.5$, as shown in Fig. 23.7b:

$$(x', y', 1) = (x, y, 1) \begin{pmatrix} 1 & 0 & 0 \\ 0.5 & 1 & 0 \\ 0 & 0 & 1 \end{pmatrix} .$$

Or, to shear by $d = 0.5$ and by $b = 0.5$, as shown in Fig. 23.7c:

$$(x', y', 1) = (x, y, 1) \begin{pmatrix} 1 & 0.5 & 0 \\ 0.5 & 1 & 0 \\ 0 & 0 & 1 \end{pmatrix} .$$

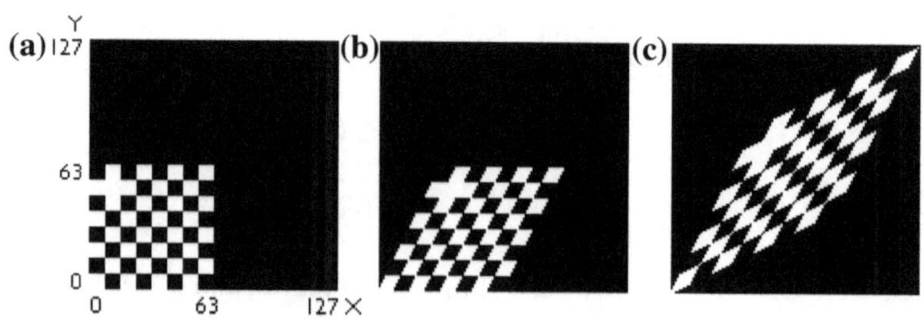

Fig. 23.7 Shearing. **a** 1-bit image of 64 × 64 px displayed in a 128 × 128 working window. **b** Image sheared by $0.5Y$ in x-direction. **c** Image sheared by $0.5y$ in x-direction and by $0.5x$ in y-direction

Graphically, a GUI editor shears by selecting a sub-window, and *skewing* it horizontally (in x) and/or vertically (in y) by so many degrees set from the keyboard. The skews are about the centre-lines of the selected sub-window, so may take some of the image outside the working window.

23.9 Inversion

Inversion [3] about a circle with centre C and radius K moves each pixel Q of an image to a location Q′ so that

$$CQ * CQ' = K^2$$

as shown in Fig. 23.8. Parallel straight lines become circles through C. Centre C remains fixed. Angles are preserved, but sense is reversed. Inversion can be thought of as reflection in a circle.

Figure 23.9a shows a working window and an image within it; Fig. 23.9b shows the image inverted about a circle with centre C = (0, 0) and radius $K = 32$ px; and

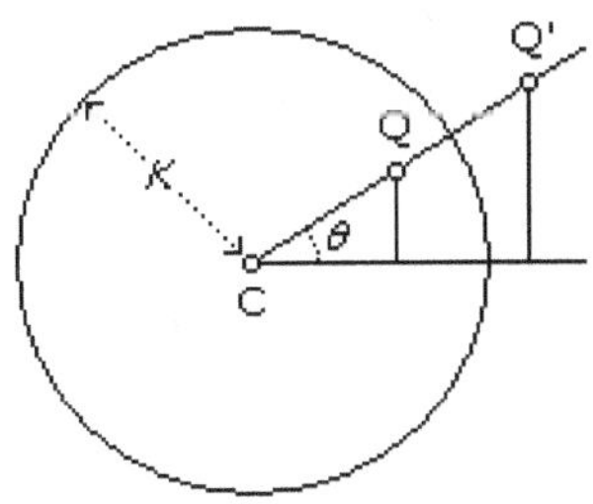

Fig. 23.8 Inversion about circle with centre C and radius K. Pixel Q goes to Q′, so that CQ * CQ′ = K^2

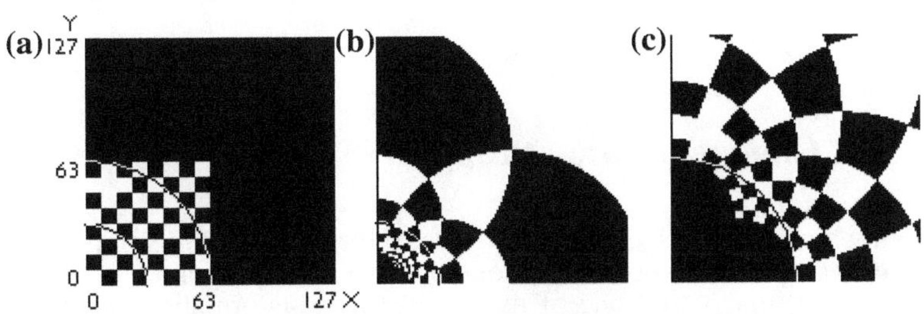

Fig. 23.9 Inversion about circle centred at origin. **a** 1-bit image of 64 × 64 px displayed in a 128 × 128 working window. **b** Image inverted about circle with centre C = (0, 0) and radius $K = 32$ px. **c** Image inverted about circle with centre C = (0, 0) and radius $K = 64$ px

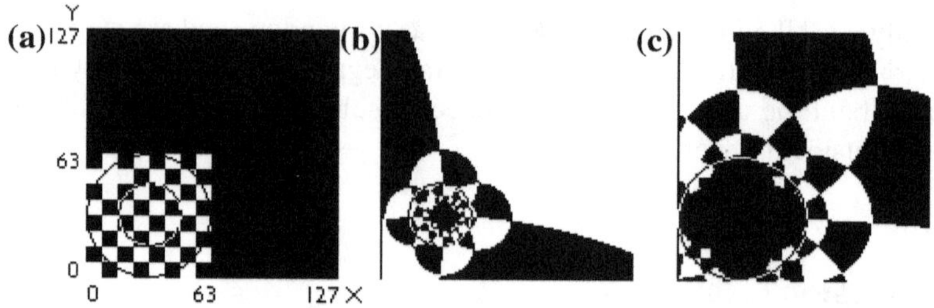

Fig. 23.10 Inversion about *circle* centred anywhere. **a** 1-bit image of 64 × 64 px displayed in a 128 × 128 working window. **b** Image inverted about circle with centre C = (32, 32) and radius $K = 16$ px. **c** Image inverted about circle with centre C = (32, 32) and radius $K = 32$ px

Fig. 23.9c shows the image inverted about a circle with centre C = (0, 0) and radius $K = 64$ px.

Figure 23.10a shows the same source image; Fig. 23.10b shows the image inverted about a circle with centre C = (32, 32) and radius $K = 16$ px; and Fig. 23.10c shows the image inverted about a circle with centre C = (32, 32) and radius $K = 32$ px.

Computationally, an algorithm which inverts each pixel (x, y) of an image to (x', y') is

```
comment CX, CY are coords of given circle centre
comment K is given circle radius
comment QX, QY are coords of source pixel
comment QQX, QQY are coords of required inverse pixel
integer CX, CY, K, QX, QY, QQX, QQY
begin
    THETA = ATAN(QY/QX)
    CQ = QX / COS(THETA)
    CQQ = K^2 / CQ
    QQX = (COS(THETA) * CQQ) + CX
    QQY = (SIN(THETA) * CQQ) + CY
end
```

23.10 1-bit Change-location Program

A program which performs all of the above transformations of location is described below (see SpringerLink web-site for source code). This program lets the user pick any 1-bit source image, and one of the geometric transformations of it. It uses a matrix equation for each transformation except inversion where it uses the algorithm given above.

```
REM  **  Program CHANLOC1 to change 1-bit image
REM  **     locations. User enters filenames.
```

```
REM  **     Program reads given "before" image from
REM  **     1-bit .BMP file #1 into array IMAGEA.
REM  **     User picks transformation and enters
REM  **     constants. Program processes pixel by
REM  **     pixel from IMAGEA to array IMAGEB,
REM  **     setting old intensity at new location.
REM  **     Program writes "after" image from
REM  **     IMAGEB to 1-bit .BMP file #2.
REM  **  Main program
REM  **  Enter old file and new file names
         CALL RHAWHB1
         CALL RDATA1
         CALL GEOMTRAN
         CALL WDATB1
         CALL REVHB1
END
         SUB GEOMTRAN
REM  **  Sub-program to enter constants for transformation
REM  **  wanted, set up transformation matrix, and
REM  **  multiply coordinates of each pixel from
REM  **  IMAGEA to IMAGEB.
REM  **  Reflection
REM  **  Translation
REM  **  Rotation
REM  **  Dilation
REM  **  Shear
REM  **  Inversion
END SUB
         SUB PUTBYTE (INTE%)
REM  **  Subprogram to put 1 byte in file #2.
END SUB
         SUB PUTI (INTE)
REM  **  Puts integer value passed into binary file as
REM  **  2 bytes and increments pointer.
END SUB
         SUB PUTL (INTL&)
REM  **  Puts long integer into binary file #2 as 4 bytes
END SUB
         SUB RDATA1
REM  **  Subprogram to read 1-bit .BMP file #1 and load
REM  **  old image data into array IMAGEA. The data
REM  **  are 0 or 1 packed in 8s.
END SUB
         SUB REVHB1
REM  **  Revises header of file #2 with values in bytes.
```

```
END SUB
              SUB RHAWHB1
REM   **      Subprogram to read header and colour-table of
REM   **      1-bit .BMP input file #1, and write same
REM   **      header and colour-table of output file
REM   **      #2 (to be revised later).
END SUB
              SUB WDATB1
REM   **      Subprogram to write data from array IMAGEB to
REM   **      file #2. Array locations are converted as groups
REM   **      of 8 to byte. Finish row on a long (4-byte)
REM   **      boundary.
END SUB
```

23.11 Sequential Transformation

Elementary transformations may be performed in sequence, the second on the result of the first, the third on the result of the second, and so on. Sequential transformations are not, in general, commutative. For example, Fig. 23.11a shows source image; Fig. 23.11b after rotation by 30° followed by translation by 32 px in x; and Fig. 23.11c after translation by 32 px followed by rotation by 30°.

23.12 Symmetry Group

A *symmetrical* image is one which is congruent with itself under one or more *symmetry operations*. In the plane, there are four symmetry operations:

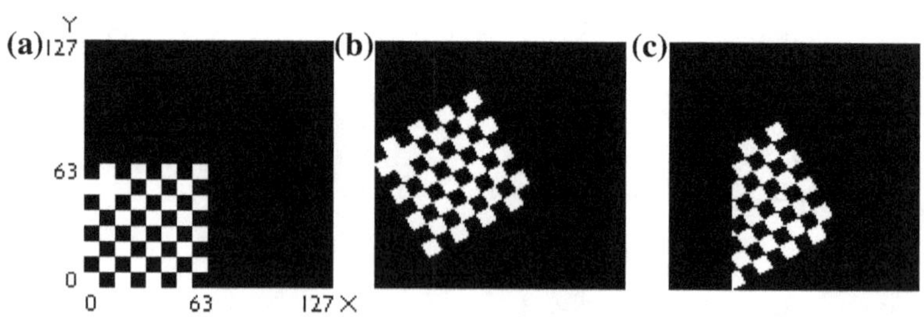

Fig. 23.11 Sequential transformation. **a** Source image. **b** Image (**a**) first rotated by 30° about origin, then translated by 32 px in X. **c** Image (**a**) first translated by 32 px in X, then rotated by 30° about origin

- N-fold rotation about a given pole by $360/N°$,
- reflection about a given mirror-line,
- translation in a given direction by a given distance,
- glide reflection, a combination of translation along and reflection about a given glide mirror-line.

For example, an equilateral triangle has three-fold rotation symmetry about its centre (by $360/3 = 120°$), and reflection symmetry about three mirror-lines (its three altitudes).

A *group* is a set of elements E and an operation $\oplus$ with the following properties:

- *closure*: if E_1 and E_2 are elements, then $E_1 \oplus E_2$ is an element,
- *associativity* $(E_1 \oplus E_2) \oplus E_3 = E_1 \oplus (E_2 \oplus E_3)$,
- *identity*: there is an element I such that, for any element E, $I \oplus E = E \oplus I = E$,
- *invertibility*: for every element E there is an inverse element E_{-1} such that $E \oplus E_{-1} = E_{-1} \oplus E = I$.

A plane *symmetry group* contains all the images which can be formed from an initial (asymmetrical) seed image by any number of applications of one or more of the plane symmetry operations. All the images belonging to a symmetry group have the same symmetry properties, though their detailed appearances vary according to the seed used.

23.13 Rotation Group

A plane *rotation symmetry group* (also called a cyclic group, or rosette group) is specified by

- a fixed pole,
- an integer N,
- and possibly a fixed mirror-line through the fixed pole.

The symmetry *cell* is a sector with angle $360/N°$ at the pole, and arbitrary radius. If there is no mirror-line, the initial *seed* is any (asymmetrical) image occupying the cell. The symmetry pattern is then the circle formed by N rotations of the cell about the pole by $360/N°$. For example, the upper row of Fig. 23.12 shows a seed image rotated by 360/1, 360/2, 360/3, and $360/4°$. A convenient notation is r for rotation, followed by integer N. If there is a mirror-line, it bisects the cell. The seed is then any (asymmetrical) image occupying the half-cell, that is to say the sector with angle $360/(2N)°$. For example, the lower row of Fig. 23.12 shows a seed reflected then rotated by 360/1, 360/2, 360/3, and $360/4°$. The notation adds m for mirror.

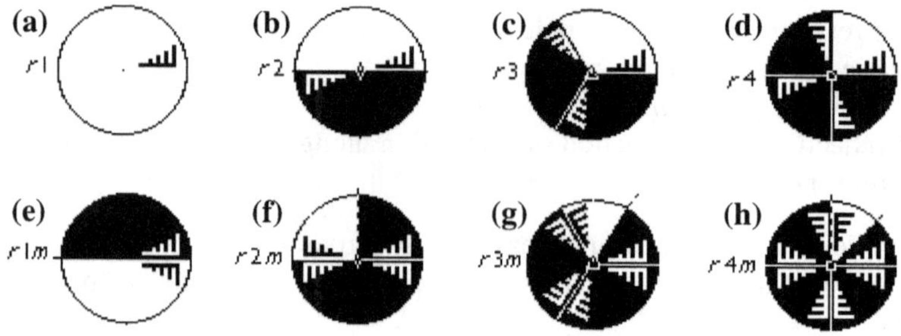

Fig. 23.12 Rotation symmetry group. In the upper row, the seed is the Black figure on White ground; and the cell is the same as the seed. **a** $r1$ rotation by $360/1 = 360°$ (the identity). **b** $r2$ rotation by $360/2 = 180°$. **c** $r3$ rotation by $360/3 = 120°$. **d** $r4$ rotation by $360/4 = 90°$. In the lower row, the seed is the White figure on Black ground. The seed is reflected about a mirror-line (*dashed*) at $360/(2N)°$, then rotated by **e** 360°, **f** 180°, **g** 120°, and **h** ≈90°

23.14 Single-Translation Group

A plane *single-translation symmetry group* (also called a frieze group) is specified by

- a given direction and distance of translation,
- and possibly a mirror-line parallel with the direction of translation,
- and possibly a mirror-line orthogonal to the direction of translation,
- and possibly a glide-line parallel to the direction of translation,
- and possibly a two-fold rotation.

The initial seed is any (asymmetrical) image occupying a rectangle with width the distance of translation and arbitrary height. The cell is formed by applying the possible reflections, glide reflections and rotation to the seed. The symmetry pattern is then the band formed by repeated translations of the cell. It turns out that there are just seven such patterns [6], as shown in Fig. 23.13. A convenient symbol is

- t for translation; followed by,
- 1 for no rotation (identity) or 2 for rotation by $360/2 = 180°$; followed by,
- 1 for no reflection parallel to direction of translation or m for mirror-line or g for glide-line parallel to translation,
- 1 for no reflection orthogonal to direction of translation or m for mirror-line or g for glide-line orthogonal to translation.

The seven single-translation symmetry groups are,

(1) $t111$: No reflection or rotation. Cell is same as seed.
(2) $t1m1$: Reflection parallel to translation. Cell is twice the size of seed.
(3) $t11m$: Reflection orthogonal to translation. Cell is twice the size of seed.
(4) $t1mm$: Reflections parallel and orthogonal to translation. Cell is four times the size of seed.
(5) $t211$: Rotation by 180°. Cell is twice the size of seed.

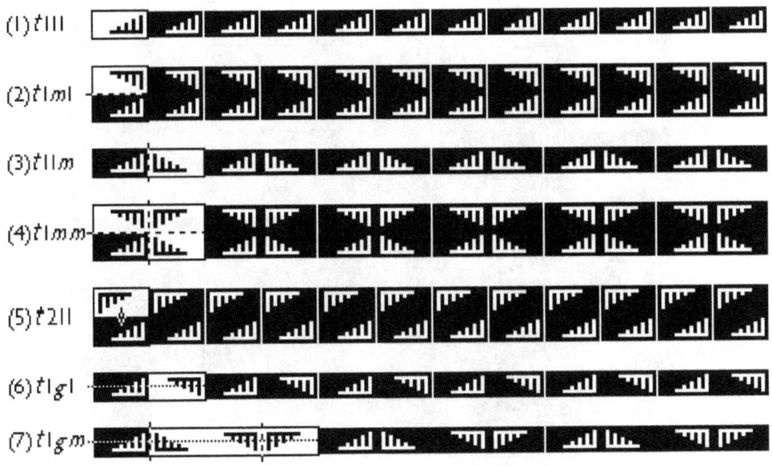

Fig. 23.13 Single-translation symmetry groups. The initial seed image is the White figure on Black ground. The symmetry cell is the White rectangle at the left

(6) $t1g1$: Glide reflection parallel to translation. Cell is twice the size of seed.
(7) $t1gm$: Glide reflection parallel to and reflection orthogonal to translation. Cell is four times the size of seed.

23.15 Double-Translation Group

A plane *double-translation symmetry group* (often called a wallpaper group) is specified by

- two given directions and distances of translations,
- and possibly a mirror-line parallel to a translation,

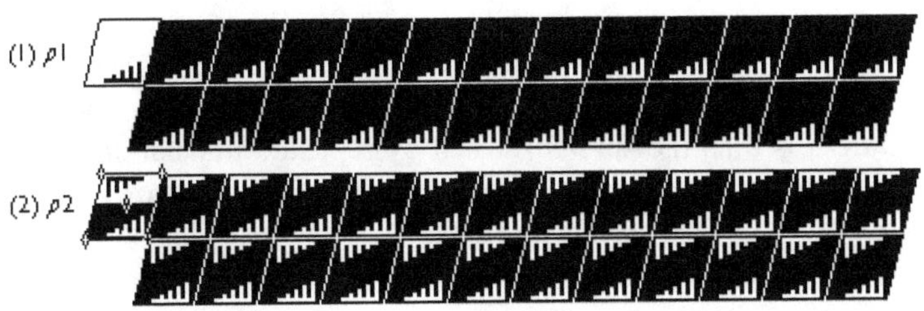

Fig. 23.14 Double-translation symmetry groups with oblique lattice. In (*1*) the initial seed image is the Black figure on White ground, and the cell is the same as the seed. In (*2*) the seed is the White figure, and the cell is the White parallelogram at the left

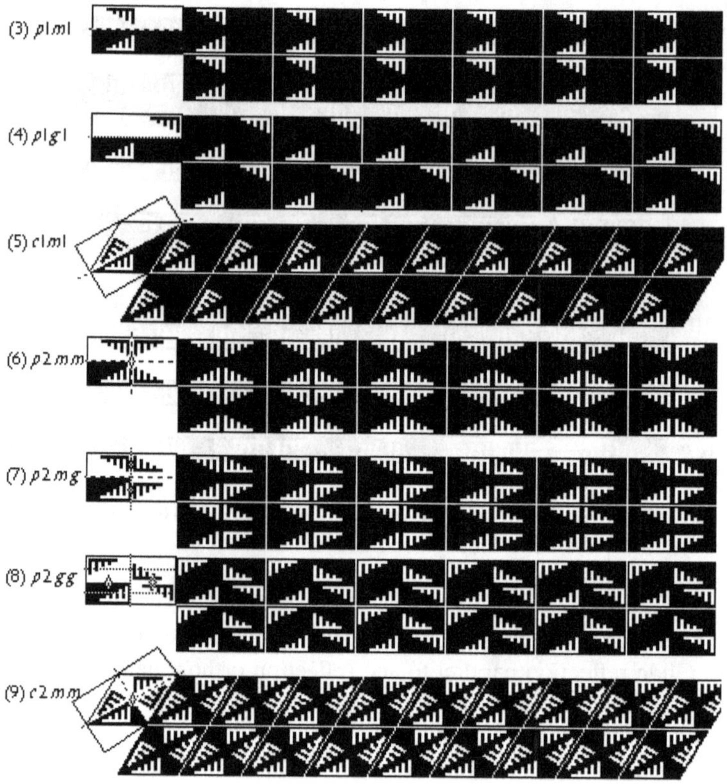

Fig. 23.15 Double-translation symmetry groups with rectangular lattice. The initial seed image is the White figure on Black ground. In (3), (4), (6), (7), and (8) the symmetry cell is the White rectangle *at the left*. In (5) and (9) the cell is the White rhomb, equivalent to the rectangle

- and possibly a mirror-line orthogonal to a translation,
- and possibly a glide-line parallel to a translation,
- and possibly a glide-line orthogonal to a translation,
- and possibly a two- or three- or four- or six-fold rotation.

The initial seed is any (asymmetrical) image. The cell is formed by applying the specified reflections, glide reflections and rotations to the seed. There are four possible cell shapes: oblique (parallelogram), rectangular, square, and hexagonal. The symmetry pattern is then the spread formed by repeated translations of the cell. It turns out that there are just seventeen such patterns [5, 7], as shown in Figs. 23.14, 23.15, 23.16 and 23.17. The standard symbols,[1] as set out in [2], are

[1]Devised by Carl Hermann (1898–1961), Professor of Crystallography, University of Marburg, and Charles-Victor Mauguin (1878–1958) Professor of Mineralogy, University of Paris.

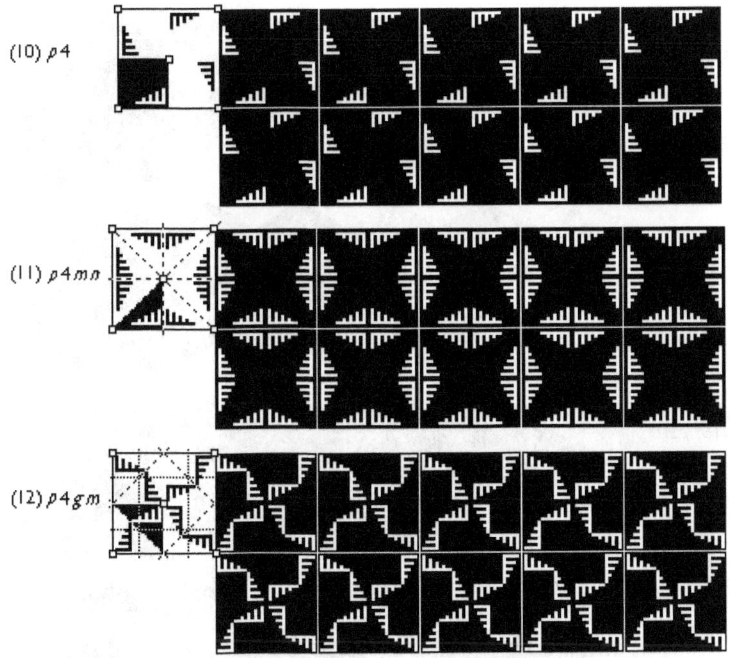

Fig. 23.16 Double-translation symmetry groups with square lattice. The initial seed image is the White figure on Black ground. The symmetry cell is the White square *at the left*

- *p* for primitive cell, or *c* for face-centred (rhombic) cell; followed by,
- 1 for no rotation (identity), or 2 for rotation by 180°, or 3 for rotation by 120°, or 4 for rotation by 90°, or 6 for rotation by 60°; followed by,
- 1 for no reflection parallel to direction of first translation, or *m* for mirror-line, or *g* for glide-line parallel to first translation,
- 1 for no reflection parallel to direction of second translation, or *m* for mirror-line, or *g* for glide-line parallel to second translation.

The 17 double-translation groups are

(1) *p*1: No reflection or rotation. Cell is a parallelogram, same as seed.
(2) *p*2: Rotation by 180°. Cell is parallelogram twice the size of seed.
(3) *p*1*m*1: No rotation, reflection horizontal. Cell is rectangle twice the size of seed.
(4) *p*1*g*1: No rotation, glide reflection horizontal. Cell is rectangle twice the size of seed.
(5) *c*1*m*1: No rotation, reflection diagonal. Cell is rhomb (special case of rectangle) twice the size of seed
(6) *p*2*mm*: Rotation by 180°, reflections horizontal and vertical. Cell is rectangle four times the size of seed.

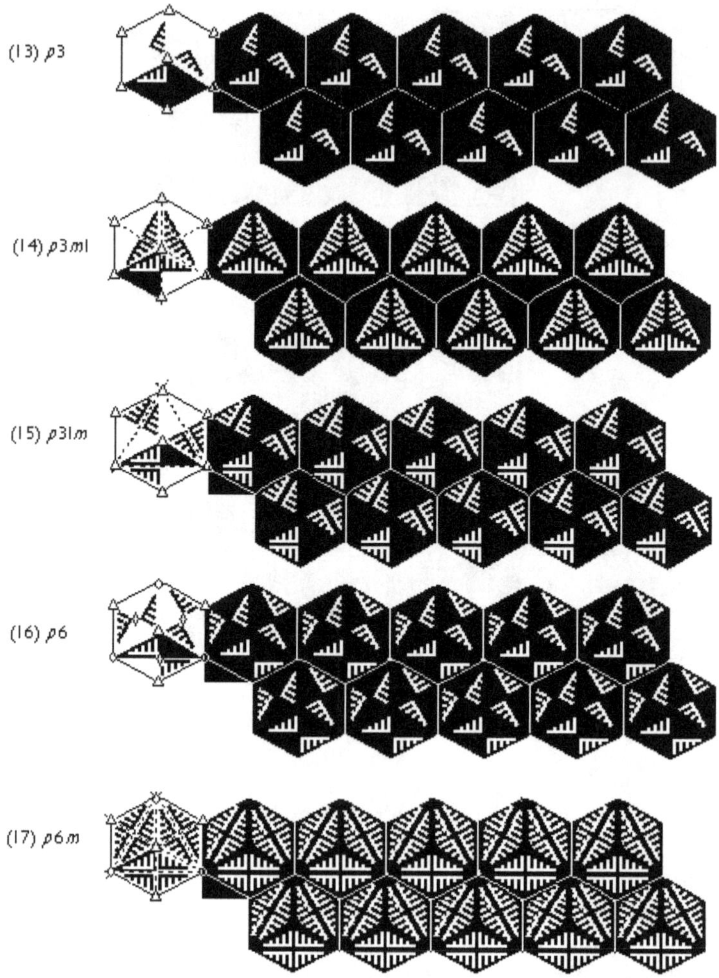

Fig. 23.17 Double-translation symmetry groups with hexagonal lattice. The initial seed image is the White figure on Black ground. The symmetry cell is the White hexagon *at the left*

(7) *p2mg*: Rotation by 180°, reflection horizontal, glide reflection vertical. Cell is rectangle four times the size of seed.
(8) *p2gg*: Rotation by 180°, glide reflection horizontal, glide reflection vertical. Cell is rectangle twice the size of seed.
(9) *c2mm*: Rotation by 180°, reflections both diagonals. Cell is rhomb (special case of rectangle) four times the size of seed.
(10) *p4*: Rotation by 90°, no reflection. Cell is square four times the size of seed.
(11) *p4mm*: Rotation by 90°, reflections horizontal, vertical, and both diagonals. Cell is square eight times the size of seed.

(12) $p4gm$: Rotation by 90°, glide reflections horizontal and vertical, reflections both diagonals. Cell is square eight times the size of seed.
(13) $p3$: Rotation by 120°, no reflection. Cell is hexagon three times the size of seed.
(14) $p3m1$: Rotation by 120°, reflections at 60-degree intervals. Cell is hexagon six times the size of seed.
(15) $p31m$: Rotation by 120°, reflections at 60-degree interals. Cell is hexagon six times the size of seed.
(16) $p6$: Rotations by 180, 120, and 60°, no reflection. Cell is hexagon six times the size of seed.
(17) $p6m$: Rotations by 180, 120, and 60°, reflections at 60-degree intervals. Cell is hexagon twelve times the size of seed.

From here, the way is open to write programs to generate any kind of repeating pattern, using seeds created by geometry, or scanner or camera.

23.16 1-bit Grating

A *1-bit grating* belongs to the single-translation group $t111$ where the seed (and the cell) is a White bar and a Black bar of equal width and arbitrary height, oriented at angle α to the x-axis, as shown in Fig. 23.18. The intensity profile of a 1-bit grating is a square wave alternating between intensity 0 and intensity 1, so we may measure the width of the cell as *wavelength* λ px, where λ is an even integer.

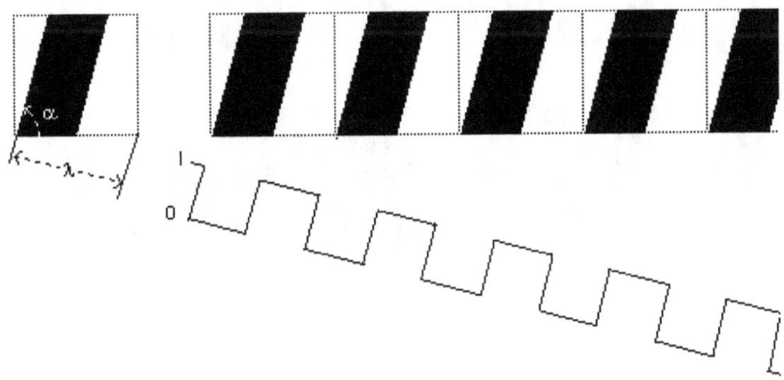

Fig. 23.18 1-bit grating. The cell, wavelength λ px, contains equal White and Black bars each $\lambda/2$ px wide, set at angle α to the x-axis. The cell, translated repeatedly in the x-direction, makes a grating. The intensity profile of a grating is a square wave

23.17 Parallel Gratings with Same Wavelength

Given two 1-bit gratings where

- wavelength $\lambda_1 = \lambda_2$,
- angle $\alpha_1 = \alpha_2$,

we may overlay transparently one on the other. If the initial cells coincide, the *phase difference* $\phi = 0$ and the two gratings are said to be *in phase*. The combination looks the same as each of the component gratings, as shown in Fig. 23.19a. If the initial cell of one grating is offset by $\phi = \lambda/N$ px (or $360/N°$) relative to the other, the combination has bars of different widths, as shown in Fig. 23.19b. If the offset is $\lambda/2$, gratings are said to be in *anti-phase* and the combination becomes a solid, as shown in Fig. 23.19c.

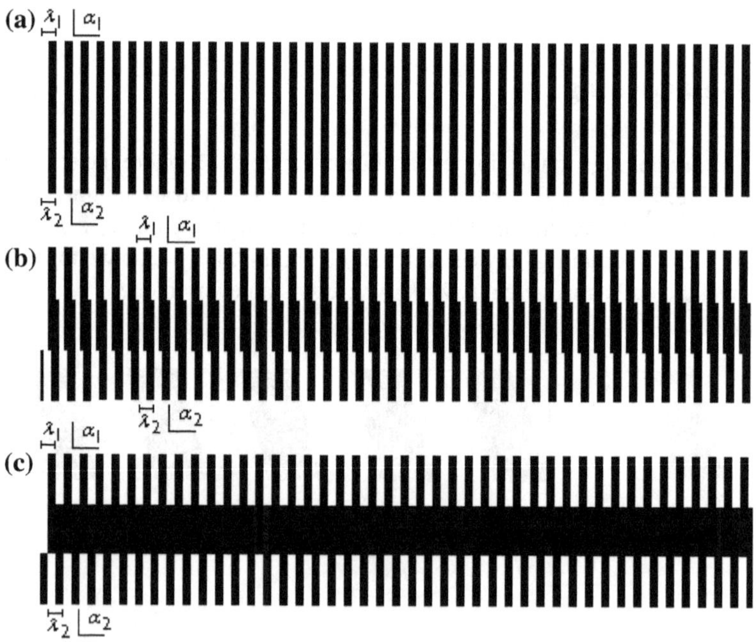

Fig. 23.19 Parallel gratings with same wavelength. **a** Gratings in phase, $\phi = 0$. **b** Gratings with phase difference $\phi = \lambda/5$ or 72°. **c** Gratings with phase difference $\phi = \lambda/2$ or 180°

Fig. 23.20 Parallel gratings with different wavelengths. Grating 1 has wavelength 10 px; grating 2 has wavelength 12 px, and the Moiré has wavelength 60 px

23.18 Parallel Gratings with Different Wavelengths

If we overlay two 1-bit gratings where

- wavelength $\lambda_1 <> \lambda_2$,
- angle $\alpha_1 = \alpha_2$,

their combination shows a *Moiré*[2] pattern [1, 4] of light and dark bands. The wavelength of the Moiré is

$$\lambda_M = (\lambda_1 \times \lambda_2) / \mid \lambda_1 - \lambda_2 \mid$$

and the angle of the Moiré is the same as the angle of the component gratings:

$$\alpha_M = \alpha_1 = \alpha_2 .$$

For example, Fig. 23.20 shows grating 1 with wavelength $\lambda_1 = 10$ px and angle $\alpha_1 = 90°$; grating 2 with wavelength $\lambda_2 = 12$ px and angle $\alpha_2 = 90°$. The Moiré has wavelength $\lambda_M = (10 \times 12) / \mid 10 - 12 \mid = 120/2 = 60$ px, and angle $\alpha_M = 90°$.

23.19 Gratings with Different Angles

If we overlay two 1-bit gratings where

- wavelength $\lambda_1 = \lambda_2$,
- angle $\alpha_1 <> \alpha_2$,

[2] The word entered English by several channels from an Arabic word meaning fine or selected applied to a type of cloth. It settled as *mohair*, which the French turned into *moaire* and later *moiré*, applied to a silk or other cloth with an impressed "watered" texture.

and set,

- angular difference $\alpha_D = \mid \alpha_1 - \alpha_2 \mid$,

then the wavelength of the Moiré is, in px:

$$\lambda_M = \lambda_1 / \sin(\alpha_D)$$

and the angle of the Moiré is, in degrees:

$$\alpha_M = \alpha_1 + 90 - (\alpha_D/2)\,.$$

For example, Fig. 23.21a shows grating 1 with wavelength $\lambda_1 = 10$ px and angle $\alpha_1 = 90°$; grating 2 with wavelength $\lambda_2 = 10$ px and angle $\alpha_2 = 75°$. The angular difference is $\alpha_D = 90 - 75 = 15°$. The wavelength of the Moiré is $\lambda_M = 10/\sin(15) = 10/0.2588 = 38$ px, and the angle of the Moiré is $\alpha_M = 90 + 90 - 15/2 = 172.5°$.

If we overlay two 1-bit gratings where

- wavelength $\lambda_1 <> \lambda_2$
- angle $\alpha_1 <> \alpha_2$

and set,

- angular difference $\alpha_D = \mid \alpha_1 - \alpha_2 \mid$,

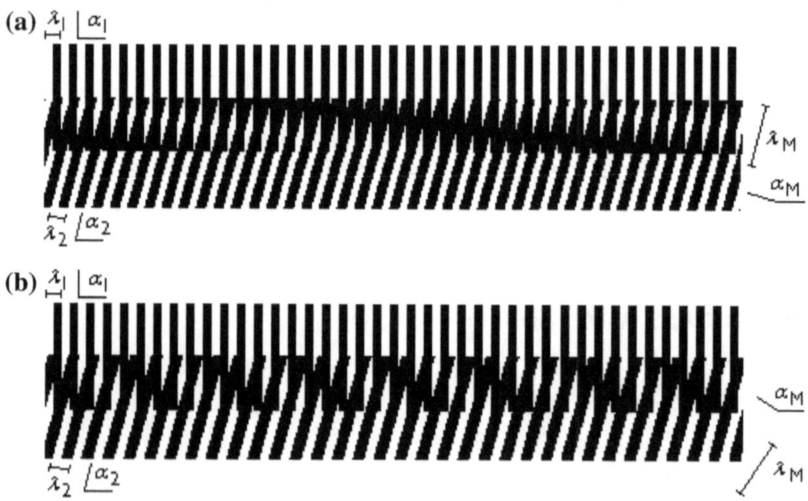

Fig. 23.21 Gratings with different angles. **a** Grating 1 has wavelength 10 px and angle 90°; grating 2 has wavelength 10 px and angle 75°. The Moiré has wavelength 39 px and angle 172.5°. **b** Grating 1 has wavelength 10 px and angle 90°; grating 2 has wavelength 12 px and angle 75°. The Moiré has wavelength 34 px and angle 145.3°

then the wavelength of the Moiré is, in px:

$$\lambda_{\mathrm{M}} = (\lambda_1\lambda_2)/\surd((\lambda_2^2 \sin^2(\alpha_{\mathrm{D}})) + (\lambda_2^2 \cos(\alpha_{\mathrm{D}}) - \lambda_1)^2)$$

and the angle of the Moiré is, in degrees:

$$\alpha_{\mathrm{M}} = 90 + \arctan(\tan(\alpha_{\mathrm{D}}/2)((\lambda_1 + \lambda_2)/(\lambda_2 - \lambda_1)))\,.$$

For example, Fig. 23.21b shows grating 1 with wavelength $\lambda_1 = 10$ px and angle $\alpha_1 = 90°$; grating 2 with wavelength $\lambda_2 = 12$ px and angle $\alpha_2 = 75°$. The angular difference is $\alpha_D = 90 - 75 = 15°$. The wavelength of the Moiré is

$$\begin{aligned}\lambda_M &= (10 \times 12)/\surd\left(12^2 \sin^2(15) + (12 \times \cos(15) - 10)^2\right)\\ &= 120/\surd(144 \times 0.0670 + (12 \times 0.9659 - 10)^2\\ &= 120/\surd(9.648 + 2.966)\\ &= 120/\surd 12.614\\ &= 120/3.552\\ &= 33\,\mathrm{px}\,.\end{aligned}$$

And the angle of the Moiré is

$$\begin{aligned}\alpha_{\mathrm{M}} &= 90 + \arctan(\tan(15/2) \times (10 + 12)/(12 - 10))\\ &= 90 + \arctan(0.1316 \times (22/2))\\ &= 90 + \arctan(1.4476)\\ &= 90 + 55.3\\ &= 145.3°\,.\end{aligned}$$

23.20 Orthogonal Gratings

If we overlay two 1-bit gratings where the angular difference is 90°, there is no Moiré, whatever the wavelengths of the component gratings. For example, Fig. 23.22a shows grating 1 with wavelength $\lambda_1 = 10$ px and angle $\alpha_1 = 90°$; grating 2 with wavelength $\lambda_2 = 10$ px and angle $\alpha_2 = 0°$. Figure 23.22b shows grating 1 with wavelength $\lambda_1 = 10$ px and angle $\alpha_1 = 90°$; grating 2 with wavelength $\lambda_2 = 12$ px and angle $\alpha_2 = 0°$.

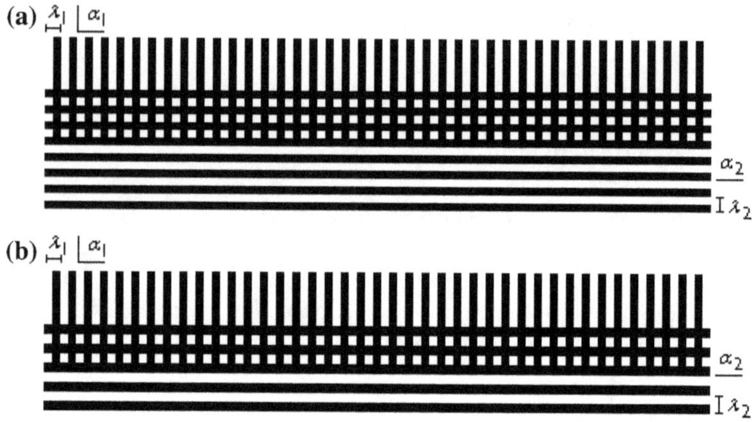

Fig. 23.22 Orthogonal gratings. **a** Grating 1 has wavelength 10 px and angle 90°; grating 2 has wavelength 10 px and angle 0°. **b** Grating 1 has wavelength 10 px and angle 90°; grating 2 has wavelength 12 px and angle 0°. There is no Moiré

23.21 Repeated Dilation

A dilation may also be repeated. Figure 23.23 shows a *repeated dilation* where (a) is the the unit image; and (b) is (a) dilated five times in the x-direction (that is, stretched to double, triple, fourfold, and fivefold width), and three times in the y-direction (that is, stretched to double and triple height). The greatest dilation is done first, then the next smaller overlaid on it, and so on.

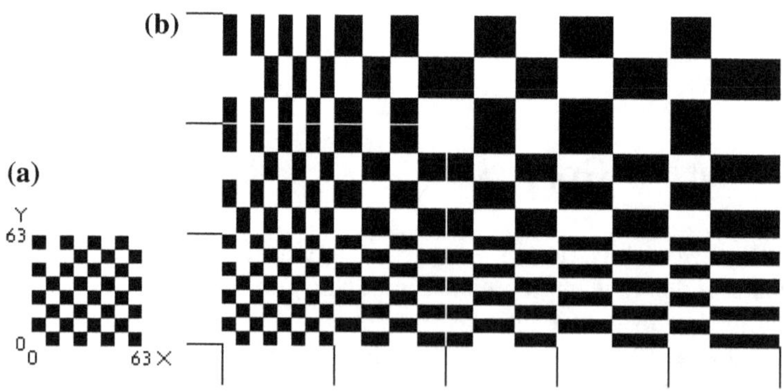

Fig. 23.23 Repeated dilation. **a** 1-bit image of 64 × 64 px in a 64 × 64 window. **b** Image (**a**) dilated five times in the x-direction (including the source unit as the zero-th repeat) and three times in the y-direction

Fig. 23.24 Combine locations. **a** 1-bit image of circle, 96 × 96 px. **b** 1-bit image of cross, 96 × 96 px **c** 1-bit combination of (**a**) and (**b**) by alternating columns, 96 × 96 px

23.22 Combine 1-bit Locations

Two 1-bit images A and B may be combined as one C by taking pixels alternately from A and from B to construct C. The alternation may be by columns or by rows or both. For example, Fig. 23.24 shows circle A, cross B, and combination C, where the alternation is by columns. Compare with Sect. 24.4 Boolean function A XNOR B. A program which combines two 1-bit images by location is described briefly below (see SpringerLink web-site for source code):

```
REM  **  Program COMBLOC1 to combine two 1-bit images
REM  **     with same pixel-count, by alternating
REM  **     locations. User enters file-names for
REM  **     input images A=#1, B=#3, and output
REM  **     image C=#2. Program reads image A data
REM  **     into IMAGEA, B data into IMAGEB, then
REM  **     processes alternate pixels to IMAGEC.
REM  **     Program writes IMAGEC to 1--bit .BMP
REM  **     file #2.
REM  **  Main program
REM  **  Enter two input and one output file-names
         CALL RHAWHB1
         CALL RDATA1
         CALL RDATC1
         CALL CALCALTE
         CALL WDATB1
         CALL REVHB1
END
         SUB CALCALTE
REM  **  Subprogram CALCALTE to take alternate pixels
REM  **  from IMAGEA and IMAGEC, and put into IMAGEB.
END SUB
         SUB PUTBYTE (INTE%) [as CHANLOC1]
         SUB PUTI (INTE%) [as CHANLOC1]
         SUB PUTL (INTL&) [as CHANLOC1]
         SUB RDATA1 [as CHANLOC1]
```

```
          SUB RDATC1
REM   **  Subprogram RDATC1 to read image data from
REM   **     1-bit .BMP file #3 and load into array
REM   **     IMAGEC. The data are 0 or 1 packed 8 to
REM   **     a byte.
END SUB
          SUB REVHB1 [as CHANLOC1]
          SUB RHAWHB1 [as CHANLOC1]
          SUB WDATB1 [as CHANLOC1]
```

23.23 Reduce 1-bit Locations

Given a 1-bit image, we may reduce the count (and increase the size) of pixels by taking every Nth pixel and repeating it N times. The reduced image has $1/N^2$ as many pixels, each N^2 the size of the original. A 1-bit image with Q px conveys Q bits of information, and a reduced image conveys Q/N^2 bits. For example, Fig. 23.25 shows a 1-bit scanner image, and a camera image saved-as 1-bit, thus reduced by $N = 2$, 4, and 8.

A program which reduces 1-bit images by location is described briefly below (see SpringerLink web-site for source code):

```
REM   **  Program REDULOC1 to reduce count of pixels
REM   **     for 1-bit image (width a multiple of 32).
```

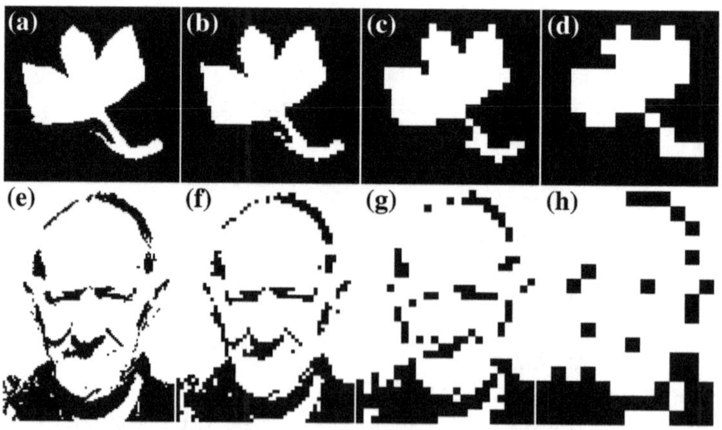

Fig. 23.25 Reduce locations. **a** Scanner image of leaf 96×96 px: 9,216 bits. **b** Reduction of (**a**) by $N = 2$, to 48×48 px: 2,304 bits. **c** Reduction of (**a**) by $N = 4$, to 24×24 px: 576 bits. **d** Reduction of (**a**) by $N = 8$, to 12×12 px: 144 bits. **e** Camera image saved-as 1-bit 96×128 px: 12,288 bits. **f** Reduction of (**e**) by $N = 2$ to 48×64 px: 3,072 bits. **g** Reduction of (**e**) by $N = 4$, to 24×32 px: 768 bits. **h** Reduction of **e** by $N = 8$, to 12×16 px: 192 bits

```
REM  **     User enters filenames. Program reads
REM  **     input image from 1-bit .BMP file #1 into
REM  **     array IMAGEA. User picks reduction.
REM  **     Program takes every Nth pixel from array
REM  **     IMAGEA and repeats it N times to array
REM  **     IMAGEB, with same intensity. Program
REM  **     writes reduced image from IMAGEB to 1-bit
REM  **     .BMP file #2.
REM  **   Main program
REM  **   Enter input and output file-names
          CALL RHAWHB1
          CALL RDATA1
          CALL LOC1REDU
          CALL WDATB1
          CALL REVHB1
END

          SUB LOC1REDU
REM  **   Sub-program LOC1REDU to enter reduction wanted,
REM  **     and perform reduction for each pixel from
REM  **     IMAGEA to IMAGEB.
END SUB
          SUB PUTBYTE (INTE%) [as CHANLOC1]
          SUB PUTI (INTE) [as CHANLOC1]
          SUB PUTL (INTL&) [as CHANLOC1]
          SUB RDATA1 [as CHANLOC1]
          SUB REVHB1 [as CHANLOC1]
          SUB RHAWHB1 [as CHANLOC1]
          SUB WDATB1 [as CHANLOC1]
```

References

1. Amidror I (2000) The theory of the moiré phenomenon. Kluwer, Dordrecht
2. IUCr (2006) International tables for crystallography, vol a, space group symmetry. http://it.iucr.org/. Accessed 1 Feb 2014
3. Morley F, Morley FV (1933) Inversive geometry. Ginn, Boston. http://babel.hathitrust.org/cgi/pt?id=uc1.b4074019;view=1up;seq=7. Accessed 1 Feb 2014
4. Oster G, Nishijuna Y (1963) Moiré patterns. Sci Am 208:54–63. doi:10.1038/scientificamerican0563-54
5. Pólya G, Niggli P (1924) Über die Analogie der Symmetrie in der Ebene. Z Kristall 60:278–282
6. Speiser A (1924) Theorie der Gruppen von endlicher Ordnung. Springer, Berlin Reissued 1980 Springer, Basel
7. Weyl H (1952) Symmetry. Princeton University Press, Princeton NJ. Available at: https://archive.org/details/Symmetry_482. Accessed 1 Feb 2014

Chapter 24
1-bit Intensity Process

A discrete intensity transformation changes all the intensities of an image in some systematic way, leaving locations unchanged. In a 1-bit image the intensities are just Black or White. Reversal changes Black to White and White to Black, and can be done in a GUI. 1-bit images with the same pixel-count constitute a Boolean algebra, and hence have 16 Boolean functions. A program to combine two images by Boolean functions is given, with examples. Images with bit-depth 4, 8, or 24 can be reduced to 1-bit by standard operations.

24.1 Flowchart

The flowchart (Fig. 24.1) is for general guidance in this chapter.

24.2 Intensity Transformation

An *intensity transformation* is an operation on a discrete image which changes the intensities of all the pixels in some systematic way, leaving the locations unchanged. A 1-bit image has but two intensities, 0 Black and 1 White, so the only intensity changes are from 0 to 1 and from 1 to 0.

24.3 Reverse Intensities

A GUI editor usually has a command which reverses the intensities of an entire image, or of a selected sub-image. In the MS PAINT menu it is found as Image-Invert Colors. In 1-bit mode, this will change every pixel of an image from Black to White or from White to Black, as shown in Fig. 24.2.

A. Parkin, *Digital Imaging Primer*, DOI 10.1007/978-3-540-85619-1_24

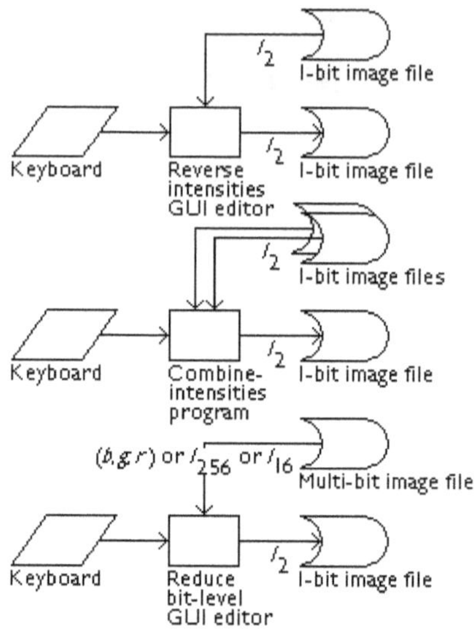

Fig. 24.1 1-bit intensity process. Given a 1-bit image file, a GUI editor may reverse the intensities Black to White. Given two 1-bit image files, a program may combine them by calculating any of their Boolean functions, and write a new image file. Given a 24-bit or 8-bit or 4-bit image file, a GUI editor may reduce the image to 1-bit

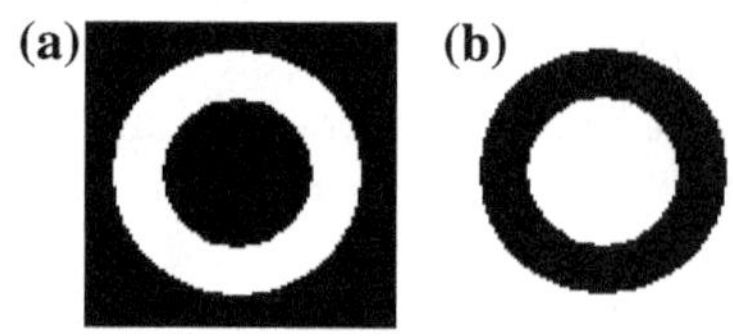

Fig. 24.2 Reverse intensities. **a** Given image. **b** Reversed-intensity image

24.4 Boolean Function

1-bit images with the same pixel-count constitute a Boolean algebra (see Sect. 3.17), since they satisfy the conditions:

- For any 1-bit image A, each pixel has intensity 0 Black or 1 White.
- For any 1-bit image A, there is a pixel-wise unary transformation, *negation* $\neg A$, which changes intensity 0 to 1, and 1 to 0.
- For any two images A and B with the same pixel-count, there is a pixel-wise binary transformation, *conjunction* $A \wedge B$, which delivers intensity 0 if both A and B are 0, else delivers 1.
- For any two images A and B with the same pixel-count, there is a pixel-wise binary transformation, *disjunction* $A \vee B$, which delivers intensity 1 if either A or B or both are 1, else delivers 0.

For example Fig. 24.3 shows two 1-bit images A and B with the same pixel-count, and the negation of A, the disjunction of A and B, and the conjunction of A and B.

A *Boolean function* is any combination of negations, disjunctions and conjunctions. For any two 1-bit images with the same pixel-count, A and B, there are 16 distinct Boolean functions, as shown in Table 24.1. For example, Fig. 24.4 shows the 16 Boolean functions of a White-on-Black ring A and a White-on-Black cross B,

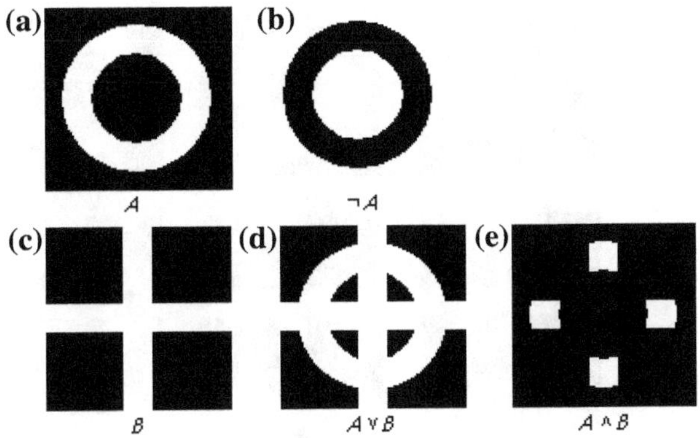

Fig. 24.3 Example 1-bit images as elements of a Boolean algebra. **a** Image A. **b** Negation $\neg A$. **c** Image B. **d** Disjunction $A \vee B$. **e** Conjunction $A \wedge B$

Table 24.1 The 16 Boolean functions of two elements A and B

Symbol	Value	Expansion	Name
A	0011		Element
B	0101		Element
0	0000	$A \wedge \neg A$	Contradiction
$A \wedge B$	0001	$A \wedge B$	Conjunction
$A \neg \rightarrow B$	0010	$A \wedge \neg B$	Non-implication
A	0011	A	Left projection
$A \neg \leftarrow B$	0100	$\neg A \wedge B$	Converse non-implication
B	0101	B	Right projection
A XOR B	0110	$(A \vee B) \wedge \neg(A \wedge B)$	Exclusive disjunction
$A \vee B$	0111	$A \vee B$	(Inclusive) Disjunction
A NOR B	1000	$\neg A \wedge \neg B$	Joint denial
A XNOR B	1001	$(A \rightarrow B) \wedge (B \rightarrow A)$	Equivalence
$\neg B$	1010	$\neg B$	Right complementation
$A \leftarrow B$	1011	$A \vee \neg B$	Converse implication
$\neg A$	1100	$\neg A$	Left complementation
$A \rightarrow B$	1101	$\neg A \vee B$	(Material) implication
A NAND B	1110	$\neg A \vee \neg B$	Non-conjunction
1	1111	$A \vee \neg A$	Tautology

When A and B have the values shown in the heading, each function takes the values shown in the body

corresponding to Table 24.1. These are combinations by intensity, as against combinations by location (see Sect. 23.22).

MS PAINT can perform some of the Boolean functions. It can "invert" a 1-bit image (negation), or overlay it on another by a "paste transparent" command. Paste transparent is the conjunction: White where both A and B are White, else Black.

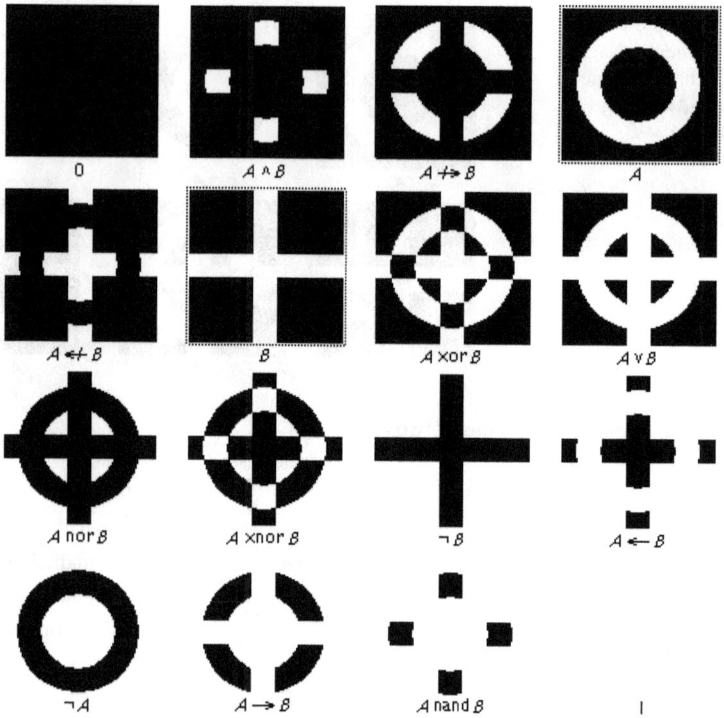

Fig. 24.4 The 16 Boolean functions of two 1-bit images. Given two 1-bit images, A and B, 16 Boolean functions can be formed, as listed in Table 24.1

Now De Morgan's law tells us that the negation of the disjunction is equivalent to the conjunction of the negations:

$$\neg(A \vee B) \equiv (\neg A) \wedge (\neg B)\,.$$

So to get the disjunction $A \vee B$ in PAINT, we invert A, invert B, paste transparent $\neg B$ on $\neg A$, then invert the result.

24.5 Boolean Intensity Function of Gratings

If A and B are parallel gratings (see Sect. 23.18), with different wavelengths λ_A and λ_B but the same angle α, their combination shows a Moiré with wavelength λ_C:

$$\lambda_{\mathrm{C}} = (\lambda_{\mathrm{A}} \times \lambda_{\mathrm{B}}) / \mid \lambda_{\mathrm{A}} - \lambda_{\mathrm{B}} \mid\,.$$

Fig. 24.5 The 16 Boolean functions of two parallel gratings. Given two gratings, A with wavelength $\lambda_A = 10\,\text{px}$, and B with wavelength $\lambda_B = 12\,\text{px}$; there are 16 Boolean functions as listed in Table 24.1, each with wavelength $\lambda_C = 60\,\text{px}$

For example, Fig. 24.5 shows grating A with wavelength $\lambda_\text{A} = 10\,\text{px}$; grating B with wavelength $\lambda_\text{B} = 12\,\text{px}$; and their Boolean functions when combined parallel. The Moiré wavelength is $(10 \times 12)/ \mid 10 - 12 \mid = 120/2 = 60\,\text{px}$.

For another example, Fig. 24.6 shows orthogonal gratings, A at angle $\alpha_\text{A} = 0°$ and B at angle $\alpha_\text{B} = 90°$, both with wavelength 10 px; and their Boolean functions.

24.6 Combine-Intensities Program

A program which combines two 1-bit images by any required Boolean function is described below (see SpringerLink web-site for source code).

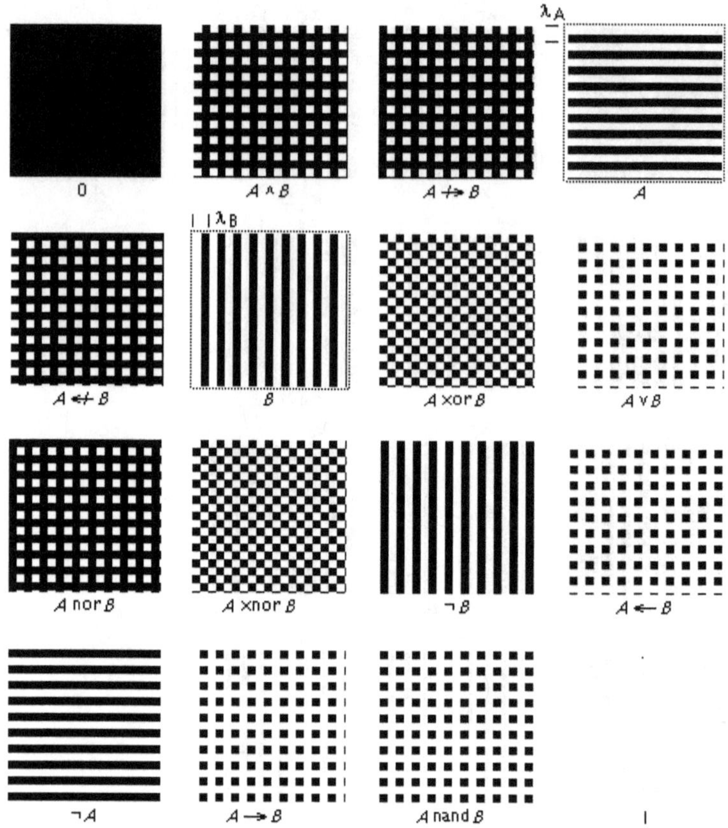

Fig. 24.6 The 16 Boolean functions of two orthogonal gratings. Given two gratings, A with angle $\alpha_A = 0°$ and wavelength $\lambda_A = 10\,\text{px}$, and B with angle $\alpha_B = 90°$ and wavelength $\lambda_B = 10\,\text{px}$; there are 16 Boolean functions as listed in Table 24.1

```
REM  **  Program COMBINT1 to combine two 1-bit images
REM  **     as one, by Boolean functions of intensities.
REM  **     User enters filenames for given images A
REM  **     and B and new image C, and picks function
REM  **     required. Program reads image data from files
REM  **     #1 and #3 into arrays IMAGEA and IMAGEB, and
REM  **     sets-up output array IMAGEC. Program steps
REM  **     through pixels of IMAGEA and IMAGEB, and puts
REM  **     resulting intensities in IMAGEC. Program
REM  **     writes 1-bit .BMP file #2 containing new image.
REM  **  Main program
REM  **  Enter two input and one output file-names
```

```
            CALL RHAWHB1
            CALL INCONSTA
            CALL RDATA1
            CALL RDATC1
            CALL CALCIMAB
            CALL WDATB1
            CALL REVHB1
END

            SUB CALCIMAB
REM   **    Sub-program CALCIMAB to scan IMAGEA and IMAGEC,
REM   **       calculate required function of intensities
REM   **       and put result in IMAGEB.
END SUB
            SUB INCONSTA
REM   **    Subprogram INCONSTA to enter Boolean function
REM   **       required.
END SUB
            SUB PUTBYTE (INTE%) [as CHANLOC1]
            SUB PUTI (INTE) [as CHANLOC1]
            SUB PUTL (INTL&) [as CHANLOC1]
            SUB RDATA1 [as CHANLOC1]
            SUB RDATC1 [as COMBLOC1]
            SUB REVHB1 [as CHANLOC1]
            SUB RHAWHB1 [as CHANLOC1]
            SUB WDATB1 [as CHANLOC1]
```

From here, the way is open to explore Boolean functions of any 1-bit images, created by geometry or by scanner or camera.

24.7 Outline of 1-bit Figure

To get the outline of a 1-bit figure A, first make B as a small dilation of A, then form the exclusive disjunction A xor B (the difference). For example, Fig. 24.7 shows a White ring, a slight dilation of the White ring, and the difference of the two images.

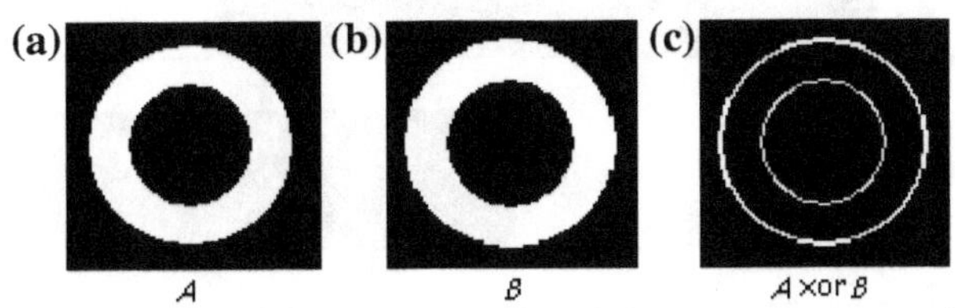

Fig. 24.7 Outline of 1-bit figure. **a** Image A: 96×96 px. **b** Image B: A stretched by 105 %, cropped to 96×96 px. **c** The difference A or B

24.8 Reduce Bit-Depth to 1-bit

If we want to reduce the bit-level of an image with 24 or 8 or 4 b/px to 1 b/px we can use the *Save-as* command in an image editor, such as MS PAINT:

- display the source file in its 24- or 8- or 4-bit mode,
- click on menu File—Save-as,
- enter new file name (or keep old file name),
- enter new file type (monochrome bitmap),
- click on Save.

The operating system will then

- create new 1-bit file with two-entry colour-table 1 BW,
- keep all pixel location values (x, y) unchanged,
- per-pixel, compare the source intensity 0-255 to fixed threshold 128,
- set 1-bit intensity index 0 or 1 accordingly.

For example, Fig. 24.8a shows a 24-bit scanned image of a test card with a line-pair strip, an inch/cm rule, six colour patches, and a step-wedge. Figure 24.8b shows Fig. 24.8a saved-as 1-bit.

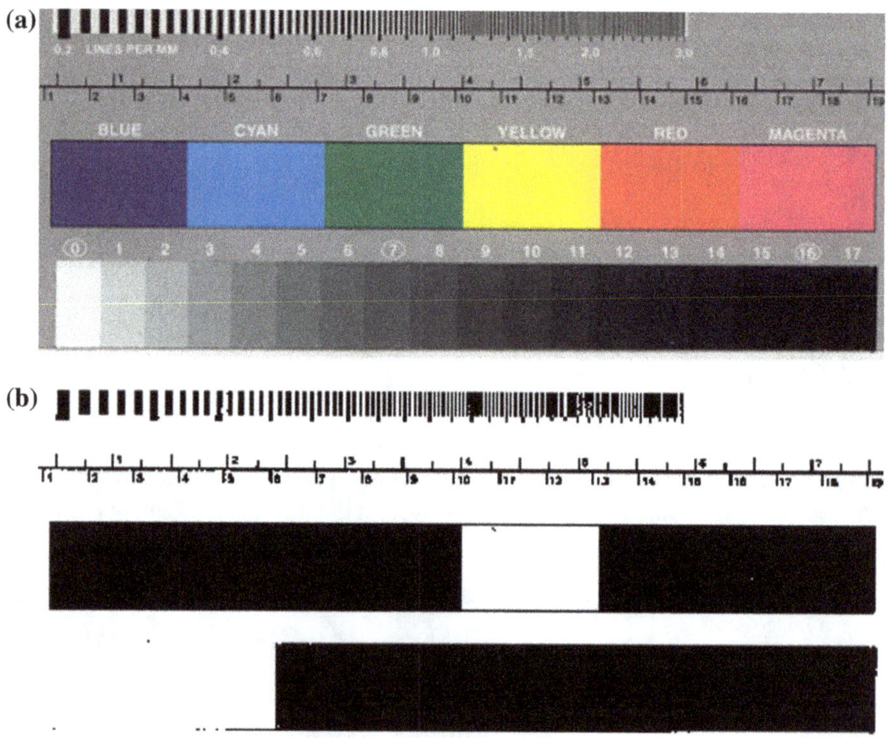

Fig. 24.8 24-bit image saved-as 1-bit. **a** 24-bit scanned image of test card. **b** (**a**) saved-as 1-bit image

Save-as gives slightly different results according to the bit-level and colour-table of the source. Appendix B shows all the reductions.

Save-as 1-bit is a very harsh reduction which loses all gradations. However, a multi-bit image can be reduced to a 1-bit image which keeps gradations by the important process of half-toning (see Sect. 39.16).

24.9 Statistics of 1-bit Image

Statistically (see Sect. 18.3), a digital image is a *finite population* of pixels, with a *population count* of N px. An individual pixel of the population is identified by its location (x, y). The population has one discrete *variate*, intensity I.

For example, a 1-bit digital image has pixels showing one of two intensities, so $I = 0$ or 1. Figure 24.9 shows an example of a camera image saved-as 1-bit. Compare with Fig. 18.2. The statistics were calculated by program DOSTATS1 (listed on the Springer web-site), quite similar to program DOSTATS8 described in Sect. 39.26.

Fig. 24.9 1-bit camera image. **a** Image 128×128 px, with two intensities. **b** Histogram. **c** Statistics

Image file is AP1.BMP

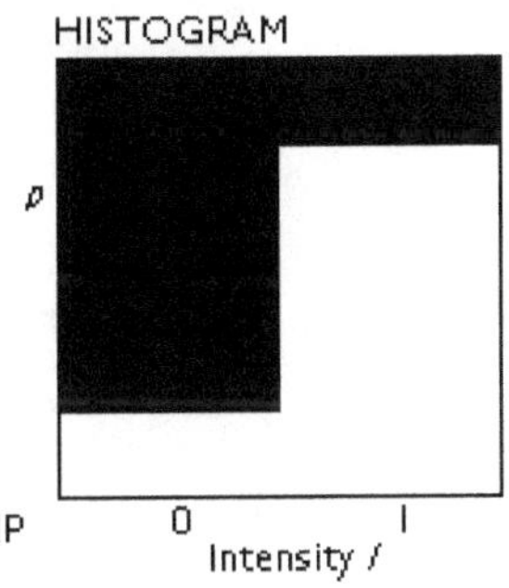


```
STATISTICS OF 1-BIT IMAGE

Image file     AP1.BMP
This file is   AP1T.TXT

------------------------
 Image width, px      128
Image height, px      128
  Image total px    16384
 Black count, px     3240
 White count, px    13144
            Mean    0.802
            Mode        1
        Variance    0.159
   Standard dev.    0.398
        Skewness   -1.518
  Int.complexity        1
  Loc.complexity       14
  Tot.complexity       14
         Entropy    0.717
```

Chapter 25
1-bit Fourier Process

A Fourier transformation changes all the locations of an image to frequencies of gratings, and back again. This allows certain frequencies to be attenuated or amplified, for one purpose or another. The process for 8-bit images is described in Chap. 40. This chapter shows examples of 1-bit images thus processed.

25.1 Flowchart

The flowchart (Fig. 25.1) shows Fourier processing for a 1-bit image, using the 8-bit programs from Chap. 40.

25.2 Fourier Process for 1-bit Image

Given an image P px wide, a Fourier analysis captures the visual information of the image as a frequency spectrum, the coefficients of a sum of sinusoidal gratings belonging to a harmonic set. Full capture uses frequencies from 0 to $(P/2) - 1$. Partial capture uses fewer frequencies. A Fourier synthesis restores an image from the frequency spectrum. For example, Fig. 25.2 shows a 1-bit source image and five restorations, where ceilings of 4, 8, 16, 32, and 63 have been placed on the spectrum. Low ceilings lose much information; high ceilings lose little.

25.3 Frequency Filtering

If we analyze an image at the highest available frequency, we have the opportunity to change the coefficients of the spectrum, then synthesize a new image. There is thus a class of *frequency transformations*, quite distinct from the location transformations of Chap. 23 and the intensity transformations of Chap. 24.

A. Parkin, *Digital Imaging Primer*, DOI 10.1007/978-3-540-85619-1_25

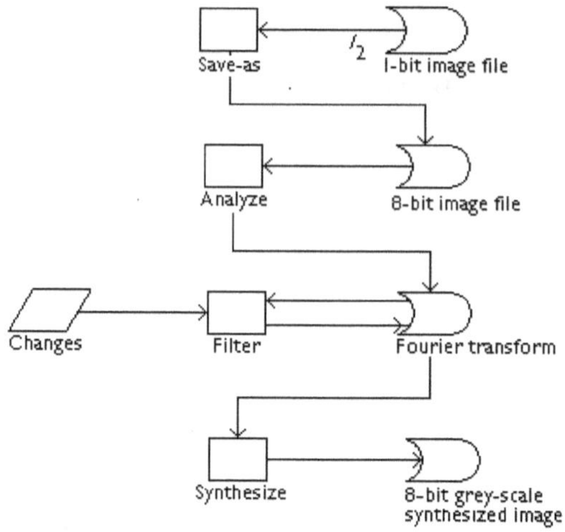

Fig. 25.1 1-bit Fourier process. Given a 1-bit image file, it can be saved-as an 8-bit file, then an analysis program reads the file, calculates the frequency transform, and outputs it to a transform file. A synthesis program reads the transform file, possibly applies frequency filters, calculates a reconstructed image, and outputs it to an 8-bit image file

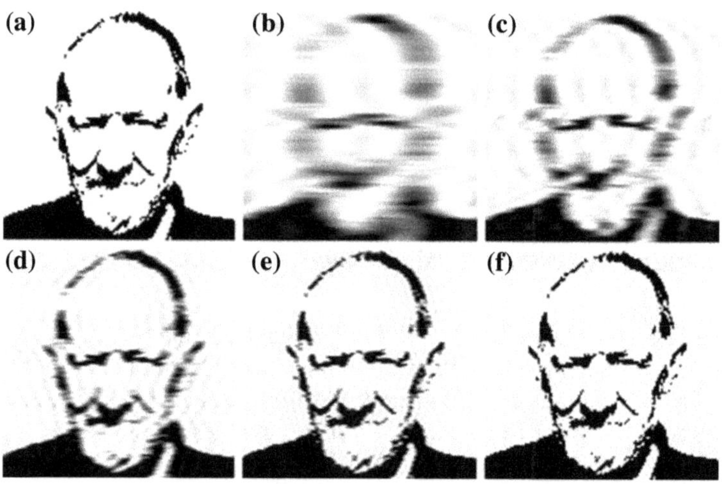

Fig. 25.2 1-bit image analysed and restored with different frequency ceilings. **a** Source image of 128×128 px. **b** Restored image using highest frequency 4. **c** Restored image using highest frequency 8. **d** Restored image using highest frequency 16. **e** Restored image using highest frequency 32. **f** Restored image using highest frequency 63

Changing a frequency spectrum can be done by setting filter factors for the various frequencies or groups of frequencies, and multiplying each frequency coefficient by the corresponding filter factor.

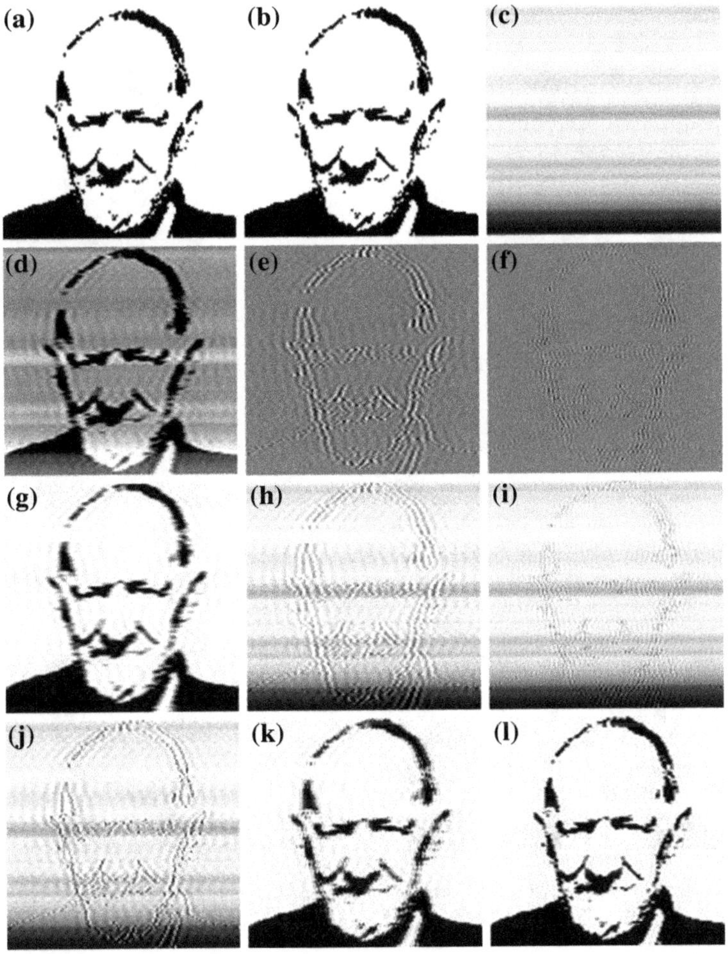

Fig. 25.3 1-bit frequency filtering. **a** Source image of 128 × 128 px. **b** Restored image with all frequencies 0–63 %. **c** Filter passes frequency 0 only (the DC component). **d** Filter stops frequency 0 and passes low frequencies. **e** Filter stops 0 and passes mid frequencies. **f** Filter stops 0 and passes high frequencies. **g** Filter passes 0 and low frequencies. **h** Filter passes 0 and mid frequencies. **i** Filter passes 0 and high frequencies. **j** Filter stops low frequencies. **k** Filter stops mid frequencies. **l** Filter stops high frequencies

- A *binary filter* has factor values 0 or 1, so either stops (0) or passes unchanged (1) the corresponding frequency.
- An *integer filter* has integer factor values 0 to M, so stops (0) or passes unchanged (1) or amplifies (>1) the corresponding frequency.
- A *real filter* has real factor values 0–m, so stops (0) or attenuates (<1) or passes unchanged (1) or amplifies (>1) the corresponding frequency.

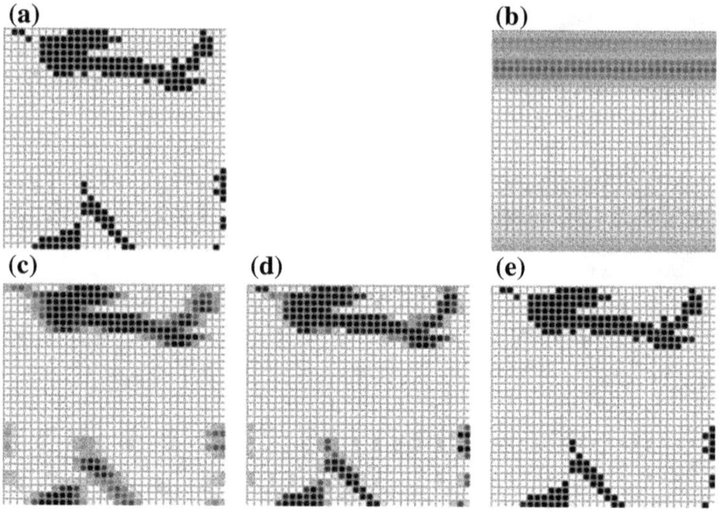

Fig. 25.4 Detail of 1-bit Fourier filtering. **a** Source image of 32 × 32 px, enlarged 4 times linear. **b** Filter passes zero frequency only. **c** Filter passes zero and low frequencies. **d** Filter passes zero low and mid frequencies. **e** Filter passes all frequencies

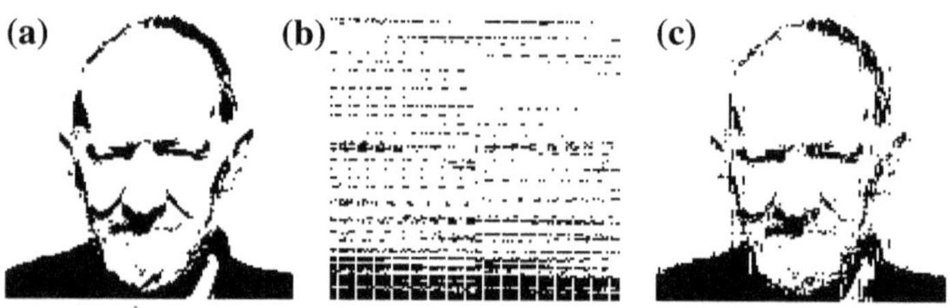

Fig. 25.5 Source image, Fourier transform, and restored image. **a** Source image of 128 × 128 px. **b** DCT transform. **c** IDCT restored image using all eight frequency components

For example, Fig. 25.3 shows a source image and the restorations after passing or stopping different frequencies. The zero-frequency component[1] is just the average intensity of the block, with no internal detail at all. The first, second, and subsequent components modify the zero-frequency component to give increasing detail. In most images the low frequencies carry most of the visual information.

Figure 25.4 shows an enlargement of the source image and syntheses from various groups of frequencies, pixel by pixel.

From here, the way is open to explore all kinds of frequency manipulations of 1-bit images, and to combine frequency operations with Boolean functions.

[1]Often called the *DC* component, following a usage from electrical signal-processing.

25.4 Alternative 1-bit Fourier Process

Fourier processing is normally done in an 8-bit context, where the frequency gratings take 256 grey values. However, it is possible to analyze and synthesize using only 1-bit values 0 and 1, with an 8-bit file for the intervening transform. A pair of such programs which analyse and synthesize 1-bit images are briefly described below, and listed on the Springer web-site. The examples which follow were produced by these programs. The results show imperfections (no doubt consequences of stretching the 0,1 source domain to the $-1, 0, +1$ frequency domain), but the alternative approach may occasionally be helpful in grasping the sometimes puzzling aspects of Fourier programming.

The given target image may be any integral multiple (zero-padded if necessary) of the block size $N \times N$ set for the DCT, and the DCT is applied again and again across the image. For example, Fig. 25.5 shows a source image of 128×128 px, the DCT transform resulting from $16 \times 16 = 256$ blocks of 8×8 px, and the IDCT

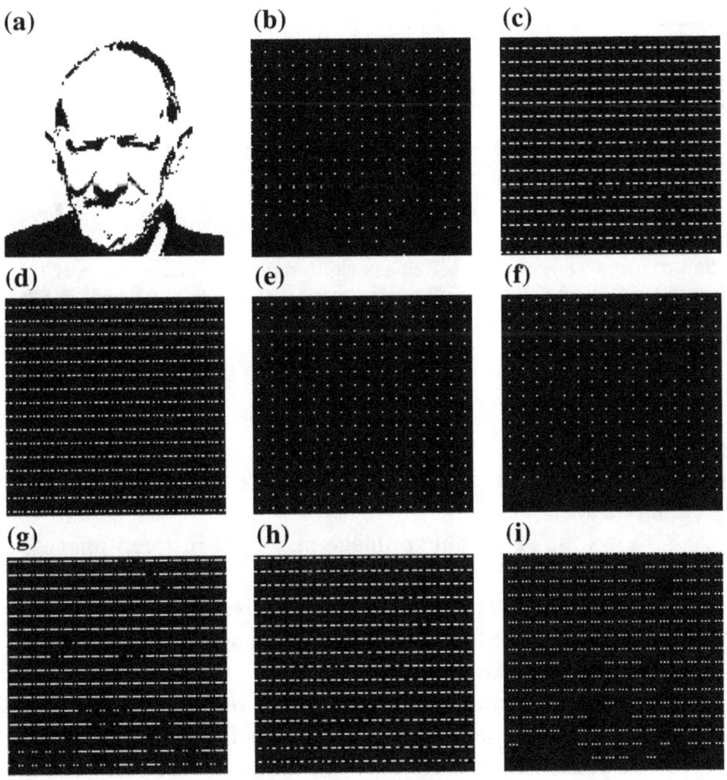

Fig. 25.6 Source image and single-frequency restorations. **a** Source image of 128×128 px. **b** Restoration using only the zero-frequency component. **c** Restoration using only the 1-frequency component. **d–f** Restorations using the 2-, 3-, 4-frequency components, singly. **g–i** Restorations using the 5-, 6-, 7-frequency components, singly

Fig. 25.7 Source image and cumulative-frequency restorations. **a** Source image of 128 × 128 px. **b** Restoration using frequency 0 only. **c** Restoration using frequencies 0 and 1. **d–f** Restorations using frequencies 0–2, 0–3, and 0–4. **g–i** Restorations using frequencies 0–5, 0–6, and 0–7

restored image using all 8 frequency components. The restored image is fairly close to the source.

Figure 25.6 shows a source image and eight restored images, each using a single frequency component.

Figure 25.7 shows the same source image and eight restored images, using an increasing set of frequency components from 0 to 7.

Figure 25.8 shows several examples of 1-bit Fourier transforms and restorations by programs DCT1 and IDCT1. In each case, the image size is 128 × 128 px, and the processing is done with block size 8 × 8 px: eight blocks per row and 8 rows per image. The source images were created by computer program, GUI editor, scanner, and camera. In each case, the source image is at the left, the frequency transform in the middle, and the image restored from the transform at the right.

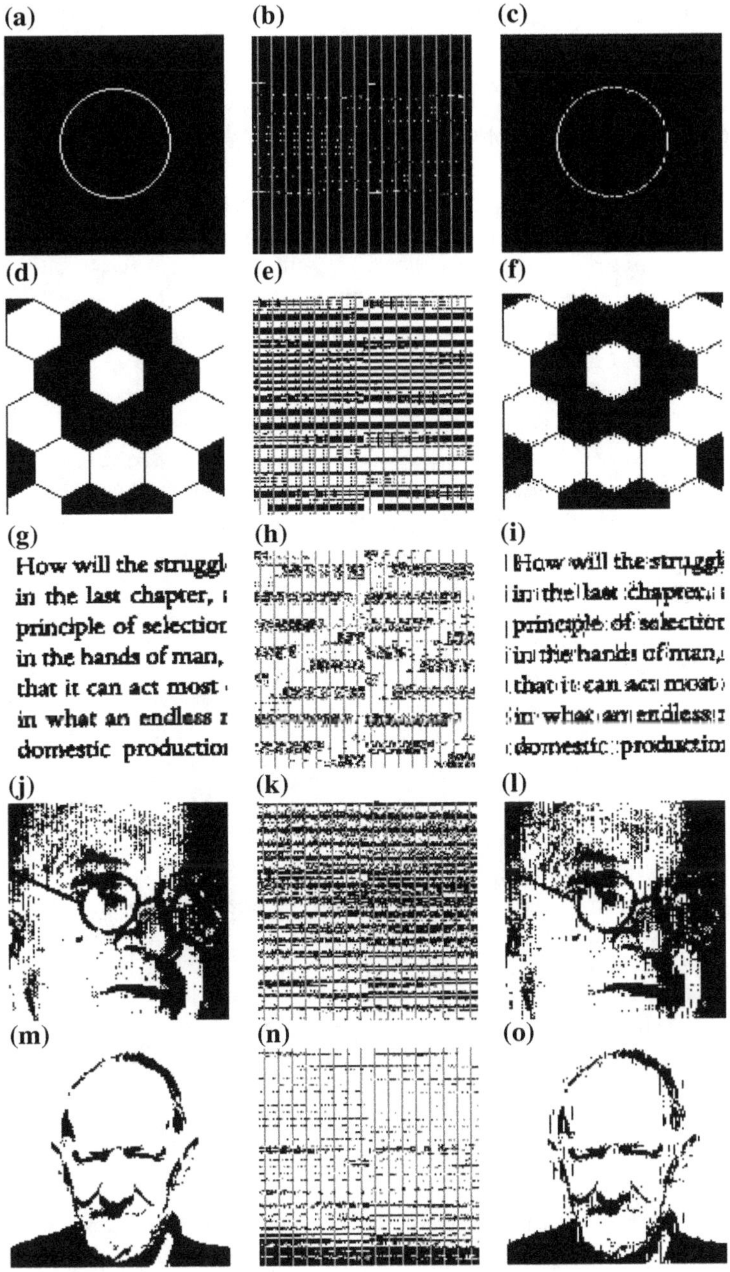

Fig. 25.8 Examples of Fourier transforms and restorations of 1-bit images. **a–c** Source image created by computer (program MKCIRCLE), frequency transform of source image, and image restored from transform. **d–f** Source image created by GUI. **g–i** Source scan of text at sample frequency $F_S = 100$ spi. **j–l** Source scan of of newspaper half-tone at sample frequency $F_S = 96$ spi. **m–o** Source 24-bit camera image reduced to 1-bit

25.5 1-bit DCT Program

A program to produce the frequency transform of a 1-bit image is briefly described below (see source code on the SpringerLink web-site). It will process an image of 160×160 px or less, with a block size of 8×8 px.

```
REM  **  Program DCT1 to read 1-bit image and write
REM  **  8-bit DCT transform, in 8x8 blocks. User
REM  **  enters filenames #1 for .BMP image to be
REM  **  transformed, #2 for .BMP transform. Program
REM  **  reads headers of file #1, and writes headers
REM  **  of file #2. Program dimensions arrays, and
REM  **  generates DCT half-cosine basis and its transpose.
REM  **  Program reads block of 8x8 pixels of image into
REM  **  target buffer; multiplies basis by target;
REM  **  multiplies products by transpose of basis, to
REM  **  get transform array. Program writes transform
REM  **  image to 8-bit file #2.
REM  **  Main program
REM  **  Enter input and output file-names, and set zero
         CALL RWHEADS
         CALL CALCCOSX
         CALL CALCBASIS
         CALL CALCTRAN
         CALL REVHEADS
END

         SUB CALCBASIS
REM  **  Subprogram to assign 1-bit cosine harmonic
REM  **     values into array ABASIS, and its transpose
REM  **     ABASIT.
END SUB
         SUB CALCCOSX
REM  **  Subprogram CALCCOSX to calculate angles in
REM  **     radians and real cosine values for harmonic
REM  **     set into array ACOSTAB.
END SUB
         SUB CALCTRAN
REM  **  Subprogram to find DCT frequency transform
REM  **    for given image. Read N complete rows of
REM  **    1-bit image from file #1 into buffer. Take
REM  **    successive NxN blocks into target array;
REM  **    premultiply by basis into products array;
REM  **    postmultiply products by transpose of basis
REM  **    into transform array; write 8-bit .BMP of
```

```
REM   **     transform to file #2.
           SUB PUTI (INTE%) [as CHANLOC1]
           SUB PUTL (INTL&) [as CHANLOC1]
           SUB REVHEADS
REM   **   Revises header of file #2 with values in bytes.
END SUB
           SUB RWHEADS
REM   **   Subprogram to read header of 1-bit input file
REM   **      #1, and write 8-bit header and colour-table
REM   **      of output file #2.
END SUB
```

25.6 1-bit IDCT Program

The following program is the inverse partner of DCT1.

```
REM   **   Program IDCT1 to read 8-bit transform file
REM   **   written by DCT1 and write 1-bit restored image,
REM   **   in 8x8 blocks. User enters filenames #1 for
REM   **   transform to be processed, #2 for image
REM   **   output, and sets lowest and highest frequencies
REM   **   wanted. Program reads headers of file #1, and
REM   **   writes headers of file #2. Program dimensions
REM   **   arrays, and generates DCT half-cosine basis
REM   **   and its transpose. Program reads 8 rows of
REM   **   input transform into buffer, and takes 8x8
REM   **   target block; multiplies basis by target;
REM   **   multiplies product by transpose of basis, to
REM   **   get restored image. Program writes image to
REM   **   file #2.
REM   **   Main program
REM   **   Enter input and output file-names, and
REM   **   frequencies wanted.
           CALL RWHEA81
           CALL CALCCOSX
           CALL CALCBASIS
           CALL CALCIMAGE
           CALL REVHEADS
END

           SUB CALCBASIS [as DCT1]
           SUB CALCCOSX [as DCT1]
           SUB CALCIMAGE
```

```
REM  **  Subprogram to restore 1-bit image from given
REM  **     8-bit DCT transform. Read N complete rows
REM  **     of transform from file #1 into buffer; take
REM  **     NxN block into target array; premultiply
REM  **     by basis-transpose into products array;
REM  **     postmultiply products by basis into image
REM  **     array; write blocks of image to 1-bit file #2.
END SUB
         SUB PUTI (INTE%) [as CHANLOC1]
         SUB PUTL (INTL&) [as CHANLOC1]
         SUB REVHEADS [as DCT1]
         SUB RWHEA81
REM  **  Subprogram to read 8-bit headers and colour-
REM  **     table from input file #1, and write 1-bit
REM  **     headers and colour-table to output file #2.
END SUB
```

Chapter 26
1-bit .BMP File

In this book, standard bitmap (BMP) files are used throughout as interfaces between computer inputs, processes, and outputs. The characteristics of computer files are described. The format of a 1-bit BMP file is defined, with examples. A 1-bit file has a colour-table with two entries, activated by a 1-bit index per pixel, value 0 or 1. The colour-table translates the index value into a (Blue, Green Red) triple. A program to edit 1-bit files is given.

26.1 Flowchart

The flowchart (Fig. 26.1) is for general guidance in this chapter.

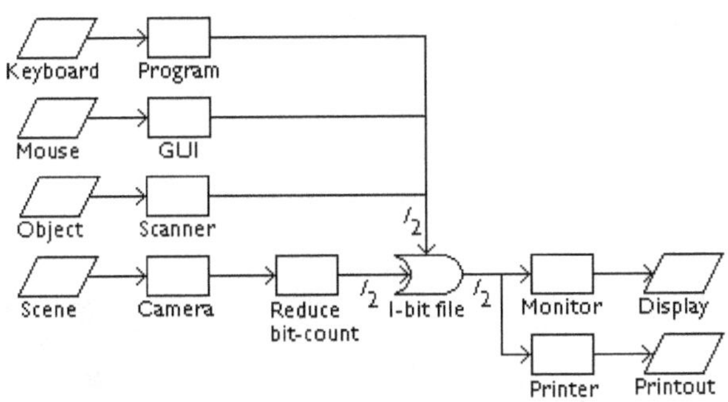

Fig. 26.1 Flowchart for 1-bit .BMP file. A 1-bit .BMP file can be created by a program or a GUI editor or a scanner or a camera (bit-reduced). It can be processed by a program. It can be displayed by a monitor or printed out by a printer. It acts as an interface between inputs, processes, and outputs

A. Parkin, *Digital Imaging Primer*, DOI 10.1007/978-3-540-85619-1_26

26.2 Memory

See Sect. 4.2.

26.3 File

See Sect. 4.4.

26.4 Data Representation

See Sect. 4.4.

26.5 Bitmap and Scan

A digital image is a rectangular array of *width*×*height* px, often called a *bitmap*. In a 1-bit image each pixel has intensity 0 or intensity 1. Figure 26.2 shows a 1-bit image of 8 × 8 px, as seen (tiny) in the top left corner of a PAINT screen. In order to write this rectangular image as a sequence of bits in a file, the array is scanned, starting at

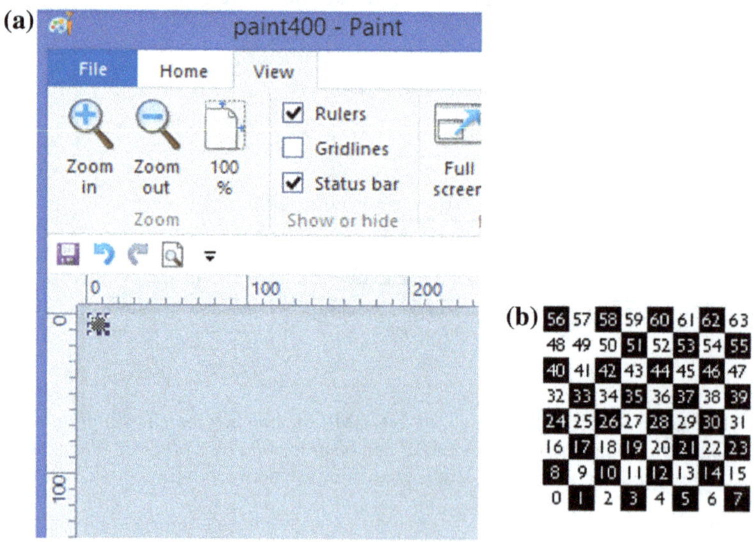

Fig. 26.2 Example of image bitmap. **a** 1-bit image of 8 × 8 px, as seen in the *top left* corner of a PAINT screen. **b** The bitmap enlarged, with the scan sequence

the left end of the bottom row (the origin of a first-quadrant window), going to the right end of the bottom row, then to the left end of the next row up, ..., to finish at the right end of the top row, as shown in the Figure. Each pixel is represented by one bit, and each sequence of 8 px by 1 B, written to the file.

Conversely, when reading the file, each byte specifies the intensities of 8 px which are displayed according to the scanning scheme.

26.6 .BMP File

Microsoft defines several *device-independent bitmap* (DIB) file structures for images, with the extension .BMP [1]. There are four distinct formats, according to how many bits are used for the colour of each pixel:

- 1 b/px, allowing 2 colours (monochrome or bi-level),
- 4 b/px, allowing 16 colours,
- 8 b/px, allowing 256 colours,
- 24 b/px, allowing 16,777,216 colours.

26.7 1-bit .BMP File

Table 26.1 shows the structure of a 1-b .BMP file. A .BMP file contains four groups of fields:

- File header,
- Info header,
- Colour-table,
- Image data.

The file header is a fixed-length group, and contains particulars of the file. The first field is the so-called Magic Cookie, ASCII BM = 66 77 decimal = 42 4D hexadecimal. Without this the file is rejected as not a .BMP file. The info header contains particulars of the image. In a 1-b .BMP file, the colour-table has two entries, indexed 0 and 1. Each entry occupies a 4-B *rgbquad*. The first byte is the blue intensity, the second byte is the green intensity, the third byte is the red intensity, and the fourth byte is empty. Each entry determines the colour intensities which are displayed or printed for an image pixel with that index. The default setting for the 1-bit colour-table is index 0 = (0, 0, 0) black, and index 1 = (F, F, F) white; but we can set it to any BGR values we please by using a .BMP file editor. The image data is a variable-length group containing, in scan-sequence, the index value 0 or 1 of each pixel in the image. Each index value occupies one bit, packed eight to the byte. A scan line must contain a whole number of 32-bit (4-B) *segments*, and it may be necessary to pad out a row with zeros to achieve this. We have in the file header:

Table 26.1 1-bit .BMP file format

Group	Field	Type	Byte	Byte	Remarks
FileHeader			1–14	01–0E	file type, size, layout
	bfType	integer	1–2	01–02	always 'BM'
	bfSize	long	3–6	03–06	file length in bytes
	bfReserved1	integer	7–8	07–08	always 0
	bfReserved2	integer	9–10	09–0A	always 0
	bfOffBits	long	11–14	0B–0E	header length in bytes
InfoHeader			15–54	0F–36	image size and colours
	biSize	long	15–18	0F–12	infoheader length in bytes
	biWidth	long	19–22	13–16	image width in pixels
	biHeight	long	23–26	17–1A	image height in pixels
	biPlanes	integer	27–28	1B–1C	always 0100
	biBitCount	integer	29–30	1D–1E	bits per pixel = 1
	biCompression	long	31–34	1F–22	not compr. = 0
	biSizeImage	long	35–38	23–26	image data length in bytes
	biXPelsPerMeter	long	39–42	27–2A	printer ppm, horizontal
	biYPelsPerMeter	long	43–46	2B–2E	printer ppm, vertical
	biClrUsed	long	47–50	2F–32	nr of colours actually used
	biClrImportant	long	51–54	33–36	nr of colours important
ColorTable			55–62	37–3E	array of 2 colours
	aColors[0]	rgbquad	55–58	37–3A	BGRZ for colour[0]
	aColors[1]	rgbquad	59–62	3B–3E	BGRZ for colour[1]
Image data			63–	3F–	in scan sequence
	aBitmapBits[0]	byte	63–	3F–	one bit per pixel
	aBitmapBits[1]	byte	–	–	
	...				
	aBitmapBits[*n*]	byte	–	–	

The groups contain fields of given types, which are represented physically in sequentially-numbered bytes of the file (here shown in decimal and hexadecimal notation). Field types are: *byte* occupying eight bits; *integer* occupying two bytes; *long* occupying four bytes; *rgbquad* occupying four bytes. Notice that rgbquad holds the colour values in the first three bytes, in sequence BGR, and the fourth byte is zero Z

- Rows per image = biHeight,
- Pixels per row = biWidth,
- Bits per pixel = biBitCount.

Hence:

- Bits per row = biWidth × biBitCount,
- Rounded-up 32-bit segments per row = ((biWidth × biBitCount) +31) \ 32,
- Bytes per row = (((biWidth * biBitCount) +31) \ 32) × 4.

26.8 Example of 1-bit .BMP File

As an example of a 1-bit .BMP file, we take the small 8 × 8-pixel image shown again in Fig. 26.3. This image is held in a file called EXAMPLE.BMP. The contents of this file can be seen in the hexadecimal dump, a utility available in many image editors, such as IRFANVIEW.

An annotated file listing is shown in Table 26.2. Bytes 39–42 and 43–46 hold values, horizontal and vertical, for printing the image. When these are 0, the printer defaults to 3189 ppm (= 3.189 ppmm = 81 ppi).

Calculation of BytesPerRow:

- Rows per image = biHeight = 8,
- Pixels per row = biWidth = 8,
- Bits per pixel = biBitCount = 1.

(a)

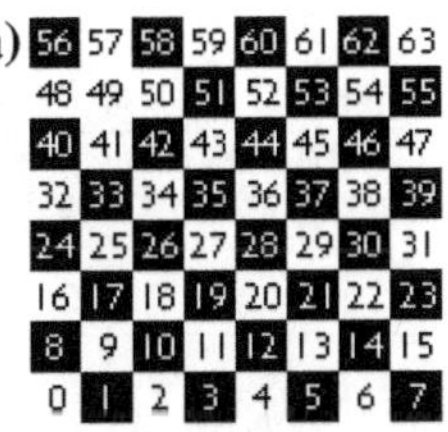

(b)

```
000: 42 4D|5E 00  00 00|00 00 |00 00|3E 00  00 00|28 00
010: 00 00|08 00  00 00|08 00  00 00|01 00 |01 00|00 00
020: 00 00|20 00  00 00|00 00  00 00|00 00  00 00|00 00
030: 00 00|00 00  00 00|00 00  00 00|FF FF  FF 00|AA 00
040: 00 00|55 00  00 00|AA 00  00 00|55 00  00 00|AA 00
050: 00 00|55 00  00 00|EA 00  00 00|55 00  00 00|
```

(c)

```
BM^........>...(.
..▮...▮...▮.▮...
.. ............
..........ÿÿÿ.ª.
..U...ª...U...ª.
..U...ê...U...
```

Fig. 26.3 Image and file EXAMPLE.BMP. **a** 8 × 8-pixel image (enlarged), with scan sequence. **b** Hexadecimal dump of file EXAMPLE.BMP, with field boundaries marked. Each pair of hexadecimal digits is the value of a byte. Starting at the *top* row, we have the 2-byte magic cookie field containing 42 4D (ASCII BM), the 4-byte file size field containing 5E, two 2-B fields containing 0, the 4-B header length 3E, the 4-B InfoHeader length 28, the 4-B image width 8, the 4-B image height 8, the 2-B planes 1, the 2-B bpp 1, the 4-B compression 0, the image data length 20 (that is 32 decimal = 8 rows of 4 B), four 4-B fields all 0, then the 4-B B G R Z for colour [0], all 0, and the 4-B B G R Z for colour [1] FF FF FF 0. Finally we have eight 4-B fields, one for each row of the image starting with the *bottom* row of the image. The rows are alternately AA = binary 10101010 and 55 = binary 01010101 zero-padded to fill 4 B; the exception is the *seventh row* EA = binary 11101010. **c** Byte-by-byte interpretation to ASCII printable characters

Table 26.2 Example of file EXAMPLE.BMP

Byte dec.	Field	Type	Value dec.	Value hex.	Remarks
1–14	FileHeader				
1–2	bfType	integer	66 77	42 4D	magic cookie
3–6	bfSize	long	94	5E	file length in bytes
7–8	bfReserved1	integer	0	0	
9–10	bfReserved2	integer	0	0	
11–14	bfOffBits	long	62	3E	header length in bytes
15–54	InfoHeader				
15–18	biSize	long	40	28	infoheader length, bytes
19–22	biWidth	long	8	8	image width in px
23–26	biHeight	long	8	8	image height in px
27–28	biPlanes	integer	1	1	
29–30	biBitCount	integer	1	1	bits per px
31–34	biCompression	long	0	0	not compressed
35–38	biSizeImage	long	32	20	image data len., bytes
39–42	biXPelsPerMeter	long	0	0	printer ppm, horiz.
43–46	biYPelsPerMeter	long	0	0	Printer ppm, vert.
47–50	biClrUsed	long	0	0	nr colours used
51–54	biClrImportant	long	0	0	nr colours important
55–62	ColorTable				
55–58	aColors[0]	rgbquad	0000	0000	BGRZ Black
59–62	aColors[1]	rgbquad	255..0	FFFFFF0	BGRZ White
63–94	image data				
63–66	aBitmapBits[0]	byte	10101010	AA	ColorTable values
67–70	aBitmapBits[1]	byte	01010101	55	
71–74	aBitmapBits[2]	byte	10101010	AA	
75–78	aBitmapBits[3]	byte	01010101	55	
79–82	aBitmapBits[4]	byte	10101010	AA	
83–86	aBitmapBits[5]	byte	01010101	55	
87–90	aBitmapBits[6]	byte	11101010	EA	
91–94	aBitmapBits[7]	byte	01010101	55	

Hence:

- Bits per row = biWidth $\times$ biBitCount = $8 \times 1 = 8$.
- Rounded-up 32-bit segments per row = ((biWidth $\times$ biBitCount) +31) \ 32=((8 $\times$ 1) + 31) \ 32 = 1.
- Bytes per row = (((biWidth * biBitCount) +31) \ 32) $\times$ 4 = 4.

26.9 1-bit .BMP File Editor

If we wish to have interactive access to BMP files to view or change header, colour-table, or image data values, we need a .BMP file editor. Such an editor is briefly described below, with source code at the SpringerLink web-site.

```
REM  **  Program EDBMP1.BAS to edit 1-bit BMP file
REM  **  (width a multiple of 32 pixels). User
REM  **  enters filename to edit. Program gets
REM  **  file and offers menu to edit headers
REM  **  or colour-table or image-array.
REM  **  Data
REM  **  Read data
REM  **  Main program
REM  **  Enter file-name to edit
REM  **  Show menu
            CALL HEADER(FILENAME$)
            CALL COLTAB(FILENAME$)
            CALL IMAARR(FILENAME$)
FINIS:   END

         FUNCTION BITS8$ (H AS STRING)
REM  **  Function BITS8$(H AS STRING) to convert a
REM  **  hexacimal byte to a binary byte. String H
REM  **  contains either one or two ASCII characters
REM  **  "0" to "FF". This function replaces each nibble
REM  **  by a four-character ASCII string "0000" to
REM  **  "1111", and returns the eight-character string
REM  **  of "0"s and "1"s corresponding to H.
END FUNCTION
         SUB COLTAB (FILENAME$)
REM  **  Sub-program COLTAB to edit 1-bit colour-table.
END SUB
         SUB HEADER (FILENAME$)
REM  **  Sub-program HEADER to edit header values of
REM  **     BMP file. Reads each field of header, shows
REM  **     old value, offers next or change or exit.
END SUB
         SUB IMAARR (FILENAME$)
REM  **  Sub-program IMAARR to edit 1-bit image-array.
REM  **  User has  option to start at any row Y, from
REM  **  0 to (height - 1). Program shows groups of 8
REM  **  pixel values 0 or 1, and offers options of
REM  **  next or change or exit.
END SUB
```

Reference

1. Microsoft (2005) DIBs and Their Use. http://msdn.microsoft.com/en-us/library/ms969901. Accessed 1 Feb 2014

Chapter 27
1-bit Display

An image composed of pixels can be displayed on an additive-light device, which has Blue, Green, and Red sub-pixels in each pixel. For a 1-bit image, when all three sub-pixels are off the screen shows Black, and when all three sub-pixels are full-on the screen shows White. Display size variables are defined, and a measure for display resolving power is given, with examples. 1-bit display intensity variables and colours are defined.

27.1 Flowchart

To view an image we can display it on a monitor screen. The flowchart (Fig. 27.1) shows a 1-bit image file sent to a display by an image editor.

27.2 Display Hardware

A *monitor* is a digital-to-analogue device which displays additive-colour images by emitting light with varying intensity per pixel. It accepts a scan-sequence of digital pixel locations (x, y), with, for each pixel, three integer intensities of Blue, Green, and Red (B, G, R). It displays the pixels on a screen, each in its proper location and each showing its proper light intensity. The functional parts of a monitor are,

- scanning circuitry,
- digital-to-analogue converter (DAC),
- display screen.

Much of the functionality is provided in a graphics card [4] between the computer and the monitor itself.

A. Parkin, *Digital Imaging Primer*, DOI 10.1007/978-3-540-85619-1_27

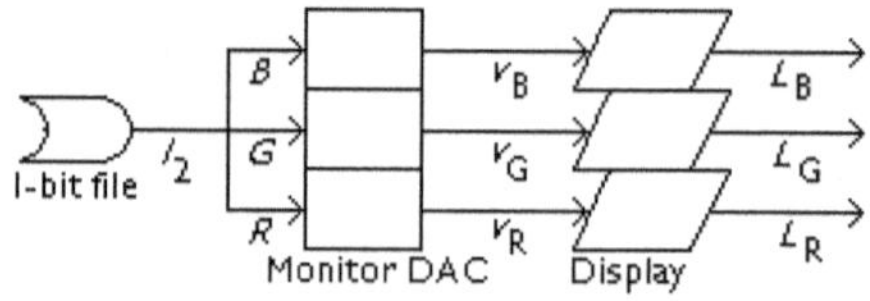

Fig. 27.1 Flowchart for 1-bit display. A 1-bit image file has, per-pixel, intensity index $I_2 = 0$ or 1. Index value 0 sends (Blue, Green, Red) values (0, 0, 0), and index value 1 sends (Blue, Green, Red) values (255, 255, 255) to the three channels in a colour monitor. In each channel, a digital-to-analogue converter sends a low or high voltage v to the appropriate screen sub-pixels, which emit low or high luminance L_B, L_G, L_R

27.3 Scanning Circuitry

Scanning circuitry uses the input image location coordinates (x, y) to activate the appropriate screen pixels, and passes the integer intensities (B, G, R) into three distinct channels.

27.4 Digital-to-Analogue Converter (DAC)

A DAC is a device which takes an integer-valued (digital) input I and smooths it to a continuous real-valued (analogue) voltage v. It is thus opposite in function to an ADC (see Sect. 22.7). A colour monitor has one such DAC in each channel, to convert incoming 8-bit-per-pixel intensities (B, G, R) to outgoing continuous voltages (v_B, v_G, v_R).

27.5 Display Screen

Display screens for personal computers have a remarkable history [3], and are still in active development. From the 1980s, a cathode ray tube (CRT) was used. In a CRT display, the screen has an array of Blue Green and Red phosphor spots. A BGR triad constitutes one screen pixel. A perforated mask is placed between the three electron guns and the screen, so that as the the Blue gun scans it activates only Blue spots, the Green gun Green spots, and the Red gun Red spots, in each triad. Figure 27.2a shows the arrangement.

In the 1990s, liquid crystal displays (LCD) were developed, and as thin film transistor liquid crystal displays (TFT-LCD) are now standard, virtually replacing CRTs. In an LCD display the sub-pixels are triads of liquid crystal bars, activated by back-light to show varying intensities of light through Blue, Green, or Red filters [1]. Figure 27.2b shows the arrangement.

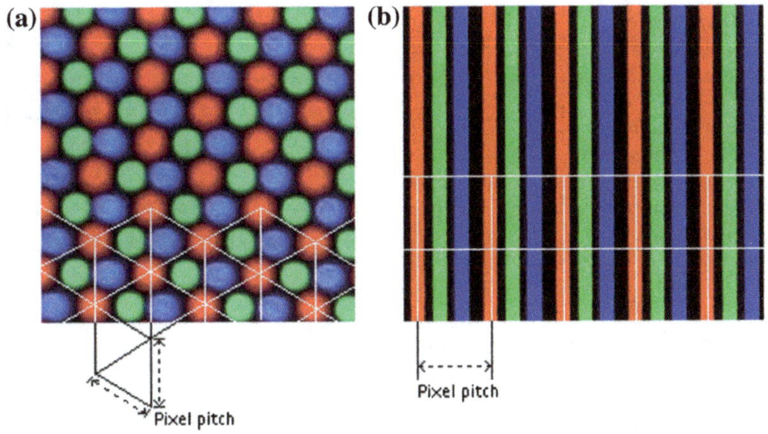

Fig. 27.2 Sub-pixels in a display screen. **a** Phosphor spots in a CRT screen. **b** Liquid crystal bars in an LCD screen. Dot pitch measures the resolution of a screen

In the 2000s, organic light-emitting diodes (OLED) became available, working without a back-light.

CRTs are cheap and offer excellent contrast, response time, and variable pixel resolution, compared with LCDs [5], but are much more bulky. LCDs are flat, require little power, and have fixed pixel resolution [6]. OLEDs are currently expensive and have relatively short working life, but will no doubt continue to improve.

27.6 Display Size Variables

Figure 27.3 shows sizes[1] on a display screen, and Table 27.1 shows their relations. Given the asterisked values, the rest can be calculated [7].

For a particular monitor, screen width w, height h, and aspect ratio $w : h$ are fixed by the manufacturer. For a CRT, count of pixels P and Q are settable to one of the available standards [2, 8] by the operating system. For example, a typical CRT display might have

- screen width $w = 11.26\,\text{in} = 286\,\text{mm}$,
- screen height $h = 8.45\,\text{in} = 215\,\text{mm}$,
- count of pixels in width $P = 1024\,\text{px}$ (XGA standard)
- count of pixels in height $Q = 768\,\text{px}$ (XGA standard),
- pixel width $a = 11.26/1024 = 0.011\,\text{in} = 0.28\,\text{mm}$,
- pixel frequency $F_{\text{P}} = 1/0.011 = 91\,\text{ppi} = 3.58\,\text{ppmm}$,
- screen aspect ratio $w : h = 4 : 3$.

[1]Display length unit is either inch or metre. Metres = inches ×0.0254; inches = metres ×39.37; millimetres = inches ×25.4; inches = millimetres ×0.0394.

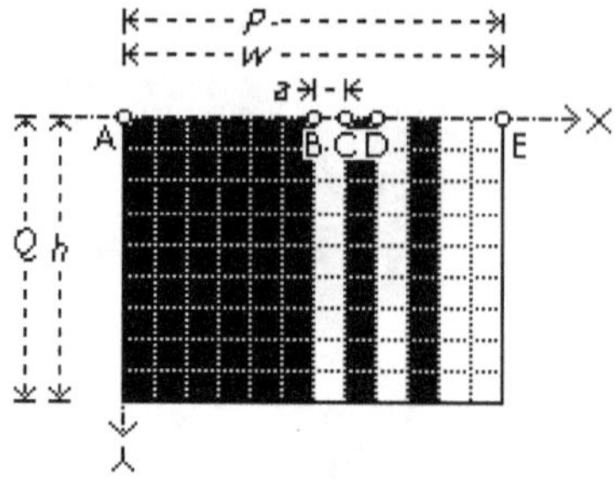

Fig. 27.3 Display size variables. Screen coordinate origin is at top left; x-axis is to the right; y-axis is downwards. w is width of screen; h is height of screen; P is count of pixels in width of screen; Q is count of pixels in height of screen; a is width of pixel

Table 27.1 Display size variables, as shown in Fig. 27.3

DISPLAY			
Screen pixel width	a	$= w/P$	Length units
Screen pixel freq.	F_P	$= 1/a$	Pixels per unit len.
Screen height *	h		Length units
Count pixels in width *	P		Pixels
Count pixels in height *	Q		Pixels
Screen width *	w		Length units
Screen aspect ratio	$w : h$		

Given values for the asterisked variables, the rest can be calculated

For an LCD, on the other hand, the native count of pixels P and Q are set by the manufacturer, to one of the many available standards. A high-end LCD display might have

- screen width $w = 20.35\,\text{in} = 517\,\text{mm}$,
- screen height $h = 12.72\,\text{in} = 323\,\text{mm}$,
- count of pixels in width $P = 1920\,\text{px}$ (WUXGA standard),
- count of pixels in height $Q = 1200\,\text{px}$ (WUXGA standard),
- pixel width $a = 20.35/1920 = 0.0106\,\text{in} = 0.27\,\text{mm}$,
- pixel frequency $F_P = 1/0.0106 = 94.3\,\text{ppi} = 3.71\,\text{ppmm}$,
- screen aspect ratio $w : h = 16 : 10$.

However, images are usually displayed in a GUI editor, such as PAINT, in a window of settable size. And an editor usually offers enlargements of the pixel width a, such as $\times 2$, $\times 4$, $\times 6$, and $\times 8$, thus reducing the pixel frequency F_P by factor 1/2, 1/4, 1/6, and 1/8.

27.7 Resolving Power of Display

The resolving power, or resolution limit, of a display is the finest detail which it can render. The resolving power can be measured as the highest-frequency grating which can be correctly displayed. A grating is an object with alternating Black and

White lines, as at BC, CD, ... in Fig. 27.3; so the finest detail is a line-pair, that is one Black line and one White line, BC and CD. At the resolution limit, successive line-pairs of the grating occupy successive pairs of pixels of the display. Clearly, the line-pair frequency of an object grating at the resolution limit of a display, F_{LP}, is half the pixel frequency of the display. So we have the linear resolution limit of a display:

$$F_{\mathrm{LP}} = F_{\mathrm{P}}/2 \,\mathrm{lppul}\,.$$

For example, if pixel frequency F_{P} is 9 ppi (or 3.58 ppmm), the resolution limit is line-pair frequency $F_{\mathrm{LP}} = 91/2 = 45.5$ lppi (or $3.58/2 = 1.79$ lppmm).

Until about 2010, most LCD display screens had native resolution between 72 and 100 ppi. Apple introduced the Retina Display range [10] for the iPad (and also for iPhones, iPod, and MacBook), which has between 220 and 326 ppi. Very many resolutions are now available [9].

27.8 Display Sizes from Program/GUI Image

Suppose a (virtual) object is created as a space graph (wire-frame model)(see Sect. 7.7) then projected to an image (see Chap. 8), or is directly constructed as an image array by a program, or by hand in a GUI image editor window. The program or GUI outputs a file of $P \times Q$ px of unspecified size. If we know the pixel frequency F'_{P} of the display, we can calculate other relevant variables. For example, Table 27.2 shows calculations for a constructed grating 100×10 px, displayed at two different pixel frequencies as in Fig. 27.4. In Fig. 27.4a, the PAINT display pixel frequency $F_{\mathrm{P}} = 91$ ppi (3.58 ppmm). In Fig. 27.4b, PAINT is operating at four-times enlargement (with pixels marked), so the effective pixel frequency is $91/4 = 22.75$ ppi (0.90 ppmm).

Table 27.2 Size variables for image by program/GUI and display, as shown in Fig. 27.4

PROGRAM/GUI			(a)	(b)	
Count of pixels in width *	P		100	100	px
Count of pixels in height *	Q		10	10	px
DISPLAYED IMAGE					
Display pixel width	a'	$= 1/F'_{\mathrm{P}}$	0.011	0.088	in
Display pixel frequency *	F'_{P}		91	22.75	ppi
Display resolution limit	F'_{LP}	$= F'_{\mathrm{P}}/2$	45.5	11.375	lppi
Image height	h'	$= Q/F'_{\mathrm{P}}$	0.11	0.44	in
Image width	w'	$= P/F'_{\mathrm{P}}$	1.10	4.40	in

Asterisked variables are given, the rest calculated

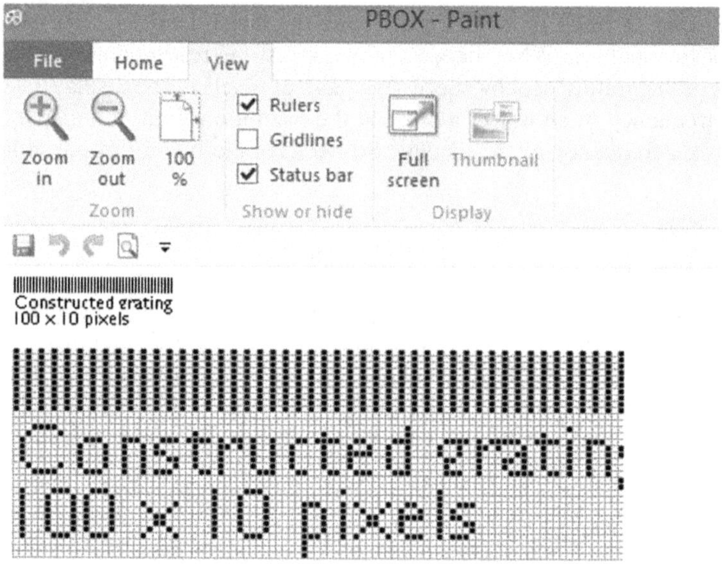

Fig. 27.4 Constructed grating 100 × 10 px. **a** Displayed at 91 ppi, **b** Displayed at four-times enlargement, 22.75 ppi

The size of a displayed image is limited upwards only by the size of memory available. The displayed image size $w' \times h'$ may well be greater than the display window size, and the aspect ratio $P : Q$ different from the display screen aspect ratio; but then the image can be panned and scrolled within the screen window.

The resolution limit of an image created by a program or GUI then displayed is determined solely by the display resolution F'_{LP}.

For another example, suppose we wish to construct a millimetre scale in a GUI editor, to display at true size:

- With unit length = mm
- and given count of pixels in screen width $P_{\mathrm{w}} = 1024\,\mathrm{px}$
- and given screen width $w = 286\,\mathrm{mm}$.
- Then pixel width $j = w/P_{\mathrm{w}} = 286/1024 = 0.28\,\mathrm{mm}$.
- Required displayed length of scale $= 50\,\mathrm{mm}$.
- Then required pixel count of scale $= 50/0.28 = 179\,\mathrm{px}$.

The (real) spacing of the marks for a mm scale 50 mm long would thus be multiples of $179/50 = 3.58\,\mathrm{px}$. But pixels are indivisible, so the best that can be done is to round the spacing to integers 0, 4, 7, 11, 14, …, 180 px. Figure 27.5 shows such a scale.

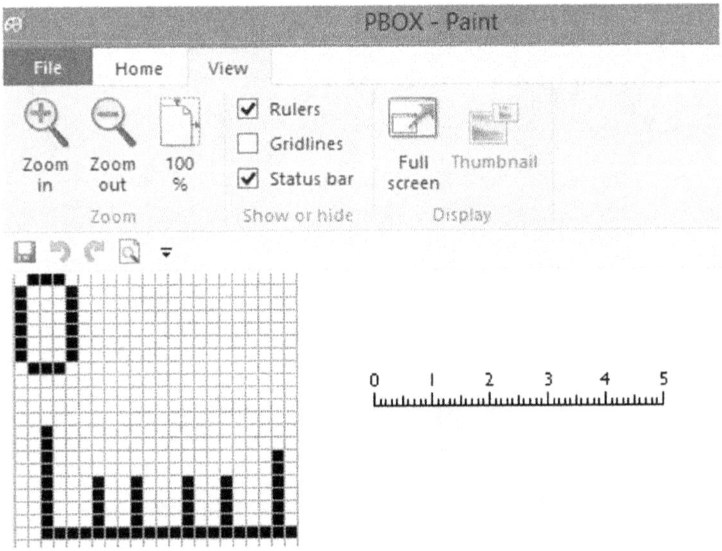

Fig. 27.5 Millimetre scale constructed for display at true size. **a** Enlarged view of part of a millimetre scale with marks spaced at rounded pixel distances 0, 4, 7, 11, 14, ... **b** Display of 50-millimetre scale

27.9 Display Sizes from Scanner Image

Suppose an actual object is small, accessible, and more or less flat. It can be scanned in 1-bit mode (see Chap. 22), with a settable sampling frequency F_S, typically in the range 30–600 spi (1.18–23.62 spmm). High-performance scanners can go to much higher sampling frequencies which exceed the acuity of even the best eye. The size of the scanned object is limited by the size of the scanner window $w \times h$. The magnification of a displayed image relative to the object is determined as $m = F_S / F'_P$. If the scan frequency is lower than the display frequency, the image is diminished; if the frequencies are equal the image is same-size; and if the scan frequency is higher than the display frequency the image is enlarged. The displayed image size $w' \times h'$ may well be greater than the display window size, and the aspect ratio different from the display screen aspect ratio; but the image can be panned and scrolled within the screen window.

The resolution limit of the displayed image which is presented to the eye is the lower of the sampling resolution F_{LP} and the display resolution F'_{LP}, that is, $(\text{MIN}(F_S, F'_P))/2$.

A convenient object is a test card produced commercially for photographers, with a line-pair gradient, an in/cm scale, 6-colour patches, and a grey-scale step-wedge. For example, Fig. 27.6a shows the test object scanned in 1-bit mode at 30 spi and displayed in PAINT at 91 ppi. The scan frequency is less than the display frequency, so the image is diminished, and the resolution limit is half the scan frequency. In

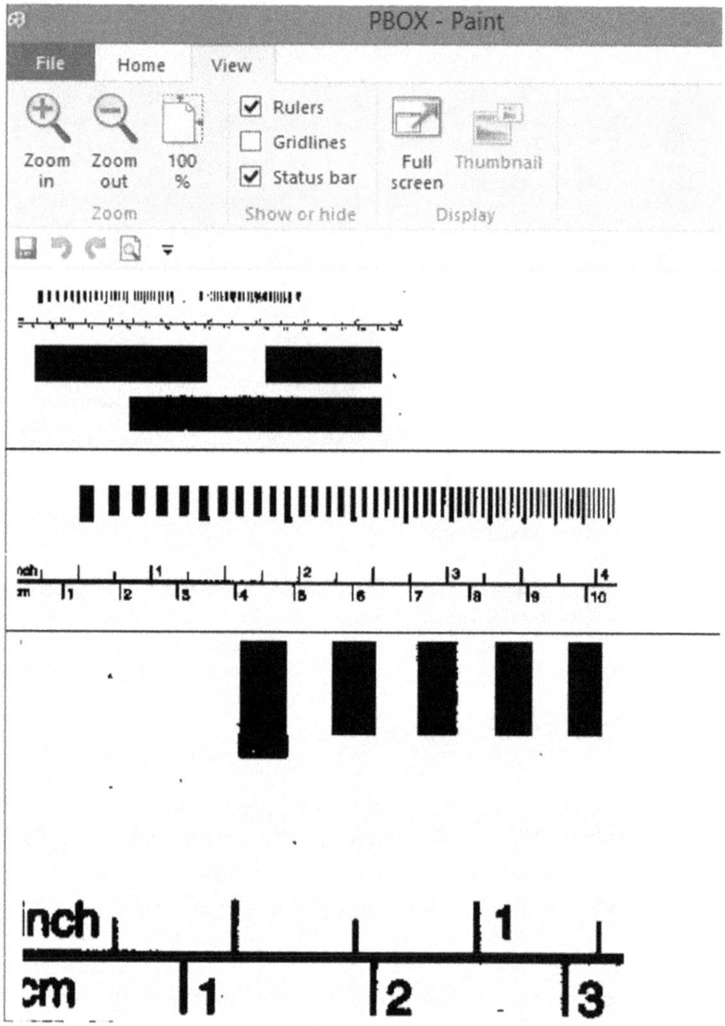

Fig. 27.6 Three scans displayed. **a** Test card scanned at $F_S = 30$ spi, displayed at $F_P = 91$ ppi, so magnification $m = 30/91 = 0.33$. **b** Scanned at $F_S = 91$ spi, displayed at $F_P = 91$ ppi, so magnification $m = 91/91 = 1$. **c** Scanned at $F_S = 300$ spi, displayed at $F_P = 91$ ppi, so magnification $m = 300/91 = 3.30$

Fig. 27.6b the object is scanned at 91 spi and displayed at 91 ppi, so the image is the same size as the object, and the resolution limit is half the scan and display frequency. In Fig. 27.6c the object is scanned at 300 spi and displayed at 91 ppi. The scan frequency is greater than the display frequency, so the image is magnified, but the resolution limit is still half the display frequency. Table 27.3 shows the values.

For displaying a camera image, see Sect. 49.4.

Table 27.3 Examples of size variables for scanned and displayed images in Fig. 27.6

SCANNED OBJECT			(a)	(b)	(c)	
Sample width	a	$= 1/F_S$	0.033	0.011	0.0033	in
Sampling frequency *	F_S		30	91	300	spi
Resolution limit	F_{LP}	$= F_S/2$	15	45.5	150	lppi
Object height *	h		6	6	6	in
Object width *	w		8	8	8	in
COMMON						
Magnification image/object	m	$= F_S/F_P$	0.33	1	3.3	
Count of samples in width	S	$= w/a$	240	728	2424	Samples
Count of samples in height	T	$= h/a$	90	546	1818	Samples
DISPLAYED IMAGE						
Image pixel width	a'	$= 1/F_P$	0.011	0.011	0.011	in
Image pixel frequency *	F'_P		91	91	91	spi
Image resolution limit	F'_{LP}	$= F'_P/2$	15.5	45.5	45.5	lppi
Image height	h'	$= m \times h$	1.98	6	19.8	in
Image width	w'	$= m \times w$	2.64	8	26.4	in

Given values for the asterisked variables, the rest can be calculated

27.10 1-bit Display Intensity Variables

When a monitor is used to display a 1-bit image, we have the following per-pixel intensity variables:

- Intensity index I_2,
- Colour-table intensity triple (B, G, R),
- Voltage triple (v_B, v_G, v_R),
- Luminance L.

Intensity index is $I_2 = 0$ or 1, held in the image section of the .BMP file (see Chap. 26).

Colour-table intensity triple is $B = G = R = 0$ or 255 (other settings are possible), held in the colour-table of the .BMP file. It defines the Blue, Green, and Red intensities to be displayed.

Voltage triple is the (normalized) analogue voltage output by the DAC of each channel, proportional to the colour-table intensity triple.

Luminance is the (normalized) light intensity output by the screen. For the measurement of luminance, see Chap. 15. The actual luminance of a CRT display can be varied by a control for White luminance (oddly labelled "Contrast") and a control for Black luminance (labelled "Brightness"). When properly adjusted a good CRT has a maximum White luminance of about 170 cd/m^2, and a minimum Black luminance of about 0.01 cd/m^2. An LCD display has a maximum luminance of about 160 cd/m^2, and a minimum luminance of about 0.72 cd/m^2 (with minimum back-light) [5].

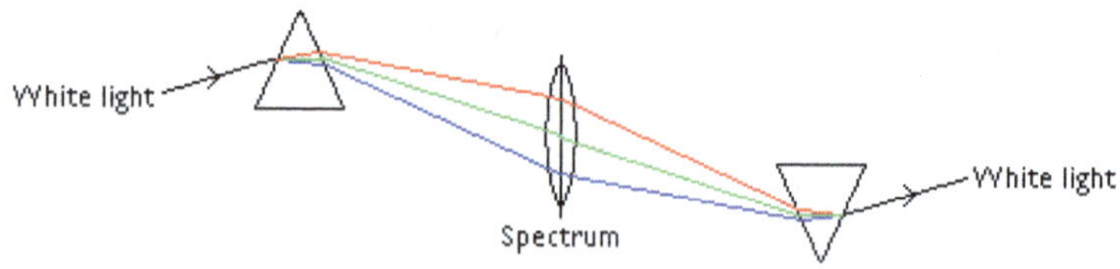

Fig. 27.7 Additive light. White light is refracted by a prism to a spectrum of colours. A spectrum of colours is converged by a lens to White light

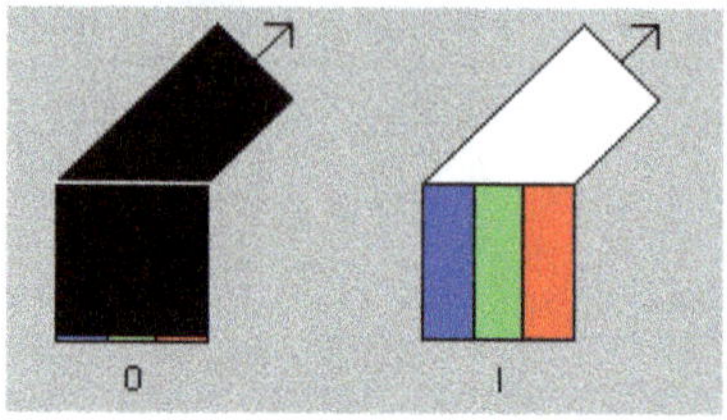

Fig. 27.8 1-bit display colours. **a** Pixel receives three low voltages, all three sources are off, so looks Black. **b** Pixel receives three high voltages, all three sources are full on, so looks White

27.11 Additive Light

Figure 27.7 shows Newton's clinching demonstration that White light, refracted by a prism, separates to a continuous spectrum of colours, Red at one end, Orange Yellow and Green in the middle, and Blue Indigo and Violet at the other end. And conversely, a spectrum of colours, converged by a lens, combine as White light.

27.12 1-bit Display Colours

Figure 27.8a shows a display pixel with intensity index $I_2 = 0$, (B, G, R) triple $= (0, 0, 0)$, all three voltages zero, all three sub-pixel sources off, so the pixel shows Black (the natural colour of the screen). Figure 27.8b shows a display pixel with intensity index $I_2 = 1$, (B, G, R) triple $= (255, 255, 255)$, all three voltages high, all three sub-pixel sources on, so the pixel shows White by additive mixing.

References

1. Case (2004) Introduction to LCDs. http://plc.cwru.edu/tutorial/enhanced/files/textbook.htm. Accessed 1 Feb 2014
2. Equasys (2014) Standard resolutions. http://www.equasys.de/standardresolution.html. Accessed 1 Feb 2014

3. History of the PC Display (2008). http://gadgets.softpedia.com/news/History-of-the-PC-Display-041-01.html. Accessed 1 Feb 2014
4. How Graphics Cards Work (2014). http://www.howstuffworks.com/graphics-card.htm. Accessed 1 Feb 2014
5. Soneira RM (2005) Shoot-out. http://www.displaymate.com/ShootOut. Accessed 1 Feb 2014
6. Steer WA (2003) LCDs for professionals. http://www.techmind.org/lcd/index.html. Accessed 1 Feb 2012
7. Sven (2014) Dots-per-inch calculator. http://www.sven.de/dpi/. Accessed 1 Feb 2014
8. Wikipedia (2014) Computer display standards. http://en.wikipedia.org/wiki/Computer_display_standard. Accessed 1 Feb 2014
9. Wikepedia (2014) List of displays by pixel density. http://en.wikipedia.org/wiki/List_of_displays_by_pixel_density. Accessed 1 Feb 2014
10. Wikipedia (2014) Retina display. http://en.wikipedia.org/wiki/Retina_Display. Accessed 1 Feb 2014

Chapter 28
1-bit Printout

A printer is a subtractive-light output device which puts coloured ink on a suitable White surface. For a 1-bit image only Black ink is needed. Black ink absorbs nearly all of the incident light, and unprinted paper reflects nearly all of the (normally White) incident light. Print size variables are defined, and a measure for printer resolving power is given, with examples. 1-bit print intensity variables and colours are defined.

28.1 Flowchart

To view an image we can print it on paper. The flowchart (Fig. 28.1) shows a 1-bit image file sent to a printer by an image editor.

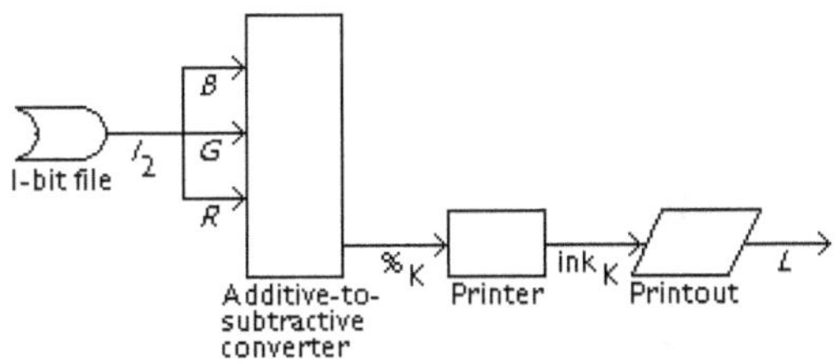

Fig. 28.1 Flowchart for 1-bit printout. Per-pixel, a 1-bit image file has intensity index $I_2 = 0$ or 1, pointing to colour-table (B, G, R) triple (0, 0, 0) or (255, 255, 255). Given triple (0, 0, 0) an additive-to-subtractive converter outputs $K = 100\,\%$, or given (255, 255, 255) outputs $K = 0\,\%$. A printer places *Black* ink on paper for a $K = 100\,\%$ pixel, or no ink for a $K = 0\,\%$ pixel. A printout reflects incident light with luminance L, low for a *Black* pixel or high for a *White* pixel

A. Parkin, *Digital Imaging Primer*, DOI 10.1007/978-3-540-85619-1_28

28.2 1-bit Printer Hardware

The first printers for personal computers were meant to output text (see Sect. 5.5). They worked rather like a typewriter, where single characters, arranged on a daisy-wheel or formed by a matrix of protruding pins, struck through an inked ribbon onto paper. Current PC printers are meant to output any digital image composed of pixels by placing tiny dots of ink on paper or a similar substrate. The dots are much smaller than the pixels of a digital image. The two leading technologies are ink-jet [3] and laser [2].

A *bubble-jet* printer[1] ejects ink droplets, formed under software control so that they land on the paper as correctly-located dots. A *continuous-jet* printer ejects a stream of droplets which are steered electro-statically either to land on the paper in the required locations or to hit a gutter for recycling. Ink-jet printers process and print stripe-by-stripe across, then down, a page. A typical fundamental dot pitch for an ink-jet printer is 300–600 dpi.

A *laser* printer[2] scans a laser beam across an electrically-charged drum to form a latent electro-static image of a page. The drum then picks up resinous ink powder, transfers it to paper, and fixes it by heat. Laser printers process and print one page at a time. A typical fundamental dot pitch is 600–1200 dpi.

Much of the functionality for a printer is provided in driver software [4] between the computer and the printer itself.

A digital image file can also be sent to other types of printer, for special purposes:

- Dye-sublimation [5], for very good colour control, permanence, and printing on fabrics,
- Photo-setter, for very high quality (2400 dpi) imaging on photo-sensitive paper roll,
- Plate-setter, for very high quality imaging directly on offset-lithography plates,
- Electronic-ink [1] printer, for erasable images.

28.3 Print Size Variables

Figure 28.2 shows sizes on a printout sheet, and Table 28.1 shows their relations.[3] The fundamental constants of a printer, fixed by the manufacturer, are sheet size, and print dot width d. Sheet size limits the largest printed image size $w \times h$ (that is, sheet size less margins). Print dot width limits the smallest print pixel width a.

Print pixel frequency F_P maps the fixed width of print dot d to the required width of print pixel a, thus determining printed image width w. In ordinary use it is calculated

---

[1]Invented at Canon laboratories around 1957.

[2]Invented at the Xerox Palo Alto Research Center around 1970.

[3]Printer sizes are commonly in inch units. Metres = inches ×0.0254; inches = metres ×39.37; millimetres = inches ×25.4; inches = millimetres ×0.0394.

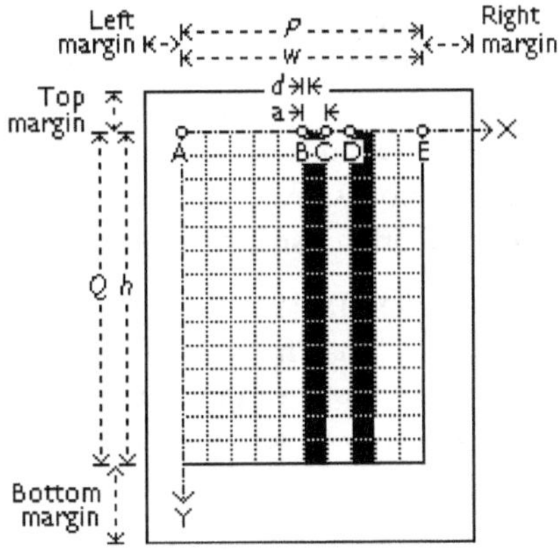

Fig. 28.2 Print size variables. Printed image coordinate origin is A; x-axis is to the right; y-axis is downwards. P is count of pixels in width and Q in height of image; w is width and h is height of printed image; d is dot width; a is pixel width

Table 28.1 Printer size variables, as shown in Fig. 28.2

PRINTOUT			
Print pixel width	a	$= w/P$	length units
Print dot width *	d		length units
Print dot frequency	F_D	$= 1/d$	dots per unit length
Print dot/pixel ratio		$= F_D/F_P$	dots per pixel
Print pixel freq. *	F_P		pixels per unit length
Print resolution lim.	F_{LP}	$= F_P/2$	line-pairs per unit length
Print height	h	$= P/Q \times w$	length units
Count pixels in width *	P		pixels
Count pixels in height *	Q		pixels
Print width	w	$= P/F_P$	length units

Given values for the asterisked variables, the rest can be calculated

by the software supplied with a printer, which manages margins and required printed image width. In a .BMP image file, F_P is held in the header fields *biXPelsPerMeter* and *biYPelsPerMeter* (see Sect. 26.7), normally both set to the same value, in ppm. When MS PAINT displays an image it shows the print pixel frequency as Image—Attributes—"Resolution $n \times m$ dpi", meaning in the terms used here ppi. But PAINT does not offer a way of changing the file value. If we want to inspect or change the file values directly we need a bitmap file editor (see Sect. 26.9). A scanner usually sets the print pixel frequency equal to the scan sample frequency, so that the image prints same-size as the scanned object. If the file has default value 0, MS PAINT treats it as 81 × 81 dpi, meaning ppi in the terms used here.

Setting print pixel frequency F_P in the .BMP file gives precise control of print size. For example, suppose we have a file image of a Snellen chart (see Sect. 10.6) where the biggest letter has pixel count $P = 99$ px wide, and we want to print it for standard 6-m viewing so that the biggest letter has width $w = 87$ mm. Say the printer has a fixed print dot frequency $F_D = 300$ dpi $= 11{,}811$ dpm. We calculate:

- Given pixel count $P = 99\,\text{px}$
- and required width of printout $w = 87\,\text{mm}$ (or 3.42 in),
- then print pixel frequency $F_P = 99/87 = 1.138\,\text{ppmm} = 1138\,\text{ppm}$ (or 29 ppi).
- Set header fields *biXPelsPerMeter* and *biYPelsPerMeter* in image file to 1138.
- PAINT now shows "Resolution 29 × 29 dpi".
- Tell the printer software to print as "100 % normal size" or similar command.

For example, Table 28.2 shows a range of print sizes in inch units for an ink-jet printer which has fixed print dot frequency $F_D = 300\,\text{dpi}$. For a given pixel count $P = 100\,\text{px}$ and variable print pixel frequency F_P, the printout has width w in. Table 28.3 shows the same range of print sizes in mm units.

Table 28.2 Example of ink-jet print sizes, in inches

F_D*	P*	F_P*	w	a	DPP	F_{LP}
300	100	200	0.5	0.005	1.5	100
300	100	100	1	0.010	3	50
300	100	91	1.10	0.011	3.3	45.5
300	100	81	1.23	0.012	3.8	40.5
300	100	50	2	0.02	6	25
300	100	25	4	0.04	12	12.5
300	100	12.5	8	0.08	24	6.25

Asterisked variables are given, and the rest are calculated. F_D is fixed dot frequency in dpi; P is given width of image in px; F_P is settable print pixel frequency in ppi. Then $w = P/F_P$ is width of printout in in; $a = w/P$ is pixel width in in; DPP $= F_D/F_P$ is dpp; and $F_{LP} = F_P/2$ is resolution limit in lppi. $F_P = 91\,\text{ppi}$ is a typical display screen frequency. $F_P = 81\,\text{ppi}$ is the MS PAINT default

Table 28.3 Example of ink-jet print sizes, in mm

F_D	P	F_P	w	a	DPP	F_{LP}
11.81	100	7874	12.70	0.127	1.5	3.9
11.81	100	3937	25.40	0.254	3	2.0
11.81	100	3583	27.91	0.279	3.3	1.8
11.81	100	3189	31.35	0.313	3.8	1.6
11.81	100	1968	50.80	0.508	6	1.0
11.81	100	984	101.60	1.016	12	0.5
11.81	100	492	203.20	2.032	24	0.25

F_D is fixed dot frequency in dpmm; P is given width of image in px; F_P is variable print pixel frequency in ppm. Then $w = P/F_P$ is width of printout in mm; $a = w/P$ is pixel width in mm; DPP $= F_D/F_P$ is dpp, and $F_{LP} = F_P/2$ is resolution limit in lppmm

28.4 Resolving Power of Printer

The resolving power, or resolution limit, of a printer is the finest detail which it can render. The resolution limit is usually measured as the line-pair frequency F_{LP} of a grating of alternating White and Black lines, when one line-pair occupies a pair of print pixels, as BC, CD, ... in Fig. 28.2. The resolution limit is half the print pixel frequency (see Sect. 10.5):

$$F_{\mathrm{LP}} = F_{\mathrm{P}}/2 \,.$$

For example, if pixel frequency F_{P} is 300 ppi (= 11.81 ppmm = 11,811 ppm), the resolution limit F_{LP} is 300/2 = 150 lppi (= 5.9 lppmm = 5905 lppm). The printer resolution limit is shown in Tables 28.2 and 28.3 for each case.

28.5 1-bit Image by Program/GUI and Printer

Given the pixel counts $P \times Q$ produced by a program or GUI, and the pixel frequency F'_{P} set for a printer, we can calculate other relevant variables. For example, Table 28.4 shows calculations for a constructed grating 100 × 10 px, printed at two different pixel frequencies as in Fig. 28.3. In Fig. 28.3a, the printer pixel frequency $F_{\mathrm{P}} = 30$ ppi (1.18 ppmm). In Fig. 28.3b, $F_{\mathrm{P}} = 300$ ppi (11.81 ppmm) (which loses all detail in this reproduction).

The size of a printed image is limited upwards only by the size of memory available. The printed image size $w' \times h'$ may well be greater than the printer paper size, and the aspect ratio $P : Q$ different from the paper aspect ratio; but then the image can be tiled across several sheets.

The resolution limit of the printed image presented to the eye is the chosen printer resolution F'_{LP}.

Table 28.4 Size variables for program/GUI image and printer, as shown in Fig. 28.3

PROGRAM/GUI			(a)	(b)	
Count of pixels in width *	P		100	100	pixels
Count of pixels in height *	Q		10	10	pixels
PRINTED IMAGE					
Printer pixel width	a'	$= 1/F'_{\mathrm{P}}$	0.033	0.0033	in
Printer pixel frequency *	F'_{P}		30	300	ppi
Printer resolution limit	F'_{LP}	$= F'_{\mathrm{P}}/2$	15	150	lppi
Image height	h'	$= Q/F'_{\mathrm{P}}$	0.33	0.033	in
Image width	w'	$= P/F'_{\mathrm{P}}$	3.33	0.33	in

Asterisked variables are given, the rest calculated

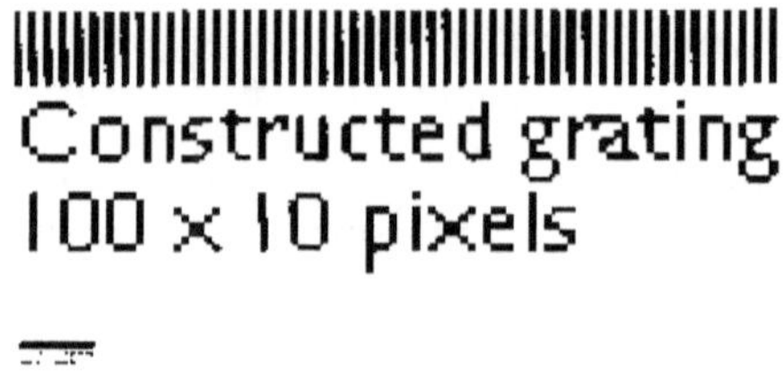

Fig. 28.3 Constructed grating 100 × 10 px. **a** Printed at $F_P = 30$ ppi. **b** Printed at $F_P = 300$ ppi

For another example, suppose we wish to construct a mm scale in a GUI editor, to print at correct size. We make the scale of alternating Black and White pixels, and print it so that 1 px prints as 1 mm wide. We calculate:

- Unit = m,
- Required printed pixel width $a = 0.001$ m,
- Hence print pixel frequency $F_P = 1/a = 1/0.001 = 1000$ ppm.

Fig. 28.4 shows such a scale.

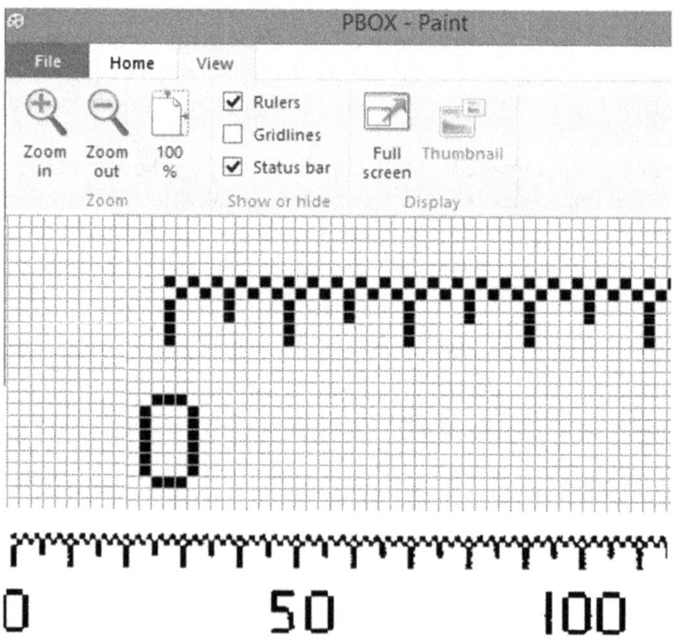

Fig. 28.4 Construction of mm scale for print. **a** Enlarged view of part of scale in PAINT editor, with marks 1 px wide. **b** Scale printed with print pixel frequency $F_P = 1000$ ppm. In the actual printout the pixels are 1 mm wide

28.6 1-bit Image by Scanner and Printer

This case is similar to the scanner-and-display case, with the difference that a printer is capable of much higher resolutions than a display. A printer has a pixel frequency F'_P, usually measured in ppi, settable over a wide range either directly in the image file (see above, Sect. 28.3), or more conveniently by printer software at print time. A typical ink-jet printer offers pixel frequencies from say 30 to 300 ppi (1.18–11.8 ppmm). A laser printer or an image-setter can go to much higher pixel frequencies, exceeding the acuity of even the best eye. The size of the scanned object is limited by the size of the scanner window $w \times h$.

The magnification of the printed image relative to the object is determined as $m = F_S/F'_P$. If the scan frequency is lower than the printer frequency, the image is diminished; if the frequencies are equal the image is same-size; and if the scan frequency is higher than the printer frequency the image is enlarged. The printed image size $w' \times h'$ may well be greater than the printer sheet size, and the aspect ratio different from the sheet aspect ratio; but the image can be tiled across several sheets.

The resolution limit of the printed image which is presented to the eye is the lower of the object resolution F_{LP} and the printer resolution F'_{LP}, that is, $(\mathrm{MIN}(F_S, F'_P))/2$.

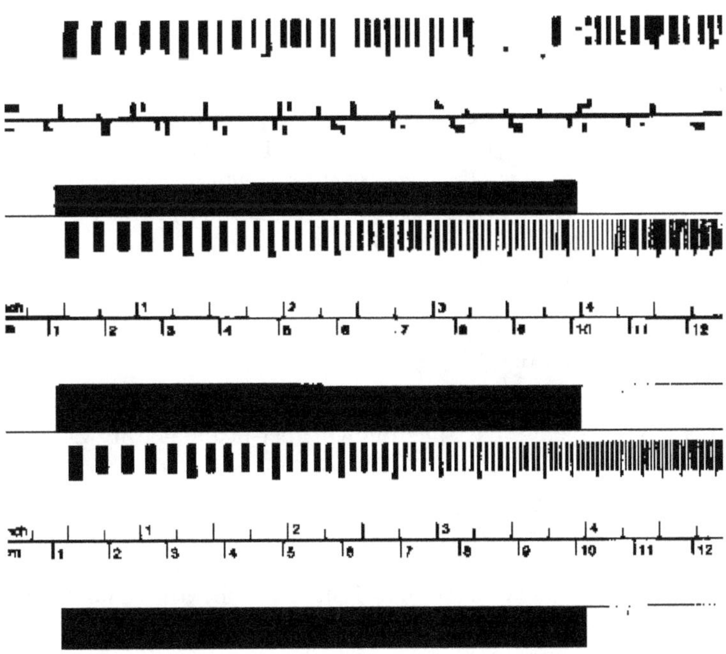

Fig. 28.5 Three scans printed. **a** Test card scanned at $F_S = 30$ spi, printed at $F_P = 30$ ppi, so magnification $m = 30/30 = 1$. **b** Scanned at $F_S = 91$ spi, printed at $F_P = 91$ ppi, so magnification $m = 91/91 = 1$. **c** Scanned at $F_S = 300$ spi, printed at $F_P = 300$ ppi, so magnification $m = 300/300 = 1$

Table 28.5 Examples of size variables for scanned and printed images as shown in Fig. 28.5

SCANNED OBJECT			(a)	(b)	(c)	
Sample width	a	$= 1/F_S$	0.033	0.011	0.0033	mm
Sampling frequency *	F_S		30	91	300	spi
Resolution limit	F_{LP}	$= F_S/2$	15	45.5	150	lppi
Object height *	h		6	6	6	in
Object width *	w		8	8	8	in
COMMON						
Magnification image/object	m	$= F_S/F_P$	1	1	1	
Count of samples in width	S	$= w/a$	242	727	2424	samples
Count of samples in height	T	$= h/a$	182	545	1818	samples
PRINTED IMAGE						
Image pixel width	a'	$= 1/F_P$	0.033	0.011	0.0033	in
Image pixel frequency *	F'_P		30	91	300	spi
Image resolution limit	F'_{LP}	$= F'_P/2$	15	45.5	150	lppi
Image height	h'	$= m \times h$	6	6	6	in
Image width	w'	$= m \times w$	8	8	8	in

For example, Fig. 28.5a shows a test object scanned in 1-bit mode at 30 spi (1.18 spmm) and printed at 30 ppi. The scan frequency is equal to the printer frequency, so the image is same-size. In Fig. 28.5b the object is scanned at 91 spi (3.58 spmm) and printed at 91 ppi, so the image is again same-size but the resolution is higher. In Fig. 28.5c the object is scanned at 300 spi (11.81 spmm) and printed at 300 ppi, so again the image is same-size but the resolution is higher still. Table 28.5 shows the values. The actual printouts are much degraded by being reproduced here. For printing a camera image, see Sect. 50.4.

28.7 1-bit Print Intensity Variables

In printing a 1-bit image, we have the following per-pixel intensity variables:

- 1-bit intensity index I_2,
- 8-bit colour-table intensity triple (B, G, R),
- 8-bit ink coverage percent K.

Intensity index is 0 or 1, held in the image section of the .BMP file (see Chap. 26). Colour-table intensity triple is $B = G = R = 0$ or 255, held in the colour-table of the .BMP file. Ink coverage percent is calculated as $K = 100$ for Black or 0 for White, as in Table 28.6.

Table 28.6 1-bit additive-to-subtractive conversion

Index	B	G	R	Name	K %
0	0	0	0	Black	100
1	255	255	255	White	0

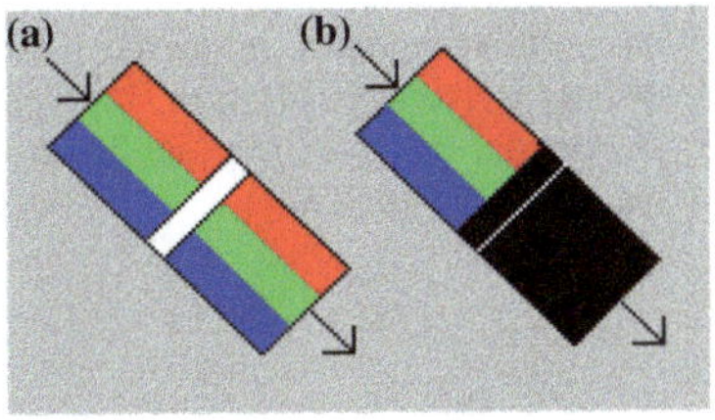

Fig. 28.6 Subtractive light. **a** A clear filter stops none of incident *White* light, passing all of *Blue Green* and *Red* wavebands. **b** A *Black* filter stops all of incident *White* light, passing none

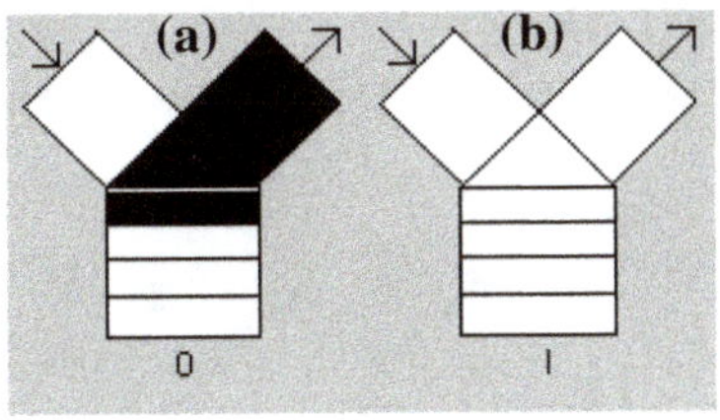

Fig. 28.7 1-bit print colours. **a** Pixel receives three low voltages, prints 100 % *Black*, so reflects almost no light. **b** Pixel receives three high voltages, prints nothing, so reflects almost all the light

28.8 Subtractive Light

Whereas a source emits light, and multiple sources combine by addition, a filter absorbs light, and multiple filters combine by subtraction.[4] Figure 28.6 shows the effect of (a) clear filter, (b) Black filter.

28.9 Additive-to-Subtractive Converter

An additive-to-subtractive converter operating in 1-bit mode changes, per-pixel, an additive triple (B, G, R) to a subtractive percentage K. Table 28.6 shows the conversion.

28.10 1-bit Print Colours

In printing, transparent inks act as filters, and White paper or another substrate acts as a reflector. Figure 28.7 shows (a) when White light falls on Black ink, it is nearly all absorbed, but (b) when it falls on White paper, it is nearly all reflected.

[4]More correctly, by fractional multiplication; but subtraction is the established usage.

References

1. How Stuff Works (2014) E-ink. http://electronics.howstuffworks.com/gadgets/high-tech-gadgets/e-ink.htm. Accessed 1 Feb 2014
2. How Stuff Works (2014) Laser Printer. http://computer.howstuffworks.com/laser-printer.htm. Accessed 1 Feb 2014
3. PC Magazine (2014) Inkjet Printer. http://www.pcmag.com/encyclopedia/term/58062/inkjet-printer#encyc_entry. Accessed 1 Feb 2014
4. Printer Drivers (2014). http://www.printer-driver.org.uk/. Accessed 1 Feb 2014
5. Wikipedia (2014) Dye-sublimation. http://en.wikipedia.org/wiki/Dye-sublimation_printer. Accessed 1 Feb 2014

Chapter 29
Viewing 1-bit Image

Viewing a 1-bit displayed image and viewing a 1-bit printed image is compared with viewing a scene directly, for scope, location resolution, luminance range, and colour resolution. An additive display and a subtractive printout behave differently under varying ambient illumination levels.

29.1 Flowchart

The flowchart (Fig. 29.1) is for general guidance in this chapter. It shows an eye viewing a scene, and by comparison viewing various input/output types of 1-bit image.

29.2 Viewing Variables

If we want to compare direct viewing of a scene with viewing a digital image of a scene, four variables emerge:

- Location range, or scope, measured as angular or linear extent of view,
- Location resolution within scope, measured as Snellen acuity, or as line-pairs per unit angle, or as line-pairs per unit length (see Sect. 10.6),
- Luminance range, measured as dynamic operating range from lowest to highest luminance in cd/m^2 (see Sect. 15.8),
- Colour resolution within luminance, measured as gamut on a CIE chromaticity chart(see Sect. 16.3).

A. Parkin, *Digital Imaging Primer*, DOI 10.1007/978-3-540-85619-1_29

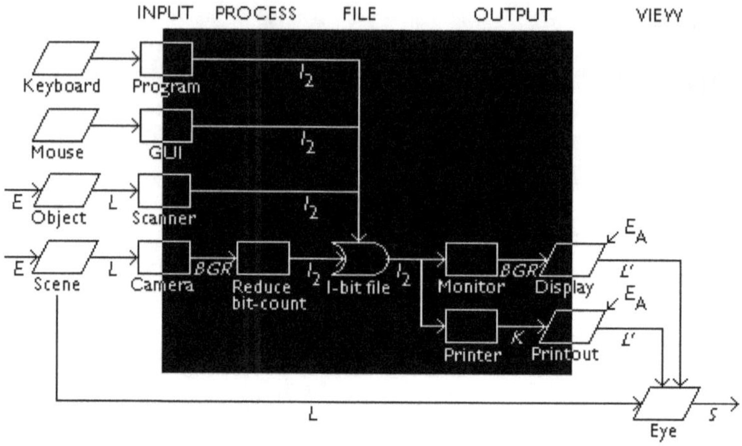

Fig. 29.1 Comparison of viewing 1-bit image with viewing scene. E is illuminance in lx, from source onto object or scene. L is luminance in cd/m^2, from scene to eye and from scene to input device. I_2 is intensity index 0 or 1 from input to file, and from file to output. BGR is intensity triple, 3× (eight bits 0–255), from camera to bit-reducer, and from monitor to display. K is Black ink coverage, from printer to paper. E_A is ambient illuminance in lx, from surroundings onto display or printout. L' is luminance from output device to eye, and S is viewer's Space/Colour sensation from eye to perception

29.3 Scope of Eye and 1-bit Image

Figure 29.2 compares the scope of the eye with the scope of various types of digital image, in angular or linear measure. A displayed or printed image has the lesser scope of input and output.

One fixation of the eye projects an angular scope of about 2° of sharp foveal vision, surrounded by about 10° of less sharp vision, surrounded by nearly 180° of peripheral vision (see Sect. 10.2). Binocular fixation by the two eyes produces an immediate sensation of depth (see Sect. 12.5). The perception of visual space is a sphere with an elliptic geometry, maintained by continually-changing fixations (see Sect. 14.2).

A digital image input by camera is a (perspective) central projection of a scene, typically of about 42° scope, which corresponds roughly to A4 scope at near viewing distance. An image input by scanner is a (contact) parallel projection of a flat object, typically of A4 scope. An image output by display is typically of about A4 scope at near viewing distance (though a 16:10 screen is much wider) An image output by printer is also typically of A4 scope at near viewing distance. An image input by scanner or camera and output by display or printer has the lesser scope of the input/output devices.

Notice that a flat image corresponds closely to the elliptic geometry of visual space at near viewing distance within about 45°, but fails to correspond beyond 90° (see Sect. 6.9).

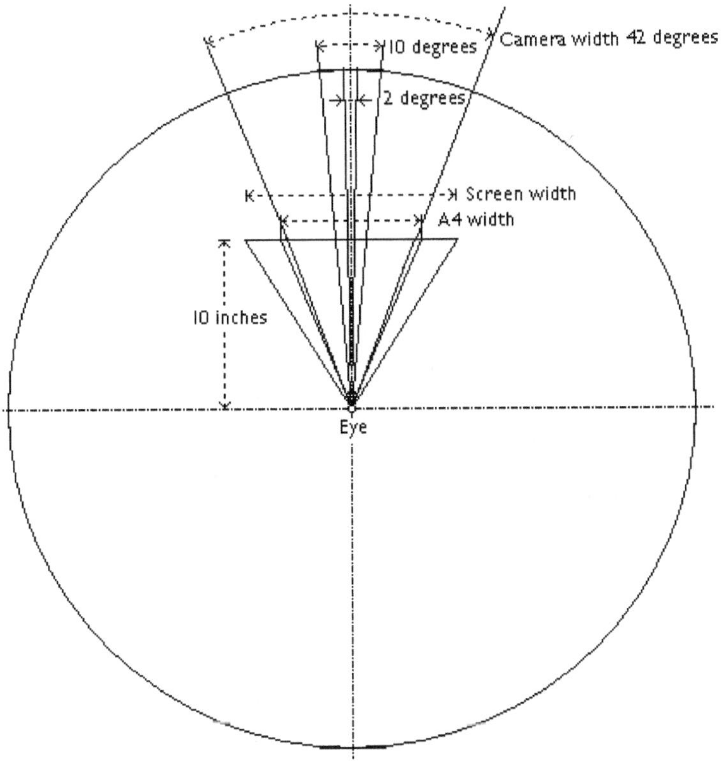

Fig. 29.2 Scope of eye and scope of digital images. One fixation of eye samples about 2° very finely, about 10° less finely, and up to nearly 180° peripherally. Near viewing distance is about 10 in. A typical scanner samples a flat object of A4 width. A typical camera samples a scene of about 42° width. A typical display has a screen width of about 12 in. A typical printer has a sheet of A4 width. A displayed or printed image has the lesser scope of input and output

29.4 Location Resolution of Eye and 1-bit Image

Figure 29.3 compares the resolution of location by the eye, with the resolution of various types of digital image, in acuity or angular or linear measure. A displayed or printed image has the lower resolution of input and output.

Eyes vary in resolution limit from very poor (Snellen 6/60, or 6 spd, or 35 spi at 10 in viewing distance), through normal (Snellen 6/6, or 60 spd or 30,lppd, or 350 spi at 10 in), to very good (Snellen 6/3, or 120 spd, or 700 spi at 10 in). See Sect. 10.6.

The resolution limit of a particular image may be compared with that of the eye by noting the input-to-output path of the image. An image created by program or GUI has pixel-counts but no sizes, so imposes no resolution limit. A scanner can be set to operate at various sampling frequencies: at 600 spi it has the resolution of a good eye (Snellen 6/4.5). A 1.3 mega-pixel camera operates at about 25 spd; a 6 Mpx

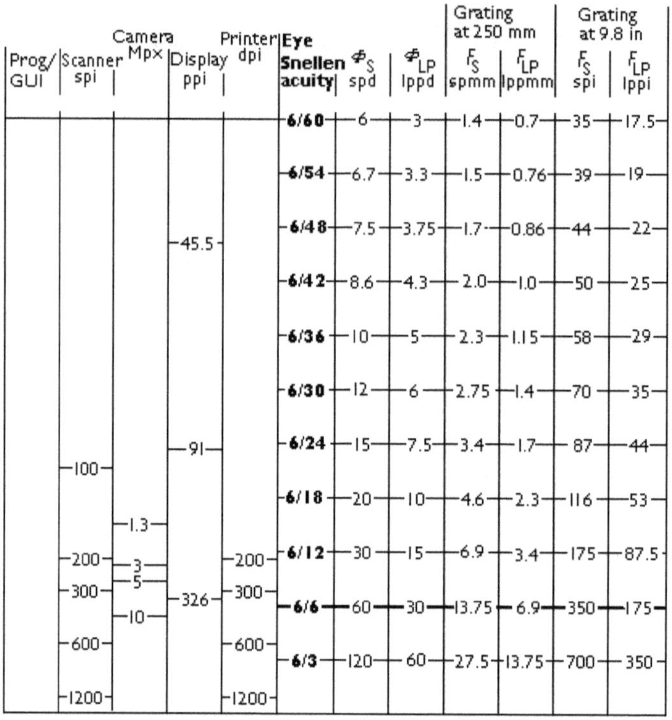

Fig. 29.3 Location resolution of eye, compared with resolution of digital images, measured as Snellen acuity and the corresponding angular and linear frequencies. A displayed or printed image has the lower resolution of input and output

camera at about 60 spd, equivalent to a normal eye; and a 16 Mpx camera at about 120 spd, equivalent to a very good eye. A typical display operates at something under 100 ppi, say 15 spd [1], comparable with a 6/24 eye. The enlargements by factor 2, 4, 6, and 8 in a GUI editor thus offer around 50, 25, 17, and 12 ppi. An Apple iPad display has 326 ppi, which viewed at 10 in is 57 ppd, slightly less than a normal 6/6 eye [2].[1] A typical ink-jet printer has about the same resolution as a scanner.

The resolution of a digital image is the lower of the input device resolution and the output device resolution. For example, a scan at 300 spi or an image from a 6 Mpx camera, displayed at 91 ppi is enlarged by factor 3.3 but has resolution 45.5 lppi (see Sect. 27.9). But a scan at 300 spi printed at 300 ppi is same-size and has resolution 150 lppi (see Sect. 28.6).

[1] There has been much controversy over the resolutions of the various i-products, apparently arising from the supposed viewing distances for different sizes of screen.

29.5 External Luminance and Internal Intensity

When the eye views a scene, it detects the differential luminance L coming from a scene, and transforms it to the differential sensation of Brightness (see Chap. 17).

Figure 29.4 shows a 1-bit digital system capturing the differential luminance L coming from a scene (actual or imaginary), processing it, and outputting an image with corresponding luminance L'. In processing, the system normalizes[2] L in the (wide) range L_{min} to L_{max} to L_{norm} in the range 0–1. It then encodes $L_{\mathrm{norm}} \leq 0.5$ as intensity index $I_2 = 0$, or $L_{\mathrm{norm}} > 0.5$ as $I_2 = 1$. I_2 is stored in a file, which, with the usual colour-table, interprets $I_2 = 0$ as tone-code triple $(B, G, R) = (0, 0, 0)$ and $I_2 = 1$ as $(B, G, R) = (255, 255, 255)$. On output from the file, the system decodes I_2 as L'_{norm} in the range 0–1, and de-normalizes[3] L'_{norm} to L' in the (restricted) range L'_{min} to L'_{max} of the output device.

For example, given scene luminance $L_{\mathrm{min}} = 0.01$ and $L_{\mathrm{max}} = 1000\,\mathrm{cd/m^2}$, a camera would normalize a grey $L = 700\,\mathrm{cd/m^2}$ as $L_{\mathrm{norm}} = (700 - 0.01)/(1000 - 0.01) = 0.7$. A 1-bit thresholder would deliver this as $L_{\mathrm{norm}} = 1$, and encode it as $(1 \times (255-0))+0 = 255$. A display would decode 255 as $L'_{\mathrm{norm}} = (255-0)/(255-0) = 1$. And if the display had a luminance range of $L'_{\mathrm{min}} = 0.01$ to $L'_{\mathrm{max}} = 100\,\mathrm{cd/m^2}$, the displayed luminance would be $L' = (1 \times (100 - 0.01)) + 0.01 = 100\,\mathrm{cd/m^2}$.

Thus a processed image has luminance L' in a restricted range (depending on the input and output devices), which is a relative rendition of the scene luminance L in a (usually wider) range.

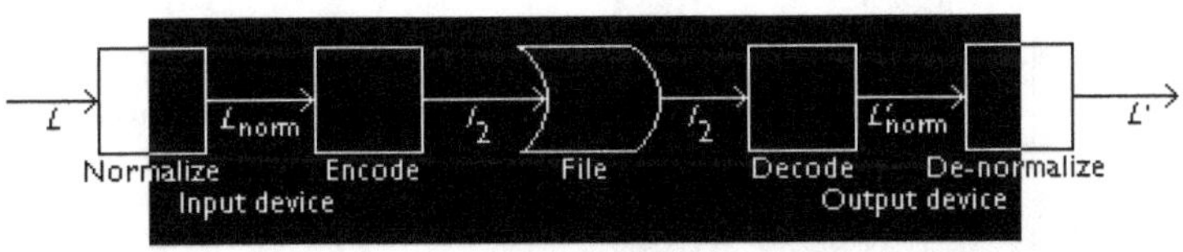

Fig. 29.4 Luminance input to luminance output in a 1-bit system. Input luminance in the (wide) range L_{min} to L_{max} is normalized in the range 0–1, then encoded in the range 0–1, and stored in a file. Output code in the range 0–1 is decoded in the range 0–1, then de-normalized in the (restricted) range L'_{min} to L'_{max}

---

[2]Given a number N in the range N_{min} to N_{max}, it is normalized in the range 0 to 1 as

$$N_{\mathrm{norm}} = (N - N_{\mathrm{min}})/(N_{\mathrm{max}} - N_{\mathrm{min}}).$$

[3]Given a number N_{norm} normalized in the range 0–1, it is de-normalized in the range N_{min} to N_{max} as

$$N = (N_{\mathrm{norm}} \times (N_{\mathrm{max}} - N_{\mathrm{min}})) + N_{\mathrm{min}}$$

29.6 Luminance Range of Eye and 1-bit Image

Luminance L, measured in cd/m^2, is the strength of light leaving a surface (see Sect. 15.8). The *contrast ratio* of a scene or an image is

$$C_R = L_{max} : L_{min} ,$$

where L_{max} is the highest luminance and L_{min} the lowest luminance in the scene or image. The *dynamic range* of a device is the greatest contrast ratio at which it operates correctly. Figure 29.5 compares the dynamic range of the eye with the dynamic ranges of various types of digital device.

The eye involuntarily adapts its dynamic range according to scene intensities, by fairly quick changes of pupil size and fairly slow retinal adaptation (see Sect. 15.3). Photopic (day) vision has a dynamic range from about 10^{-2} to 10^6 cd/m^2; scotopic (night) vision from 10^{-6} to 1 cd/m^2; and mesopic vision a mixture of the two. If the

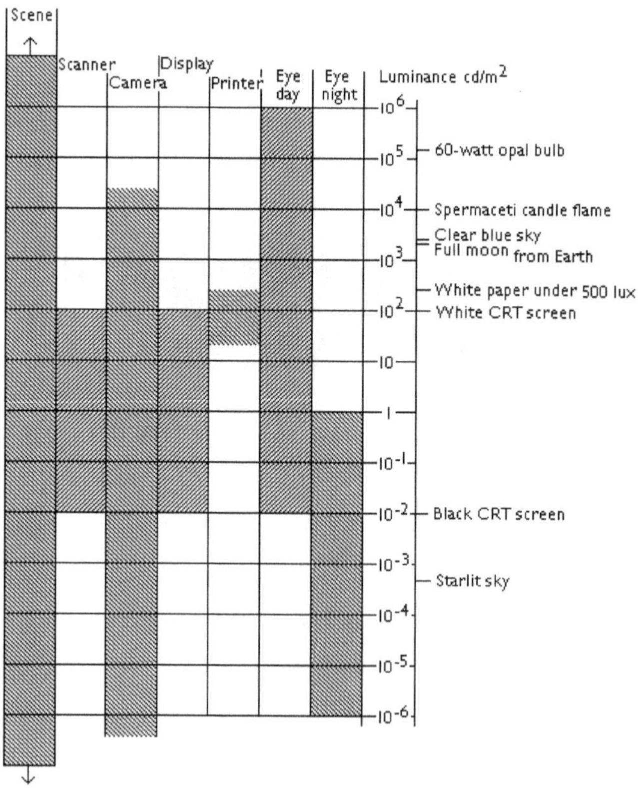

Fig. 29.5 Dynamic range of eye, compared with dynamic range of digital devices. The contrast ratio of a displayed or printed image is the lesser range of input and output

contrast ratio of a scene over-tops the dynamic range of the eye, perception suffers *glare* or *dazzle*. The eye cannot receive local differences of luminance in the too-bright extent, so cannot detect features there (see Sect. 17.13). If the contrast ratio of a scene under-shoots the dynamic range of the eye, the eye cannot receive local differences of luminance in the too-dark extent, so again cannot detect any detail.

A program or GUI in 1-bit mode specifies an imagined object with an imagined contrast ratio, which is normalized as a dynamic range in 0 to 1.

A scanner in 1-bit mode senses light from its internal light source reflected by the object (see Sect. 22.5). The contrast ratio of the object is therefore the same as the dynamic range of the scanner, typically from near zero to 100 cd/m^2, about the same as a display.

A digital camera has a dynamic range from just a few quanta (hence its wide-spread use in astronomy) to say 10,000 cd/m^2. If the contrast ratio of a scene exceeds this range the camera sensor suffers blooming, like dazzle for the eye. Images with very *high dynamic range* (HDR) can be produced by combining differently-exposed images of the same scene (see Sect. 46.15).

A CRT display has a dynamic range from about 0.01–100 cd/m^2; an LCD display rather less (see Sect. 27.10).

A printout has a dynamic range which is entirely dependent on the incident illuminance: under typical conditions say from 25–135 cd/m^2 (see Sect. 29.7 below).

29.7 Ambient Illuminance

A display screen is viewed in surroundings which add *ambient illuminance* E_A lx to the emittance M lx of the image. If we say that an average surface in the surroundings reflects about 20 % of the incident illuminance, then we have the approximate ambient illuminances listed in Table 29.1.

For a display, adding a fixed ambient illuminance to the additive image decreases the contrast, in much the same way that aerial perspective adds blue-white mist to a distant landscape to decrease contrast. To get a viewing contrast ratio for a display device, we should use

Table 29.1 Typical values of illuminance E and ambient illuminance $E_A = 0.2 \times E$

Surroundings	E lx	E_A lx
Dimly-lit room	15–50	3–10
Living room	50–200	10–40
Brightly-lit office	300–500	60–100
TV studio	1000	200
Overcast sky light	1000	200
Clear sky light	10,000–25,000	2,000–5,000

Table 29.2 Viewing contrast ratios for display with constant screen White luminance $L_W = 100\,\text{cd/m}^2$, and three screen Black luminances $L_B = 0.01$, 0.1 and $1\,\text{cd/m}^2$

Surroundings	$E_A l\times$	$C_R 0.01$	$C_R 0.1$	$C_R 1$
Dim room	5	63.67	62.55	53.17
Living room	25	13.55	13.51	13.04
Overcast sky	200	2.57	2.57	2.56
Clear sky	4000	1.08	1.08	1.08

$$C_R = (M_W + E_A) : (M_B + E_A)\,,$$

where M_W lx is the emittance of screen White, E_A lx is the ambient illuminance, and M_B is the emittance of screen Black. For example, if screen White is set at luminance $L_W = 100\,\text{cd/m}^2$, screen Black at luminance $L_B = 0.01\,\text{cd/m}^2$, and ambient illuminance at $E_A = 25\,\text{lx}$, we have

- White emittance $M_W = \pi \times 100 = 314\,\text{lx}$,
- Black emittance $M_B = \pi \times 0.01 = 0.03\,\text{lx}$,
- $M_W + E_A = 314 + 25 = 339\,\text{lx}$,
- $M_B + E_A = 0.01 + 25 = 25.01\,\text{lx}$,
- Viewing contrast ratio $C_R = 339/25.01 = 13.55$.

Table 29.2 lists display viewing contrast ratios for several ambient levels, and Fig. 29.6 plots viewing contrast ratio C_R against (logarithmic) ambient illuminance E_A. We see that the first 100 lx or so of ambient light makes a dramatic reduction in the contrast ratio, with relatively less effect thereafter. The very large contrast ratios of 10,000:1 or more quoted by manufacturers of CRTs are calculated from L_{max} and L_{min} in completely dark surroundings. For an LCD L_{max} and L_{min} are both dependent on the intensity of the back-light chosen.

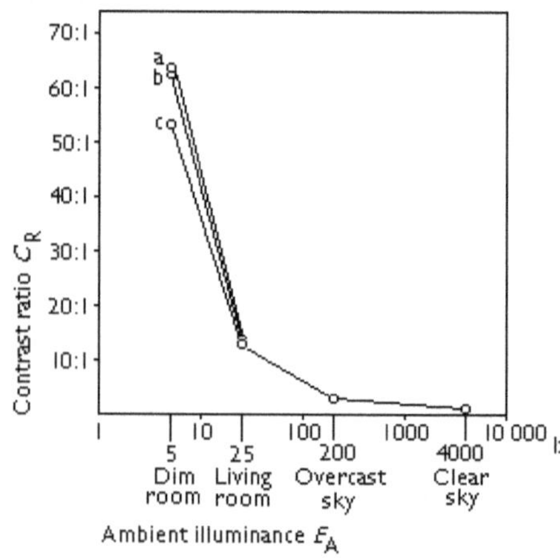

Fig. 29.6 Plot of display viewing contrast ratio against ambient illuminance (values from Table 29.2). **a** screen White luminance $L_W = 100\,\text{cd/m}^2$, screen Black luminance $L_B = 0.01\,\text{cd/m}^2$. **b** $L_W = 100$, $L_B = 0.1\,\text{cd/m}^2$. **c** $L_W = 100$, $L_B = 1\,\text{cd/m}^2$

A 1-bit printout contains White (paper) pixels and Black (ink) pixels, viewed under ambient illuminance E_{A} lx. A White pixel has a reflection factor $0 \leq rf_{\mathrm{W}} \leq 1$, say 0.85 for a good paper; and a Black pixel $0 \leq rf_{\mathrm{B}} \leq 1$, say 0.15 for a good ink. So the emittance of a White pixel is $M_{\mathrm{W}} = E_{\mathrm{A}} \times rf_{\mathrm{W}}$, and of a Black pixels is $M_{\mathrm{B}} = E_{\mathrm{A}} \times rf_{\mathrm{B}}$. The luminance contrast ratio of a printout is

$$C_{\mathrm{R}} = M_{\mathrm{W}} / M_{\mathrm{B}} .$$

For a printout, multiplying a fixed ambient illuminance by two reflection factors keeps a constant contrast.

Table 29.3 lists viewing contrast ratios for a printout under a series of ambient levels, and Fig. 29.7 plots viewing contrast ratio C_{R} against (logarithmic) ambient illuminance E_{A}. We see that varying ambient light does not change the contrast ratio. On a printout the contrast ratio is about 6:1.

Table 29.3 Viewing contrast ratios for printout with White reflection factor $rf_{\mathrm{W}} = 0.85$ and Black reflection factor $rf_{\mathrm{B}} = 0.15$, under four different illuminances $E_{\mathrm{A}} l\times$

Surroundings	$E_{\mathrm{A}} l\times$	$M_{\mathrm{W}} l\times$	$M_{\mathrm{B}} l\times$	C_{R}
Dim room	5	4.25	0.75	5.67
Living room	25	21.25	3.75	5.67
Overcast sky	200	170	30	5.67
Clear sky	4000	3400	600	5.67

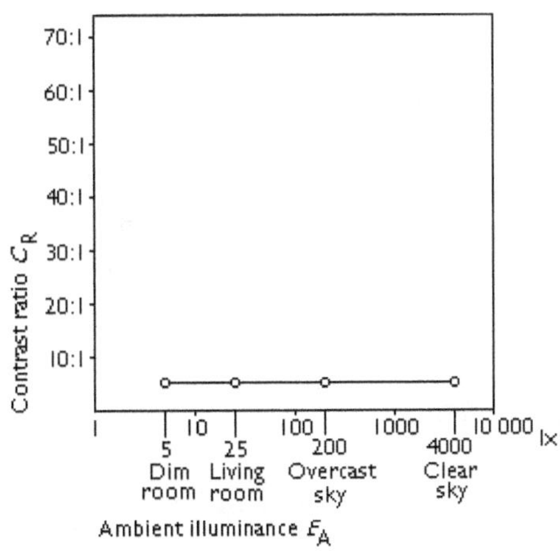

Fig. 29.7 Plot of printout viewing contrast ratio against ambient illuminance (values from Table 29.3). Contrast ratio is constant

29.8 Colour Resolution of Eye and 1-bit Image

Figure 29.8 shows the colour gamut of the eye compared with a 1-bit image. Physically-measurable power distribution in the electro-magnetic spectrum has an enormous range, from gamma rays to radio waves. Physiologically, the eye can register wavelengths from about 400 to about 800 nm, and can detect the shape of the power distribution within this visible spectrum, by combining the signals from three types of cone. Psychologically, we have the sensation of Colour, as Hue and Saturation, which can be mapped on a CIE chromaticity chart (see Sect. 16.3).

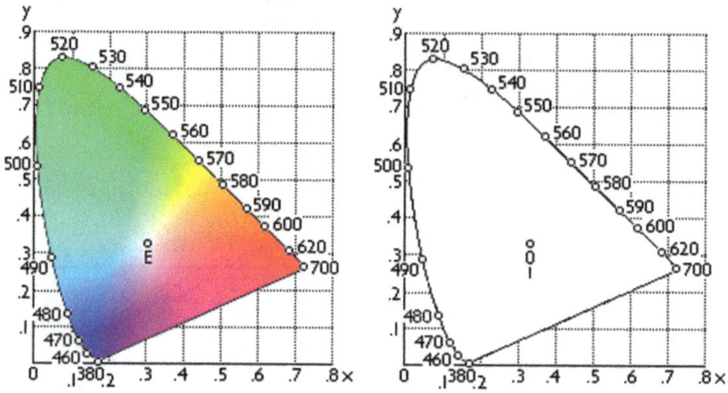

Fig. 29.8 Colour gamut of eye, compared to that of 1-bit image. **a** CIE chromaticity chart showing the colour gamut to which the eye responds. E is equal-energy illuminant. Notice that the colours printed here are merely indicative: the eye gamut extends to much more saturated colours. **b** CIE chart showing 1-bit gamut 0 Black and 1 White, differing only in luminance. On an sRGB screen White usually corresponds to illuminant D65, a little cooler than E

The colour gamut of a digital image is determined by the colour-table of the file. For a 1-bit image this is either 0 Black or 1 White. This extreme reduction of colour to a binary discrimination is the defining characteristic of a 1-bit image.

References

1. Wikepedia (2014) List of displays by pixel density. https://en.wikipedia.org/wiki/List_of_displays_by_pixel_density. Accessed 1 Feb 2014
2. Wikipedia (2014) Retina display. https://en.wikipedia.org/wiki/Retina_Display. Accessed 1 Feb 2014

Part VI
4-bit Imaging

Chapter 30
4-bit Image

A 4-bit image has the same properties of location as a 1-bit image. But it has 16 intensities available, specified by an index 0 to 15 in 4 binary digits (or one hexadecimal digit) per pixel. The index points to a 16-entry colour-table in its file, each holding a triple of (Blue, Green, Red) intensities. Alternative colour-tables hold 16 colours (4CO), or 16 grey-matches to the colours (4GM). The 4-bit colour-tables are defined. Examples are given of typical 4-bit images created by program, graphic user interface, scanner, and camera, under both of the colour-tables.

30.1 Bit-Depth 4

Sixteen intensities (B, G, R) can be specified by an intensity index I_{16} taking a 4-bit value 0 to 15 decimal, (= 0 to F hexadecimal, = 0 to 1111 binary). We then say that the plane has *bit-depth 4*, or that it is the *4-bit plane*. The intensity index I_{16} refers to a 16-entry colour-table in a 4-bit .BMP file (see Chap. 33). Each of the index values points to an intensity triple (B, G, R) used to display that pixel on a monitor screen (see Chap. 34) or print it on paper (see Chap. 35). The colour-table may be set up in any way we please with a .BMP file editor (see Sect. 33.6). Two standard settings are shown in Fig. 30.1 and Table 30.1.

30.2 4-bit Colour-table 4CO

When an additive BGR display (see Chap. 34) and/or a subtractive YMCK printer (see Chap. 35) is available, we can use a colour-table with 16 colours, which we here label 4CO.

The intensities in each of the three channels are limited to 0, 128, 192, or 255. There are then

A. Parkin, *Digital Imaging Primer*, DOI 10.1007/978-3-540-85619-1_30

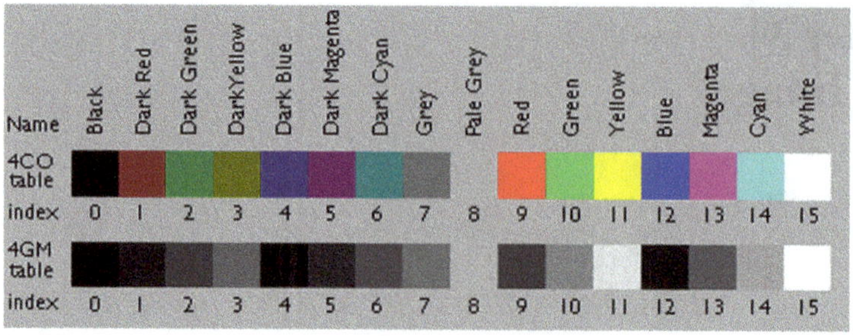

Fig. 30.1 Two 4-bit colour-tables. **a** 4CO sixteen colours. **b** 4GM sixteen *grey*-matches

Table 30.1 Two 4-bit colour-tables

Name	Index	B	4CO G	R	B	4GM G	R
Black	0	0	0	0	0	0	0
Dark Red	1	0	0	128	38	38	38
Dark Green	2	0	128	0	76	76	76
Dark Yellow	3	0	128	128	114	114	114
Dark Blue	4	128	0	0	14	14	14
Dark Magenta	5	128	0	128	52	52	52
Dark Cyan	6	128	128	0	90	90	90
Grey	7	128	128	128	128	128	128
Pale Grey	8	192	192	192	192	192	192
Red	9	0	0	255	77	77	77
Green	10	0	255	0	150	150	150
Yellow	11	0	255	255	227	227	227
Blue	12	255	0	0	28	28	28
Magenta	13	255	0	255	105	105	105
Cyan	14	255	255	0	179	179	179
White	15	255	255	255	255	255	255

- 3 *primary* colours, Blue, Green, and Red, where one channel has intensity 255 and the other two 0.
- 3 *secondary* colours, Cyan, Yellow, and Magenta, where two channels have intensity 255 and the other 0.
- 3 *primary shades*, Dark Blue, Dark Green, and Dark Red, where one channel has intensity 128 and the other two 0.
- 3 *secondary shades*, Dark Cyan, Dark Yellow, and Dark Magenta, where two channels have intensity 128 and the other 0.

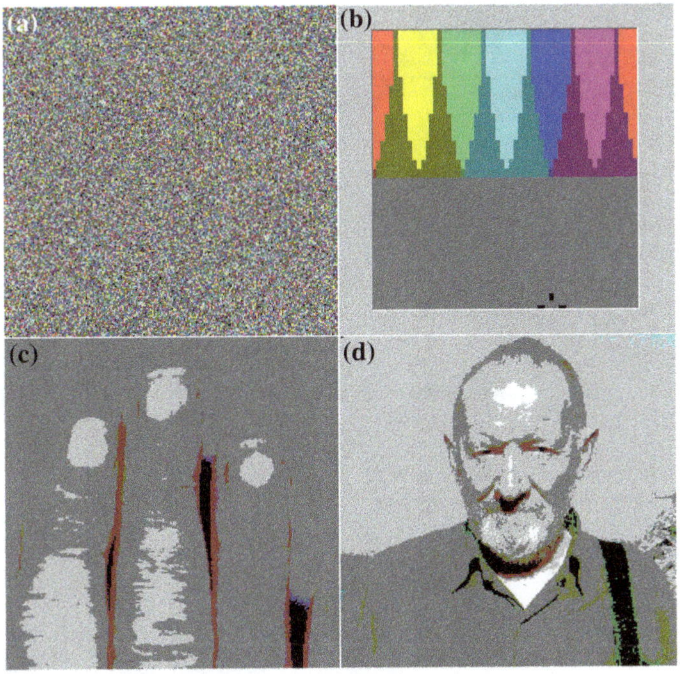

Fig. 30.2 Examples of 4-bit images with colour-table 4CO. **a** Image generated by 4-bit program: 222 × 222 px randomly set at one of 16 colours. **b** Image created by PAINT in 4-bit mode. **c** Fingers scanned in 24-bit mode then saved-as 4CO. **d** Scene photographed by 24-bit camera then saved-as 4CO

- 4 *neutral* colours, Black, Grey, Pale Grey, and White, where the three channels have equal values.

To set this colour-table, use a program, or an image editor in 4-bit mode, to create a 4-bit .BMP file. A 4-bit colour-table can also be set by reducing an 8-bit or 24-bit image to 4-bit (see Appendix A).

30.3 4-bit Grey-match Colour-table 4GM

When the available display device and/or printer can show a range of grey intensities, but not different colours, we may re-set the colour-table to deliver greys which are a reasonable Brightness match to the 4CO colours. We here label this colour-table 4GM. A grey has equal Blue Green and Red intensities, B = G = R. But Blue looks dark Grey, Green looks light Grey, and Red between the two; so we find the grey-match as the weighted sum of the three components:

Fig. 30.3 Examples of 4-bit images with colour-table 4GM. The per-pixel locations and index values are the same as Fig. 30.2, but colour-table 4CO has been converted-to 4GM

$$\text{Grey-match} = 0.11\ \text{Blue} + 0.59\ \text{Green} + 0.30\ \text{Red}\,.$$

These weights are based on a set of CRT phosphors, and are apparently used by MS PAINT. For the newer sRGB values see Sect. 16.5, and [1] and [2].

To set this colour-table, and change all the pixel index values accordingly, apply the transformation *Convert to grey-scale* to a 4CO .BMP file (see Appendix B). Convert-to is offered in IRFANVIEW and other image editors, but not in PAINT.

30.4 4-bit Digital Image

A 4-bit digital image is like a 1-bit image (see Chap. 14) as to locations, but different as to intensities. Graphically, a 4-bit image is a set of $x \times y$ px, each pixel one of 16 intensities. Numerically, it is a set of integer triples:

$$(x, y; I_{16})\,,$$

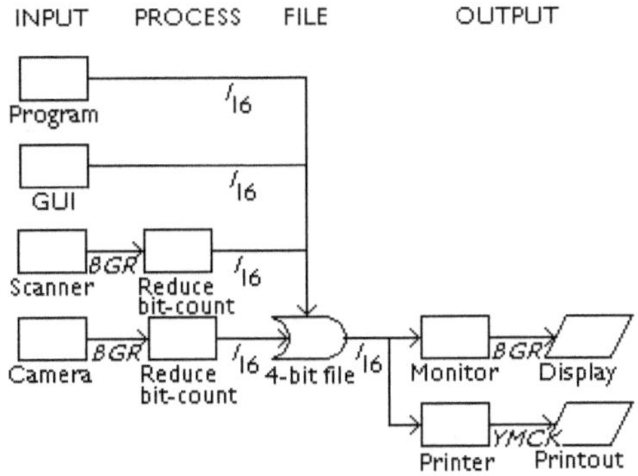

Fig. 30.4 Flowchart for 4-bit image. A 4-bit image may be constructed numerically by program, or constructed graphically by GUI image editor, or reduced from a 24-bit scanner or camera image. The image is stored in a 4-bit file. The image may be displayed on a screen or printed on paper, to be viewed by an eye. I_{16} is intensity index 0–15 from encoder to file, and from file to decoder. *BGR* is intensity triple, 3× (8 bits 0–255), from scanner/camera to encoder, and from monitor to display. *CMYK* is ink coverages, from printer to paper

where (for a first-quadrant image) $0 \leq x \leq x_{\text{max}}$, $0 \leq y \leq y_{\text{max}}$, $0 \leq I_{16} \leq 15$ decimal.

Figure 30.2 shows examples of 4-bit digital images with colour-table 4CO, comparable with the 1-bit examples in Chap. 19. In Fig. 30.2b 13 of the 16 4-bit colours are present (Dark Red, Dark Green, and Dark Blue are also available but are not seen here. In Figs. 30.2c, d nearly all the original colours have reduced to neutral Black, Grey, Pale Grey, and White.

Figure 30.3 shows examples of 4-bit digital images with colour-table 4GM. They are grey-matches of Fig. 30.2.

30.5 Flowchart

The flowchart in Fig. 30.4 is for general guidance in this Part. To make a 4-bit image, we can use a program or a GUI image editor (see Chap. 31) or reduce a 24-bit scanner or camera image to 4-bit. The image can be stored in a 4-bit .BMP file (see Chap. 33), then displayed (see Chap. 34) or printed (see Chap. 35) for viewing (see Chap. 36).

References

1. ITU-R Recommendation BT 709-5 (2008) Parameter values for the HDTV standards for program and international exchange. http://www.itu.int/rec/R-REC-BT.709/en. Accessed 1 Feb 2014
2. Poynton CA (2009) Color FAQ. http://www.poynton.com/ColorFAQ.html. Accessed 1 Feb 2014

Chapter 31
4-bit Input

A 4-bit image can be created by a 4-bit program, or a GUI in 4-bit mode, or by scanning in 8-bit or 24-bit mode and reducing to 4-bit, or by photographing in 24-bit mode and reducing to 4-bit. Examples are given of all four inputs. In each case the location properties are the same as for 1-bit images; only the intensity properties are different.

31.1 4-bit Input

A 4-bit image can be created in any of four main ways:

- From integer numeric data specified via keyboard, or via data file, or within a program.
- From graphic data specified via keyboard and mouse to a GUI.
- From a continuous flat object via a scanner and a reduction process.
- From a continuous space object via a camera and a reduction process.

31.2 4-bit Program Input

A 4-bit program is like a 1-bit program (see Chap. 20), with the differences:

- Read and write 4-bit .BMP file instead of 1-bit file.
- Set colour-table 4CO or 4GM instead of 1BW.
- Process 4-bit intensity $0 \leq I_{16} \leq 15$ instead of 1-bit $0 \leq I_2 \leq 1$.

Figure 31.1 shows the flow for a 4-bit program. A program which creates a 4-bit random image with either of the two colour-tables is briefly described below (full listing on the SpringerLink web-site). If programs are written with a modular structure,

A. Parkin, *Digital Imaging Primer*, DOI 10.1007/978-3-540-85619-1_31

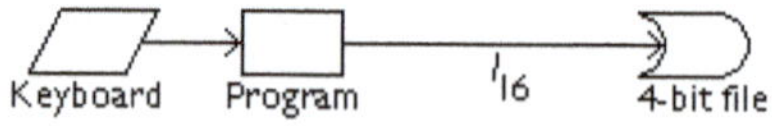

Fig. 31.1 4-bit program input. Input is from from keyboard or data file to program. Output to 4-bit file has the form $(x, y; I_{16})$ per pixel, where $0 \le x \le w_{max}$ and $0 \le y \le h_{max}$ are location coordinates, and $0 \le I_{16} \le 15$ is the index referring to a 4-bit colour-table in the .BMP file

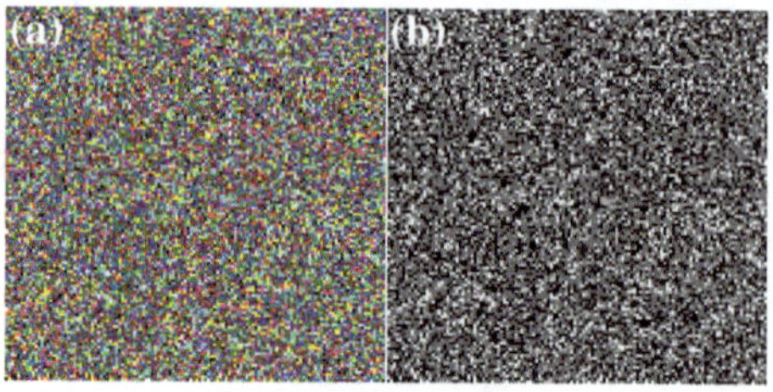

Fig. 31.2 Example of two 4-bit random images, each 128 × 128 px. **a** With colour-table 4CO. **b** With grey-match colour-table 4GM. The pixel-by-pixel intensity index values are the same for both

then read and write sub-programs (such as CREHEA4 and WRIDAT4 below) can be used over and again.

Figure 31.2 shows two random images, each 128 × 128 px, created by program GENRAND4. The two images have the same I_{16} index values, but different colour-tables.

```
REM   **   Program GENRAND4.BAS to generate 4-bit random
REM   **   image, as .BMP file. User enters output
REM   **   .BMP filename, image width and height,
REM   **   and which colour-table (CO or GM).
REM   **   Program writes 4-bit .BMP file header,
REM   **   colour-table, and image data.
REM   **   Data for colours
REM   **   Data for grey-match
REM   **   Data for gamma-encoded
REM   **   Main program
REM   **   Enter file_name, width, height and colour-table
           CALL CREHEA4
REM   **   Generate a random image:
REM   **   Set each pixel in array AIMAG4 to random
REM   **   4-bit index 0 to Fh
           CALL WRIDAT4
           CALL REVHEA1
FINIS: END

           SUB CREHEA4
REM   **   Subprogram to write header and colour-table of
REM   **   4-bit .BMP file #1.
```

```
END SUB
           SUB PUTBYTE (ASCII%)
REM   **   Subprogram to put 1 byte in file #1
END SUB
           SUB PUTI (INTE%)
REM   **   Sub-program PUTI to put integer value in binary
REM   **      file at pointer position (set outside) and
REM   **      increment pointer.
END SUB
           SUB PUTL (INTL&)
REM   **   Sub-program PUTL to put long integer into
REM   **      binary file as 4 bytes
END SUB
           SUB REVHEA1
REM   **   Revises header of .BMP file #1 with values in bytes.
END SUB
           SUB WRIDAT4
REM   **   Subprogram to write data from string array
REM   **      AIMAGE4 to .BMP file #1. Take two hex values
REM   **      N1$ AND N2$ as nibbles in a byte, convert
REM   **      byte to ASCII and write in 4-bit file.
END SUB
```

31.3 4-bit GUI Input

We can also create a 4-bit image by graphic input to a graphic user interface, such as MS PAINT, as shown in Fig. 31.3. A GUI in 4-bit mode is like in 1-bit mode (see Chap. 21), but offers 4-bit intensities $0 \le I_{16} \le 15$ instead of 1-bit $0 \le I_2 \le 1$.

To put the MS PAINT image editor into 4-bit mode, save a blank file (or already-existing file) as a 16-colour bitmap. The 16 4-bit colours 0 to 15 can be (tediously) set up as "Custom colors" in the Edit Colours window, as shown in Fig. 31.4. The current (B, G, R) values are at lower right, and can be changed to re-set the colour.

To set the foreground colour, select the Eyedropper and left-click the Custom color patch wanted. To set the background colour, right-click the patch wanted. To see or

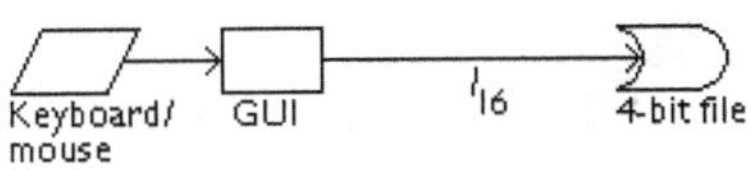

Fig. 31.3 4-bit GUI input. Input from keyboard/mouse has the form $(x, y; I_{16})$ per pixel, where $0 \le x \le w_{max}$ and $0 \le y \le h_{max}$ are the integer Cartesian coordinates of location in the first quadrant, and $0 \le I_{16} \le 15$ is the index referring to the 4-bit colour-table in the .BMP file

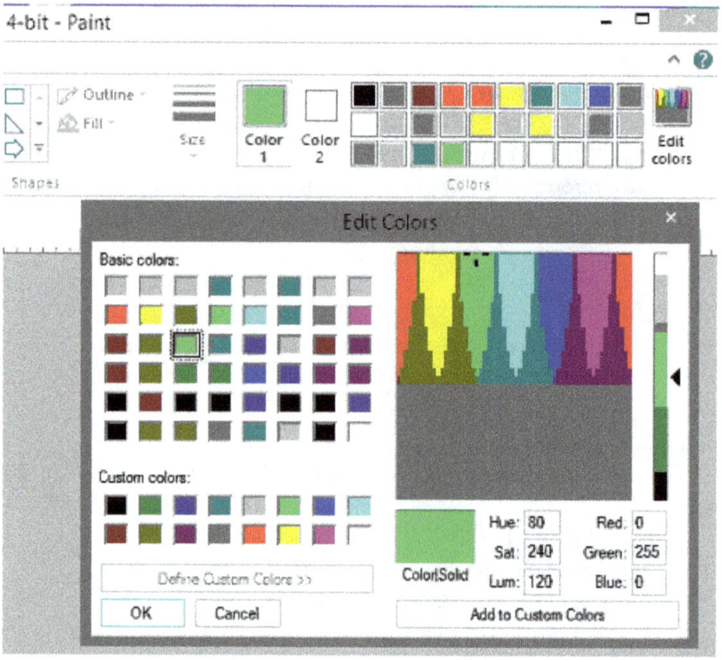

Fig. 31.4 PAINT Edit Colours window. The 16 4-bit colours have been set in "Custom colors". (B, G, R) and (Hue, Saturation, Luminance) values of current foreground colour are displayed at *lower right*

change the (B, G, R) values of any displayed image pixel, select the Eyedropper and click the pixel wanted, then go to Edit Colours.

A 4-bit plane graph has vertices, edges, and faces (see Sect. 21.4), and the faces may be filled by any of 16 colours. Text in any available font and size, and in any of 16 colours, may be set in the GUI. The colours are displayed according to the colour-table in the .BMP file. For example, Fig. 31.5 shows two 4-bit images, each containing all 16 intensities.

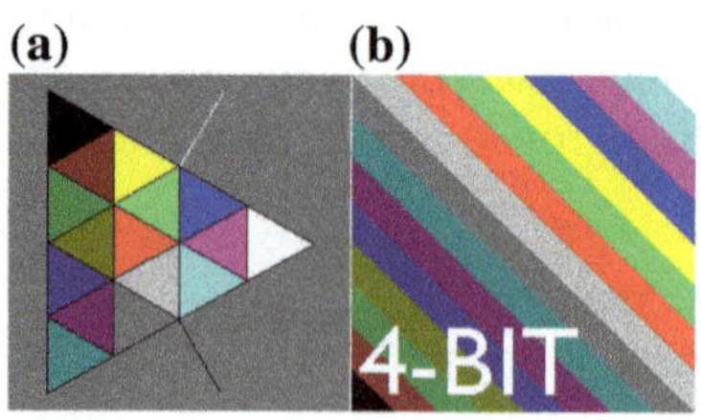

Fig. 31.5 4-bit images created by GUI. **a** Plane graph with all 16 colours. **b** 4-bit image with text. The stripes have intensity indexes from *bottom left* to *top right* 0 Black, 1 Dark red, 2 Dark green, ..., 13 Magenta, 14 Cyan, 15 White

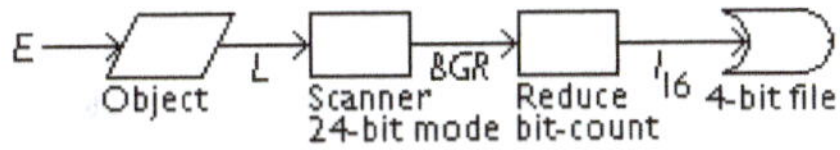

Fig. 31.6 4-bit scanner input. E is illuminance in lx, from scanner source onto object. L is luminance in cd/m^2, from object to scanner. BGR is intensity triple, 3× (eight bits 0–255), from scanner to reduction process. I_{16} is intensity index 0–15 from reduction process to file

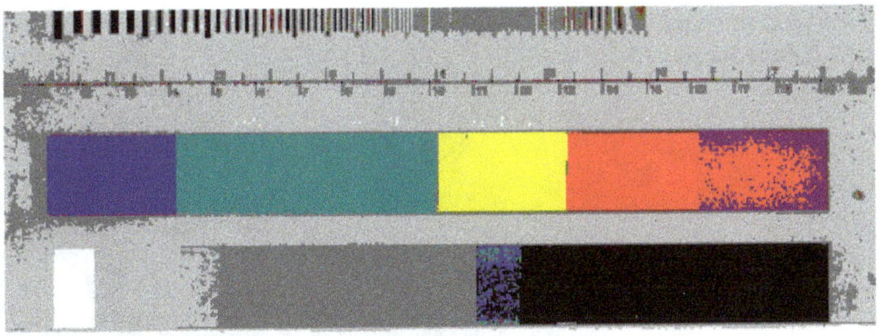

Fig. 31.7 Test card scanned in 24-bit colour mode, then saved-as 4-bit colour 4CO. 24-bit Blue reduces to 4 *Dark Blue*, Green and Cyan to 6 Dark Cyan, Yellow remains 11 Yellow, Red remains 9 Red, and Magenta equivocates between 9 Red and 5 Dark Magenta. The step-wedge reduces to 15 White, 8 Pale Grey, 7 Grey, and 0 Black, with some equivocation in the middle

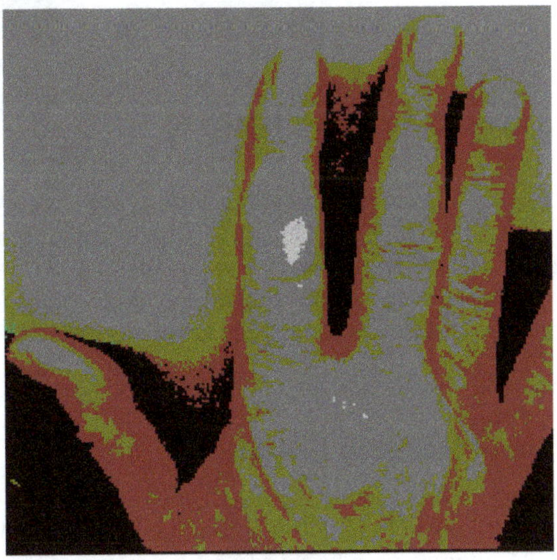

Fig. 31.8 Object scanned in 24-bit colour mode, then reduced to 4-bit colour 4CO

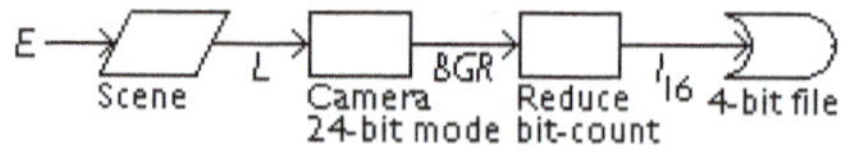

Fig. 31.9 4-bit camera input. E is illuminance in lx, from source onto scene. L is luminance in cd/m^2, from scene to camera. BGR is intensity triple, 3× (eight bits 0–255), from camera to reduction process. I_{16} is intensity index 0–15 from reduction process to file

Fig. 31.10 Image created by 24-bit camera, then reduced to 4-bit colour 4CO

31.4 Scanner Input Reduced to 4-bit

A simple scanner usually operates in 1-bit monochrome, or 8-bit grey-scale, or 24-bit colour mode. To get a 4-bit image, we can scan in 24-bit colour mode, then save-as 4-bit colour (see Fig. 31.6 and Appendix A). For example, Fig. 31.7 shows a test card with 6 colour patches and a step-wedge with 18 grey patches, and Fig. 31.8 a hand treated thus. Colours are thresholded and graduations are stepped.

31.5 Camera Input Reduced to 4-bit

A digital camera usually operates in 24-bit colour mode. To get a 4-bit image from such a camera, the 24-bit image can be saved-as 4-bit colour or grey-scale as shown in shown in Fig. 31.9 and Appendix A. For example Fig. 31.10 shows that colours are thresholded and graduations are stepped, with spectacular effect.

Chapter 32
4-bit Process

4-bit processes are like 1-bit processes for location properties, but with differences in intensity properties. A program for changing 4-bit locations is described. Examples of combining and reducing 4-bit locations, and of reducing bit-depth, are given.

32.1 Flowchart

The flowchart (Fig. 32.1) is for general guidance in this chapter. A location process is an operation on a discrete image which changes all the locations in some systematic way, keeping the intensities unchanged. An intensity process is an operation which changes all the intensities in some systematic way, keeping the locations unchanged.

32.2 Change 4-bit Image Locations

Changing 4-bit locations is like changing 1-bit locations (see Sects. 23.3–23.8) with the differences:

- Read and write 4-bit .BMP files instead of 1-bit files.
- Set colour-table 4CO or 4GM instead of 1BW.
- Process 4-bit intensity $0 \leq I_{16} \leq 15$ instead of 1-bit $0 \leq I_2 \leq 1$.

Program CHANLOC4, briefly described below and listed on the Springer web-site, performs the same transformations of 4-bit images as CHANLOC1 does of 1-bit images. For example, Fig. 32.2 shows six transformations as performed by program CHANLOC4.

```
REM  **  Program CHANLOC4 to change locations of
REM  **  4-bit image image. User enters filenames.
REM  **  Program reads given "before" image from 4-bit
```


A. Parkin, *Digital Imaging Primer*, DOI 10.1007/978-3-540-85619-1_32

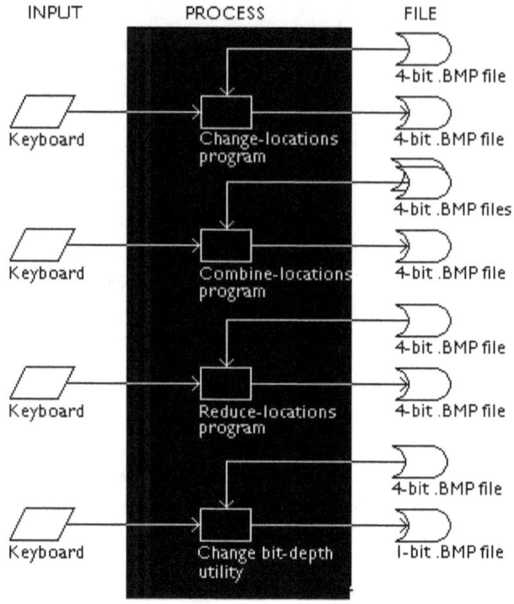

Fig. 32.1 4-bit process. A program can change the locations of a given 4-bit image (keeping the intensities unchanged). Or it can combine two 4-bit images as one, by alternating locations. Or it can reduce the pixel-count of locations. A standard utility can reduce a 4-bit image to 1-bit

```
REM  **  .BMP file #1 into array IMAGEA. User picks
REM  **  transformation and enters constants.
REM  **  Program then processes pixel by pixel
REM  **  from IMAGEA to array IMAGEB, setting old
REM  **  intensity at new location. Program writes
REM  **  "after" image from IMAGEB to 4-bit .BMP
REM  **  file #2.
REM  **  Main program
REM  **  Enter old and new file-names
         CALL RWHEA4AB
         CALL REIMAGE4
         CALL GEOMTRAN
         CALL WRIMAGE4
         CALL REVHEAD4
END
         SUB PUTBGR (INTE%) [as CHANLOC1]
         SUB PUTBYTE (INTE%) [as CHANLOC1]
         SUB PUTI (INTE) [as CHANLOC1]
         SUB PUTL (INTL&) [as CHANLOC1]
         SUB REIMAGE4
REM  **  Subprogram RE4IMAGE to read 4-bit .BMP file
```

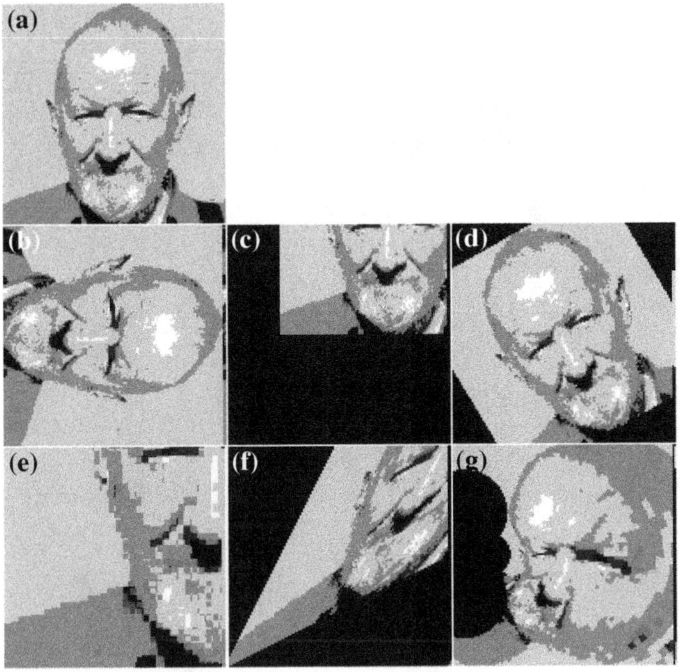

Fig. 32.2 Change 4-bit image locations. **a** Given 4-bit image of 128 × 128 px. **b** Reflected in mirror-line through the origin at 45°. **c** Translated by 32 px in X and 64 px in Y. **d** Rotated about (64, 64) by 30°. **e** Dilated about (0, 0) by factor 2 in X and Y. **f** Sheared about (0, 0) by factor 0.5 in X and 0.5 in Y. **g** Inverted in circle with centre (8, 64) and radius 56 px

```
REM  **     #1 and load image intensities into array
REM  **     IMAGEA.
END SUB
         SUB REVHEAD4
REM  **  Revises header of .BMP file #2 with values in bytes.
END SUB
         SUB RWHEA4AB
REM  **  Subprogram RWHEA4AB to read header of 4-bit
REM  **     .BMP input file #1, and write header and
REM  **     colour-table of 4-bit output file #2 (to
REM  **     be revised later).
END SUB
         SUB WRIMAGE4
REM  **  Subprogram WRIMAGE4 to write data from integer
REM  **     array IMAGEB to 4-bit .BMP file #2. Take
REM  **     two successive pixel intensities A1% and A2%,
REM  **     combine as A% and write as byte in file.
END SUB
```

Fig. 32.3 Two 4-bit double-repeating patterns, created by using parts of program CHANLOC4 in different symmetry groups

For another example, Fig. 32.3 shows two double-repeating patterns made using parts of program CHANLOC4 in the symmetry framework described in Sect. 23.15.

32.3 Combine 4-bit Images by Location

Combining 4-bit locations is like combining 1-bit locations (see Sect. 23.22) with the differences noted above. For example, Fig. 32.4 shows two 4-bit images and their combination by alternating columns.

32.4 Reduce 4-bit Image Locations

Reducing the count (and increasing the size) of pixels in a 4-bit image is like reducing 1-bit locations (see Sect. 23.23) with the differences noted above. For example, Fig. 32.5 shows a 4-bit image thus reduced by $N = 2$, 4, and 8.

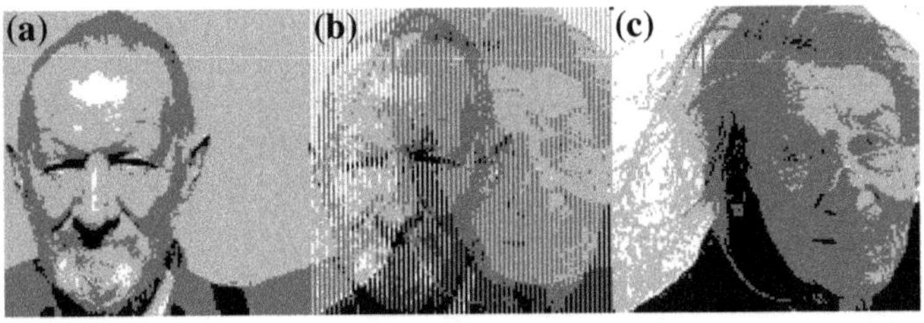

Fig. 32.4 Combine by location. **a** 4-bit image, 128 × 128 px. **b** Combination of (**a**) and (**c**) by alternating columns, 128 × 128 px. **c** 4-bit image, 128 × 128 px

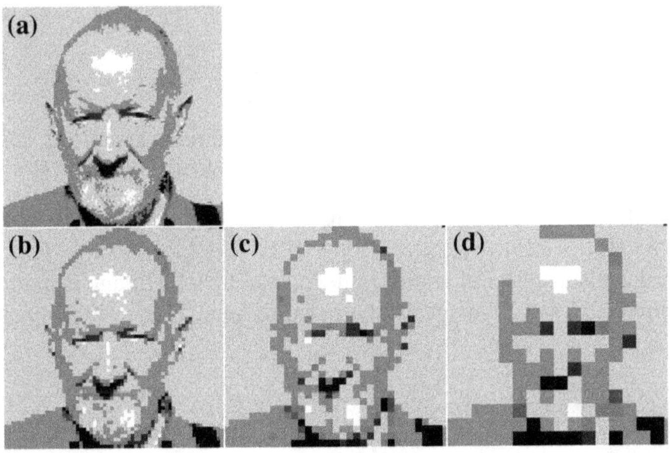

Fig. 32.5 Reduce image locations. **a** 4-bit image 128 × 128 px: 65,536 bits. **b** Reduction of (**a**) by $N = 2$ to 64 × 64 px: 16,384 bits. **c** Reduction of (**a**) by $N = 4$ to 32 × 32 px: 4,096 bits. **d** Reduction of (**a**) by $N = 8$ to 16 × 16 px: 1,024 bits

32.5 Change Bit-depth

To convert a given 4-bit file with colour-table 4CO to a 4-bit file with colour-table 4GM, use the *Convert-to* command which is offered in some GUI editors (such as IRFANVIEW, but not MS PAINT):

- Display the source file
- Click on *Convert-to* grey-scale
- Click on Save
- Enter new file name (or keep old file name).

The operating system will then

- create new 4-bit file with 16-entry grey-match colour-table 4GM
- set file header field biClrUsed = 16
- keep all pixel location values (X, Y) unchanged

- per-pixel, calculate grey-match value $gmv = (0.11B + 0.59G + 0.30R)$
- set pixel intensity index $I_{16} = gmv$.

To change a given 4-bit file to a 1-bit file use the *Save-as* command in an image editor, such as MS PAINT:

- Display the source file
- Click on Save-as
- Enter new file name (or keep old file name)
- Enter new file type (monochrome bitmap)
- Click on Save.

The operating system will then

- create new 1-bit file with two-entry colour-table 1BW
- keep all pixel location values (X, Y) unchanged
- per-pixel, compare 4-bit (B, G, R) values to fixed thresholds at 128, 128, 128
- set 1-bit intensity index 0 or 1 according to threshold intervals.

For example, Fig. 32.6 shows the two 4-bit colour-tables, and both reduced to 1-bit. See Appendix A.

A 4CO image is a subset of an 8CO image, and an 8CO image is in turn a subset of a 24CO image, so a 4CO image can be saved-as an 8CO or a 24CO image without

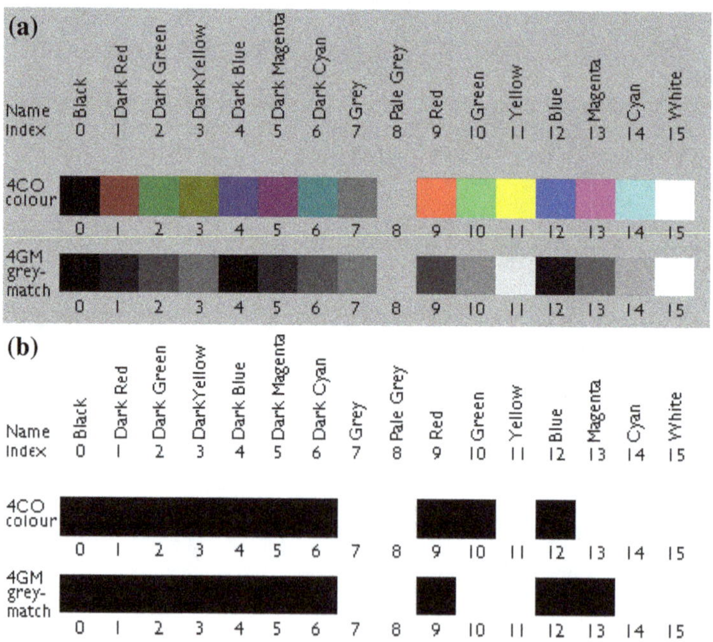

Fig. 32.6 Reduce bit-depth. **a** 4-bit image with colour-table 4CO and colour-table 4GM. **b** Image (**a**) saved-as 1-bit

change of appearance. Intensity processes belonging to the higher bit-level will then work for the lower. For example, to make 8-bit separations then 1-bit half-tones for a 4CO image, save-as 24-bit and proceed as such. This is not true for a 4GM image.

32.6 Statistics of 4-bit Image

Statistically (see Sect. 18.3), a digital image is a *finite population* of pixels, with a *population count* of N pixels. An individual pixel of the population is identified by its location (X, Y). The population has one discrete *variate*, intensity I.

For example, a 4-bit digital image has pixels showing one of 16 intensities, so $0 \leq I \leq 15$. Figure 32.7 shows an example of a camera image saved-as 4-bit. Compare with Fig. 32.7. The statistics were calculated by program DOSTATS4 (listed on the Springer web-site), quite similar to program DOSTATS8 described in Sect. 39.26.

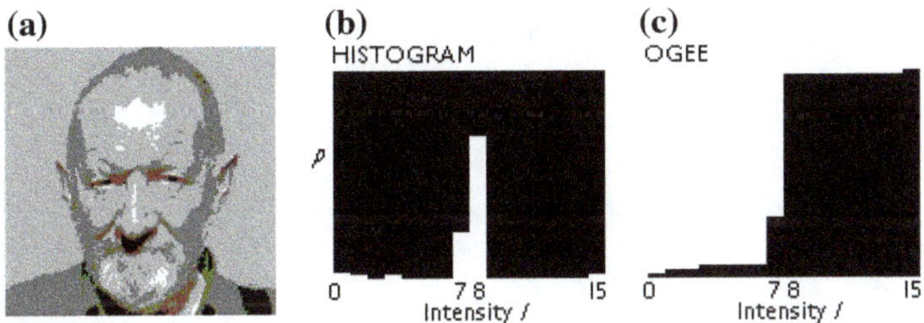

Image file is AP4.BMP

(d)

FREQUENCY TABLE

I	Freq.	Prob.	Cu.Freq.	Cu.%
0	483	.029	483	3
1	297	.018	780	5
2	61	.004	841	5
3	351	.021	1192	7
4	0	.000	1192	7
5	1	.000	1193	7
6	4	.000	1197	7
7	3682	.225	4879	30
8	11110	.678	15989	98
9	0	.000	15989	98
10	0	.000	15989	98
11	0	.000	15989	98
12	0	.000	15989	98
13	0	.000	15989	98
14	5	.000	15994	98
15	390	.024	16384	100

(e)

STATISTICS OF IMAGE

Image bit-depth	4
Image width, px	128
Image height, px	128
Image total px	16384
Mean intensity	7
Mode intensity	8
Variance	5
Standard deviation	2
Skew	-1
Kurtosis	7
Lower qu.intensity	7
Median intensity	8
Upper qu.intensity	8
Intensi.complexity	4
Locatio.complexity	14
Total complexity	56
Entropy	1.403

Fig. 32.7 4-bit camera image. **a** Image 128 × 128 px, with 16 intensities. **b** Histogram. **c** Ogee plot. **d** Frequency table. **e** Statistics

Chapter 33
4-bit .BMP File

The format of a 4-bit BMP file is defined, with examples of the two colour-tables. The colour-table has 16 entries, and the image data has a 4-bit index per pixel. A program to edit 4-bit files is given.

33.1 Flowchart

The flowchart Fig. 33.1 is for general guidance in this chapter.

33.2 Files and Data

For Memory, File, Data Representation, Bitmap Scan, and .BMP File, see Chap. 26.

33.3 4-bit .BMP File

Table 33.1 shows the structure of a 4-bit .BMP file.

A 4-bit .BMP file contains four groups of fields:

- File header
- Info header
- Colour-table
- Image data

The headers are nearly the same as a 1-bit .BMP file (see Chap. 26), but the colour-table and image data are quite different.

A. Parkin, *Digital Imaging Primer*, DOI 10.1007/978-3-540-85619-1_33

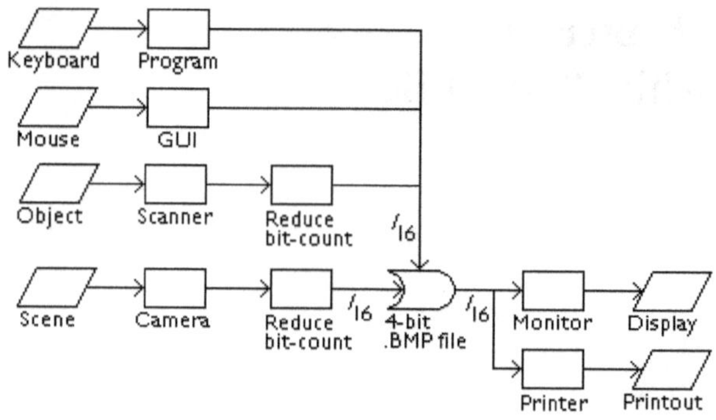

Fig. 33.1 A 4-bit .BMP file can be created or processed by a program or by a GUI editor, or reduced from a 24-bit or an 8-bit file, or displayed or printed out by a GUI editor. It acts as an interface between inputs, processes, and outputs

The file header is a fixed-length group, and contains particulars of the file. The first field is the so-called Magic Cookie, ASCII BM = 66 77 decimal = 42 4D hexadecimal. Without this the file is rejected as not a .BMP file.

The info header contains particulars of the image.

In a 4-bit .BMP file, the colour-table has 16 entries, indexed 0 to F hexadecimal. Each entry occupies a 4-byte *rgbquad*. The first byte is the Blue intensity, the second byte is the Green intensity, the third byte is the Red intensity, and the fourth byte is empty. Each entry determines the colour intensities which are displayed or printed for an image pixel with that index. There are 16 possible colour intensities, 0 to 15 (or 0 to F hexadecimal). There are two standard 4-bit colour-tables, which we here label for convenience 4CO colour and 4GM grey-match (see Chap. 30). Non-standard colour-tables may be set up in any way we please.

The image data is a variable-length group containing, in scan-sequence, the index value 0 to F of each pixel in the image. Each index value occupies one 4-bit nibble, packed two to the byte, and a multiple of 32 bytes to the row.

A field is what can be written or read by a command to the operating system. A field contains one, two, or four bytes, and a byte contains eight bits.

33.4 4-bit Colour File with Colour-table 4CO

Figure 33.2 shows an example of a 4-bit colour image and its file. The image is 16×1 px (shown enlarged). The file is seen in the hexadecimal dump. Table 33.2 is an annotated file listing for Fig. 33.2.

Table 33.1 4-bit .BMP file format

Group	Field	Type	Dec.	Hex.	Remarks
FileHeader			1–14	01–0E	File type, size, layout
	bfType	integer	1–2	01–02	Always 'BM'
	bfSize	long	3–6	03–06	File length in bytes
	bfReserved1	integer	7–8	07–08	Always 0
	bfReserved2	integer	9–10	09–0A	Always 0
	bfOffBits	long	11–14	0B–0E	Header length in bytes
InfoHeader			15–54	0F–36	Image size and colours
	biSize	long	15–18	0F–12	Infoheader length in bytes
	biWidth	long	19–22	13–16	Image width in px
	biHeight	long	23–26	17–1A	Image height in px
	biPlanes	integer	27–28	1B–1C	Always 1
	biBitCount	integer	29–30	1D–1E	Bits per pixel = 4
	biCompression	long	31–34	1F–22	Not compr. = 0
	biSizeImage	long	35–38	23–26	Image length,bytes
	biXPelsPerMeter	long	39–42	27–2A	Printer ppm, horiz.
	biYPelsPerMeter	long	43–46	2B–2E	Printer ppm, vert.
	biClrUsed	long	47–50	2F–32	Nr colours actually used
	biClrImportant	long	51–54	33–36	Nr colours important
ColorTable			55–118	37–76	Array of 16 colours
	aColors[0]	rgbquad	55–58	37–3A	BGRZ for colour[0]
	aColors[1]	rgbquad	59–62	3B–3E	BGRZ for colour[1]
	aColors[2]	rgbquad	63–66	3B–3E	BGRZ for colour[2]
	...				
	aColors[F]	rgbquad	115–118	73–76	BGRZ for colour[F]
Image data			119–	77–	In scan sequence
	aBitmapBits[0]	byte	119–	77–	One nibble per pixel
	aBitmapBits[1]	byte	–	–	
	...				
	aBitmapBits[n]	byte	–	–	

Field types are: *byte* occupying eight bits; *integer* occupying two bytes; *long* occupying four bytes; *rgbquad* occupying four bytes. Notice that rgbquad holds the colour values in the first three bytes, in sequence BGR, and the fourth byte is zero Z

33.5 4-bit Grey-match File with Colour-table 4GM

Figure 33.3 shows an example of a 4-bit grey-match image and its file. The image is 16×1 px with exactly the same index values as before, but pointing to a 4GM colour-table. The file is seen in the hexadecimal dump. Table 33.3 shows the differences for Fig. 33.3.

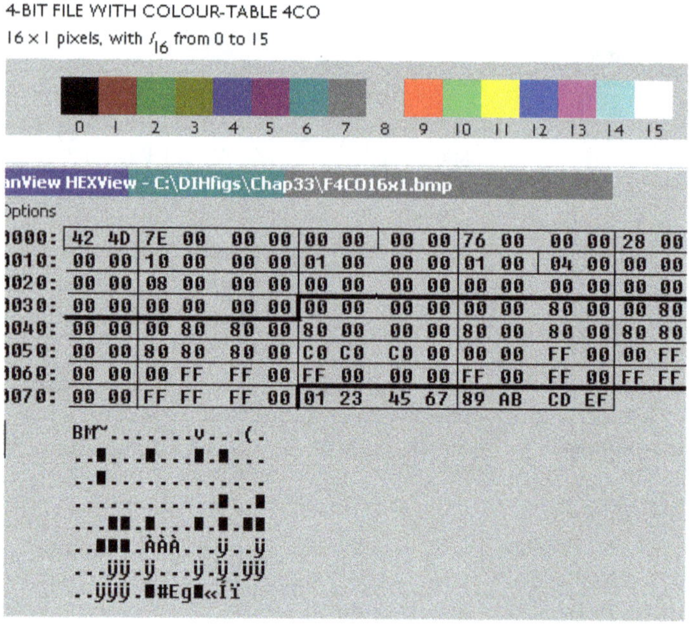

Fig. 33.2 Example of 4-bit image with 4CO colour-table, and its .BMP file shown in hexadecimal dump (with ASCII interpretation below). Each pair of hexadecimal digits is the value of a byte, and the field boundaries are marked. In a multi-byte field the hexadecimal pairs are taken as most-significant right, least-significant left. The file header fields are: a 2-byte magic cookie field containing 42 4D (ASCII BM); a 4-byte file size field containing 7 Eh = 126 d; two 2-byte fields (ignored); a 4-byte header length 76 h = 118 d; a 4-byte info-header length 28 h = 40 d; a 4-byte image width 10 h = 16 d; a 4-byte image height 1; a 2-byte planes field 1; a 2-byte bpp field 4; a 4-byte compression field 0; an image data length 8; two 4-byte ppm fields 0; a 4-byte colours used field 0; and a 4-byte colours important field 0. Between the *thick lines* are the 16 entries of the colour-table, each a 4-byte field for intensity of Blue Green Red and pad zero. Finally there are eight 1-byte fields, each containing the index values of 2 px of the image (origin *bottom right*)

33.6 4-bit BMP File Editor

If we wish to have interactive access to .BMP files, including the ability to change header values, we need a BMP file editor. A 4-bit BMP file editor is much the same as the 1-bit editor in Chap. 26 but for the detailed handling of the 4-bit values. A brief description of a 4-bit editor EDBMP4.BAS follows (with source code at the SpringerLink web-site).

```
REM  **  Program EDBMP4.BAS to edit 4-bit BMP file. User
REM  **     enters filename to edit. Program gets file and
REM  **     offers menu to edit headers or color-table or
REM  **     image-array.
REM  **  Declarations
REM  **  Data
```

Table 33.2 File listing for Fig. 33.2

Byte			Dec.	
Dec.	Field	Type	Value	Remarks
1–14	FileHeader			
1–2	bfType	integer	66 77	'BM'
3–6	bfSize	long	126	File length in bytes
7–8	bfReserved1	integer	0	
9–10	bfReserved2	integer	0	
11–14	bfOffBits	long	118	Header length in bytes
15–54	InfoHeader			
15–18	biSize	long	40	Infoheader length, bytes
19–22	biWidth	long	16	Image width in px
23–26	biHeight	long	1	Image height in px
27–28	biPlanes	integer	1	
29–30	biBitCount	integer	4	Bits per pixel
31–34	biCompression	long	0	Not compressed
35–38	biSizeImage	long	8	Image data len., bytes
39–42	biXPelsPerMeter	long	0	Printer ppm., horiz.
43–46	biYPelsPerMeter	long	0	Printer ppm., vert.
47–50	biClrUsed	long	0	Nr colours used
51–54	biClrImportant	long	0	Nr colours important
55–118	ColorTable			
55–58	aColors[0]	rgbquad	0 0 0 0	BGRZ Black
59–62	aColors[1]	rgbquad	0 0 128 0	BGRZ Dark Red
63–65	aColors[2]	rgbquad	0 128 0 0	BGRZ Dark Green
...				
115–118	aColors[F]	rgbquad	255 255 255 0	BGRZ White
119–126	Image data			
119	aBitmapBits[0]	byte		First two pixels
120	aBitmapBits[1]	byte		Next two pixels
...				
126	aBitmapBits[7]	byte		Last two pixels

```
REM  **  Main program
REM  **     Enter filename
REM  **     Read headers
            CALL HEADER(FILENAME$)
            CALL COLTAB(FILENAME$)
            CALL IMAARR(FILENAME$)
END $
         SUB COLTAB (FILENAME$)
```

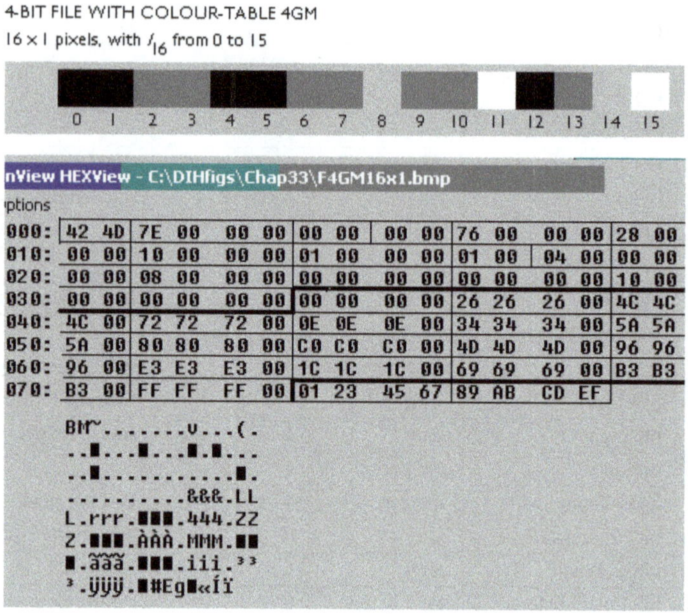

Fig. 33.3 Example of 4-bit image with 4GM colour-table, and its .BMP file shown in hexadecimal dump (with ASCII interpretation below). The header fields and the image fields have exactly the same values as in Fig. 33.2. But here the 16 colour-table fields between the *thick lines* have weighted-sum Greys 0, 26 h = 38 d, 4Ch = 76, ..., FFh = 255 d, which give tonal matches to the 4CO colours

Table 33.3 File differences for Fig. 33.3

Byte			Dec.	
Dec.	Field	Type	Value	Remarks
55–118	ColorTable			
55–58	aColors[0]	rgbquad	0 0 0 0	BGRZ Grey-match 0
59–62	aColors[1]	rgbquad	38 38 38 0	BGRZ Grey-match 1
63–65	aColors[2]	rgbquad	76 76 76 0	BGRZ Grey-match 2
...				
115–118	aColors[F]	rgbquad	255 255 255 0	BGRZ Grey-match F

```
REM   **   Sub-program COLTAB to edit 1-bit colour-table.
END  SUB
           SUB HEADER (FILENAME$)
REM   **   Sub-program HEADER to edit header values of
REM   **      BMP file. Reads each field of header, shows
REM   **      old value, offers next or change or exit.
END  SUB
```

```
          SUB IMAARR (FILENAME$)
REM   **  Sub-program IMAARR to edit 4-bit image-array. User
REM   **  has option to start at any row Y, 0 to (height-1).
REM   **  Program shows for each pixel its colour-table
REM   **  index, and offers options of next or change or exit
END SUB
```

Chapter 34
4-bit Display

A 4-bit image needs a display with 16 intensity values. Display gamma should be compensated in the image, but very little difference occurs if it is ignored. 4-bit display intensity variables and colours are defined for the two colour-tables.

34.1 Flowchart

The flowchart (Fig. 34.1) shows a 4-bit image file sent to a display. The flow is like the 1-bit case (see Chap. 27), but with 16 intensities. Since there are now intermediate intensities, display gamma comes into play, but can usually be ignored.

34.2 Display Hardware

4-bit display hardware is the same as 1-bit hardware, described in Chap. 27.

34.3 Display Size Variables

4-bit display size variables are the same as 1-bit size variables, described in Chap. 27.

34.4 4-bit Display Intensity Variables

When a monitor is used to display a 4-bit image, we have the following per-pixel intensity variables:

A. Parkin, *Digital Imaging Primer*, DOI 10.1007/978-3-540-85619-1_34

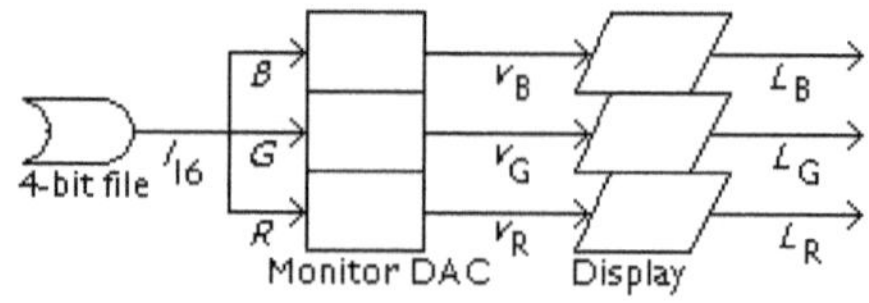

Fig. 34.1 Flowchart for 4-bit display. Per-pixel, a 4-bit image file has intensity index $0 \leq I_{16} \leq 15$, pointing to a colour-table. The colour-table sends (B, G, R) values to the three channels in a colour monitor. In each channel, a digital-to-analogue converter sends a voltage v to the appropriate screen sub-pixel. The screen emits normalized luminance $L_{\text{norm}} = v_{\text{norm}}^{\gamma}$

Table 34.1 From file index to display luminance, with colour-table 4CO

	File				v			L		
I_{16}	B	G	R	Name	B	G	R	B	G	R
0	0	0	0	Black	0	0	0	0	0	0
1	0	0	128	Dark Red	0	0	0.5	0	0	0.22
2	0	128	0	Dark Green	0	0.5	0	0	0.22	0
3	0	128	128	Dark Yellow	0	0.5	0.5	0	0.22	0.22
4	128	0	0	Dark Blue	0.5	0	0	0.22	0	0
5	128	0	128	Dark Magenta	0.5	0	0.5	0.22	0	0.22
6	128	128	0	Dark Cyan	0.5	0.5	0	0.22	0.22	0
7	128	128	128	Grey	0.5	0.5	0.5	0.22	0.22	0.22
8	192	192	192	Pale Grey	0.75	0.75	0.75	0.54	0.54	0.54
9	0	0	255	Red	0	0	1	0	0	1
10	0	255	0	Green	0	1	0	0	1	0
11	0	255	255	Yellow	0	1	1	0	1	1
12	255	0	0	Blue	1	0	0	1	0	0
13	255	0	255	Magenta	1	0	1	1	0	1
14	255	255	0	Cyan	1	1	0	1	1	0
15	255	255	255	White	1	1	1	1	1	1

The v columns are the corresponding v_{norm} values, as shown in Fig. 34.2. The L columns are the corresponding $v_{\text{norm}}^{2.2}$ values, somewhat darker for colours 1 to 8

- Intensity index I_{16}
- Colour-table intensity triple (B, G, R)
- Voltage triple ($v_{\text{B}}, v_{\text{G}}, v_{\text{R}}$)
- Display gamma γ
- Luminance L.

Intensity index is $0 \leq I_{16} \leq 15$, held in the image section of a 4-bit .BMP file (see Chap. 33).

Colour-table intensity triple is $0 \leq \text{B} \leq 255, 0 \leq \text{G} \leq 255, 0 \leq \text{R} \leq 255$, held in the colour-table of a 4-bit .BMP file. It defines the Blue, Green, and Red intensities to be displayed. Colour-tables for 4-bit colour 4CO and 4-bit grey-match 4GM are shown in Chap. 30.

Fig. 34.2 4-bit display colours, with colour-table 4CO. Each square is a display pixel with intensity index I_{16} from 0 Black to 15 White, showing the relative intensities of sub-pixel sources B, G, and R, and the result by additive mixing (values in Table 34.1)

Voltage triple is the (normalized) analogue voltage output by the DAC of each channel, proportional to the colour-table intensity triple.

Display gamma is the darkening of intensity imposed by the display device (see Sect. 22.12). In 4-bit imaging the effect is slight, and is swamped by the thresholding inherent in 4-bit images, so can be ignored.

In each channel, the normalized luminance $L_{\text{norm}} = v_{\text{norm}}^{\gamma}$. For the measurement of luminance, see Chap. 15. The actual luminance of a CRT display can be varied by a control for White luminance (labelled "Contrast") and a control for Black luminance (labelled "Brightness"). When properly adjusted a good CRT has a maximum White luminance of about 170 cd/m^2, and a minimum Black luminance of about 0.01 cd/m^2. An LCD display has a maximum luminance of about 160 cd/m^2, and a minimum luminance of about 0.72 cd/m^2 (with minimum back-light) [1].

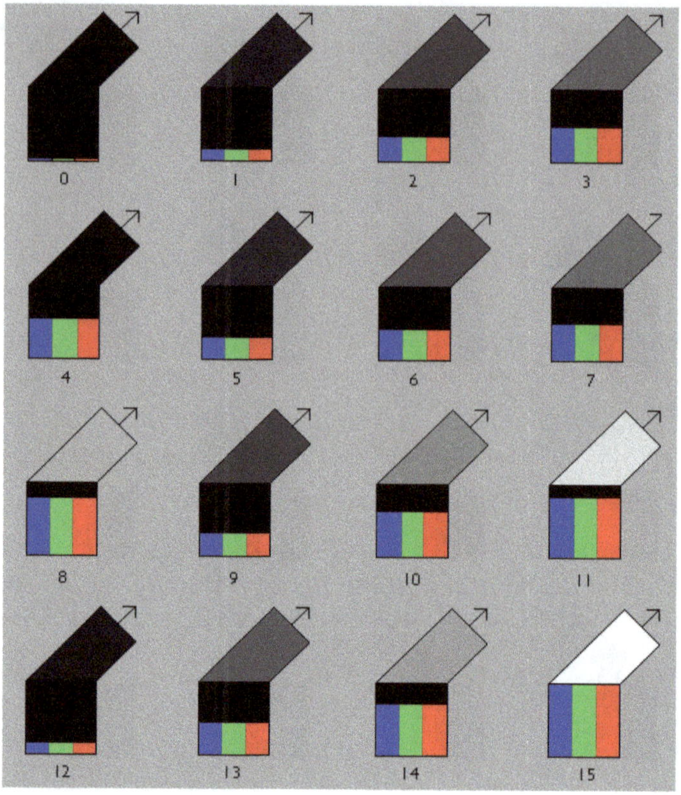

Fig. 34.3 4-bit display colours, with colour-table 4GM. Each square is a display pixel with intensity index I_{16} from 0 Black to 15 White, showing the relative intensities of sub-pixel sources B, G, and R, and the result by additive mixing (values in Table 34.2)

34.5 4-bit Display Colours

Figure 34.2 shows the 16 possible colours of a 4-bit display pixel, where the file colour-table is 4CO. In each case the B G and R sub-pixels are off, half-on, three-quarters on (index 8), or full-on. The resulting colour is by additive mixing (see Sect. 27.11). A monitor receives, per-pixel, an integer triple (B, G, R) from a file. In each channel, the monitor converts to analogue voltage, and displays luminance with gamma = 2.2. Table 34.1 shows the values for Fig. 34.2.

Figure 34.3 shows the 16 possible colours of a 4-bit display pixel, where the file colour-table is 4GM. Table 34.2 shows the values for Fig. 34.3.

Table 34.2 From file index to display luminance with colour-table 4GM

	File				v			L		
I_{16}	B	G	R	Name	B	G	R	B	G	R
0	0	0	0	Black	0	0	0	0	0	0
1	38	38	38	Dark Red	0.15	0.15	0.15	0.01	0.01	0.01
2	76	76	76	Dark Green	0.30	0.30	0.30	0.07	0.07	0.07
3	114	114	114	Dark Yellow	0.45	0.45	0.45	0.17	0.17	0.17
4	14	14	14	Dark Blue	0.05	0.05	0.05	0.00	0.00	0.00
5	52	52	52	Dark Magenta	0.20	0.20	0.20	0.03	0.03	0.03
6	90	90	90	Dark Cyan	0.35	0.35	0.35	0.10	0.10	0.10
7	128	128	128	Grey	0.50	0.50	0.50	0.22	0.22	0.22
8	192	192	192	Pale Grey	0.75	0.75	0.75	0.54	0.54	0.54
9	77	77	77	Red	0.30	0.30	0.30	0.07	0.07	0.07
10	150	150	150	Green	0.59	0.59	0.59	0.31	0.31	0.31
11	227	227	227	Yellow	0.89	0.89	0.89	0.77	0.77	0.77
12	28	28	28	Blue	0.11	0.11	0.11	0.01	0.01	0.01
13	105	105	105	Magenta	0.41	0.41	0.41	0.14	0.14	0.14
14	179	179	179	Cyan	0.70	0.70	0.70	0.46	0.46	0.46
15	255	255	255	White	1	1	1	1	1	1

The v columns are the corresponding v_{norm} values, as shown in Fig. 34.3. The L columns are the corresponding $v_{\mathrm{norm}}^{2.2}$ values, somewhat darker for colours 1 to 8

Reference

1. Soneira RM (2005) Shoot-out. http://www.displaymate.com/ShootOut. Accessed 1 Feb 2014

Chapter 35
4-bit Printout

Printer size variables are the same for 4-bit as for 1-bit images. Ideally, a printer should convert the three additive colours (Blue, Green, Red) to the complementary subtractive inks (Yellow, Magenta, Cyan). But deficiencies in the available inks means that Black ink has to be used also, and the Black component removed from the other three inks. A calculation for this conversion is defined. Printing then requires separation of the four ink components as 8-bit grey-scale sub-images, and then half-toning each to 1-bit sub-images. The 4-bit intensity variables are defined, with examples of colours, separation, and half-toning. The spreading of ink when a half-tone is printed causes dot gain, which acts in a similar manner to display gamma. The flowchart (Fig. 35.1) shows a 4-bit image file sent to a printer by an image editor.

35.1 Flowchart

A colour printout is a very complex construction which is difficult to control. The image is made of four sets of ink dots on a White ground. In viewing, dots which overlap combine by subtractive filtering, but dots which are separate combine additively. Different papers cause different spreading of dots. Compromises are forced by the imperfections of the inks, variations of paper, and differences between make of printer. Desktop colour printing is normally a closed process within the software supplied with a printer. More elaborate equipment, or commercial printing services, offer proper colour management (see Sect. 16.6).

35.2 Printer Hardware

4-bit printer hardware is the same as 1-bit hardware, described in Chap. 28.

A. Parkin, *Digital Imaging Primer*, DOI 10.1007/978-3-540-85619-1_35

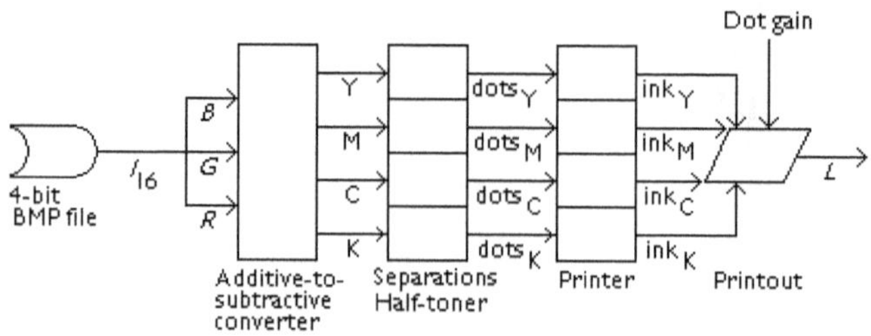

Fig. 35.1 Flowchart for 4-bit printout. A 4-bit .BMP file has, per pixel, location coordinates (X, Y), and intensity index I_{16} pointing to colour-table intensities (B, G, R). An additive-to-subtractive converter has three input channels for (B, G, R) and four output separation channels for (Y, M, C, K). In each separation channel a half-toner converts 8-bit intensities to 1-bit dot densities, which a printer applies as ink on paper. Dot gain is caused by ink spread on paper

35.3 Print Size Variables

4-bit print size variables and resolving power are the same as for 1-bit print, described in Chap. 28.

35.4 Ideal YMC Printing

Printing a digital image is the inverse of displaying it. Whereas a display emits light from three sources which combine additively, a printout reflects light through transparent inks which combine as subtractive filters.

Suppose we have three transparent inks: Yellow Y which absorbs Blue light, Magenta M which absorbs Green light, and Cyan C which absorbs Red light.[1] Figure 35.2 shows the eight possible combinations of the three inks.

But in going from additive display to subtractive print there are two big differences. First, available inks do not transmit perfectly in their pass bands, do not absorb perfectly in their stop bands, and have overlaps of bands. Consequently the Black produced by YMC is a disappointing purple-brown. To get a good Black, we can use a fourth ink K which absorbs equally in all three wavebands.[2] Second, whereas a display can vary the intensities of its three sources by varying the analogue voltages supplied, a printer has only a binary choice between ink or no-ink for each of its primaries. To get variation of ink intensity, a printer uses half-toning, in effect diluting the ink with White paper.

[1]Y is sometimes called a *minus-Blue* ink; M *minus-Green*; C *minus-Red*.

[2]In the printing trades K stood for key block or plate or ink.

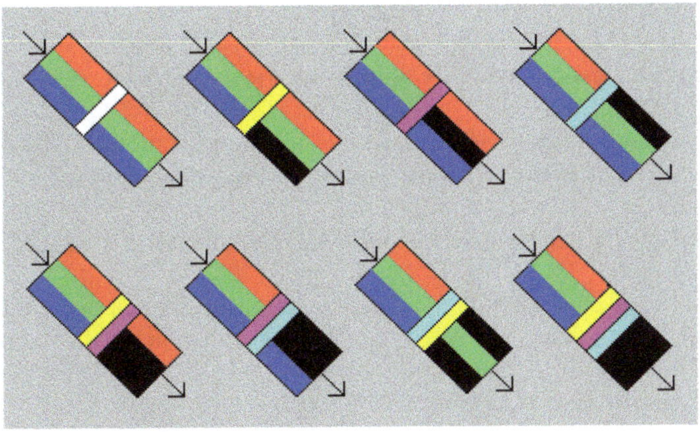

Fig. 35.2 YMC filters. **a** Clear filter stops nothing, passes White light. **b** Yellow filter stops Blue, passes Green and Red light. **c** Magenta filter stops Green, passes Blue and Red light. **d** Cyan filter stops Red, passes Blue and Green light. **e** Yellow and Magenta filters stop Blue and Green, pass Red light. **f** Magenta and Cyan filters stop Green and Red, pass Blue light. **g** Cyan and Yellow filters stop Red and Blue, pass Green light. **h** Yellow and Magenta and Cyan filters should stop all light, pass none

35.5 Practical YMCK Printing

Given three additive intensities (B, G, R) each in the range 0 to 255, first we get the 255-complements Y, M, C, then find the intensity of Black boost K as the minimum of Y, M, C:

$$\begin{aligned} Y &= 255 - B \\ M &= 255 - G \\ C &= 255 - R \\ K &= \text{MIN}\ (Y, M, C) \end{aligned}$$

This overprints the K contribution on the YMC. But heavy overprinting is unwelcome because the paper gets too wet (and in successive-run commercial production one ink picks up another), so it is usual to *remove under-colour*,[3] reducing each of Y, M, C by K:

$$\begin{aligned} Yur &= Y - K \\ Mur &= M - K \\ Cur &= C - K \end{aligned}$$

[3]Sometimes only part of the under-colour is removed.

For example:

- Suppose we have 4-bit additive colour, index $I_{16} = 1$ Dark Red, specified as (B, G, R) = (0, 0, 128).
- In the Blue band, to transmit 0 we must set the Yellow ink to absorb $255 - 0 = 255$.
- In the Green band, to transmit 0 we must set the Magenta ink to absorb $255 - 0 = 255$.
- In the Red band, to transmit 128 we must set the Cyan ink to absorb $255 - 128 = 127$.
- But the three inks have a common Black component K equal to the minimum of (255, 255, 127), namely 127. We assign this absorption to the Black ink:
- $\mathrm{Y} = 255 - 0 = 255$,
- $\mathrm{M} = 255 - 0 = 255$,
- $\mathrm{C} = 255 - 128 = 127$,
- $\mathrm{K} = \mathrm{MIN}\ (255, 255, 127) = 127$,
- If we overprint the Black on the other inks, (Y, M, C, K) = (255, 255, 127, 127) is the subtractive equivalent of additive (B, G, R) = (0, 0, 128).
- If we remove the under-colour we have,
- $\mathrm{Yur} = 255 - 127 = 128$,
- $\mathrm{Mur} = 255 - 127 = 128$,
- $\mathrm{Cur} = 127 - 127 = 0$,
- $\mathrm{Kur} = 127$.
- So then (Yur, Mur, Cur, Kur) = (128, 128, 0, 127) is the under-colour-removed subtractive equivalent of additive (B, G, R) = (0, 0, 128).

The inverse conversion from four subtractive ink intensities (Y, M, C, K) (under-colour-removed) to three additive intensities (B, G, R), is

$$
\begin{aligned}
\mathrm{Y} &= \mathrm{Yur} + \mathrm{Kur} \\
\mathrm{M} &= \mathrm{Mur} + \mathrm{Kur} \\
\mathrm{C} &= \mathrm{Cur} + \mathrm{Kur} \\
\mathrm{B} &= 255 - \mathrm{Y} \\
\mathrm{G} &= 255 - \mathrm{M} \\
\mathrm{R} &= 255 - \mathrm{C}
\end{aligned}
$$

35.6 Grey-scale Separations

To print a given 4-bit colour image, we need to separate the 4-bit image to four 8-bit grey-scale images, one for each of the Y M C K channels. The Y-separation contains only the intensity of Yellow ink to be printed, and similarly the M C and

Fig. 35.3 4-bit source images. Top is a 24-bit camera image saved-as 4-bit. Bottom is 16 stripes with index I_{16} from 0 Black at *bottom left* to 15 White at *top right*

K separations. The grey-scale intensities are the 255-complements of the (under-colour-removed) intensities (since grey-scale White, to print no ink, is (255, 255, 255), and grey-scale Black, to print solid ink, is (0, 0, 0)). The calculation is

$$\begin{aligned} \text{Ygs} &= 255 - \text{Yur} \\ \text{Mgs} &= 255 - \text{Mur} \\ \text{Cgs} &= 255 - \text{Cur} \\ \text{Kgs} &= 255 - \text{K} \end{aligned}$$

For example, Fig. 35.3 shows two source 4-bit images with colour-table4CO, and Fig. 35.4 shows the corresponding Y M C K grey-scale separations.

Separation is normally done by the printer software. Some of the more elaborate image editors offer YMCK separation. IRFANVIEW has BGR separation, which can be inverted to YMC, but not YMCK. Those who wish to dabble can save a 4-bit image as 24-bit and do the separating by a program like that described in Chap. 47. But there appears to be no simple way to send own-produced separations to a desk-top printer.

35.7 Separation Half-tones

Finally we need to make a 1-bit half-tone from each of the Y M C K 8-bit separations, as shown in Fig. 35.5. The ink percent density applied by the printer will be $100/255\times$ the 255-complement of the grey-scale intensity:

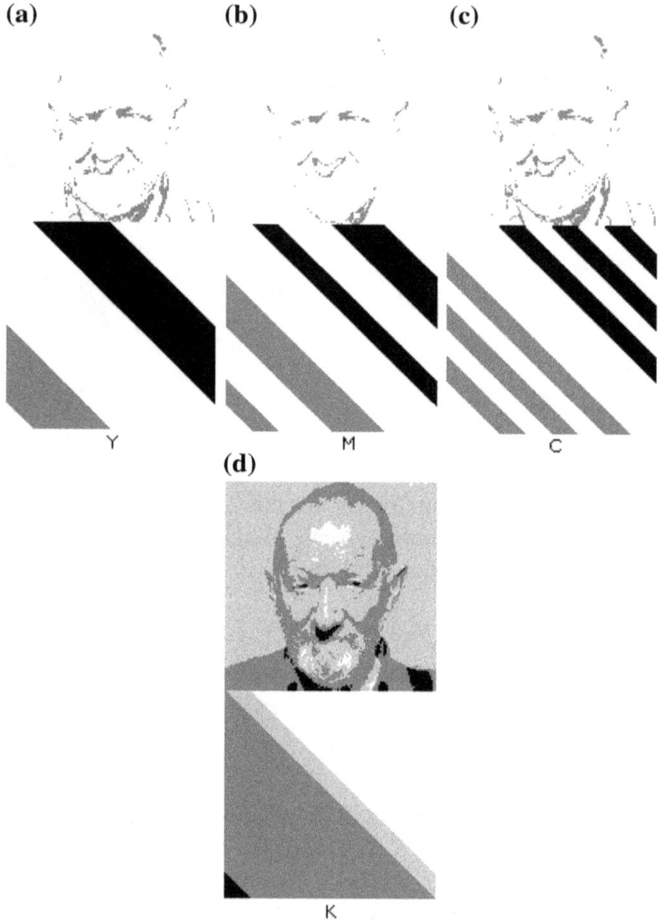

Fig. 35.4 8-bit grey-scale separations. **a** Yellow separation. Stripes 1, 2, 3 show mid-grey 127, so the printer will place 50 % Yellow. **b** Magenta separation. Stripe 1 shows mid-grey 127, so the printer will place 50 % Magenta; stripes 2 and 3 will have no Magenta. **c** Cyan separation. Stripe 2 shows mid-grey 127, so the printer will place 50 % Cyan; stripes 1 and 3 will have no Cyan. **d** Black separation. Stripe 0 shows Black 0, so the printer will place solid Black. Stripes 1–7 show mid-grey 127, so the printer will place 50 % Black. Stripe 8 shows pale grey 192, so the printer will place 25 % Black. Stripe 1 will end up with 50 % Yellow, 50 % Magenta, and 50 % Black, making Dark Red

$$\begin{aligned} \text{Ypc} &= (255 - \text{Ygs})/2.55 \\ \text{Mpc} &= (255 - \text{Mgs})/2.55 \\ \text{Cpc} &= (255 - \text{Cgs})/2.55 \\ \text{Kpc} &= (255 - \text{Kgs})/2.55 \end{aligned}$$

Half-toning is normally done by the printer software. Several half-toning schemes are described in Chap. 39.

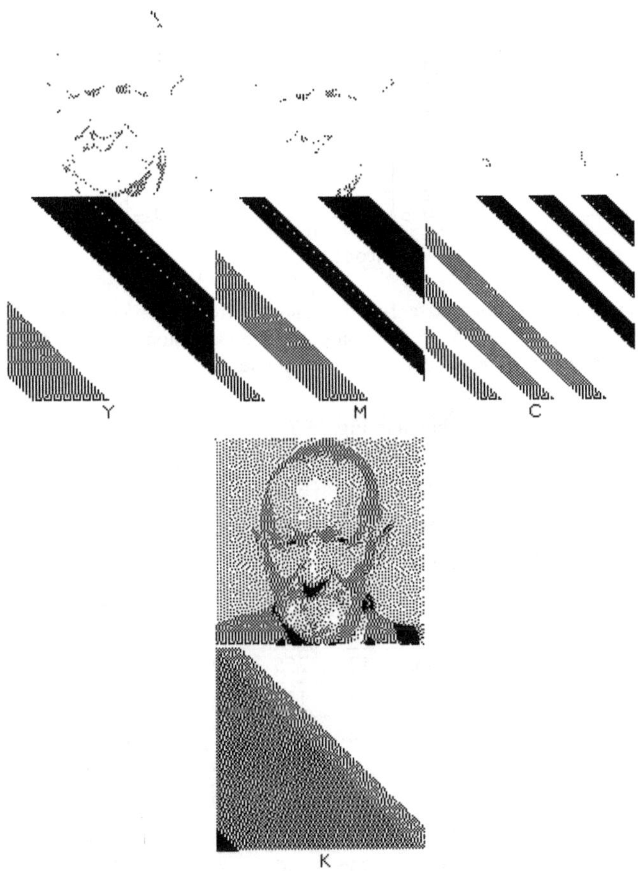

Fig. 35.5 1-bit half-toned separations from Fig. 35.4

35.8 4-bit Print Intensity Variables

In printing a 4-bit image, we have the following per-pixel intensity variables:

- 4-bit intensity index I_{16},
- 8-bit colour-table intensity triple (B, G, R),
- 8-bit separation intensity quadruple (Y, M, C, K),
- 8-bit separation grey-scale quadruple (Ygs, Mgs, Cgs, Kgs),
- 1-bit ink percent density quadruple (Ypc, Mpc, Cpc, Kpc),
- Dot gain δ.

Intensity index is $0 \leq I_{16} \leq 15$, held in the image section of a 4-bit .BMP file (see Chap. 33).

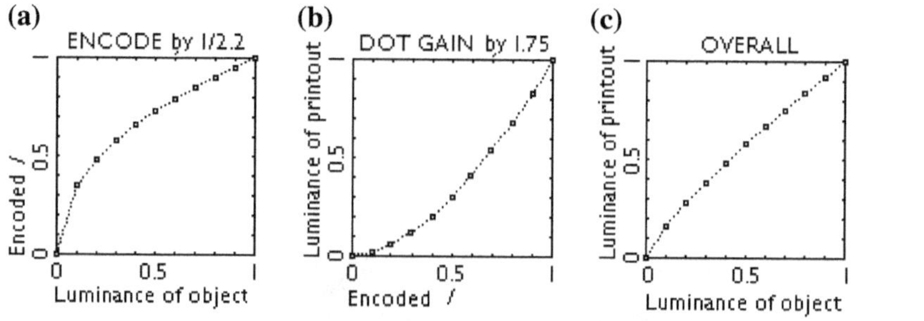

Fig. 35.6 Encoding and dot gain transfer functions. **a** Input from creating program, GUI, or scanner is encoded with $1/\gamma = 1/2.2 = 0.4545$, then output to file. **b** Printout has dot gain (say) $\delta = 1.75$. **c** When (**a**) is followed by (**b**), printout is somewhat lighter than input

Table 35.1 4-bit print colours, as shown in Fig. 35.7

Index	B	G	R	Name	Y	M	C	K	Yur	Mur	Cur	Kur
0	0	0	0	Black	255	255	255	255	0	0	0	255
1	0	0	128	Dk Red	255	255	127	127	128	128	0	127
2	0	128	0	Dk Green	225	127	255	127	128	0	128	127
3	0	128	128	Dk Yellow	255	127	127	127	128	0	0	127
4	128	0	0	Dk Blue	127	255	255	127	0	128	128	127
5	128	0	128	Dk Magenta	127	255	127	127	0	128	0	127
6	128	128	0	Dk Cyan	127	127	255	127	0	0	128	127
7	128	128	128	Grey	127	127	127	127	0	0	0	127
8	192	192	192	Pl Grey	63	63	63	63	0	0	0	63
9	0	0	255	Red	255	255	0	0	255	255	0	0
10	0	255	0	Green	255	0	255	0	255	0	255	0
11	0	255	255	Yellow	255	0	0	0	255	0	0	0
12	255	0	0	Blue	0	255	255	0	0	255	255	0
13	255	0	255	Magenta	0	255	0	0	0	255	0	0
14	255	255	0	Cyan	0	0	255	0	0	0	255	0
15	255	255	255	White	0	0	0	0	0	0	0	0

Colour-table intensity triple is $0 \le \mathrm{B} \le 255, 0 \le \mathrm{G} \le 255, 0 \le \mathrm{R} \le 255$, held in the colour-table of a 4-bit .BMP file. It defines the proportions of additive primaries in the pixel. Colour-tables for 4-bit colour 4CO and 4-bit grey-match 4GM are shown in Chap. 30.

Separation intensity quadruple is $\le \mathrm{Y} \le 255$, $0 \le \mathrm{M} \le 255$, $0 \le \mathrm{C} \le 255$, $0 \le \mathrm{K} \le 255$. It defines the proportions of subtractive primaries in the pixel. In a 4-bit image the intensities are limited to 0, 128, 192, and 255. The conversion is normally done by the printer software. The calculation is shown below.

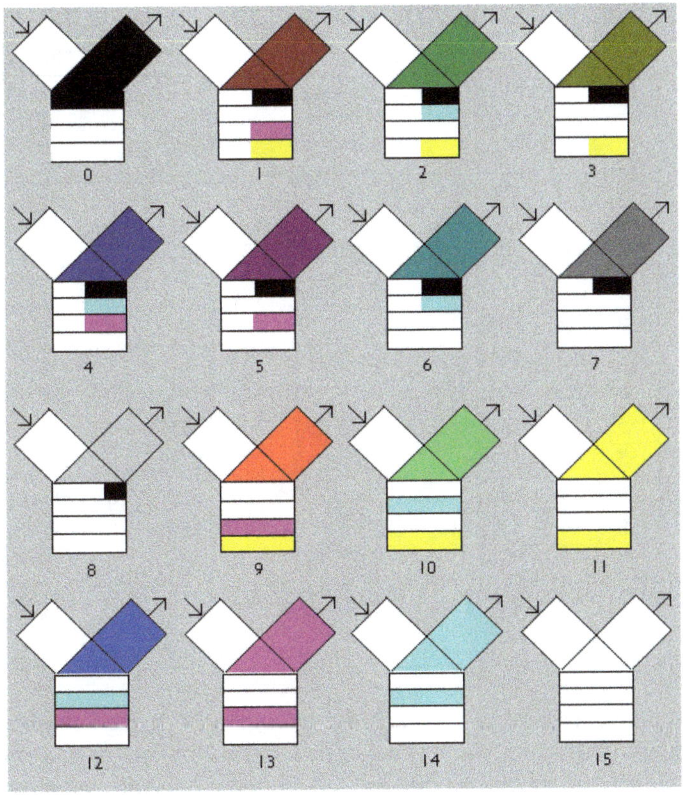

Fig. 35.7 4-bit print colours, as listed in Table 35.1. White light incident on a printout is reflected differentially according to the coverage of White paper by Black, Cyan, Magenta, and Yellow inks

Separation grey-scale intensity is the 255-complement of the separation intensity quadruple; that is to say 0 becomes 255, 128 becomes 127, and 255 becomes 0.

Ink percent density is $100/255\times$ the 255-complement of the grey-scale intensity; that is to say, 0 becomes 100 %, 127 becomes 50 %, 192 becomes 25 %, and 255 becomes 0 %.

In printing, *dot gain* is the spreading of ink on the paper. It occurs at the peripheries of ink dots. For a large inked area it is negligible, but for a half-toned area the dots become larger than intended and the White area smaller, thus darkening the image. Dot gain can be represented reasonably well by a power-law transfer function with exponent δ:

$$\text{Actual} = \text{calculated}^{\delta}$$

where *Actual* is the normalized actual ink coverage, and *calculated* is the normalized calculated ink coverage. The numerical value of δ varies according to the printing

Table 35.2 4-bit print colours in 8-bit grey-scale separations, and half-toned percent coverages

Index	B	G	R	Name	Ygs	Mgs	Cgs	Kgs	Ypc	Mpc	Cpc	Kpc
0	0	0	0	Black	255	255	255	0	0	0	0	100
1	0	0	128	Dk Red	127	127	255	128	50	50	0	50
2	0	128	0	Dk Green	127	255	127	128	50	0	50	50
3	0	128	128	Dk Yellow	127	255	255	128	50	0	0	50
4	128	0	0	Dk Blue	255	127	127	128	0	50	50	50
5	128	0	128	Dk Magenta	255	127	255	128	0	50	0	50
6	128	128	0	Dk Cyan	255	255	127	128	0	0	50	50
7	128	128	128	Grey	255	255	255	128	0	0	0	50
8	192	192	192	Pl Grey	255	255	255	192	0	0	0	25
9	0	0	255	Red	0	0	255	255	100	100	0	0
10	0	255	0	Green	0	255	0	255	100	0	100	0
11	0	255	255	Yellow	0	255	255	255	100	0	0	0
12	255	0	0	Blue	255	0	0	255	0	100	100	0
13	255	0	255	Magenta	255	0	255	255	0	100	0	0
14	255	255	0	Cyan	255	255	0	255	0	0	100	0
15	255	255	255	White	255	255	255	255	0	0	0	0

process and the paper used: it will be near 1 (no gain) for a high-precision image-setter on resin-coated paper, somewhat larger for a desktop ink-jet printer on special paper, and larger still for ordinary paper. A widely-used average value is $\delta = 1.75$, as shown in Fig. 35.6, which is comparable with the display $\gamma = 2.2$ (see Sect. 34.4). Such a function has no effect on the extremes 0 (solid ink) and 1 (White paper); only on half-tones.

35.9 4-bit Print Colours

Table 35.1 lists the 16 4-bit colours with their additive display intensities (B, G, R) and subtractive print intensities (Y, M, C, K) and (Yur, Mur, Cur, Kur). Figure 35.7 shows the appearance.

35.10 4-bit Separations and Half-tones

Table 35.2 lists the 16 4-bit colours with their additive display intensities (B, G, R), 8-bit grey-scale intensities (Ygs, Mgs, Cgs, Kgs) and half-toned percent coverage (Ypc, Mpc, Cpc, Kpc).

Chapter 36
Viewing 4-bit Image

Viewing a 4-bit displayed image and a 4-bit printed image is compared with viewing a scene directly. Scope, location resolution, and luminance range are the same for 4-bit as for 1-bit images, but colour resolution is quite different.

36.1 Flowchart

The flowchart (Fig. 36.1) is for general guidance in this Chapter. It shows an eye viewing various types of 4-bit image, and, by comparison, viewing a scene. The computing flow is like that in Chap. 29, but with 16 intensities instead of two.

36.2 Viewing Variables

If we want to compare viewing a scene with viewing a digital image of a scene, four variables emerge:

- Location range, or scope, measured as angular or linear extent of view
- Location resolution within scope, measured as Snellen acuity, or as line-pairs per unit angle, or as line-pairs per unit length (see Sect. 10.6)
- Luminance range, measured as dynamic operating range from lowest to highest luminance in cd/m^2 (see Sect. 15.8)
- Colour resolution within luminance, measured as gamut on a CIE chromaticity chart(see Sect. 16.3).

A. Parkin, *Digital Imaging Primer*, DOI 10.1007/978-3-540-85619-1_36

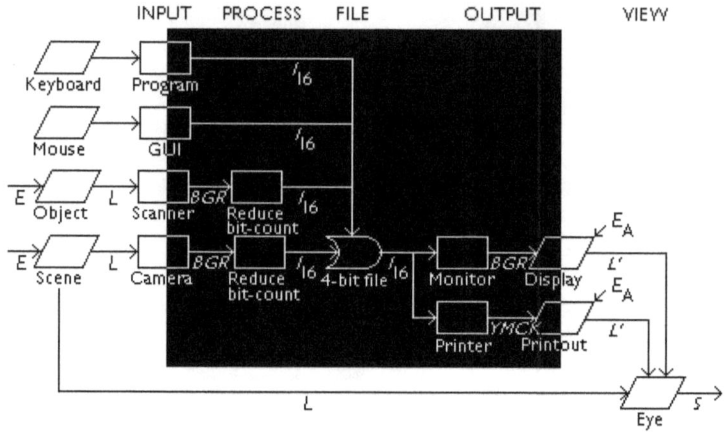

Fig. 36.1 Comparison of viewing 4-bit image with viewing scene. E is illuminance in lx, from source onto object or scene. L is luminance in cd/m^2, from scene to input device, from output device to eye, and from scene to eye. I_{16} is intensity index 0 or 1 from input to file, and from file to output. BGR is intensity triple, 3× (eight bits 0–255), from camera to bit-reducer, and from monitor to display. CMYK are ink coverages, from printer to paper. E_A is ambient illuminance in lux, from surroundings onto display or printout. L' is luminance from output device to eye, and S is viewer's Space/Colour sensation from eye to perception

The location range and resolution for 4-bit images are the same as for 1-bit images in Chap. 29.

The luminance range for 4-bit images is the same as for 1-bit images in Chap. 29, if we ignore display gamma and dot gain effects.

The ambient illuminance effects for 4-bit are the same as for 1-bit in Chap. 29.

36.3 Colour Resolution of Eye and 4-bit Image

Figure 36.2 shows the colour gamut of the eye compared with 4-bit images. In the chromaticity chart, Brightness is imagined as a third dimension, from low below the chart to high above the chart; thus Dark Red 1 and Red 9 have the same chromaticity but different Brightnesses, and neutrals 0 7 8 and 15 likewise. In Fig. 36.2b, a display has additive mixture: for example primaries Red 9 and Green 10 mix to make secondary Yellow 11. In Fig. 36.2c, a printout has subtractive mixture: primaries Yellow 11 and Magenta 13 mix to make secondary Red 9, and the saturations are lower. In Fig. 36.2d, the colours are all neutrals.

The colour gamut of a digital image is determined by the colour-table of the file, and the output device used.

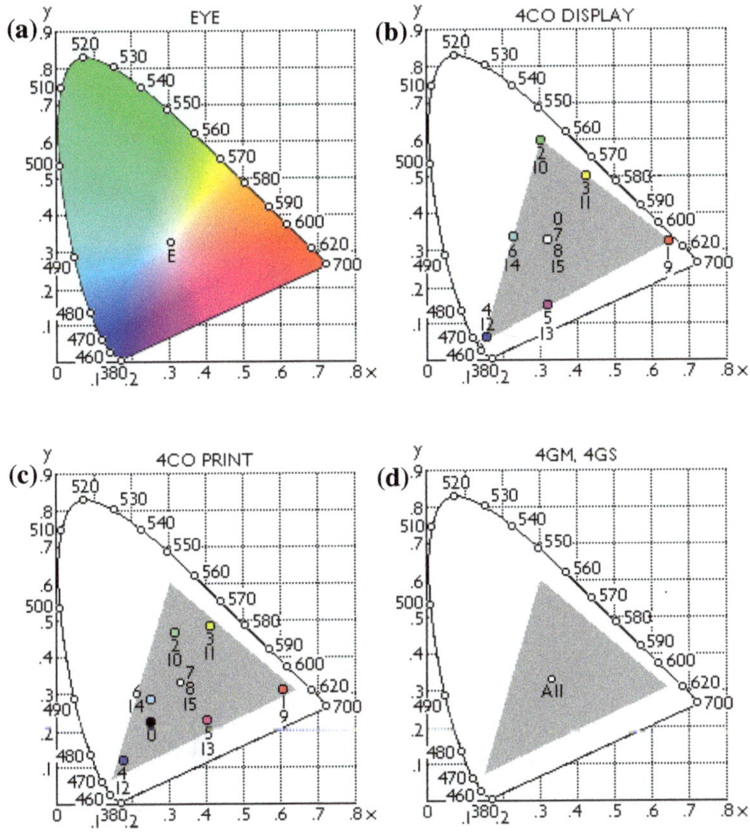

Fig. 36.2 Colour gamut of eye, compared to that of 4-bit image. **a** CIE chromaticity chart showing the colour gamut to which the eye responds. E is equal-energy illuminant. Notice that the colours printed here are merely indicative: the eye gamut extends to much more saturated colours. **b** Discrete 16-colour gamut of 4-bit display with colour-table 4CO. The numbers are the intensity index I_{16}. The sRGB triangle is added for guidance. **c** Gamut of 4-bit printout with colour-table 4CO. **d** Gamut of 4-bit image with colour-table 4GM

Part VII
8-bit Imaging

Chapter 37
8-bit Image

An 8-bit image has the same properties of location as a 1-bit image. But it has 256 intensities available, specified by an index 0 to 255 in eight binary digits (or two hexadecimal digits) per pixel. The index points to a 256-entry colour-table in its file, each holding a triple of (Blue, Green, Red) intensities. There are three alternative 8-bit colour-tables: 8CO holding 256 colours, 8GM holding 256 grey-matches to the colours, and 8GS holding a complete grey-scale of 256 greys between Black and White. The 8-bit colour-tables are defined. Examples are given of typical 8-bit images created by program, graphic user interface, scanner, and camera, under each of the colour-tables.

37.1 Bit-depth 8

256 intensities (B, G, R) can be specified by an intensity index I_{256} taking an 8-bit value 0 to 255 decimal, (= 0 to FF hexadecimal, = 0 to 11111111 binary). We then say that the plane has *bit-depth 8*, or that it is the *8-bit plane*. The intensity index I_{256} refers to a 256-entry colour-table in an 8-bit .BMP file (see Chap. 41). Each of the index values points to an intensity triple (B, G, R) used to display that pixel on a monitor screen (see Chap. 42) or print it on paper (see Chap. 43). The colour-table may be set up in any way we please by using a file editor such as EDBMP8.BAS described in Chap. 41. There are two standard 8-bit colour-tables:

- 256 colours, which we call here 8CO, including the 16 4-bit colours (see Chap. 30), and the 1-bit Black and White (see Chap. 19);
- 256 greys, which we call here 8GS, which form a smooth gradation from Black to White.

Table 37.1 shows just the 28-colour subset which is offered by MS PAINT in 8-bit mode (see Chap. 38). The 16-colour subset with names are the 4CO 4-bit colours (see Chap. 30).

A. Parkin, *Digital Imaging Primer*, DOI 10.1007/978-3-540-85619-1_37

Table 37.1 Colour-table 8CO (28-colour subset), shown in Fig. 37.1

8CO				
Name	Index	B	G	R
Black	0	0	0	0
Dark Red	1	0	0	128
Dark Green	2	0	128	0
Dark Yellow	3	0	128	128
Dark Blue	4	128	0	0
Dark Magenta	5	128	0	128
Dark Cyan	6	128	128	0
Pale Grey	7	192	192	192
	9	240	202	166
	20	0	64	128
	80	64	64	0
	100	64	128	128
	103	64	128	224
	135	128	0	224
	144	128	64	0
Grey	164	128	128	128
	184	128	224	0
	191	128	224	224
	196	192	0	128
	224	192	128	0
	228	192	128	128
Red	249	0	0	255
Green	250	0	255	0
Yellow	251	0	255	255
Blue	252	255	0	0
Magenta	253	255	0	255
Cyan	254	255	255	0
White	255	255	255	255

37.2 8-bit Colour-table 8CO

When an additive B G R display (see Chap. 42) and/or a subtractive Y M C K printer (see Chap. 43) is available, we use the colour-table 8CO, as shown in Table 37.1 and Fig. 37.1. In the full table, entries between 10 and 239 have Blue increasing by steps of 64, Green by steps of 32, and Red by steps of 32; entries 8 and 9, and 240 to 247, are oddities. To make an image with this gamut (see Appendix A):

- Create an image by program or GUI in 8-bit mode (see Chap. 38).
- Or create an image by scanner or camera in 24-bit mode, then save-as 8-bit with 256 colours.

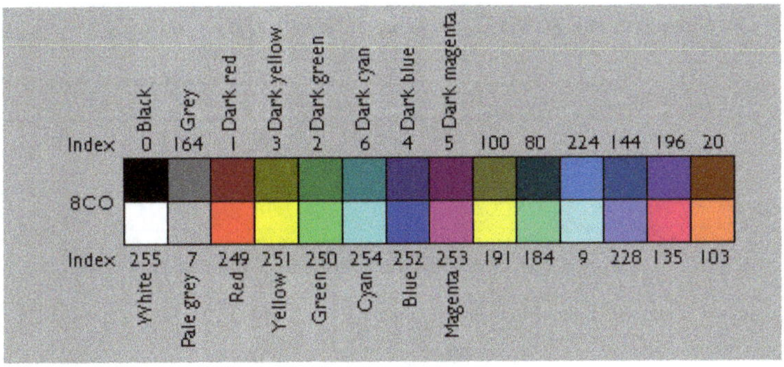

Fig. 37.1 Colour gamut 8CO (28-colour subset). The index numbers point to the (B, G, R) triples in the 8CO colour-table used to display or print, as listed in Table 37.1. For example, Dark Red has index 1 and (B, G, R) = (0, 0, 128); Dark Green has index 2 and (B, G, R) = (0, 128, 0)

37.3 8-bit Grey-match Colour-table 8GM

When the available display device and/or printer can show a range of grey intensities, we use a colour-table to deliver greys which are a reasonable Brightness match to the 8CO colours, as shown in Table 37.2 and Fig. 37.2. A grey has equal intensities B = G = R. But Blue looks dark, Green looks light, and Red between the two; so we find the grey-match as the weighted sum of the three components:

$$\text{Grey} - \text{match} = 0.11 \text{ Blue} + 0.59 \text{ Green} + 0.30 \text{ Red}$$

For example Dark Red (B, G, R) $=$ (0, 0, 128) becomes grey-match $(0.11 \times 0 + 0.59 \times 0 + 0.30 \times 128) = (38, 38, 38)$. The intensity index I_{256} is set as 38, which points to the 8GM colour-table entry (38, 38, 38). To make an image with this gamut (see Appendix A):

- Create an 8CO image by program or GUI in 8-bit mode, or by scanner or camera in 24-bit mode saved-as 8CO; then convert-to grey-scale.

37.4 8-bit Grey-scale 8GS

The 8GS colour-table has 256 equal-step greys, each with equal intensities B $=$ G $=$ R, and equal to the index I_{256}. Figure 37.3 and Table 37.3 show the first, middle, and last of these greys. 8GS is of great importance because it supports grey-scale

Table 37.2 Grey-match gamut 8GM (28-grey-match subset), shown in Fig. 37.2

8GM

Name	Index	B	G	R
Black	0	0	0	0
Dark Red	38	38	38	38
Dark Green	76	76	76	76
Dark Yellow	114	114	114	114
Dark Blue	14	14	14	14
Dark Magenta	52	52	52	52
Dark Cyan	90	90	90	90
Pale Grey	192	192	192	192
	195	195	195	195
	76	76	76	76
	45	45	45	45
	121	121	121	121
	150	150	150	150
	81	81	81	81
	52	52	52	52
Grey	128	128	128	128
	146	146	146	146
	213	213	213	213
	60	60	60	60
	97	97	97	97
	135	135	135	135
Red	77	77	77	77
Green	150	150	150	150
Yellow	227	227	227	227
Blue	28	28	28	28
Magenta	105	105	105	105
Cyan	179	179	179	179
White	255	255	255	255

In the .BMP file, the colour-table is actually 8GS with has 256 greys from (0, 0, 0) Black to (255, 255, 255) White, but in 8GM some greys are used more than once and some are not used at all

image gradients which appear smooth, and which can be half-toned to 1-bit smooth gradients (see Chap. 39). Smooth gradients are not possible in 8CO nor in 8GM (nor in 4-bit images) because of jumps in the colour-tables; (but smooth gradients can be made in 1-bit images by half-toning). To make an image with this gamut (see Appendix A):

- Create an 8GS image by program in 8-bit grey-scale mode (see Chap. 38).
- Or create an image by scanner in 8-bit grey-scale mode.

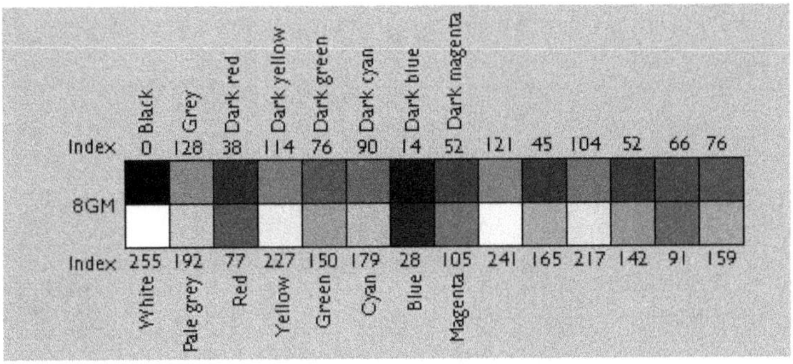

Fig. 37.2 Colour gamut 8GM (28-colour subset). The index numbers point to the (B, G, R) triples in the 8GM colour-table used to display or print, as listed in Table 37.2. For example, Dark Red has index 38 and (B, G, R) = (38, 38, 38); Dark Green has index 76 and (B, G, R) = (76, 76, 76)

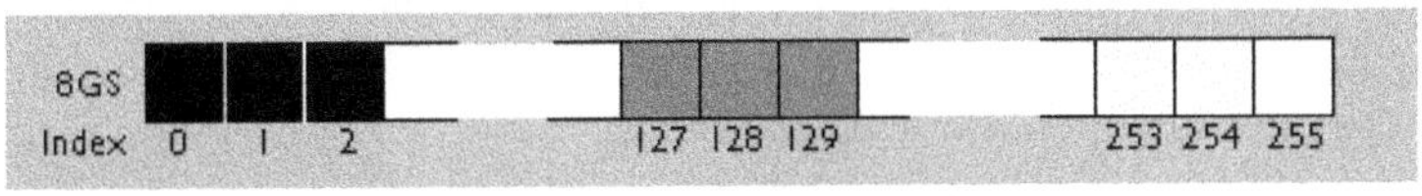

Fig. 37.3 First middle and last greys from 8GS. The index values point to the (B, G, R) colours in the 8GS colour-table. For example, index 1 points to (B, G, R) = (1, 1, 1); index 2 points to (B, G, R) = (2, 2, 2); and so on. Neighbouring greys are indistinguishable to the eye

Table 37.3 Grey-scale gamut 8GS

8GS				
Name	Index	B	G	R
Black	0	0	0	0
	1	1	1	1
	2	2	2	2
	3	3	3	3
	4	4	4	4
	...	...	...	...
Grey	128	128	128	128
	...	...	...	...
Pale Grey	192	192	192	192
	...	...	...	...
	252	252	252	252
	253	253	253	253
	254	254	254	254
White	255	255	255	255

- Or create a 24-bit .BMP file by by program, GUI, scanner, or camera, then convert-to 8-bit grey-scale.

37.5 8-bit Digital Image

An 8-bit digital image is like a 4-bit image (see Chap. 30) as to location, but different as to intensity. Graphically, an 8-bit image is a set of $X \times Y$ px, each pixel one of 256 intensities. Numerically, it is a set of integer triples:

$$(X, Y; I_{256}),$$

where $\leq X \leq x_{\max}, 0 \leq Y \leq y_{\max}, 0 \leq I \leq 255.$

Figure 37.4 shows examples of 8-bit digital images with colour-table 8CO, comparable with the 4-bit examples in Chap. 30. In Fig. 37.4b–d we see that smooth gradients between colours are not possible.

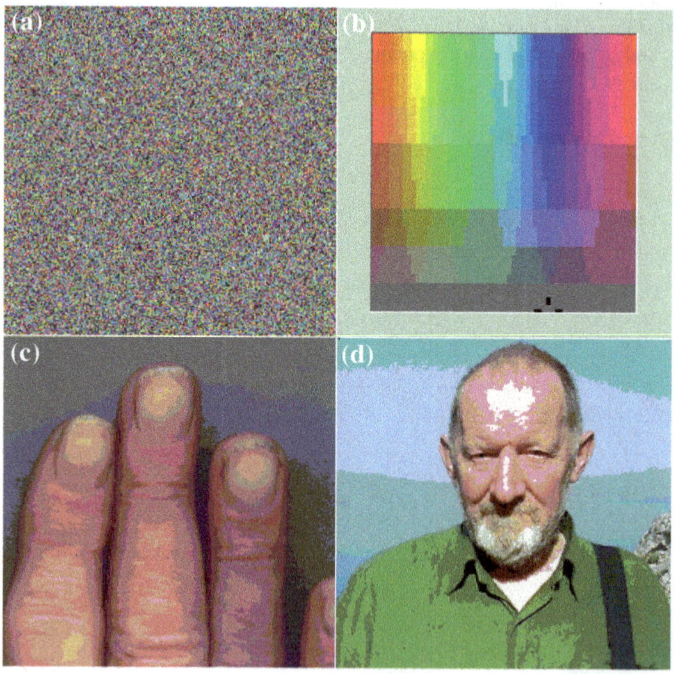

Fig. 37.4 Examples of 8-bit images with colour-table 8CO. **a** Image generated by 8-bit program: 222 × 222 px randomly set at one of 256 colours. **b** Image created by PAINT in 8-bit mode. **c** Fingers scanned in 24-bit mode then saved-as 8CO. **d** Scene photographed by 24-bit camera then saved-as 8CO

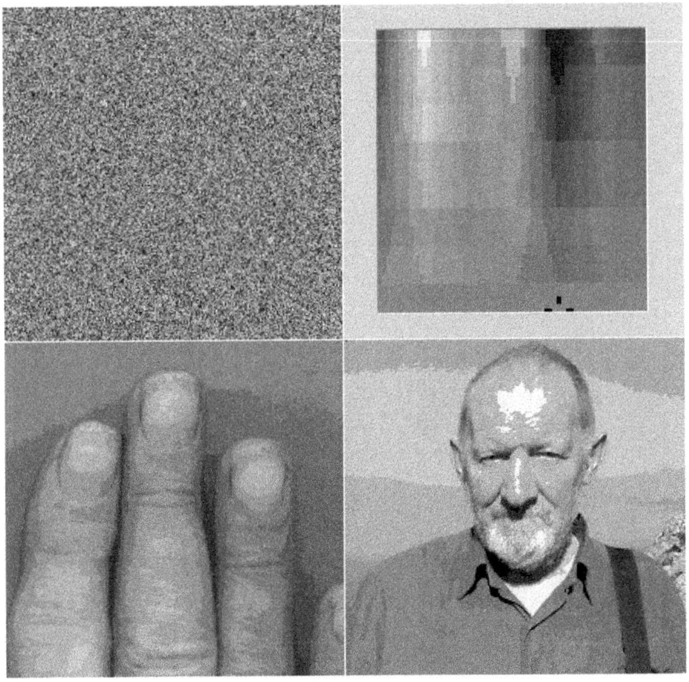

Fig. 37.5 Examples of 8-bit images with colour-table 8GM. Figure 37.4 converted-to 8GM

Figure 37.5 shows examples of 8-bit digital images with colour-table 8GM. In Fig. 37.5b–d we see again that smooth gradients between greys are not possible.

Figure 37.6 shows examples of 8-bit digital images with colour-table 8GS. In Fig. 37.6b–d we see that the 256 greys form a smooth gradient.

37.6 Flowchart

The flowchart in Fig. 37.7 is for general guidance in this Part. To make an 8-bit image, we can use a program or a GUI image editor or a scanner in 8-bit grey-scale mode (see Chap. 38), or reduce a 24-bit camera image to 8-bit (see Chap. 46). The image can be stored in an 8-bit .BMP file (see Chap. 41), then displayed (see Chap. 42) or printed (see Chap. 43) for viewing (see Chap. 44).

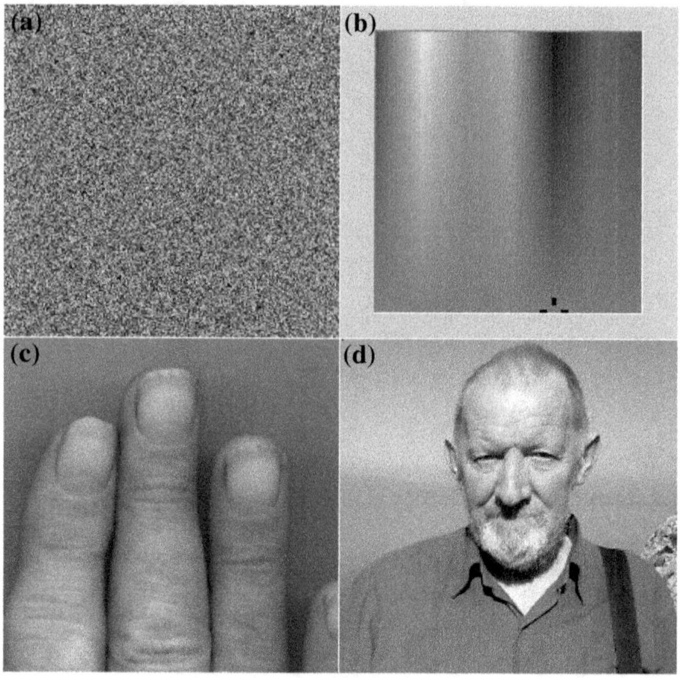

Fig. 37.6 Examples of 8-bit images with colour-table 8GS. **a** Image generated by 8-bit program: 222 × 222 px randomly set at one of 256 greys. **b** Image created by PAINT in 24-bit mode, then converted-to 8GS. **c** Fingers scanned in 8-bit grey-scale mode 8GS. **d** Scene photographed by 24-bit camera, then converted-to 8GS

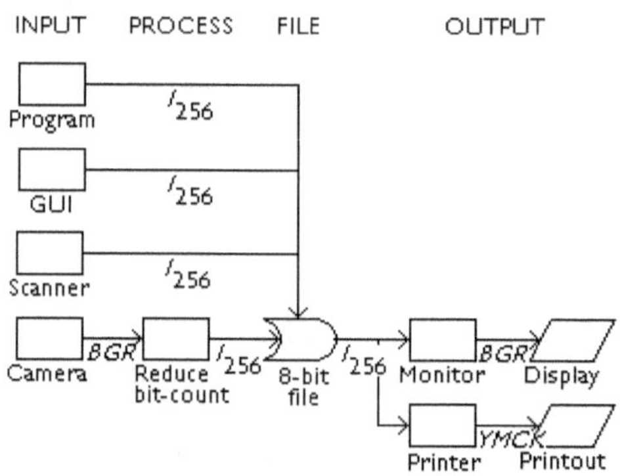

Fig. 37.7 Flowchart for 8-bit image. An 8-bit image may be input alpha-numerically from keyboard to program, or graphically from mouse/keyboard to GUI editor, or scanned in 8-bit mode, or reduced from a 24-bit image. The image is stored in an 8-bit file. The image may be displayed on a screen or printed on paper, to be viewed by an eye. I_{256} is intensity index 0–255 from encoder to file, and from file to decoder. BGR is intensity triple, 3× (eight bits 0–255), from scanner/camera to encoder, and from monitor to display. CMYK is ink coverages, from printer to paper

Chapter 38
8-bit Input

An 8-bit image can be created by an 8-bit program, or a GUI in 8-bit mode, or by scanning in 8-bit mode, or by photographing in 24-bit mode and reducing to 8-bit. In each case the location properties are the same as for 1-bit images. Smooth gradients of intensity are possible using the 8GS grey-scale colour-table, but not the other colour-tables. Gamma-encoding is significant in 8-bit input. Examples are given of all four inputs.

38.1 8-bit Input

An 8-bit image can be created in any of four main ways, broadly similar to 4-bit input (see Chap. 31):

- From integer numeric data via a program.
- From graphic data via a GUI.
- From a continuous flat object via a scanner.
- From a continuous space object via a camera and a reduction process.

38.2 8-bit Program Input

An 8-bit program is like a 4-bit program (see Chap. 31), with the differences:

- Read and write 8-bit .BMP file instead of 4-bit file.
- Set colour-table 8CO or 8GM or 8GS (see Chap. 37) instead of 4CO or 4GM or 4GS.
- Process 8-bit intensity $0 \leq I_{256} \leq 255$ instead of 4-bit $0 \leq I_{16} \leq 15$.

Figure 38.1 shows the flow for an 8-bit program. The 8GS grey-scale colour-table is easy to set up within a program (whereas the 8CO and 8GM colour-tables need a tedious 768 data numbers to be supplied). For example, a program which creates an

A. Parkin, *Digital Imaging Primer*, DOI 10.1007/978-3-540-85619-1_38

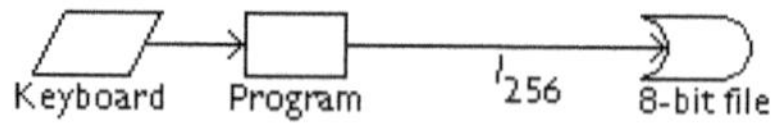

Fig. 38.1 8-bit program input. Input is from from keyboard or data file to program. Output to 8-bit file has the form $(X, Y; I_{256})$ per pixel, where $0 \leq X \leq w_{max}$ and $0 \leq Y \leq h_{max}$ are integer location coordinates, and $0 \leq I_{256} \leq 255$ is an integer index referring to an 8-bit colour-table in the .BMP file

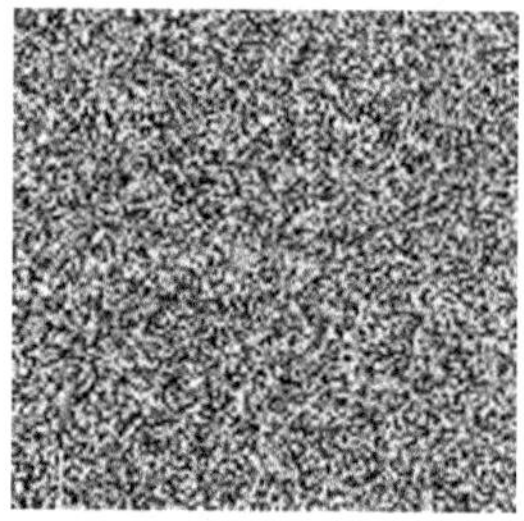

Fig. 38.2 Example of 8-bit random image, 128×128 px, with grey-scale colour-table 8GS

8-bit random image with colour-table 8GS is briefly described below (full program is on SpringerLink web-site). Figure 38.2 shows an image created by this program.

```
REM  **  Program GENRAND8.BAS to generate 8-bit random
REM  **     image, as .BMP file with grey-scale colour-
REM  **     table 8GS. User enters output .BMP filename,
REM  **     image width and height. Program writes 8-bit
REM  **     .BMP file header, colour-table, and image data.
REM  **  Main program
REM  **  Enter output file-name, width, height
         CALL CREHEA8
REM  **  Generate a random image:
REM  **     Set each pixel in array AIMAGE to random
REM  **     8-bit intensity 0 to 255
         RANDOMIZE
         FOR Y% = 0 TO (HIGH% - 1)
            FOR X% = 0 TO (WIDE% - 1)
               INTEN% = RND * 255
               AIMAGE(X%, Y%) = INTEN%
            NEXT X%
         NEXT Y%
         CALL WRIDAT8
         CALL REVHEA1
FINIS: END PROGRAM

         SUB CREHEA8
REM  **  Subprogram CREHEA8 to write header and
REM  **     colour-table of 8-bit .BMP file #1.
```

```
          SUB WRIDAT8
REM  **   Subprogram WRIDAT8 to write data from array
REM  **     AIMAGE to .BMP file #1. Each pixel has
REM  **     intensity 0-255: write as one byte in file.
          SUB REVHEA1
REM  **   Revises header of .BMP file #1 with values in bytes.
```

38.3 Encoding Gamma

If the colour-table is set as 8GS, we can create a smooth gradient as a sequence of 256 px with intensity index I_{256} running from 0 Black to 255 White. (A smooth gradient is not possible if the colour-table is set as 8CO or 8GM, because of jumps in the successive (B, G, R) values for the index (see Sect. 37.2).) When such a gradient is displayed, it remains smooth, but the luminance is governed by the power-law transfer function of the display (see Sect. 22.12). To counteract this, scanner and camera inputs are routinely delivered gamma-encoded, that is after applying encoding gamma = 1/2.2 = 0.4545. A program should do the same.

Figure 38.3 shows two gradients, one gamma-un-encoded ($\gamma = 1$) and one gamma-encoded ($\gamma = 0.4545$), produced by the program briefly described below and with source code at the SpringerLink web-site. At the right a patch of the middle grey of each gradient is put between patches of Black and White. In isolation either can be accepted as mid-Brightness. The relation between physical luminance and perceived brightness is examined in some depth in Chap. 17.

```
REM  **   Program GENGR8GS.BAS to generate 8-bit
REM  **     gradient with colour-table 8GS.
REM  **     User enters .BMP filename, width and
REM  **     height wanted, and gamma-encoded or
REM  **     un-encoded. Program generates 8-bit
REM  **     .BMP file header, colour-table, and
REM  **     image data, then outputs file.
```

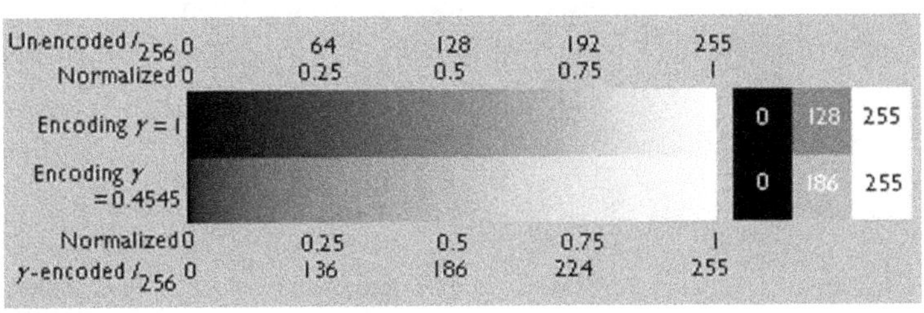

Fig. 38.3 Gradients produced by program, with colour-table 8GS. The upper gradient is gamma-unencoded ($\gamma = 1$), and the lower is gamma-encoded ($\gamma = 0.4545$). At the right, which grey looks midway between black and white? The background is very slightly paler than a 50 % grey

```
REM    **   Main program
REM    **   Enter file-name, width, height, gamma-encoded
REM    **   or un-encoded
            CALL CREHEA
            CALL CRECT8
            CALL CREIMD
            CALL REVHEA
FINIS: END PROGRAM

            SUB CREHEA
REM    **   Sub-program CREHEA to create new 8-bit .BMP
REM    **      file and write header.
            SUB CRECT8
REM    **   Sub-program CRECT8 to create 8GS 8-bit grey-
REM    **   scale colour-table. The colour table has
REM    **   256 entries, indexed 00 to FF. Each 4-byte
REM    **   entry contains the (equal) intensities
REM    **   (0-255) of Blue Green Red, and an empty byte 0.
            SUB CREIMD
REM    **   Sub-program CREIMD to put intensity values
REM    **      from array IMAR as bytes in data part
REM    **      of 8-bit file.
            SUB REVHEA
REM    **   Sub-program REVHEA to revise header with byte
REM    **      lengths.
```

38.4 8-bit Graphic Input by GUI

Figure 38.4 shows the flow for 8-bit imaging by GUI. To put the PAINT image editor into 8-bit colour mode, save a file as a 256-colour bitmap, or open a file which is already a 256-colour bitmap. Figure 38.5 shows the window. It is like 4-bit mode (see Sect. 31.3), but the screen now displays a colour box containing 20 patches, in two rows of 10. To set the foreground colour, select the Eyedropper and left-click the patch wanted. To set the background colour, right-click the patch wanted.

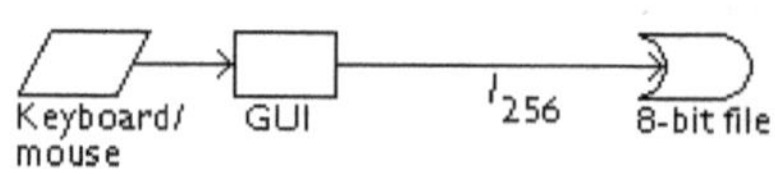

Fig. 38.4 Flowchart for 8-bit graphic imaging. Input from mouse and/or keyboard to GUI has the form $(X, Y; I_{256})$ per pixel, where $0 \le X \le w_{\max}$ and $0 \le Y \le h_{\max}$ are the integer Cartesian coordinates of location in the first quadrant, and $0 \le I_{256} \le 255$ is the index referring to the 8-bit colour-table in the .BMP file

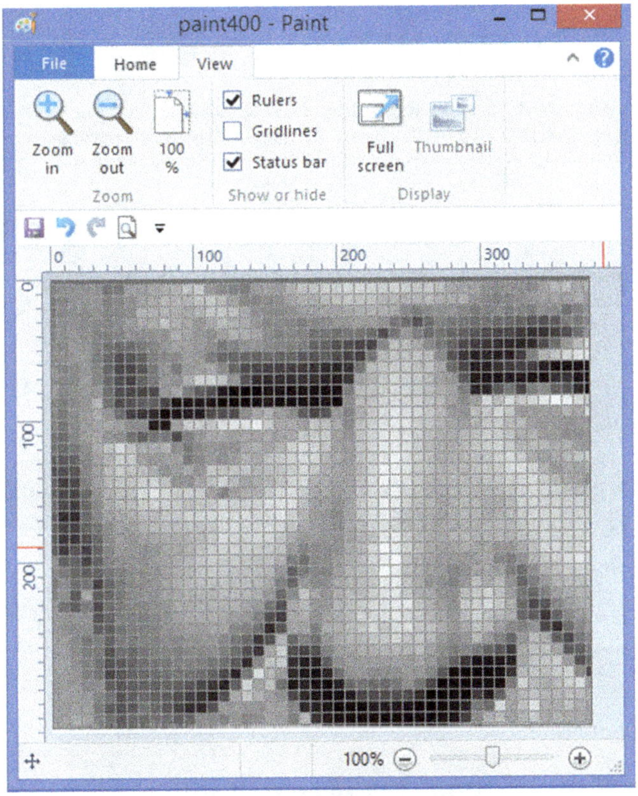

Fig. 38.5 PAINT window in 8-bit mode. The displayed image is a 24-bit camera image converted-to 8-bit grey-scale, and enlarged ×8 in the window to discriminate pixels

To see or change the (B, G, R) values of any displayed image pixel, select the Eyedropper and click the pixel wanted, select Colours on the top menu, then Edit Colours. The Edit Colours window appears. The (B, G, R) values are at lower right, and can be changed to re-set the current colour for the graphic tool.

38.5 8-bit Scanner Input

A scanner can usually be set to 8-bit grey-scale mode, to deliver an image which is like 1-bit mode (see Chap. 22) as to location, but with 256 grey intensities and colour-table 8GS (see Sect. 37.5) instead of two intensities and colour-table 1BW. The difference lies in the thresholder. Figure 38.6 shows input from object via scanner to file.

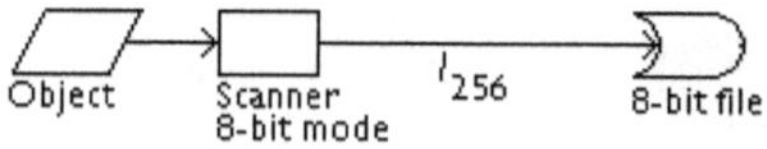

Fig. 38.6 Flowchart for 8-bit scanner input. Input from scanner has the form $(X, Y; I_{256})$ per pixel, where $0 \leq X \leq w_{max}$ and $0 \leq Y \leq h_{max}$ are the integer Cartesian coordinates of location, and $0 \leq I_{256} \leq 255$ is the index referring to the 8-bit colour-table in the .BMP file

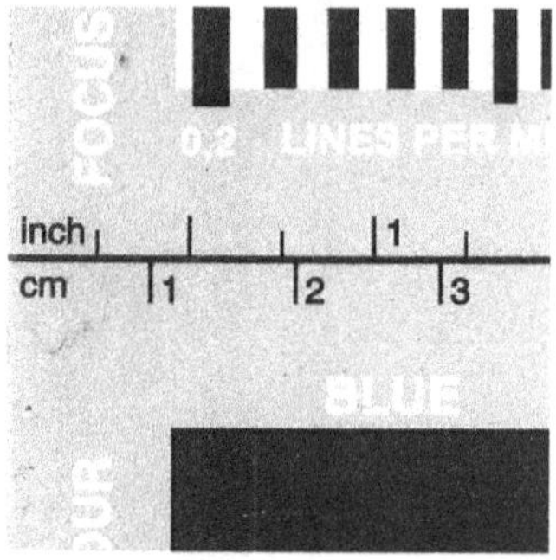

Fig. 38.7 8-bit scanned image. Test card scanned in 8-bit grey-scale mode at 300 spi

For example, Fig. 38.7 shows an 8-bit grey-scale 8GS scan of a test card at 300 spi. The precision is high, gradients of intensity are smooth, but of course there is no colour.

38.6 24-bit Camera Input Reduced to 8-bit

To get an 8-bit grey-scale image 8GS from a 24-bit camera image, convert-to grey-scale. For example, Fig. 38.5 shows such a camera image part-displayed in a PAINT window, enlarged 8 times. To get an 8-bit colour image 8CO from a 24-bit camera image, save-as 256-colour. See Appendix A.

Chapter 39
8-bit Process

8-bit processes are like 1-bit processes for location properties, but with differences in intensity properties. Programs for changing 8-bit locations, for combining two 8-bit images by location, and for reducing location-count, are described, with examples. The 8GS grey-scale colour-table has a complete range of index values 0 to 255, so some interesting intensity transformations can be performed by doing arithmetic on the index values. Six elementary intensity transformations are defined, with a program and examples. A program for combining two images by averaging intensities is given, with examples. Bit-depth can be changed by standard operations. Digital half-toning is an important process, whereby an 8-bit grey-scale image with intensities in the range 0 to 255 is transformed into an image with 1-bit intensities Black or White. Seven half-toning schemes which have been used in digital imaging are defined, with programs and examples. A program for 8-bit statistical analysis is given.

39.1 Flowchart

The flowchart (Fig. 39.1) is for general guidance in this chapter. It is noteworthy that an 8-bit image file with colour-table 8GS has a complete range of index values from 0 to 255 corresponding to the complete range of grey-scale intensities. Thus a 24-bit colour image file is a combination of three such 8GS files, one for each of the Blue, Green, and Red channels. This complete range allows us to do arithmetic on the intensity values of a grey-scale image, in a way which is not possible for images with colour-tables 8CO, 8GM, 4CO, 4GM, 4GS, 1BW, or 1GS, which threshold intensities into large and visible steps (see Appendix A).

A. Parkin, *Digital Imaging Primer*, DOI 10.1007/978-3-540-85619-1_39

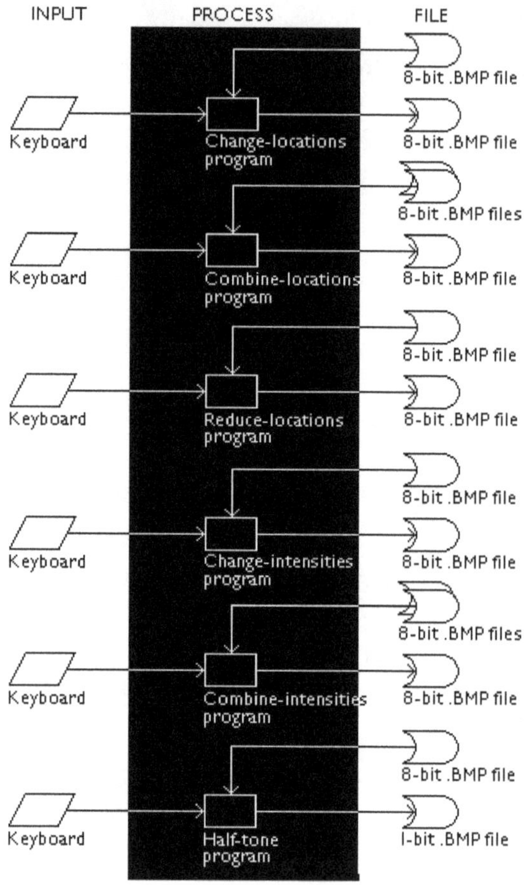

Fig. 39.1 8-bit process. A program can change the locations of a given 8-bit image (keeping the intensities unchanged). Or it can combine two 8-bit files as one, by alternating locations. Or it can reduce the pixel-count of locations. A program can change the intensities of a given 8-bit image (keeping the locations unchanged). Or it can combine two 8-bit files as one, by averaging intensities. A half-tone program can reduce an 8-bit grey-scale image to a 1-bit image which keeps graduations

39.2 Change 8-bit Image Locations

Changing 8-bit locations is like changing 1-bit locations (see Sects. 23.3–23.8) with the differences:

- Read and write 8-bit .BMP files instead of 1-bit files.
- Set colour-table 8CO or 8GM or 8GS instead of 1BW.
- Process 8-bit intensity $0 \leq I_{256} \leq 255$ instead of 1-bit $0 \leq I_2 \leq 1$.

Six elementary transformations are

- *Reflection* of an image about a mirror-line,
- *Translation* of an image along a direction,
- *Rotation* of an image about a pole,

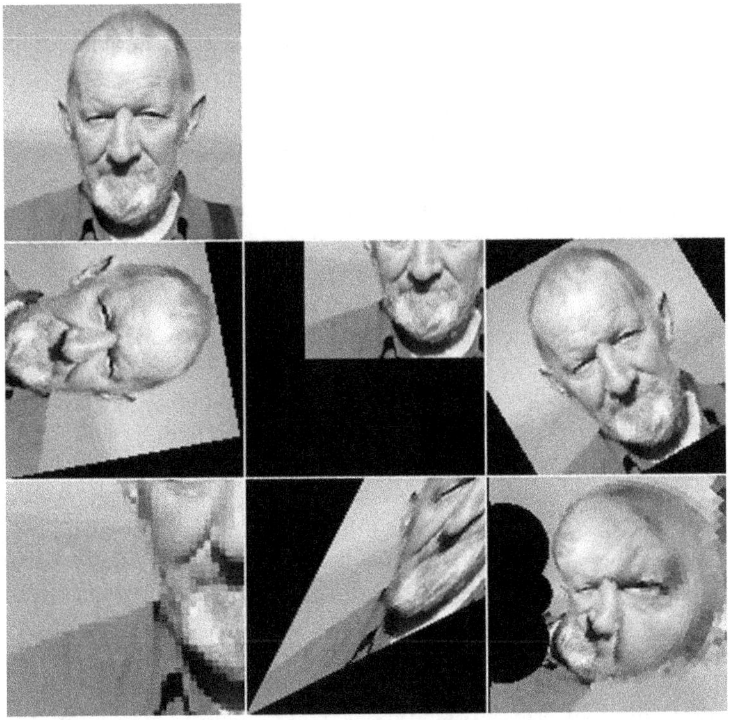

Fig. 39.2 Change 8-bit image locations. **a** Given 8-bit image of 128 × 128 px. **b** Reflected in mirror-line through the origin at 45°. **c** Translated by 32 px in X and 64 px in Y. **d** Rotated about (64, 64) by 30°. **e** Dilated about (0, 0) by factor 2 in X and Y. **f** Sheared about (0, 0) by factor 0.5 in X and 0.5 in Y. **g** Inverted in circle with centre (8, 64) and radius 56 px

- *Dilation* of an image about an origin,
- *Shearing* of an image along an axis,
- *Inversion* of an image about a circle.

Figure 39.2 shows a given 8-bit image, and an example of each of the six transformations.

Reflection moves each pixel of an image from one side of a mirror-line to the other side. Pixels on the mirror-line remain fixed. The sense of the image (left/right, clockwise/anticlockwise) is reversed. Figure 39.2b shows the image Fig. 39.2a reflected in a mirror-line through the origin at 50°. Two successive reflections in the same mirror-line give the identity transformation.

Translation moves each pixel of an image by a certain distance in a certain direction. No pixel remains fixed. The sense of the image is preserved. Two successive reflections in parallel mirror-lines give a translation by twice the distance between the mirror-lines. Figure 39.2c shows the image translated by 32 px to the right and 64 px upwards.

Rotation moves each pixel of an image through a certain angular arc of a circle whose centre is the pole of the rotation. The pole remains fixed. The sense of the image is preserved. Angles are conventionally measured anticlockwise from the x-axis. Two successive reflections in mirror-lines which meet give a rotation by twice the angle between the mirror-lines, and the meet of the mirror-lines is the pole of the rotation. Figure 39.2d shows the the image rotated about (64, 64) by 30°.

Dilation about the origin (0, 0) moves each pixel of an image in the x-direction by a certain proportion of its original x-distance, and in the y-direction by a certain proportion of its original y-distance. The origin remains fixed. If the dilation factor is positive, it is a magnification or enlargement or stretch or expansion. If the dilation factor is negative, it is a diminution or reduction or shrink or compression. If the dilation is equal in x and y it is a similarity: the sense and angles of the image are preserved, but all lengths are increased or decreased. Figure 39.2e shows the image dilated by a factor of 2 (that is stretched to twice size) in both x- and y-directions.

Shearing about the origin (0, 0) moves each pixel of an image in the x-direction by a certain proportion of its original y-distance, and in the y-direction by a certain proportion of its original x-distance. The effect is a slanting of the image. The origin remains fixed. Straight lines remain straight lines, and the sense of the image is preserved, but not lengths or angles. Figure 39.2f shows the image sheared by a factor of $0.5Y$ in the X-direction and by a factor of $0.5X$ in the y-direction.

Inversion is like reflection in a mirror-circle. Figure 39.2g shows the image inverted about a circle with centre (8, 64) and radius 56 px.

A program which performs all of the above transformations of location is described below (see SpringerLink web-site for source code). This program lets the user pick any 8-bit source image, and one of the geometric transformations of it. It is like the 1-bit change-location program in Chap. 23, but with different sub-programs for reading and writing 8-bit intensities.

```
REM  **  Program CHANLOC8 to change locations of
REM  **     8-bit image. User enters filenames.
REM  **     Program reads given "before" image
REM  **     from 8-bit.BMP file #1 into array
REM  **     IMAGEA. User picks transformation and
REM  **     enters constants. Program then processes
REM  **     pixel by pixel from IMAGEA to array
REM  **     IMAGEB, setting old intensity at new
REM  **     location. Program writes "after" image
REM  **     from IMAGEB to 8-bit .BMP file #2.
REM  **  Main program
REM  **  Enter file-names
         CALL RWHEA8AB
         CALL REIMAGE8
         CALL GEOMTRAN
         CALL WRIMAGE8
         CALL REVHEAD8
```

```
END

          SUB RWHEA8AB
REM  **   Sub-program to read header and colour-table of
REM  **      8-bit .BMP input file #1, and write same
REM  **      header and colour-table of output file
REM  **      #2 (to be revised later).
          SUB REIMAGE8
REM  **   Sub-program to read 8-bit .BMP file #1 and
REM  **      load old image data (intensities) into
REM  **      array IMAGEA.
          SUB GEOMTRAN
REM  **   Sub-program to enter constants for transformation
REM  **      wanted, set up transformation matrix, and
REM  **      multiply coordinates of each pixel from
REM  **      IMAGEA to IMAGEB.
REM  **   As program CHANLOC4
          SUB WRIMAGE8
REM  **   Subprogram WRIMAGE8 to write data from array
REM  **      IMAGEB to 8-bit .BMP file #2, one byte
REM  **      per pixel.
          SUB REVHEAD8
REM  **   Sub-program REVHEAD8 to revise header of .BMP
REM  **      file #2 with values in bytes.
```

39.3 Combine 8-bit Images by Location

Two images A and B with the same pixel-count may be combined as one C by taking pixels alternately from A and from B to construct C. The alternation may be by columns or by rows or both. For example, Fig. 39.3 shows two 8-bit images and their combination by alternating columns. A program which combines 8-bit images either by alternating locations or by averaging intensities is described briefly below (see SpringerLink web-site for source code). It is like the 4-bit combine-locations program in Chap. 32, but with different sub-programs for reading and writing 8-bit intensities.

```
REM  **   Program COMALAV8 to combine two 8-bit images
REM  **      with same pixel-count. User picks scheme
REM  **      either alternating or weighted averaging,
REM  **      and enters 8-bit filenames for images A, B,
REM  **      and C, Program sets up array IMAGEC, and
REM  **      processes pixel by pixel from file #1 and #2,
```

```
REM  **      calculating new intensity for each location
REM  **      into IMAGEC. Program writes IMAGEC to
REM  **      8-bit .BMP file #3.
REM  **   Main program
REM  **   Enter file-names
          CALL RWHEDABC
          CALL INCONSTA
          CALL CALCAVER
          CALL WRIMAGEC
          CALL REVHEADC
END

          SUB CALCAVER
REM  **   Subprogram CALCAVER to read image data from 8-bit
REM  **      .BMP file #1 and file #2, and calculate
REM  **      weighted average into array IMAGEC.
          IF AVAL$ = "AV" THEN
REM  **   Read 8-bit image data from files #1 and #2,
REM  **   calculate weighted average into array IMAGEC.
          IF AVAL$ = "AL" THEN
REM  **   Read 8-bit image data from files #1 and #2, put
REM  **   alternating pixels into array IMAGEC.
          SUB INCONSTA
REM  **   Subprogram INCONSTA to enter weights W and V
REM  **      for average.
          SUB REVHEADC
REM  **   Revises header of file #3 with values in bytes.
          SUB RWHEDABC
REM  **   Subprogram RWHEDABC to read header and colour-table
REM  **   of 8-bit .BMP input file #1, and preliminary-write
REM  **   same header and colour-table of output file
REM  **   #3 (to be revised later).
          SUB WRIMAGEC
REM  **   Subprogram WRIMAGEC to write data from array
REM  **      IMAGEC to 8-bit .BMP file #3.
```

39.4 Reduce 8-bit Image Locations

Given an 8-bit image, we may reduce the count (and increase the size) of pixels by taking every Nth pixel and repeating it N times. The reduced image has $1/N^2$ as many pixels, each N^2 the size of the original. An 8-bit image with Q px conveys

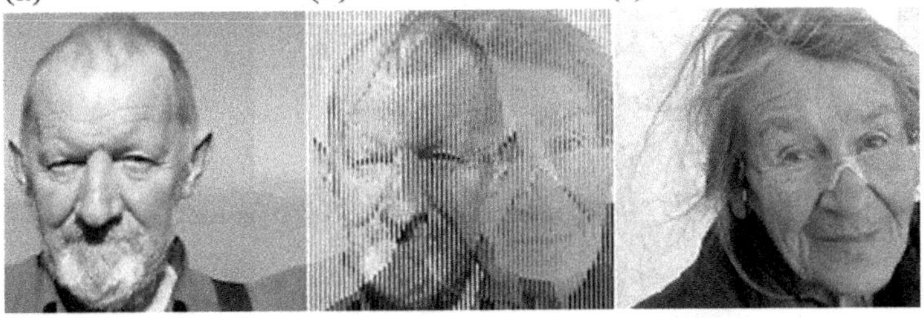

Fig. 39.3 Combine by location. **a** 8-bit image, 128 × 128 px. **b** Combination of (**a**) and (**c**) by alternating columns, 128 × 128 px. **c** 8-bit image, 128 × 128 px

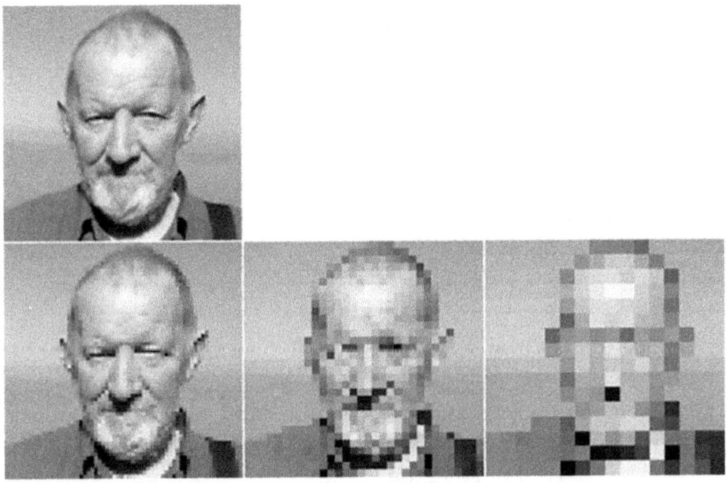

Fig. 39.4 Reduce 8-bit image locations. **a** 8-bit image 128 × 128 px: 65,536 b. **b** Reduction of (**a**) by $N = 2$ to 64 × 64 px: 16,384 b. **c** Reduction of (**a**) by $N = 4$ to 32 × 32 px: 4,096 b. **c** Reduction of (**a**) by $N = 8$ to 16 × 16 px: 1,024 b

$8Q$ bits of information, and a reduced image conveys $(8Q)/N^2$ bits. For example, Fig. 39.4 shows an 8-bit image thus reduced by $N = 2$, 4, and 8.

A program which reduces 8-bit image locations is described briefly below (see SpringerLink web-site for source code):

```
REM  **  Program REDULOC8 to reduce count of pixels
REM  **     for 8-bit image. User enters filenames.
REM  **     Program reads input image from 8-bit .BMP
REM  **     file #1 into array IMAGEA. User picks
REM  **     reduction. Program takes every Nth pixel
REM  **     from array IMAGEA and repeats it N times
REM  **     to array IMAGEB, with same intensity.
```

```
REM  **     Program writes reduced image from IMAGEB
REM  **     to 8-bit .BMP file #2.
REM  **   Main program
REM  **   Enter file-names
          CALL RWHEA8AB
          CALL REIMAGE8
          CALL LOC8REDU
          CALL WRIMAGE8
          CALL REVHEAD8
END

          SUB LOC8REDU
REM  **   Sub-program LOC8REDU to enter reduction wanted,
REM  **      and perform reduction for each pixel from
REM  **      IMAGEA to IMAGEB.
          SUB REIMAGE8
REM  **   Subprogram to read 8-bit .BMP file #1 and
REM  **      load old image data (intensities) into
REM  **      array IMAGEA.
          SUB REVHEAD8
REM  **   Sub-program REVHEAD8 to revise header of .BMP
REM  **      file #2 with values in bytes.
          SUB RWHEA8AB
REM  **   Subprogram to read header and colour-table of
REM  **      8-bit .BMP input file #1, and write same
REM  **      header and colour-table of output file
REM  **      #2 (to be revised later).
          SUB WRIMAGE8
REM  **   Subprogram WRIMAGE8 to write data from array
REM  **      IMAGEB to 8-bit .BMP file #2, one byte
REM  **      per pixel.
```

39.5 8-bit Intensity Process

An intensity transformation is an operation on a discrete image which changes the intensities of all the pixels in some systematic way, leaving the locations unchanged. An 8-bit grey-scale image has intensities $0 \leq I_{256} \leq 255$, and colour-table 8GS where each BGR triple has B = G = R = I_{256}. In an 8-bit grey-scale image the tone-codes in the table are a complete integer sequence from 0 to 255, and since the difference between successive tones is less than the discrimination of the eye, the sequence appears as a smooth intensity gradient from Black to White. (This is not the case for colour-table 8CO or 8GM, nor for 4-bit colour-tables 4CO or 4GM, because of gaps in the table values.) It follows that we can transform the intensities of an 8-bit grey-scale image freely into any other grey-scale intensities by a numerical process:

- Given an 8-bit grey-scale image with per-pixel intensity I_{256}.
- Normalize intensity as $I_{\text{norm}} = I_{256}/255$.
- Apply a numerical function $f(I_{\text{norm}})$.
- Restore transformed intensity $I'_{256} = (f(I_{\text{norm}})) \times 255$.

Some elementary intensity transformations are:

- Reverse intensity.
- Add/subtract intensity.
- Multiply/divide intensity.
- Exponentiate intensity.
- Re-scale intensity range.
- Combine intensities of two images.

Examples of each of these transformations follow. In each case the given source is a camera image 128 × 128 px, a gamma-un-encoded gradient 128 × 64 px and a gamma-encoded gradient 128 × 64 px (see Sect. 38.3), and a scale showing the normalized intensities of the gradients. The transformations are reversible and combinable. Some GUI image editors offer ready-made intensity transformations. In general, a program is needed.

39.6 Reverse Intensity

Numerically, the reversal function for each pixel is

$$I'_{256} = (1 - I_{\text{norm}}) \times 255\,,$$

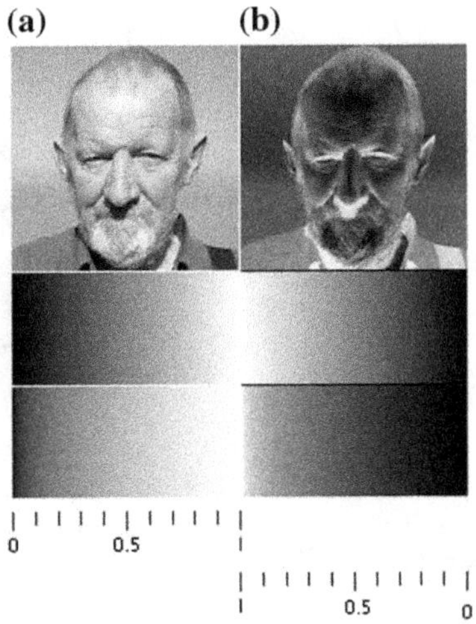

Fig. 39.5 Reverse intensity. **a** Given 8-bit grey-scale images: camera image 128 × 128 px; gamma-un-encoded gradient 128 × 64 px; gamma-encoded gradient 128 × 64 px. **b** Intensity-reversed images. *Below the images* are the corresponding scales of normalized intensities

where $0 \leq I \leq 255$ is the given intensity, and $I_{\text{norm}} = I/255$. Figure 39.5a shows three given 8-bit grey-scale images: a camera image, a gamma-un-encoded gradient, and a gamma-encoded gradient (see Sect. 38.3); Fig. 39.5b shows the corresponding intensity-reversals.

Graphically, reversing the intensities of an image is like making a photographic negative, where Black becomes White, dark greys become pale greys, and White becomes Black. Notice that the un-encoded gradients are symmetrical, but the encoded are not. MS PAINT offers reversal as Image-Invert Colors.

39.7 Add/Subtract Intensity

Numerically, the add/subtract function for each pixel is

$$I'_{256} = (I_{\text{norm}} \pm k) \times 255\,,$$

where $0 \leq I \leq 255$ is the given intensity, $0 \leq I_{\text{norm}} \leq 1 = I/255$, and $0 \leq k \leq 1$ is a given constant. Figure 39.6 shows in the middle the given images; at the left after adding $k = 0.5$; and at the right after subtracting $k = 0.5$. Adding is a linear shift of image intensities up the scale; subtracting a linear shift down the scale.

Graphically, adding a constant intensity to an image loses high-intensity distinctions by pushing them to White. Subtracting loses low-intensity distinctions by pulling them to Black.

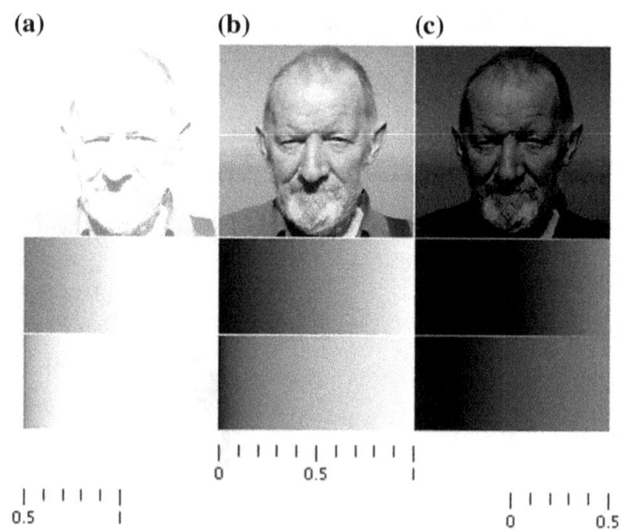

Fig. 39.6 Add/subtract intensity. **a** After adding $k = 0.5$ to (**b**). **b** Given images. **c** After subtracting $k = 0.5$ from (**b**)

39.8 Multiply/Divide Intensity

Numerically, the multiply/divide function for each pixel is

$$I'_{256} = (I_{\text{norm}} \times k) \times 255$$

or

$$I'_{256} = (I_{\text{norm}} \div k) \times 255\,,$$

where $0 \leq I \leq 255$ is the given intensity, $0 \leq I_{\text{norm}} \leq 1 = I/255$, and $0 \leq k \leq 1$ is a given constant. Figure 39.7 shows in the middle the given images; at the left after multiplying by $k = 0.5$; and at the right after dividing by $k = 0.5$. Multiplying is a linear compression of image intensities from 0 upwards on the scale; dividing is a linear expansion from 0 upwards on the scale.

Graphically, multiplying image intensity by a constant between 0 and 1 keeps all intensities, and hardens transitions. Dividing loses high intensities, and softens the surviving transitions.

39.9 Exponentiate Intensity

Numerically, the exponentiate function for each pixel is

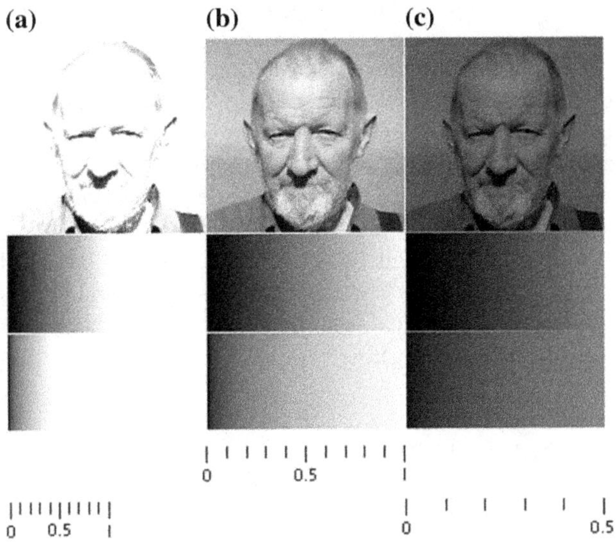

Fig. 39.7 Multiply/divide intensity. **a** After multiplying (**b**) by $k = 0.5$. **b** Given images. **c** After dividing (**b**) by $k = 0.5$

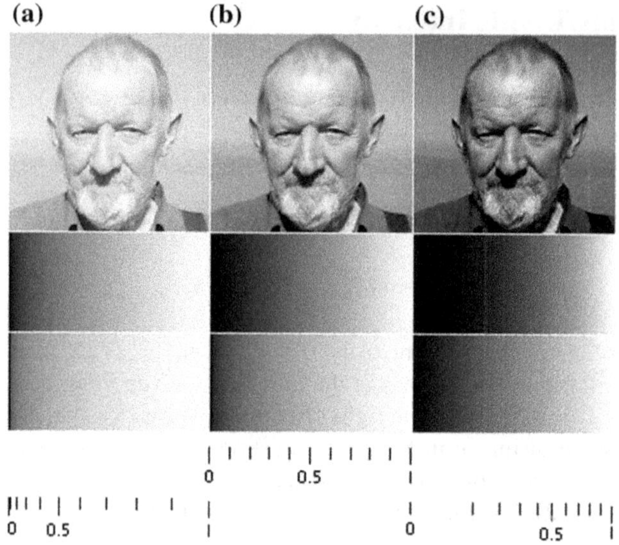

Fig. 39.8 Exponentiate intensity. **a** After raising (**b**) to power $k = 0.5$. **b** Given images. **c** After raising to power $1/k = 1/0.5 = 2$

$$I'_{256} = (I_{\text{norm}} \wedge k) \times 255$$

or

$$I'_{256} = (I_{\text{norm}} \wedge 1/k) \times 255\,,$$

where $0 \le I \le 255$ is the given intensity, $0 \le I_{\text{norm}} \le 1 = I/255$, and $0 \le k \le 1$ is a given constant. Figure 39.8 shows in the middle the given images; at the left after applying power $k = 0.5$; and at the right after applying power $1/k = 1/0.5 = 2$. Raising to a power less than 1 gives a scale with exponentially-increasing steps from 0 upwards; raising to a power greater than 1 a scale with exponentially-decreasing steps from 0 upwards.

Graphically, raising image intensity by a power less than 1 hardens low-intensity transitions and softens high-intensity transitions (as in gamma-encoding). Raising by a power greater than 1 softens low-intensity transitions and hardens high-intensity transitions (as in gamma-decoding).

39.10 Re-scale Intensity

Numerically, to re-scale from full to restricted range, the function for each pixel is

$$I'_{256} = (I_{\text{norm}} \times (I'_{\max} - I'_{\min})) + I'_{\min}\,,$$

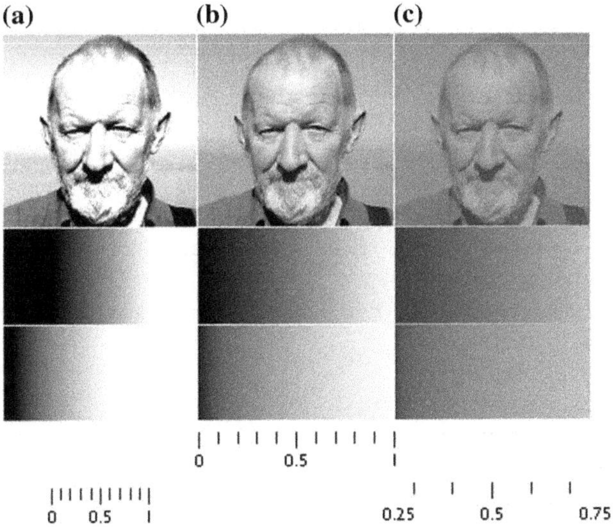

Fig. 39.9 Re-scale intensity. **a** After re-scaling full range of (**b**) to restricted range between 0.25 and 0.75. **b** Given images. **c** After re-scaling restricted range of (**b**) from 0.25 to 0.75 to full range

where $0 \leq I \leq 255$ is the given intensity, $0 \leq I_{\text{norm}} \leq 1 = I/255$, and $0 \leq I'_{\text{min}} \leq 1$ and $0 < I'_{\text{max}} < 1$ is the new restricted range. To re-scale from restricted to full range, the function is:

$$I'_{256} = (I_{\text{norm}} - I_{\text{min}})/(I_{\text{max}} - I_{\text{min}}),$$

where $0 \leq I \leq 255$ is the given intensity, and $0 \leq I_{\text{min}} \leq 1$ and $0 \leq I_{\text{max}} \leq 1$ is the old restricted range. Figure 39.9b shows the given images. At the left the full range of Fig. 39.9b between 0 and 1 is re-scaled to restricted range between 0.25 and 0.75: a linear compression of the entire scale. At the right the restricted range of Fig. 39.9b between 0.25 and 0.75 is re-scaled to full range 0 to 1: linear expansion of part of the scale.

Graphically, re-scaling from full to part range keeps all intensities, and hardens transitions. Re-scaling from part to full range loses high and low intensities, and softens the surviving transitions.

39.11 8-bit Transform Intensity Program

A program which does any of these intensity transformations is briefly described below (see SpringerLink web-site for full listing).

```
REM  **  Program TRANPIN8 to perform transformations of
```

```
REM  **     8-bit plane image intensities.
REM  **     User enters filenames and transformation
REM  **     constants. Program reads given "before" image
REM  **     from 8-bit .BMP file #1 into array IMAGEA.
REM  **     Program then processes pixel by pixel from
REM  **     IMAGEA to array IMAGEB, calculating new
REM  **     intensity for each location. Program writes
REM  **     "after" image from IMAGEB to 8-bit .BMP file #2.
REM  **  Main program
REM  **  Enter filenames for input file and output file
         CALL RWHEADAB
         CALL INCONSTA
         CALL REIMAGEA
         CALL TRANSFOR
         CALL WRIMAGEB
         CALL REVHEADB
END PROGRAM
         SUB RWHEADAB
REM  **  Subprogram to read header and colour-table of
REM  **     8-bit .BMP input file #1, and preliminary-write
REM  **     same header and colour-table of output file
REM  **     #2 (to be revised later).
         SUB INCONSTA
REM  **  Subprogram INCONSTA to enter intensity
REM  **     transformation constants.
REM  **  Reverse intensity
REM  **  Add/subtract intensity
REM  **  Multiply/divide intensity
REM  **  Re-scale intensity full to short range
REM  **  Re-scale intensity short to full range
         SUB REIMAGEA
REM  **  Subprogram to read 8-bit .BMP file #1 and
REM  **     load old image data (intensities) into
REM  **     array IMAGEA.
         SUB TRANSFOR
REM  **  Sub-program TRANSFOR to calculate new intensities.
REM  **     Initializes array IMAGEB; for each pixel from
REM  **     IMAGEA finds new intensity in IMAGEB.
         SUB WRIMAGEB
REM  **  Subprogram to write data from array IMAGEB to
REM  **     8-bit .bmp file #2.
         SUB REVHEADB
REM  **  Revises header of file #2 with values in bytes.
```

Considerable insight into intensity transformations can be gained by taking the before-and- after statistics of transformed images (see Sect. 39.26 onwards). From here, the way is open to explore intensity transformations of images created by gui or scanner or camera, and to combine such transformations with bit-depth reductions.

39.12 Combine Images by Averaging Intensities

Two images A and B with the same pixel-count may be combined as C by taking the weighted average of intensities per pixel. Figure 39.10 shows two images combined with varying weights.

Numerically, the weighted average is, for each pixel

$$I_C = (W \times I_{\mathrm{A}}) + (V \times I_{\mathrm{B}}) \backslash (W + V),$$

where W and V are integer weights, I_{A} is the intensity of the pixel in image A, I_{B} is the intensity of the corresponding pixel in image B, and I_C is the weighted average in the combined image C.

Graphically, when $W > V$, image A dominates; when $W = V$, images A and B are equally strong; when $W < V$, B dominates.

Program COMALAV8 described above in Sect. 39.3 produces the weighted average intensities of two given 8 bit images.

39.13 Change 8-bit Colour-table

To change a given 8-bit file with colour-table 8CO to an 8-bit file with colour-table 8GM, use the *Convert-to* command of an image editor such as IRFANVIEW:

(a) (b) (c)

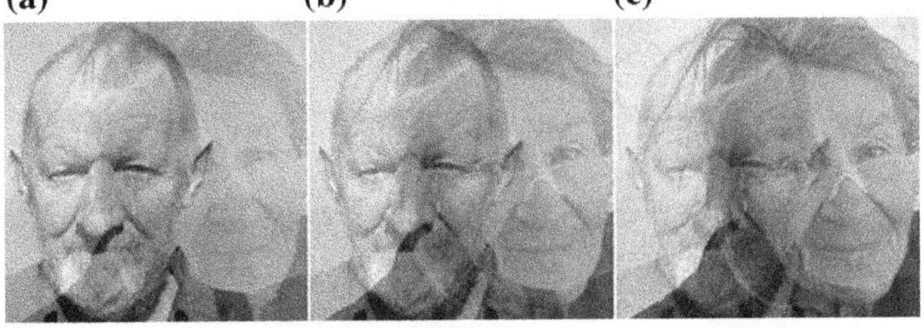

Fig. 39.10 Combine by averaging intensities. Given images A and B each have 128 × 128 px. **a** Intensity-averaged image C, where A has weight 2, B weight 1. **b** Intensity-averaged image C, where A has weight 1, B weight 1. **c** Intensity-averaged image C, where A has weight 1, B weight 2

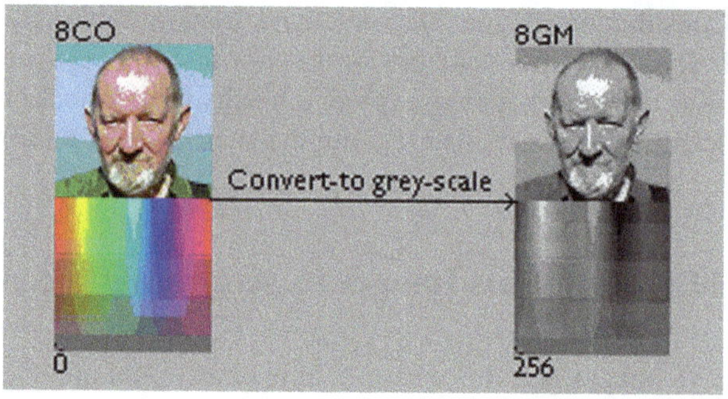

Fig. 39.11 Image with colour-table 8CO converted-to 8GM

- Display the source file.
- Click on *Convert-to* grey-scale.
- Click on Save.
- Enter new file name (or keep old file name).

The operating system will then

- Create new 8-bit file with 256-entry grey-match colour-table 8GM.
- Set file header field $biClrUsed = 256$.
- Keep all pixel location values (X, Y) unchanged.
- Per-pixel, calculate grey-match value gmv $= (0.11\text{B} + 0.59\text{G} + 0.30\text{R})$.
- Set pixel intensity index $I_{256} = \text{gmv}$.

Figure 39.11 shows an 8CO colour image converted-to 8GM grey-match. The index values I_{256} are unchanged, but the colour-table is 8GM.

39.14 Change Bit-depth

To change a given 8-bit file to a 4-bit file, use the *Save-as* command in an image editor, such as MS PAINT:

- Display the source file.
- Click on Save-as.
- Enter new file name (or keep old file name).
- Enter new file type (16-colour .BMP).
- Click on Save.

The operating system will then

- Create new 4-bit file with 16-entry colour-table.
- Keep all pixel location values (X, Y) unchanged.
- Per-pixel, compare 8-bit (B, G, R) values to 4-bit thresholds.
- Set 4-bit intensity index I_{16} according to threshold intervals.

Or to change a given 8-bit file to a 1-bit file:

- Display the source file.
- Click on Save-as.
- Enter new file name (or keep old file name).
- Enter new file type (monochrome .BMP).
- Click on Save.

The operating system will then

- Create new 1-bit file with two-entry colour-table.
- Keep all pixel location values (X, Y) unchanged.
- Per-pixel, compare 8-bit (B, G, R) values to 1-bit thresholds.
- Set 1-bit intensity index I_2 according to threshold intervals.

For example, Fig. 39.12 shows several reductions of bit-depth. (see Appendix A.)

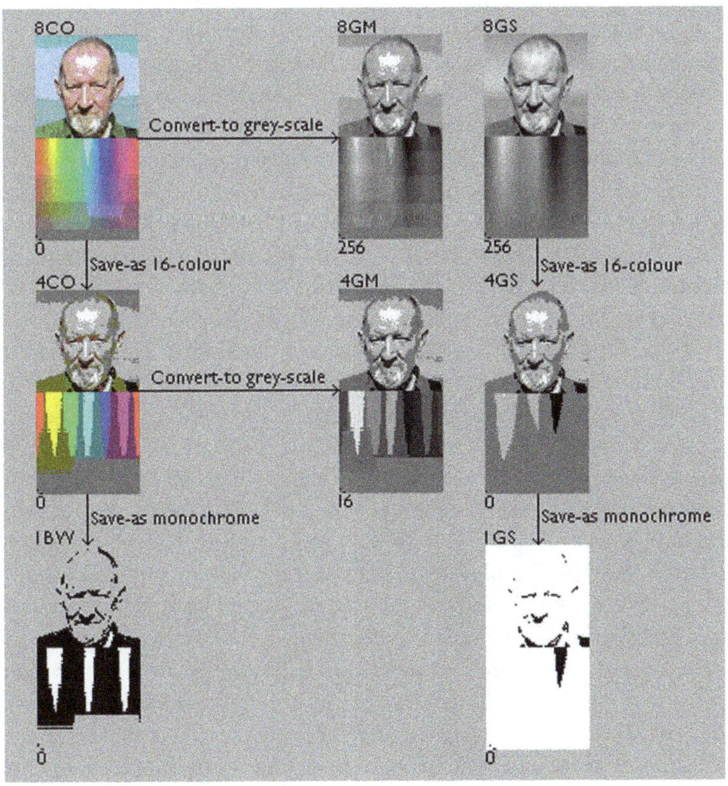

Fig. 39.12 8-bit images saved-as 4-bit and 1-bit

39.15 Half-toning

A white substance reflects nearly all of the light incident upon it, and a colourant absorbs some of the light. So the intensity of light reflected by a coloured substance is controlled inversely by the *concentration* of the colourant: the greater the concentration, the more the absorption and the less the reflection. Concentration can be varied by diluting a colourant, as for example in textile dyeing or in watercolour painting. Alternatively, concentration can be varied by area-division, that is to say the percent area covered by a colourant. In print *half-toning*, dots of a fixed-strength ink part-cover a white paper: in effect, the ink is diluted with white paper. When viewed, the print seems to show varying intensities.

Half-toning by manual or photographic techniques has a long and interesting history in the graphic arts and the printing trades. For *letterpress* printing, at an early date Holbein, Durer, and many others cut black-printing lines and cross-hatchings on the side grain of a type-high plank of wood, which was then locked up with the type and imposed as an illustrated page [4]. Around 1775 Bewick began cutting white-printing dots, lines, and cross-hatchings on the end-grain of a type-high boxwood block, and this became the commercial standard for illustrated letterpress for much of the nineteenth century [8]. Various attempts were made to convert a photograph to a printable surface, culminating in F.E. Ives' second process, exploiting a curious optical phenomenon whereby a graduated image, projected through a screen of closely-spaced crossed parallel opaque lines, separates into an array of fuzzy spots of varying brightness on a surface a small distance from the screen. When a negative image is thus projected onto a photosensitive resist coated onto a zinc plate, chemical development hardens the resist where the intensity of light received is above a certain threshold. After washing away the unhardened resist, the plate is masked by an array of spots of resist, small in the shadows and large in the lights. The plate is then bitten by an acid, which lowers the white-printing spots but leaves the black-printing spots high. The plate is mounted type-high on a block to form part of the letterpress page.

For *intaglio* printing [6], a copper or steel plate is engraved by hand, with fine black-printing lines and cross-hatching. The plate is then inked, wiped to leave ink in the lines only, and printed with great pressure on damp paper. An etching plate is first coated with a wax resist, through which black-printing lines are scratched with a needle, then bitten into the plate with acid. A dry-point plate holds ink in the burr raised by a needle, without acid biting. A mezzo-tint plate is first given an all-over burr, to print black, which is then selectively burnished down to print greys and white. An aqua-tint plate is first given a granular resist, bitten by acid, then progressively stopped out and re-bitten area by area. A photo-gravure plate has engraved dots and lines produced by projecting a positive image through a screen then biting with acid.

For *lithographic* printing [7], a stone surface which can be wetted is drawn upon by hand, with an oily crayon or ink. The stone is inked with an oily printing ink which is accepted by the drawn marks only, and pressed to paper. Photo-lithography transfers a positive image photographically, via a screen, to a metal plate. The plate is wetted, inked with an oily ink and pressed to paper. Offset lithography transfers the inked image to an intermediate rubber surface and thence to paper.

39.16 Digital Half-toning

Digital half-toning transforms intensities in the range 0–255 into area-coverages in the range 0–100 %. It replaces multi-bit intensity-division by 1-bit area-division. Much work has gone into the development of digital half-toning methods [3, 9]. Here we examine in detail seven schemes:

- *Single threshold per pixel*
  - Fixed threshold
  - Random threshold
  - Fixed threshold, distributed error
- *Vector of thresholds per pixel*
  - Clustered vector
  - Dispersed vector
- *Matrix of thresholds per pixel*
  - Clustered matrix
  - Dispersed matrix

For each scheme we explain the principle, then transform the 8-bit images shown in Fig. 39.13 to 1-bit half-tones with the same pixel-count.

39.17 Fixed Threshold

Set threshold $0 \leq T \leq 255$. For each pixel (X, Y) of the input image, compare its intensity $0 \leq I_{\text{in}} \leq 255$ to the threshold T. If $I_{\text{in}} < T$, output pixel (X, Y) with 1-bit intensity index $I_{\text{out}} = 0$ Black, else output $I_{\text{out}} = 1$ White:

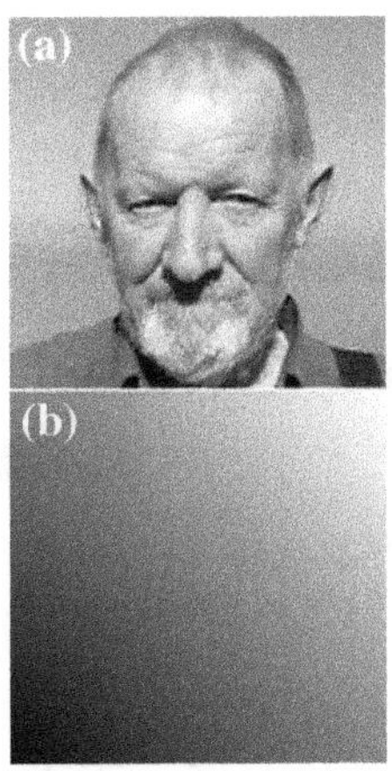

Fig. 39.13 8-bit grey-scale images used as input for the following examples of half-toning schemes. **a** Photograph 128 × 128 px, with 8GS intensities $0 \leq I_{256} \leq 255$. **b** Gradient 128 × 128 px, with 8GS intensities from 0 at *bottom left* to 255 at *top right*

$$I_{out} = 0 \quad \text{if } I_{in} < T$$
$$I_{out} = 1 \quad \text{if } I_{in} \geq T \,.$$

Figure 39.14 shows the principle of fixed thresholding. At the top are three patches of 8 × 8 px (enlarged), before half-toning, with 4-bit grey-scale intensities $I_{in} = 0$, 128, and 192. Underneath are the patches after half-toning with fixed threshold $T = 160$. Figure 39.15 shows examples of images half-toned by three different fixed thresholds. Fixed thresholding is what the Save-as 1-bit command does, with threshold 128. It is equivalent to reducing a continuous-tone original to a line image in traditional print technology.

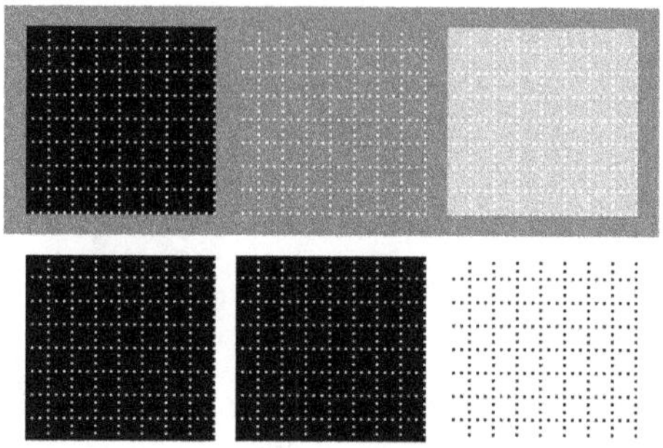

Fig. 39.14 Principle of half-toning by fixed threshold. **a** 8 × 8-pixel input patch, uniform intensity $I_{in} = 0$ Black. **b** Input patch, intensity 128 Grey. **c** Input patch, intensity 192 Pale grey. Each input patch is half-toned with threshold $T = 160$. **d** For each pixel, $I_{in} < T$, so the whole patch has $I_{out} = 0$ Black. **e** For each pixel, $I_{in} < T$, so the whole patch has $I_{out} = 0$ Black. **f** For each pixel, $I_{in} \geq T$, so the whole patch has $I_{out} = 1$ White

Fig. 39.15 Examples of half-toning by fixed threshold. (**a**) + (**d**) Fixed threshold $T = 64$. (**b**) + (**e**) Fixed threshold $T = 128$. (**c**) + (**f**) Fixed threshold $T = 192$

39.18 Random Threshold

For each pixel, set threshold $0 \leq T \leq 255$ as a (pseudo-)random integer. If the input intensity index of that pixel $0 \leq I_{\text{in}} \leq 255$ is low, it has a small chance of exceeding T, and so is likely to output $I_{\text{out}} = 0$ black. If the input intensity in the middle of the range, it is equally likely to output black or white. If the input intensity is high, it is likely to output white. The thresholding level can be altered by an adjusting value A added to or subtracted from I_{in} for every pixel.

Figure 39.16 shows the principle of random thresholding. At the top are three input patches with intensities $I_{\text{in}} = 0$, 128, and 192. Underneath are the patches after half-toning with random threshold $0 \leq T \leq 255$ and $A = 0$. Figure 39.17 shows examples of images half-toned by random threshold, with three different adjustments A.

Fig. 39.16 Principle of half-toning by random threshold. **a** Input patch, intensity $I_{\text{in}} = 0$. Input patch, intensity $I_{\text{in}} = 128$. **c** Input patch, intensity $I_{\text{in}} = 192$. Each pixel is half-toned with random threshold $0 \leq T \leq 255$ and adjustment $A = 0$. **d** For most pixels, $(I_{\text{in}} \pm A) < T$, so the patch is mainly black. **e** About half the pixels, go black, and about half white. **f** For most pixels, $(I_{\text{in}} \pm A) \geq T$, so the patch is mainly white

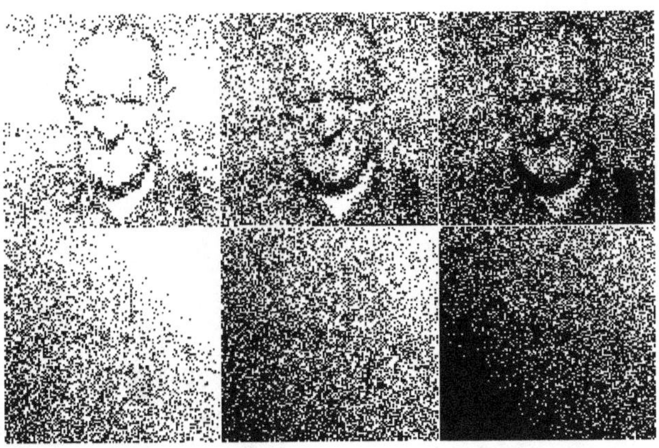

Fig. 39.17 Examples of half-toning by random threshold. (**a**) + (**d**) Random threshold $0 \leq T \leq 255$ and adjustment $A = -64$. (**b**) + (**e**) Random threshold $0 \leq T \leq 255$ and adjustment $A = 0$. (**c**) + (**f**) Random threshold $0 \leq T \leq 255$ and adjustment $A = 64$

Roberts [5] proposed adding pseudo-random noise to a six-bit signal to transmit over a three-bit PCM channel.

39.19 Fixed Threshold, Distributed Error

This clever and pleasing method [2] is variously called Floyd-Steinberg, stochastic, error-distributed, or error-diffused half-toning. Figure 39.18 shows the principle of error-distributed half-toning. Figure 39.19 shows examples of error-distributed half-toning.

Set threshold $T = 128$. For each input pixel (X, Y), compare its intensity $0 \leq I_{\text{in}} \leq 255$ to the threshold T. If $I_{\text{in}} < T$, output pixel (X, Y) with 1-bit intensity index $I_{\text{out}} = 0$ Black, else output $I_{\text{out}} = 1$ White:

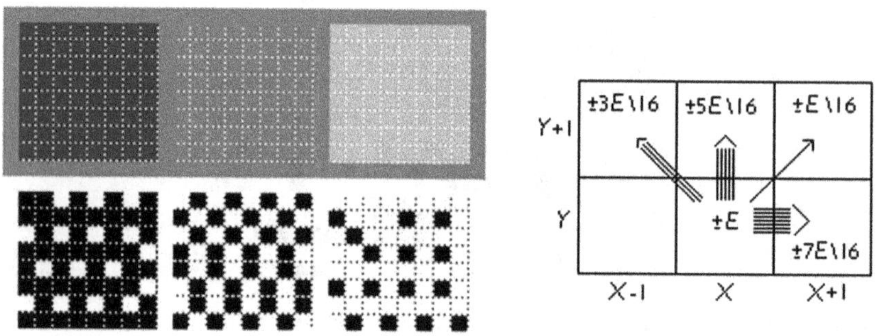

Fig. 39.18 Principle of error-distributed half-toning. **a–c** Input patches as above. **d** Each pixel is half-toned with threshold $T = 128$. The error E is the difference, positive or negative, between the input and the output intensities. The error is distributed proportionately to four neighbouring pixels yet to be processed. **e** The half-tone of (**a**) **f** The half-tone of (**b**). **g** The half-tone of (**c**)

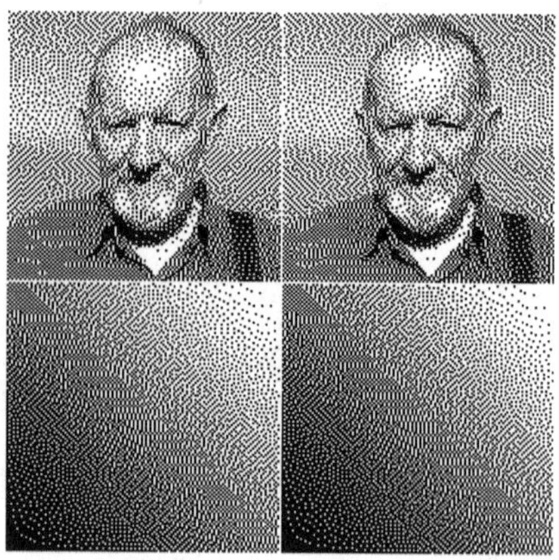

Fig. 39.19 Examples of error-distributed half-toning. (**a**) + (**c**) Half-toned with all rows scanned *left* to *right*. (**b**) + (**d**) Half-toned with rows scanned alternately *left* to *right* and *right* to *left*

$$I_{\text{out}} = 0 \quad \text{if } I_{\text{in}} < T$$
$$I_{\text{out}} = 1 \quad \text{if } I_{\text{in}} \geq Tts.$$

The error E incurred at this pixel is the difference between the input intensity I_{in} and the thresholded 1-bit output I_{out}:

$$E = I_{\text{in}} - (255 \times I_{\text{out}}) .$$

That is, when $I_{\text{out}} = 0$ black, the error is positive; and when $I_{\text{out}} = 1$ white, the error is negative. The error at this pixel is distributed to the four neighbouring pixels yet to be processed, in proportions $E\backslash 16$, $3E\backslash 16$, $5E\backslash 16$, and $7E\backslash 16$, as shown in Fig. 39.18. The proportional weights are such that a mid-grey area $I_{\text{in}} = 128$ half-tones to about 50 % Black and 50 % White. Clearly, changing the weights gives slightly different results.

The scanning of the given image may be done in the usual way, taking the rows from bottom to top and the columns from left to right; or taking the columns of alternate rows left to right and right to left.[1] Figure 39.19a is scanned in one direction, and Fig. 39.19b in alternating directions: the difference is slight.

39.20 Clustered Vector

The pixels in each row of an image may be grouped into *cells* of width M. When $M = 2$, each cell in the 1-bit output image has three possible states: both pixels black, one pixel white, or both pixels white. When $M = 4$, each cell has five possible states: all black, one white, two white, three white, or all white. And in general, a cell of width M has $(M + 1)$ possible states.

We associate a cell of width M with two vectors, each of width M:

- A *sequence vector* $\mathrm{S} = [S_1, S_2, \ldots, S_{\mathrm{M}}]$,
- a *threshold vector* $\mathrm{T} = [T_1, T_2, \ldots, T_{\mathrm{M}}]$,

where S contains ordinals $1, 2, \ldots, M$, and T contains threshold intensities $0 \leq T_{\mathrm{J}} \leq 255$ calculated as

$$T_{\mathrm{J}} = (255 \times S_{\mathrm{J}})\backslash(M + 1)$$

for $J = 1, 2, \ldots, M$.

Each pixel (X, Y) of the input image maps to threshold $T_{\mathrm{X\,mod\,M}}$; that is to say, the cell is tiled across each row of the image. Now compare the input intensity $0 \leq I_{\text{in}} \leq 255$ to the threshold $T_{\mathrm{X\,mod\,M}}$. If $I_{\text{in}} < T_{\mathrm{X\,mod\,M}}$, output pixel (X, Y) with 1-bit intensity $I_{\text{out}} = 0$ Black, else output 1 White:

[1] The ancient practice of writing alternate lines from left to right and from right to left was called by the Greeks *boustrophedon*—ox-turning—since the pattern is like ploughing successive furrows.

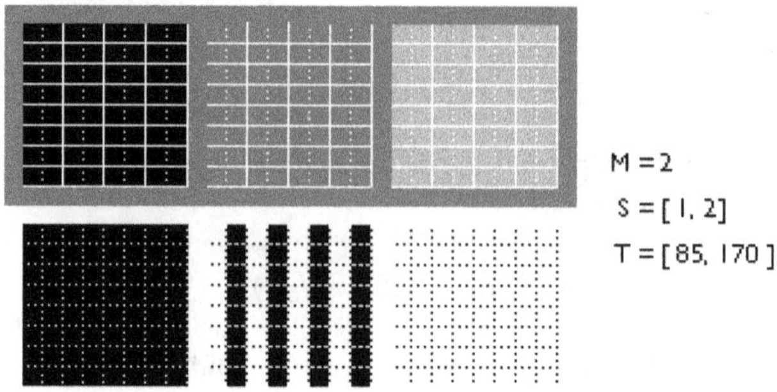

Fig. 39.20 Principle of half-toning by clustered threshold vector. **a–c** Input patches as above. **d** The half-tone of (**a**) **e** The half-tone of (**b**). **f** The half-tone of (**c**)

$$I_{\text{out}} = 0 \quad \text{if } I_{\text{in}} < T_{\text{X mod M}}$$
$$I_{\text{out}} = 1 \quad \text{if } I_{\text{in}} \geq T_{\text{X mod M}}\,.$$

Figure 39.20 shows the principle of half-toning with cell width $M = 2$. The sequence vector S contains 1 and 2. The threshold vector T contains $(255 \times 1)\backslash 3 = 85$, and $(255 \times 2)\backslash 3 = 170$. For the patch $I_{\text{in}} = 0$, both pixels of the cell go black. For the patch $I_{\text{in}} = 128$, the first pixel of the cell goes white and the second black. For the patch $I_{\text{in}} = 192$, both pixels of the cell go white.

Figure 39.21 shows examples of images half-toned with three different cell widths M. Half-toning by threshold vector separates the input intensities into $M+1$ distinct levels. There is an inverse relation between the fineness of intensity discrimination and the coarseness of the cell structure.

Fig. 39.21 Examples of half-toning by clustered threshold vector. (**a**) + (**d**) Cell width $M = 2$. (**b**) + (**e**) Cell width $M = 4$. (**c**) + (**f**) Cell width $M = 8$

39.21 Dispersed Vector

We can get a somewhat different half-tone by changing the set-up of the sequence vector S, thus dispersing the pixels in each cell. For example, with $M = 8$, a clustered sequence vector contains 1, 2, 3, 4, 5, 6, 7, 8 in that order. We may replace it by a dispersed sequence vector containing 1, 5, 3, 7, 2, 6, 4, 8 in that order. Figure 39.22 shows dispersed half-toning, to be compared with Fig. 39.21.

39.22 Clustered Matrix

The pixels of an image may be grouped into rectangular cells of width M and height N. Each cell then has $(M \times N) + 1$ possible states. (The standard Adobe PostScript page description language used for this book has cell size 16×16px, giving 257 8-bit grey levels.)

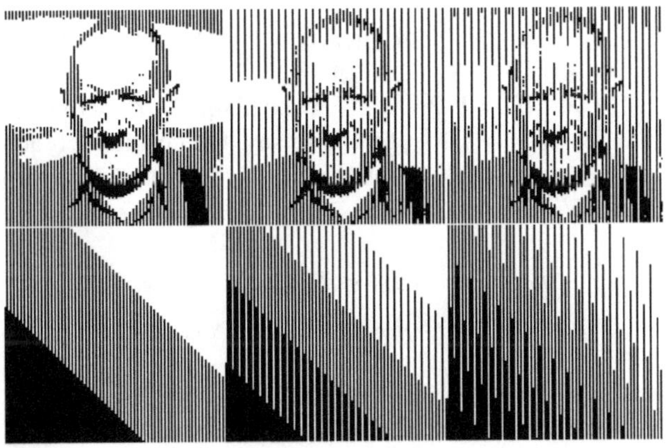

Fig. 39.22 Examples of half-toning by dispersed vector. (**a**) + (**d**) Cell width $M = 2$. (**b**) + (**e**) Cell width $M = 4$. (**c**) + (**f**) Cell width $M = 8$

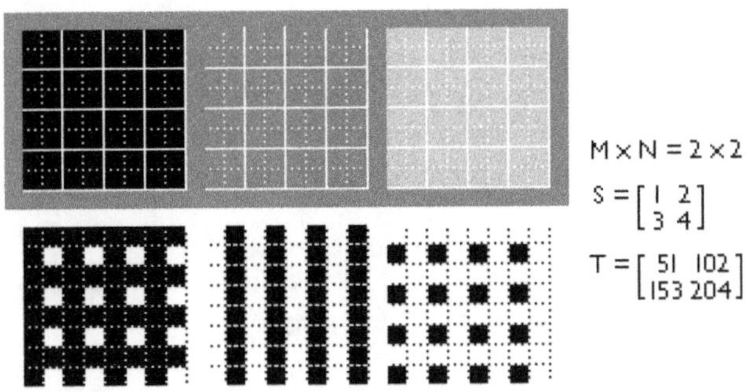

Fig. 39.23 Principle of half-toning by clustered threshold matrix with cell 2×2. **a–c** Input patches as above. **d** The half-tone of (**a**). **e** The half-tone of (**b**). **f** The half-tone of (**c**)

We associate with a cell an $M \times N$ sequence matrix:

$$\mathbf{S} = \begin{pmatrix} S_{11} & S_{12} & \ldots & S_{1N} \\ S_{21} & S_{22} & \ldots & S_{2N} \\ \vdots & \vdots & \vdots & \vdots \\ S_{M1} & S_{M2} & \ldots & S_{MN} \end{pmatrix}$$

and an $M \times N$ threshold matrix:

$$\mathbf{T} = \begin{pmatrix} T_{11} & T_{12} & \ldots & T_{1N} \\ T_{21} & T_{22} & \ldots & T_{2N} \\ \vdots & \vdots & \vdots & \vdots \\ T_{M1} & T_{M2} & \ldots & T_{MN} \end{pmatrix}$$

where **S** contains ordinals $1, 2, \ldots, (M \times N)$, and **T** contains threshold intensities $0 \leq T_{JH} \leq 255$ calculated as

$$T_{JH} = (255 \times S_{JH}) \backslash ((M \times N) + 1)$$

for $J = 1, 2, \ldots, M$ and $H = 1, 2, \ldots, N$. Each pixel (X, Y) of the input 8-bit image maps to threshold $T_{X \bmod M, Y \bmod N}$; that is to say, the cell is tiled across the entire image. If $I_{in} < T_{X \bmod M, Y \bmod N}$, output pixel (X, Y) with 1-bit intensity $I_{out} = 0$ Black, else output 1 White:

$$\begin{aligned} I_{out} &= 0 \quad \text{if } I_{in} < T_{X \bmod M, Y \bmod Y} \\ I_{out} &= 1 \quad \text{if } I_{in} \geq T_{X \bmod M, Y \bmod N} \,. \end{aligned}$$

Figure 39.23 shows the principle of half-toning with cell size $M \times N = 2 \times 2$. The sequence vector **S** contains 1, 2, 3, 4 arranged so that the next always abuts a previous, forming a cluster. The threshold vector **T** contains 3, 6, 9, 12 in the same arrangement. For the patch $I_{in} = 0$, all but one pixel of each cell go Black. For the patch $I_{in} = 128$, two pixels go White and two Black. For the patch $I_{in} = 192$, all but one pixel go white.

Figure 39.24 shows examples of images half-toned with three different clustered cell sizes $M \times N$. Half-toning by threshold matrix separates the input intensities into $((M \times N) + 1)$ distinct levels. There is an inverse relation between the number of grey levels distinguished and the prominence of the cell structure.

Fig. 39.24 Examples of half-toning by clustered threshold matrix. (**a**) + (**d**) Cell $M \times N = 2 \times 2$; 5 grey levels, 64 cells. (**b**) + (**e**) Cell $M \times N = 4 \times 4$; 17 grey levels, 32 cells. (**c**) + (**f**) Cell $M \times N = 8 \times 8$; 65 grey levels, 16 cells

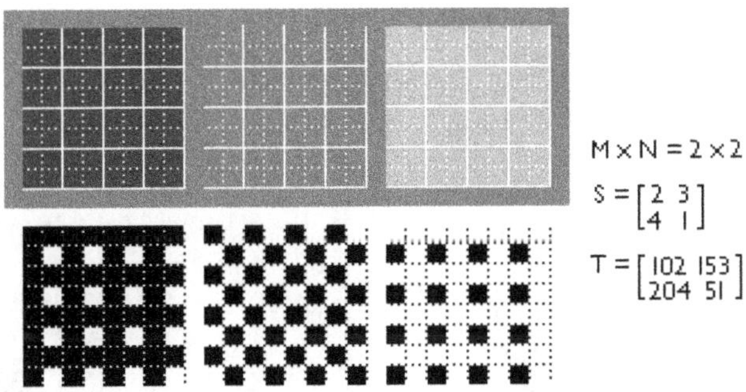

Fig. 39.25 Principle of half-toning by dispersed matrix, with cell 2×2. **a** Patch with 4-bit intensity $I_{\text{in}} = 0$, and cells marked. **b** Patch with intensity $I_{\text{in}} = 128$. **c** Patch with intensity $I_{\text{in}} = 192$. **d** The half-tone of (**a**). **e** The half-tone of (**b**). **f** The half-tone of (**c**)

39.23 Dispersed Matrix

Marked improvement in the half-toning process can be obtained by arranging the ordinals in the sequence matrix so that successive pixels are dispersed as far apart as possible in the cell. Bayer has proved [1] that an optimal sequence matrix for cell size $M \times M$, M a power of 2, is generated by the following recursion:

$$S_2 = \begin{pmatrix} 2 & 3 \\ 4 & 1 \end{pmatrix}, \quad S_{2\text{N}} = \begin{pmatrix} 4S_N + 1 & 4S_N + 2 \\ 4S_N + 3 & 4S_N \end{pmatrix},$$

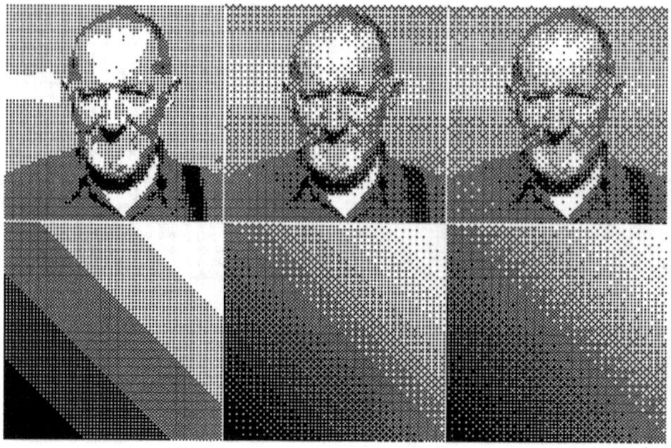

Fig. 39.26 Examples of half-toning by dispersed matrix. (**a**) + (**d**) Cell $M \times N = 2 \times 2$; 5 grey levels, 64 cells. (**b**) + (**e**) Cell $M \times N = 4 \times 4$; 17 grey levels, 32 cells. (**c**) + (**f**) Cell $M \times N = 8 \times 8$; 65 grey levels, 16 cells

so we get

$$S_4 = \begin{pmatrix} 6 & 10 & 7 & 11 \\ 14 & 2 & 15 & 3 \\ 8 & 12 & 5 & 9 \\ 16 & 4 & 13 & 1 \end{pmatrix}, \quad S_8 = \begin{pmatrix} 22 & 38 & 26 & 42 & 23 & 39 & 27 & 43 \\ 54 & 6 & 58 & 10 & 55 & 7 & 59 & 11 \\ 30 & 46 & 18 & 34 & 31 & 47 & 19 & 35 \\ 62 & 14 & 50 & 2 & 63 & 15 & 51 & 3 \\ 24 & 40 & 28 & 44 & 21 & 37 & 25 & 41 \\ 56 & 8 & 60 & 12 & 53 & 5 & 57 & 9 \\ 32 & 48 & 20 & 36 & 29 & 45 & 17 & 33 \\ 64 & 16 & 52 & 4 & 61 & 13 & 49 & 1 \end{pmatrix}$$

Figure 39.25 shows the principle of half-toning by a dispersed sequence matrix. It is like the clustered case, but here the middle patch $I = 128$ shows a chequer instead of bars. The difference between clustering and dispersing is greater with larger cell sizes. Figure 39.26 shows examples of images half-toned by dispersed matrices, with three different cell sizes. Half-toning by dispersed matrix separates the input intensities into $((M \times N) + 1)$ distinct levels, but the cell structure is much less obtrusive than it is in the clustered case.

39.24 8-bit Half-tone Programs

The following programs implement the bare bones of all seven of the half-tone methods described above. Brief descriptions are given below, and source code at the SpringerLink web-site. The reading and writing sub-programs are the same for all three programs.

```
REM  **  Program HT8STHRE to half-tone 8-bit image by
REM  **     single threshold. User enters filenames.
REM  **     Program reads 8-bit .BMP input file #1
REM  **     heads and writes 1-bit output file #2 heads.
REM  **     User chooses fixed or random or error-
REM  **     distributed threshold, and sets variables.
REM  **     Program reads input image intensities into
REM  **     array IMAGEA. Program then processes pixel
REM  **     by pixel from array IMAGEA to array IMAGEB,
REM  **     calculating 1-bit intensity 0 or 1 for each
REM  **     pixel. Program writes half-tone image from
REM  **     IMAGEB to 1-bit .BMP output file.
REM
REM  **  Main program
REM  **  Enter file-names
         CALL R8W1HEAD
         CALL HT8STVAR
         CALL REIMAGE8
         CALL CALC8STH
         CALL WRIMAGE1
         CALL REVHEAD1
END

         SUB CALC8STH
REM  **  Subprogram CALC8STH to scan IMAGEA and calculate
REM  **     half-tone into IMAGEB by method requested.
REM  **  Fixed threshold
REM  **  Random threshold
REM  **  Error-distributed threshold, forward
REM  **  Error_distributed threshold, forward and back
END SUB
         SUB HT8STVAR
REM  **  Subprogram HT8STVAR to enter half-tone
REM  **     method and variable.
END SUB
         SUB R8W1HEAD
REM  **  Subprogram RWHEADS to read header and 8-bit
REM  **     colour-table of input file #1, and
REM  **     preliminary-write header and 1-bit colour-
REM  **     table of output file #2 (to be revised later).
END SUB
         SUB REIMAGE8
REM  **  Subprogram REIMAGE8 to read 8-bit image data
REM  **  into array IMAGEA.
END SUB
```

```
        SUB REVHEAD1
REM  **  Sub-program REVHEAD1 to revise header of
REM  **     output file #2 with values in bytes.
END SUB
        SUB WRIMAGE1
REM  **  Sub-program WRIMAGE1 to write data from array
REM  **     IMAGEB to 1-bit file #2. Packs array locations
REM  **     as groups of 8 to a byte. Must finish row
REM  **     on a long (4-byte) boundary.
END SUB
```

```
REM  **  Program HT8VTHRE to make 1-bit halftone
REM  **     of 8-bit image, by vector of thresholds.
REM  **     User enters 8-bit .BMP filename #1 for
REM  **     input, 1-bit .BMP filename #2 for output,
REM  **     and size of vector (=cell). Program reads
REM  **     file #1 header and writes file #2 header,
REM  **     and dimensions image arrays. Program puts
REM  **     image data into array IMAGEA, then
REM  **     processes pixel by pixel into array IMAGEB,
REM  **     using thresholds in vector. Program
REM  **     writes data from IMAGEB to .BMP file #2.
REM
DECLARE SUB REVHEAD1 ()
DECLARE SUB WRIMAGE1 ()
DECLARE SUB REIMAGE8 ()
DECLARE SUB R8W1HEAD ()
DECLARE SUB CALCHTOV ()
DECLARE SUB CALCULVT ()
DECLARE SUB PUTBYTE (I%)
DECLARE SUB PUTL (INTL&)
DECLARE SUB PUTI (INTE%)
REM **  Set up sequence and threshold vectors for M=2
REM **  Set up sequence and threshold vectors for M=4
REM **  Set up dispersed sequence and thr vectors for M=4
REM **  Set up sequence and threshold vectors for M=8
REM **  Set up dispersed sequence and thr vectors for M=8
REM **  Main program
REM **  Enter file-names and cell size
        CALL R8W1HEAD
        CALL REIMAGE8
        CALL CALCULVT
        CALL CALCHTOV
```

```
          CALL WRIMAGE1
          CALL REVHEAD1
END

          SUB CALCHTOV
REM  **   Subprogram CALCHTOV to scan IMAGEA and calculate
REM  **      halftone into IMAGEB using vector thresholds.
END SUB
          SUB CALCULVT
REM  **   Subprogram CALCULVT to calculate threshold vector
END SUB
          SUB R8W1HEAD
          [as HT8STHRE]
          SUB REIMAGE8
          [as HT8STHRE]
          SUB REVHEAD1
          [as HT8STHRE]
          SUB WRIMAGE1
          [as HT8STHRE]
```

```
REM  **   Program HT8MTHRE to make 1-bit halftone
REM  **      of 4-bit image, by matrix of thresholds.
REM  **      User enters 4-bit .BMP filename #1 for
REM  **      input, 1-bit .BMP filename #2 for output,
REM  **      size of (square) cell, and clustered/
REM  **      dispersed. Program reads file #1 header
REM  **      and writes file #2 header, and dimensions
REM  **      image arrays. Program puts image data into
REM  **      array IMAGEA, then processes pixel by
REM  **      pixel into array IMAGEB, using thresholds
REM  **      in matrix. Program writes data from
REM  **      IMAGEB to 1-bit .BMP file #2.
REM
DECLARE SUB REVHEAD1 ()
DECLARE SUB WRIMAGE1 ()
DECLARE SUB REIMAGE8 ()
DECLARE SUB R8W1HEAD ()
DECLARE SUB CALCHTOM ()
DECLARE SUB CALCULMT ()
DECLARE SUB PUTBYTE (I%)
DECLARE SUB PUTL (INTL&)
DECLARE SUB PUTI (INTE%)
REM **   Set up 2x2 clustered sequence array
```

```
REM **   Set up 4x4 clustered sequence array
REM **   Set up 8x8 0 deg. clustered sequence array
REM **   Set up 2x2 dispersed sequence array
REM **   Set up 4x4 dispersed sequence array
REM **   Set up 8x8 dispersed sequence array
REM **   Main program
REM **   Enter file-names and cell size
         CALL R8W1HEAD
         CALL REIMAGE8
         CALL CALCULMT
         CALL CALCHTOM
         CALL WRIMAGE1
         CALL REVHEAD1
END
         SUB CALCHTOM
REM  **  Subprogram CALCHTOM to scan IMAGEA and calculate
REM  **     halftone into IMAGEB using matrix thresholds.
END SUB
         SUB CALCULMT
REM  **  Subprogram to calculate threshold arrays MT.
END SUB
         SUB R8W1HEAD
         [as HT8STHRE]
         SUB REIMAGE8
         [as HT8STHRE]
         SUB REVHEAD1
         [as HT8STHRE]
         SUB WRIMAGE1
         [as HT8STHRE]
END SUB
```

39.25 Half-toning for Print

Digital half-toning is needed whenever a 4-bit, 8-bit, or 24-bit image is sent to a printer, because a printer cannot vary the intensity of its inks, only their area-distribution. Half-toning for print is routinely performed by software supplied by a printer manufacturer. The fundamental process is from an 8-bit grey-scale image to a 1-bit image. For a colour image, first separate to four 8-bit grey-scale sub-images. For a 4-bit image, first save-as a 24-bit image.

39.26 Program for 8-bit Statistics

Figure 39.27 shows the processing flow for a program which calculates statistics from a given 8-bit image file. The program is described briefly below, and listed on the Springer web-site. Examples of the output are in Sect. 18.3.

```
\section{DOSTATS8}
\label{sec:b13}
\begin{small}\begin{verbatim}
REM   **  Program DOSTATS8 to read 8-bit .BMP file #1,
REM   **     scan image data and count frequencies of
REM   **     intensities I = 0-255 into array FCOUNT(I);
REM   **     then write cumulative frequencies into
REM   **     array CUFCOUNT(I); then write .TXT file #2
REM   **     containing statistics and frequency tables;
REM   **     then if asked write 1-bit .BMP file #3
REM   **     containing histogram 256 x 100 pixels; and
REM   **     if asked write 1-bit .BMP file #4 containing
REM   **     ogee plot 256 x 100 pixels.
REM   ** Main program
REM   **  Enter file_names, request histogram and/or ogee plot
             CALL FCOUNT8
             CALL CALCMVS
             CALL WRISTATI
             IF HIS$ = "Y" THEN
                CALL WRIHIHEA
                CALL WRIHICOL
                CALL WRIHIIMA
                CALL REVHIHEA
             END IF
             IF OGE$ = "Y" THEN
                CALL WRIOGHEA
```

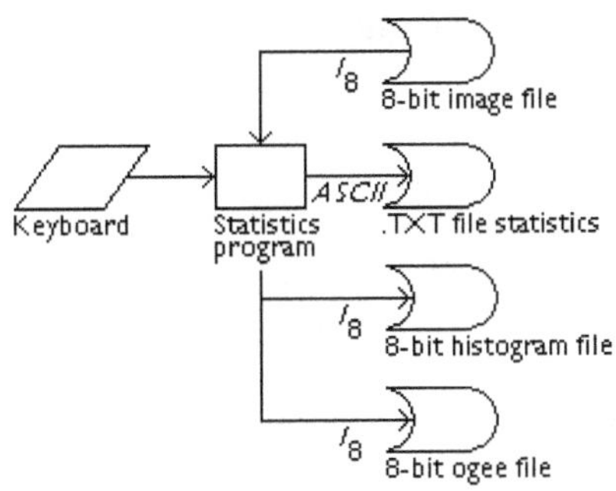

Fig. 39.27 Flowchart for 8-bit statistics program. Program reads an 8-bit image held in a bitmap file, calculates statistics to a text file, and calculates a histogram to a bitmap file (which can be displayed or printed by a GUI editor)

```
                CALL WRIOGCOL
                CALL WRIOGIMA
                CALL REVOGHEA
            END IF
    END

FUNCTION BITS8$ (H AS STRING)
         SUB CALCMVS
REM  **  Subprogram to calculate mean, variance, and
REM  **  standard deviation from frequency table, and lower
REM  **  quartile, median, upper quartile, and skew from
REM  **  cumulative frequency table.
END SUB
         SUB FCOUNT8
REM  **  Subprogram to count frequencies of intensities
REM  **  of pixels for image bitdepth 8, into FCOUNT(I),
REM  **  and cumulative frequencies into CUFCOUNT(I)
END SUB
     SUB IMAARR (FILENAME$)
PRINT "........................................
............"
PRINT "     EDIT IMAGE ARRAY FOR BMP FILE "; FILENAME$
PRINT " "
REM FIRST THE 1-BIT CASE
IF BDEPTH% = 1 THEN
    PRINT "     The image array is "; WPIX&; " pixels
wide and "; HPIX&; " pixels high"
    PRINT "     Each pixel (x,y) has 0 <= x <= "; WPI
X&; ", 0 <= y <= "; HPIX&
    PRINT "     and colour index 0 or 1"
    PRINT " "
    Y = 0
    INPUT "     Start at row y = "; Y
    PRINT "...................................
................."
    PRINT " "
    X = 0
    BYPERRO% = (((WPIX& * BDEPTH%) + 31) \ 32) * 4
    BIPERRO% = BYPERRO% * 8
    FOR YP% = Y TO (HPIX& - 1)
        FBP% = OFFSET& + (YP% * BYPERRO%)
        FOR XP% = X TO (WPIX& - 1) STEP 8
            FBP% = FBP% + 1
            GET #1, FBP%, BYTE
            HBY$ = HEX$(INT(ASC(BYTE)))
```

```
            B$ = BITS8$(HBY$)
            B0$ = MID$(B$, 1, 1)
            B1$ = MID$(B$, 2, 1)
            B2$ = MID$(B$, 3, 1)
            B3$ = MID$(B$, 4, 1)
            B4$ = MID$(B$, 5, 1)
            B5$ = MID$(B$, 6, 1)
            B6$ = MID$(B$, 7, 1)
            B7$ = MID$(B$, 8, 1)
            PRINT " "
            PRINT "File "; FILENAME$; " -- Image ar
ray"
            PRINT "Y (row) = "; YP%
            PRINT "X (col) = ", XP%; XP% + 1; XP% +
 2; XP% + 3; XP% + 4; XP% + 5; XP% + 6; XP% + 7
            PRINT"--------------------------------
----------------------------"
            PRINT "Colour index"," "; B0$;"  "; B
1$;"  "; B2$;"  "; B3$;"  "; B4$;"  "; B5$;"  ";
B6$;"  "; B7$
            PRINT " "
            INPUT "Pick action: 1=continue 2=change
3=exit "; AC%
            IF AC% = 2 THEN
                PRINT "Replace "; B$
                INPUT "     by "; B$
                REM  CONVERT 8-BIT STRING B$ TO AS
CII A
                A = VAL(MID$(B$, 1, 1)) * 128
                A = A + VAL(MID$(B$, 2, 1)) * 64
                A = A + VAL(MID$(B$, 3, 1)) * 32
                A = A + VAL(MID$(B$, 4, 1)) * 16
                A = A + VAL(MID$(B$, 5, 1)) * 8
                A = A + VAL(MID$(B$, 6, 1)) * 4
                A = A + VAL(MID$(B$, 7, 1)) * 2
                A = A + VAL(MID$(B$, 8, 1))
                BYTE = CHR$(A)
                PUT #1, FBP%, BYTE
            END IF
            IF AC% = 3 THEN GOTO ENDIMA
        NEXT XP%
    NEXT YP%
END IF
REM NEXT THE 24-BIT CASE
```

```
IF BDEPTH% = 24 THEN
     PRINT "     The image array is "; WPIX&;" pixels wi
de and "; HPIX&;" pixels high"
     PRINT "     Each pixel (x,y) has BGRZ quad"
     PRINT " "
     Y = 0
     INPUT "     Start at row y = "; Y
     PRINT ".........................................
..............."
     PRINT " "
     X = 0
     BYPERRO% = (((WPIX& * BDEPTH%) + 31) \ 32) * 4
     BIPERRO% = BYPERRO% * 8
     FOR YP% = Y TO (HPIX& - 1)
          FBP% = OFFSET& + (YP% * BYPERRO%)
          FOR XP% = X TO (WPIX& - 1) STEP 1
               FBP% = FBP% + 1
               GET #1, FBP%, BYTE
               B% = ASC(BYTE)
               FBP% = FBP% + 1
               GET #1, FBP%, BYTE
               G% = ASC(BYTE)
               FBP% = FBP% + 1
               GET #1, FBP%, BYTE
               R% = ASC(BYTE)
               FBP% = FBP% + 1
               GET #1, FBP%, BYTE
               Z% = ASC(BYTE)
               PRINT " "
               PRINT "File "; FILENAME$;" -- Image ar
ray"
               PRINT "Y (row) = "; YP%
               PRINT "X (col) = ", XP%
               PRINT "---------------------------------
---------------------------"
               PRINT "BGRZ "," "; B%;"  "; G%;"  ";
 R%;"  "; Z%
               PRINT " "
               INPUT "Pick action: 1=continue 2=change
 3=exit "; AC%
               IF AC% = 2 THEN
                    PRINT "Replace "; B%
                    INPUT "     by "; B%
                    PRINT "Replace "; G%
                    INPUT "     by "; G%
```

```
                    PRINT "Replace "; R%
                    INPUT "     by "; R%
                    BYTE = CHR$(B%)
                    FBP% = FBP% - 3
                    PUT #1, FBP%, BYTE
                    BYTE = CHR$(G%)
                    FBP% = FBP% + 1
                    PUT #1, FBP%, BYTE
                    BYTE = CHR$(R%)
                    FBP% = FBP% + 1
                    PUT #1, FBP%, BYTE
                    BYTE = CHR$(Z%)
                    FBP% = FBP% + 1
                    PUT #1, FBP%, BYTE
               END IF
               IF AC% = 3 THEN GOTO ENDIMA
          NEXT XP%
     NEXT YP%
END IF
ENDIMA:
PRINT " "
PRINT "..........................................
..........."
INPUT "End of image array: hit any key to go on ", OK
PRINT " "
PRINT " "
PRINT " "
PRINT " "
CLOSE #1
END SUB
DEFINT A-Z
          SUB PUTI (INTE)
REM  **   Puts integer value into binary file at next po
sition as 2 bytes
          PUT$ = MKI$(INTE)
          PUT #2, SEEK(2), PUT$
END SUB
          SUB PUTL (INTL&)
REM  **   Puts long integer into binary file as 4 bytes
          PUT$ = MKL$(INTL&)
          PUT #2, SEEK(2), PUT$
END SUB
          SUB REVHIHEA
REM  **   Subprogram to revise output file 3 header with b
```

```
yte lengths
REM  **  Update bfSize length of file in bytes, as long
REM  **  and keep file length for end message
         INTL& = LOF(3)
         PUT$ = MKL$(INTL&)
         PUT #3, 3, PUT$
         FILELENGTH& = LOF(3)
REM  **  Update biSizeImage length of image data in bytes
, as long
         INTL& = LOF(3) - 62
         PUT$ = MKL$(INTL&)
         PUT #3, 35, PUT$
END SUB
         SUB REVOGHEA
REM  **  Subprogram to revise output file 4 header for oge
e with byte lengths
END SUB
         SUB WRIHICOL
REM  **  Subprogram to write 1-bit colourtable to binary
file, for histogram
END SUB
         SUB WRIHIHEA
REM  **  Subprogram to create histogram output file
and write .BMP header
END SUB
        SUB WRIHIIMA
REM  **  Subprogram WRIHIIMA to generate histogram 100
REM  **     rows high. Set up each row as string HISROW
REM  **     of "0" or "1", and write histogram row to
REM  **     output file #3.
END SUB
         SUB WRIOGCOL
REM  **  Subprogram to write 1-bit colourtable to binary
 file, for ogee plot
END SUB
         SUB WRIOGHEA
REM  **  Subprogram to create ogee plot output file and
 write .BMP header
         SUB WRIOGIMA
REM  **  Subprogram to generate ogee image from CUP(I),
 row 0 to row 100,
REM  **     as string OGEROW of "0" or "1", and call sub
program WRIOGROW
REM  **     to write ogee plot row to output file.
bunching counts
```

```
END SUB
        SUB WRIOGROW
REM  **  Subprogram to take the 256-char string OGEROW$ 8
 chars at a time,
REM  **     32 times. Each 8-char string is converted to
ASCII A, then to
REM  **     byte, then written to binary file 4 as ogee p
lot data
END SUB
        SUB WRIOGROWH
REM  **  Subprogram to take the 128-char string OGEROWH$
 8 chars at a time,
REM  **     16 times. Each 8-char string is converted to
 ASCII A, then to
REM  **     byte, then written to binary file 4 as ogee
plot data
END SUB
        SUB WRIOGTXT
REM  **  Subprogram to write text file 2 for statistics
and frequency
REM  **  tables.
END SUB
        SUB WRISTATI
REM  **  Subprogram to write text file 2 for statistics a
nd frequency
REM  **  tables.
END SUB
        SUB WRITEFILE
REM  **  Subprogram to write filename.FRQ with header and
frequency data
END SUB
        SUB WRITEROW
REM  **  Subprogram to take the 256-char string HISRO
W$ 8 chars at a time,
REM  **     32 times. Each 8-char string is converted
 to ASCII A, then to
REM  **     byte, then written to binary file 3 as hi
stogram image data.
END SUB
        SUB WRITEROWH
REM  **  Subprogram to take the 128-char string HISRO
WH$ 8 chars at a time,
REM  **     16 times. Each 8-char string is converted
to ASCII A, then to
```

```
REM  **      byte, then written to binary file 3 as his
togram image data
END SUB
```

References

1. Bayer BE (1973) An optimal method for two-level rendition of continuous-tone pictures. IEEE Int Conf Commun 1:11–15
2. Floyd RW, Steinberg L (1976) An adaptive algorithm for spatial greyscale. J Soc Inf Disp 17:75–77
3. Lau Consulting (2014) Digital halftoning bibliography. http://www.drhalftone.com/Lau_Consulting_Inc./Halftoning/Entries/2009/4/1_Bibliography.html
4. Library of Congress (2014) Heavenly craft: the woodcut in early printed books. http://www.loc.gov/exhibits/heavenlycraft/heavenly-15th.html. Accessed 1 Feb 2014
5. Roberts LG (1961) Picture Coding Using Pseudo-random Noise. SM thesis, MIT. Available at: http://www.packet.cc/files/pic-code-noise.html. Accessed 1 Feb 2014
6. Twyman M (1970) Printing 1770–1970. Eyre and Spottiswoode, London
7. Twyman M (1970) Lithography 1800–1850. Oxford University Press, London
8. Uglow J (2006) Nature's Engraver: a life of Thomas Bewick. Faber and Faber, London
9. Ulichney R (1987) Digital halftoning. MIT Press, Cambridge

Chapter 40
8-bit Fourier Process

A Fourier transformation changes all the locations of an image to frequencies of gratings, and back again. This allows certain frequencies to be attenuated or amplified, for one purpose or another. The theory is developed here for continuous curves, then for 8-bit grey-scale images and 8-bit sinusoidal gratings. A pair of programs for analysis and synthesis is given, with examples.

40.1 Flowchart

The flowchart (Fig. 40.1) shows the general arrangement for Fourier processing of 8-bit grey-scale images (see programs in Sect. 40.16).

40.2 Fourier Transformation

The fundamental idea of Fourier transformation is to pass from a given set of values in a space (or time) domain to an equivalent set of values in a frequency domain, and back again. This enables certain frequencies to be filtered out or attenuated or amplified, for one purpose or another. The classical theory deals with continuous curves varying in amplitude, as summarized in Sect. 40.3 below. We then apply the same ideas to 8-bit grey-scale images in Sects. 40.4–40.15, with a pair of programs described in Sect. 40.16.

40.3 Fourier Transformation of Real Curve

A real curve can be specified by an equation in two variables, x and y. We have two forms of periodic *sinusoidal curve*:

A. Parkin, *Digital Imaging Primer*, DOI 10.1007/978-3-540-85619-1_40

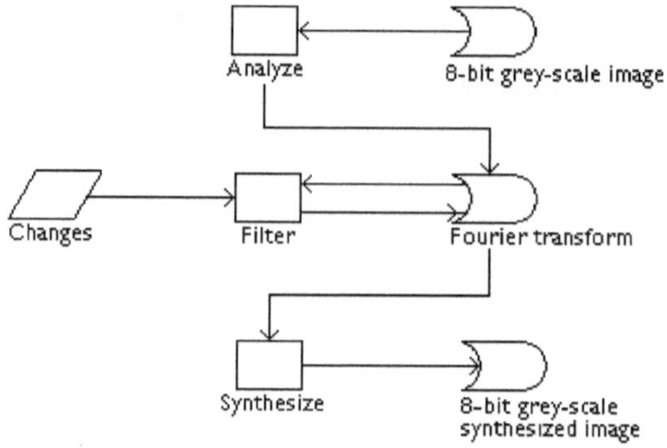

Fig. 40.1 8-bit Fourier process. Given an 8-bit grey-scale image file, an analysis program reads the file, calculates the frequency transform, and outputs it to a transform file. A filter program reads the transform file, changes it, and writes a new transform file. A synthesis program reads the transform file, calculates a reconstructed image, and outputs it to an 8-bit grey-scale image file

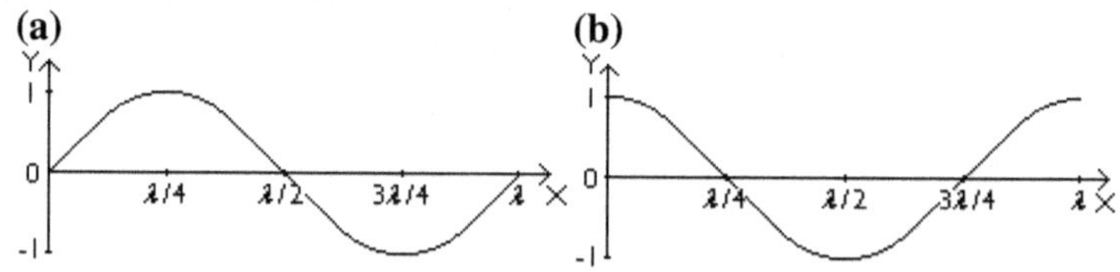

Fig. 40.2 Sinusoidal curves. **a** Sine curve with wavelength λ and amplitude 1. The curve has rotation symmetry by a half-turn about poles at $x = 0$ and $x = \lambda/2$, and reflection symmetry about vertical axes at $x = \lambda/4$ and $x = 3\lambda/4$. **b** Cosine curve with wavelength λ and amplitude 1. The cosine curve is the same shape as the sine curve, but shifted a quarter-wavelength *to the left*. It has rotation symmetry by a half-turn about poles at $x = \lambda/4$ and $x = 3\lambda/4$, and reflection symmetry about vertical axes at $x = 0$ and $x = \lambda/2$

- *Sine curve* $y = \sin(x)$,
- *Cosine curve* $y = \cos(x)$,

where the x-axis is calibrated in angle $0 \le x \le 360°$, or $0 \le x \le 2\pi$ radians; and the y-axis is calibrated in length. Figure 40.2 shows one cycle of a sine curve and one cycle of a cosine curve, understood to be continuous. The *wavelength* λ of a sinusoid is the length of one complete cycle. The *amplitude* a is the maximum height reached by the curve.

We may take any sine curve:

$$y = \sin(x)$$

as a *fundamental* or *first harmonic*, and derive from it a *harmonic set* of further sine curves. The fundamental has integer *frequency* $F = 1$ and normalized wavelength $\lambda = 1/F = 1$. The *second harmonic*:

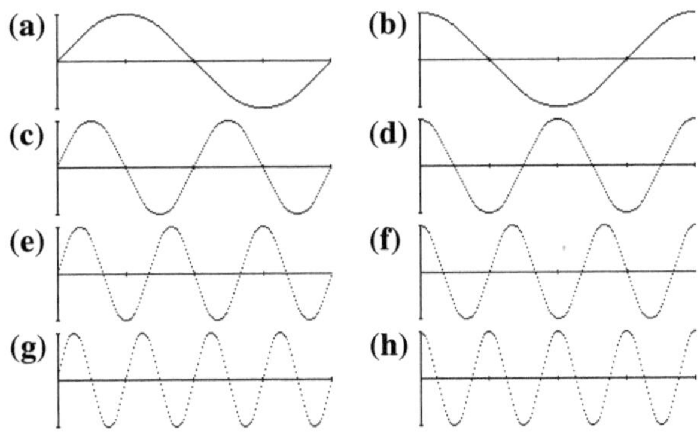

Fig. 40.3 Harmonic sets of sinusoids. On the *left*, **a** Fundamental sine curve with frequency $F = 1$ and wavelength $\lambda = 1$. **b** Second sine harmonic with frequency $F = 2$ and wavelength $\lambda = 1/2$. **c** Third sine harmonic with frequency $F = 3$ and wavelength $\lambda = 1/3$. **d** Fourth sine harmonic with frequency $F = 4$ and wavelength $\lambda = 1/4$. On the *right*, **e** Fundamental cosine curve with frequency $F = 1$ and wavelength $\lambda = 1$. **f** Second cosine harmonic with frequency $F = 2$ and wavelength $\lambda = 1/2$. **g** Third cosine harmonic with frequency $F = 3$ and wavelength $\lambda = 1/3$. **h** Fourth cosine harmonic with frequency $F = 4$ and wavelength $\lambda = 1/4$

$$y = \sin(2x)$$

has frequency 2 and wavelength 1/2: that is, exactly two cycles for one cycle of the fundamental. The *third harmonic*:

$$y = \sin(3x)$$

has frequency 3 and wavelength 1/3: that is, exactly three cycles for one cycle of the fundamental. And so on, to the *Nth harmonic*:

$$y = \sin(Nx)$$

has frequency N and wavelength $1/N$: that is, exactly N cycles for one cycle of the fundamental. Figure 40.3 shows *harmonic sets* of sinusoids.

We may change the amplitude of a sinusoid by multiplying each ordinate of the curve by a real *coefficient* c:

$$y = c\sin(x)$$
$$y = c\cos(x)$$

For example, Fig. 40.4 shows fundamentals with coefficient $c = 1$, second harmonics with $c = 0.6$, third harmonics with $c = 0.3$, and fourth harmonics with $c = 0$. (Also -1, -0.6, -0.3).

We may add any two or more sinusoids (sine with sine, sine with cosine, or cosine with cosine) to get a *sum curve*, by algebraically-adding corresponding ordinates:

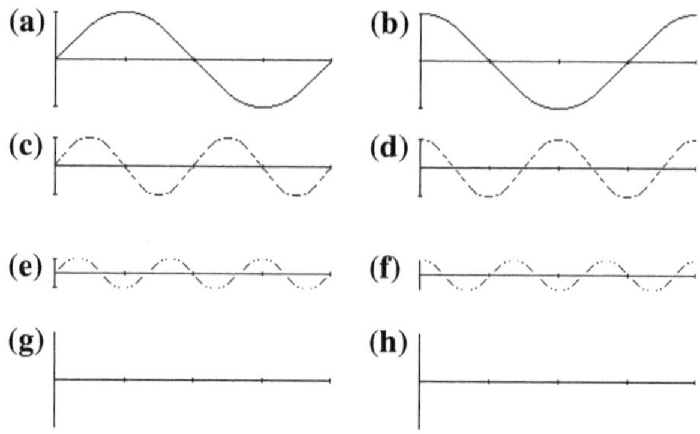

Fig. 40.4 Various amplitudes of sinusoids. On the *left*, **a** Fundamental sine curve with amplitude 1. **b** Second sine harmonic with amplitude 0.6. **c** Third sine harmonic with amplitude 0.3. **d** Fourth sine harmonic with amplitude 0. On the *right*, **e** Fundamental cosine curve with amplitude 1. **f** Second cosine harmonic with amplitude 0.6. **g** Third cosine harmonic with amplitude 0.3. **h** Fourth cosine harmonic with amplitude 0

$$y_3 = \sin(x_1) + \sin(x_2)\,.$$

For example, Fig. 40.5 shows the sum of two fundamental sine curves with amplitude 1, and the sum of a fundamental sine curve with amplitude 1 and a second sine harmonic with amplitude 0.6.

We may multiply any two or more sinusoids (sine with sine, sine with cosine, or cosine with cosine) to get a *product curve*, by multiplying corresponding ordinates:

$$y_3 = \sin(x_1) \times \sin(x_2)\,.$$

For example, Fig. 40.6 shows the product of two fundamental sine curves with amplitude 1, and the product of a fundamental sine curve with amplitude 1 and a second sine harmonic with amplitude 0.6.

The *mean* of a sinusoidal curve is the average of the ordinates. In practice it is found as the algebraic sum of N equally-spaced ordinates, divided by N. In a harmonic set of sinusoids:

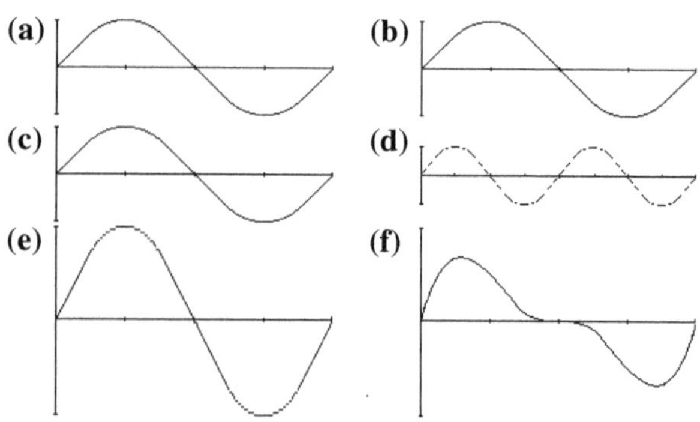

Fig. 40.5 Sum of sinusoids. On the *left*, **a** Fundamental sine curve with amplitude 1. **b** Another fundamental sine curve with amplitude 1. **c** Sum curve with amplitude 2. On the *right*, **d** Fundamental sine curve with amplitude 1. **e** Second sine harmonic with amplitude 0.6. **f** Sum curve with amplitude 1.3

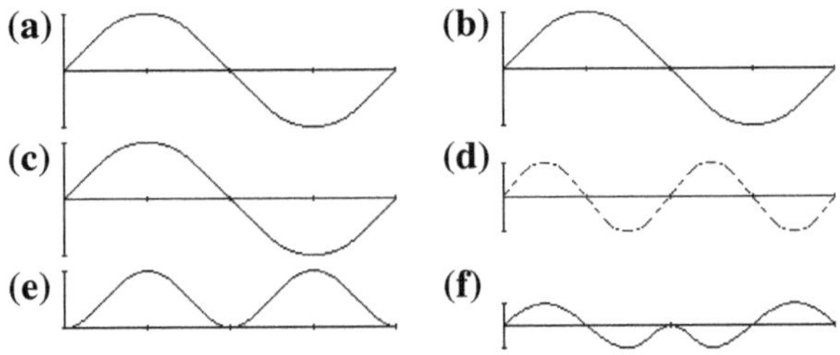

Fig. 40.6 Product of sinusoids. On the *left*, **a** Fundamental sine curve with amplitude 1. **b** Another fundamental sine curve with amplitude 1. **c** Product curve with amplitude 1. On the *right*, **d** Fundamental sine curve with amplitude 1. **e** Second sine harmonic with amplitude 0.6. **f** Product curve with amplitude 0.4

- The mean of each harmonic curve is zero (as is evident from the symmetries).
- The mean of the sum of any two harmonics is zero.
- The mean of the product of any two harmonics is zero, with the sole exception of the square of a harmonic.
- The mean of the square of a harmonic is positive, and is half the square of the amplitude.

This exceptional property of the square of a harmonic is the key to Fourier analysis.

Fourier[1] proposed [3][2] that

- any periodic single-valued real function of x can be specified exactly as the sum of an infinite series of harmonic cosine and sine curves of various amplitudes; and further that any such function can be approximated as closely as desired by the sum of a finite series of N terms, by taking larger N.

This surprising theorem also holds for non-periodic functions. Thus an arbitrary curve can be increasingly closely matched by adding harmonic cosine and sine curves, each with its own coefficient.

A continuous waveform (such as the amplitude of a telephone or other signal varying in the time dimension) can be reduced to a set of *discrete samples* according to the *Sampling Theorem*[3]:

- If a signal is limited to bandwith B (that is, contains no frequencies higher than B), then it can be completely specified by uniformly-spaced discrete samples taken at a rate higher than $2B$.

Or, put differently, if the sampling rate is S samples per unit time, the signal bandwidth which can be captured is $B \leq S/2$.

---

[1] J.B.J. Fourier (1768–1830) Governor of Lower Egypt 1801, Prefect of Isere 1802, created Baron 1808 (by Napoleon, annulled at his fall), Secretary of Academie des Sciences 1823, Academie Francaise 1826.

[2] This was doubted by Poisson and others, but proved under increasingly rigorous conditions by Sturm and by Dirichlet.

[3] Originated by Whitaker, with further contributions by Nyquist, Kotelniakov, Shannon, and others.

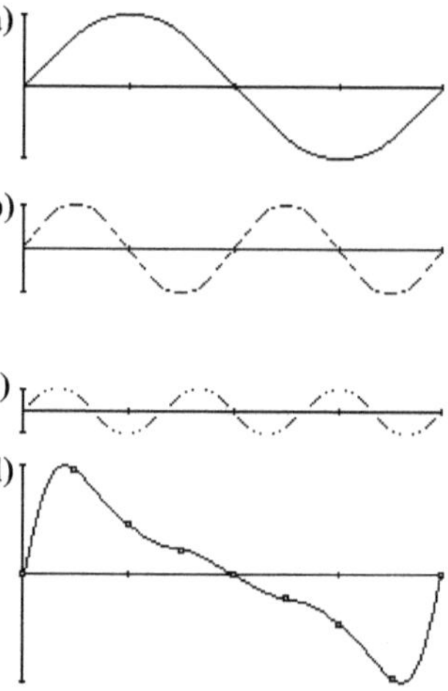

Fig. 40.7 Fourier series of sines. **a** Fundamental sine curve with amplitude $b_1 = 1$. **b** Second sine harmonic with amplitude $b_2 = 0.6$. **c** Third sine harmonic with amplitude $b_3 = 0.3$. **d** Sum curve with amplitude 1.5

A *finite Fourier series* is the sum of all the sine and cosine harmonic curves in a harmonic set up to the Nth, each with an independent coefficient:

$$f(x) = a_0 + a_1 \cos(x) + a_2 \cos(2x) + \cdots + a_N \cos(Nx) \\ + b_1 \sin(x) + b_2 \sin(2x) + \cdots + b_N \sin(Nx)\,,$$

where a_0 is the amplitude of the *zeroth harmonic*, a constant value which raises or lowers the entire curve. Terms with a zero coefficient are omitted. For example, Fig. 40.7 shows the component curves and the sum curve for the Fourier series:

$$f(x) = \sin(x) + 0.6 \sin(2x) + 0.3 \sin(3x)\,,$$

which has only sine harmonics.

A Fourier series may be specified as a *frequency spectrum* of the amplitude of each sine harmonic and of each cosine harmonic in the harmonic set. For such a frequency spectrum, the horizontal axis is integer frequency F, and the vertical axis is amplitude a or b. For example, Fig. 40.8 shows the sine frequency spectrum for the Fourier series of Fig. 40.7.

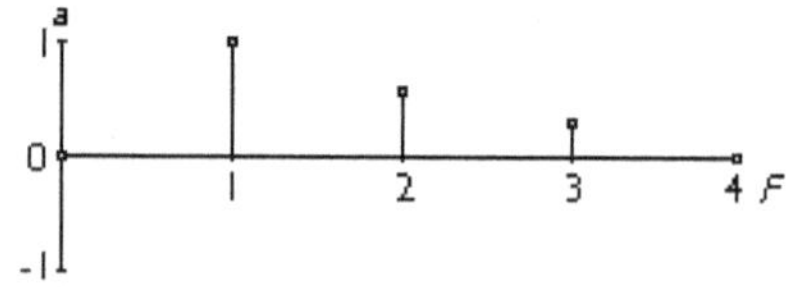

Fig. 40.8 Frequency spectrum for Fig. 40.7. The sine harmonics have amplitudes $a_0 = 0$, $a_1 = 1$, $a_2 = 0.6$, $a_3 = 0.3$, $a_4 = 0$

Given any curve (the *target*) with wavelength λ, the process of *Fourier analysis* is

- Sample the target at $2N$ equally-spaced ordinates

Table 40.1 Fourier analysis

	0	1	2	3	4	5	6	7	Sum	Mean	2× Mean
(a)	0	1.52	0.70	0.32	0	−0.32	−0.70	−1.52	0	0	
(b)	0	0.71	1	0.71	0	−0.71	−1	−0.71			
(c)	0	1	0	−1	0	1	0	−1			
(d)	0	0.71	−1	0.71	0	−0.71	1	−0.71			
(e)	0	0	0	0	0	0	0	0			
(f)	0	1.10	0.70	0.23	0	0.23	0.70	1.10	4.06	0.51	1
(g)	0	1.52	0	−0.32	0	−0.32	0	1.52	2.40	0.30	0.60
(h)	0	1.10	−0.70	0.23	0	0.23	−0.70	1.10	1.26	0.16	0.31

Row (a) Target curve, given as $2N = 8y$-values, which finds coefficient $a_0 = 0$. Row (b) Fundamental sine curve $y = \sin(x)$. Row (c) Second sine harmonic $y = \sin(2x)$. Row (d) Third sine harmonic $y = \sin(3x)$. Row (e) Fourth sine harmonic $y = \sin(4x)$. Row (f) Product of (a) and (b), which finds coefficient $b_1 = 1$. Row (g) Product of (a) and (c), which finds coefficient $b_2 = 0.6$. Row (h) Product of (a) and (d), which finds coefficient $b_3 = 0.3$. The final column contains the frequency spectrum 1, 0.6, 0.3

- Multiply the target successively by each cosine harmonic and each sine harmonic in the set with wavelength λ, frequencies 1 to N, and save the products
- The mean of each product is positive if and only if that harmonic is present in the target (since in that case the product is a square); otherwise the mean intensity is zero. Coefficient $a_1, a_2, \ldots, b_1, b_2, \ldots$ is twice the mean of its product sequence
- Coefficient a_0 for the zeroth harmonic (constant 1) is found directly as the mean of the target.

For example, Table 40.1 shows the analysis of the target curve built up in Fig. 40.7.

Given a spectrum of cosine and sine frequencies 0 to N, the process of *Fourier synthesis* is just the inverse of Fourier analysis:

- Construct the cosine and sine harmonics with frequencies $0, 1, 2, \ldots, N$ at $2N$ equally-spaced ordinates
- Multiply the zeroth harmonic (constant 1) by coefficient a_0 across the entire wavelength, and save the products
- Multiply each cosine harmonic, frequencies 1 to N, by the corresponding coefficient a_1 to a_N, and save the products
- Multiply each sine harmonic, frequencies 1 to N, by the corresponding coefficient b_1 to b_N, and save the products
- Add the products across the entire wavelength, to reconstruct the target curve.

For example, Table 40.2 shows the synthesis of the curve built up in Fig. 40.7.

Given a spectrum of amplitudes of cosine and sine frequencies 0 to N, we may filter the spectrum by changing one or more of the amplitude coefficients, then synthesizing a new curve from the changed coefficients. For example, Fig. 40.9 shows an original spectrum and a changed spectrum, and the original curve and the modified curve. The effect is to raise the entire curve by height 0.5, and to halve and invert the fundamental

Table 40.2 Fourier synthesis of Fig. 40.7

	0	1	2	3	4	5	6	7
(a)	0	0.71	1	0.71	0	−0.71	−1	−0.71
(b)	0	1	0	−1	0	1	0	−1
(c)	0	0.71	−1	0.71	0	−0.71	1	−0.71
(d)	0	0	0	0	0	0	0	0
(e)	0	0.71	1	0.71	0	−0.71	−1	−0.71
(f)	0	0.60	0	−0.60	0	0.60	0	−0.60
(g)	0	0.21	−0.30	0.21	0	−0.21	0.30	−0.21
(h)	0	1.52	0.70	0.11	0	−0.32	−0.70	−1.52

Row (a) Fundamental sine curve $y = \sin(x)$. Row (b) Second sine harmonic $y = \sin(2x)$. Row (c) Third sine harmonic $y = \sin(3x)$. Row (d) Fourth sine harmonic $y = \sin(4x)$. Row (e) Product of (a) and coefficient $b_1 = 1$. Row (f) Product of (b) and coefficient $b_2 = 0.6$. Row (g) Product of (c) and coefficient $b_3 = 0.3$. Row (h) Sum of (e)–(g), the synthesized curve

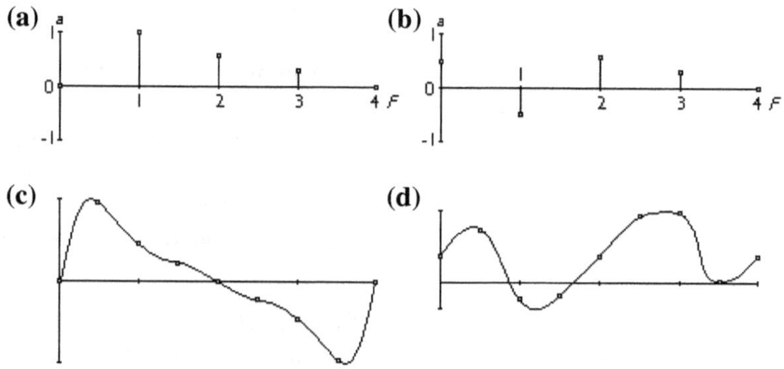

Fig. 40.9 Filtering of frequency spectrum. **a** Original spectrum: the harmonics have amplitudes $a_0 = 0$, $b_1 = 1$, $b_2 = 0.6$, $b_3 = 0.3$, $a_4 = 0$. **b** Changed spectrum: the harmonics have amplitudes $a_0 = 0.5$, $b_1 = -0.5$, $b_2 = 0.6$, $b_3 = 0.3$, $a_4 = 0$. **c** Original curve. **d** Modified curve

component. Table 40.3 shows the synthesis of the modified curve built up from the changed amplitude coefficients.

Table 40.3 Fourier synthesis of modified curve in Fig. 40.9

	0	1	2	3	4	5	6	7
(a)	1	1	1	1	1	1	1	1
(b)	0	0.71	1	0.71	0	−0.71	−1	−0.71
(c)	0	1	0	−1	0	1	0	−1
(d)	0	0.71	−1	0.71	0	−0.71	1	−0.71
(e)	0	0	0	0	0	0	0	0
(f)	0.50	0.50	0.50	0.50	0.50	0.50	0.50	0.50
(g)	0	−0.35	−0.50	−0.35	0	0.35	0.50	0.35
(h)	0	0.60	0	−0.60	0	0.60	0	−0.60
(i)	0	0.21	−0.30	0.21	0	−0.21	0.30	−0.21
(j)	0.50	0.96	−0.30	−0.24	0.50	1.24	1.30	0.04

Row (a) Zeroth harmonic (constant). Row (b) Fundamental sine curve $y = \sin(x)$. Row (c) Second sine harmonic $y = \sin(2x)$. Row (d) Third sine harmonic $y = \sin(3x)$. Row (e) Fourth sine harmonic $y = \sin(4x)$. Row (f) Product of (a) and coefficient $a_0 = 0.5$. Row (g) Product of (b) and coefficient $b_1 = -0.5$. Row (h) Product of (c) and coefficient $b_2 = 0.6$. Row (i) Product of (d) and coefficient $b_3 = 0.3$. Row (j) Sum of (f)–(i), the modified curve

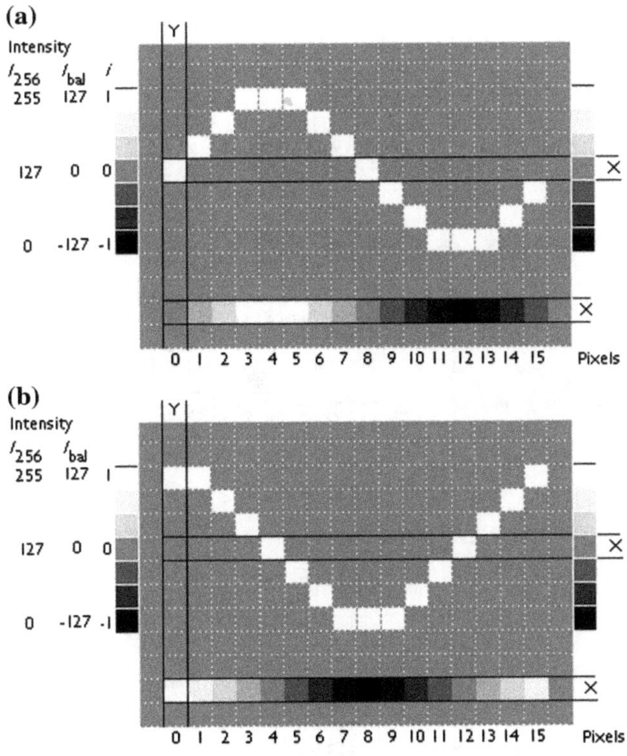

Fig. 40.10 Discrete sinusoidal curves and gratings. **a** Discrete sine curve with amplitude $R = 3$ and wavelength $W = 16$px, and the corresponding sine grating using the intensities shown. **b** Discrete cosine curve with amplitude $r = 3$ and wavelength $W = 16$px, and the corresponding cosine grating using the intensities shown

Table 40.4 Calculated values for Fig. 40.10

X	0	1	2	3	4	5	6	7	8	9	10	11	12	13	14	15
i	0.0	0.38	0.71	0.92	1.0	0.92	0.71	0.38	0.0	−0.38	−0.71	−0.92	−1.0	−0.92	−0.71	−0.38
I_{bal}	0	48	90	116	127	116	90	48	0	−48	−90	−116	−127	−116	−90	−48
I_{256}	127	175	217	243	254	243	217	175	127	79	37	11	0	11	37	79
i	1.0	0.92	0.71	0.38	0.0	−0.38	−0.71	−0.92	−1.0	−0.92	−0.71	−0.38	0.0	0.38	0.71	0.92
I_{bal}	127	116	90	48	0	−48	−90	−116	−127	−116	−90	−48	0	48	90	116
I_{256}	254	243	217	175	127	79	37	11	0	11	37	79	127	175	217	243

Each grating has wavelength $W = 16\,\text{px}$, $X = 0, 1, 2, \ldots, 15$. $i = \sin(360 \times (X/W))$ is normalized intensity between -1 and 1. $I_{bal} = i \times 127$ is integer balanced intensity between -127 and 127; $I_{256} = (i \times 127) + 127$ is 8-bit intensity index between 0 and 255

40.4 8-bit Sinusoidal Grating

We now turn to the Fourier analysis and synthesis of 8-bit grey-scale images.

In the discrete (integer-value) Cartesian domain (X, Y) we can generate a discrete sine curve and a discrete cosine curve as described in Sect. 20.15. Figure 40.10a shows a sine curve generated by program MKSINCUR, where the amplitude is 3 px (the radius of the generating circle) and the wavelength is 16 px. If we normalize the Y-values between -1 and 1, and replace each by a normalized intensity i between Black -1 and White 1, we get a discrete sine grating, as shown. Normalized intensity i corresponds to an integer balanced intensity $I_{\mathrm{bal}} = i \times 127$; and to the 8-bit intensity index $I_{256} = (i \times 127) + 127$. Figure 40.10b shows a cosine curve and cosine grating.

Table 40.4 shows the i, I_{bal}, and I_{256} intensities calculated for Fig. 40.10.

40.5 Harmonic Set of Gratings

We may take any sine grating with wavelength W px and

$$i(X) = \sin(360 \times X/W)$$

as a *fundamental* or *first harmonic*, and derive from it a *harmonic set* of further sine gratings. The fundamental has *frequency* $F = 1$ and normalized wavelength $W = 1/F = 1$.

The *second harmonic*:

$$i(X) = \sin(2 \times 360 \times X/W)$$

has frequency 2 and wavelength 1/2: that is, exactly two cycles for one cycle of the fundamental.

The *third harmonic*:

$$i(X) = \sin(3 \times 360 \times X/W)$$

has frequency 3 and wavelength 1/3: that is, exactly three cycles for one cycle of the fundamental.

And so on, to the *Nth harmonic*:

$$i(X) = \sin(N \times 360 \times X/W)$$

where N is an integer, which has frequency N and wavelength $1/N$.

To complete the harmonic set, we adjoin the *zeroth harmonic*, which has frequency 0, and constant values $i = 1$, $I_{\mathrm{bal}} = 127$, $I_{256} = 255$.

Table 40.5 Harmonic set of sine gratings

X	0	1	2	3	4	5	6	7	8	9	10	11	12	13	14	15
$F0$	1.0	1.0	1.0	1.0	1.0	1.0	1.0	1.0	1.0	1.0	1.0	1.0	1.0	1.0	1.0	1.0
$F1$	0.0	0.38	0.71	0.92	1.0	0.92	0.71	0.38	0.0	−0.38	−0.71	−0.92	−1.0	−0.92	−0.71	−0.38
$F2$	0.0	0.71	1.0	0.71	0.0	−0.71	−1.0	−0.71	0.0	0.71	1.0	0.71	0.0	−0.71	−1.0	−0.71
$F3$	0.0	0.92	0.71	−0.38	−1.0	−0.38	0.71	0.92	0.0	−0.92	−0.71	0.38	1.0	0.38	−0.71	−0.92
$F4$	0.0	1.0	0.0	−1.0	0.0	1.0	0.0	−1.0	0.0	1.0	0.0	−1.0	0.0	1.0	0.0	−1.0
$F5$	0.0	0.92	−0.71	−0.38	1.0	−0.38	−0.71	0.92	0.0	−0.92	0.71	0.38	−1.0	0.38	0.71	−0.92
$F6$	0.0	0.71	−1.0	0.71	0.0	−0.71	1.0	−0.71	0.0	0.71	−1.0	0.71	0.0	−0.71	1.0	−0.71
$F7$	0.0	0.38	−0.71	0.92	−1.0	0.92	−0.71	0.38	0.0	−0.38	0.71	−0.92	1.0	−0.92	0.71	−0.38
$F8$	0.0	0.0	0.0	0.0	0.0	0.0	0.0	0.0	0.0	0.0	0.0	0.0	0.0	0.0	0.0	0.0

The fundamental has wavelength $W = 16$ px. Each row shows normalized intensities i for the harmonic of that frequency

Table 40.6 Harmonic set of cosine gratings

X	0	1	2	3	4	5	6	7	8	9	10	11	12	13	14	15
$F0$	1.0	1.0	1.0	1.0	1.0	1.0	1.0	1.0	1.0	1.0	1.0	1.0	1.0	1.0	1.0	1.0
$F1$	1.0	0.92	0.71	0.38	0.0	−0.38	−0.71	−0.92	−1.0	−0.92	−0.71	−0.38	0.0	0.38	0.71	0.92
$F2$	1.0	0.71	0.0	−0.71	−1.0	−0.71	0.0	0.71	1.0	0.71	0.0	−0.71	−1.0	−0.71	0.0	0.71
$F3$	1.0	0.38	−0.71	−0.92	0.0	0.92	0.71	−0.38	−1.0	−0.38	0.71	0.92	0.0	−0.92	−0.71	0.38
$F4$	1.0	0.0	−1.0	0.0	1.0	0.0	−1.0	0.0	1.0	0.0	−1.0	0.0	1.0	0.0	−1.0	0.0
$F5$	1.0	−0.38	−0.71	0.92	0.0	−0.92	0.71	0.38	−1.0	0.38	0.71	−0.92	0.0	0.92	−0.71	−0.38
$F6$	1.0	−0.71	0.0	0.71	−1.0	0.71	0.0	−0.71	1.0	−0.71	0.0	0.71	−1.0	0.71	0.0	−0.71
$F7$	1.0	−0.92	0.71	−0.38	0.0	0.38	−0.71	0.92	−1.0	0.92	−0.71	0.38	0.0	−0.38	0.71	−0.92
$F8$	1.0	−1.0	1.0	−1.0	1.0	−1.0	1.0	−1.0	1.0	−1.0	1.0	−1.0	1.0	−1.0	1.0	−1.0

The fundamental has wavelength $W = 16$ px. Each row shows intensities i for the harmonic of that frequency

For example, Table 40.5 shows the intensities i for harmonic sine gratings with frequencies from 0 to 8, where the fundamental wavelength is $W = 16$ px. Figure 40.11 shows the corresponding gratings. Similarly, Table 40.6 shows the intensities i for harmonic cosine gratings with frequencies from 0 to 8, where the fundamental wavelength is $W = 16$ px. Figure 40.12 shows the corresponding gratings.

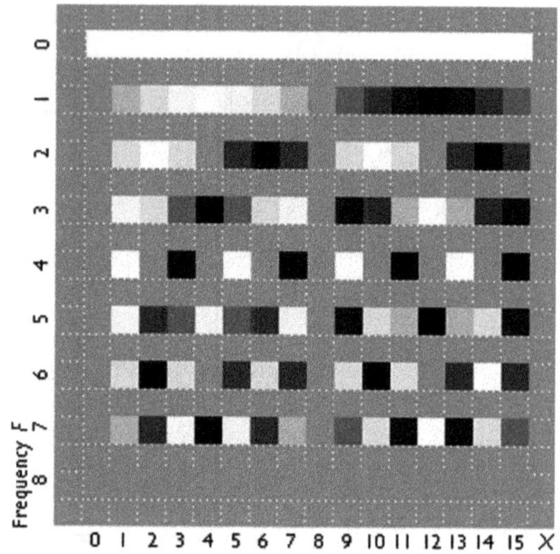

Fig. 40.11 Harmonic set of sine gratings corresponding to Table 40.5. The fundamental wavelength W is 16 px

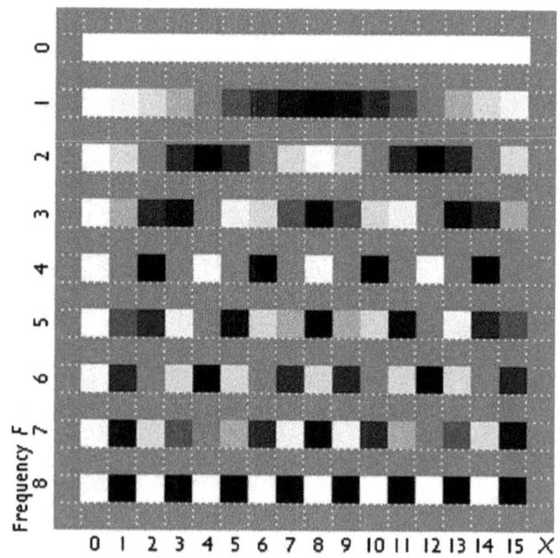

Fig. 40.12 Harmonic set of cosine gratings corresponding to Table 40.6. The fundamental wavelength W is 16 px

Table 40.7 Coefficient of grating

X	0	1	2	3	4	5	6	7	8	9	10	…	15
$c = 2$	0.0	0.76	1.42	1.84	2.0	1.84	1.42	0.76	0.0	−0.76	−1.4	…	−0.76
$c = 1$	0.0	0.38	0.71	0.92	1.0	0.92	0.71	0.38	0.0	−0.38	−0.71	…	−0.38
$c = 0.5$	0.0	0.19	0.35	0.46	0.5	0.46	0.35	0.19	0.0	−0.19	−0.35	…	−0.19
$c = 0$	0.0	0.0	0.0	0.0	0.0	0.0	0.0	0.0	0.0	0.0	0.0	…	0.0
$c = -1$	0.0	−0.38	−0.71	−0.92	−1.0	−0.92	−0.71	−0.38	0.0	0.38	0.71	…	0.38

Intensities i for a sine grating (row 2), with coefficients 2, 1, 0.5, 0, and −1 applied

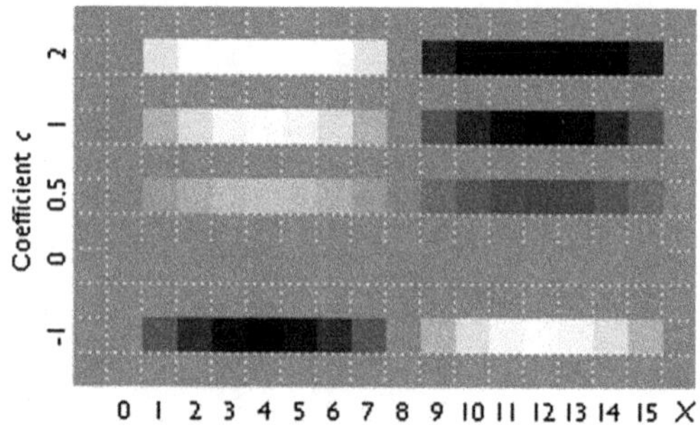

Fig. 40.13 Coefficient of grating, corresponding to Table 40.7. **a** Grating (b) amplified by coefficient 2: contrast is increased. **b** Identity grating, coefficient 1. **c** Grating (b) attenuated by coefficient 0.5: contrast is reduced. **d** Grating (b) nullified by coefficient 0: no contrast at all. **e** Grating (b) inverted by coefficient -1: all the positive intensities become negative and all the negative intensities positive, contrast unchanged

40.6 Coefficient of Grating

We may apply a real *coefficient* c to any discrete grating by multiplying each intensity i, across the entire wavelength, by the coefficient:

$$i(X) = c\sin(X)\,, \quad \text{or,} \quad i(X) = c\cos(X)\,.$$

Evidently,

- when $c > 1$ the intensities are amplified positively and negatively
- when $c = 1$ the intensities are unchanged
- when $1 > c > 0$ the intensities are attenuated positively and negatively
- when $c = 0$ the intensities are all nullified to mid-Grey
- when $c = -1$ the dark intensities are inverted to light and light to dark.

For a display device which has limits $-127 \le I_{\text{bal}} \le 127$, that is $0 \le I_{256} \le 255$, a coefficient may take parts of the grating outside the device limits. In such cases the grating is capped at the device limits.

For example, Table 40.7 shows the intensities i for a sine grating with coefficients 2, 1, 0.5, 0, and -1. Figure 40.13 shows the corresponding gratings.

40.7 Sum of Gratings

We may add any two or more discrete gratings by algebraically-adding corresponding intensities, across the entire wavelength, to get a *sum grating*:

$$i(X) = \sin(X_a) + \sin(X_b)$$

Table 40.8 Sum of gratings

X	0	1	2	3	4	5	6	7	8	9	10	…	15
a	0.0	0.38	0.71	0.92	1.0	0.92	0.71	0.38	0.0	−0.38	−0.71	…	−0.38
b	1.0	0.92	0.71	0.38	0.0	−0.38	−0.71	−0.92	−1.0	−0.92	−0.71	…	0.92
$a+a$	0.0	0.76	1.42	1.84	2.0	1.84	1.42	0.76	0.0	−0.76	−1.42	…	−0.76
$a+b$	1.0	1.30	1.42	1.30	1.0	0.54	0.0	−0.54	−1.0	−1.30	−1.42	…	0.54
$b+b$	2.0	1.84	1.42	0.76	0.0	−0.76	−1.42	−1.84	−2.0	−1.84	−1.42	…	1.84

Intensities i for a sine grating (a), its twin cosine grating (b), and their sums

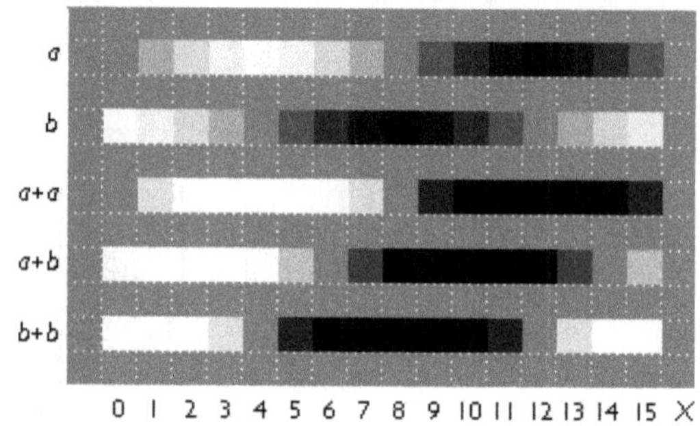

Fig. 40.14 Sum of gratings, corresponding to Table 40.8. **a** Sine grating. **b** Cosine grating. **c** Sum grating (a) + (a) (the same as 2× (a)). **d** Sum grating (a) + (b). **e** Sum grating (b) + (b) (the same as 2× (b)

or with cos in place of either or both sin. For a display device which has limits $0 \leq I_{256} \leq 255$, the sum is capped at the device limits.

For example, Table 40.8 shows the intensities i for a sine grating, its twin cosine grating, and three sum gratings. Figure 40.14 shows the corresponding gratings.

40.8 Product of Gratings

We may multiply any two or more discrete gratings by multiplying corresponding intensities, across the entire wavelength, to get a *product grating*:

$$i(X) = \sin(X_\mathrm{a}) \times \sin(X_\mathrm{b}))$$

or with cos in place of either or both sin. For a display device which has limits $0 \leq I_{256} \leq 255$, the product is capped at the device limits.

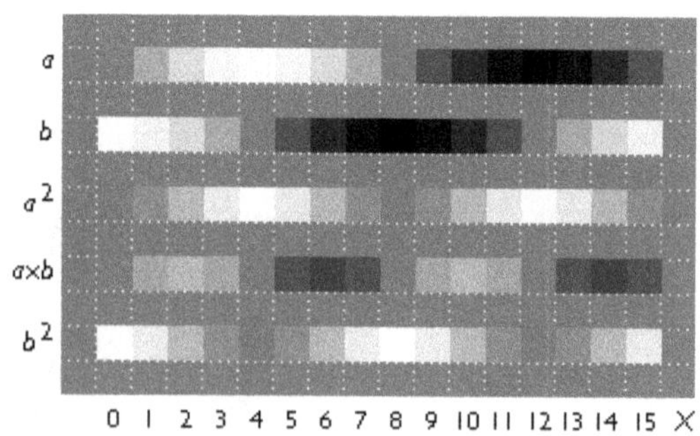

Fig. 40.15 Product of gratings, corresponding to Table 40.9. **a** Sine grating. **b** Cosine grating. **c** Product grating (a)^2: all intensities positive. **d** Product grating (a) + (b). **e** Product grating (b)^2: all intensities positive

Table 40.9 Product of gratings

X	0	1	2	3	4	5	6	7	8	9	10	...	15
a	0.0	0.38	0.71	0.92	1.0	0.92	0.71	0.38	0.0	−0.38	−0.71	...	−0.38
b	1.0	0.92	0.71	0.38	0.0	−0.38	−0.71	−0.92	−1.0	−0.92	−0.71	...	0.92
a^2	0.0	0.14	0.50	0.85	1.0	0.85	0.50	0.14	0.0	0.14	0.50	...	0.14
$a \times b$	0.0	0.35	0.50	0.35	0.0	−0.35	−0.50	−0.35	0.0	0.35	0.50	...	−0.35
b^2	1.0	0.85	0.50	0.14	0.0	0.14	0.50	0.85	1.0	0.85	0.50	...	0.85

Intensities i for a sine grating (a), its twin cosine grating (b), and their products

For example, Table 40.9 shows the intensities i for a sine grating, its twin cosine grating, and three product gratings. Figure 40.15 shows the corresponding gratings.

40.9 Mean Intensity of Grating

The *mean intensity* m of a grating is the algebraic sum of its intensities across the entire fundamental wavelength, divided by the number of pixels in the fundamental wavelength:

$$m = (\Sigma i)/W\,,$$

where W is the wavelength in px, and Σi is the sum of the intensities from $X = 0$ to $X = (W - 1)$.

In a harmonic set of sinusoidal gratings,

- the mean intensity of *each* harmonic grating is zero (as is evident from the symmetries),
- the mean intensity of the *sum* of any two harmonic gratings is zero,
- the mean intensity of the *product* of any two harmonic gratings is zero, with the sole exception of the square of a grating,
- the mean intensity of the *square* of a grating is positive, and is half the square of the amplitude: that is to say, ($2\times$ sum of intensities)/W = amplitude^2.

This exceptional property of the square of a harmonic grating is the key to Fourier analysis.

For example, Table 40.10 shows the mean intensities of the gratings and products in Fig. 40.15.

40.10 Fourier Series of Gratings

A *finite Fourier series* of gratings is the sum grating of all the sine and cosine harmonic gratings in a harmonic set up to the Nth, each amplified, attenuated, nullified or inverted by an independent coefficient:

$$\begin{aligned} i(X) = a_0 + a_1\cos(X) + a_2\cos(2X) + \cdots + a_{\mathrm{N}}\cos(NX) \\ +b_1\sin(X) + b_2\sin(2X) + \cdots + b_{\mathrm{N}}\sin(NX)\,, \end{aligned}$$

where

- $\cos(X)$ is the fundamental cosine grating
- $\cos(2X)$ the second harmonic cosine grating
- ...

Table 40.10 Mean intensity of grating

X	0	1	2	3	4	5	6	7	8	9	10	…	15	Σ	m
a	0.0	0.38	0.71	0.92	1.0	0.92	0.71	0.38	0.0	−0.38	−0.71	…	−0.38	0	0
b	1.0	0.92	0.71	0.38	0.0	−0.38	−0.71	−0.92	−1.0	−0.92	−0.71	…	0.92	0	0
a^2	0.0	0.14	0.50	0.85	1.0	0.85	0.50	0.14	0.0	0.14	0.50	…	0.14	8	0.5
$a \times b$	0.0	0.35	0.50	0.35	0.0	−0.35	−0.50	−0.35	0.0	0.35	0.50	…	−0.35	0	0
b^2	1.0	0.85	0.50	0.14	0.0	0.14	0.50	0.85	1.0	0.85	0.50	…	0.85	8	0.5

Σ is the sum of the intensities; and m is the mean intensity. Row 1 is a sine grating, with intensity sum 0 and mean intensity 0. Row 2 is a cosine grating, with intensity sum 0 and mean intensity 0. Row 3 is a sine grating squared, with intensity sum 8 and mean intensity 0.5 (half the square of the amplitude). Row 4 is the product of a sine grating and a cosine grating, with intensity sum 0 and mean intensity 0. Row 5 is a cosine grating squared, with intensity sum 8 and mean intensity 0.5 (half the square of the amplitude)

- $\cos(NX)$ is the Nth harmonic cosine grating
- a_0 is the coefficient applied to the zeroth (constant) harmonic
- a_1 is the coefficient applied to the first (fundamental) cosine harmonic
- a_2 is the coefficient applied to the second cosine harmonic
- ...
- a_N is the coefficient applied to the Nth cosine harmonic
- $\sin(X)$ is the fundamental sine grating
- $\sin(2X)$ the second harmonic sine grating
- ...
- $\sin(NX)$ is the Nth harmonic sine grating
- b_1 is the coefficient applied to the first (fundamental) sine harmonic
- b_2 is the coefficient applied to the second sine harmonic
- ...
- b_N is the coefficient applied to the Nth sine harmonic.

In writing a Fourier series it is customary to put the cosine terms first, and also to omit from the series any term with coefficient 0.

For example, Table 40.11 shows the intensities i of component and sum gratings of the Fourier series:

$$i(X) = 0.4 + 0.3\cos(X) + 0.2\cos(2X) + 0.1\sin(X)\,.$$

Figure 40.16 shows the corresponding gratings.

40.11 Discrete Fourier Analysis

In the discrete plane of pixels, Fourier's Theorem says that any integer intensity function of X can be approximated as closely as desired by the sum of a finite series of gratings. Thus an arbitrary sequence of pixel intensities can be increasingly closely matched by adding higher and higher harmonics, each with its own coefficient.

The Sampling Theorem says that, if one row of an image contains N px, then the highest frequency discernible is no greater than $N/2$ (called the *Nyquist frequency*[4] or folding frequency). For example, in a given sequence of $N = 16$ px intensities, we may detect frequencies (harmonics) up to 8, but no higher.

The aim of discrete Fourier analysis of a given sequence of N discrete values is to find a Fourier series which matches the given sequence. The Fourier series will have $N+1$ coefficients: $a_0, a_1, \ldots, a_{N/2}, b_1, \ldots, b_{N/2}$ (some of which may be zero). The process is:

- Call the given sequence of values the *target*, with wavelength $W = N$.
- Multiply the target successively by each of the harmonic set of cosines and the harmonic set of sines, wavelength W, frequencies 1 to $W/2$, and save the *products*.

[4]Harry T. Nyquist (1889–1976), ATandT Labs 1917–1934, Bell Telephone Labs 1934–1954.

Table 40.11 Fourier series example

X	0	1	2	3	4	5	6	7	8	9	10	11	12	13	14	15
(0)	0.4	0.4	0.4	0.4	0.4	0.4	0.4	0.4	0.4	0.4	0.4	0.4	0.4	0.4	0.4	0.4
(1)	0.30	0.28	0.21	0.11	0.0	−0.11	−0.21	−0.28	−0.30	−0.28	−0.21	−0.11	0.0	0.11	0.21	0.28
(2)	0.20	0.14	0.0	−0.14	−0.20	−0.14	0.0	0.14	0.20	0.14	0.0	−0.14	−0.20	−0.14	0.0	0.14
(3)	0.0	0.04	0.07	0.09	0.10	0.09	0.07	0.04	0.0	−0.04	−0.07	−0.09	−0.10	−0.09	−0.07	−0.04
(4)	0.90	0.86	0.68	0.46	0.30	0.24	0.26	0.30	0.30	0.22	0.12	0.06	0.10	0.28	0.54	0.78

Intensities i of: row (0): zeroth harmonic grating (constant 1) multiplied by coefficient $a_0 = 0.4$; row (1): fundamental cosine grating multiplied by coefficient $a_1 = 0.3$; row(2): second cosine harmonic multiplied by coefficient $a_2 = 0.2$; row (3) fundamental sine multiplied by coefficient $b_1 = 0.1$. Row (4) shows the sum of the four component gratings in the series

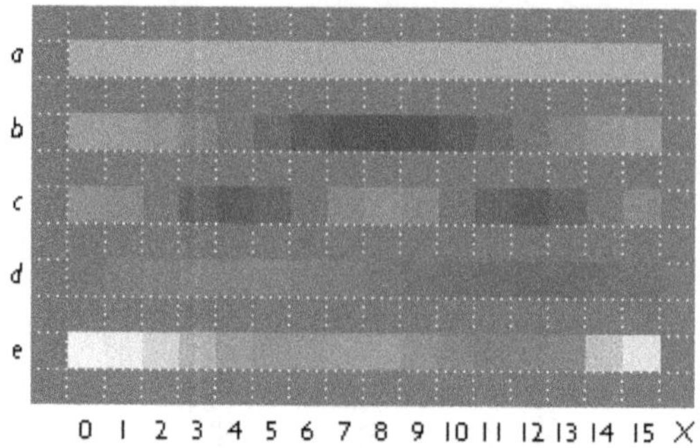

Fig. 40.16 Fourier series example (from Table 40.11). **a** Zeroth harmonic grating (constant 1) multiplied by coefficient $a_0 = 0.4$. **b** Fundamental cosine grating multiplied by coefficient $a_1 = 0.3$. **c** Second cosine harmonic multiplied by coefficient $a_2 = 0.2$. **d** Fundamental sine multiplied by coefficient $b_1 = 0.1$. **e** Sum of the four component gratings in the series

- Coefficient a_0 for the zeroth harmonic (constant 1) is found directly from the target as the mean intensity of the target, that is Σ/N.
- The mean intensity of each product is positive if and only if that harmonic is present in the target (since in that case the product is a square); otherwise the mean intensity is zero. Coefficient $a_1, a_2, \ldots, b_1, b_2, \ldots$ is the twice the mean intensity of its product sequence, that is $2(\Sigma/N)$.

The coefficients $a_0, a_1, \ldots, a_{N/2}$ are the *cosine spectrum* of the given target, specifying the amplitudes of the respective frequencies 0 to $N/2$. Similarly, the coefficients $b_1, \ldots, b_{N/2}$ are the *sine spectrum* of the target.

For example, let us take the sum grating constructed in Table 40.11 row (4) as target, and calculate the coefficients. Table 40.12 shows the first part of the analysis, and Table 40.13 shows the second part of the analysis. The coefficients found by the analysis agree with the construction coefficients.

40.12 Discrete Fourier Synthesis

The aim of discrete Fourier synthesis is, given a spectrum of cosine and sine coefficients, to construct the corresponding grating. If the spectrum has $N+1$ coefficients: $a_0, a_1, \ldots, a_{N/2}, b_1, \ldots, b_{N/2}$ (some of which may be zero), the grating will have up to $N/2$ cosine and $N/2$ sine components. The process is just the inverse of Fourier analysis:

- Multiply the zeroth harmonic (constant 1) by coefficient a_0 across the entire wavelength, and save the products.
- Multiply each cosine harmonic, frequencies 1 to $N/2$, by the corresponding coefficient a_1 to $a_{N/2}$, and save the products.

Table 40.12 Fourier analysis part 1: target and products

	0	1	2	3	4	5	6	7	8	9	10	11	12	13	14	15
a	0.90	0.86	0.68	0.46	0.30	0.24	0.26	0.30	0.30	0.22	0.12	0.06	0.10	0.28	0.54	0.78
b	0.90	0.79	0.48	0.17	0.0	−0.09	−0.43	−0.28	−0.30	−0.20	−0.08	−0.02	0.0	0.11	0.38	0.72
c	0.90	0.61	0.0	−0.33	−0.30	−0.17	0.0	0.21	0.30	0.16	0.0	−0.04	−0.10	−0.20	0.0	0.55
d	0.0	0.33	0.48	0.42	0.30	0.22	0.18	0.11	0.0	−0.08	−0.08	−0.05	−0.10	−0.26	−0.38	−0.30

Row (a) shows the i-values of the target grating. Row (b) shows the products of target multiplied by fundamental cosine grating $i(X) = \cos(X)$. Row (c) shows the products of target multiplied by second cosine harmonic $i(X) = \cos(2X)$. Row (d) shows the products of target multiplied by fundamental sine grating $i(X) = \sin(X)$

Table 40.13 Fourier analysis part 2: calculation of coefficients

	Σ	Mean	2× Mean	*a*	*b*
a	6.40	0.40		0.4	
b	2.15	0.13	0.27	0.3	
c	1.59	0.10	0.20	0.2	
d	0.79	0.05	0.10		0.1

Σ is the sum of the products in the corresponding row of Table 40.12; Mean = Σ/N (here, $N = 16$); 2× Mean is twice the mean; a and b are the rounded coefficients

- Multiply each sine harmonic, frequencies 1 to $N/2$, by the corresponding coefficient b_1 to $b_{N/2}$, and save the products.
- Add the products across the entire wavelength, to get the grating wanted.

For example, see Table 40.11.

40.13 One-Dimensional Discrete Cosine Transformation

Fourier transformation has been widely developed and applied in several fields of science, particularly since the invention of the *Fast Fourier Transform* (FFT)[5] [2] which dramatically reduces computation time. In image processing *Discrete Cosine Transforms* (DCTs) [1, 6] have been of great importance. DCTs use only cosine harmonics, and transform N image pixels into N frequency coefficients. In this Section we show how a one-dimensional DCT works, and in the next Section expand to two dimensions.

A DCT assumes that a given sequence of N px occupies the left half of a cosine grating, full wavelength $2N$, with reflection axes at both ends. Figure 40.17 shows the left half of a cosine curve, with the full range of 256 intensity levels superimposed; and two gratings where $N = 8$ px, one odd and the other even. In an *odd* scheme, the pixel boundaries are at odd multiples of $1/(2N)$, and consequently the intensity levels are at even multiples of $1/(2N)$. In an *even* scheme, the pixel boundaries are at even multiples of $1/(2N)$, and consequently the intensity levels are at odd multiples of $1/(2N)$. Since the half-cosine has reflection axes at each end, in the odd scheme the reflection axes go through the centres of pixels, while in the even scheme they go between pixels.

The so-called *Type II DCT* uses the even scheme. An algorithm to calculate the i-values of a harmonic set of half-cosines with this spacing is

[5]Published in 1965 by Cooley and Tukey; it was later found that Gauss had used essentially the same process in 1805 for astronomical calculations. James W. Cooley (born 1926) was a programmer with von Neumann at Princeton 1953–1956, and with IBM 1962–1991. John W. Tukey (1915–2000) was a statistician at Princeton, Professor 1950–1985; coined the term "bit" 1946, and the term "software" 1958.

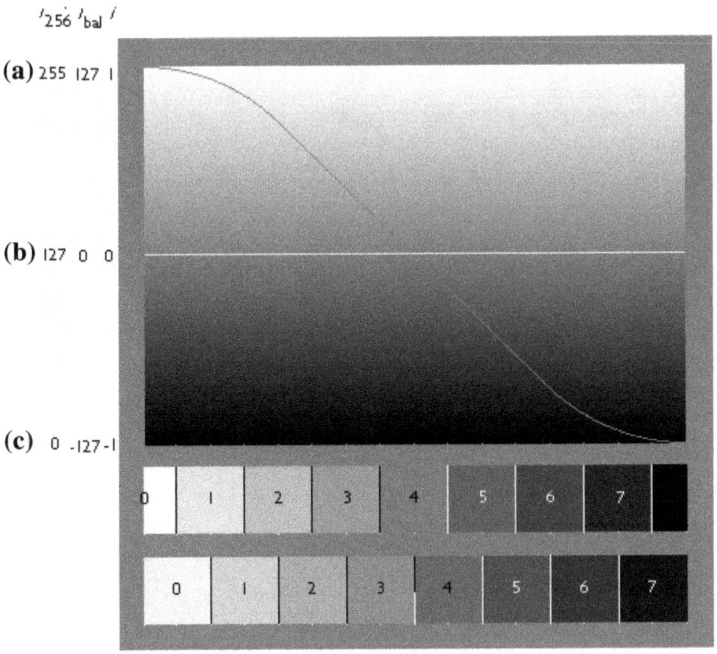

Fig. 40.17 Odd and even schemes of pixels in a DCT. **a** Cartesian domain, with half-cosine curve, full wavelength 720 px and amplitude 127 px; and superimposed on it the 256 8-bit intensity levels (so, here, one pixel per level). **b** Odd half-cosine grating of 8 px. **c** Even half-cosine grating of 8 px. The intensities here are only approximate: for the accurate construction of DCT gratings see the text

```
real array axcosx(0:720)
comment given table of x-to-cos(x) values
real array acoshar(0:n-1, 0:maxf)
comment to hold calculated harmonic gratings i-values
integer maxf,n,q,w
comment maxf is the given highest frequency wanted
comment n is the given pixel count in fundamental
comment    half-cosine
comment q is the given length of quadrant
comment w is the given wavelength of full cosine
comment first initialize
w = 720
q = 180
comment next find dbal values
for k = 1 to maxf
begin
   for j = 0 to n-1
```

```
        begin
           x = cint((q *  ((2*j)+1)  * k)  \ n )
           x = x mod w
           acoshar(j,k)  = axcosx(x)
        end
    end
end
```

To complete the harmonic set, we adjoin the zeroth harmonic, which has frequency 0, and constant intensity $i = 1$.

For example, Table 40.14 shows the intensity values i for $N = 8$-pixel half-cosine harmonics with frequencies $F = 0, 1, 2, \ldots, 7$. This $x_{\max} \times F_{\max}$ array is the *cosine basis function* for the DCT where $N = 8$. Figure 40.18 shows the corresponding gratings.

Table 40.14 Harmonic set of half-cosine gratings

X	0	1	2	3	4	5	6	7
$F = 0$	1.000	1.000	1.000	1.000	1.000	1.000	1.000	1.000
$F = 1$	0.984	0.843	0.535	0.173	−0.173	−0.535	−0.843	−0.984
$F = 2$	0.937	0.354	−0.354	−0.937	−0.937	−0.354	0.354	0.937
$F = 3$	0.843	−0.173	−0.984	−0.535	0.535	0.984	0.173	−0.843
$F = 4$	0.709	−0.709	−0.709	0.709	0.709	−0.709	−0.709	0.709
$F = 5$	0.535	−0.984	0.173	0.843	−0.843	−0.173	0.984	−0.535
$F = 6$	0.354	−0.937	0.937	−0.354	−0.354	0.937	−0.937	0.354
$F = 7$	0.173	−0.535	0.843	−0.984	0.984	−0.843	0.535	−0.173

The gratings have $N = 8$ px, so the fundamental has wavelength $W = 2N = 16$ px, in an even scheme. Values are intensity i

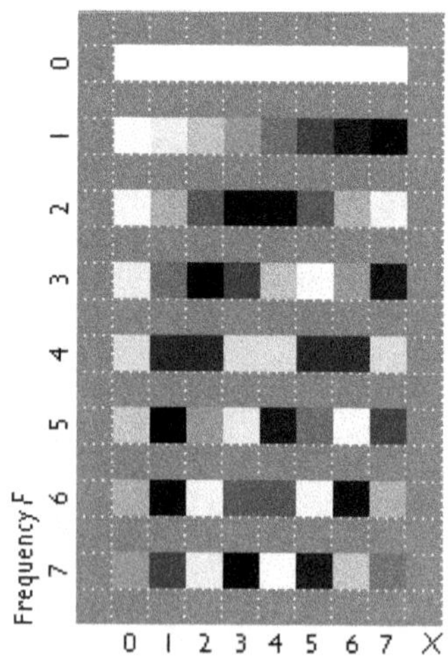

Fig. 40.18 Harmonic set of half-cosine gratings. Values are taken from Table 40.14 for a sequence of $N = 8$ px

The (one-dimensional Type II) *Discrete Cosine Transform* of a given sequence of *N* px with intensities $i_0, i_1, \ldots, i_{N-1}$ is a sequence of *N* px with intensities $i'_0, i'_1, \ldots, i'_{N-1}$. It is produced by the cosine half of the general Fourier analysis:

- Call the given sequence of pixels the *target.*
- Multiply the target successively by each of the harmonic set of cosines, frequencies 0 to $N-1$, and save the *products.*
- Coefficient a_0 for the zeroth harmonic (constant 1) is the mean intensity of the target, that is $\Sigma/N-1$.
- The mean intensity of each product is positive if and only if that harmonic is present in the target (since in that case the product is a square); otherwise the mean intensity is zero. Coefficient $a_1, a_2, \ldots, a_{N-1}$ is the twice the mean intensity of its product sequence, that is $2(\Sigma/N-1)$.

The coefficients $a_0, a_1, \ldots, a_{N-1}$ are the *cosine spectrum* of the given target, specifying the amplitudes of the respective frequencies 0 to $N-1$.

An algorithm for this forward transformation is

```
begin
   real array atarget(0:maxx)
   comment holds the given target image as i-values
   real array acoshs(0:maxx,0:maxf)
   comment holds the calculated cosine harmonic set
   real array aprods(0:maxx,0:maxf)
   comment holds the calculated products
   real array coeff(0:maxf)
   comment holds calculated frequency coeffs of transform
   comment next multiply target by each harmonic
   for f = 0 to maxf
   begin
      for x = 0 to maxx
      begin
         aprods(x,f) = atarget(x) * acoshs(x,f)
      end
   end
   comment next calculate coefficients
   sum = 0
   for x = 0 to maxx
   begin
      sum = sum + aprods(x,0)
   end
   coef = sum / n
   coeff(0) = coef
   for f = 1 to maxf
   begin
      sum = 0
```

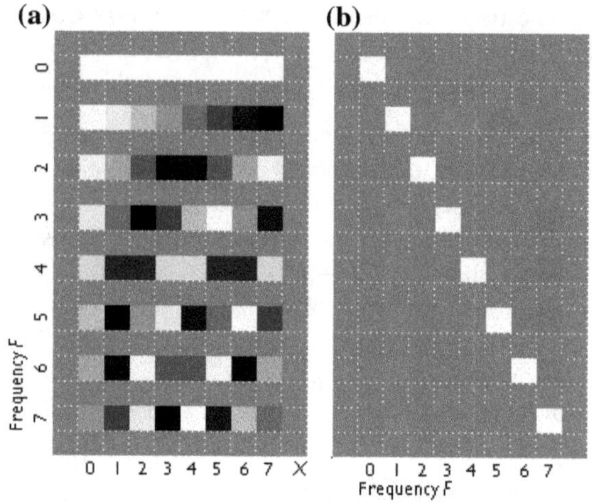

Fig. 40.19 Target images and their transforms. **a** Target images 0–7; chosen as the cosine harmonic gratings. **b** Transforms 0–7; each shows intensity 1 (*White*) in one pixel, and intensity 0 (*Mid-Grey*) elsewhere

```
        for x = 0 to maxx
        begin
            sum = sum + aprods(x,f)
        end
        coef = (2 * sum) / n
        acoeff(f) = coef
    end
end
```

For example, Fig. 40.19 shows eight 8-pixel target images and their transforms. The target images are precisely the eight cosine harmonics, so their respective coefficients in the transform are a single 1 and seven zeros.

For another example, Fig. 40.20 also shows eight 8-pixel target images and their transforms. Here the target images are the Black-to-White inverses of the eight cosine harmonics, so their respective coefficients in the transform are a single -1 and seven zeros.

The (one-dimensional) *Inverse Discrete Cosine Transform* (IDCT) works the DCT backwards. Given a transform spectrum of N px with intensities $i'_0, i'_1, \ldots, i'_{N-1}$, an IDCT restores the original image sequence of N px with intensities $i_0, i_1, \ldots, i_{N-1}$. For a Type II forward DCT the corresponding IDCT is Type III, executing the following steps:

- Call the given sequence of pixels the *spectrum.*
- Multiply the given spectrum successively by each of the harmonic set of cosines, frequencies 0 to $N - 1$, and save the *products.*
- The restored intensity of jth pixel is the sum of the intensities in the jth column of the products, $0 \leq j \leq N - 1$.

An algorithm for a one-dimensional IDCT is

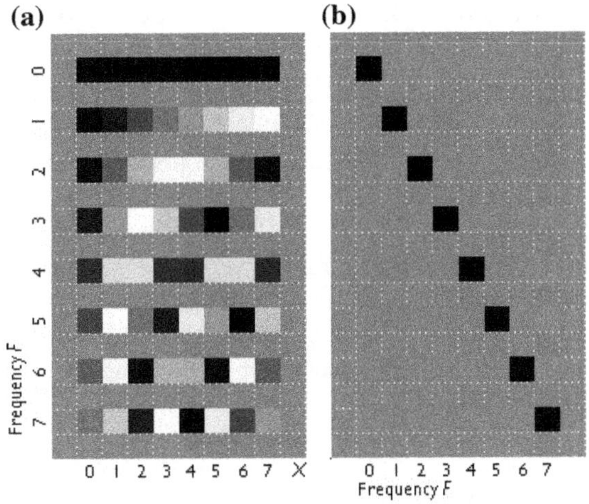

Fig. 40.20 Target images and their transforms. **a** Target images 0–7; chosen as the *Black-to-White* inverses of the cosine harmonic gratings. **b** Transforms 0–7; each shows intensity −1 (*Black*) in one pixel, and intensity 0 (*Mid-Grey*) elsewhere

```
begin
   real array aspectrum(0:maxf)
   comment holds the given transform as i-values
   real array acoshs(0:maxx,0:maxf)
   comment holds the calculated cosine harmonic set
   real array aprods(0:maxx,0:maxf)
   comment holds the calculated products
   real array arestore(0:maxx)
   comment holds calculated intensities of restored image
   comment next multiply spectrum coeff by each harmonic
   for f = 0 to maxf
   begin
      for x = 0 to maxx
      begin
         aprods(x,f) = aspectrum(f) * acoshs(x,f)
      end
   end
   comment next restored intensities
   for x = 1 to maxx
   begin
      sum = 0
      for f = 0 to maxf
      begin
         sum = sum + aprods(x,f)
      end
      arestore(x) = sum
   end
end
```

40.14 Two-Dimensional Discrete Cosine Transformation

A *two-dimensional DCT* is essentially a one-dimensional DCT applied once horizontally and again vertically. If the one-dimensional process transforms a row of N image pixels to a row of N frequency coefficients, the two-dimensional process transforms a block of $N \times N$ image pixels to a block of $N \times N$ frequency coefficients. For example, Fig. 40.21 shows some elementary 8×8-pixel target images, and Fig. 40.22 shows their transforms. The target images are two-dimensional products of the one-dimensional cosine harmonics in Fig. 40.19, so their respective transforms show a single pixel with intensity $i = 1$, and 63 px with intensity zero. But any $N \times N$-pixel image block whatsoever can be matched by combining N horizontal cosine harmonics and N vertical cosine harmonics with intensities specified by the $N \times N$ coefficients in the transform.

A two-dimensional DCT is usually implemented as a matrix calculation (see Sect. 20.10 onwards). For a forward DCT:

- Treat a given $N \times N$ block of image pixels as an $N \times N$ integer *target matrix* **A**.
- Treat the one-dimensional cosine basis function (Sect. 8.13) as a real $N \times N$ *basis matrix* **B**.

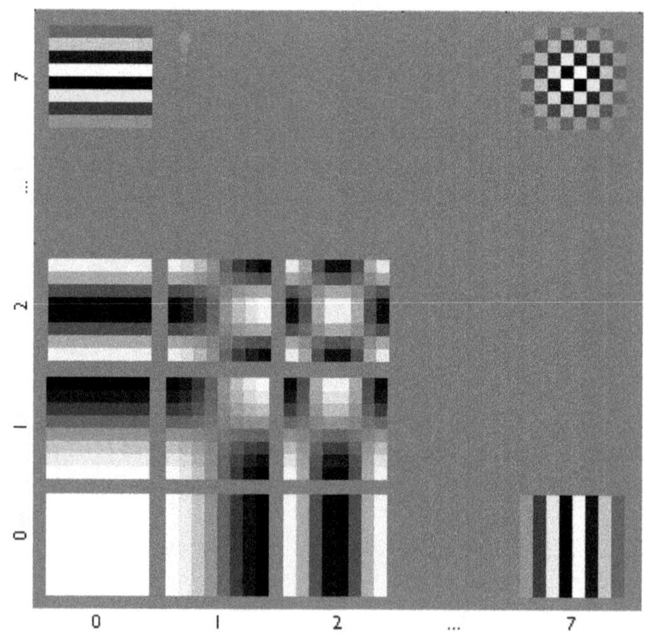

Fig. 40.21 Some elementary target images. Each target image has 8×8 px. Image (0, 0) is the product of cosine harmonic zero (constant intensity 1) with itself. Image (0, 1) is the product of harmonic zero and harmonic one (the fundamental). Image (1, 0) is the product of harmonic zero and harmonic one set vertically. And so on. Figure 40.22 shows the corresponding transforms

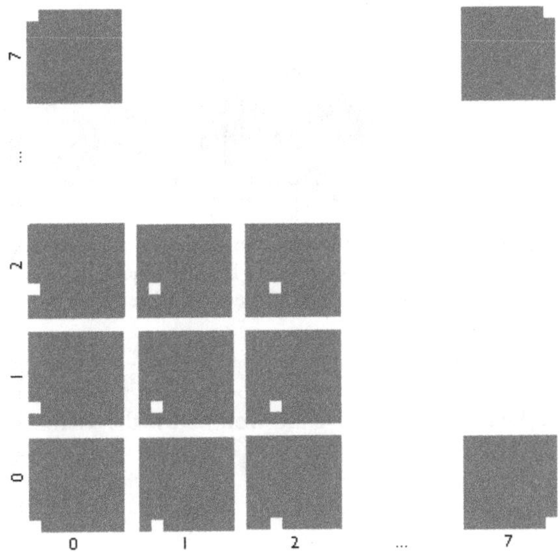

Fig. 40.22 Some elementary transforms. Each transform has 8 × 8 px. Transform (0, 0) corresponds to image (0, 0) in Fig. 40.21, transform (0, 1) to image (0, 1), and so on

- Divide row 1 of **B** (the zeroth harmonic, constant 1) by $1/\surd N$, and rows 2 to N by $\surd(2/N)$, rather than later dividing the products to get the coefficients.
- Pre-multiply target matrix by basis matrix to get products matrix $\mathbf{BA} = \mathbf{P}$.
- Transpose rows and columns of basis matrix to get transposed basis matrix $\mathbf{B}^{\mathrm{T}}$.
- Post-multiply products matrix by transposed basis matrix to get coefficients matrix $\mathbf{PB}^{\mathrm{T}} = \mathbf{C}$.

The $N \times N$ coefficients in **C** are the (two-dimensional) DCT of the given target image. The coefficients are in the range -1023 to 1023. If we wish to display the DCT as an 8-bit spectrum we must divide by 8 to get D_{bal} values in the range -128 to 127, then add 127 to get D values in the range 0 to 255.

A two-dimensional backward IDCT is the inverse of a two-dimensional forward DCT:

- The $N \times N$ coefficients matrix **C** is now the target.
- The basis matrix **B** and the transposed basis matrix $\mathbf{B}^{\mathrm{T}}$ are the same as in the forward process.
- Pre-multiply coefficients matrix by transposed basis matrix to get products matrix $\mathbf{B}^{\mathrm{T}}\mathbf{C} = \mathbf{P}$.
- Post-multiply products matrix by basis matrix to get restored image matrix $\mathbf{PB} = \mathbf{A}$.

The $N \times N$ values in the restored image matrix **A** are I_{bal} values in the range -128–127: add 127 to get I_{256} values in the range 0–255.

For a very simple example, the adjoining printouts show the matrix values for an 8 × 8-pixel image, and its DCT transform, and the IDCT restored image. Notice that there is a small dithering between integer values at successive stages, because

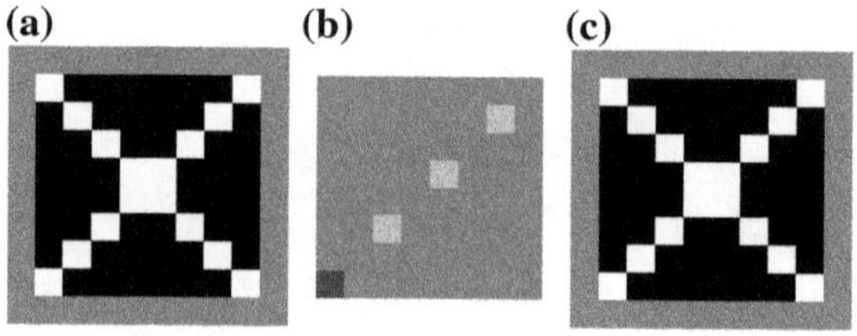

Fig. 40.23 Example of image, transform, and restored image corresponding to adjoining printout. **a** Target image, 8 × 8 px (enlarged 16 times linear). **b** DCT spectrum (where the transform i-values have been divided by 8 then 127 added, to get $D_{\text{bal}} = -64$ and 64, $D = 63$ and 191). **c** IDCT restored image

of the roundings: these are well below the visible threshold. Figure 40.23 shows the corresponding graphics.

```
DCT FROM 8-BIT IMAGE FILE

TARGET MATRIX  Dbal VALUES FROM FILE X8.BMP

         0      1      2      3      4      5      6      7

0      128   -127   -127   -127   -127   -127   -127    128
1     -127    128   -127   -127   -127   -127    128   -127
2     -127   -127    128   -127   -127    128   -127   -127
3     -127   -127   -127    128    128   -127   -127   -127
4     -127   -127   -127    128    128   -127   -127   -127
5     -127   -127    128   -127   -127    128   -127   -127
6     -127    128   -127   -127   -127   -127    128   -127
7      128   -127   -127   -127   -127   -127   -127    128

BASIS MATRIX

         0      1      2      3      4      5      6      7

0   0.354  0.354  0.354  0.354  0.354  0.354  0.354  0.354
1   0.490  0.416  0.278  0.098 -0.098 -0.278 -0.416 -0.490
2   0.462  0.191 -0.191 -0.462 -0.462 -0.191  0.191  0.462
3   0.416 -0.098 -0.490 -0.278  0.278  0.490  0.098 -0.416
4   0.354 -0.354 -0.354  0.354  0.354 -0.354 -0.354  0.354
5   0.278 -0.490  0.098  0.416 -0.416 -0.098  0.490 -0.278
6   0.191 -0.462  0.462 -0.191 -0.191  0.462 -0.462  0.191
7   0.098 -0.278  0.416 -0.490  0.490 -0.416  0.278 -0.098

TRANSPOSED BASIS MATRIX
```

	0	1	2	3	4	5	6	7
0	0.354	0.490	0.462	0.416	0.354	0.278	0.191	0.098
1	0.354	0.416	0.191	-0.098	-0.354	-0.490	-0.462	-0.278
2	0.354	0.278	-0.191	-0.490	-0.354	0.098	0.462	0.416
3	0.354	0.098	-0.462	-0.278	0.354	0.416	-0.191	-0.490
4	0.354	-0.098	-0.462	0.278	0.354	-0.416	-0.191	0.490
5	0.354	-0.278	-0.191	0.490	-0.354	-0.098	0.462	-0.416
6	0.354	-0.416	0.191	0.098	-0.354	0.490	-0.462	0.278
7	0.354	-0.490	0.462	-0.416	0.354	-0.278	0.191	-0.098

TRANSFORM MATRIX i VALUES TO FILE X8T.BMP

	0	1	2	3	4	5	6	7
0	-506	0	-0	0	-0	0	-0	0
1	0	-0	0	-0	0	-0	0	-0
2	-0	0	510	-0	0	-0	0	-0
3	0	-0	-0	0	0	-0	0	-0
4	-0	0	0	0	510	-0	0	-0
5	0	-0	-0	-0	-0	0	0	-0
6	-0	0	0	0	0	0	510	-0
7	0	-0	-0	-0	-0	-0	-0	0

BACKWARD IDCT OF 8-BIT TRANSFORM FILE

TRANSFORM MATRIX i VALUES FROM FILE X8T.BMP

Col=	0	1	2	3	4	5	6	7
0	-504	0	0	0	0	0	0	0
1	0	0	0	0	0	0	0	0
2	0	0	512	0	0	0	0	0
3	0	0	0	0	0	0	0	0
4	0	0	0	0	512	0	0	0
5	0	0	0	0	0	0	0	0
6	0	0	0	0	0	0	512	0
7	0	0	0	0	0	0	0	0

RESTORED IMAGE MATRIX Dbal VALUES TO FILE X8R.BMP

Col= 0 1 2 3 4 5 6 7

0	129	-127	-127	-127	-127	-127	-127	129
1	-127	129	-127	-127	-127	-127	129	-127
2	-127	-127	129	-127	-127	129	-127	-127
3	-127	-127	-127	129	129	-127	-127	-127
4	-127	-127	-127	129	129	-127	-127	-127
5	-127	-127	129	-127	-127	129	-127	-127
6	-127	129	-127	-127	-127	-127	129	-127
7	129	-127	-127	-127	-127	-127	-127	129

The given target image may be any integral multiple (zero-padded if necessary) of the block size $N \times N$ set for the DCT, and the DCT is applied again and again across the image. For example, Fig. 40.24 shows a target image of 16×16 px, the DCT spectrum resulting from four blocks of 8×8 px, and the IDCT restored image. In comparison, Fig. 40.25 shows the same target image of 16×16 px, the DCT spectrum resulting from a single block of 16×16 px, and the IDCT restored image. The restored images from the two processes are indistinguishable.

The given target image may be any integral multiple (zero-padded if necessary) of the block size $N \times N$ set for the DCT, and the DCT is applied again and again across the image. When N is small relative to target size, the breaks between adjoining blocks are usually not visible. For example, Fig. 40.26 shows a target image of 128×128 px, the DCT spectrum resulting from 256 blocks of 8×8 px, and the IDCT

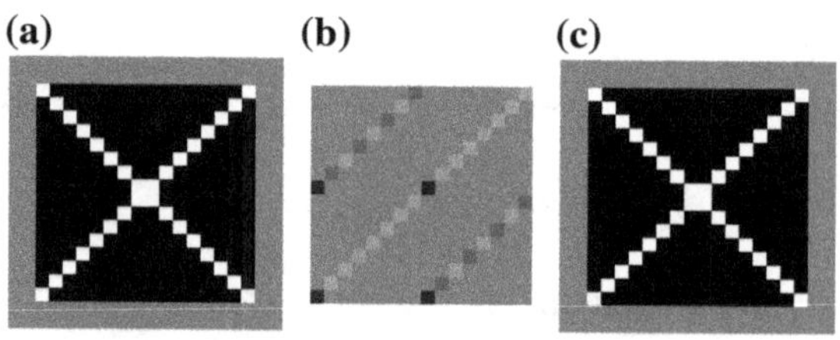

Fig. 40.24 Example of image, fourfold DCT, and restored image. **a** Target image of 16×16 px (enlarged eight times linear). **b** DCT spectrum resulting from four blocks of 8×8 px (where the transform values have been divided by 8 then 127 added, to get $D_{\text{bal}} = -64$ and 64, $D = 63$ and 191). **c** IDCT restored image

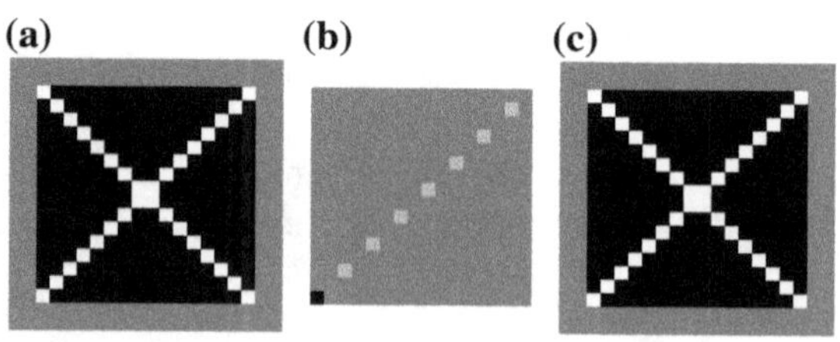

Fig. 40.25 Example of image, single DCT, and restored image. **a** Target image of 16×16 px, the same as in Fig. 40.24. **b** DCT spectrum resulting from a single block of 16×16 px. **c** IDCT restored image, indistinguishable from Fig. 40.24

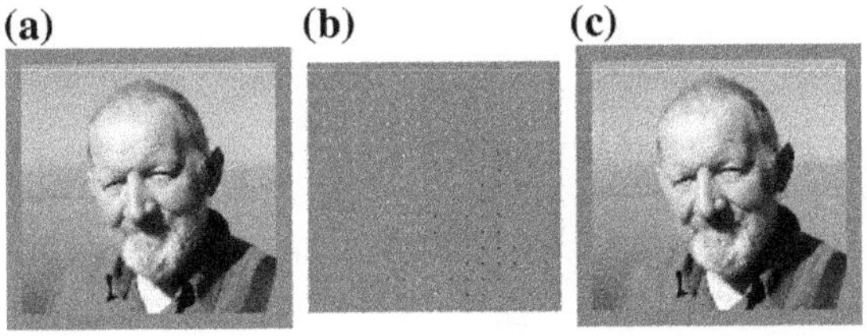

Fig. 40.26 Example of image, small-block DCT, and restored image. **a** Target image of 128 × 128 px (natural size). **b** DCT spectrum resulting from 256 blocks of 8 × 8 px. **c** IDCT restored image

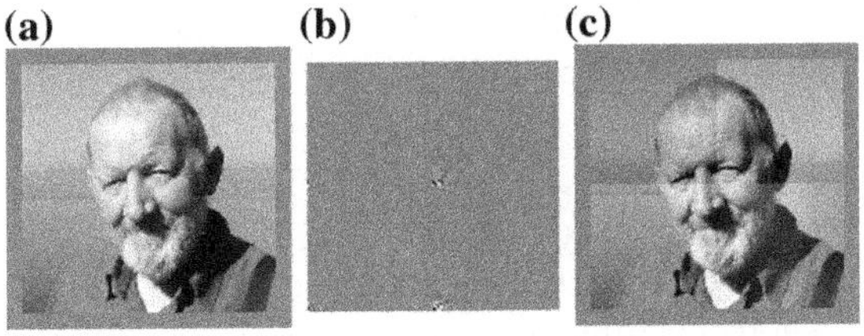

Fig. 40.27 Example of image, large-block DCT, and restored image. **a** Target image of 128 × 128 px, the same as in Fig. 40.26. **b** DCT spectrum resulting from a four blocks of 64 × 64 px. **c** IDCT restored image, showing breaks between blocks

restored image. The restored image is very close to the target. However, when N is large relative to target size, the breaks between blocks will generally become visible. Figure 40.27 shows the same target image of 128 × 128 px, the DCT spectrum resulting from four blocks of 64 × 64 px, and the IDCT restored image. The restored image shows visible breaks between the four blocks.

The quality of the synthesis depends on the highest frequency of component in the analysis. Figure 40.28 shows a source image, analysed then synthesised with highest frequency 4 and 16 and 64.

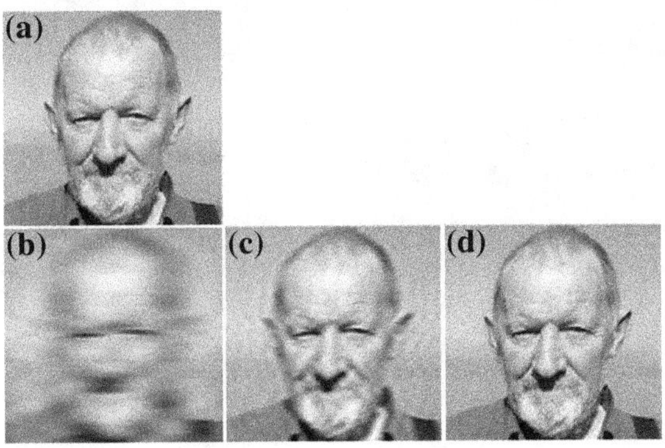

Fig. 40.28 Example of image analysed and restored with different frequency ceilings. **a** Target image of 128 × 128 px. **b** Restored image using highest frequency 4. **c** Restored image using highest frequency 16. **d** Restored image using highest frequency 63

40.15 Frequency Transformation

A Fourier analysis captures nearly all of the visual information of an image as a frequency spectrum, the coefficients of a sum of sinusoidal gratings belonging to a harmonic set, with frequencies from 0 to N. We then have the opportunity to change the coefficients of the spectrum, and synthesize a new image. We thus have a class of *frequency transformations*, quite distinct from the location transformations and the intensity transformations of Chap. 39.

A straightforward way to do this is to construct a *filter block* containing $N \times N$ factor values, and apply it to the successive $N \times N$ blocks of the DCT by multiplying each frequency coefficient by the corresponding filter factor.

- A *binary filter* has factor values 0 or 1, so either stops (0) or passes unchanged (1) the corresponding frequency.
- An *integer filter* has integer factor values 0 to M, so stops (0) or passes unchanged (1) or amplifies (>1) the corresponding frequency.
- A *real filter* has real factor values 0 to m, so stops (0) or attenuates (<1) or passes unchanged (1) or amplifies (>1) the corresponding frequency.

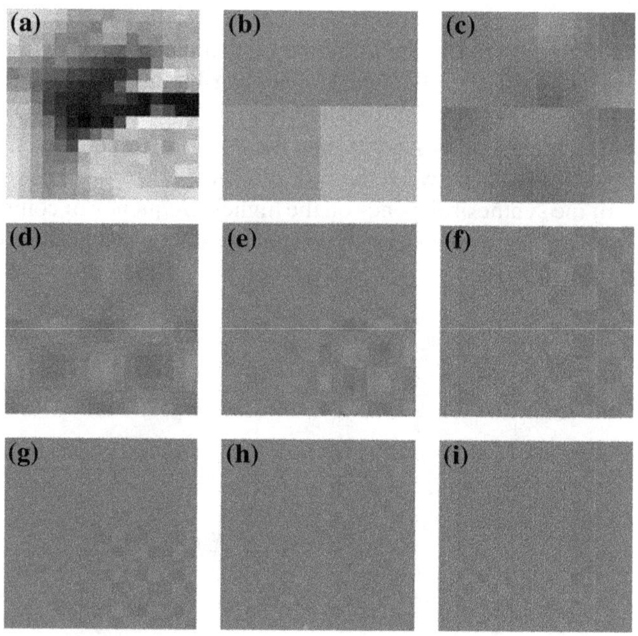

Fig. 40.29 An image and its separate frequency components. **a** Image of 16×16 px, enlarged 8 times linear. **b** The single zero-frequency component of the DCT where $N = 8$; four blocks each showing the block-average intensity. **c** The single component of frequency 1. **d–f** The single components of frequencies 2, 3, and 4. **g–i** The single components of frequencies 5, 6, and 7

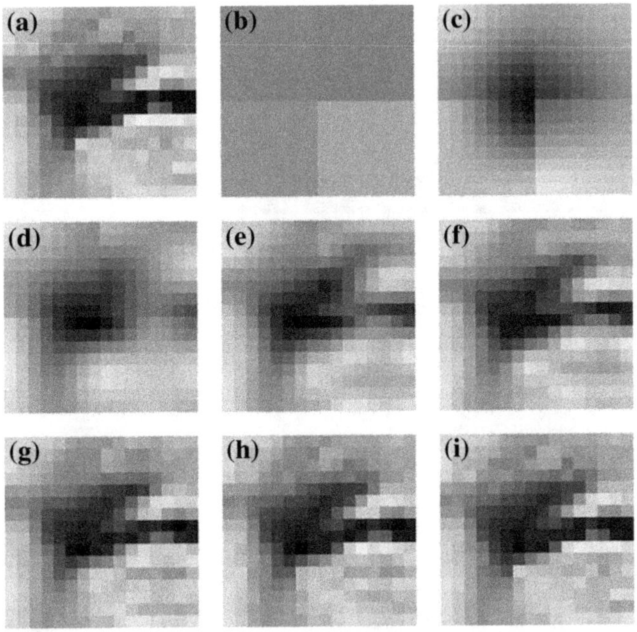

Fig. 40.30 An image and its cumulative frequency components. **a** Image of 16 × 16 px, enlarged 8 times linear. **b** The single zero-frequency component of the DCT where $N = 8$; four blocks each showing the block-average intensity. **c** The combined components of frequency 0 and frequency 1. **d–f** The combined components of frequencies 0–2, 0–3, and 0–4. **g–i** The combined components of frequencies 0–5, 0–6, and 0–7

We can then separate out one or more frequency components of an image. For example, Fig. 40.29 shows an image of 16 × 16 px taken from the middle of the source image, enlarged 8 times linear for clarity. The DCT of this image contains four blocks of 8 × 8 px, and each block contains eight components, with frequencies 0–7. The IDCT image of each frequency component is shown separately.

Figure 40.30 shows the corresponding cumulation of the frequency components. The zero-frequency component[6] is just the average intensity of the block, with no internal detail at all. The first, second, and subsequent components modify the zero-frequency component to give increasing detail. In most images the low frequencies carry most of the visual information.

Figure 40.31 shows syntheses from various groups of frequencies.

[6]Often called the *DC* component, following a usage from electrical signal-processing.

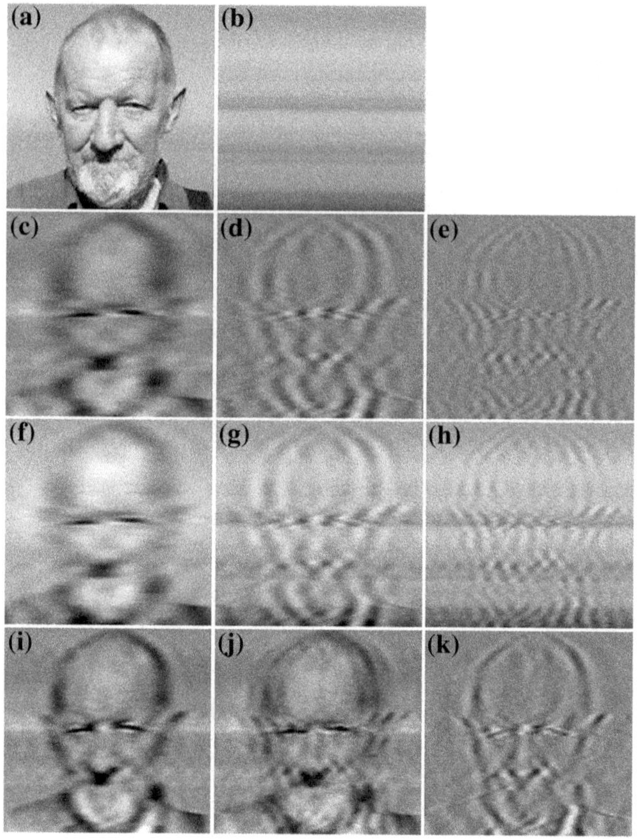

Fig. 40.31 An image and groups of frequency components. **a** Image of 128 × 128 px. **b** Synthesis from zero-frequency component alone. **c** Synthesis from low-frequency components. **d** Synthesis from mid-frequency components. **e** Synthesis from high-frequency components. **f** Synthesis from zero- plus low-frequency components. **g** Synthesis from zero- plus mid-frequency components. **h** Synthesis from zero- plus high-frequency components. **i** Synthesis from low- plus mid-frequency components. **j** Synthesis from low- plus high-frequency components. **k** Synthesis from mid- plus high-frequency components

40.16 Programs for Fourier Analysis and Synthesis

A pair of programs which analyse and synthesize 8-bit grey-scale images are briefly described below, with source code on the Springer web-site. Figure 40.1 is a flowchart for these programs. All the examples in this chapter, and some in Chap. 25, were produced by these programs.

```
REM  **  Program FOURIANA to analyze 8-bit image as one
REM  **  Fourier series per row. User inputs filenames
```

```
REM   **   #1 for image to be analyzed, #2 for coefficients
REM   **   binary output. Program reads headers of
REM   **   file #1, shows particulars, asks highest
REM   **   frequency to process. Program dimensions
REM   **   arrays, and generates COS and SIN harmonic
REM   **   sets. Then for each row, program  reads
REM   **   image intensities into target, multiplies
REM   **   target by each harmonic into product,
REM   **   calculates coefficients, and writes row
REM   **   of file #2.
REM   **   Main program
REM   **   Enter input and output file-names
           CALL REAHEADS
           CALL WRIHEADS
           CALL CALCFUND
           CALL CALCHSET
           CALL ANALYZE
END

           SUB ANALYZE
REM   **   Subprogram ANALYZE  to find Fourier frequency
REM   **      coefficients A0,A1,.,B1,B2,.. For each row
REM   **      of image, repeat: read row from file #1 into
REM   **      ATARGET; multiply by each cos and sin harmonic
REM   **      into ACOSPROD and ASINPROD; calculate coeffs
REM   **      into AACOEFF and ABCOEFF; write coeffs to
REM   **      binary file #2.
END SUB
           SUB CALCFUND
REM   **   Bresenham's circle algorithm adapted to generate
REM   **      fundamental cosine grating into array
REM   **      ACOSHS(0 TO MAXX%, 1) and fundamental sine
REM   **      grating into array ASINHS(0 TO MAXX%, 1)
END SUB
           SUB CALCHSET
REM   **   Subprogram CALCSET to calculate sine harmonics
REM   **      2 to MAXF from fundamental in array ASINHS,
REM   **      and cosine harmonics from 2 to MAXF from
REM   **      fundamental in array ACOSHS. FREPTR% points
REM   **      to frequency F; FUNPTR% points to X of
REM   **      fundamental; HARPTR% points to X of F-harmonic.
END SUB
           SUB PUTI (INTE%)
REM   **   Puts integer value passed into binary file
REM           #2 as 2 bytes and increments  pointer.
```

```
          PUT$ = MKI$(INTE%)
          PUT #2, SEEK(2), PUT$
END SUB
          SUB REAHEADS
REM  **   Subprogram REAHEADS to read header of input
REM  **      image file #1.
END SUB
          SUB WRIBINOUT
REM  **   Subprogram  WRIBINOUT to write real coefficients
REM  **      for row to binary file #2.
END SUB
          SUB WRIHEADS
REM  **   Subprogram  WRIHEADS to write header of output
REM  **      file #2 (binary .FFC).
REM  **   Wavelength in pixels, as integer
END  SUB $
```

```
REM  **   Program FOURISYN to synthesize 8-bit image from
REM  **      one Fourier series per row. User inputs
REM  **      filenames #1 for .FFC file to be synth-
REM  **      esized, #2 for .BMP image file output,
REM  **      #3 for text output. Program reads headers
REM  **      of file #1, shows particulars, asks for
REM  **      transformation parameters. Program dimen-
REM  **      sions arrays, and generates COS and SIN
REM  **      harmonic sets. Then for each row, program
REM  **      reads coefficients into AACOEFF and
REM  **      ABCOEFF, multiplies harmonics each by coeffs,
REM  **      finds sums for image, and writes image row
REM  **      to file #2.
REM  **   Main program
REM  **   Enter input and output file-names
          CALL REAFFCHD
          CALL WRIBMPHD
          CALL CALCFUND
          CALL CALCHSET
             CALL SYNTHESI
          CALL REVBMPHD
END

          SUB CALCFUND
REM  **   Bresenham's circle algorithm adapted to generate
REM  **      fundamental cosine grating into array
REM  **      ACOSHS(0 TO MAXX%, 1) and fundamental sine
```

```
REM  **     grating into array ASINHS(0 TO MAXX%, 1)
END SUB
        SUB CALCHSET
REM  **  Subprogram CALCHSET to calculate sine harmonics
REM  **     2 to MAXF from fundamental in array ASINHS,
REM  **     and cosine harmonics from 2 to MAXF from
REM  **     fundamental in array ACOSHS. FREPTR% points
REM  **     to frequency F; FUNPTR% points to X of
REM  **     fundamental; HARPTR% points to X of F-harmonic.
END SUB
        SUB REAFFCHD
REM  **  Subprogram REAFFCHD to read header of .FFC input
REM  **     file #1. END SUB
        SUB REVBMPHD
REM  **  Subprogram REVBMPHD to revise header of .BMP
REM  **     file #2 with values in bytes.
REM  **  Next update bfSize length of file in bytes, as long
END SUB
        SUB SYNTHESI
REM  **  Subprogram SYNTHESI to find image intensities
REM  **     from Fourier frequency coefficients A0,A1,..,
REM  **     B1,B2,... For each set of coeffs (row), repeat:
REM  **     read coeffs from file #1 into AACOEFF and
REM  **     ABCOEFF; multiply by corresponding cos and
REM  **     sin harmonics into ACOSPROD and ASINPROD;
REM  **     calculate column sums into AIMROW; write
REM  **     image row to .BMP file #2.
END SUB
        SUB WRIBMPDA
REM  **  Subprogram WRIBMPDA to write 8-bit image row
REM  **     data to .BMP output file #2.
END SUB
        SUB WRIBMPHD
REM  **  Subprogram to write header and 8-bit colour-table
REM  **  of output file #2 (to be revised later).
END SUB
```

From here, the way is open to explore frequency manipulations of all kinds of 8-bit images created by geometry or scanner or camera, and to combine such manipulations with intensity transformations, pixel-count reductions, and half-toning.

40.17 JPG Compression and Coding

One of the great technological achievements in digital imaging is the recommendation of the Joint Photographic Experts Group [5] for digital compression and coding of images [4]. This combines Fourier analysis, psycho-physical observations, and information coding theory to provide a method of compressing an image file to much smaller size with remarkably little loss.

References

1. Ahmed N, Natarajan T, Rao KR (1974) Discrete Cosine Transform. IEEE Trans Comput C-3(1):90–93. http://ieeexplore.ieee.org/xpl/articleDetails.jsp?arnumber=1672377. Accessed 1 Feb 2014
2. Cooley JW, Tukey JW (1965) An algorithm for the machine calculation of complex Fourier series. Math Comput 19:297–301
3. Fourier JBJ (1822) Théorie Analytique de la Chaleur. Didot, Paris
4. ISO Standard 10918-1 (1994) Digital compression and coding of continuous-tone still images. http://www.iso.org/. Accessed 1 Feb 2014
5. Joint Photographic Experts Group (2014) http://www.jpeg.org/index.html. Accessed 1 Feb 2014
6. Rao KR, Yip P (1990) Transform algorithms, advantages, applications. Academic Press, Boston

Chapter 41
8-bit File

In an 8-bit BMP file, each pixel has an 8-bit index 0–255. The index points to a colour-table with 256 entries. An 8CO colour-table holds 256 colours, an 8GM colour-table 256 grey-matches to colours, and an 8GS colour-table a complete range of 256 greys between Black and White. The format of an 8-bit BMP file is defined for the three colour-tables, with examples. A program to edit 8-bit files is given.

41.1 Flowchart

The flowchart Fig. 41.1 is for general guidance in this chapter.

41.2 Files and Data

For Memory, File, Data Representation, Bitmap Scan, and .BMP File, see Chap. 26.

41.3 8-bit .BMP File with Colour-table 8CO

Table 41.1 shows the structure of an 8-bit .BMP file with colour-table 8CO.

In a .BMP file the fields are in four groups. The FileHeader and the InfoHeader groups give the particulars of the file. The first field is bfType, ASCII BM = 66 77 decimal = 42 4D hexadecimal. Without this the file is rejected as improper. The biXPelsPerMeter and biYPelsPerMeter fields are used for printing (see Sect. 28.3). The biClrUsed field = 0 for an 8CO colour-table.

A. Parkin, *Digital Imaging Primer*, DOI 10.1007/978-3-540-85619-1_41

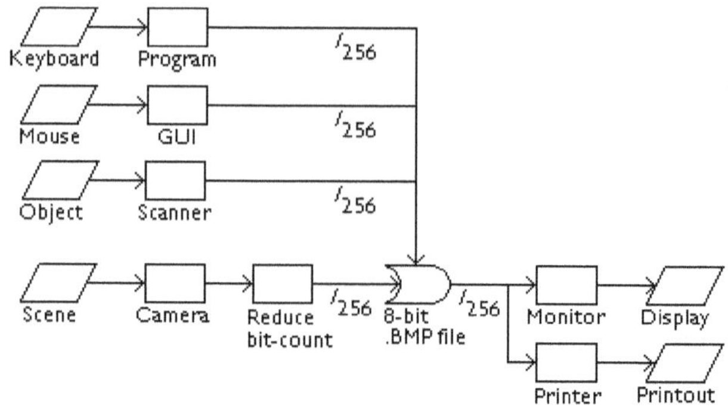

Fig. 41.1 An 8-bit .BMP file can be created, or changed, by a program or by a GUI image editor. It can be reduced from a 24-bit file (which has been previously created by program, GUI editor, scanner or camera. It can be displayed or printed-out by a GUI editor. It acts as an interface between inputs, processes, and outputs

In an 8-bit file the ColorTable group contains 256 fields, specifying the Blue, Green, Red intensities for colours 0–255. A display screen has three colour elements per pixel, Blue, Green, and Red; and each element has 256 possible intensities, from 0 off to 255 full on. The screen shows Black when all three are off, and shows White when all three are full on. This is customarily called an RGB colour system, though the sequence is reversed to BGR in a .BMP file. Each of the 256 fields occupies a four-byte *rgbquad*. The first byte is the Blue intensity, the second byte is the Green intensity, the third byte is the Red intensity, and the fourth byte is empty. An 8CO colour-table (see Chap. 36 and Sects. 38.3–38.5) contains Black, Dark red, Dark green, ..., White.

The image data group contains the colour index values, 0 to FF hexadecimal, of the pixels taken in scan sequence bottom left to top right of the image. The index values occupy one 8-bit byte each, and point to the colour-table, which sets the actual colours.

Figure 41.2 shows an example of an 8-bit .BMP file with colour-table 8CO. The image is 14 × 2 px, shown here much enlarged. The decimal index values and colour names have been added. The file contents are seen in the hexadecimal dump.

41.4 8-bit .BMP File with Colour-table 8GM

Table 41.2 shows the structure of an 8-bit .BMP file with colour-table 8GM.

In an 8GM file the FileHeader is the same as a CO8 file. The InfoHeader is the same, excepting the biClrUsed field which contains 100 h = 256 d.

An 8GM colour-table (see Chap. 36 and Sects. 38.3–38.5) contains a complete grey-scale from Black (0, 0, 0) to White (FF, FF, FF)h = (255, 255, 255)d.

Table 41.1 8-bit 8CO file format

Group	Field	Type	dec.	hex.	Remarks
FileHeader			1–14	01–0E	File type, properties
	bfType	integer	1–2	01–02	Always ‘BM’
	bfSize	long	3–6	03–06	File length, bytes
	bfReserved1	integer	7–8	07–08	Always 0
	bfReserved2	integer	9–10	09–0A	Always 0
	bfOffBits	long	11–14	0B–0E	Header length, bytes
InfoHeader			15–54	0F–36	Image size, properties
	biSize	long	15–18	0F–12	Infoheader leng, bytes
	biWidth	long	19–22	13–16	Image width, pixels
	biHeight	long	23–26	17–1A	Image height, pixels
	biPlanes	integer	27–28	1B–1C	Always 1
	biBitCount	integer	29–30	1D–1E	Bits per pixel = 8
	biCompression	long	31–34	1F–22	Not compr. = 0
	biSizeImage	long	35–38	23–26	Image data leng, bytes
	biXPelsPerMeter	long	39–42	27–2A	Printer p.p.m., horiz.
	biYPelsPerMeter	long	43–46	2B–2E	Printer p.p.m., vert.
	biClrUsed	long	47–50	2F–32	0 for CO8
	biClrImportant	long	51–54	33–36	Colours important
ColorTable			55–1078	37–436	256 colours for CO8
	aColors[0]	rgbquad	55–58	37–3A	Black: 0 0 0 0
	aColors[1]	rgbquad	59–62	3B–3E	Dark red: 0 0 128 0
	aColors[2]	rgbquad	63–66	3B–3E	Dark green: 0 128 0 0
	...				
	aColors[255]	rgbquad	1075–1078	433–436	White: 255 255 255 0
Image data			1079–	437–	In scan sequence
	aBitmapBits[0]	byte	1079–	437–	One byte per pixel
	aBitmapBits[1]	byte	–	–	
	...				
	aBitmapBits[*n*]	byte	–	–	

Field types are: *byte* occupying eight bits; *integer* occupying two bytes; *long* occupying four bytes; *rgbquad* occupying four bytes. Notice that rgbquad holds the colour values in the first three bytes, in sequence BGR, and the fourth byte is zero Z

Table 41.2 8-bit 8GM file format

Group	Field	Type	dec.	hex.	Remarks
FileHeader			1–14	01–0E	File type, properties
	bfType	integer	1-2	01–02	Always 'BM'
	bfSize	long	3–6	03–06	File length, bytes
	bfReserved1	integer	7–8	07–08	Always 0
	bfReserved2	integer	9–10	09–0A	Always 0
	bfOffBits	long	11–14	0B–0E	Header length, bytes
InfoHeader			15–54	0F–36	Image size, properties
	biSize	long	15–18	0F–12	Infoheader leng, bytes
	biWidth	long	19–22	13–16	Image width, pixels
	biHeight	long	23–26	17–1A	Image height, pixels
	biPlanes	integer	27–28	1B–1C	Always 1
	biBitCount	integer	29–30	1D–1E	Bits per pixel = 8
	biCompression	long	31–34	1F–22	Not compr. = 0
	biSizeImage	long	35–38	23–26	Image data leng,bytes
	biXPelsPerMeter	long	39–42	27–2A	Printer p.p.m., horiz.
	biYPelsPerMeter	long	43–46	2B–2E	Printer p.p.m., vert.
	biClrUsed	long	47–50	2F–32	256 for 8GM
	biClrImportant	long	51–54	33–36	Colours important
ColorTable			55–1078	37–436	256 colours for 8GM
	aColors[0]	rgbquad	55–58	37–3A	Black: 0 0 0 0
	aColors[1] .	rgbquad	59–62	3B–3E	Grey-m 1: 38 38 38 0
	aColors[2]	rgbquad	63–66	3B–3E	Grey-m 2: 76 76 76 0
	...				
	aColors[255]	rgbquad	1075–1078	433–436	White: 255 255 255 0
Image data			1079–	437–	In scan sequence
	aBitmapBits[0]	byte	1079–	437–	One byte per pixel
	aBitmapBits[1]	byte	–	–	
	...				
	aBitmapBits[*n*]	byte	–	–	

Field types are: *byte* occupying eight bits; *integer* occupying two bytes; *long* occupying four bytes; *rgbquad* occupying four bytes. Notice that rgbquad holds the colour values in the first three bytes, in sequence BGR, and the fourth byte is zero Z

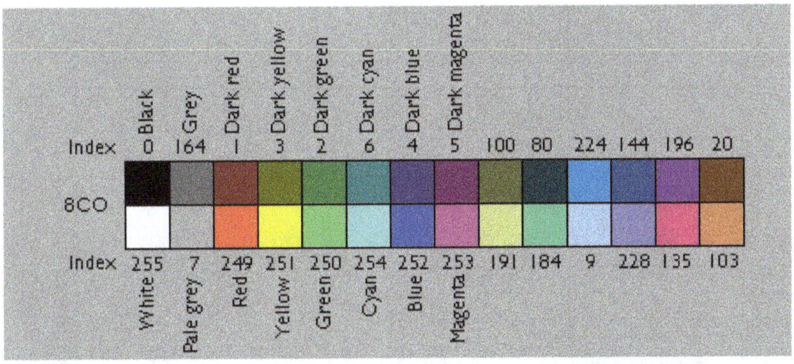

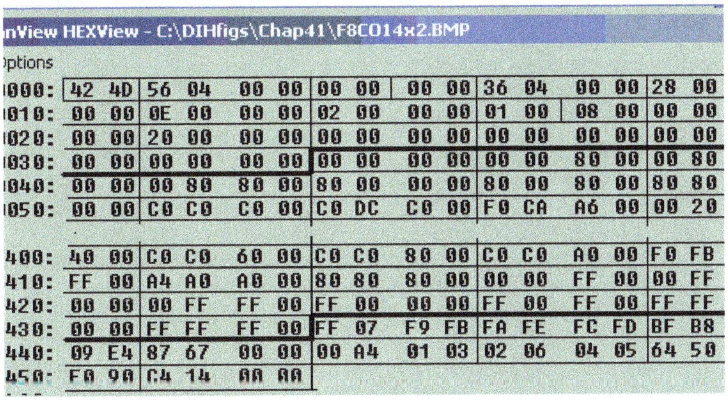

nView HEXView - C:\DIHfigs\Chap41\F8CO14x2.BMP

ptions

```
000: 42 4D 56 04 00 00 00 00 00 00 36 04 00 00 28 00
010: 00 00 0E 00 00 00 02 00 00 00 01 00 08 00 00 00
020: 00 00 20 00 00 00 00 00 00 00 00 00 00 00 00 00
030: 00 00 00 00 00 00 00 00 00 00 00 00 80 00 00 80
040: 00 00 00 80 80 00 80 00 00 00 80 00 80 00 80 80
050: 00 00 C0 C0 C0 00 C0 DC C0 00 F0 CA A6 00 00 20

400: 40 00 C0 C0 60 00 C0 C0 80 00 C0 C0 A0 00 F0 FB
410: FF 00 A4 A0 A0 00 80 80 80 00 00 00 FF 00 00 FF
420: 00 00 00 FF FF 00 FF 00 00 00 FF 00 FF 00 FF FF
430: 00 00 FF FF FF 00 FF 07 F9 FB FA FE FC FD BF B8
440: 09 E4 87 67 00 00 00 A4 01 03 02 06 04 05 64 50
450: F0 90 C4 14 00 00
```

Fig. 41.2 8-bit 8CO file. The image is 14 × 2 px (shown here with decimal index values and some colour names added). The file is shown as a hexadecimal dump, where each pair of digits is the hex value of a byte. In a multi-byte field the hex pairs are taken as most-significant right, least-significant left. The file-header fields are: a 2-byte magic cookie field containing 42 4D = ASCII BM; a 4-byte file-size field containing 456 h = 1110 d; two 2-byte fields (ignored); a 4-byte header-length 436 h = 1028 d; a 4-byte info-header-length 28 h = 40 d; a 4-byte image-width Eh = 14 d; a 4-byte image-height 2; a 2-byte planes field = 1; a 2-byte bpp field = 8; a 4-byte compression field 0; an image data-length 20 h = 32 d; two 4-byte ppm fields = 0; a 4-byte colours-used field = 0; and a 4-byte colours-important field = 0. After the *thick line* is the start of the 8CO colour-table: 4-byte fields for intensities of Blue Green Red and pad zero. After the second *thick line* are 32 1-byte fields: the first 14 (FFh = 255 d, 7h = 7 d, F9h = 249 d, ...) for the lower row of the image followed by two zero-padding fields, and the second 14 for the upper row followed by two zero-padding fields. The padding is to make the row-length a multiple of 4 bytes

The image data group contains the colour index values, 0 to FF hexadecimal, of the pixels taken in scan sequence bottom left to top right of the image. The index values occupy one 8-bit byte each, and point to the colour-table, which sets the actual colours.

Figure 41.3 shows an example of an 8-bit .BMP file with colour-table 8GM. The image is 14 × 2 px, shown here hugely enlarged. The decimal index values are added. The file contents are seen in the hexadecimal dump.

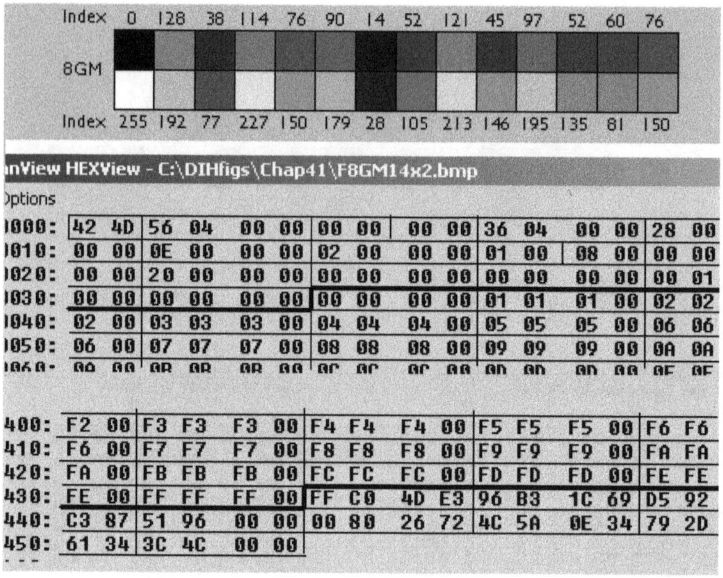

Fig. 41.3 8-bit 8GM file. The image is 14 × 2 px, as shown (with 8CO decimal index values and some colour names added). The file is shown as a hexadecimal dump, where each pair of digits is the hex value of a byte. In a multi-byte field the hex pairs are taken as most-significant right, least-significant left. The file header fields are: a 2-byte magic cookie field containing 42 4D (ASCII BM); a 4-byte file size field containing 456 h = 1110 d; two 2-byte fields (ignored); a 4-byte header length 436 h = 1028 d; a 4-byte info-header length 28 h = 40 d; a 4-byte image width Eh = 14 d; a 4-byte image height 2; a 2-byte planes field = 1; a 2-byte bpp field = 8; a 4-byte compression field 0; an image data length 20 h = 32 d; two 4-byte ppm fields = 0; a 4-byte colours used field 100 h = 256 d; and a 4-byte colours important field = 0. After the *thick line* is the start of the 8GM colour-table: 4-byte fields for intensities of Blue Green Red and pad zero. The table is a complete grey-scale, from 0 to FFh = 255 d. After the second *thick line* are 16 1-byte fields for the indexes of the lower row of the image (FFh = 255 d, C0h = 192 d, 4D h = 77 d, …, and two zero pads); then 16 fields for the upper row (00 h = 0 d, 80 h = 128 d, 26 h = 38 d, …, and two zero pads)

41.5 8-bit .BMP File with Colour-table 8GS

Table 41.3 shows the structure of an 8-bit .BMP file with colour-table 8GS.

In an 8GS file the FileHeader and the InfoHeader is the same as in an 8CO file.

An 8GS colour-table (see Sect. 37.4) contains Black (0, 0, 0), Grey 1 (1, 1, 1), Grey 2 (2, 2, 2), …, White (FF, FF, FF)h = (255, 255, 255)d.

The image data group contains the colour index values, 0 to FF hexadecimal, of the pixels taken in scan sequence bottom left to top right of the image. The index values occupy one 8-bit byte each, and point to the colour-table, which sets the actual colours.

Figure 41.4 shows an example of an 8-bit .BMP file with colour-table 8GS. The image is the first three, middle three, and last three of the 256 grey tones. The file contents are seen in the hexadecimal dump.

Table 41.3 8-bit 8GS file format

Group	Field	Type	dec.	hex.	Remarks
FileHeader			1–14	01–0E	File type, properties
	bfType	integer	1–2	01–02	Always 'BM'
	bfSize	long	3–6	03–06	File length, bytes
	bfReserved1	integer	7–8	07–08	Always 0
	bfReserved2	integer	9–10	09–0A	Always 0
	bfOffBits	long	11–14	0B–0E	Header length, bytes
InfoHeader			15–54	0F–36	Image size, properties
	biSize	long	15–18	0F–12	Infoheader leng, bytes
	biWidth	long	19–22	13–16	Image width, pixels
	biHeight	long	23–26	17–1A	Image height, pixels
	biPlanes	integer	27–28	1B–1C	Always 1
	biBitCount	integer	29–30	1D–1E	Bits per pixel = 8
	biCompression	long	31–34	1F–22	Not compr. = 0
	biSizeImage	long	35–38	23–26	Image data leng,bytes
	biXPelsPerMeter	long	39–42	27–2A	Printer p.p.m., horiz.
	biYPelsPerMeter	long	43–46	2B–2E	Printer p.p.m., vert.
	biClrUsed	long	47–50	2F–32	256 for 8GS
	biClrImportant	long	51–54	33–36	Colours important
ColorTable			55–1078	37–436	256 colours for 8GS
	aColors[0]	rgbquad	55–58	37–3A	Black: 0 0 0 0
	aColors[1]	rgbquad	59–62	3B–3E	Grey 1: 1 1 1 0
	aColors[2]	rgbquad	63–66	3B–3E	Grey 2: 2 2 2 0
	...				
	aColors[255]	rgbquad	1075–1078	433–436	White: 255 255 255 0
Image data			1079–	437–	In scan sequence
	aBitmapBits[0]	byte	1079–	437–	One byte per pixel
	aBitmapBits[1]	byte	–	–	
	...				
	aBitmapBits[*n*]	byte	–	–	

Field types are: *byte* occupying eight bits; *integer* occupying two bytes; *long* occupying four bytes; *rgbquad* occupying four bytes. Notice that rgbquad holds the colour values in the first three bytes, in sequence BGR, and the fourth byte is zero Z

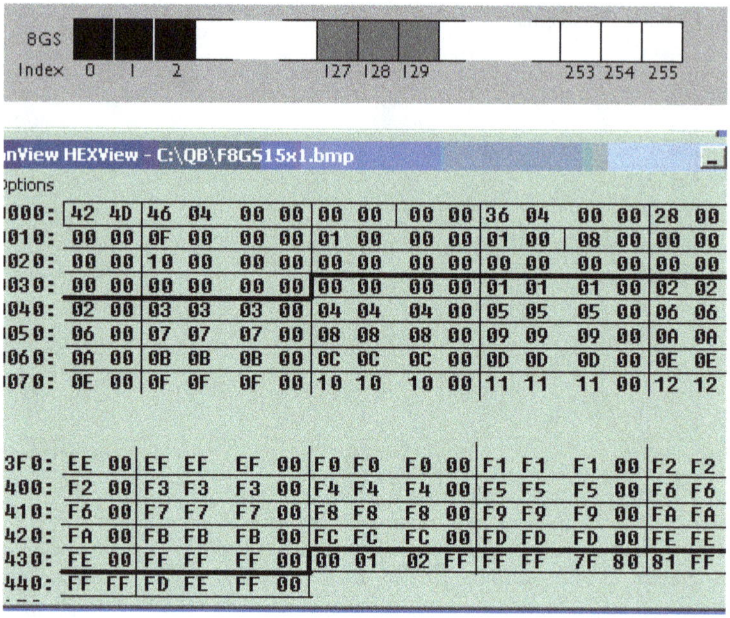

Fig. 41.4 8-bit 8GS file. The image is 15 px: the first three tones 0–2 in the 8GS *colour-table*, three blanks the middle three tones 127–129, three blanks, and and the last three tones 253–255. Neighbouring tones are not distinguished by the eye. The decimal indexes have been added. The file is shown as a hexadecimal dump, where each pair of digits is the hex value of a byte. In a multi-byte field the hex pairs are taken as most-significant right, least-significant left. The file header fields are: a 2-byte magic cookie field containing 42 4D (ASCII BM); a 4-byte file size field containing 446 h = 1094 d; two 2-byte fields (ignored); a 4-byte header length 436 h = 1028 d; a 4-byte info-header length 28 h = 40 d; a 4-byte image width Fh = 15 d; a 4-byte image height 1; a 2-byte planes field = 1; a 2-byte bpp field = 8; a 4-byte compression field 0; an image data length 10 h = 16 d; two 4-byte ppm fields = 0; a 4-byte colours used field = 0; and a 4-byte colours important field = 0. After the *thick line* is the start of the 8GS *colour-table*: 4-byte fields for intensities of Blue Green Red and pad zero. The table is a complete grey-scale, from 0 to FFh = 255 d. After the second *thick line* are 16 1-byte fields for the indexes of the single row of the image: 0, 1, 2, (three blanks), 7Fh = 127 d, 80h = 128 d, 81h = 129 d, (three blanks), FDh = 253 d, FEh = 254 d, FFh = 255 d, and a zero pad to make the row-length a multiple of 4 bytes

41.6 8-bit .BMP File Editor

If we wish to have interactive access to .BMP files, to view or change header, colour-table, and image-data values, we need a .BMP file editor. A brief description of such an editor follows, with source code at the SpringerLink web-site.

```
REM  **  Program EDBMP8.BAS to edit 8-bit BMP file. User
REM  **        enters filename to edit. Program gets file and
REM  **        offers menu to edit headers or colour-table or
REM  **        image-array.
```

```
REM
REM   **   Main program
REM   **   Enter filename to edit
                 CALL HEADER8(FILENAME$)
                 CALL COLTAB8(FILENAME$)
                 CALL IMAARR8(FILENAMES$)
END PROGRAM

           SUB HEADER8 (FILENAME$)
REM   **   Sub-program HEADER8 to edit header values of 8-bit
REM   **   BMP file. Reads each field of header, shows
REM   **   old value, offers next or change or exit.
           SUB COLTAB8 (FILENAME$)
REM   **   Sub-program COLTAB8 to edit 8-bit colour-table.
           SUB IMAARR8 (FILENAMES$)
REM   **   Sub-program IMAARR8 to edit 8-bit image-array.
REM   **   User has option to start at any row Y, from 0 to
REM   **   (height - 1). Program shows pixel value 0 to 255,
REM   **   and offers options of next or change or exit.
```

Chapter 42
8-bit Display

8-bit images are routinely input to a computer with an encoding gamma applied to intensities which is equal and opposite to the current display gamma. If this is not done the display darkens the image significantly. 8-bit display location variables are the same as for 1-bit images. 8-bit display intensity variables are defined.

42.1 Flowchart

The flowchart (Fig. 42.1) shows an 8-bit image file sent to a display. The flow is like the 4-bit case (see Chap. 34), but with 256 intensities. Display gamma is significant.

42.2 Display Hardware

8-bit display hardware is the same as 1-bit hardware, described in Chap. 27.

42.3 Display Size Variables

8-bit display size variables are the same as 1-bit size variables, described in Chap. 27.

42.4 8-bit Display Intensity Variables

When a monitor is used to display a 8-bit image, we have the following per-pixel intensity variables:

- Intensity index I_{256}.
- Colour-table intensity triple (B, G, R).
- Voltage triple (v_B, v_G, v_R).

A. Parkin, *Digital Imaging Primer*, DOI 10.1007/978-3-540-85619-1_42

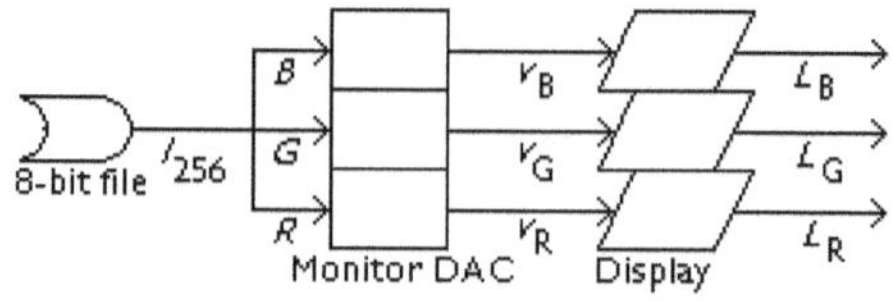

Fig. 42.1 Flowchart for 8-bit display. Per-pixel, an 8-bit image file has intensity index $0 \leq I_{256} \leq 255$, pointing to a colour-table. The colour-table sends (B, G, R) values to the three channels in a colour monitor. In each channel, a digital-to-analogue converter sends a voltage v to the appropriate screen sub-pixel, which emits luminance L_B, L_G, L_R. Normalized luminance $L_{norm} = v_{norm}^{\gamma}$

- Display gamma γ.
- Luminance L.

Intensity index is $0 \leq I_{256} \leq 255$, held in the image section of an 8-bit .BMP file (see Chap. 41).

Intensity triple is $0 \leq B \leq 255$, $0 \leq G \leq 255$, $0 \leq R \leq 255$, held in the colour-table of the .BMP file, defines the Blue, Green, and Red intensities to be displayed. Colour-tables for colour 8CO, grey-match 8GM, and grey-scale 8GS are shown in Chap. 37.

Voltage triple is the (normalized) analogue voltage output by the DAC of each channel, proportional to the colour-table intensity triple.

Display gamma is the darkening of intensity imposed by the display device, standard value 2.2 (see Sect. 22.12).

In each channel, the normalized luminance $L_{norm} = v_{norm}^{\gamma}$. For the measurement of luminance, see Chap. 15. The actual luminance of a CRT display can be varied by a control for White luminance (usually mis-labelled "Contrast") and a control for Black luminance ("Brightness"). When properly adjusted a good CRT has a maximum White luminance of about 170 cd/m^2, and a minimum Black luminance of about 0.01 cd/m^2. An LCD display has a maximum luminance of about 160 cd/m^2, and a minimum luminance of about 0.72 cd/m^2 (with minimum back-light) [1].

42.5 8-bit Display Colours

An 8CO colour-table offers 256 display colours, produced by additive mixing of the Blue Green and Red sub-pixel light sources of the display, like the 16 4CO colours shown in Sect. 34.5.

An 8GM colour-table offers 256 grey-matches, where the B G and R sub-pixels have equal intensities, like the 16 4GM colours shown in Sect. 34.5. Some greys are used more than once in the colour-table, and some are not used at all.

An 8GS colour-table offers a complete grey-scale of 256 greys, each used once.

Reference

1. Soneira RM (2013) http://www.displaymate.com/ShootOut. Cited 30 Sept 2013

Chapter 43
8-bit Printout

Printer size variables are the same for 8-bit as for 1-bit images. 8-bit colour images are printed with four inks: Yellow, Magenta, Cyan, and Black (YMCK). Each ink requires an 8-bit grey-scale separation, which is half-toned to a 1-bit sub-image. 8-bit grey-scale images are printed with Black ink only, so the image is already a K separation which requires half-toning to 1-bit. 8-bit print intensity variables are defined, with examples for colour and grey-scale.

43.1 Flowchart

The flowchart (Fig. 43.1) shows an 8-bit image file sent to a printer by an image editor.

43.2 Printer Hardware

8-bit printer hardware is the same as 1-bit hardware, described in Chap. 28.

43.3 Print Size Variables

8-bit print size variables and resolving power are the same as for 1-bit print, described in Chap. 28.

A. Parkin, *Digital Imaging Primer*, DOI 10.1007/978-3-540-85619-1_43

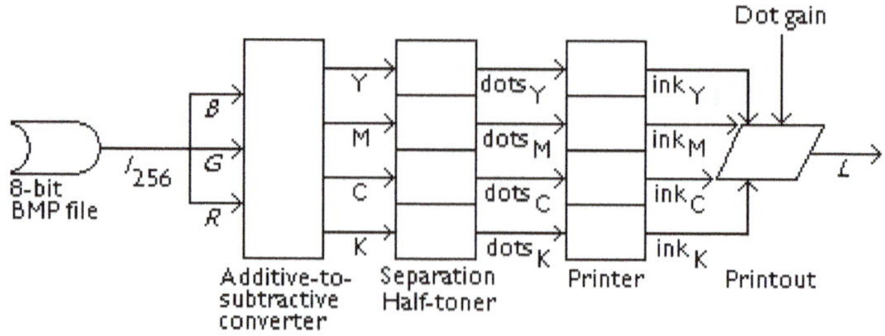

Fig. 43.1 Flowchart for 8-bit printout. An 8-bit .BMP file has, per pixel, location coordinates (X, Y), and intensity index I_{256} pointing to colour-table intensities (B, G, R). An additive-to-subtractive converter has three input channels for (B, G, R) and four output separation channels for (Y, M, C, K). In each separation channel a half-toner converts 8-bit intensities to 1-bit dot densities, which a printer applies as ink on paper. Dot gain is caused by ink spread on paper

43.4 YMCK Printing

Printing an 8-bit colour image is done by converting additive (B, G, R) intensities to subtractive (Y, M, C, K) intensities, making grey-scale separations, and 1-bit half-tones, as described in Chap. 35.

For example, Fig. 43.2 shows two source images with colour-table 8CO; Fig. 43.3 shows the corresponding Y M C K separations with grey-scale colour-table 8GS; and Fig. 43.4 shows the 1-bit half-tones made from each of the separations. These may be compared with the 4-bit examples in Chap. 35.

Fig. 43.2 8CO source images. Top is a 24-bit camera image saved-as 8CO. Bottom is a 24-bit gamut saved-as 8CO

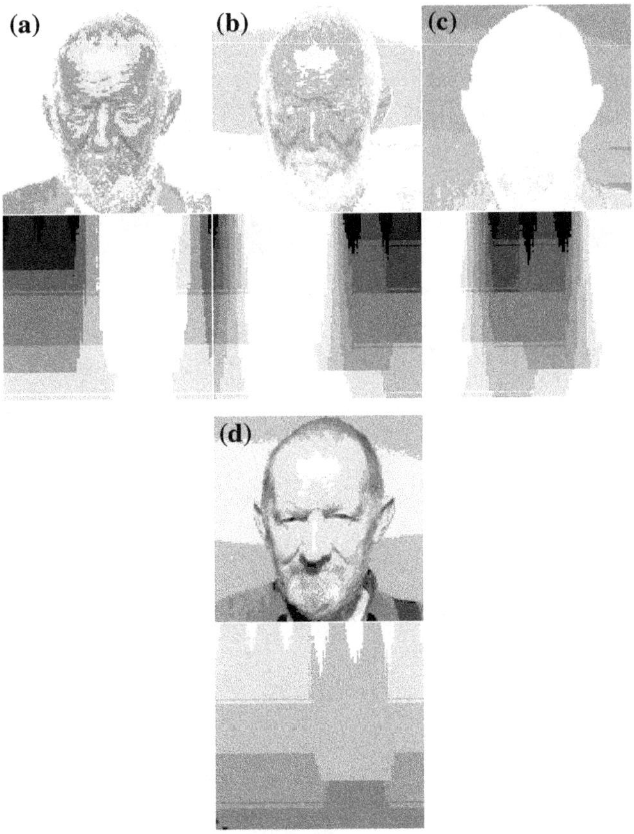

Fig. 43.3 8-bit grey-scale separations from 8CO images in Fig. 43.2. **a** Yellow separation. **b** Magenta separation. **c** Cyan separation. **d** Black separation

43.5 8-bit Grey-scale Printing

An 8-bit grey-scale image with colour-table 8GS has, per-pixel, $I_{256} = \mathrm{B} = \mathrm{G} = \mathrm{R}$. It is thus already a K-separation, with Black intensity Kgs $= I_{256}$. The separation is half-toned to 1-bit, and the final Black ink percent coverage is Kpc = (255−Kgs)/2.55. For example, Fig. 43.5a shows two grey-scale source images, and Fig. 43.5b shows the 1-bit half-toned image for printing.

Fig. 43.4 1-bit half-tones of separations in Fig. 43.3

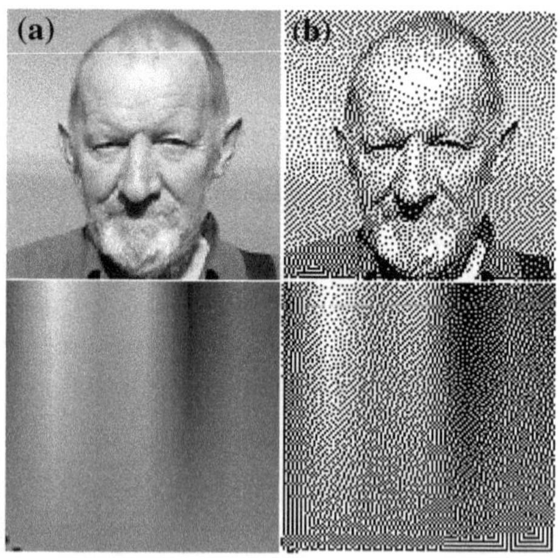

Fig. 43.5 8-bit grey-scale source and half-tone. **a** Source is 24-bit camera image and 24-bit gamut converted-to 8GS. Source is its own K-separation. **b** 1-bit half-tone

43.6 8-bit Print Intensity Variables

In printing an 8-bit image, we have the following per-pixel intensity variables:

- 8-bit intensity index I_{256}.
- 8-bit colour-table intensity triple (B, G, R).
- 8-bit separation intensity quadruple (Y, M, C, K).
- 8-bit separation grey-scale quadruple (Ygs, Mgs, Cgs, Kgs).
- 1-bit ink percent density quadruple (Ypc, Mpc, Cpc, Kpc).
- Dot gain δ

Table 43.1 28 of the colours, with their additive B G R intensities and subtractive Ypc Mpc Cpc Kpc percentages (under-colour removed)

Index	B	G	R	Name	Ypc	Mpc	Cpc	Kpc
0	0	0	0	Black	0	0	0	100
1	0	0	128	Dark Red	50	50	0	50
2	0	128	0	Dark Green	50	0	50	50
3	0	128	128	Dark Yellow	50	0	0	50
4	128	0	0	Dark Blue	0	50	50	50
5	128	0	128	Dark Magenta	0	50	0	50
6	128	128	0	Dark Cyan	0	0	50	50
7	128	128	128	Pale Grey	0	0	0	25
9	240	202	166		0	15	29	6
20	0	64	128		50	25	0	50
80	64	64	0		0	0	25	75
100	64	128	128		25	0	0	50
103	64	128	224		63	38	0	12
135	128	0	224		38	38	0	12
144	128	64	0		0	25	50	50
164	128	128	128	Grey	0	0	0	50
184	128	224	0		50	12	0	0
191	128	224	224		38	0	0	12
196	192	0	128		0	75	25	25
224	192	128	0		0	25	75	25
228	192	128	128		0	25	25	25
249	0	0	255	Red	100	100	0	0
250	0	255	0	Green	100	0	100	0
251	0	255	255	Yellow	100	0	0	0
252	255	0	0	Blue	0	100	100	0
253	255	0	255	Magenta	0	100	0	0
254	255	255	0	Cyan	0	0	100	0
255	255	255	255	White	0	0	0	0

Table 43.2 Beginning, middle, and end of grey-scale colour-table 8GS, with the additive B G R intensities and subtractive Kpc percentage

8GS					
Name	Index	B	G	R	Kpc
Black	0	0	0	0	100
	1	1	1	1	100
	2	2	2	2	99
	3	3	3	3	99
	4	4	4	4	98
	...	...	...	...	...
Grey	128	128	128	128	50
	...	...	...	...	...
Pale Grey	192	192	192	192	25
	...	...	...	...	...
	252	252	252	252	1
	253	253	253	253	1
	254	254	254	254	0
White	255	255	255	255	0

Intensity index is $0 \leq I_{256} \leq 255$, held in the image section of an 8-bit .BMP file (see Chap. 41).

Colour-table intensity triple is $0 \leq \mathrm{B} \leq 255, 0 \leq \mathrm{G} \leq 255, 0 \leq \mathrm{R} \leq 255$, held in the colour-table of an 8-bit .BMP file. It defines the proportions of additive primaries in the pixel.

Separation intensity quadruple is $\leq \mathrm{Y} \leq 255$, $0 \leq \mathrm{M} \leq 255$, $0 \leq \mathrm{C} \leq 255$, $0 \leq \mathrm{K} \leq 255$. It defines the proportions of subtractive primaries in the pixel.

Separation grey-scale intensity is the 255-complement of the separation intensity quadruple; that is to say 0 becomes 255, 128 becomes 127, and 255 becomes 0.

Ink percent density is $100/255\times$ the 255-complement of the grey-scale intensity; that is to say, 0 becomes 100 %, 127 becomes 50 %, 192 becomes 25 %, and 255 becomes 0 %.

Dot gain is the spreading of ink on the paper (see Sect. 35.8).

43.7 8-bit Print Colours

Table 43.1 shows a subset of 28 of the colours in colour-table 8CO, with their final Ypc Mpc Cpc Kpc ink percent coverages. Table 43.2 shows the beginning, middle, and end of the 8GS colour-table, and the subtractive 1-bit Kpc percent coverages.

Chapter 44
Viewing 8-bit Image

Viewing an 8-bit displayed image and an 8-bit printed image is compared with viewing a scene directly. Scope, location resolution, and luminance range are the same for 8-bit as for 1-bit images, but colour resolution is quite different.

44.1 Flowchart

The flowchart Fig. 44.1 is for general guidance in this chapter. It shows an eye viewing various types of 8-bit image, and, by comparison, viewing a scene. The computing flow is like that in Chap. 36, but with 256 intensities instead of 16.

44.2 Viewing Variables

If we want to compare viewing a scene with viewing a digital image of a scene, four variables emerge:

- location range, or scope, measured as angular or linear extent of view,
- location resolution within scope, measured as Snellen acuity, or as line-pairs per unit angle, or as line-pairs per unit length (see Sect. 10.6),
- luminance range, measured as dynamic operating range from lowest to highest luminance in cd/m^2 (see Sect. 15.8),
- colour resolution within luminance, measured as gamut on a CIE chromaticity chart (see Sect. 16.3).

The location range and resolution for 8-bit images are the same as for 1-bit images in Chap. 29.

A. Parkin, *Digital Imaging Primer*, DOI 10.1007/978-3-540-85619-1_44

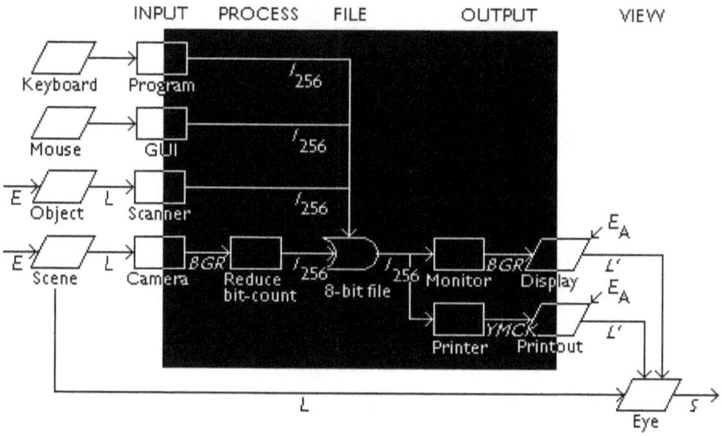

Fig. 44.1 Viewing a scene and an 8-bit image. E is illuminance in lx, from source onto object or scene. L is luminance in cd/m^2, from scene to eye or to input device, and from output device to eye. I_{256} is intensity index 0–255 from encoder to file, and from file to decoder. BGR is intensity triple, each eight bits 0–255, from monitor to display. CMYK are ink coverages, from printer to paper. E_A is ambient illuminance in lx, from surroundings onto display or printout. L' is luminance from output device to eye, and S is viewer's colour or *grey-scale* brightness sensation from eye to consciousness

The luminance range for 8-bit images is the same as for 1-bit images in Chap. 29, but taking notice of display gamma and dot gain effects.

The ambient illuminance effects for 8-bit are the same as for 1-bit in Chap. 29.

44.3 Colour Resolution of Eye and 8-bit Image

Figure 44.2 shows the colour gamut of the eye compared with 8-bit images. In the chromaticity chart, Brightness is imagined as a third dimension, from low below the chart to high above the chart; thus a dark Red and a light Red may have the same chromaticity but different Brightnesses, and neutral greys likewise. In Fig. 44.2b, colour-table 8CO has 256 colours, with gaps. A display has additive mixture: for example primaries R and G mix to make secondary Y. In Fig. 44.2c, a printout has subtractive mixture: primaries Y and M mix to make secondary R, with saturations lower than a display. In Fig. 44.2d, the 8GM gamut has gaps between the grey-matches, whereas the 8GS gamut has a smooth grey-scale.

The colour gamut of a standard eye is specified by the CIE chromaticity chart. The colour gamut of a digital image is determined by the colour-table of the file, and the output device used.

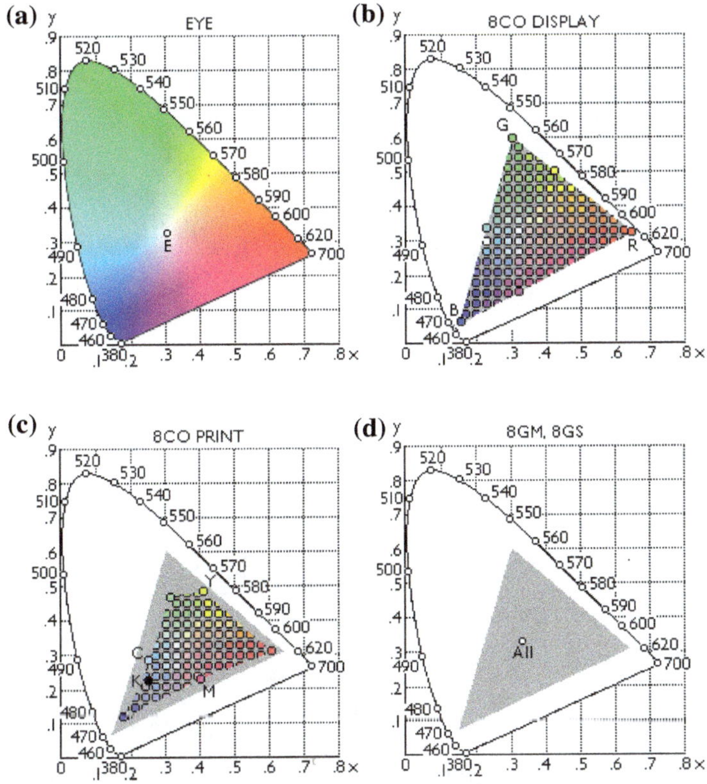

Fig. 44.2 Colour gamut of eye, compared to 8-bit images. **a** CIE chromaticity chart showing the colour gamut to which the eye responds. E is equal-energy illuminant. Notice that the colours printed here are merely indicative: the eye gamut extends to much more saturated colours. **b** Discrete 256-colour gamut of 8-bit display with colour-table 8CO, with gaps between colours. **c** Discrete 256-colour gamut of 8-bit printout with colour-table 8CO. The sRGB *triangle* is added for guidance. **d** Discrete 256-grey-match gamut of 8-bit image with colour-table 8GM, or smooth 256-*grey-scale* gamut of 8-bit image with colour-table 8GS

Part VIII
24-bit Imaging

Chapter 45
24-bit Image

A 24-bit image has any combination of 256 intensities of Blue, 256 intensities of Green, and 256 intensities of Red available. It has the same properties of location as an 8-bit image, but different properties of intensity. In the image file, each pixel carries its (B, G, R) triple as 3×8 binary digits = 24 bits (or 3 bytes), and no index and no colour-table is used. Examples are given of typical 24-bit images created by program, graphic user interface, scanner, and camera.

45.1 Bit-depth 24

If a computer system has three intensity variables B, G, R, each taking an 8-bit value 0–255 decimal, (= 0–FF hexadecimal, = 0–11 111 111 binary), then the intensity of each pixel can be represented by an integer triple:

$$(\mathrm{B}, \mathrm{G}, \mathrm{R}),$$

where $0 \leq \mathrm{B} \leq 255$, $0 \leq \mathrm{G} \leq 255$, and $0 \leq \mathrm{R} \leq 255$. Each of the three components is specified in 8 bits, and $3 \times 8 = 24$ bits, so we say that the plane now has *bit-depth 24*, or that it is the *24-bit plane*. There are $256 \times 256 \times 256 = 16\,777\,216$ possible intensity triples, far more than the eye can discriminate (see Sects. 17.11 and 17.12). In a 24-bit file pixel intensities are stored as (B, G, R) triples: there is no index and no colour-table.

45.2 24-bit Digital Image

A 24-bit digital image is like an 8-bit image (see Chap. 37) as to location, but different as to intensity. Graphically, a 24-bit image is a set of $X \times Y$ px, each pixel one of 16,777,216 intensities. Just as an 8-bit grey-scale forms a visually-smooth continuum

A. Parkin, *Digital Imaging Primer*, DOI 10.1007/978-3-540-85619-1_45

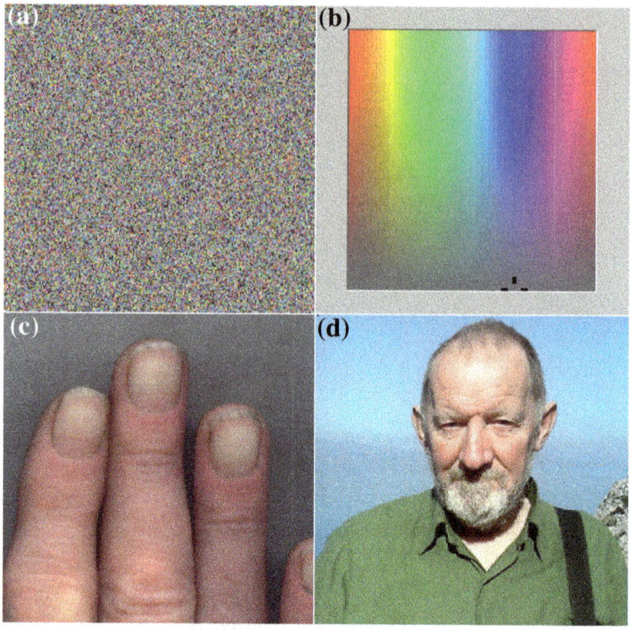

Fig. 45.1 Examples of 24-bit images. **a** Image generated by 24-bit program: 222 × 222 px randomly set at one of 16,777,216 colours. **b** Image created by PAINT in 24-bit mode. **c** Fingers scanned in 24-bit mode. **d** Scene photographed by 24-bit camera

from Black to White, so do the three 8-bit intensity components form a visually-smooth three-dimensional continuum from Black to White via Blue, Green, and Red, that is to say, the sRGB gamut (see Sect. 16.5). Numerically, it is a set of integer quintuples:

$$(\mathrm{X}, \mathrm{Y}; \mathrm{B}, \mathrm{G}, \mathrm{R}),$$

where $\leq X \leq x_{\max}$, $0 \leq Y \leq y_{\max}$; $0 \leq \mathrm{B} \leq 255$, $0 \leq \mathrm{G} \leq 255$, $0 \leq \mathrm{R} \leq 255$.

Figure 45.1 shows examples of 24-bit digital images, comparable with the 8-bit examples in Chap. 37. In Fig. 45.1b–d we see that smooth gradients between any two colours are possible.

45.3 Flowchart

The flowchart in Fig. 45.2 is for general guidance in this Part. To make a 24-bit image, we can use a program or a GUI image editor or a scanner or a camera (see Chap. 46). The image can be stored in a 24-bit .BMP file (see Chap. 48), then displayed (see Chap. 49) or printed (see Chap. 50) for viewing (see Chap. 51).

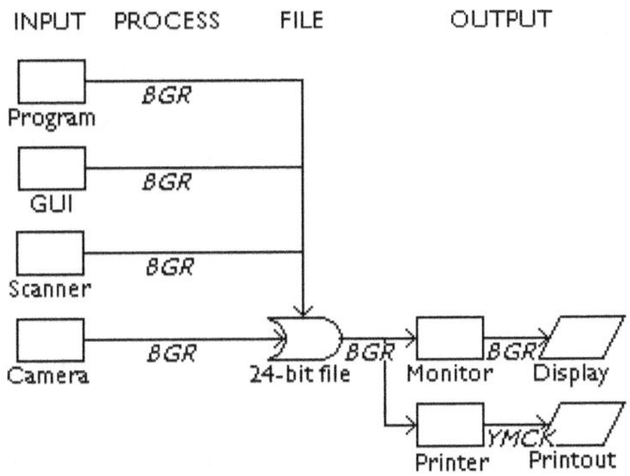

Fig. 45.2 Flowchart for 24-bit image. A 24-bit image may be input alpha-numerically from keyboard to program, or graphically from mouse/keyboard to GUI editor, or scanned in 24-bit mode, or photographed by camera. The image is stored in a 24-bit file. The image may be displayed on a screen or printed on paper, for viewing. BGR is intensity triple, 3× (eight bits 0–255), from input device to file, and from file to display. CMYK is ink coverages, from printer to paper

Chapter 46
24-bit Input

A 24-bit image can be created by a 24-bit program, or by a GUI in 24-bit mode, or by a scanner in 24-bit mode, or by a camera in 24-bit mode. In each case the location properties are the same as for 8-bit images. Smooth gradients of intensity are possible between any two colours. Gamma-encoding is significant in 24-bit input. Examples are given of a 24-bit program, GUI input, and scanner input. The operation of a digital camera is described. Camera size variables and a measure of camera resolving power are defined, and the acuity of a camera is compared with the acuity of the eye. Depth of field for a camera is defined. Camera intensity variables are defined, and a camera colour test is described.

46.1 24-bit Input

A 24-bit image can be created in any of four main ways, broadly similar to 8-bit input (see Chap. 38):

- From integer numeric data via a program.
- From graphic data via a GUI.
- From a continuous flat object via a scanner.
- From a continuous space object via a camera.

46.2 24-bit Program Input

A 24-bit program is like an 8-bit program (see Sect. 38.2), with the differences:

- Read and write 24-bit .BMP file instead of 8-bit file.
- Process 24-bit (B, G, R) intensities instead of 8-bit indexed intensities I_{256}.

A. Parkin, *Digital Imaging Primer*, DOI 10.1007/978-3-540-85619-1_46

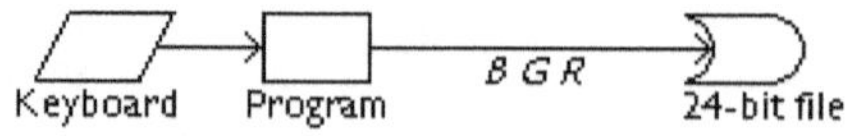

Fig. 46.1 Flowchart for 24-bit program. Input is from from keyboard or data file to program. Output to 24-bit file has the form (X, Y; B, G, R) per pixel, where $0 \leq X \leq w_{\max}$ and $0 \leq Y \leq h_{\max}$ are integer coordinates of location, and $0 \leq \mathrm{B} \leq 255$, $0 \leq \mathrm{G} \leq 255$, $0 \leq \mathrm{R} \leq 255$ are integer Blue Green and Red intensities

Fig. 46.2 Example of 24-bit random image, 256 × 256 px

Figure 46.1 shows the flow for a 24-bit program. For example, a program which creates a 24-bit random image is briefly described below (source code at the SpringerLink web-site). Figure 46.2 shows an image created by this program.

```
REM  **  Program GENRAN24.BAS to generate 24-bit random
REM  **     image, as .BMP file. User enters output
REM  **     .BMP filename, image width and height.
REM  **     Program writes 24-bit .BMP file header
REM  **     and image data.
REM  **  Main program
REM  **  Enter file-name for output, image width and height
         CALL CREHEA24
REM  **  Generate a random image:
REM  **  Write three bytes B G R each randomly 0-255
REM  **  to file #1
         CALL REVHEA24
FINIS: END

         SUB CREHEA24
REM  **  Subprogram CREHEA24 to write header of
REM  **     24-bit .BMP file #1.
END SUB
         SUB REVHEA24
REM  **  Revises header of 24-bit .BMP file #1 with
REM  **  values in bytes.
END SUB
```

46.3 Encoding Gamma

In the 24-bit plane we can create a smooth gradient between any two colours, since each of the components of the intensity triple (B, G, R) can take any integer value from 0 to 255. When such a gradient is displayed, it remains smooth, but the luminance is governed by the power-law transfer function of the display (see Sect. 22.12). To counteract this, scanner and camera inputs are routinely delivered gamma-encoded, that is after applying encoding gamma $= 1/2.2 = 0.4545$. A program should do the same. For example, Fig. 46.3 shows three pairs of gradients, one of each pair gamma-un-encoded ($\gamma = 1$) and one gamma-encoded ($\gamma = 0.4545$). At the right a patch of the middle colour of each gradient is put between patches of the extremes. In isolation either can be accepted as mid-colour. The relation between physical luminance and perceived brightness is examined in Chap. 17.

The gradients in Fig. 46.3 were produced by the program briefly described below (source code at the SpringerLink web-site):

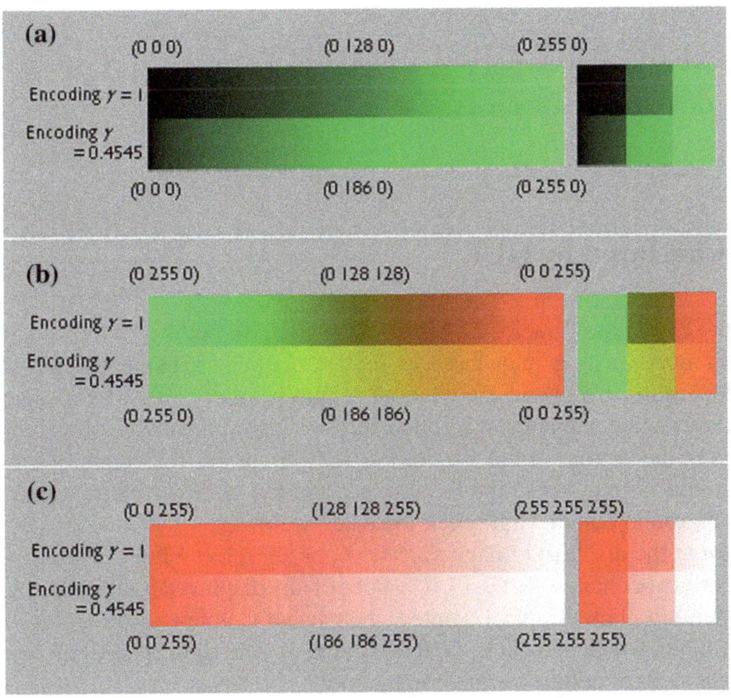

Fig. 46.3 Gradients produced by 24-bit program, with (B, G, R) values added. **a** Shade from *Black* to *Green*. **b** Mix from *Green* to *Red*. **c** Tint from *Red* to *White*. In each pair the upper gradient is gamma-unencoded ($\gamma = 1$), and the lower is gamma-encoded ($\gamma = 0.4545$). At the *right*, which colour looks midway between the extremes? The background is just slightly paler than a 50 % grey

```
REM   **   Program GENGRA24.BAS to generate 24-bit
REM   **      gradient between any two chosen colours.
REM   **      User enters .BMP file-name, chosen
REM   **      colours, width and height wanted, and
REM   **      gamma-encoded or un-encoded. Program creates
REM   **      24-bit .BMP file header, generates image
REM   **      data, revises header, then outputs file.
REM   **   Main program
REM   **   Enter file-name, BGR from, BGR to, width, height,
REM   **   and gamma-encoded or un-encoded
           CALL CREHEA24
REM   **   Generate un-gamma-encoded gradient
REM   **   Generate gamma-encoded gradient
FINIS: END

           SUB CREHEA24
REM   **   Subprogram CREHEA24 to write header of
REM   **      24-bit .BMP file #1.
           SUB REVHEA24
REM   **   Revises header of 24-bit .BMP file #1 with
REM   **   values in bytes.
```

46.4 24-bit Input by GUI

Figure 46.4 shows the flow for 24-bit input by GUI. To put the PAINT image editor into 24-bit mode, save an image as a 24-bit file, or open an image which is already a 24-bit file. Figure 46.5 shows the window displaying part of a 24-bit camera image. The window can be panned and scrolled across a large image. 28 colours are shown in the box at the bottom of the window, as in 8-bit mode. To set any of these 28 as the foreground colour, select the Eyedropper and left-click the patch wanted. To set as the background colour, right-click the patch wanted.

To enlarge the displayed image x2, x4, x6, or x8, select View—Zoom—Custom.

To ascertain or change the (B, G, R) values of any displayed image pixel, select the Eyedropper and click the pixel wanted, then select Colours—Edit Colors—Define Custom Colors, which shows (B, G, R) values at bottom right. These can be changed to re-set the current colour for the graphic tool.

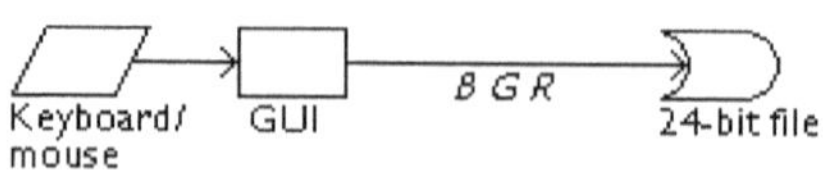

Fig. 46.4 Flowchart for 24-bit GUI input. (B, G, R) are per-pixel integer intensities

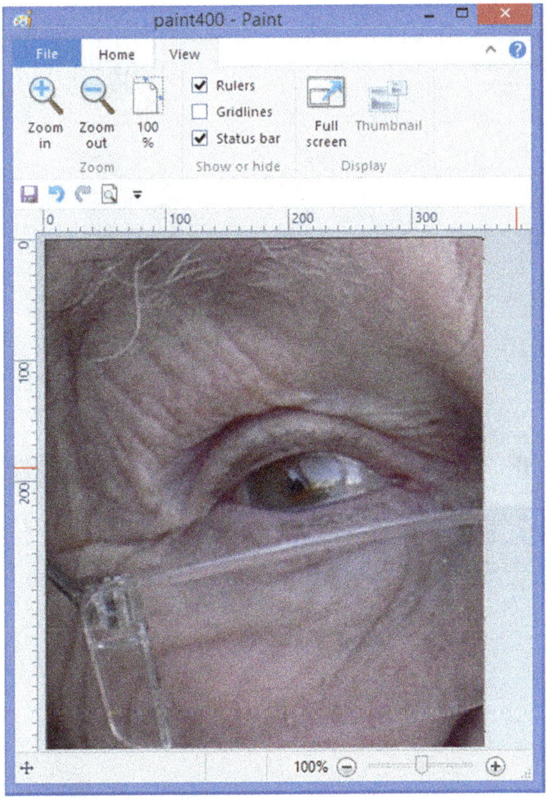

Fig. 46.5 PAINT window in 24-bit mode. The window shows a 24-bit camera image. The window can be panned and scrolled across the image

46.5 24-bit Input by Scanner

Figure 46.6 shows the flow for 24-bit input by scanner. The scanner converts continuous (real) values of point location to discrete (integer) pixel coordinates, and real values of spectral intensity to 24-bit integer (B, G, R) intensities. The output to file is usually gamma-encoded.

24-bit scanner input is like 1-bit scanner input (see Chap. 22) as to location. For example, Fig. 46.7a shows part of a 24-bit scan of a line-pair test strip.

The scanning frequency is $F_S = 300\,\text{spi}$ (=11.8 spmm), so the resolution limit is $F_{LP} = F_S/2 = 150\,\text{lppi}$ (=5.9 lppmm). The highest frequency of the test strip is $F_{LP} = 3\,\text{lppmm}$. When the scanning frequency is 11.8 lppmm, we calculate that

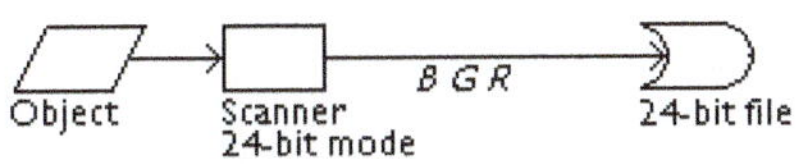

Fig. 46.6 Flowchart for 24-bit scanner input. (B, G, R) are per-pixel integer intensities

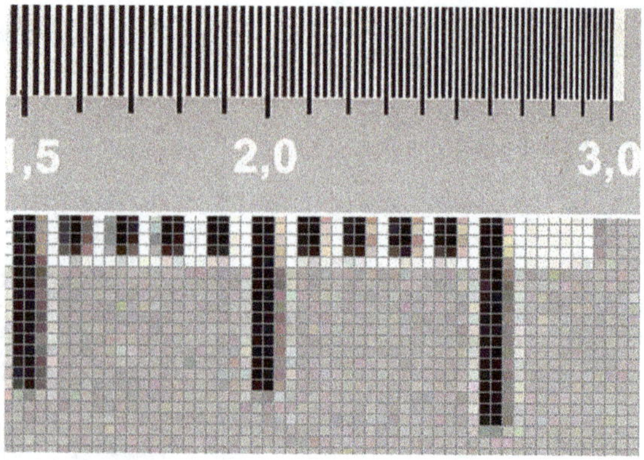

Fig. 46.7 24-bit scanner location test. **a** Line-pair test strip scanned in 24-bit mode at 300 spi. **b** Enlargement ×8 of (**a**) to show samples

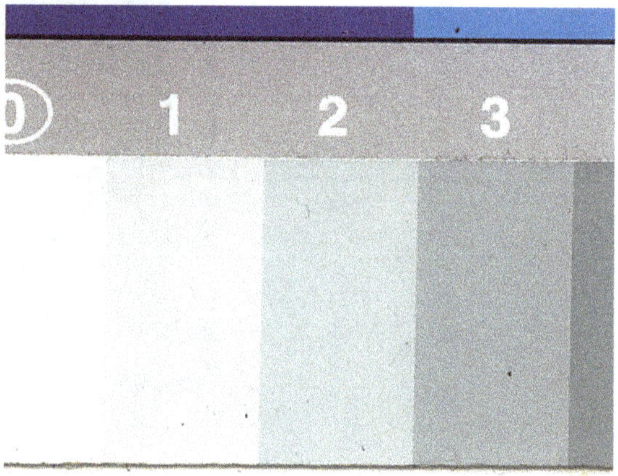

Fig. 46.8 24-bit scanner intensity test: step-wedge scanned in 24-bit mode. Table 46.1 shows calculated and actual intensity values

a line-pair of frequency 3 lppmm should occupy $11.8/3 = 3.94$ samples, and that a line-pair occupying four samples has a frequency of $11.8/4 = 2.95$ lppmm. The enlargement in Fig. 46.7b shows that the line-pair frequency actually occupying four samples is 2.87 lppmm, which agrees fairly well with the calculation.

24-bit scanner input is unlike 1-bit scanner input as to intensity. For example, Fig. 46.8 shows part of a 24-bit scan of a standard step-wedge. Table 46.1 shows calculated values for a perfect scan of a perfect step-wedge (see Sect. 22.15), compared with actual values for this scan of this step-wedge. The actual values have considerable variation from sample to sample within each patch, because of minute variations in the step-wedge and the scanner. In aggregate they agree fairly well with the calculated perfect case.

Table 46.1 24-bit scanner intensity test

d	r	v	Perfect			Actual		
			B	G	R	B	G	R
0	1	1	255	255	255	255	255	255
0.1	0.79	0.90	229	229	229	243	246	248
0.2	0.63	0.81	207	207	207	223	227	213
0.3	0.50	0.73	186	186	186	195	195	186
0.4	0.40	0.66	168	168	168	175	165	171

The columns are: Optical density d of step-wedge patch; normalized reflectance $r = 1/10^d$; encoded voltage $v = r^{0.4545}$; intensities (B, G, R) $= 255 \times v$ calculated for perfect scan; actual intensities (B, G, R) for Fig. 46.8

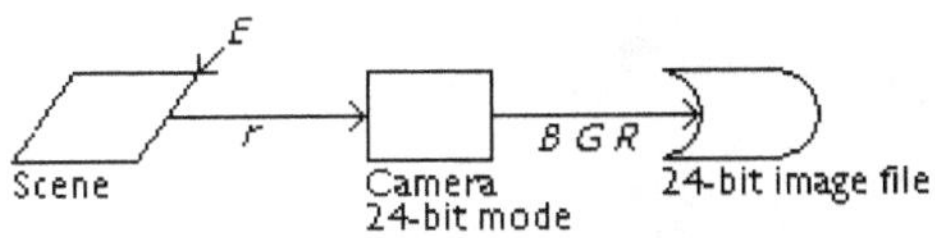

Fig. 46.9 Flowchart for 24-bit camera input. Illuminance E falling on a scene is differentially reflected in all directions. Within its scope, a camera receives per-sample reflectance r, and sends encoded intensities (B, G, R) to a 24-bit file

46.6 24-bit Input by Camera

A *digital camera* is a device which accepts light from a continuous scene of objects in space, and delivers a discrete plane image of the scene. It is like a scanner with a lens, or a single unblink of an immobile eye.[1] The functional parts of a camera are,

- lens,
- sensor,
- scanning system,
- analogue-to digital converter,
- internal image file.

Figure 46.9 shows the flow for 24-bit input by camera. Figure 46.10 shows the functional parts (compare Sect. 22.2). We take a space reference frame with origin at the centre of the camera lens, z-axis on the lens axis, x-axis to the right, and y-axis upwards (a right-handed set). The lens forms an inverted relief perspective image of the object at back focal distance f' (see Sect. 9.4). In a low-end digital camera the lens is fixed, with short focal length and small diameter. A high-end camera has an adjustable-focus lens.

A camera sensor is a photo-electric device which contains an array of square sample buckets (see scanner sensor Sect. 22.5). Each bucket independently registers the light reflectance r which falls upon it, and outputs a voltage v proportional to the reflectance. Figure 46.10 shows a sensor fixed inside the camera at the back focal

[1] Turn towards a scene and close your eyes. Open and shut your eyes as fast as possible, once.

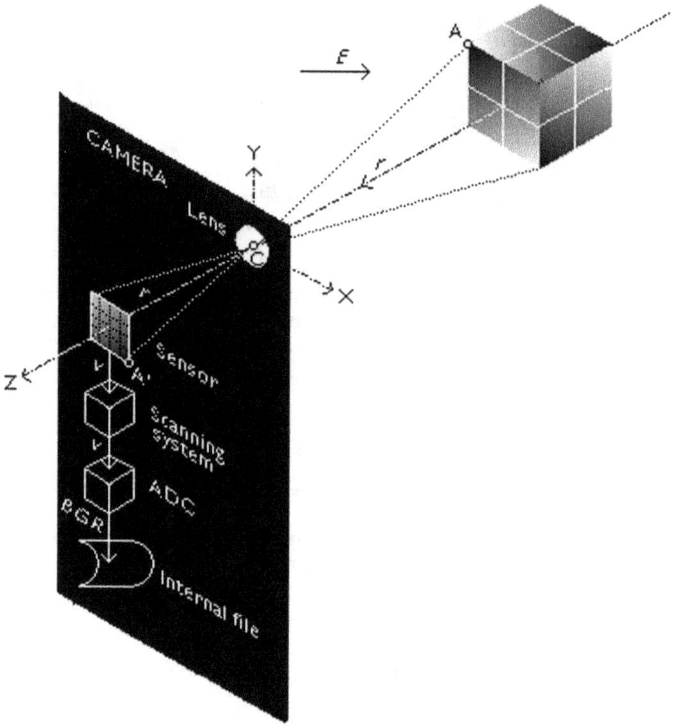

Fig. 46.10 Functional parts of a digital camera. Incident light E falls on a continuous object, which reflects differential intensities r per point location. The lens forms a flat perspective image on the sensor. The sensor is an array of discrete sampling buckets, each of which registers the intensity falling on it as a voltage v. The scanning system orders the samples as a sequence, and the analogue-to-digital converter turns the analogue voltages into 8-bit intensities B, G, R. The digital image is held in camera memory as a file, usually in .JPG format

point F′, focal length f from the lens. The sensor thus receives an inverted flat perspective image, continuous in location and in intensity; and it registers the image as voltages proportional to the intensity in discrete sample buckets. Almost all digital cameras register intensity through a Bayer[2] filter [1], where in each group of 4 px two respond to mid-wavelength Green light, one to short-wavelength Blue light, and one to long-wavelength Red light. The separate intensities are then interpolated to three values for each pixel.

A scanning system turns the simultaneous registration of intensities across the sensor array into a sequence of voltages $(X, Y; v)$, sample-by-sample across each row, and row-by-row across the array.

An 8-bit ADC is a device which converts a real input $0 \leq v \leq 1$ to an integer output $0 \leq I \leq 255$ corresponding to a grey-scale of discrete intensities between 0

[2]Bryce Edward Bayer (1929–2012), Eastman Kodak 1951–1986.

Black and 255 White (see Sect. 22.7). A sensor with Bayer filter delivers three colour channels, passing Blue, Green or Red intensities. Each channel has an 8-bit ADC to digitize the respective colour intensity, giving a 24-bit output.

A digital camera stores images internally, gamma-encoded, in fixed memory or in a removable memory card, in (usually) 24-bit .JPG format. Some cameras have an option to store images in RAW, or in 8-bit grey-scale, or in another format.

A 24-bit .JPG file can be transferred from camera to computer by standard software. In a computer image editor, a .JPG file can be opened then saved as a .BMP file.

46.7 Camera Size Variables

Figure 46.11 shows the size variables for a camera, in plan and elevation. A camera lens projects incoming rays from a continuous space object ABCDE at distance v to a real image on a discrete sensor A′B′C′D′E′ at distance v' (=focal length f), where the centre of projection is N (the centre of a simple lens or the nodal points of a compound lens). We may also say that it projects outgoing rays from a discrete sensor array to a virtual grid on a continuous object. Thus at a given object distance v there is a grid of virtual samples of calculable size on the object.

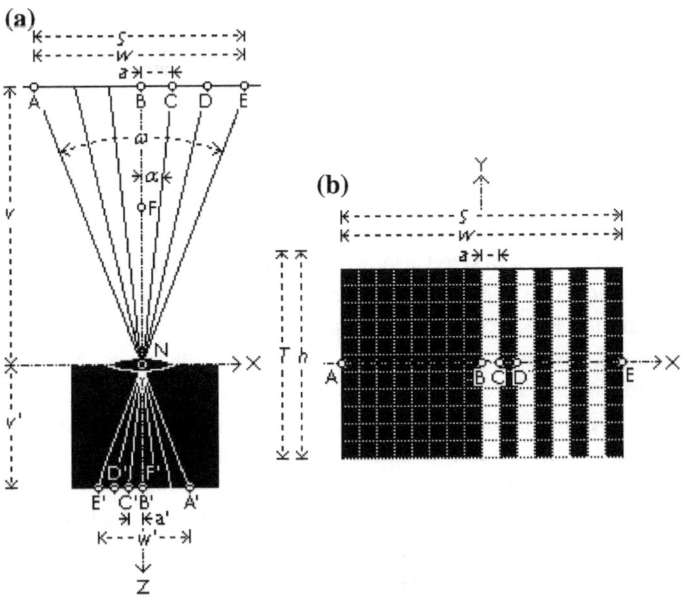

Fig. 46.11 Camera size variables, as listed in Table 46.2. **a** A camera projects an integer sampling grid from sensor A′B′C′D′E′ via centre N to a continuous space object ABCDE. The coordinate frame is a right-handed set with origin at centre of lens N. **b** Elevation of object ABCDE, with projected sample grid

Table 46.2 Camera size variables, as shown in Fig. 46.11

OBJECT			
Object sample width	a	$= 1/F_S$	Length units
Object sampling freq.	F_S	$= S/w$	Samples per unit len.
Object resolution lim.	F_{LP}	$= F_S/2$	Line-pairs per unit len.
Object distance*	v		Length units
Object width*	w		Length units
COMMON			
Lens diameter*	d		Length units
Focal length*	f		Length units
F-number	F/n	$n = f/d = f/d$	
Width/distance const.	k	$= w/v$	
Magnification im./ob.	m	$= v'/v$	
Count samples in width*	S		Samples
Count samples in height*	T		Samples
Angular sample width	α	$= \omega/S$	Angle units
Angular sampling freq.	Φ_S	$= S/\omega$	Samples per unit angle
Angular resolution lim.	Φ_{LP}	$= \Phi_S/2$	Line-pairs per unit angle
Angular width	ω	$= 2 \times \arctan(w/2v)$	Angle units
CAMERA IMAGE			
Image sample width	a'	$= 1/F'_S$	Length units
Image sampling frequency	F'_S	$= S/w'$	Samples per unit len.
Image resolution limit	F'_{LP}	$= F'_S/2$	Line-pairs per unit len.
Image distance	v'	$= f$	Length units
Image width	w'	$= m \times w$	Length units

Given values for the asterisked variables, the rest can be calculated

Table 46.2 lists the camera size variables. A particular camera has certain constants which can be noted once and for all. The manufacturer fixes lens diameter d, focal length f, F-number F/n, count of samples in width S, and count of samples in height T. Further constants can be got by experiment with an accurate bench set-up: photograph a millimetre rule at distance v from the centre of the camera lens, then examine the photograph in an image editor, as shown in Fig. 46.12, to read the object width w and the sample counts S and T. The further constants are calculated as shown in COMMON of Table 46.2.

For example, Table 46.3 shows sizes in millimetres and inches calculated for a typical low-end (4 Mpx) camera, compared with a normal (Snellen 6/6) eye (see Sect. 10.6). Notice that sample counts S and T are fundamental for a camera, whereas angular resolution Φ_{LP}, from cone diameter a', is fundamental for the eye. There is a fast-developing range of digital cameras which have much better performances than this example, and much higher prices (see Sect. 46.14 below).

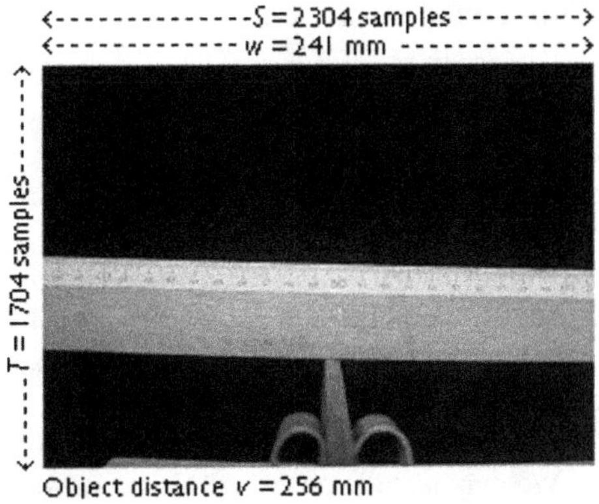

Fig. 46.12 Example of bench test for camera constants. Camera is set at distance *v* from rule. An image editor shows object width *w* on the rule, and sample counts *S* and *T*

Table 46.3 Example of sizes calculated for a typical 4 Mpx camera, compared with a normal Snellen 6/6 eye

		4 Mpx camera		6/6 eye		
		mm	in.	mm	in.	
OBJECT						
Sample width	a	0.10	0.004	0.073	0.0029	lu
Sampling freq.	F_S	9.56	243	13.1	333	spul
Resolution limit	F_{LP}	4.8	121	6.5	166	lppul
Object distance	v	256*	10.1	250*	10	lu
Object width	w	241*	9.5	200*	8	lu
COMMON						
Pupil diameter	d	3.1*	0.122	4*	0.16	lu
Focal length	f	9*	0.354	22.38*	0.88	lu
F-number	F/n	F/2.9	F/2.9	F/5.6	F/5.6	
Width/distance constant	k	0.941	0.941	0.80	0.80	
Magnification image/object	m	0.0352	0.0352	0.0895	0.0895	
Count of samples in width	S	2304*	2304	2616	2616	Samples
Count of samples in height	T	1704*	1704	–	–	Samples
Angular sample width	α	0.0219	0.0219	0.0167	0.0167	Degrees
Angular sampling freq.	Φ_S	45.7	45.7	60*	60	spd
Angular resolution limit	Φ_{LP}	22.8	22.8	30	30	lppd
Angular width	ω	50.4	50.4	43.6	43.6	Degrees
IMAGE						
Sample width	a'	0.0037	0.0001	0.0068	0.00025	lu
Sampling frequency	F'_S	272	6898	147	3902	spul
Resolution limit	F'_{LP}	136	3449	73.5	1951	lppul
Image distance	v'	9*	0.354	22.38*	0.881	lu
Image width	w'	8.5	0.33	17.9	0.70	lu

The asterisked variables are given, and the rest are calculated

46.8 Resolving Power of Camera

The resolving power or resolution limit of a digital camera can be measured as the highest angular frequency of object grating which it can register correctly. The resolution limit is reached at half the sampling frequency. So the constant angular resolution limit for a digital camera is $\Phi_{LP} = \Phi_S/2$ lppd. The constant angular resolution limit of a camera converts to a linear resolution limit measured on an object at a specified distance v, which of course varies with v. Given a camera's sample count S and width/distance constant k, and an object's distance v, we can calculate object width $w = k \times v$, object sample width $a = w/S$, object sampling frequency $F_S = 1/a$, and linear resolution limit $F_{LP} = F_S/2$.

We may test the actual resolving power of a camera by photographing a line-pair test strip (see Sect. 22.10) at a given distance. For example, Fig. 46.13 shows a test of the 4 Mpx camera specified in Table 46.3. The line-pair test strip is at given distance $v = 2000$ mm, and the camera has given sample count $S = 2304$ and width/distance constant $k = 0.941$. We calculate view width $w = 0.941 \times 2000 = 1882$ mm; object sample width $a = 1882/2304 = 0.817$ mm; object sampling frequency $F_S =$

(a)
(b)

Fig. 46.13 Line-pair test of camera resolving power. **a** Test strip photographed at two metres: full frame 2304 × 1704 px. **b** Enlargement of strip from (**a**), with object sampling grid marked. Resolution limit is where line-pairs coincide with adjacent samples, about 0.6 lppmm

Table 46.4 Typical camera resolution limits in millimetres

S	k	v (mm)	w (mm)	a (mm)	F_S (lpmm)	F_{LP} (lppmm)
2304	0.941	250	235	0.10	10	5
2304	0.941	500	470	0.20	5	2.5
2304	0.941	1000	941	0.41	2.44	1.22
2304	0.941	2000	1882	0.82	1.22	0.61
2304	0.941	4000	3764	1.63	0.61	0.31
2304	0.941	8000	7528	3.27	0.31	0.15
2304	0.941	16,000	15056	6.53	0.15	0.08
2304	0.941	32,000	29248	12.69	0.08	0.04

S is given sample count; k is given width/distance constant; v is variable object distance; $w = k \times v$ is object width; $a = w/S$ is object sample width; $F_S = 1/a$ is linear sampling frequency; and $F_{LP} = F_S/2$ is linear resolution limit

Table 46.5 Typical camera resolution limits in inches

S	k	v (in.)	w (in.)	a (in.)	F_S (lpi)	F_{LP} (lppi)
2304	0.941	10	9.41	0.004	245	122
2304	0.941	20	18.82	0.008	122	61
2304	0.941	40	37.64	0.016	61	31
2304	0.941	80	75.28	0.033	31	15
2304	0.941	160	150.56	0.065	15	7.6
2304	0.941	320	301.12	0.131	7.6	3.8
2304	0.941	640	602.24	0.261	3.8	1.9
2304	0.941	1280	1204.48	0.523	1.9	1.0

S is constant sample count; k is width/distance constant; v is variable object distance; $w = k \times v$ is object width; $a = w/S$ is object sample width; $F_S = 1/a$ is linear sampling frequency; and $F_{LP} = F_S/2$ is linear resolution limit

$1/0.817 = 1.224$ spmm; and linear resolution limit at 2000 mm $F_{LP} = 1.224/2 = 0.6$ lppmm. In Fig. 46.13b the bars of the test strip do not coincide with pixels, except by fluke, and pixels show varying greys due to manifold tiny imperfections of lens, sensor, object, lighting, and set-up. However, the linear resolution limit at 2,000 mm is indeed seen to be about 0.6 lppmm.

Table 46.4 shows the calculated linear resolution limit F_{LP} for increasing object distance v, in millimetres, using the same 4 Mpx camera as in Table 46.3 above. Table 46.5 shows the same, in inches.

46.9 Acuity of Camera

We may also test the resolving power of a camera by a Snellen chart, as if it were a human eye (see Sect. 10.6). For example, Fig. 46.14 shows a photograph of a Snellen chart at six metres, taken by the typical 4 Mpx camera specified in Table 46.3. (a) is the full frame, 2304 × 1704 px; (b) is an enlargement of the chart, and (c) is a further enlargement to show the sampling grid projected onto the object. The

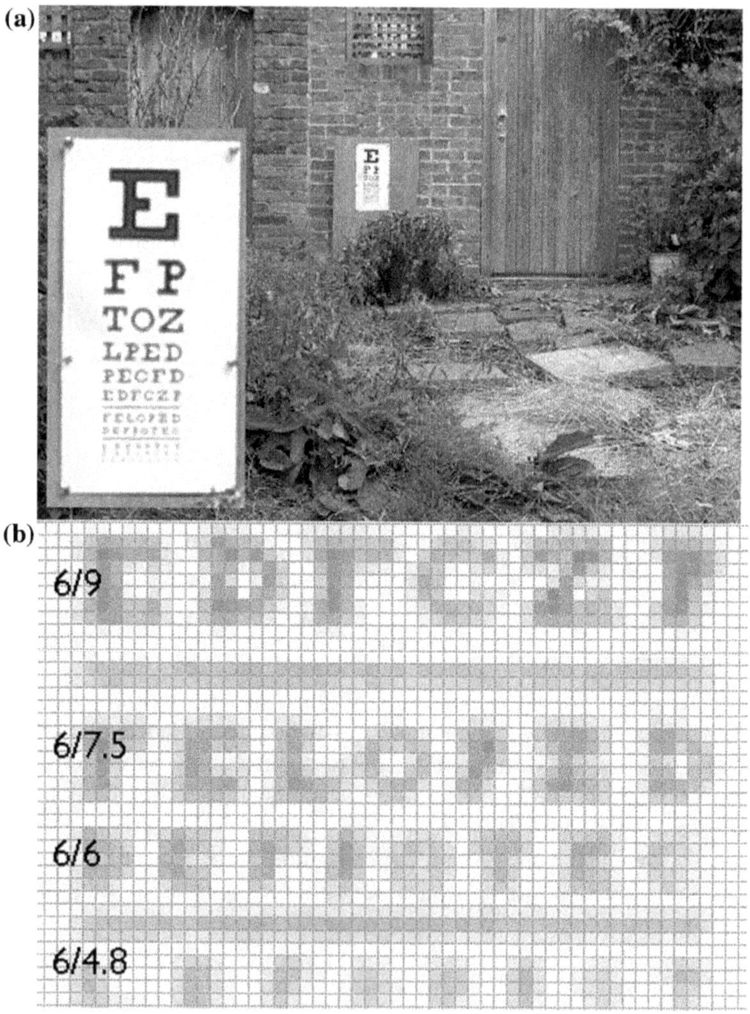

Fig. 46.14 Snellen test of camera acuity. **a** Snellen chart photographed at 6 m by camera with 2304 × 1704 px. **b** Enlargement of chart from (**a**). **c** Further enlargement with object sampling grid marked. Resolution limit is where letter stroke-width occupies one sample, between Snellen row 6/7.5 and 6/9

chart is at given distance $v = 6000$ mm, and the camera has given sample count $S = 2304$ and width/distance constant $k = 0.941$. We calculate view width $w = 0.941 \times 6000 = 5646$ mm; object sample width $a = 5646/2304 = 2.45$ mm; object sampling frequency $F_S = 1/2.45 = 0.41$ spmm, and linear resolution limit at 6000 mm $F_{LP} = 0.41/2 = 0.204$ lppmm. We see from Sect. 10.6 that this camera has an acuity between Snellen 6/7.5 and 6/9, as is evident in Fig. 46.14b. More expensive cameras match or exceed the acuity of the eye.

46.10 Depth of Field

For any camera image, the usual measure of sharpness of focus is the diameter of a *circle of confusion* c. It is variously estimated as (diagonal of sensor image)/1500, or (focal length of lens)/1720, or 2 minutes of arc. For a low-end digital camera we may take $c = 0.005$ mm. For any camera lens focussed at infinity, the *hyper-focal distance* H is that beyond which all objects are deemed sharp, calculated as

$$H = f^2/Nc \quad \text{length units,}$$

where f is the focal length of the lens, N is the F-number of the lens, and c is the diameter of the circle of confusion. For example:

$$H = (9^2)/(2.9 \times 0.005) = 5586\,\text{mm}.$$

But this may be inconveniently large. To bring nearer objects to sharp focus in the image we can use a supplementary lens with focal length $f_2 < H$ touching or very close to the camera lens. Figure 46.15 shows that rays from an object at the front focus F_2 of the supplementary lens emerge from that lens parallel to the axis, and are then focussed by the camera lens at back focus F'_1 as if from infinity. A zoom lens changes the focal length by different means.

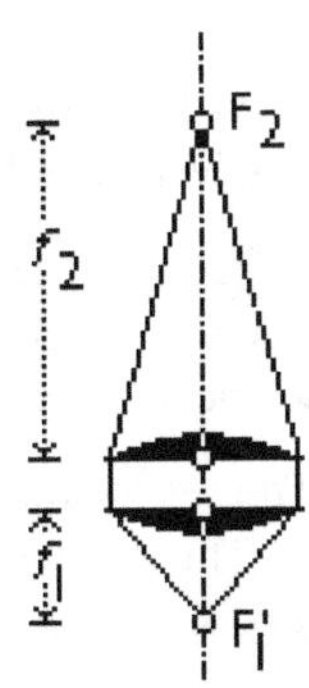

Fig. 46.15 Supplementary lens. Lens 2 is placed very close to camera lens 1. Rays from an object at front focal point F_2 are refracted to parallel to the lens axis, and are then refracted as if from infinity to focus at back focal point F'_1

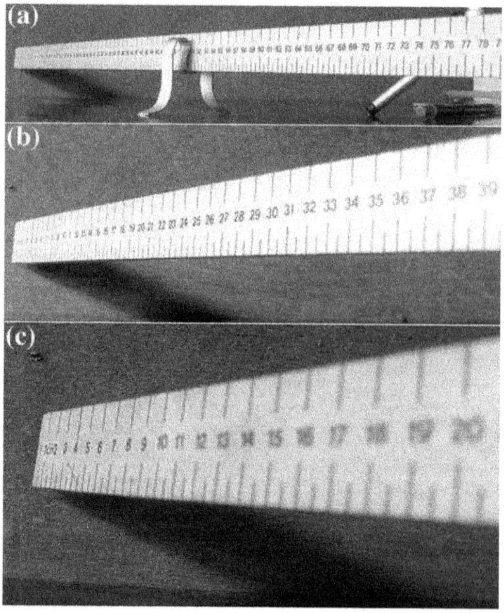

Fig. 46.16 Test of camera depth of field. **a** End of metre stick $v = 1000$ mm from camera: focal length 500 mm. **b** $v = 500$ mm; focal length 250 mm. **c** $v = 250$ mm; focal length 170 mm

The *depth of field* (see Sect. 9.5) is the distance between the farthest and nearest objects which are in focus in the image. At the hyper-focal distance, depth of field goes out to infinity. At closer distances, depth of field becomes increasingly shallow. For example, Fig. 46.16 shows three photographs of a metre stick taken with the same camera with different depths of field. In Fig. 46.16a the camera is one metre from the far end of the stick, with focal length 500 mm. In Fig. 46.16b the camera is 500 mm from the end, with focal length 250 mm. In Fig. 46.16c, the camera is 250 mm from the end, with focal length 170 mm.

46.11 Camera Intensity Variables

A 24-bit camera has, per sample, the following intensity variables in each of the B G R wavebands:

- fractional reflectance r from object,
- voltage v from sensor,
- encoded intensity triple (B, G, R) from ADC to file.

The sensor receives analogue reflectance r and outputs voltage v proportional to r.

The ADC receives voltage v, calculates $i_{\text{norm}} = v_{\text{norm}}^{0.4545}$, and outputs 8-bit integer tone-code between 0 and 255 $I = i_{\text{norm}} \times 255$.

46.12 Own Test Image

In a standard photographic test card the colour patches are subtractive inks, usually of unknown spectral characteristics. In digital imaging the fundamental colour space is additive sRGB (B, G, R) (see Sect. 16.5). We can construct our own additive test image, using the devices in current use. For example, Fig. 46.17a shows (of course somewhat degraded by subtractive reproduction here) an additive test image displaying eight colour patches,with their sRGB (B, G, R) values. (It also includes a line-pair strip down to 2 px per line-pair on the current display, a subtractive step-wedge half-toned by hand, and an additive step-wedge with reflectances, grey-scale tone-codes, and optical densities.)

46.13 Camera Colour Test

Figure 46.17b shows a camera image of Fig. 46.17a, shot at distance 333 mm from the screen with a 4 Mpx camera. The camera image has an overall Blue cast, and general loss of saturation. Evidently this camera is set for recording colours over a far greater luminance range than the 1–100 cd/m^2 of a display screen, and for a warmer colour balance, so the White of Fig. 46.17a becomes a bluish grey in Fig. 46.17b, and all the colours are greyed accordingly. We can put the camera image in a GUI editor to examine the (B, G, R) values of the patches, and compare them with the test image, as in Fig. 46.18. Loss of saturation is equivalent to replacing equal intensities of B G and R by grey (like under-colour removal in subtractive printing).

A camera (like a scanner) introduces considerable variation from sample to sample, not present in the original image.

46.14 Camera State of the Art

The digital camera with a CCD sensor [9] was invented in 1975 by Steven Sasson (born 1950) of Eastman Kodak. In the 1990s digital cameras overtook film cameras [14].The CMOS or Active-Pixel Sensor [3] was invented in 1993 by Eric Fossum (born 1957) of NASA Jet Propulsion Laboratory at California Institute of Technology. In the 2010s smartphones [15] incorporating CMOS cameras [13] overtook simple compact digital cameras.

A compact (or point-and-shoot) digital camera has a small fixed lens, a CCD sensor of about 10 Mpx, a preview/review display screen, and .JPG format output. It has various pre-set photographic modes combining focus, aperture, exposure, and white balance [12]. It couples to a computer by a USB cable, or a wireless connection, or an interchangeable SD memory card.

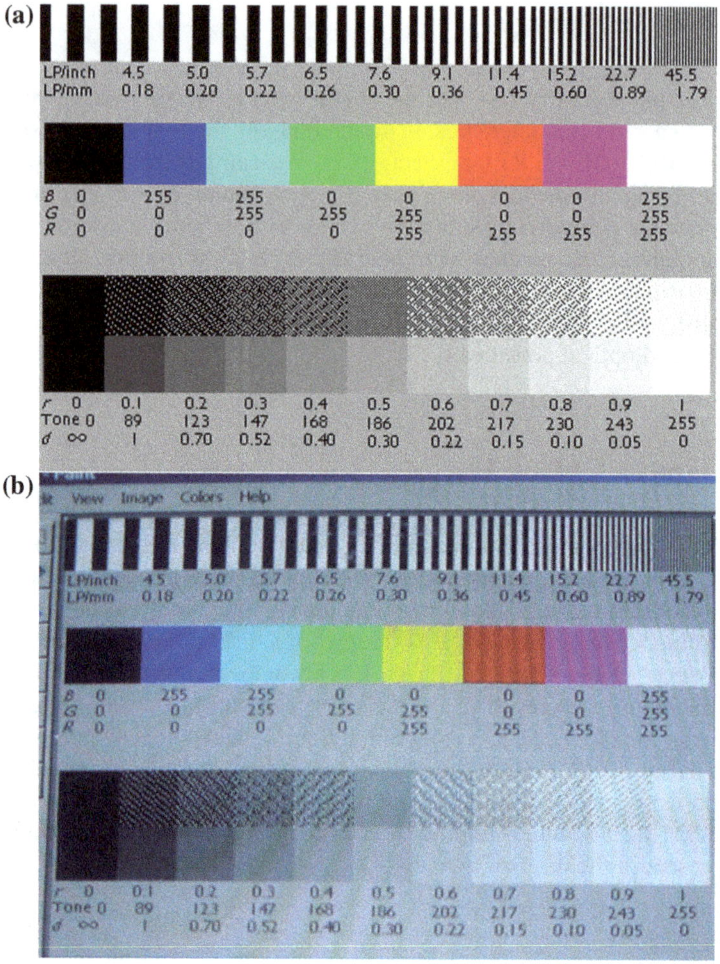

Fig. 46.17 Camera colour test. **a** Own test image displayed. **b** Camera image of (**a**), displayed. (B, G, R) values are compared in Fig. 46.18

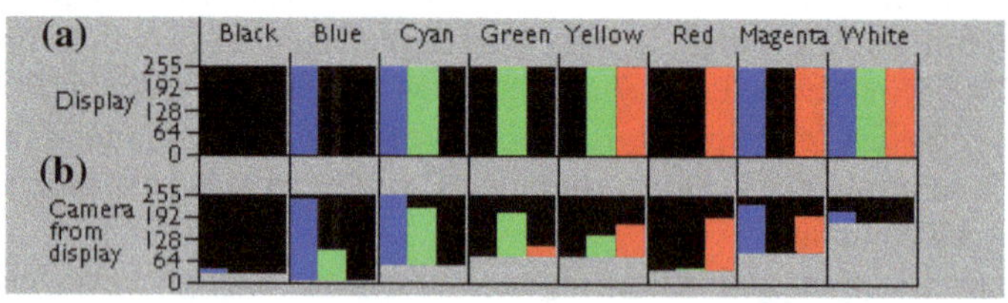

Fig. 46.18 Camera colour test. **a** Histograms of (B, G, R) relative intensities for Fig. 46.17a. **b** Histograms of (B, G, R) relative intensities for Fig. 46.17b, with equal intensities replaced by *grey*

A digital bridge camera [11] has a large zoom lens, a CCD or CMOS sensor of about 16 Mpx, a preview/review display screen, and .JPG or RAW output. It has more adjustable photographic modes, and is bulkier, than a compact.

A digital single lens reflex (DSLR) camera has interchangeable lenses, a CMOS sensor of 16 Mpx or more, through-the-lens view-finding, and adjustable modes throughout.

Mirrorless interchangeable-lens cameras, with near-DSLR quality but compact size, are likely to become increasingly popular [7].

A smartphone camera has a fixed-focus lens, CMOS sensor up to perhaps 40 Mpx, no physical shutter (so large shutter lag), a preview/review display screen, .JPG output, and cable or wireless coupling to a computer.

46.15 Computational Photography State of the Art

Current topics [2] include panoramic imaging, high dynamic range imaging, tone-mapping, and various special-camera techniques.

A panoramic image is one where the view width is much greater than the view height. It can be produced by central projection from a viewpoint to a cylindrical surface, when it is more properly called a dioramic image (see Sect. 8.22. The image should be presented as a (part-) cylinder: if presented flat it looks peculiar. Since the 19th century cameras have been constructed for taking school photographs and similar, with progressive exposure left to right over a restricted sector of view. Recently cameras have been developed which rotate, up to 360°, to capture a very wide image [6]. Alternatively, several images from a standard camera can be stitched together as a single wide image [10]. Image-stitching is available in many photo-manipulation packages.

A natural scene may emit or reflect a very large range of intensities from darkest dark to lightest light: it is then said to have a *high dynamic range* (HDR). Different digital devices, and the eye, have characteristically different dynamic ranges within which they function correctly (see Sect. 29.6). Where a scene exceeds the dynamic range of a camera it is possible to combine several differently-exposed images of the same scene as one HDR image, by techniques originated by Mann[3] in 1994 [4, 5], using Fourier transforms in the amplitude domain. HDR processing is now incorporated in high-end cameras and photo-processing packages.

When an image is displayed or printed-out the dynamic range is usually compressed or curtailed, sometimes severely. *Tone-mapping* is the adjustment of the source range to the presentation range, where the main aim is to preserve the most significant local contrasts. In painting, the subject is discussed in great depth by Ruskin [8]. In film photography, it is achieved by dodging and burning when printing from one or more negatives. In digital photography [16] it is an elaboration of the elementary intensity transformations described in Sect. 49.5.

[3] Steve Mann (born 1962), Professor of Electrical and Computer Engineering, University of Toronto.

References

1. Bayer BE (1976) Color imaging array. US Patent 3,971,065, 20 July 1976
2. Durand F, Szeliki R (2007) Computational photography. IEEE Comput Graph Appl 27(2):21-22. http://www.computer.org/csdl/mags/cg/2007/02/mcg2007020021.pdf. Accessed 1 Feb 2014
3. Fossum ER (1998) CMOS active pixel sensor type imaging on a chip. US Patent 5,841,126, 24 Nov 1998
4. Mann S, Picard RW (1994) On being 'undigital' with digital cameras: extending dynamic range by combining differently exposed pictures. http://wearcam.org/is_t95_myversion.pdf. Accessed 1 Feb 2014
5. Mann S (1998) Method and apparatus for producing digital images having extended dynamic ranges. US Patent 5,828,793 A, 27 Oct 1998
6. Open Directory (2013) Panoramic and 360 degree. http://www.dmoz.org/Arts/Photography/Techniques_and_Styles/Panoramic_and_360. Accessed 1 Feb 2014
7. Photography Life (2014) The future of digital cameras. http://photographylife.com/future-digital-cameras. Accessed 1 Feb 2014
8. Ruskin J (1843) Modern painters: of truth of tone, vol I, pt II, Sect. II, Chap. I. http://www.gutenberg.org/files/29907/29907-h/29907-h.htm#page140. Accessed 1 Feb 2014
9. Sasson SA (1978) Electronic still camera. US Patent 4,131,919, 26 Dec 1978
10. Szeliski R (2005) Image augmentation and stitching. http://www.caam.rice.edu/~zhang/caam699/p-files/Im-Align2005.pdf. Accessed 1 Feb 2014
11. Wikipedia (2014) Bridge camera. http://en.wikipedia.org/wiki/Bridge_camera. Accessed 1 Feb 2014
12. Wikipedia (2014) Camera modes. http://en.wikipedia.org/wiki/List_of_digital_camera_modes. Accessed 1 Feb 2014
13. Wikipedia (2014) Camera phone. http://en.wikipedia.org/wiki/Camera_phone. Accessed 1 Feb 2014
14. Wikipedia (2014) Digital camera. http://en.wikipedia.org/wiki/Digital_camera. Accessed 1 Feb 2014
15. Wikipedia (2014) Smartphone. http://en.wikipedia.org/wiki/Smartphone. Accessed 1 Feb 2014
16. Wikipedia (2013) Tone mapping. http://en.wikipedia.org/wiki/Tone_mapping#Purpose_and_methods. Accessed 1 Feb 2014

Chapter 47
24-bit Process

A program which changes the locations of a 24-bit image (keeping the intensities unchanged) is given, with examples. Programs to combine two 24-bit files as one by alternating locations, and to reduce the pixel-count of locations, are given, with examples. Programs to change the intensities of a 24-bit image (keeping the locations unchanged), and to combine two 24-bit files as one by averaging intensities, are given, with examples. A separation program to replace a 24-bit image by four 8-bit images ready for half-toning is also given.

47.1 Flowchart

The flowchart (Fig. 47.1) is for general guidance in this chapter.

47.2 Change 24-bit Image Locations

Changing 24-bit locations is like changing 1-bit locations as described in Sects. 23.3–23.11. Six elementary transformations are

- *Reflection* of an image about a mirror-line,
- *Translation* of an image along a direction,
- *Rotation* of an image about a pole,
- *Dilation* of an image about an origin,
- *Shearing* of an image along an axis,
- *Inversion* of an image about a circle.

Figure 47.2 shows a given 24-bit image, and an example of each of the six transformations.

Reflection moves each pixel of an image from one side of a mirror-line to the other side. Pixels on the mirror-line remain fixed. The sense of the image (left/right,

A. Parkin, *Digital Imaging Primer*, DOI 10.1007/978-3-540-85619-1_47

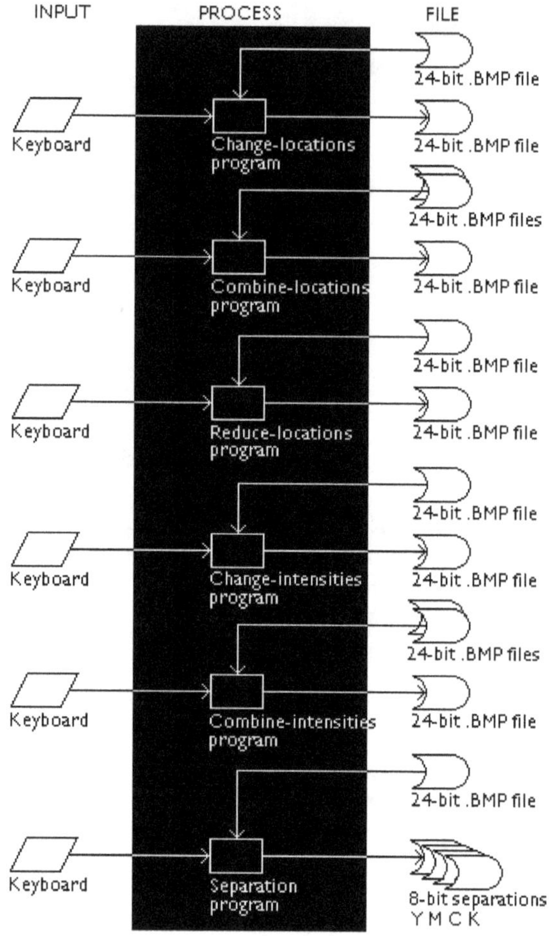

Fig. 47.1 24-bit process. A program can change the locations of a given 24-bit image (keeping the intensities unchanged). Or it can combine two 24-bit files as one, by alternating locations. Or it can reduce the pixel-count of locations. A program can change the intensities of a given 24-bit image (keeping the locations unchanged). Or it can combine two 24-bit files as one, by averaging intensities. A separation program can replace a 24-bit image by four 8-bit images ready for half-toning

clockwise/anticlockwise) is reversed. Figure 47.2b shows the image Fig. 47.2a reflected in a mirror-line (marked) through the origin at 45°; the pale neck goes to the other side. Two successive reflections in the same mirror-line give the identity transformation.

Translation moves each pixel of an image by a certain distance in a certain direction. No pixel remains fixed. The sense of the image is preserved. Two successive reflections in parallel mirror-lines give a translation by twice the distance between the mirror-lines. Figure 47.2c shows the image translated by 16 px to the right and 32 px upwards.

Rotation moves each pixel of an image through a certain angular arc of a circle whose centre is the pole of the rotation. The pole remains fixed. The sense of the image is preserved. Angles are conventionally measured anticlockwise from the x-axis. Two successive reflections in mirror-lines which meet give a rotation by twice the angle between the mirror-lines, and the meet of the mirror-lines is the pole

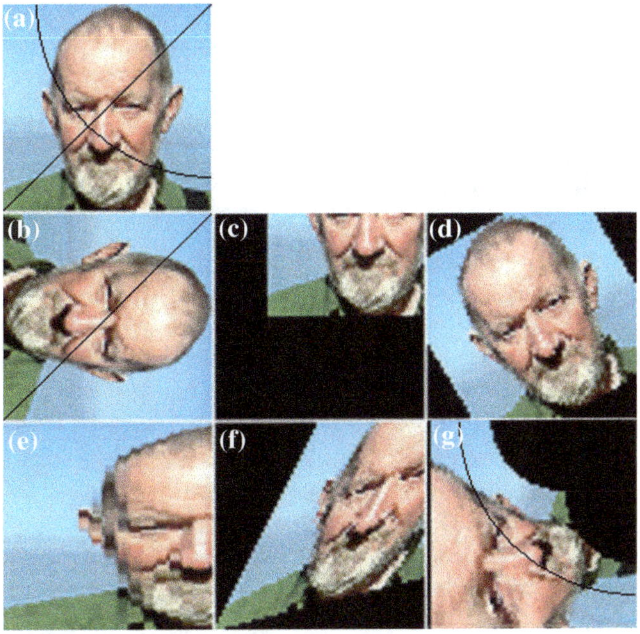

Fig. 47.2 Change 24-bit image locations. **a** Given 24-bit image of 64 × 64 px. **b** Reflected in mirror-line through the origin at 45°. **c** Translated by 16 px in X and 32 px in Y. **d** Rotated about (32, 32) by 30°. **e** Dilated about (0, 0) by factor 2 in X and 0 in Y. **f** Sheared about (0, 0) by factor 0.5 in X and 0.25 in Y. **g** Inverted in circle with centre (64, 64) and radius 54 px

of the rotation. Figure 47.2d shows the image rotated about pole at centre (32, 32) by 30°.

Dilation about the origin (0, 0) moves each pixel of an image in the x-direction by a certain proportion of its original X-distance, and in the y-direction by a certain proportion of its original Y-distance. The origin remains fixed. If the dilation factor is positive, it is a magnification or enlargement or stretch or expansion. If the dilation factor is negative, it is a diminution or reduction or shrink or compression. If the dilation is equal in X and Y it is a similarity: the sense and angles of the image are preserved, but all lengths are increased or decreased. Figure 47.2e shows the image dilated by a factor of 2 (that is stretched to twice size) in the x-direction and unchanged in the y-direction.

Shearing about the origin (0, 0) moves each pixel of an image in the x-direction by a certain proportion of its original Y-distance, and in the y-direction by a certain proportion of its original X-distance. The effect is a slanting of the image. The origin remains fixed. Straight lines remain straight lines, and the sense of the image is preserved, but not lengths or angles. Figure 47.2f shows the image sheared by a factor of $0.5Y$ in the x-direction and $0.5X$ in the y-direction.

Inversion is reflection in a mirror-circle. Parts originally inside the circle go outside and are magnified; parts originally outside the circle go inside and are diminished.

The sense of the image (left/right, clockwise/anticlockwise) is reversed. Figure 47.2g shows the image inverted about a circle (marked) with centre (64, 64) and radius 54 px. The pale neck goes to the other side of the circle.

47.3 24-bit Change-location Program

A program which performs all of the above transformations of location is described below (see the SpringerLink web-site for source code). This program lets the user pick any 24-bit source image, and one of the geometric transformations of it. It is like the 1-bit change-location program in Chap. 28, but with different sub-programs for reading and writing 24-bit intensities.

```
REM  **  Program CHANLO24 to perform geometric
REM  **     transformations of plane 24-bit image
REM  **     locations. User enters filenames.
REM  **     Program reads given "before" image from
REM  **     24-bit .BMP file #1 into array IMAGEA.
REM  **     User picks transformation and enters
REM  **     constants. Program then processes pixel
REM  **     by pixel from IMAGEA to array IMAGEB,
REM  **     setting old intensity at new location.
REM  **     Program writes "after" image from IMAGEB
REM  **     to 24-bit .BMP file #2.
REM  **  Main program
REM  **  Enter file-names
         CALL RWHEAD24
         CALL REIMAG24
         CALL GEOMTR24
         CALL WRIMAG24
         CALL REVHEADB
END $
REM $STATIC
         SUB GEOMTR24
REM  **  Sub-program GEOMTR24 to enter constants for
REM  **     transformation wanted, set up transformation
REM  **     matrix, and multiply coordinates of each
REM  **     pixel from IMAGEA to IMAGEB.
REM  **  Reflection
REM  **  Translation
REM  **  Rotation
REM  **  Dilation
REM  **  Shear
REM  **  Inversion
```

```
END SUB
DEFSNG A-Z
         SUB REIMAG24
END SUB
         SUB REVHEADB
REM  **  Revises header of file #2 with values in bytes.
END SUB
         SUB RWHEAD24
REM  **  Subprogram RWHEAD24 to read header of 24-bit
REM  **     .BMP input file #1, and preliminary-write
REM  **     same header to output file #2 (to be
REM  **     revised later).
END SUB
         SUB WRIMAG24
REM  **  Subprogram WRIMAG24 to write data from array
REM  **     IMAGEB to 24-bit file #2. Each intensity
REM  **     B G R is one byte in the file.
END SUB
```

47.4 Combine 24-bit Images by Location

Two images A and B with the same pixel-count may be combined as one C by taking pixels alternately from A and from B to construct C. The alternation may be by columns or by rows or both. For example, Fig. 47.3 shows two 24-bit images and their combination by alternating columns.

A program which combines two given 24-bit images by location or by averaging intensity is briefly described below (see the SpringerLink web-site for source code).

```
REM  **  Program COMBIN24 to combine two 24-bit images
REM  **     with same pixel-count. User enters
REM  **     file-names for images A, B, and C, alternating
REM  **     or averaging with integer weights.
REM  **     Program processes pixel by pixel from
REM  **     file #1 and file #2, and puts new intensity
REM  **     for each location into file #3.
REM  **  Main program
REM  **  Enter file-names for A, B, C, and alternating or
REM  **  averaging with weights.
         CALL RWHEAD24
         CALL CALCCOMB
END
         SUB CALCCOMB
```

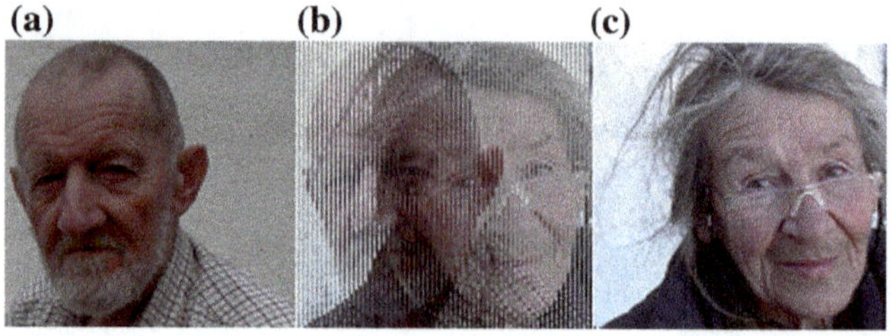

Fig. 47.3 Combine by location. **a** 24-bit image, 128 × 128 px. **b** Combination of (**a**) and (**c**) by alternating columns, 128 × 128 px. **c** 24-bit image, 128 × 128 px

```
REM   **   Sub-program CALCCOMB to read image data from
REM   **   24-bit .BMP file #1 and file #2, and put either
REM   **   alternate or weighted-average intensities into
REM   **   file  #3.
END SUB
           SUB RWHEAD24
REM   **   Sub-program RWHEAD24 to read header of 24-bit
REM   **  .BMP input file #1, and write same header to
REM   **   output file #3.
END SUB
```

47.5 Reduce 24-bit Image Locations

Given a 24-bit image, we may reduce the count (and increase the size) of pixels by taking every Nth pixel and repeating it N times. The reduced image has $1/N^2$ as many pixels, each N^2 the size of the original. A 24-bit image with Q px conveys $24 \times Q$ bits of information, and a reduced image conveys $(24Q)/N^2$ bits. For example, Fig. 47.4 shows a 24-bit image thus reduced by $N = 2, 4$, and 8. The reduced images show the full range of colours in the fewer pixels.

A program which reduces 24-bit image locations is described briefly below (see the SpringerLink web-site for source code):

```
REM   **   Program REDULO24 to reduce count of pixels
REM   **      for 24-bit image. User enters filenames.
REM   **      Program reads input image from 24-bit .BMP
REM   **      file #1 into array IMAGEA. User picks
REM   **      reduction. Program takes every Nth pixel
REM   **      from array IMAGEA and repeats it N times
REM   **      to array IMAGEB, with same intensity.
REM   **      Program writes reduced image from IMAGEB
REM   **      to 24-bit .BMP file #2.
```

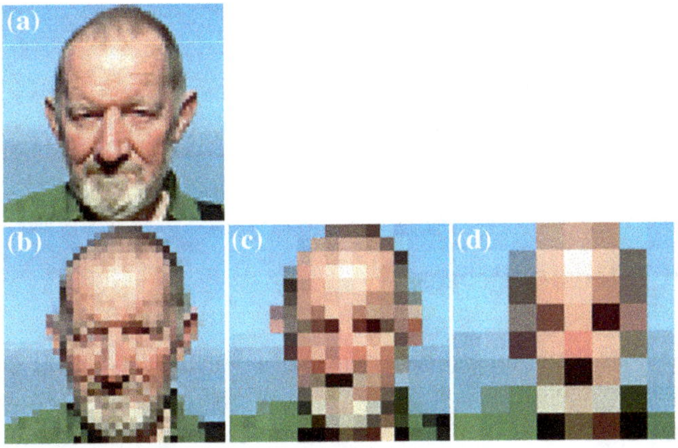

Fig. 47.4 Reduce 24-bit image locations. **a** 24-bit image 64 × 64 px: 98,304 bits. **b** Reduction of (**a**) by $N = 2$ to 32 × 32 px: 24,576 bits. **c** Reduction of (**a**) by $N = 4$ to 16 × 16 px: 6,144 bits. **d** Reduction of (**a**) by $N = 8$ to 8 × 8 px: 1,536 bits

```
REM   **   Main program
REM   **   Enter file-names
           CALL RWHEAD24
           CALL REIMAG24
           CALL LO24REDU
           CALL WRIMAG24
           CALL REVHEADB
END
           SUB LO24REDU
REM   **   Sub-program LO24REDU to enter reduction wanted,
REM   **      and perform reduction for each pixel from
REM   **      IMAGEA to IMAGEB.
END SUB
           SUB REIMAG24
REM   **   Subprogram REIMAG24 to read 24-bit .BMP file
REM   **      #1 and load old image data into array IMAGEA.
REM   **      The data are B, G, R each 0-255.
END SUB
           SUB REVHEADB
REM   **   Revises header of file #2 with values in bytes.
END SUB
           SUB RWHEAD24
REM   **   Subprogram RWHEAD24 to read header of 24-bit
REM   **      .BMP input file #1, and preliminary-write
REM   **      same header to output file #2 (to be
REM   **      revised later).
```

```
END SUB
         SUB WRIMAG24
REM  **  Subprogram WRIMAG24 to write data from array
REM  **     IMAGEB to 24-bit file #2. Each intensity
REM  **     B G R is one byte in the file.
```

47.6 24-bit Intensity Process

A 24-bit image has three independent ranges of intensity, each a full sequence 0 to 255 (and no colour-table). In such an image a smooth gradient from any colour to any colour can be formed, and various numerical operations on 24-bit intensities are possible. The 24-bit processes are like the 8-bit grey-scale case (see Sect. 39.5) done three times over.

Given a 24-bit image with per-pixel intensities (B, G, R), to perform an intensity transformation:

- Normalize each intensity as $I_{\text{norm}} = I_{256}/255$.
- Apply a numerical function $f(I_{\text{norm}})$.
- Restore transformed intensity $I'_{256} = (f(I_{\text{norm}})) \times 255$.

Some elementary intensity functions are,

- reverse intensity,
- add/subtract intensity,
- multiply/divide intensity,
- exponentiate intensity,
- re-scale intensity range,
- combine intensities of two images.

The functions are reversible and combinable. Some GUI image editors offer ready-made intensity transformations. In general, a program is needed.

47.7 Reverse Intensity

Numerically, the reversal for the Blue intensity of each pixel is

$$\text{B}'_{256} = (1 - \text{B}_{\text{norm}}) \times 255,$$

where $0 \leq \text{B} \leq 255$ is the given intensity, and $\text{B}_{\text{norm}} = \text{B}/255$. And similarly for the Green and Red intensities. Figure 47.5a shows a given 24-bit colour image, and (b) the corresponding intensity-reversal.

The transformation can also be represented geometrically by a scale of tone-codes 0 to 255 (for all three channels), normalized between 0 and 1. That is to say,

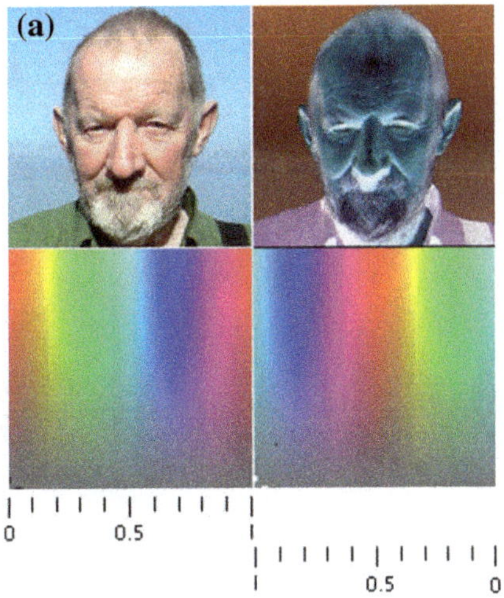

Fig. 47.5 Reverse intensity. **a** Given 24-bit image: camera image 128 × 128 px and 24-bit colour gamut 128 × 128 px. **b** Intensity-reversal of (**a**). Below the images are the corresponding scales of intensities of B, G, and R normalized between 0 and 1

scale 0 represents tone 0, scale 0.5 tone 128, and scale 1 tone 255. In this case the transformation reverses the scale linearly, so 0 becomes 1, 0.5 becomes 0.5, and 1 becomes 0.

Visually, reversing intensity is like making a colour negative of a photograph. MS PAINT offers reversal as Image—Invert Colors.

47.8 Add/Subtract Intensity

Numerically, the function for Blue for each pixel is

$$B'_{256} = (B_{norm} \pm k) \times 255,$$

where $0 \leq B \leq 255$ is the given intensity, $0 \leq B_{norm} \leq 1 = B/255$, and $0 \leq k \leq 1$ is a given constant. And similarly for Green and Red. Figure 47.6 shows in the middle the given image; at the left after adding $k = 0.5$; and at the right after subtracting $k = 0.5$.

Geometrically, adding shifts intensities up the scale, so lower values become higher values and higher values merge at the ceiling. Subtracting shifts intensities down the scale, so upper values become lower values and lower values merge at the floor.

Visually, adding makes an image paler; subtracting makes an image darker. In the gamut, adding makes the primary B G R bands narrower and the secondary Y C M bands wider; while subtracting makes the primaries wider and the secondaries narrower.

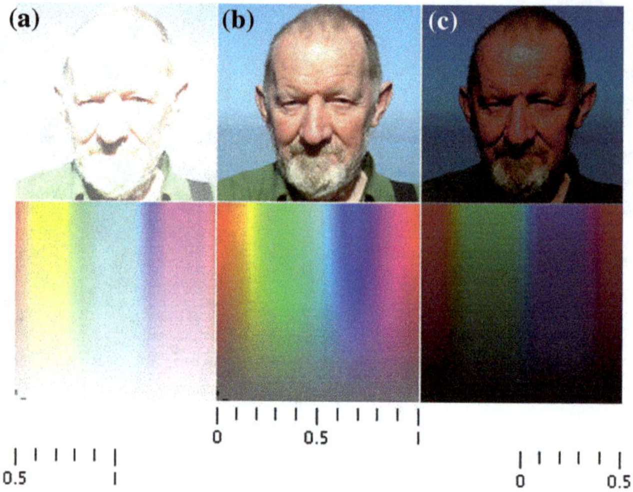

Fig. 47.6 Add/subtract intensity. **a** After adding $k = 0.5$ to (**b**). **b** Given images. **c** After subtracting $k = 0.5$ from (**b**)

47.9 Multiply/Divide Intensity

Numerically, the function for Blue for each pixel is

$$\mathrm{B}'_{256} = (\mathrm{B}_{\mathrm{norm}} \times k) \times 255$$

or

$$\mathrm{B}'_{256} = (\mathrm{B}_{\mathrm{norm}} \div k) \times 255,$$

where $0 \leq \mathrm{B} \leq 255$ is the given Blue intensity, $0 \leq \mathrm{B}_{\mathrm{norm}} \leq 1 = \mathrm{B}/255$, and $0 \leq k \leq 1$ is a given constant. Figure 47.7 shows in the middle the given images; at the left after multiplying by $k = 0.5$; and at the right after dividing by $k = 0.5$.

Geometrically, multiplying by a factor less than 1 compresses all intensities linearly into the lower part of the scale, so gradients are steeper but none are pushed below the floor. Dividing by a factor less than 1 expands lower intensities into the whole scale, so gradients are shallower but upper intensities are pushed above the ceiling.

Visually, multiplying by a factor less than 1 makes an image paler and gradients steeper; dividing by a factor less than 1 makes an image darker and gradients shallower. In the gamut, multiplying by less than 1 makes the primary bands narrower and the secondary bands wider; while dividing by less than 1 makes the primaries wider and the secondaries narrower.

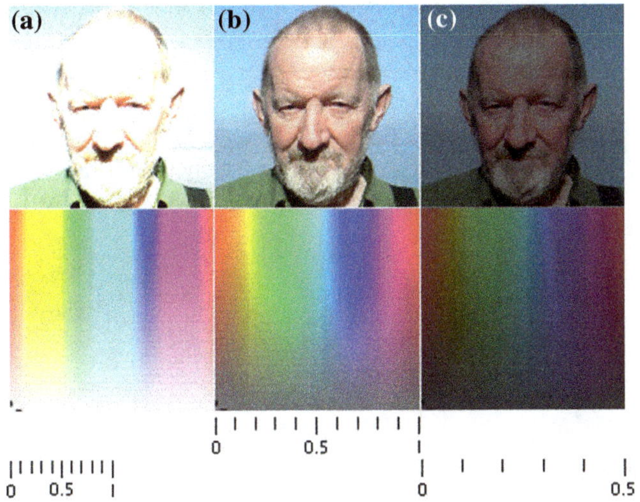

Fig. 47.7 Multiply/divide 24-bit intensity. **a** After multiplying (**b**) by $k = 0.5$. **b** Given image. **c** After dividing (**b**) by $k = 0.5$

47.10 Exponentiate Intensity

Numerically, the function for Blue for each pixel is

$$B'_{256} = (B^{k}_{\text{norm}}) \times 255$$

or

$$B'_{256} = (B^{1/k}_{\text{norm}}) \times 255,$$

where $0 \leq B \leq 255$ is the given blue intensity, $0 \leq B_{\text{norm}} \leq 1 = B/255$, and $0 \leq k \leq 1$ is a given constant. Figure 47.8 shows in the middle the given images; at the left after applying power $k = 0.4545$; and at the right after applying power $1/k = 1/0.4545 = 2.2$. These are indeed the standard gamma functions for encoding and decoding intensities in a digital system (see Sect. 22.12).

Geometrically, raising to a power less than 1 compresses lower intensities and expands upper intensities, so low-intensity gradients are steeper and high-intensity gradients shallower, but none are lost. Raising to a power greater than 1 does just the opposite.

Visually, raising to a power less than 1 makes an image paler, because high intensities dominate. Raising to a power greater than 1 makes an image darker, because low intensities dominate. In the gamut, exponent < 1 makes the primary bands narrower and the secondary bands wider; while exponent > 1 makes the primaries wider and the secondaries narrower.

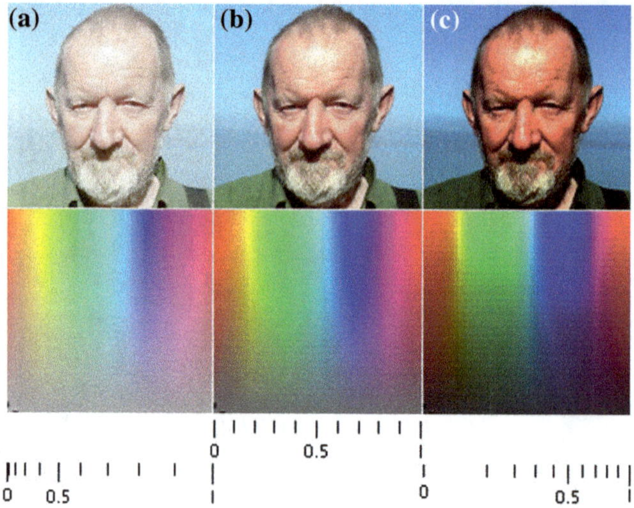

Fig. 47.8 Exponentiate intensity. **a** After raising (**b**) to power $k = 0.5$. **b** Given images. **c** After raising to power $1/k = 1/0.5 = 2$

47.11 Re-scale Intensity

Numerically, to re-scale from full to restricted range the function for Blue for each pixel is

$$B'_{256} = (B_{norm} \times (I'_{max} - I'_{min})) + I'_{min}$$

where $0 \leq B \leq 255$ is the given intensity, $0 \leq B_{norm} \leq 1 = B/255$, and $0 \leq I'_{min} \leq 1$ and $0 \leq I'_{max} \leq 1$ is the new restricted range. To re-scale from restricted to full range the function is:

$$B'_{256} = (B_{norm} - I_{min})/(I_{max} - I_{min}),$$

where $0 \leq B \leq 255$ is the given intensity, and $0 \leq I_{min} \leq 1$ and $0 \leq I_{max} \leq 1$ is the old restricted range. Figure 47.9b shows the given images. In (c) the full range of (b) between 0 and 1 is re-scaled to restricted range between 0.25 and 0.75. In (a) the restricted range of (b) between 0.25 and 0.75 is re-scaled to full range 0–1.

Geometrically, re-scaling from restricted to full range expands the middle tones and loses the extremes, so all gradients are shallower. Re-scaling from full to restricted range compresses all tones without loss, so all gradients are steeper.

Visually, re-scaling from restricted to full range makes an image less saturated, in effect imposing a grey mist. Re-scaling from full to restricted range makes an image more saturated, removing most of the grey. In the gamut, the differences of saturation are clearly displayed.

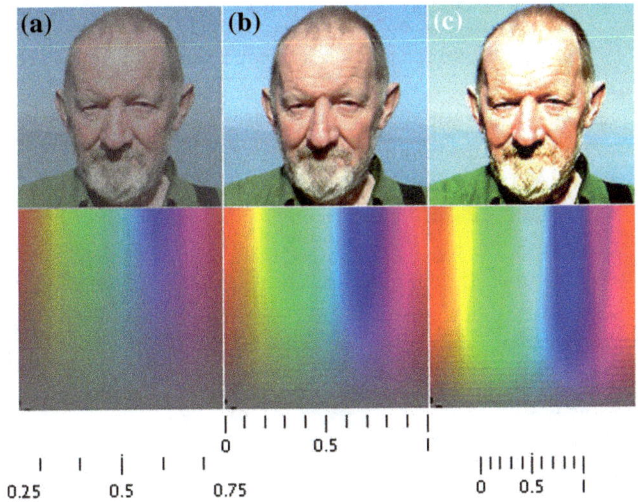

Fig. 47.9 Re-scale intensity. **a** Restricted range 0.25 to 0.75 re-scaled to full range. **b** Given images. **c** Full range re-scaled to restricted range 0.25 to 0.75

47.12 24-bit Transform-intensity Program

A program which does any of these intensity transformations is briefly described below (see the SpringerLink web-site for source code).

```
REM  **  Program TRANPI24 to perform transformations
REM  **     of 24-bit plane image intensities.
REM  **     User enters filenames and transformation
REM  **     wanted, and constants. Program reads pixel
REM  **     by pixel old B G R intensities from 24-bit
REM  **     .BMP file #1, and calculates new intensities
REM  **     back to same file.
REM  **  Main program
REM  **  Enter file-name for image
         CALL REHEAD24
         CALL TRANIN24
END
         SUB REHEAD24
REM  **  Subprogram REHEAD24 to read some fields of
REM  **     24-bit .BMP file #1.
END SUB
         SUB TRANIN24
REM  **  Sub-program TRANIN24 to calculate new intensities.
REM  **     User picks case wanted and enters constant.
```

```
REM  **     For each pixel in file, program calculates new
REM  **     B G R intensities and replaces in same file.
REM  **  Pick case
REM **   Enter constant
REM  **  Process blue intensity of each pixel
REM  **  from file #1 and write back to file
REM  **  #1; then green, then red.
REM  **  Invert
REM  **  Shift up
REM  **  Compress
REM  **  Expand
REM  **  Power up
REM  **  Power down
REM  **  Compress range
REM  **  Expand range
END SUB
```

47.13 Combine Images by Averaging Intensities

Two images A and B with the same pixel-count may be combined as C by taking the weighted average of intensities per pixel.

Numerically, the weighted average is, for each pixel:

$$I_{\mathrm{C}} = (W \times I_{\mathrm{A}}) + (V \times I_{\mathrm{B}}) \backslash (W + V),$$

where W and V are integer weights, I_{A} is the intensity of the pixel in image A, I_{B} is the intensity of the corresponding pixel in image B, and I_{C} is the weighted average in the combined image C.

Visually, when $W > V$, image A dominates; when $W = V$, images A and B are equally strong; when $W < V$, B dominates.

For example, Fig. 47.10 shows two 24-bit images and their combination by averaging intensities.

Program COMBIN24 in Sect. 47.4 above will do this combination.

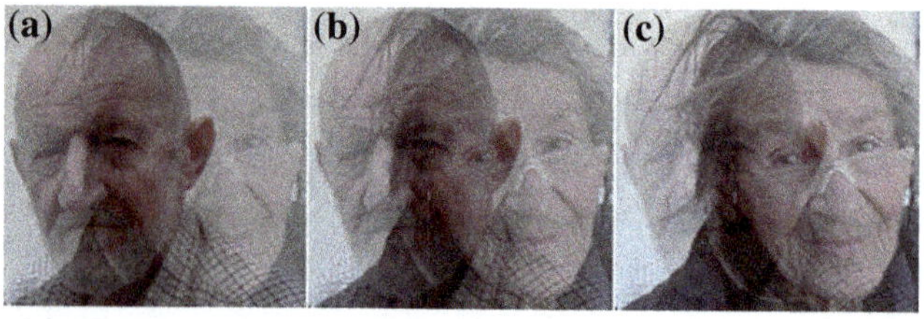

Fig. 47.10 Combine by averaging intensities. **a** Image A weight 2, image B weight 1. **b** Image A weight 1, image B weight 1. Image A weight 1, image B weight 2

47.14 Separate 24-bit Image to 8-bit

Printing a colour image requires separation to four grey-scale sub-images which are then half-toned to four 1-bit sub-images, as described in Chaps. 35, 43, and 50. Separation and half-toning is usually done within closed proprietary software supplied with a printer. BGR-separation is a standard operation in some image editors (but not in MS PAINT). The program briefly described below and listed in the Springer Link web-site produces one separation per run, and can be run four times over on any 24-bit .BMP file.

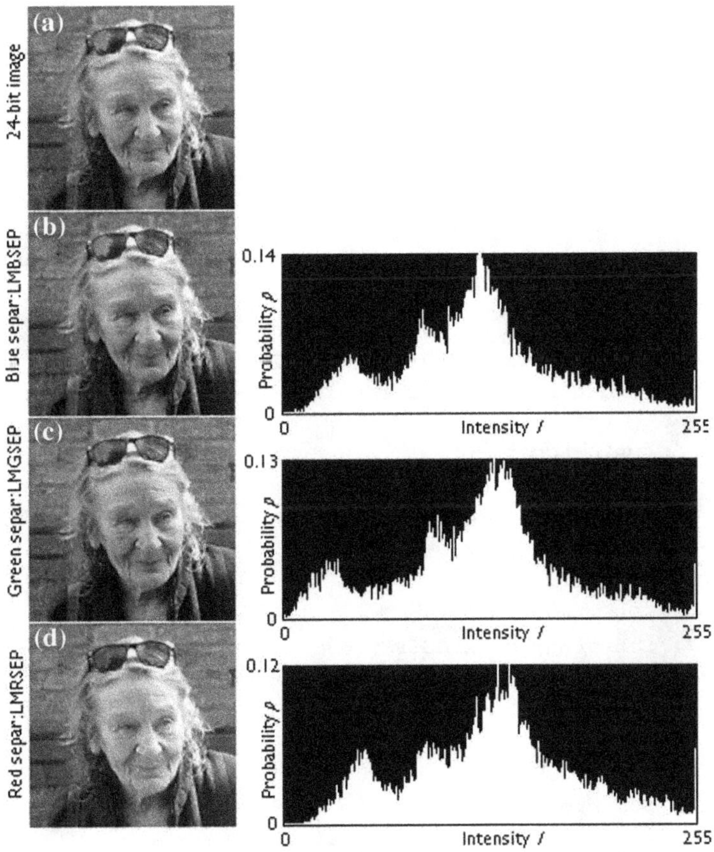

Fig. 47.11 Image, separations, and histograms. **a** 24-bit camera image. **b** Blue separation and histogram. **c** Green separation and histogram. **d** Red separation and histogram. The original camera image has mainly muted greys, with a slight predominance of Red over Green, and of Green over Blue. The separations are closely similar, with Red somewhat lighter than Green, and Green somewhat lighter than Blue

```
REM  **  Program SEP24TO8 to make an 8-bit separation
REM  **  image from a 24-bit image (need 4 runs for
REM  **  Y M C K). User enters filename for 24-bit
REM  **  and for separation file. Program reads
REM  **  24-bit file heads and writes 8-bit file
REM  **  heads. Program reads each 24-bit pixel,
REM  **  calculates YMCK values and writes whichever
REM  *   to output file.
REM  **  Main program
REM  **  Enter 24-bit file-name, pick Y or M or C or K,
REM  **  and output file-name
         CALL REA24HEA
            CALL CRE8HEAD
            CALL CALCSEPY
            CALL CALCSEPM
            CALL CALCSEPC
            CALL CALCSEPK
         CALL REVHSEP
END

         SUB CALCSEPC
REM  **  Sub-program CALCSEPM to read 3 bytes B G R
REM  **     from file #2, calculate C-separation
REM  **     and write to file #1
END SUB
         SUB CALCSEPK
REM  **  Sub-program CALCSEPM to read 3 bytes B G R
REM  **     from file #2, calculate K-separation
REM  **     and write to file #1
END SUB
         SUB CALCSEPM
REM  **  Sub-program CALCSEPM to read 3 bytes B G R
REM  **     from file #2, calculate M-separation
REM  **     and write to file #1
END SUB
         SUB CALCSEPY
REM  **  Sub-program CALCSEPY to read 3 bytes B G R
REM  **     from file #2, calculate Y-separation
REM  **     and write to file #1
END SUB
         SUB CRE8HEAD
REM  **  Subprogram CREHEAD to write header and
REM  **  colour-table of 8-bit .BMP file #1.
END SUB
         SUB PUTI (INTE%)
```

```
REM   **   Sub-program PUTI to put integer value in binary
REM   **      file at pointer position (set outside) and
REM   **      increment pointer.
END SUB
           SUB PUTL (INTL&)
REM   **   Sub-program PUTL to put long integer into
REM   **      binary file #1 as 4 bytes
END SUB
           SUB REA24HEA
REM   **   Subprogram REA24HEA to read some fields of
REM   **      24-bit .BMP file #2.
END SUB
           SUB REVHSEP
REM   **   Sub-program REVHSEP to revise header of
REM   **      separation file #1 with values in bytes.
END SUB
```

47.15 24-bit Statistics

A 24-bit colour image may be decomposed into three 8-bit grey-scale separations, Blue, Green, and Red, as in Sect. 47.14. We may then take statistics for each of the three separations as for any other 8-bit grey-scale images. Figure 47.11 shows a 24-bit colour image, its Blue Green and Red grey-scale separations, and their histograms. Figure 47.12 shows the statistics of the separations.

```
Image file is LMBSEP.BMP        is LMGSEP.BMP      is LMRSEP.BMP

STATISTICS OF IMAGE INTENSITY
-------------------------  -----------------  -----------------
 Image bit-depth        8  bit-depth        8  bit-depth        8
 Image width, px      128  width, px      128  width, px   -  128
Image height, px      128  eight, px      128  eight, px      128
  Image total px    16384   total px    16384   total px    16384
            Mean      119       Mean      120       Mean      128
            Mode      122       Mode      131       Mode      140
Lowest intensity        0  intensity        0  intensity        0
Highest intensit      255   intensit      255   intensit      255
        Variance     2434   Variance     2661   Variance     2710
Standard deviatn       49  d deviatn       52  d deviatn       52
            Skew        0       Skew       -0       Skew        0
        Kurtosis       -0   Kurtosis       -0   Kurtosis       -0
  Lower quartile       87   quartile       91   quartile       91
          Median      119     Median      125     Median      130
  Upper quartile      144   quartile      148   quartile      158
Inten.complexity        8  omplexity        8  omplexity        8
Locat.complexity       14  omplexity       14  omplexity       14
Total complexity      112  omplexity      112  omplexity      112
         Entropy    7.549    Entropy    7.580    Entropy    7.628
```

Fig. 47.12 Statistics for separations in Fig. 47.11. Left columns are blue separation, centre are green, right are red

Chapter 48
24-bit File

In a 24-bit BMP file, each pixel has three 8-bit intensity values 0 to 255 (B, G, R). There is no index and no colour-table. The format of a 24-bit BMP file is defined, with an example. A program to edit 24-bit files is given.

48.1 Flowchart

The flowchart Fig. 48.1 is for general guidance in this chapter.

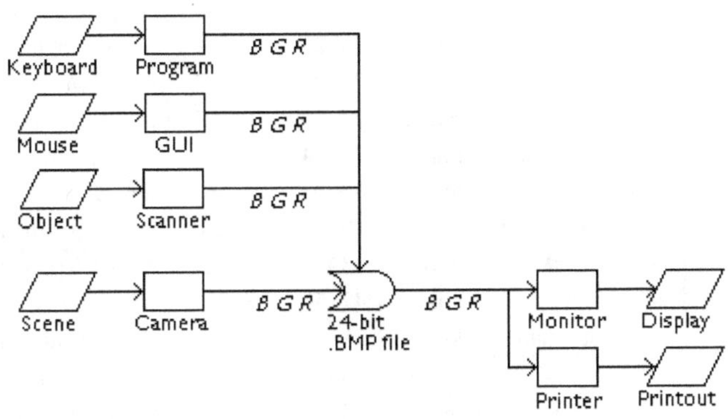

Fig. 48.1 A 24-bit .BMP file can be created by a program, a GUI editor, a scanner, or a camera. A .BMP file can be processed by a program or by a GUI editor. It can be displayed or printed-out by a GUI editor. It acts as an interface between inputs, processes, and outputs

A. Parkin, *Digital Imaging Primer*, DOI 10.1007/978-3-540-85619-1_48

48.2 Files and Data

For Memory, File, Data Representation, Bitmap Scan, and .BMP File, see Chap. 26.

48.3 24-bit .BMP File

Table 48.1 shows the structure of a 24-bit .BMP file.

The fields are in three groups. The FileHeader and the InfoHeader groups give the particulars of the file. The first field is bfType, ASCII BM = 66 77 decimal = 42 4D hexadecimal: without this the file is rejected as improper. The biXPelsPerMeter and

Table 48.1 24-bit file format

Group	Field	Type	dec.	hex.	Remarks
FileHeader			1–14	01–0E	File type, properties
	bfType	integer	1–2	01–02	Always 'BM'
	bfSize	long	3–6	03–06	File length, bytes
	bfReserved1	integer	7–8	07–08	Always 0
	bfReserved2	integer	9–10	09–0A	Always 0
	bfOffBits	long	11–14	0B–0E	Header length, bytes
InfoHeader			15–54	0F–36	Image size, properties
	biSize	long	15–18	0F–12	Infoheader len., bytes
	biWidth	long	19–22	13–16	Image width, pixels
	biHeight	long	23–26	17–1A	Image height, pixels
	biPlanes	integer	27–28	1B–1C	Always 1
	biBitCount	integer	29–30	1D–1E	Bits per pixel = 24
	biCompression	long	31–34	1F–22	Not compr. = 0
	biSizeImage	long	35–38	23–26	Image data len., bytes
	biXPelsPerMeter	long	39–42	27–2A	Printer p.p.m., horiz.
	biYPelsPerMeter	long	43–46	2B–2E	Printer p.p.m., vert.
	biClrUsed	long	47–50	2F–32	Always 0
	biClrImportant	long	51–54	33–36	Clrs considered imp.
Image data			55–	37–	In scan sequence
	aBitmapBits[0]	byte	55	37	Pixel 0, blue value
	aBitmapBits[1]	byte	56	38	Pixel 0, green value
	aBitmapBits[2]	byte	57	39	Pixel 0, red value
	aBitmapBits[3]	byte	58	3A	Pixel 1, blue value
	aBitmapBits[4]	byte	59	3B	Pixel 1, green value
	aBitmapBits[5]	byte	60	3C	Pixel 1, red value
	etc. . . .				

Field types are: *byte* occupying eight bits; *integer* occupying two bytes; *long* occupying four bytes

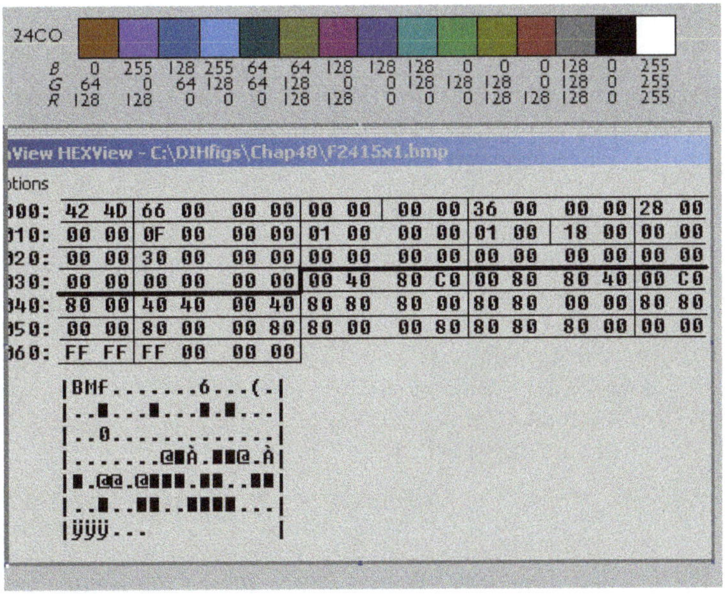

Fig. 48.2 24-bit file. The image is 15 px, shown with decimal (B, G, R) values. In the hexadecimal dump each pair of digits is the hex value of a byte. In a multi-byte field the hex pairs are taken as most-significant right, least-significant left. The file header fields are: a 2-byte magic cookie field containing 42 4D (ASCII BM); a 4-byte file size field containing 66h = 102d; 2 2-byte fields (ignored); a 4-byte header length 36h = 54d; a 4-byte info-header length 28h = 40d; a 4-byte image width Fh = 15d; a 4-byte image height = 1; a 2-byte planes field = 1; a 2-byte bpp field 18h = 24d; a 4-byte compression field = 0; an image data length 30h = 48d; 2 4-byte ppm fields = 0; a 4-byte colours used field = 0; and a 4-byte colours important field = 0. After the thick line are 15 triples of 1-byte fields for the (B, G, R) values of the image. The last 3 bytes 0 0 0 are a pad to make the file length a multiple of 4 bytes

biYPelsPerMeter fields are used for printing (see Sect. 28.3). The biClrUsed field = 0 for a 24-bit file.

There is no colour-table in a 24-bit .BMP file.

The image data group contains the (B, G, R) colour values of the pixels taken in scan sequence bottom left to top right of the image. The values occupy one 8-bit byte each, 3 bytes to the pixel.

Figure 48.2 shows an example of a 24-bit .BMP file with 15 px. The file contents are seen in the hexadecimal dump.

48.4 24-bit .BMP File Editor

If we wish to have interactive access to .BMP files, to view or change header, colour-table, and image-data values, we need a .BMP file editor. A brief description of such an editor follows (see the SpringerLink web-site for source code).

```
REM   **  Program EDBMP24.BAS to edit 24-bit BMP file. User
REM   **     enters filename to edit. Program gets file and
REM   **     offers menu to edit headers or color-table or
REM   **       image-array.
REM   **  Main program
REM   **  Enter file-name to edit
             CALL HEADER24
             CALL IMAARR24
FINIS:    END

          SUB HEADER24
REM   **  Sub-program HEADER24 to edit header values
REM   **     of 24-bit .BMP file. Reads each field of
REM   **     header, shows old value, offers next or
REM   **     change or exit.
END SUB
          SUB IMAARR24
REM   **  Sub-program IMAARR24 to edit 4-bit image-array.
REM   **  User has option to start at any row Y, from
REM   **  0 to (height - 1). Program shows for each
REM   **  pixel its Blue Green and Red intensities, and
REM   **  offers options of next or change or exit.
END SUB
```

Chapter 49
24-bit Display

Display of 24-bit images is like display of 8-bit images for location properties. A procedure for displaying part of a camera or scanner image in a window of chosen size is described. 24-bit intensity variables are defined, allowing display of the whole sRGB gamut.

49.1 Flowchart

The flowchart (Fig. 49.1) shows a 24-bit image file sent to a display. The flow is like the 8-bit case (see Chap. 42), but with any (B, G, R) intensities. Display gamma is significant.

49.2 Display Hardware

24-bit display hardware is the same as 1-bit hardware, described in Chap. 27.

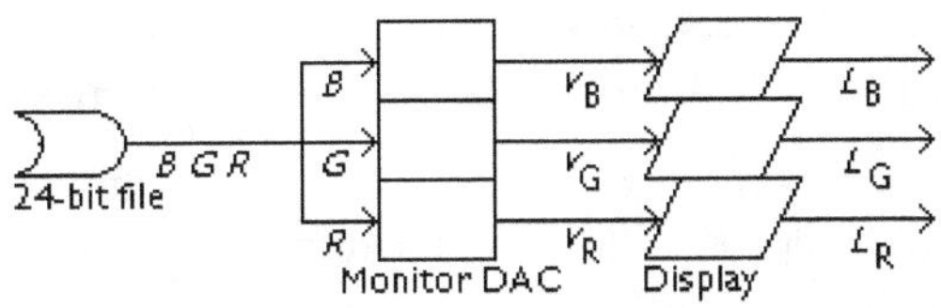

Fig. 49.1 Flowchart for 24-bit display. Per-pixel, the intensity triple (B, G, R) goes to the three channels in a colour monitor. In each channel, a digital-to-analogue converter sends a voltage v to the appropriate screen sub-pixel, which emits luminance L_B, L_G, L_R

A. Parkin, *Digital Imaging Primer*, DOI 10.1007/978-3-540-85619-1_49

49.3 Display Size Variables

24-bit display size variables are the same as 1-bit size variables, described in Chap. 27.

49.4 Display of Camera Image

A common task is to display part of a camera (or scanner) image in a window of a chosen size. For example, Fig. 49.2a shows a source image and the part we want to display. Say the source image is 1280 × 960 px, and we want to display the part shown in a window 128 × 128 px. We need to both diminish and crop the source image. One way to do this is by the following PAINT[1] procedure. Figure 49.2 shows some of the steps:

1. Make an empty window of the required final size (File—New, Properties—(set Width, set Height))
2. Set a thin box all round the edge of the window (Home—Shapes—Rectangle); and save (File—Save-as (BOX.BMP, 24-bit))
3. Make a transparent copy of the box on the clipboard (Home—Image—Select—Select All—Transparent. Right-click—Copy)
4. Open the source image (File—Open—SOURCE.BMP)
5. Put the transparent box on the source (Home—Clipboard—Paste)
6. Drag the box to near the part wanted and estimate the diminution percent DPC of source image needed to get the part into the box
7. Remove the box to the clipboard (Click outside the box, then Undo)
8. Diminish the source image (Home—Image—Resize (set Width and Height to DPC))
9. Bring back the box (Home—Clipboard—Paste) and drag to get the best framing. If no good, go back to 7
10. When good, copy filled box to clipboard (Click outside the box, then Image—Select—Not transparent—Crop. Right-click—Select All—Copy)
11. Open the empty box (File—Open—BOX.BMP)
12. Paste the filled box (Home—Clipboard—Paste)
13. Save the filled box (File—Save-as (FULLBOX.BMP))
14. Check that FULLBOX.BMP is as wanted (File—Properties)

There are of course other ways to get the same result,[2] but this method allows visual judgement, and gives precision without calculation.

---

[1]Using Windows 8 PAINT. Older versions of PAINT do the same steps with somewhat different screen commands.

[2]There should always be at least two ways to do anything in Windows—alleged dictum of Bill Gates.

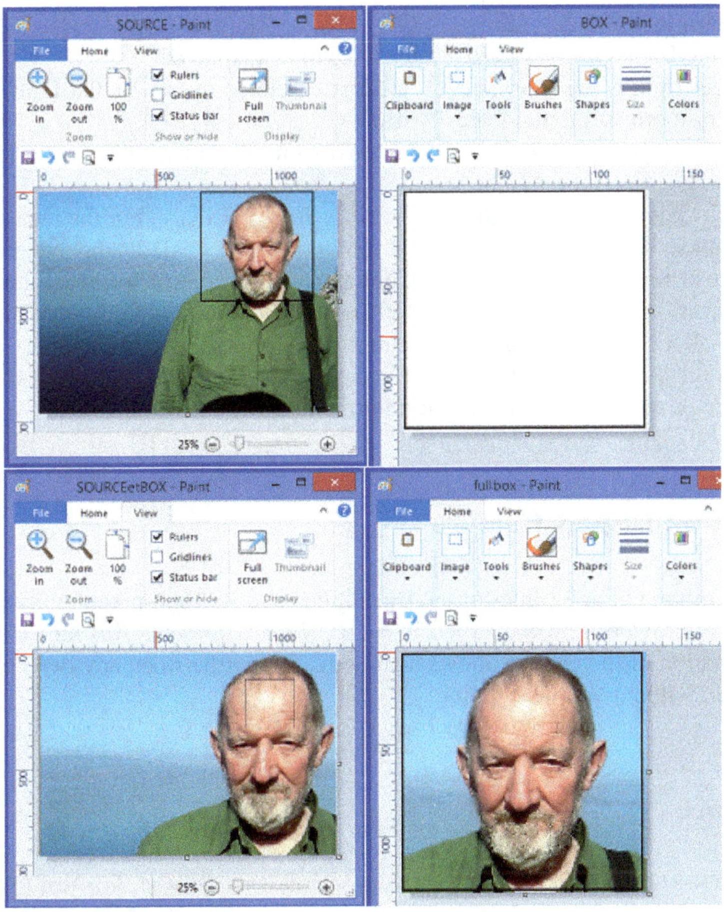

Fig. 49.2 Display of part-image. **a** Source image 1280 × 960 px and approximate required display window. The task is to display the part shown in a window 128 × 128 px. **b** Empty box, exactly 128 × 128 px. **c** Source image at native size with initially-pasted box. Estimated source diminution is 25 %. **d** After diminution of source, box contains part wanted

49.5 24-bit Display Intensity Variables

When a monitor is used to display a 24-bit image, we have the following per-pixel intensity variables:

- Intensity triple (B, G, R)
- Voltage triple (v_B, v_G, v_R).
- Display gamma γ.
- Luminance L.

Intensity triple is $0 \leq B \leq 255$, $0 \leq G \leq 255$, $0 \leq R \leq 255$, held in the .BMP file, defining the Blue, Green, and Red intensities to be displayed.

Voltage triple is the (normalized) analogue voltage output by the DAC of each channel, proportional to the intensity triple.

Display gamma is the darkening of intensity imposed by the display device, standard value 2.2 (see Sect. 22.12).

In each channel, the normalized luminance $L_{\text{norm}} = v_{\text{norm}}^{\gamma}$. For the measurement of luminance, see Chap. 15. The actual luminance of a CRT display can be varied by a control for White luminance ("Contrast") and a control for Black luminance ("Brightness"). When properly adjusted a good CRT has a maximum White luminance of about 170 cd/m^2, and a minimum Black luminance of about 0.01 cd/m^2. An LCD display has a maximum luminance of about 160 cd/m^2, and a minimum luminance of about 0.72 cd/m^2 (with minimum back-light) [1].

49.6 24-bit Display Colours

Additive mixing of any values of the Blue Green and Red sub-pixel light sources of the display provides $256 \times 256 \times 256 = 16{,}777{,}216$ colours. This far exceeds the discrimination of the eye, so allows a smooth gradient between any two colours in the gamut, without visible stepping.

Reference

1. Soneira RM (2005) Shoot-out. http://www.displaymate.com/ShootOut. Accessed 1 Feb 2014

Chapter 50
24-bit Printout

Printer size variables are the same for 24-bit as for 1-bit images. A procedure for printing part of a camera or scanner image in a print area of chosen size is described. 24-bit colour images are printed with four inks: Yellow, Magenta, Cyan, and Black (YMCK). Each ink requires an 8-bit grey-scale separation, which is half-toned to a 1-bit sub-image. Examples are shown. 24-bit print intensity variables are defined.

50.1 Flowchart

The flowchart (Fig. 50.1 shows a 24-bit image file sent to a printer by an image editor.

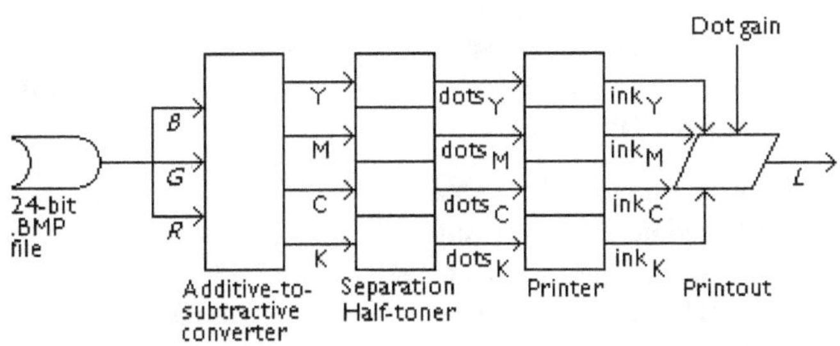

Fig. 50.1 Flowchart for 24-bit printout. A 24-bit .BMP file has, per pixel, location coordinates (X, Y), and intensities (B, G, R). An additive-to-subtractive converter has three input channels for (B, G, R) and four output separation channels for (Y, M, C, K). In each separation channel a half-toner converts grey-scale intensities to 1-bit dot densities, which a printer applies as ink on paper. Dot gain is caused by ink spread on paper

A. Parkin, *Digital Imaging Primer*, DOI 10.1007/978-3-540-85619-1_50

50.2 Printer Hardware

24-bit printer hardware is the same as 1-bit hardware, described in Chap. 28.

50.3 Print Size Variables

24-bit print size variables and resolving power are the same as for 1-bit print, described in Chap. 28.

50.4 Printout of Camera Image

A common task is to print part of a camera (or scanner) image in a print area of a chosen size. For example, Fig. 50.2a shows a source image and the part we want to print. We can find the native printout size of the source image in PAINT; for example:

- File—Properties—pixels 1200 × 1600, resolution 72 dpi.
- File—Properties—cm 42.33 × 56.44, resolution 72 dpi.
- File—Properties—inches 16.67 × 22.22, resolution 72 dpi.

The Resolution is held in the image file as BiXPelsPerMeter = BiYPelsPerMeter = 2835 = 72 × 39.375, and cannot be changed in PAINT, only by a suitable bitmap editor (see Sect. 28.3, print pixel frequency).

Say we want to print the part shown on A6 paper, 105 × 148 mm = 4.13 × 5.84 in, with margins 5 mm = 0.2 in. The required print area is thus 95 × 138 mm = 3.74 × 5.45 in. We need to crop and diminish the source image as held in the file. One way to do this is by the following Windows 8 PAINT procedure, working in inches. Figure 50.2 shows some of the steps:

1. Display the source image (File—Open—BYEBYE).
2. Re-set the width and height to the required print area at the same print resolution as the source (File—Properties—Inches—Width = 3.74 Height = 5.45).
3. Clear the image (Home—Image—Select—Select all. Right-click—Cut).
4. Draw a line all round the edge of the empty box (Home—Shapes—Rectangle), and save (File—Save as—PBOX 24-bit).
5. Make a transparent copy of the box on the clipboard (Home—Image—Select—Select all—Transparent. Right-click—Copy).
6. Display the source image again (File—Open—BYEBYE).
7. Put the transparent box on the source (Home—Clipboard—Paste).
8. Drag the box to near the part wanted and estimate the diminution percent DPC of source image needed to get the part into the box.
9. Remove the box to the clipboard (Undo).

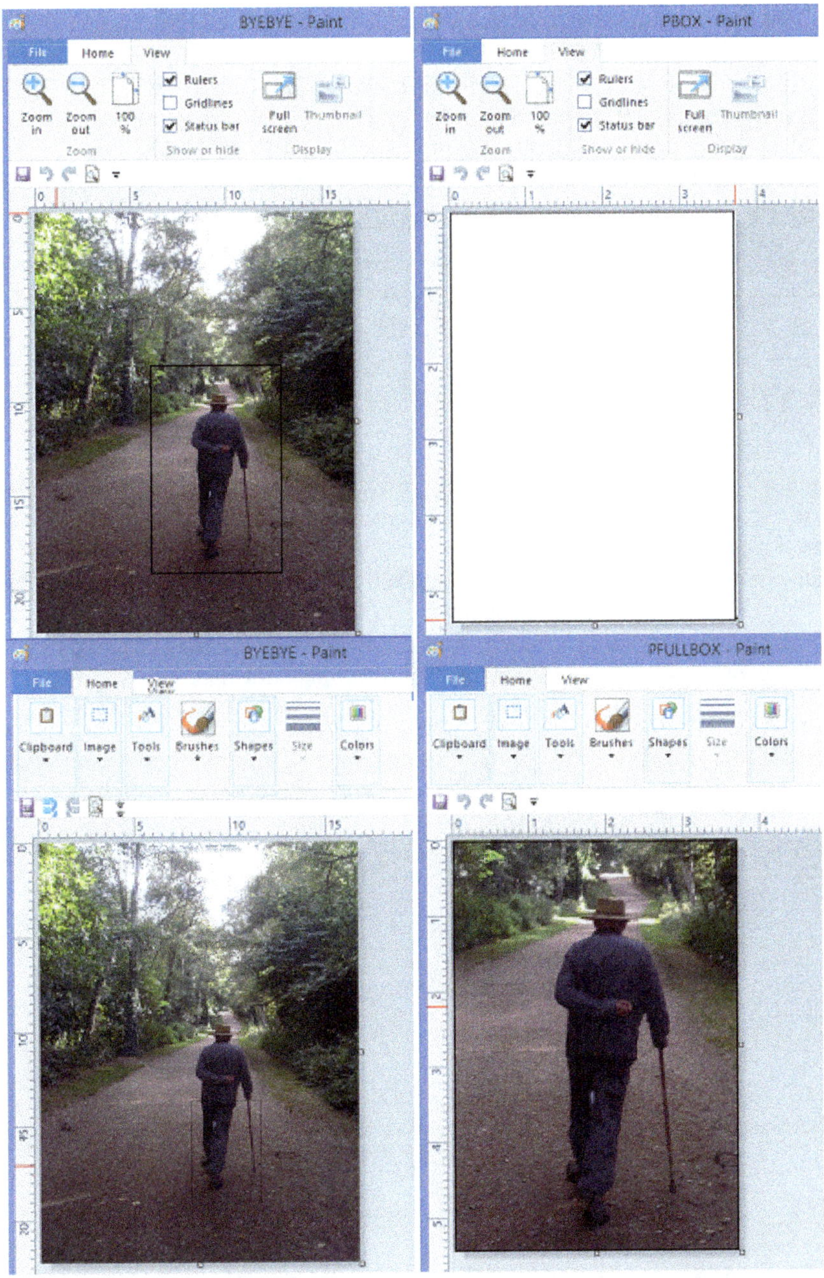

Fig. 50.2 Printout of part-image. **a** The native printout size of the source image is 16.67 × 22.22 in at print resolution 72 dpi. The task is to print the part roughly shown, on an A6 sheet of paper with 0.2-inch margins, that is 3.74 × 5.45 in. **b** Empty box exactly 3.74 × 5.45 in. **c** Source image at native size with initially-pasted box. Estimated source diminution to 58 %. **d** After diminution of source, box contains part wanted

10. Diminish the source image (Home—Image—Resize—horizontal = DPC vertical = DPC).
11. Bring back the box (Home—Clipboard—Paste) and drag to get the best framing. If no good, go to 9.
12. When good, save the filled box (File—Save as—PFULLBOX).
13. Check that PFULLBOX is as wanted (File—Properties—Inches—Width = 3.74 Height = 5.45 at 72 dpi).

There are of course other ways to get the same result, but this method allows visual judgement, and gives precision without calculation.

50.5 YMCK Printing

Printing a 24-bit colour image is done by converting additive (B, G, R) intensities to subtractive (Y, M, C, K) intensities, making grey-scale separations, and 1-bit half-tones, as described in Chap. 35.

For example, Fig. 50.3 shows two 24-bit source images; Fig. 50.4 shows the corresponding Y M C K separations with grey-scale colour-table 8GS; and Fig. 50.5 shows the 1-bit half-tones made from each of the separations. These may be compared with the 4-bit examples in Chap. 35 and the 8-bit examples in Chap. 43.

Fig. 50.3 8CO source images. *Top* is a 24-bit camera image saved-as 8CO. *Bottom* is a 24-bit gamut saved-as 8CO

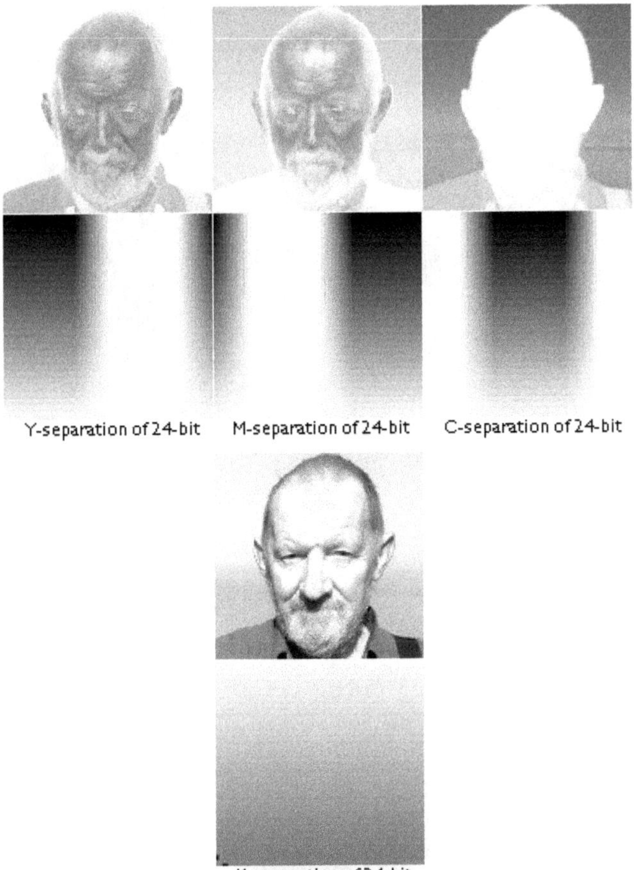

Fig. 50.4 8-bit *grey-scale* separations from images in Fig. 50.3. **a** Yellow separation. **b** Magenta separation. **c** Cyan separation. **d** Black separation

50.6 24-bit Print Intensity Variables

In printing a 24-bit image, we have the following per-pixel intensity variables:

- 8-bit file intensity triple (B, G, R).
- 8-bit separation intensity quadruple (Y, M, C, K).
- 8-bit separation grey-scale quadruple (Ygs, Mgs, Cgs, Kgs).
- 1-bit ink percent density quadruple (Ypc, Mpc, Cpc, Kpc).
- Dot gain δ

Intensity triple is $0 \leq B \leq 255$, $0 \leq G \leq 255$, $0 \leq R \leq 255$, held in a 24-bit .BMP file. It defines the proportions of additive primaries in the pixel.

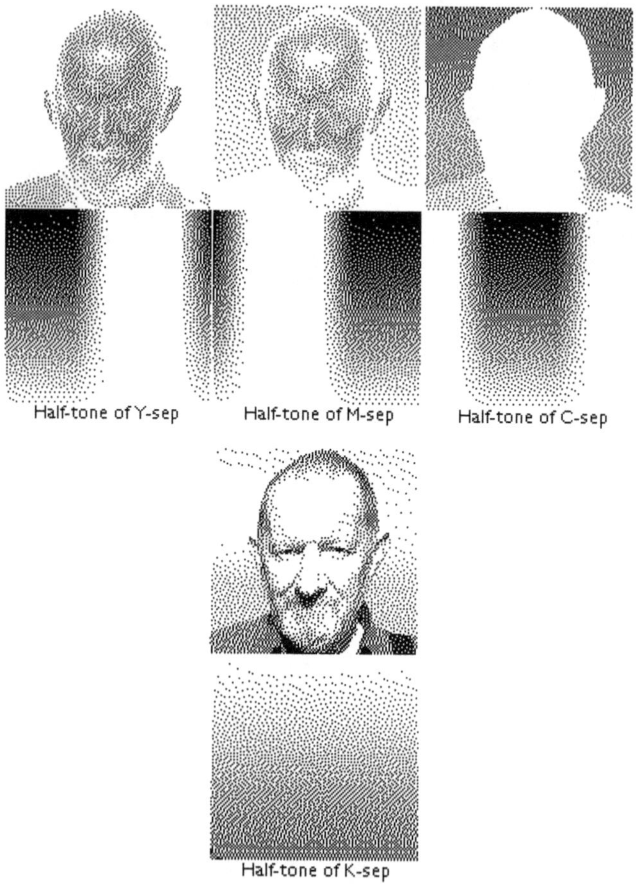

Fig. 50.5 1-bit half-tones of separations in Fig. 50.4

Separation intensity quadruple is $\leq$ Y $\leq$ 255, 0 $\leq$ M $\leq$ 255, 0 $\leq$ C $\leq$ 255, 0 $\leq$ K $\leq$ 255. It defines the proportions of subtractive primaries in the pixel.

Separation grey-scale intensity is the 255-complement of the separation intensity quadruple; that is to say 0 becomes 255, 128 becomes 127, and 255 becomes 0.

Ink percent density is $100/255\times$ the 255-complement of the grey-scale intensity; that is to say, 0 becomes 100 %, 127 becomes 50 %, 192 becomes 25 %, and 255 becomes 0 %.

Dot gain is the spreading of ink on the paper (see Sect. 35.8).

50.7 24-bit Print Colours

Converting from three given additive intensities (B, G, R) each in the range 0–255, to four subtractive ink coverages (Y, M, C, K) each in the range 100–0, is like converting from 4-bit intensities, as in Sect. 35.5.

In printing a 24-bit image, the YMCK inks correspond to the BGR display colours, but the printout is systematically duller, and prone to manifold imperfections of inks, paper, and printer.

Chapter 51
Viewing 24-bit Image

Viewing a 24-bit displayed image and a 24-bit printed image is compared with viewing a scene directly. Scope, location resolution, and luminance range are the same for 24-bit as for 1-bit images, but colour resolution is quite different.

51.1 Flowchart

The flowchart Fig. 51.1 is for general guidance in this chapter. It shows an eye viewing various types of 24-bit image, and, by comparison, viewing a scene. The computing flow is like that in Chap. 44, but with a full range of intensities instead of 256.

51.2 Viewing Variables

If we want to compare viewing a scene with viewing a digital image of a scene, four variables emerge:

- location range, or scope, measured as angular or linear extent of view,
- location resolution within scope, measured as Snellen acuity, or as line-pairs per unit angle, or as line-pairs per unit length (see Sect. 10.6),
- luminance range, measured as dynamic operating range from lowest to highest luminance in cd/m^2 (see Sect. 15.8),
- colour resolution within luminance, measured as gamut on a CIE chromaticity chart (see Sect. 16.3).

The location range and resolution for 24-bit images are the same as for 1-bit images in Chap. 29.

A. Parkin, *Digital Imaging Primer*, DOI 10.1007/978-3-540-85619-1_51

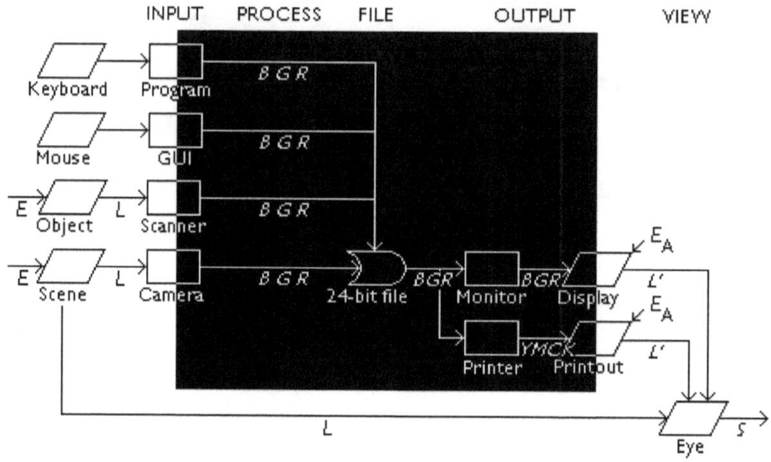

Fig. 51.1 Viewing a scene and a 24-bit image. E is illuminance in lx, from source onto object or scene. L is luminance in cd/m^2, from scene to eye or to input device, and from output device to eye. BGR is intensity triple, each eight bits 0–255, from encoder to file, from file to decoder, and from monitor to display. CMYK are ink coverages, from printer to paper. E_A is ambient illuminance in lux, from surroundings onto display or printout. L' is luminance from output device to eye, and S is viewer's colour or grey-scale brightness sensation from eye to consciousness

The luminance range for 24-bit images is the same as for 1-bit images in Chap. 29, but taking notice of display gamma and dot gain effects.

The ambient illuminance effects for 24-bit images are the same as for 1-bit in Chap. 29.

51.3 Colour Resolution of Eye and 24-bit Image

Figure 51.2a shows the colour gamut of the eye. In the chromaticity chart, Brightness is imagined as a third dimension, from low below the chart to high above the chart; thus a dark Red and a light Red may have the same chromaticity but different Brightnesses, and neutral greys likewise. Figure 51.2b shows the gamut of a sRGB display with apparently smooth colour gradients everywhere. A display has additive mixture: for example primaries R and G mix to make secondary Y. In Fig, 51.2c, a printout has subtractive mixture: primaries Y and M mix to make secondary R, with saturations lower than a display.

The colour gamut of a standard eye is specified by the CIE chromaticity chart. The colour gamut of a digital image is determined by the (B, G, R) values in the file, and the output device used.

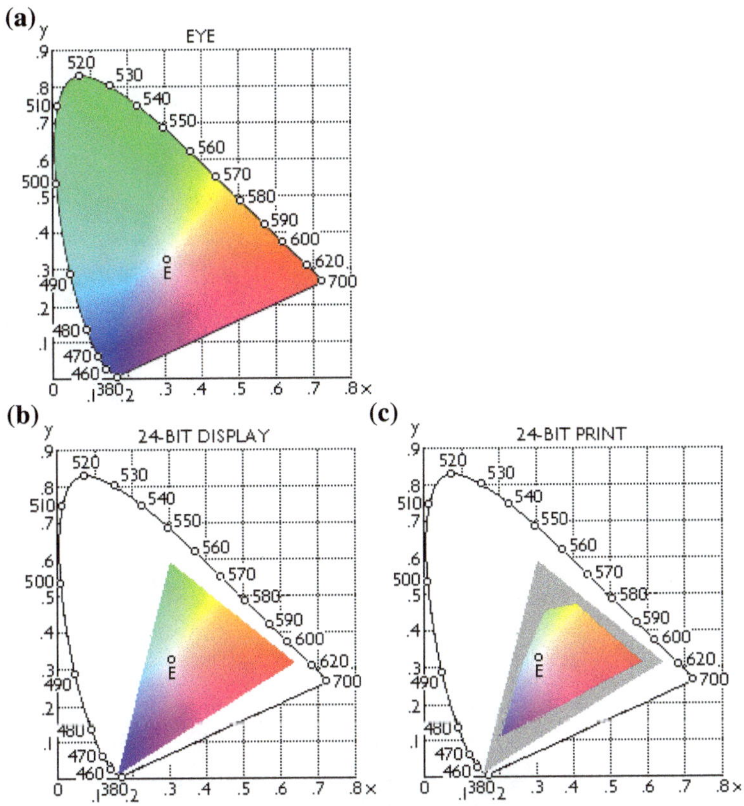

Fig. 51.2 Colour gamut of eye, compared to 24-bit images. **a** CIE chromaticity chart showing the colour gamut to which the eye responds. E is equal-energy illuminant. Notice that the colours printed here are merely indicative: the eye gamut extends to much more saturated colours. **b** Effectively-smooth gamut of 24-bit display on a sRGB monitor. **c** Effectively-smooth gamut of 24-bit printout on typical ink-jet printer

Appendix A
Summary of Image Types

Figure A.1 shows examples of the bit-depths, colour-tables, and change processes used in this book.

A 24-bit colour image (24CO) can be created by a 24-bit program, or by a GUI in 24-bit mode, or by a scanner in 24-bit mode, or by a camera in 24-bit mode. It has no colour-table. It can be saved-as an 8CO image, or converted-to an 8GS image.

An 8-bit grey-scale image with colour-table 8GS can be created by an 8-bit program, or by a scanner in 8-bit grey-scale mode. It can be saved-as a 4GS grey-scale image, which can in turn be saved-as a 1GS monochrome image alternative to 1BW.

An 8-bit colour image with colour-table 8CO can be created by an 8-bit program, or by a GUI in 8-bit mode. It can be converted-to an 8GM grey-match image, or saved-as a 4CO colour image.

A 4-bit colour image with colour-table 4CO can be created by by a 4-bit program, or by a GUI in 4-bit mode. It can be converted-to a 4GM grey-match image, or saved-as a 1BW monochrome image.

A 1-bit monochrome image with colour-table 1BW can be created by by a 1-bit program, or by a GUI in 1-bit mode, or by a scanner in 1-bit mode.

A. Parkin, *Digital Imaging Primer*, DOI 10.1007/978-3-540-85619-1

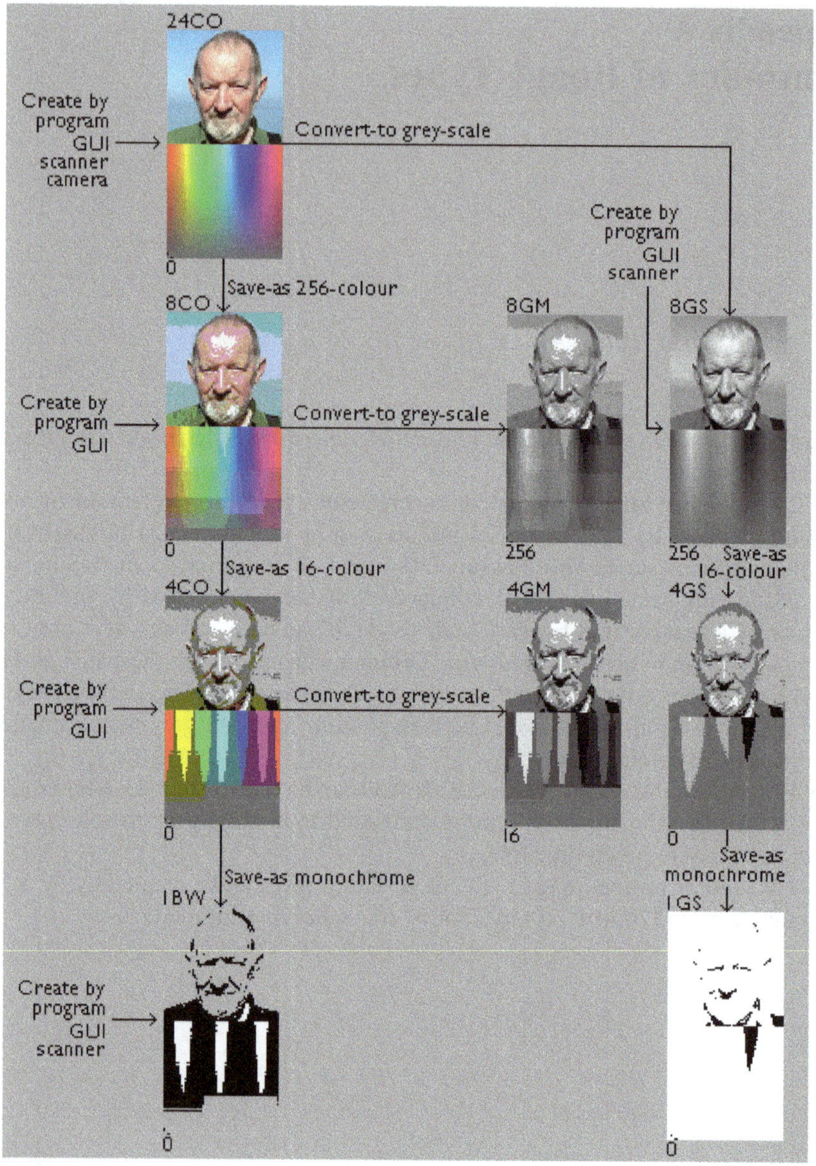

Fig. A.1 Summary of the image types used in this book. *Save-as* is a utility in MS PAINT; *Convert-to* is a utility in IRFANVIEW. The number below each example is the value of *biClrUsed* in the .BMP file

Appendix B
Source Code for Programs

Source code for the programs described in this book is available for download at the SpringerLink web-site. The programs are written in Microsoft QuickBASIC 4.5.

Microsoft QuickBASIC 4.5 is available at http://www.phatcode.net/downloads.php?id=172. Accessed 1 Feb 2014

The *.BAS files should be downloaded into the DOS directory where QuickBASIC has been installed.

(Table B.1) lists the programs: for page references to the book, see Index under Programs.

Table B.1 Programs described in this book

ASTEREO4	GENGRA24
CHANLO24	GENRAN24
CHANLOC1	GENRAND1
CHANLOC4	GENRAND4
CHANLOC8	GENRAND8
COMALAV8	HT8MTHRE
COMBIN24	HT8STHRE
COMBINT1	HT8VTHRE
COMBLOC1	IDCT1
DCT1	MKBEZIER
DOSTATS1	MKLINE
DOSTATS4	MKPGRAPH
DOSTATS8	MKSINCUR
EDBMP1	PROJECTI
EDBMP24	REDULO24
EDBMP4	REDULOC1
EDBMP8	REDULOC8
EDPGRAPH	SEP24TO8
EDSGRAPH	SRGB2ETC
ETC2SRGB	TRANPI24
FOURIANA	TRANPIN8
FOURISYN	TRANSFOS

A. Parkin, *Digital Imaging Primer*, DOI 10.1007/978-3-540-85619-1

Index

A. Parkin, *Digital Imaging Primer*, DOI 10.1007/978-3-540-85619-1

www.ingramcontent.com/pod-product-compliance
Lightning Source LLC
Chambersburg PA
CBHW071403230926
51846CB00001B/10